Alexander Koldobsky and Alexander Volberg (Eds.)
Harmonic Analysis and Convexity

Advances in Analysis and Geometry

—

Editor-in-Chief
Jie Xiao, Memorial University, Canada

Volume 9

Harmonic Analysis and Convexity

Edited by
Alexander Koldobsky and Alexander Volberg

DE GRUYTER

Mathematics Subject Classification 2020
Primary: 52A20, 46B09; Secondary: 44A12, 52A38, 46B07

Editors

Prof. Dr. Alexander Koldobsky
Department of Mathematics
University of Missouri-Columbia
Columbia, MO 65211
USA
koldobskiya@missouri.edu

Prof. Dr. Alexander Volberg
Michigan State University
Department of Mathematics
2630 Pin Oak Drive
Ann Arbor, MI 48103
USA
volberg@msu.edu

ISBN 978-3-11-077537-2
e-ISBN (PDF) 978-3-11-077538-9
e-ISBN (EPUB) 978-3-11-077543-3
ISSN 2511-0438

Library of Congress Control Number: 2023934187

Bibliographic information published by the Deutsche Nationalbibliothek
The Deutsche Nationalbibliothek lists this publication in the Deutsche Nationalbibliografie;
detailed bibliographic data are available on the Internet at http://dnb.dnb.de.

© 2023 Walter de Gruyter GmbH, Berlin/Boston
Typesetting: VTeX UAB, Lithuania
Printing and binding: CPI books GmbH, Leck

www.degruyter.com

Contents

Mark Agranovsky, Jan Boman, Alexander Koldobsky, Victor Vassiliev, and
Vladyslav Yaskin

Algebraically integrable bodies and related properties of the Radon transform

Abstract: Generalizing Lemma 28 from Newton's "Principia" [25], Arnold [10] asked for a complete characterization of algebraically integrable domains. In this chapter we describe the current state of Arnold's problems. We also consider closely related problems involving the Radon transform of indicator functions.

Keywords: Integrability, analytic continuation, monodromy, Picard–Lefschetz theory, Radon transform, convex body, Fourier transform

MSC 2020: 14D05, 42B10, 44A12, 44A99, 52A20

1 Introduction

The questions considered in this survey belong to the area of geometric tomography (see the book [17]), which lies at the crossroads between convex geometry and integral geometry and can be defined as the study of geometric properties of solids based on data about their sections and projections.

We study algebraic properties of two important volumetric characteristics in geometric tomography. For a body (compact set with non-empty interior) K in $\mathbb{R}^n$, $\xi \in S^{n-1}$, and $t \in \mathbb{R}$, the *cutoff functions* of K represent the n-dimensional volume of the parts of K cut by the hyperplane perpendicular to ξ at distance t from the origin:

Acknowledgement: The third named author was supported in part by the U. S. National Science Foundation Grant DMS-2054068. The fifth author was supported in part by NSERC. This material is partially based on the work supported by the U. S. National Science Foundation grant DMS-1929284. The third and fifth authors were in residence at the Institute for Computational and Experimental Research in Mathematics in Providence, RI, during the Harmonic Analysis and Convexity semester program. The main part of the work of the fourth author was done at the Steklov Mathematical Institute, Moscow.

Mark Agranovsky, Department of Mathematics, Bar-Ilan University, Ramat Gan, 5290002, Israel, e-mail: agranovs@math.biu.ac.il

Jan Boman, Department of Mathematics, Stockholm University, 10691 Stockholm, Sweden, e-mail: jabo@math.su.se

Alexander Koldobsky, Department of Mathematics, University of Missouri-Columbia, Columbia, MO 65211, USA, e-mail: koldobskiya@missouri.edu

Victor Vassiliev, Faculty of Mathematics and Computer Science, Weizmann Institute of Science, 7610001, Rehovot, Israel, e-mail: vavassiliev@gmail.com

Vladyslav Yaskin, Department of Mathematical and Statistical Sciences, University of Alberta, Edmonton, AB T6G2G1, Canada, e-mail: yaskin@ualberta.ca

https://doi.org/10.1515/9783110775389-001

$$V_K^+(\xi, t) = \mathrm{Vol}_n(K \cap \{x \in \mathbb{R}^n : \langle x, \xi \rangle \leq t\}) = \int\limits_{K \cap \{x \in \mathbb{R}^n:\ \langle x,\xi\rangle \leq t\}} dx,$$

$$V_K^-(\xi, t) = \mathrm{Vol}_n(K \cap \{x \in \mathbb{R}^n : \langle x, \xi \rangle \geq t\}) = \int\limits_{K \cap \{x \in \mathbb{R}^n:\ \langle x,\xi\rangle \geq t\}} dx. \tag{1.1}$$

The *section function* of K is the $(n-1)$-dimensional volume of the section of K by the same hyperplane:

$$A_K(\xi, t) = \mathrm{Vol}_{n-1}(K \cap \{x \in \mathbb{R}^n : \langle x, \xi \rangle = t\}) = \mathcal{R}(\chi_K)(\xi, t) = \int\limits_{K \cap \{x \in \mathbb{R}^n:\langle x,\xi\rangle = t\}} dx. \tag{1.2}$$

Here $\mathcal{R}$ stands for the Radon transform, χ_K is the indicator (characteristic function) of K, $\langle x, \xi \rangle$ is the inner (scalar) product in $\mathbb{R}^n$, and dx is the Lebesgue measure on $\mathbb{R}^n$ or $\{x : \langle x, \xi \rangle = t\}$, correspondingly. Clearly, the cutoff functions and the section function are related via differentiation in t.

Most of our problems take root in Lemma 28 about ovals from Newton's Principia [25]; see also the discussion in [8, 9, 34]. Newton proved that if K is a convex infinitely smooth domain in $\mathbb{R}^2$, then the cutoff function of K cannot appear as the solution of a polynomial equation involving the parameters of the cutting hyperplane. Formalizing the question and extending it to higher dimensions, Arnold [10] asked whether there exist domains with smooth boundaries in $\mathbb{R}^n$ (apart from ellipsoids for odd n) for which the cutoff functions $V_K^\pm$ are branches of an algebraic function. Recall that a function $f(\xi, t)$ is *algebraic* if there exists a non-zero polynomial $\Phi(\xi, t, w)$ of $n+2$ variables such that

$$\Phi(\xi, t, f(\xi, t)) \equiv 0.$$

Definition 1.1 (cf. [10, 8, 29]). A domain K is *algebraically integrable* if the two-valued cutoff function $V_K^\pm(\xi, t)$ coincides with some branches of an algebraic function.

In Section 2, we present the current state of Arnold's problems. In particular, it was proved in [29] that there are no algebraically integrable bodies with infinitely smooth boundaries in even dimensions. However, the odd-dimensional case is still open.

In Sections 3–7, we consider similar questions that are motivated by Arnold's problem and address the single-valued section function $A_K(\xi, t)$ rather than the multi-valued cutoff function $V_K(\xi, t)$. Therefore, we study geometric properties of bodies K from the point of view of algebraic properties of their Radon transform A_K. The following definition is similar to Definition 1.1.

Definition 1.2. Let K be a body in $\mathbb{R}^n$. We say that K has *algebraic Radon transform* if there exists a function $\Psi \in C(S^{n-1})[t, w]$ which is an element of the polynomial ring of two variables over the algebra $C(S^{n-1})$ (i. e., it is a polynomial with respect to t, w with coefficients which are continuous functions of ξ) and satisfies the equation

$$\Psi(\xi, t, A_K(\xi, t)) = 0$$

for every t such that the hyperplane $\langle \xi, x \rangle = t$ intersects K.

The essential difference between the two definitions is that in Definition 1.2 we do not assume that Ψ is a polynomial in ξ, as we do in Definition 1.1, so the section function $A_K(\xi, t)$ is algebraic only with respect to the variable t.

Note that if K is algebraically integrable (i. e., the cutoff function $V_K^{\pm}$ is algebraic), then the section function $A_K(\xi, t)$ is also algebraic as the derivative

$$A_K(\xi, t) = \pm \frac{d}{dt} V_K^{\pm}(\xi, t)$$

of an algebraic function. Thus, the class of domains with algebraic Radon transform contains algebraically integrable domains.

Our basic example is the unit ball B^n in $\mathbb{R}^n$. In this case

$$A_{B^n}(\xi, t) = \frac{\pi^{\frac{n-1}{2}}}{\Gamma(\frac{n+1}{2})} \left(1 - t^2\right)^{\frac{n-1}{2}}.$$

If n is odd, then $A_{B^n}(\xi, t)$ is a polynomial in t. Applying an affine transformation to B^n we obtain that $A_K(\xi, t)$ is a polynomial in t if n is odd and K is an ellipsoid.

In this chapter, we consider classes of bodies K satisfying Definition 1.2 with the defining polynomial Ψ of a certain form. The property of ellipsoids in odd-dimensional spaces mentioned above gives rise to the following.

Definition 1.3 ([1]). Let K be a domain in $\mathbb{R}^n$. We call K *polynomially integrable* if the Radon transform $A_K(\xi, t)$ of χ_K is a polynomial with respect to t when the corresponding hyperplane intersects K.

In the case of polynomially integrable domains, the equation $\Psi(\xi, t, w) = 0$ in Definition 1.2 has the form $\Psi(\xi, t, w) = w - \sum_{k=0}^{N} a_k(\xi) t^k = 0$. Another example is given by *rationally integrable domains* where $A_K(\xi, t)$ is the ratio of polynomials in t, $A_K(\xi, t) = \frac{P(\xi, t)}{Q(\xi, t)}$, and, correspondingly, $\Psi(\xi, t, w) = Q(\xi, t)w - P(\xi, t)$.

A polynomially integrable body is algebraically integrable if we additionally demand that $A_K(\xi, t)$ extends from the unit sphere $|\xi| = 1$ to $\mathbb{R}^n$ as a polynomial when t is fixed. However, in Definition 1.3 no essential condition is imposed on the behavior of $A_K(\xi, t)$ with respect to ξ, and hence the two classes are different, although they intersect.

In Section 3, we describe the result of [1, 23] that the only polynomially integrable bodies are ellipsoids in odd dimensions. In Section 4, we extend this result to the case where the section function is real analytic; in particular, it can be a rational function without real poles. In Section 5, a relation between polynomial integrability and finite stationary phase expansions of certain Fourier integrals is established. This relation is used to characterize locally polynomially integrable hypersurfaces. In Section 6, domains with algebraic X-ray transform (chord length functions) are studied. Finally, in Section 7 we present Theorem 7.2, showing that the Radon transform of a compactly supported distribution can be supported in the set of tangent planes to the boundary ∂D of a bounded convex domain $D \subset \mathbb{R}^n$ only if ∂D is an ellipsoid. This result gives a new proof of the fact that polynomially integrable bodies must be ellipsoids (Theorem 3.2).

2 Algebraically integrable bodies in Euclidean space

2.1 Problems and main results

By a theorem of Archimedes (see [6, 33]), spheres in $\mathbb{R}^3$ are algebraically integrable. Indeed, the volume cut from the unit ball in $\mathbb{R}^3$ by a hyperplane at distance $t < 1$ from the origin is a polynomial in t. It is easy to check that the same is true for arbitrary ellipsoids in odd-dimensional spaces. On the contrary, Newton's result [25, Lemma 28] mentioned in the introduction asserts that there are no convex algebraically integrable bodies with smooth boundaries in $\mathbb{R}^2$.

V. Arnold [10, Problems 1987-14, 1988-13, and 1990-27] conjectured that there are no algebraically integrable bodies with smooth boundaries in even-dimensional spaces and asked whether there exist such bodies other than ellipsoids in odd-dimensional spaces. The even-dimensional conjecture was confirmed in [29].

Theorem 2.1 ([29]). *There are no algebraically integrable bodies with C^∞-smooth boundaries in even-dimensional spaces.*

The odd-dimensional case is still open; see Theorem 2.3 and Remark 2.5 below for some partial results towards it.

Remarks. 1. By projective duality and the Tarski–Seidenberg theorem, if a body in $\mathbb{R}^n$ is algebraically integrable, then its boundary is semialgebraic. Therefore, it is enough to consider the case when our body is bounded by a smooth component of a hypersurface defined by a polynomial equation $F(x) = 0$.

2. The condition of infinite smoothness is essential in this problem: for an arbitrary natural N there exist algebraically integrable bodies with C^N-smooth boundaries in even-dimensional spaces.

3. In fact, we prove even more: under the conditions of Theorem 2.1 the analytic continuation of the volume function to the space of complex hyperplanes in $\mathbb{C}^n$ cannot be even algebroid, because it necessarily takes infinitely many different values at the same hyperplanes.

Conjecture 2.2. *For any odd number k, even number m, and $\varepsilon \in (0, 1)$, the body in $\mathbb{R}^{k+m}$ bounded by the hypersurface*

$$\left(\sqrt{x_1^2 + \cdots + x_k^2} - 1\right)^2 + y_1^2 + \cdots + y_m^2 = \varepsilon^2 \tag{2.1}$$

(i. e., the ε-neighborhood of the unit sphere $S^{k-1} \subset \mathbb{R}^k \subset \mathbb{R}^{k+m}$) is algebraically integrable.

It is encouraging that the obstruction to algebraic integrability mentioned in the third remark (the infinite ramification of the analytic continuation of the volume function) fails for this body.

Theorem 2.3 ([32]). *The body introduced in Conjecture 2.2 is algebroidally integrable (i. e., its cutoff functions are algebroid). In particular, the analytic continuation of this function from any domain of the space of real hyperplanes where this function is regular is finite-valued.*

So, to prove Conjecture 2.2 it suffices to check that this analytic continuation has only power growth at its singular points. Even if this conjecture was confirmed, one more of Arnold's questions would remain unsolved, namely, Problem 1990-27 of [10], asking whether there are *convex* algebraically integrable bodies in $\mathbb{R}^{2k+1}$ except for ellipsoids.

In any case, algebraically integrable bodies in $\mathbb{R}^{2k+1}$ are very rare. In particular, the local geometry of their boundaries satisfies very strong conditions.

Theorem 2.4 ([30, 28]). *If a body $K \subset \mathbb{R}^{2k+1}$ is algebraically integrable, then:*
(1) *the inertia indices of the second fundamental form of its boundary are even at all points where this form is non-degenerate;*
(2) *the algebraic closure of this boundary in $\mathbb{C}^n$ has no tame parabolic points.*

Recall that a regular point of an affine hypersurface is *parabolic* if the second fundamental form is degenerate at this point; a parabolic point is called *tame* if the tangent hyperplane at this point has no other tangencies with the hypersurface in a neighborhood of this point.

Remark 2.5. If either of the two conditions of Theorem 2.4 is not satisfied, then not only the cutoff function (1.1) is not algebraic, but also the section function (1.2) is not algebraic.

For additional restrictions on the geometry of algebraically integrable bodies, see [28, Chapter III] and [33, Chapter 7].

In addition, we consider the property of *local* algebraic integrability. The volume function is regular analytic on the set of hyperplanes transversal to the boundary of the body; the set of tangent hyperplanes divides this set into several connected components. For example, the body bounded by the surface (2.1) with arbitrary k and m has four such components: the hyperplanes from them intersect the hypersurface (2.1) along manifolds diffeomorphic to (a) the empty set, (b) S^{k+m-2}, (c) $S^{k-1} \times S^{m-1}$, and (d) $S^{k-2} \times S^m$.

We call such a component a *lacuna* if the volume function coincides in it with an algebraic function. A trivial example of a lacuna is the domain consisting of hyperplanes not intersecting the body. We show that this example is not unique even in the even-dimensional case.

Proposition 2.6 (See [31]). *If m is even (and k is arbitrary), then the component of the set of generic hyperplanes containing the hyperplane $x_1 = 0$ is a lacuna of the body bounded by the hypersurface*

$$\left(x_1^2 + \cdots + x_k^2 - 1\right)^2 + \left(y_1^2 + \cdots + y_m^2\right) = \varepsilon^2. \tag{2.2}$$

In the case $k = 3$ a more general class of examples is given in [30].

Remark 2.7. There is a deep analogy between this set of problems and Petrovsky's theory of lacunas of hyperbolic partial differential equations (PDEs) and systems (developed further by Leray, Gårding, Atiyah, Bott, and others; see, e. g., Chapter 4 of [28]). In particular, the radical difference in the behavior of both volume functions and solutions of hyperbolic PDEs in spaces of different parity of dimensions is explained by the fact that the intersection form in middle homology groups of complex varieties (which is the main part of *Picard–Lefschetz formulas* controlling the ramification of integrals) is symmetric or antisymmetric depending on the parity of the dimension.

2.2 Integrability and Picard–Lefschetz theory

Let K be a body in $\mathbb{R}^n$, the boundary of which is a smooth component of the hypersurface defined by a polynomial equation $F(x) = 0$. Let $A \subset \mathbb{C}^n$ be the set of complex zeros of this polynomial F. A complex affine hyperplane $X \subset \mathbb{C}^n$ is *generic* if its closure $\bar{X} \subset \mathbb{C}P^n$ is transversal to the stratified variety $A \cup \mathbb{C}P_\infty^{n-1}$, where $\mathbb{C}P_\infty^{n-1} \equiv \mathbb{C}P^n \setminus \mathbb{C}^n$. Denote by P_n the space of all affine hyperplanes in $\mathbb{C}^n$ and by Σ its subset consisting of non-generic hyperplanes.

By Thom's isotopy lemma (see, e. g., [18]), pairs $(\mathbb{C}^n, A \cup X)$ form a locally trivial fiber bundle over the space $P_n \setminus \Sigma$ of generic hyperplanes X. In particular, there is a vector bundle over $P_n \setminus \Sigma$, whose fiber over a point $\{X\}$ is the relative homology group

$$\mathcal{H}(X) \equiv H_n(\mathbb{C}^n, A \cup X; \mathbb{C}). \tag{2.3}$$

Moreover, the latter bundle admits a natural local trivialization (called *the Gauss–Manin connection*) defined by covering homotopy of relative cycles in the fibers $(\mathbb{C}^n, A \cup X)$ of the former bundle. The group

$$\pi_1(P_n \setminus \Sigma, \{X\}) \tag{2.4}$$

acts on the group (2.3) by monodromy operators defined by this connection. Explicit formulas for this action are provided by *the Picard–Lefschetz theory*; see, e. g., [7, 28].

It is easy to see that integrals of the volume form

$$dx_1 \wedge \cdots \wedge dx_n \tag{2.5}$$

along the elements of the group (2.3) are well-defined and form a linear function on this group for any X.

For any hyperplane $\{X\} \in P_n \setminus \Sigma$ and any element $\gamma \in \mathcal{H}(X)$, define a function on any simply connected neighborhood of the point $\{X\}$ in P_n as follows: Its value at the point $\{X'\}$ is equal to the integral of the form (2.5) along the element of the group $\mathcal{H}(X')$ obtained from γ by the Gauss–Manin continuation over an arbitrary path connecting $\{X\}$ and $\{X'\}$ in our neighborhood. This function is holomorphic and hence can be continued to a (multivalued) analytic function on the entire set $P_n \setminus \Sigma$.

If a hyperplane X is *real* (i. e., its intersection with $\mathbb{R}^n \subset \mathbb{C}^n$ is a hyperplane in $\mathbb{R}^n$) and γ is the homology class of one of the parts cut by X from our body K, then this analytic function coincides in the set of neighboring *real* hyperplanes $\{X'\} \approx \{X\}$ with one of the two branches of the volume function participating in the definition of algebraic integrability. If our body is algebraically integrable, then this analytic function is algebraic and, in particular, its analytic continuation to the space P_n of complex hyperplanes is finitely valued.

Therefore, to prove the non-integrability of a body it is enough to present a real generic hyperplane X such that the integrals of the volume form take infinitely many values on the orbit of this element $\gamma \in \mathcal{H}(X)$ under the monodromy action of the group (2.4).

2.2.1 Example: convex case

Let n be even, let $F : \mathbb{R}^n \to \mathbb{R}$ be a polynomial, and let K be a bounded *convex* connected component of the subset in $\mathbb{R}^n$ where $F \leq 0$; suppose that its boundary ∂K is smooth. The restriction of any linear function $L : \mathbb{R}^n \to \mathbb{R}$ to ∂K has exactly two critical points. By Sard's lemma we can choose L in such a way that these critical points will be Morse. Denote by m and M the minimal and maximal values of this restriction, respectively. For any generic value $t \in (m, M)$ denote by $[K]_-(t)$ and $[K]_+(t)$ two elements of the group $\mathcal{H}(L^{-1}(t))$ defined by the positively oriented domains $K \cap \{x \mid L(x) \leq t\}$ and $K \cap \{x \mid L(x) \geq t\}$, respectively. Fix a generic point t_0 of the interval (m, M) so that the hyperplane $L^{-1}(t_0)$ does not belong to Σ. Let α and β be two elements of the group $\pi_1(P_n \setminus \Sigma, L^{-1}(t_0))$ defined by *pinches* related to the segments $[m, t_0]$ and $[t_0, M]$ (that is, loops consisting of hyperplanes $L^{-1}(t)$, where $t \in \mathbb{C}^1$ goes from t_0 to a very small neighborhood of the point m or M along the segment and then turns in the positive direction around this point and comes back to t_0 along the same path).

Lemma 2.8. *Monodromy along the loop α moves the class $[K]_-(t_0)$ to $-[K]_-(t_0)$. Monodromy along β moves $[K]_+(t_0)$ to $-[K]_+(t_0)$.*

This lemma easily follows from the Picard–Lefschetz formula; see, e. g., [28].

Of course, these loops (and arbitrary elements of $\pi_1(P_n \setminus \Sigma, L^{-1}(t_0))$) do not change the cycle $[K] \equiv [K]_-(t_0) + [K]_+(t_0)$ which defines an element of the groups $\mathcal{H}(X)$ for all X simultaneously.

Denote by $v(t)$ the volume of the domain $[K]_-(t)$.

Corollary 2.9. *The analytic continuation along the loop α (respectively, β) moves the function $v(t)$ to $-v(t)$ (respectively, to $2V - v(t)$, where V is the volume of the entire domain K). In particular, for any integer p the analytic continuation along the loop $(\alpha\beta)^p$ moves $v(t)$ to $v(t) + 2pV$.*

So, these continuations take infinitely many values at one and the same point $L^{-1}(t_0) \in P_n \setminus \Sigma$, and the function v cannot be algebraic.

An explicit construction of the loop in $P_n \setminus \Sigma$ increasing the volume function by twice the volume of the body (and hence proving the non-algebraicity of this function) can be presented also for arbitrary bodies with smooth boundaries in $\mathbb{R}^2$; see [29]. For greater even n and general (non-convex) bodies, we have only a non-constructive proof of Theorem 2.1, based on the theory of reflection groups; see the next section.

2.3 Outline of the proof of Theorem 2.1

2.3.1 General scheme

Let K be an arbitrary domain in $\mathbb{R}^n$ (n even) bounded by a C^∞-smooth component ∂K of the set $\{x \mid F(x) = 0\}$. Again, let $L : (\mathbb{C}^n, \mathbb{R}^n) \to (\mathbb{C}, \mathbb{R})$ be a real linear function, the restriction of which to ∂K is strictly Morse.

Starting from these data, we will construct an integer lattice $\mathbb{Z}^r$, a $\mathbb{Z}$-valued bilinear form $\langle \cdot, \cdot \rangle$ on it, and a system of generators of $\mathbb{Z}^r$ (corresponding to all critical points of L on ∂K). If K is algebraically integrable, then the subgroup of the orthogonal group of the space $\mathbb{Z}^r \otimes \mathbb{R}$ generated by reflections in hyperplanes orthogonal (in the sense of our bilinear form) to these generators should be finite, i. e., be a Weyl group. All Weyl groups are well known (see, e. g., [16]); it is known, in particular, that they do not admit non-trivial elements of the lattice which are invariant under all reflections. On the other hand, we will present such an invariant element and hence get an obstruction to integrability of K.

2.3.2 Lattice

Let $m = m_1 < m_2 < \cdots < m_q = M$ be all critical values of the function $L|_{\partial K}$ and let $t_0 \in [m, M]$ be a generic value so that the hyperplane $L^{-1}(t_0)$ does not belong to $\Sigma \subset P_n$. Let $O_1, \ldots, O_q \in \mathbb{R}^n$ be corresponding critical points. By the Morse lemma, for a small ball $B_j \subset \mathbb{C}^n$ centered at any of these points O_j and a sufficiently small (compared with the size of B_j) positive number ε, all groups $H_n(B_j, B_j \cap (A \cup L^{-1}(m_j + \tau)))$, $\tau \in (0, \varepsilon)$, are isomorphic to $\mathbb{Z}$ and are generated by some relative cycles $\Delta_j(\tau)$ called *vanishing cycles*. Let us fix arbitrarily an orientation of these vanishing cycles which depends continuously on τ and consider the function v_j on the interval $(m_j, m_j + \varepsilon)$, whose value at the point $m_j + \tau$ is equal to the integral of the form (2.5) along the cycle $\Delta_j(\tau)$. This function is analytic there; its values on the interval are real or purely imaginary depending on the parity of the Morse index of the critical point O_j of the function $L|_{\partial K}$. The rotation of τ around the origin in $\mathbb{C}^1$ moves the vanishing cycle $\Delta_j(\tau)$ to minus itself, and therefore the function $w_j(\tau) \equiv v_j(m_j + \tau)$ splits on the interval $(0, \varepsilon)$ into a power series in half-integer (but not integer) powers of τ.

Let us connect a distinguished point $m_j + \tau_j$ of each interval $(m_j, m_j + \varepsilon)$ and the non-critical value t_0 by a path in $\mathbb{C}^1$ going along the real line in the upper half-plane. Let $\bar{v}_j$ be the germ at the point t_0 of the analytic continuation of the function v_j along this path: its value at t_0 is equal to the integral of the form (2.5) along the element $\bar{\Delta}_j \in \mathcal{H}(L^{-1}(t_0))$ obtained from the vanishing cycle $\Delta_j(\tau_j)$ (considered as an element of the group $\mathcal{H}(L^{-1}(m_j + \tau_j))$) by the Gauss–Manin connection along this path.

Consider the group $\mathbb{Z}^q$ of formal linear combinations of germs $\bar{v}_j$ with integer coefficients. The obvious evaluation homomorphism maps this group into the space of germs of holomorphic functions at t_0. The lattice $\mathbb{Z}^r$ promised in Section 2.3.1 is the image of this homomorphism.

2.3.3 Bilinear form and reflection group

Define first a bilinear form on the lattice $\mathbb{Z}^q$ of formal linear combinations of germs $\bar{v}_j$. Consider the chain of homomorphisms

$$\mathbb{Z}^q \to H_n(\mathbb{C}^n, A \cup L^{-1}(t_0)) \to H_{n-1}(A \cup L^{-1}(t_0)) \to H_{n-2}(A \cap L^{-1}(t_0)), \qquad (2.6)$$

the first of which maps any formal sum $\sum a_j \bar{v}_j$ to the homology class of the cycle $\sum a_j \bar{\Delta}_j$, the second is the boundary operator, and the third is the differential of the Mayer–Vietoris exact sequence. The bilinear form in the lattice $\mathbb{Z}^q$ is lifted by this composite map from the intersection form in the (smooth part of the) $(n-2)$-dimensional complex variety $A \cap L^{-1}(t_0)$.

Lemma 2.10. *This bilinear form can be lowered to the lattice $\mathbb{Z}^r$.*

Proof. Suppose that a linear combination $\sum a_j \bar{v}_j$, $a_j \in \mathbb{Z}$, defines the zero germ at t_0, but its pairing $\langle \sum a_j \bar{v}_j, \bar{v}_l \rangle$ with some element $\bar{v}_l$ is a non-zero number C. Consider the "pinch" loop in $\mathbb{C}^1$ starting and ending at t_0, embracing the critical value m_l, and running twice along our path connecting the points t_0 and $m_l + \tau_l$. This loop defines an element of the group $\pi_1(P_n \setminus \Sigma, L^{-1}(t_0))$: any point $t \in \mathbb{C}^1$ is associated with the hyperplane $L^{-1}(t)$. According to the Picard–Lefschetz formula, the analytic continuation of our *zero function* $\sum a_j \bar{v}_j$ along this loop adds to it the (definitely non-zero) function $\bar{v}_l$ with coefficient $\pm C \neq 0$. $\qquad \square$

Such analytic continuations of the functions $\bar{v}_j$ along all q pinch loops preserve the lattice $\mathbb{Z}^r$. By the Picard–Lefschetz formula, they act on this lattice as reflections in hyperplanes orthogonal to corresponding elements $\bar{v}_l$ with respect to our bilinear form; in particular, the pinch corresponding to m_l moves $\bar{v}_l$ to $-\bar{v}_l$. Consider the subgroup of the orthogonal group of $\mathbb{Z}^r \otimes \mathbb{R}$ generated by these q reflections.

2.3.4 If K is integrable, then this reflection group is finite

Lemma 2.11 (See [29]). *The class of the domain $K \cap \{x \mid L(x) \leq t_0\}$ (respectively, $K \cap \{x \mid L(x) \geq t_0\}$) in the group $\mathcal{H}(L^{-1}(t_0))$ is equal to the sum of the (appropriately oriented) vanishing cycles $\bar{\Delta}_j$ over all j such that $m_j < t_0$ (respectively, $m_j > t_0$).*

If K is algebraically integrable, then the volume of the domain $K \cap \{x \mid L(x) \leq t\}$ should be an algebraic function of t, and hence the sum $\sum \bar{v}_j$ over j such that $m_j < t_0$ should have a finite orbit under the action of our reflection group in the space $\mathbb{Z}^r$. Replacing t_0 with a point t_0' from another interval of non-critical values in the segment $[m, M]$ we prove the analogous statement for the sum of similar germs $\bar{v}_j'$ at the point t_0' over all j such that $m_j < t_0'$. Identifying then spaces of germs at points t_0 and t_0' by analytic continuation along a path between these points in the upper half-plane of $\mathbb{C}^1$ we prove that all sums $\sum_{j \leq s} \bar{v}_j$ for arbitrary $s = 1, \ldots, q$ have finite orbits under our reflection group in $\mathbb{Z}^r$.

Therefore, also the orbits of all particular generators $\bar{v}_j$ of this lattice should be finite, which implies the finiteness of the entire reflection group.

Remark 2.12. The group $\pi_1(P_n \setminus \Sigma, L^{-1}(t_0))$ acts transitively on the set of all vanishing cycles $\bar{\Delta}_j$ (although the action of only its subgroup generated by our pinch loops may be not sufficient for this); see [29].

2.3.5 Invariant element

By Lemma 2.11, the sum of all q function germs $\bar{v}_j$ is the constant function equal to the volume of the entire body K. This volume is positive, and therefore this sum is a non-zero element of the lattice $\mathbb{Z}^r$. On the other hand, it is invariant under all our reflections: indeed, the homomorphism (2.6) is obviously trivial on it; moreover, a non-trivial action on it of some reflection would imply a non-trivial ramification of the constant function. Therefore, our reflection group cannot be finite; otherwise it would be one of the (well-known) Weyl groups that do not admit non-trivial invariant lattice elements.

2.4 On proofs of other statements

The proof of Theorem 2.3 consists of an explicit calculation of the monodromy action of the group $\pi_1(P_n \setminus \Sigma, \{X\})$ on the space $\mathcal{H}(X)$: the common orbit of all vanishing cycles which can participate in domains cut by hyperplanes from the body (2.1) consists of exactly four elements. Theorem 2.4 and Remark 2.5 follow from the *local* monodromy theory of isolated function singularities: the violation of either of its two conditions at a point of A implies a logarithmic ramification of the analytic continuation of the volume function in an arbitrary neighborhood of the tangent hyperplane at such a point. Proposition 2.6 is proved in [31] by explicit calculation of integrals.

3 Polynomially integrable convex bodies

In this section we completely characterize infinitely smooth (having infinitely smooth boundary) polynomially integrable bodies; see Definition 1.3. Theorem 3.2 immediately implies that the only such bodies in odd dimensions are ellipsoids, as proved in [23]. On the other hand, Theorem 3.3 generalizes the result from [1] that there are no such bodies in even dimensions.

Let K be an infinitely smooth convex body in $\mathbb{R}^n$ that is polynomially integrable, i. e.,

$$A_K(\xi, t) = \sum_{k=0}^{N} a_k(\xi) t^k$$

for some integer N, all $\xi \in S^{n-1}$, and all t for which the set $K \cap \{x : \langle x, \xi \rangle = t\}$ is non-empty. Here, a_k are functions on the sphere.

Since the function $\xi \to A_K(\xi, t)$ is continuous, all the coefficients $a_k(\xi)$ are continuous functions on S^{n-1}. Without loss of generality we can assume that the origin is an interior point of K, since polynomial integrability is invariant under translations. Observe that for all $k > N$ and all $\xi \in S^{n-1}$ we have

$$\frac{\partial^k}{\partial t^k} A_K(\xi, t) \bigg|_{t=0} = 0.$$

We will use this to conclude that K is an ellipsoid in odd dimensions. First let us show that in the case of centrally symmetric bodies we need much less information.

Theorem 3.1. *Let K be an infinitely smooth origin-symmetric convex body in $\mathbb{R}^n$, where n is odd. Suppose that for some even integer $k > n$ and all $\xi \in S^{n-1}$ we have*

$$\frac{\partial^k}{\partial t^k} A_K(\xi, t) \bigg|_{t=0} = 0 \tag{3.1}$$

and

$$\frac{\partial^{k+2}}{\partial t^{k+2}} A_K(\xi, t) \bigg|_{t=0} = 0. \tag{3.2}$$

Then K is an ellipsoid.

Proof. It is known (see [22, Theorem 3.18]) that the derivatives of $A_K(\xi, t)$ with respect to t at $t = 0$ can be expressed in terms of the Fourier transform of powers of the Minkowski functional of K. Namely, if $k \geq 0$ is an even integer, $k \neq n - 1$, then

$$\frac{\partial^k}{\partial t^k} A_K(\xi, t)\bigg|_{t=0} = \frac{(-1)^{k/2}}{\pi(n - k - 1)} (\|x\|_K^{-n+1+k})^\wedge(\xi), \quad \forall \xi \in S^{n-1}. \tag{3.3}$$

Using condition (3.1) and homogeneity of the Fourier transform of $\|x\|_K^{-n+1+k}$, we get

$$(\|x\|_K^{-n+1+k})^\wedge(\xi) = 0, \quad \forall \xi \in \mathbb{R}^n \setminus \{0\}.$$

It is a well-known fact that a distribution supported at the origin is a linear combination of derivatives of the delta function (see, for example, [26, Theorem 6.25]). Therefore, the Fourier transform of $\|x\|_K^{-n+1+k}$ is a finite linear combination of derivatives of the delta function, implying that $\|x\|_K^{-n+1+k}$ is a polynomial. Denoting $m = -n + 1 + k$, we have

$$\|x\|_K^m = P(x),$$

for some homogeneous polynomial P of even degree m. Similarly, (3.2) implies

$$\|x\|_K^{m+2} = Q(x),$$

where Q is a homogeneous polynomial of degree $m + 2$.

The latter two equations yield $(P(x))^{m+2} = (Q(x))^m$ for all x. Now consider any 2-dimensional subspace H of $\mathbb{R}^n$. The restrictions of P and Q to H are again homogeneous polynomials of degrees m and $m+2$, respectively. Abusing notation, we will denote these restrictions by $P(u, v)$ and $Q(u, v)$, where $(u, v) \in \mathbb{R}^2$. Thus, we have $(P(u, v))^{m+2} = (Q(u, v))^m$ for all $(u, v) \in \mathbb{R}^2$. Since both P and Q are homogeneous, the latter is equivalent to

$$(P(u, 1))^{m+2} = (Q(u, 1))^m, \quad \forall u \in \mathbb{R}.$$

We have the equality of two polynomials of the real variable u; therefore, these polynomials are equal for all $u \in \mathbb{C}$. Let u_0 be a complex root of $P(u, 1)$ of multiplicity $\alpha \leq m$. Then u_0 is also a root of $Q(u, 1)$ of some multiplicity $\beta \leq m + 2$. Hence we have

$$\alpha(m + 2) = \beta m,$$

$$\frac{\alpha}{\beta} = \frac{m}{m + 2}.$$

Recall that m is even, say $m = 2l$, $l \in \mathbb{N}$. Thus,

$$\frac{\alpha}{\beta} = \frac{l}{l + 1}.$$

Since l and $l + 1$ are co-prime, there are only two possibilities for α and β: either $\alpha = l$ and $\beta = l + 1$ or $\alpha = 2l$ and $\beta = 2l + 2$. The latter is impossible since it implies that

$$\|(u, v)\|_{L \cap H}^m = P(u, v) = c(u - vu_0)^m,$$

for some constant c. So the remaining possibility is that $P(u, 1)$ has two complex roots, say a and b of multiplicity l. Therefore,

$$\begin{aligned}
\|(u, 1)\|_{L \cap H}^m &= P(u, 1) \\
&= c\big[(u - a)(u - b)\big]^l = c\big[u^2 - (a + b)u + ab\big]^l \\
&= c\left[u^{2l} - l(a + b)u^{2l-1} + \left(\binom{l}{2}(a + b)^2 + lab\right)u^{2l-2} + \cdots\right].
\end{aligned}$$

Since the restriction of this polynomial to $\mathbb{R}$ has real coefficients, it follows that $a + b$ and ab are real numbers. Since a and b cannot be real, we conclude that they are complex conjugates of each other. Therefore, $\|(u, v)\|_{K \cap H}^2 = \bar{c}[u^2 - (a + b)uv + abv^2]$ is a non-degenerate quadratic form. Thus, $L \cap H$ is an ellipse. Since every 2-dimensional central section of L is an ellipse, L has to be an ellipsoid. The latter is a consequence of the Jordan–von Neumann characterization of inner product spaces by the parallelogram equality; see [21]. $\qquad\square$

Proving this result for non-symmetric bodies is more involved, so we will just provide a sketch of the proof.

Theorem 3.2. *Let n be a positive odd integer and let K be an infinitely smooth convex body in $\mathbb{R}^n$ containing the origin in its interior. Suppose there exists $N \geq n$ such that for every integer $k \geq N$ and every $\xi \in S^{n-1}$ we have*

$$\frac{\partial^k}{\partial t^k} A_K(\xi, t)\bigg|_{t=0} = 0.$$

Then K is an ellipsoid.

Proof. We will use an analog of formula (3.3) for non-symmetric bodies obtained in [27]. If $k \geq 0$ is an even integer, $k \neq n - 1$, then for every $\xi \in S^{n-1}$,

$$\frac{\partial^k}{\partial t^k} A_K(\xi, t)\bigg|_{t=0} = \frac{(-1)^{k/2}}{2\pi(n - k - 1)}\big(\|x\|_K^{-n+1+k} + \|-x\|_K^{-n+1+k}\big)^{\wedge}(\xi), \tag{3.4}$$

and if $k > 0$ is an odd integer, $k \neq n - 1$, then

$$\frac{\partial^k}{\partial t^k} A_K(\xi, t)\bigg|_{t=0} = \frac{i(-1)^{(k-1)/2}}{2\pi(n - k - 1)}\big(\|x\|_K^{-n+1+k} - \|-x\|_K^{-n+1+k}\big)^{\wedge}(\xi). \tag{3.5}$$

Setting the Fourier transforms equal to zero in (3.4) and (3.5) and arguing as in the proof of Theorem 3.1, we get that $\|x\|_K^m + \|-x\|_K^m$ is a polynomial for every even $m \geq N-n+1$, and $\|x\|_K^m - \|-x\|_K^m$ is a polynomial for every odd $m \geq N - n + 1$.

Thus, for any integer $s \geq (N - n + 1)/2$ we have

$$\|x\|_K^{2s+1} - \|-x\|_K^{2s+1} = P(x)$$

and

$$\|x\|_K^{4s+2} + \|-x\|_K^{4s+2} = Q(x),$$

where P and Q are homogeneous polynomials of degrees $2s + 1$ and $4s + 2$, respectively.

Solving the latter system of equations, we get, for every $s \geq (N - n + 1)/2$,

$$\|x\|_K^{2s+1} = \frac{1}{2}\left(P_s(x) + \sqrt{Q_s(x)}\right),$$

where P_s is an odd homogeneous polynomial of degree $2s + 1$ and Q_s is an even homogeneous polynomial of degree $4s + 2$.

Theorem 3.6 from [23] allows to conclude that the Minkowski functional of K is of the form

$$\|x\|_K = R(x) + \sqrt{S(x)},$$

where R is a linear polynomial and S is a positive quadratic polynomial. From this it is easy to see that K is an ellipsoid. Indeed, if $x \in \partial K$, then $\|x\|_K = 1$ and therefore

$$1 - R(x) = \sqrt{S(x)}.$$

Squaring both sides, we get an equation of a quadric surface. Since K is compact, this surface can only be the surface of an ellipsoid. $\qquad\square$

The methods used in this section also allow us to obtain an alternative proof of the result obtained in [1], saying that there are no infinitely smooth polynomially integrable convex bodies in $\mathbb{R}^n$ for even n (see Corollary 4.5). We will prove a little more.

Theorem 3.3. *Let n be a positive even integer. There is no infinitely smooth convex body $K \subset \mathbb{R}^n$ containing the origin in its interior and satisfying*

$$\left.\frac{\partial^m}{\partial t^m} A_K(\xi, t)\right|_{t=0} = 0 \tag{3.6}$$

for some even $m \geq n$ and all $\xi \in S^{n-1}$.

Proof. Assume that there exists an infinitely smooth convex body K in $\mathbb{R}^n$ satisfying (3.6) for some $m \geq n$. Let m be even. Using (3.4) we get

$$\left(\|x\|_K^{-n+1+m} + \|-x\|_K^{-n+1+m}\right)^{\wedge}(\xi) = (-1)^{m/2} 2\pi(n - m - 1) \frac{\partial^m}{\partial t^m} A_K(\xi, t)\Big|_{t=0} = 0,$$

for every $\xi \in S^{n-1}$.

Thus, the Fourier transform of $f(x) = \|x\|_K^{-n+1+m} + \|-x\|_K^{-n+1+m}$ is zero outside of the origin, implying that $f(x)$ can only be a polynomial. This polynomial has to be even, since the function $f(x)$ is even. On the other hand, since $-n + 1 + m$ is an odd number, $f(x)$ has to be an odd polynomial. Thus, $f(x)$ is zero for all $x \in \mathbb{R}^n$, which is impossible. $\qquad\square$

We have just proved that the section function $A_K(\xi, t)$ is never a polynomial with respect to t in even dimensions. However, ellipsoids in even-dimensional spaces have a section function which in a sense is close to a polynomial, namely, this function differs from a polynomial by a simple factor. Indeed, if K is an ellipsoid centered at the origin, then $A_K(\xi, t) = C(\xi)(h_K^2(\xi) - t^2)^{\frac{n-1}{2}}$, where $h_K(\xi)$ is the support function. It follows that if n is even, then $A_K(\xi, t)$ can be represented in two ways:

$$A_K(\xi, t) = C(\xi)\sqrt{h_K^2(\xi) - t^2}\, P(\xi, t) = C(\xi)\frac{P_1(\xi, t)}{\sqrt{h_K^2(\xi) - t^2}},$$

where $P(\xi, t)$ and $P_1(\xi, t)$ are polynomials in t. It was proved in [5] that such a presentation of the section function characterizes ellipsoids in even-dimensional spaces.

Denote by $\mathcal{H}$ the Hilbert transform

$$\mathcal{H}f(t) = \frac{1}{\pi}\text{p.v.}\int_{\mathbb{R}} \frac{f(s)}{t - s}\, ds \tag{3.7}$$

of a continuous function f with sufficiently fast decay at infinity.

The main result of the article [5] is as follows.

Theorem 3.4. *Let n be an even positive integer. Let K be a bounded convex domain in $\mathbb{R}^n$ with C^{∞}-boundary ∂K. The following are equivalent:*
(i) *The section function $A_K(\xi, t)$ has the form*

$$A_K(\xi, t) = \sqrt{q(\xi, t)}\, P(\xi, t),$$

where $P(\xi, t)$ and $q(\xi, t)$ are continuous in ξ and polynomials in t with $\deg q(\xi, \cdot) = 2$, $q(\xi, t) > 0$.
(ii) *The section function $A(\xi, t)$ has the form*

$$A_K(\xi, t) = \frac{P(\xi, t)}{\sqrt{q(\xi, t)}},$$

where $P(\xi, t)$ and $q(\xi, t)$ are as in (i).
(iii) *For every fixed $\xi \in S^{n-1}$, the Hilbert transform of the function $t \to A_K(\xi, t)$ is a polynomial of t.*
(iv) *K is an ellipsoid.*

The equalities for the section function appearing above hold for values of t for which the hyperplane $\langle \xi, x \rangle = t$ meets K.

Note that the appearance of the Hilbert transform in the latter theorem is not very surprising, since $A_K(\xi, t)$ is the Radon transform of the indicator of the body K and the Hilbert transform is involved in the back-projection inversion formula for the Radon transform in even-dimensional spaces.

4 Domains with algebraic Radon transform without real singularities

In this section we consider classes of bodies with algebraic properties more general than polynomial integrability. These classes correspond to a choice of the form of the corresponding defining polynomial Ψ in Definition 1.2.

Classes of algebraic functions are characterized by the form of the corresponding defining polynomial Ψ in Definition 1.2. In the case of polynomially and rationally integrable domains (see the introduction), the defining polynomial $\Psi(\xi, t, w)$ is linear with respect to w and hence does not have multiple roots w. Starting from this observation, we consider equations $Q(\xi, t, w) = 0$ having only simple roots $w \in \mathbb{C}$, for any fixed $\xi \in S^{n-1}$ and for any real t. This means that for any fixed ξ, the algebraic function $w = w(t)$ has no real branching points. The set Br_ξ of branching points is finite for every ξ, but depends on ξ. We will also assume that the union $\bigcup_{\xi \in S^{n-1}} Br_\xi$ of all branching points when ξ runs over the unit sphere. Also it is a bounded set in the complex plane $\mathbb{C}$, i. e., the branching points do not go to infinity when the normal vector ξ runs over the sphere S^{n-1}.

We will formulate the conditions for the defining polynomial $\Psi(\xi, t, w)$ in terms of its discriminant with respect to the variable w:

$$D(\xi, t) = \psi_N(\xi, t)^{2N-1} \prod_{i<j} (w_i(\xi, t) - w_j(\xi, t)),$$

where $N = \deg_w \Psi$, $\psi_N(\xi, t)$ is the leading coefficient of $\Psi(\xi, t, w)$ as a polynomial of w and $w = w_i(\xi, t)$ are the roots of the algebraic equation $\Psi(\xi, t, w) = 0$. The discriminant is a polynomial of the coefficients of the polynomial $w \to \Psi(\xi, t, w)$ and since $\Psi(\xi, t, w)$ is a polynomial in t, the function $D(\xi, t)$ is a polynomial with respect to t.

Definition 4.1. We say that the body K has *algebraic Radon transform without real singularities* if the discriminant $D(\xi, t) \neq 0$ for all $\xi \in S^{n-1}$ and for all real t, and also the leading coefficient of the polynomial $t \to D(\xi, t)$ does not vanish for all $\xi \in S^{n-1}$. The latter means that the null-set $\{\xi \in S^{n-1}, z \in \mathbb{C} : D(\xi, z) = 0\}$ is a compact subset of $S^{n-1} \times \mathbb{C}$.

Let us explain the relation of the condition $d(\xi) \neq 0$, $\xi \in S^{n-1}$, in Definition 4.1 with the location of branching points of the algebraic function $t \to w(\xi, t)$. Since $\Psi(\xi, t, w)$ is continuous with respect to ξ, $D(\xi, t)$ and $d(\xi, t)$ are the same. Therefore, if $d(\xi)$ does not vanish on S^{n-1}, then $|d(\xi)| \geq C > 0$, $|\xi| = 1$. Write

$$D(\xi, t) = d(\xi)t^M + d_1(\xi)t^{M-1} + \cdots + d_M(\xi),$$

where t is complex. Then

$$|D(\xi, t)| \geq |t|^M \left(C - \frac{|d_1(\xi)|}{|t|} - \cdots - \frac{|d_M(\xi)|}{|t|^M} \right).$$

Since the coefficients $d_1(\xi), \ldots, d_M(\xi)$ are continuous and hence bounded on S^{n-1}, there exists $R > 0$ such that $D(\xi, t) \neq 0$ for all $\xi \in S^{n-1}$ and for all complex t with $|t| > R$.

Now, let $|t_0| > R$. Then $D(\xi, t_0) \neq 0$ for any $\xi \in S^{n-1}$. Fix such $\xi \in S^{n-1}$. Then, by definition of the discriminant, all the roots of the polynomial $w \to Q(\xi, t, w)$ are simple when t is close to t_0 and, moreover, they are continuous functions of t in a neighborhood of t_0. Therefore, t_0 is a regular point of the algebraic function $t \to w(\xi, t)$, i.e., $t \notin Br_\xi$. Thus, the union of the sets Br_ξ, $\xi \in S^{n-1}$, is contained in the disc $|t| \leq R$ and is bounded. Therefore, all the branching points (ξ, t) are contained in a compact subset of $S^{n-1} \times \mathbb{C}$.

Examples. If $A_K(\xi, t)$ is a polynomial in t, then the Radon transform of the characteristic function of any polynomially integrable body K has no real singularities since if $A_K(\xi, t) = P(\xi, t)$ is a polynomial in t, then $D(\xi, t) = 1$. Another example is any rationally integrable body K with $A_K(\xi, t) = \frac{P(\xi,t)}{Q(\xi,t)}$ and $Q(\xi, t) \neq 0$ for $(\xi, t) \in S^{n-1} \times \mathbb{R}$. In this case $D(\xi, t) = Q(\xi, t)$ satisfies the above condition.

Theorem 4.2 ([2]). *Let n be an odd integer and let K be a body in $\mathbb{R}^n$ with C^∞-boundary ∂K. Suppose that the Radon transform $A_K(\xi, t)$ of χ_K is an algebraic function, free of real singularities (Definition 4.1). Then ∂K is an ellipsoid. There are no bodies satisfying all those properties if n is even.*

Remark 4.3. In [2], Theorem 4.2 is formulated in terms of the cutoff function $V_K(\xi, t)$.

4.1 Outline of the proof

The idea of the proof is to show that K is polynomially integrable and use the result of [23].

First, we notice that the condition that the polynomial $t \to \Psi(\xi, t, w)$ has no real multiple roots implies that for any fixed ξ the algebraic function $w = w(t)$ defined by the equation $\Psi = 0$ has no branching points for real t. Hence, for every fixed $\xi \in S^{n-1}$, the function $t \to A_K(\xi, t)$ (where t is in an open interval such that $\{\langle \xi, x \rangle = t\}$ meets K) represents a germ of a real analytic branch of the algebraic function $w = w(t)$ and extends to all $t \in \mathbb{R}$ as a real analytic function. However, this situation is impossible when n is odd, since the function $A_K(\xi, t)$ does not extend analytically through the tangent plane at boundary Morse points $a \in \partial K$.

Lemma 4.4. *Let $a \in \partial K$ be a Morse point and let $\langle \xi_0, x \rangle = t_0$ be the tangent plane to ∂K at the point a. Then $A_K(\xi_0, t) = c(t - t_0)^{\frac{n-1}{2}}(1 + o(1))$, $t \to t_0$, $c \neq 0$.*

Corollary 4.5 ([1]). *Let n be even. There is no polynomially integrable body $K \subset \mathbb{R}^n$ with infinitely smooth boundary.*

Indeed, if n is even, then the exponent $\frac{n-1}{2}$ is fractional, and since ∂K contains an open set of Morse points, $A_K(\xi, t)$ cannot be a polynomial with respect to t for all $\xi \in S^{n-1}$. In Theorem 3.3 a different approach to this phenomenon is presented. Notice that both Corollary 4.5 and Theorem 3.3 are consonant with Theorem 2.1, which states that there are no infinitely smooth algebraically integrable bodies in even-dimensional spaces.

The proof of Lemma 4.4 immediately follows from the fact that in a neighborhood of the Morse point a the C^∞-hypersurface ∂K can be represented (after a suitable rotation) as the graph of the function $x_n = a_n + \sum_{j=0}^{n-1} \lambda_j^2 x_j^2 + o(x_1^2 + \cdots + x_{n-1}^2)$. Then $\xi = (0, \ldots, 0, 1)$, the tangent plane is given by $x_n = a_n = t_0$, and $A_K(\xi, t)$ is the volume of the cross-section $\{x_n = t\} \cap K$. It is equal, up to a small term of higher order, to the volume of the ellipsoid $\sum_{j=0}^{n-1} \lambda_j^2 x_j^2 = t - t_0$, which is proportional to $(t - t_0)^{\frac{n-1}{2}}$. Again, if n is even, then $\frac{n-1}{2}$ is non-integer and hence the function $A_K(\xi, a)$ is not real analytic in t near $t = t_0$. It remains to notice that on any smooth closed hypersurface there is an open set of Morse points. This proves that if n is even, then for no body K the conditions of Theorem 4.2 are fulfilled.

From now on, n is odd. Applying a translation, we can assume that 0 is an interior point of the body K. Then for any $\xi \in S^{n-1}$ the function $t \to A_K(\xi, t)$ is well-defined and real analytic in an interval $|t| < \varepsilon$. The key fact is that the Fourier coefficients of $A_K(\xi, t)$ with respect to ξ are polynomials of the variable t.

Lemma 4.6. *Let n be odd. Let $A_K(\xi, t) = \sum_{k=0}^{\infty} \sum_{a=1}^{d_k} p_{k,a}(t) Y_k^{(a)}(\xi)$ be the Fourier decomposition of the function $\xi \to A_K(\xi, t)$, $|t| < \varepsilon$, on the unit sphere. Here $\{Y_k^a\}_{a=1}^{d_k}$ is an orthonormal basis in the space H_k of all spherical harmonics of degree k. Then the Fourier coefficients $p_{k,a}(t)$ are polynomials and $\deg p_{k,a} \leq k + n$.*

Proof. Write the back-projection inversion formula for the Radon transform $A_K(\xi, t) = (\mathcal{R}\chi_K)(\xi, t)$ [19, Chapter 1, Theorem 3.1] in odd-dimensional Euclidean spaces:

$$1 = \chi_K(x) = c \int_{|\xi|=1} \frac{d^{n-1}}{dt^{n-1}} A_K(\xi, \langle x, \xi \rangle) d\xi, \tag{4.1}$$

where $d\xi$ is the normalized Lebesgue measure on S^{n-1}, when $x \in K$. Applying the Laplace operator to the Radon transform results in differentiating twice in t [19, Chapter 1, Lemma 2.1], and hence applying the Laplace operator to both sides yields

$$\int_{|\xi|=1} B(\xi, \langle x, \xi \rangle)d\xi = 0,$$

where we have denoted $B(\xi, t) = \frac{d^{n+1}}{dt^{n+1}} A_K(\xi, t)$.

The function $t \to B(\xi, t)$ is real analytic near $t = 0$:

$$B(\xi, t) = \sum_{j=0}^{\infty} b_j(\xi)t^j, \quad |t| < \varepsilon,$$

and hence

$$\sum_{j=0}^{\infty} \int_{|\xi|=1} b_j(\xi)\langle x, \xi \rangle^j d\xi = 0$$

for $|x| < \varepsilon$. Comparing homogeneous (in x) terms we get

$$\int_{|\xi|=1} b_j(\xi)\langle x, \xi \rangle^j d\xi = 0.$$

The functions $\mathbb{R}^n \ni \xi \to \langle x, \xi \rangle^j$, when x runs over an open neighborhood of $x = 0$, span the space $\mathcal{P}_j$ of all homogeneous polynomials of degree j. Since the restricted space $\mathcal{P}_j|_{S^{n-1}} = \bigoplus_{s=0}^{\frac{j}{2}} H_{j-2s}$, we conclude that each coefficient $b_j(\xi)$ is orthogonal on S^{n-1} to all spherical harmonics of degree $\leq j$, of the same parity as j.

Besides, $A_K(\xi, t) = A_K(-\xi, -t)$ implies that a_j is even when j is even and odd when j is odd. This property is inherited by the functions $b_j(t) = a_j^{(n+1)}(t)$ because $n + 1$ is even. Thus, b_j is orthogonal to all spherical harmonics of the parity which is opposite to that of j. Then b_j is orthogonal on S^{n-1} to all spherical harmonics $Y_k^{(a)}$ of degree $1 \leq k \leq j$, regardless of the parity of j, and hence

$$b_j = \sum_{k \geq j+1} \sum_{a=1}^{d_k} b_{j,k,a} Y_k^a.$$

Then

$$B(\xi, t) = \sum_{j} \sum_{k \geq j+1} \sum_{a=1}^{d_k} b_{j,k,a} Y_k^a(\xi)t^j = \sum_{k} \sum_{a=1}^{d_k} q_{k,a}(t)Y_k^a(\xi),$$

where

$$q_{k,a}(t) = \sum_{j \leq k-1} b_{j,k,a} t^j$$

is a polynomial of degree $\leq k-1$. By the construction, $q_{k,a}(t) = p_{k,a}^{(n+1)}(t)$, and hence $p_{k,a}(t)$ is also a polynomial, of degree $\leq k - 1 + n + 1 = k + n$. $\qquad\square$

The condition for the discriminant of the polynomial Ψ can be rephrased as follows: The projection of the set $\Psi(\xi, t, w) = \partial_w \Psi(\xi, t, w) = 0$ on t is a compact subset of $\mathbb{C}$ disjoint from the real axis. Therefore, if $R > 0$ is sufficiently large, then the circle $C_R = \{|z| = R\}$ encloses all complex t such that for some $\xi \in S^{n-1}$ the polynomial $w \to \Psi(\xi, t, w)$ has a multiple root.

Denote $C_R^{\pm} = [-R, R] \cup (C_R \cap \Pi^{\pm})$, where $\Pi^{\pm}$ stands for upper and lower half-planes. Since the fundamental group of $S^{n-1} \times (C_R^+ \setminus 0)$ is trivial and all w-zeros of $\Psi(\xi, t, w) = 0$, $(\xi, t) \in S^{n-1} \times C_R^+$, are simple, the lemma about covering homotopy [20, Theorem 16.2] implies that there is a leaf $w_\xi^+(t)$ of the multi-valued algebraic function $w = w_\xi(t)$ defined by $\Psi(\xi, t, w)$ which is continuous on $S^{n-1} \times (C_R^+ \setminus 0)$ and coincides with $A_K(\xi, t)$ on $S^{n-1} \times ((-\varepsilon, \varepsilon) \setminus 0)$ (and therefore is continuous on $S^{n-1} \times C_R^+$).

Fix a spherical harmonic Y_k^α. The functions $b_{k,a}^+(t) = \int_{|\xi|=1} Y_k^\alpha w_\xi^+(t) d\xi$ are real analytic in t because $w_\xi^+(t)$ is real analytic by the construction. On $(-\varepsilon, \varepsilon)$, $b_{k,a}^+(t)$ has Fourier coefficients which are polynomials by Lemma 4.6. Therefore, $b_{k,a}^+(t)$ are polynomials and by Cauchy's theorem

$$\int_{C_R^+} b_{k,a}^+(t) dt = 0.$$

This yields

$$\int_{|\xi|=1} \int_{t \in C_R^+} Y_k^\alpha(\xi) w_\xi^+(t) t^m dt d\xi = 0$$

for all $m = 0, 1, \ldots$. Since the harmonic in the integral is arbitrary, we have

$$\int_{C_r^+} w_\xi^+(t) t^m dt = 0.$$

The vanishing complex moments imply that w_ξ^+ is the boundary value of a function, analytic inside the closed contour C_R^+. Similarly, constructing an analytic extension $w_\xi^-(t)$ along $S^{n-1} \times C_R^-$ in the lower half-plane yields that this extension is a boundary value of an analytic function inside the contour C_R^-. Since there are no ramification points outside of the circle C_R, we conclude that $A_K(\xi, t)$ extends to $S^{n-1} \times \mathbb{C}$ as an entire function in t. By the great Picard theorem, entire algebraic functions are polynomials and therefore $A_K(\xi, t)$ is a polynomial in t, i. e., K is polynomially integrable. Then Theorem 1 from [23] implies that K is an ellipsoid.

5 Local polynomial integrability

5.1 Polynomial integrability and finite stationary phase expansion

Recall that the stationary phase method of asymptotic expansion of oscillatory integrals depending on a large parameter is based on the idea that the main contribution in the asymptotic is delivered by critical points of the phase function.

Generally, the expansion is presented as an infinite asymptotic series. However, such an expansion can be finite, i. e., have only a finite number of non-zero terms. This phenomenon is related to the so-called Hamiltonian maps and was studied in [11, 12]. A simple example of finite asymptotic expansion is given by the oscillatory integral on the unit sphere:

$$I(\lambda) = \int_{S^{n-1}} e^{i\lambda x_n} dS(x),$$

where the dimension n is odd. Indeed, integration by parts yields

$$I(\lambda) = c \int_{-1}^{1} e^{i\lambda x_n} (1 - x_n^2)^{\frac{n-3}{2}} dx_n = \sum_{x_n = \pm 1} e^{i\lambda x_n} Q_{x_n}\left(\frac{1}{\lambda}\right),$$

where $Q_{\pm 1}$ are polynomials.

Surprisingly, the polynomial integrability of a body K appears equivalent to the finiteness of the stationary phase expansion for oscillatory Fourier integrals on ∂K, with linear phases.

Proposition 5.1 ([3]). *Let K be a convex body in $\mathbb{R}^n$ with C^∞-boundary. Then K is polynomially integrable if and only if the family of oscillatory integrals $I_\xi(\lambda)$, $|\xi| = 1$,*

$$I_\xi(\lambda) = \int_{\partial K} e^{i\lambda \langle x,\xi \rangle} \langle \xi, n(x) \rangle dS(x),$$

where $n(x)$ is the unit outward normal vector and dS is the area measure on ∂K, possesses a finite stationary phase expansion of the form

$$I_\xi(\lambda) = \sum_{\pm} e^{i\lambda b_\pm(\xi)} Q_{\xi,\pm}\left(\frac{1}{\lambda}\right), \tag{5.1}$$

where $b_-(\xi) = \min_{x \in K} \langle x, \xi \rangle$, $b_+(\xi) = \max_{x \in K} \langle x, \xi \rangle$.

Proof. By the projection-slice theorem [19, Chapter 1, Section 2, formula (4)] the Fourier transform of χ_K equals

$$\hat{\chi}_K(\lambda\xi) = \int_{b_-(\xi)}^{b_+(\xi)} e^{i\lambda t} A_K(\xi, t)\, dt.$$

If $A_K(\xi, t)$ is a polynomial in t, then integration by parts shows that $\hat{\chi}_K(\lambda\xi)$ is represented in the form as in the right-hand side in (5.1). On the other hand, by Stokes' formula,

$$\lambda\hat{\chi}_K(\lambda\xi) = -\lambda^{-1} \int_K \Delta e^{i\langle x,\lambda\xi\rangle}\, dx = -\lambda^{-1} \int_{\partial K} \frac{\partial}{\partial n} e^{i\langle x,\lambda\xi\rangle}\, dS(x)$$

$$= -i \int_{\partial K} e^{i\lambda\langle x,\xi\rangle} \langle \xi, n(x)\rangle\, dS(x) = -i I_\xi(\lambda).$$

This shows that $I_\xi(\lambda)$ has the required form (5.1). The "only if" part is proved by a "reverse" reasoning. □

Remark 5.2. This proposition gives another argument why there are no polynomially integrable bodies in even dimensions. Indeed, if ξ is such that at least one of the points $b_\pm(\xi)$ with the normal vector $\pm\xi$ has non-zero Gaussian curvature, then the leading term of the expansion is known to be $c\lambda^{-\frac{n-1}{2}}$, and if n is even, then the expansion cannot have the form (5.1) because the exponent $\frac{n-1}{2}$ is non-integer.

5.2 Characterization of locally polynomially integrable surfaces

The relation with stationary phase expansion can be exploited for the study of a local version of the polynomial integrability property. Given a smooth strictly convex hypersurface $M \subset \mathbb{R}^n$, denote by $T_a(M)$ the tangent plane at the point $a \in M$ and by v_a the unit normal vector at a directed "inside" M, i. e., M lies in the half-space $T_a + \mathbb{R}_+ v_a$.

Definition 5.3. Let M be a smooth strictly convex hypersurface in $\mathbb{R}^n$. We say that M is locally polynomially integrable if the $(n-1)$-dimensional volume $A(a, t)$ of the $(n-1)$-dimensional domain in $T_a(M) + t v_a$, bounded by $M \cap (T_a(M) + t v_a)$, where $t > 0$ is sufficiently small, polynomially depends on t.

Examples. All strictly convex quadrics in $\mathbb{R}^{2k+1}$, i. e., ellipsoids, elliptic paraboloids, and single sheets of two-sheet hyperboloids, are locally polynomially integrable.

The same argument as in the case of global polynomial integrability shows that there are no locally polynomially integrable surfaces in even dimensions.

Conjecture 5.4. *All smooth polynomially integrable hypersurfaces are (strictly convex) quadrics in* $\mathbb{R}^{2k+1}$.

Expressing polynomial integrability in terms of the stationary phase expansion appeared useful in proving results toward this conjecture. Lemma 5.5, a local version of

Proposition 5.1, expresses the function $v(a, t)$ in terms of an asymptotic expansion of an oscillatory integral in a neighborhood of a with a linear phase with the only critical point a. Since the construction is local, we have to use a cutoff function. As in Proposition 5.1, the essential part of the expansion is finite; however, the representation now is not exact due to the presence of a remainder of fast decay with respect to the large parameter λ.

For a fixed $a \in M$ consider the oscillatory integral

$$I_a(\lambda) := \int_M \frac{\partial}{\partial v_x}[e^{i\lambda\langle x, v_a\rangle}]\rho_a(\langle x, v_x\rangle)\,dS(x),$$

where $\rho_a(u)$ is an infinitely smooth function of one variable, supported in a sufficiently small interval $(-\varepsilon_a, \varepsilon_a)$, $\rho_a(0) = 1$. The point a is a critical point of the phase function and there are no other critical points if ε is small.

Lemma 5.5. *The volume function $A(a, t)$ is a polynomial in t if and only if*

$$I_a(\lambda) = ce^{i\langle a, v_a\rangle}Q_a\left(\frac{1}{\lambda}\right) + o\left(\frac{1}{\lambda^m}\right), \quad \lambda \to \infty, \tag{5.2}$$

where Q_a is a polynomial and $m \in \mathbb{N}$ is arbitrary.

Recall that a point $a \in M$ is *elliptic* if all the principal curvatures at a are positive. Near such a point, M can be represented, after a suitable translation and rotation, as the graph $x_n = f(x_1, \ldots, x_{n-1})$, where the second differential is of the form $d_a^2 f(h) = \sum_{j=1}^{n-1} \gamma_j h_j^2$, $\gamma_j > 0$. We say that a is an *elliptic point of contact of order q* if $d^3 f_a = \cdots = d^q f_a = 0$.

The following theorem is a partial result towards Conjecture 5.4.

Theorem 5.6 ([3]). *Let M be a real analytic locally polynomially integrable hypersurface in $\mathbb{R}^n, n = 2k + 1$. Suppose that M contains an elliptic point $a_0 \in M$ of contact of order $q > 4$. Then M is an elliptic paraboloid.*

Ellipsoids and two-sheet hyperboloids do not satisfy the condition about an elliptic point of a high order and hence are out of consideration in Theorem 5.6. Getting rid of this condition (as well as the condition of real analyticity) would allow to obtain a full description of locally polynomially hypersurfaces.

The proof is based on a variation of the expansion in Lemma 5.5 with respect to a in a neighborhood of the elliptic point a_0. More precisely, we represent M near the point a_0 as the graph $x_n = f(x_1, \ldots, x_{n-1})$ and then pass to Morse coordinates u in which f has the form $f(u) = u_1^2 + \cdots + u_{n-1}^2$. Then we come up with an oscillatory integral with a quadratic phase function. We parametrize the integral $I_a(\lambda)$ by the normal vector $\xi = v_a$ and then differentiate in ξ, applying, at the point a_0, to (5.2) the Schrödinger operator $\square = i\lambda\frac{\partial}{\partial \xi_n} - \Delta'$, where Δ' is the Laplace operator in $\xi_1, \ldots, \xi_{n-1}$. The condition of the high order of tangency at a_0 yields that this operation results in reducing the degree of the polynomial Q in (5.2) and shortening the length of the expansion. Then, by applying the iterated operator $\square^k$ with sufficiently large k, we annihilate the essential part of the expansion. Furthermore, the coefficients of the stationary phase expansion with quadratic phase are expressed as powers of the Laplace operator of the density function at the point a, and in our case all of them are zero. Being translated in terms of the Morse diffeomorphism $u = u(x)$, the vanishing powers of the Laplace operator imply that the mapping $u(x)$ is affine. This in turn implies that f is a quadratic polynomial and, correspondingly, M is an elliptic paraboloid.

6 Domains with algebraic X-ray transform

As mentioned before, polynomial integrability never occurs in even dimensions. However, the Radon transforms of the indicator functions of ellipsoids in even dimensions are, in a sense, close to being polynomials. Namely, the sectional volume function $A_E(\xi, t)$, where E is an ellipsoid, is a *square root of a polynomial* in t. Thus, in both even and odd dimensions, $A_E^2(\xi, t)$ is a polynomial.

Conjecture 6.1. *Let K be a body in $\mathbb{R}^n$ with C^∞-boundary. Suppose that there exists $m \in \mathbb{N}$ such that $A_K^m(\xi, t)$ is a polynomial in t. Then ∂K is an ellipsoid and therefore we can take $m = 1$ if n is odd and $m = 2$ if n is even.*

The following theorem confirms Conjecture 6.1 for $n = 2$ and for domains K with algebraic boundary ∂K. This means that ∂K is contained in the zero set of a non-zero polynomial. In the 2-dimensional case the function $A_K(\xi, t)$ is the X-ray transform of the characteristic function χ_K and evaluates the length of the chord $K \cap (t\xi + \xi^\perp)$. This chord intersects K if and only if $A_K(\xi, t) \neq 0$.

Theorem 6.2 ([4]). *Let K be a domain in $\mathbb{R}^2$ with C^∞-algebraic boundary. If the chord length function $A_K(\xi, t)$ has the form $A_K(\xi, t) = \sqrt[m]{P_\xi(t)}$ (as long as $A_K(\xi, t) \neq 0$), where P_ξ is a polynomial, then ∂K is an ellipse.*

Here we present a sketch of the proof. The first step is to understand the degree of the polynomial P_ξ for a generic direction ξ by establishing lower and upper bounds for it.

The lower bound relies on the fact that at the Morse points $a \in \partial K$ with the unit normal ν_a the chord length function $A_K(\nu_a, t)$ behaves, near the tangent lines to the boundary, as $const\,(h_K^+(\nu_a) - t)^{\frac{1}{2}}$, where $h_K^+(\xi) = \sup_{x \in K}\langle x, \xi\rangle$ is the support function. There are two tangent lines orthogonal to a given unit vector ξ, namely, $\langle x, \xi\rangle = h_K^+(\xi)$ and $\langle x, \xi\rangle = h_K^-(\xi)$, where $h_K^-(\xi) = \inf_{x \in K}\langle x, \xi\rangle = -h_K^+(-\xi)$. Therefore, $P_\xi(t) = A_K^m(\xi, t)$ vanishes at the points $t_\pm = h_K^\pm(\pm\xi)$, to the order $\frac{m}{2}$ at each. Thus, $\deg P_\xi \geq m$ when ξ is a normal vector at Morse points. Since Morse points are dense on the boundary, the estimate holds everywhere, by continuity.

The upper bound for $\deg P_\xi$ can be obtained from the estimate of the growth of $A_K(\xi, t)$ as $t \to \infty$. However, moving t in the real direction does not help since when

t is so large that the line $t\xi + \xi^\perp$ does not meet K, then $A_K(\xi, t) = 0$ and the relation between $A_K(\xi, t)$ and $P_\xi(t)$ is lost. The idea is to move t to infinity in a complex direction. To this end, we need to analytically extend the chord length function $A_K(\xi, t)$ into the complex plane. By the uniqueness theorem, this extension must coincide with the algebraic function $\sqrt[m]{P_\xi(t)}$. The analytic extension of $A_K(\xi, t)$ relies on our assumption that ∂K is algebraic and hence is a trace on $\mathbb{R}^2$ of a complex algebraic curve in $\mathbb{C}^2$. Then we construct a path connecting a real t and ∞, providing a single-valued analytic extension $A_K(\xi, t)$ along it, and prove that the order of growth of $A_K(\xi, t)$ along this path is at most one. This results in the estimate $\deg P_\xi(t) \le m$. Thus, for generic ξ the degree of the polynomial P_ξ is exactly m and therefore $P_\xi(t)$ has the only zeros $t_\pm = h_K^\pm(\xi)$, each of multiplicity $\frac{m}{2}$. Therefore, we have $P_\xi(t) = c(\xi)(h_K^+(\xi) - t)^{\frac{m}{2}}(t - h_K^-(\xi))^{\frac{m}{2}}$.

The second step relies on using the range description of the X-ray transform $A_K(\xi, t)$ (see, e. g., [19, Chapter 1, Theorem 2.4]). This yields that the moment of degree s

$$M_s(\xi) = \int_{\mathbb{R}} A_K(\xi, t)t^s dt = c(\xi) \int_{h_K^-(\xi)}^{h_K^+(\xi)} (h_K^+(\xi) - t)^{\frac{1}{2}}(t - h_K^-(\xi))^{\frac{1}{2}} t^s dt$$

is the restriction to the circle $|\xi| = 1$ of a homogeneous polynomial of degree s. It is not difficult to obtain from here, using only the moments $M_s(\xi)$ of degrees $s = 0, 1, 2$, that, after a suitable translation of K, the support function $h_K^+(\xi)$ is the restriction to $|\xi| = 1$ of the square root of a quadratic polynomial. This yields that ∂K is an ellipse.

Remark 6.3. The obtained result can be interpreted as follows. While Newton's lemma about ovals [8] states that the cutoff function $V_K^\pm(\xi, t)$ of any domain $K \subset \mathbb{R}^2$ (with C^∞-boundary) is a transcendental function, Theorem 6.2 specifies that among those transcendental functions, the Abel integrals $\int \sqrt[m]{P_\xi(t)}dt$ (P_ξ is a polynomial) detect ellipses.

7 Radon transforms supported in hypersurfaces

7.1 The interior problem in tomography

In odd dimensions the inversion formula for the Radon transform is local in the sense that the value at x of the function f can be computed from the restriction of the Radon transform $\mathcal{R}f(L) = \int_L f\, dx$ (dx denotes the surface measure on L) to the set of hyperplanes L that intersect an arbitrarily small neighborhood of the point x. As is well known, this is not the case in even dimensions. On the contrary, the so-called "interior problem" for the even-dimensional Radon transform is not solvable. In the 2-dimensional case this statement is usually understood to mean the following. Let D be an open disc in the plane and let K be a closed subset of D. Then a function f supported in $\overline{D}$, the closure of D, cannot be determined anywhere in K from the knowledge of its integrals over all lines that intersect K. The proof of this statement runs as follows; see, e. g., [24]. Choose coordinates so that D is a unit disc centered at the origin and let D_0 be a smaller concentric disc such that $K \subset D_0$. Then look for radial functions $f(r)$, $r = |x|$ that are supported in $r \leq 1$ and satisfy $\mathcal{R}f(\xi, p) = \mathcal{R}f(p) = g(p) = 0$ for $|p| \leq p_0 < 1$. This leads to an Abel integral equation in which $f(r)$ can be solved in terms of $g(p)$, and the support of the solution f is in general equal to $\bar{D}$. By an affine transformation we see that the same is true if the disc D is replaced by the region bounded by an ellipse. However, if D is an arbitrary bounded, convex domain, this argument does not work, and as far as we know it is not known if the corresponding statement is true.

Conjecture 7.1. *Let D be an open, bounded, convex domain in the plane and let K be a closed subset of D. Then there exists a function $f \in C_c^\infty(\bar{D})$, not identically zero in K, such that its Radon transform $\mathcal{R}f(L)$ vanishes for all lines L that intersect K.*

What is the connection between the conjecture and Radon transforms supported in hypersurfaces? Let us explain. Denote by f_0 the function $\pi^{-1}(1 - |x|^2)_+^{-1/2}$, supported in the unit disc. An easy calculation shows that $\mathcal{R}f_0(\xi, p) = 1$ for $|p| < 1$ and all ξ, and obviously $\mathcal{R}f_0 = 0$ if $|p| > 1$. Hence the distribution $\partial_p \mathcal{R}f_0$ is supported in the set of (ξ, p) with $|p| = 1$, which corresponds to the set of lines that are tangent to the circle, and the same is true for $\partial_p^2 \mathcal{R}f_0$. Using the formula

$$(\mathcal{R}\Delta h)(\xi, p) = \partial_p^2 \mathcal{R}h(\xi, p), \tag{7.1}$$

we can now conclude that the Radon transform of the distribution $f = \Delta f_0$ is supported on the set of tangents to the unit circle (Δ is the Laplace operator). An affine transformation gives of course similar examples where the disc is replaced by an elliptic region. A couple of years ago one of us (J. B.) got the idea to prove Conjecture 7.1 by constructing analogous examples where D is replaced by an arbitrary convex region.

To be specific, assume for simplicity that D is symmetric, $D = -D$. Choose an open, convex region D_0 with smooth boundary such that $D_0 = -D_0$, $K \subset D_0$, and the closure $\bar{D}_0$ is contained in D. Then try to find a distribution f, supported in $\bar{D}_0$, such that its Radon transform in the distributional sense, $\mathcal{R}f = g$, is supported on the set of tangent lines to the boundary ∂D_0. A natural way to do this is to set

$$g(\xi, p) = q(\xi)\delta(p - h_{D_0}(\xi)) + q(\xi)\delta(p + h_{D_0}(\xi)),$$

where $h_{D_0}(\xi)$ is the support function for D_0,

$$h_{D_0}(\xi) = \sup\{\langle x, \xi\rangle; \ x \in D_0\}, \quad \xi \in S^1,$$

and $\delta(\cdot)$ is the Dirac measure at the origin, and try to choose the density function $q(\xi)$ so that g is equal to $\mathcal{R}f$ for some compactly supported distribution f. As is well known, the condition for $g(\xi, p)$ to be in the range of $\mathcal{R}$ is that g satisfies the moment conditions

$$\xi = (\xi_1, \xi_2) \mapsto \int_{\mathbb{R}} g(\xi, p) p^k \, dp$$

is a homogeneous polynomial in (ξ_1, ξ_2) of degree k for every $k \geq 0$.

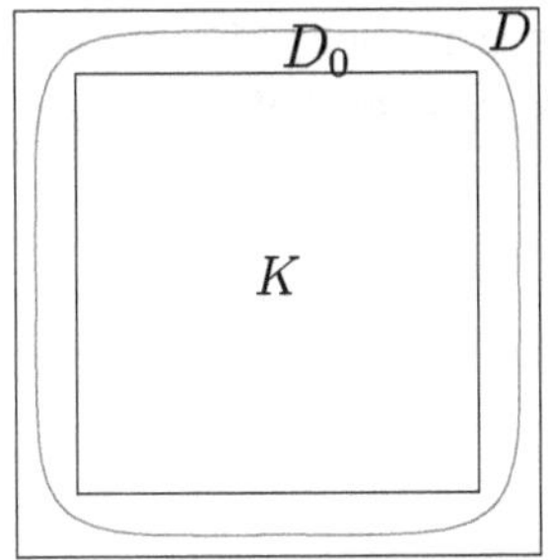

The problem then becomes to determine the function $q(\xi)$ on the circle such that the range conditions are fulfilled. When f is found, we could regularize f by convolving it with a smooth function $\phi(x)$ supported in a small neighborhood of the origin, $f_1 = \phi * f$. If the support of ϕ is sufficiently small, then the support of f_1 will be contained in D and $\mathcal{R}f_1(L) = \int_L f_1 dx$ will vanish for all lines that intersect K as desired. This idea reduces the problem to a one-variable problem just as the rotational symmetry did in the problem with two concentric discs discussed earlier.

However, surprisingly such distributions f can exist only if D_0 is an ellipse. In fact, the following theorem holds [13, 14].

Theorem 7.2. *Let D be an open, convex, bounded subset of $\mathbb{R}^n$ with boundary ∂D. Assume that there exists a distribution f, supported in $\overline{D}$, such that the Radon transform of f is supported in the set of supporting planes to D. Then the boundary of D is an ellipsoid.*

The next surprise was that this theorem gave a new proof of the result from [23] which solved a special case of Arnold's problem, as described in Section 3. In fact, assume that the domain D is polynomially integrable in the sense of Definition 1.3, and let χ_D be the characteristic function of D. The assumption that D is polynomially integrable means that for every ξ the function $p \mapsto \mathcal{R}\chi_D(\xi, p)$ is a polynomial function in the interval consisting of those p for which the line $L(\xi, p) = \{x \in \mathbb{R}^n; \langle x, \xi \rangle = p\}$ intersects D. Choose the integer s so large that $2s$ is greater than the degrees of all those polynomials. Using repeatedly formula (7.1) with $h = \chi_D$ we see that the Radon transform of the distribution $\Delta^s \chi_D$ must vanish in the open set of lines that intersect D, as well as of course in the open set of lines that are disjoint from $\overline{D}$. Hence $\mathcal{R}(\Delta^s f)$ is supported on the set of support planes to the boundary ∂D of D. But by Theorem 7.2 such distributions can exist only if ∂D is an ellipse.

There is no smoothness assumption on the boundary ∂D in Theorem 7.2; therefore, we have used the term supporting plane (to D) instead of tangent plane (to ∂D). The assumption that f is supported in $\overline{D}$ can be weakened to f being compactly supported in $\mathbb{R}^n$, because Helgason's support theorem [19, Corollary 2.8] shows that f must vanish in the complement of the convex compact set $\overline{D}$ if f is compactly supported and the Radon transform $\mathcal{R}f(L)$ vanishes for all hyperplanes L that are disjoint from $\overline{D}$.

7.2 On the proof of Theorem 7.2

Here we will prove Theorem 7.2 for the special case when the Radon transform of f is a distribution of order zero, and we will briefly indicate how the arguments given here can be modified to cover the general case.

Let us begin by writing down an expression for an arbitrary distribution $g(\xi, p)$ of order zero on the manifold $\mathcal{P}_n$ of hyperplanes in $\mathbb{R}^n$ that is supported on the set of supporting planes to D.

Since $L(\xi, p)$ and $L(-\xi, -p)$ are the same hyperplane, the distribution $g(\xi, p)$ must be even, $g(\xi, p) = g(-\xi, -p)$. A hyperplane $L(\xi, p)$ is a supporting plane for D if and only if

$$p = h_D(\xi) \quad \text{or} \quad p = \inf\{\langle x, \xi \rangle; \ x \in D\} = -\sup\{-\langle x, \xi \rangle; \ x \in D\} = -h_D(-\xi).$$

An arbitrary distribution $g(\xi, p)$ on $S^{n-1} \times \mathbb{R}$ of order zero that is supported on ∂D can therefore be written

$$g(\xi, p) = q_+(\xi)\delta(p - h_D(\xi)) + q_-(\xi)\delta(p + h_D(-\xi))$$

for some functions or distributions $q_+(\xi)$ and $q_-(\xi)$. Since $\delta(t) = \delta(-t)$ we then have

$$g(-\xi, -p) = q_+(-\xi)\delta(-p - h_D(-\xi)) + q_-(-\xi)\delta(-p + h_D(\xi))$$
$$= q_+(-\xi)\delta(p + h_D(-\xi)) + q_-(-\xi)\delta(p - h_D(\xi)).$$

The condition for $g(\xi, p)$ to be even therefore becomes

$$q_-(-\xi) = q_+(\xi) \quad \text{for all } \xi.$$

Hence it is sufficient to introduce one density function, say $q_+(\xi) = q(\xi)$, because then $q_-(\xi) = q(-\xi)$ (actually $q(\xi)$ must be a continuous function [14, Lemma 2.1]). We conclude that an arbitrary distribution of order zero on the manifold $\mathcal{P}_n$ that is supported on the set of supporting planes to D can be represented as

$$g(\xi, p) = q(\xi)\delta(p - h_D(\xi)) + q(-\xi)\delta(p + h_D(-\xi)).$$

Observing that $\int_{\mathbb{R}} \delta(p \pm h_D(\xi))p^k \, dp = (\mp h_D(\xi))^k$ we conclude that $q(\xi)$ and $h_D(\xi)$ must satisfy the infinitely many equations

$$\begin{aligned} q(\xi) + q(-\xi) &= p_0(\xi), \\ q(\xi)h_D(\xi) - q(-\xi)h_D(-\xi) &= p_1(\xi), \\ q(\xi)h_D(\xi)^2 + q(-\xi)h_D(-\xi)^2 &= p_2(\xi), \\ q(\xi)h_D(\xi)^3 - q(-\xi)h_D(-\xi)^3 &= p_3(\xi), \\ &\vdots \end{aligned} \tag{7.2}$$

for some homogeneous polynomials $p_k(\xi)$ in $\xi = (\xi_1, \ldots, \xi_n)$ of degree k for each k. We have to prove that those identities imply that the boundary of D is an ellipsoid.

If D is a ball centered at the origin, then it is clear that $h_D(\xi) = c|\xi|$ for all $\xi \in S^{n-1}$ and some constant c, and hence $h_D(\xi)^2 = c^2|\xi|^2$, which is a homogeneous quadratic polynomial. It follows from the definition of $h_D(\xi)$ that a linear transformation A that transforms D to $\tilde{D} = AD = \{Ax; x \in D\}$ transforms $h_D(\xi)$ to $h_{\tilde{D}}(\xi) = h_D(A^*\xi)$. Hence, $h_D(\xi)^2$ must be a homogeneous quadratic polynomial whenever $D = -D$ and the boundary of D is an ellipsoid. This argument can obviously be reversed. Hence we conclude that for bounded convex domains with $D = -D$ we have

the boundary of D is an ellipsoid if and only if

$h_D(\xi)^2$ is a homogeneous quadratic polynomial.

$$\tag{7.3}$$

These observations make it possible to give a very short proof of the special case of Theorem 7.2 when D is symmetric and the distribution $g(\xi, p)$ has order zero. Indeed, since $h_D(\xi)$ must be even, $h_D(\xi) = h_D(-\xi)$, we get from the first and third equations in (7.2)

$$p_2(\xi) = (q(\xi) + q(-\xi))h_D(\xi)^2 = p_0(\xi)h_D(\xi)^2.$$

But $p_0(\xi) = q(\xi) + q(-\xi)$ must be equal to some constant c. If $c \neq 0$, then the fact that $p_2(\xi)$ is a quadratic polynomial proves the assertion. If $c = 0$, then $q(\xi)$ is odd, so

$$q(\xi)h_D(\xi) - q(-\xi)h_D(\xi) = 2q(\xi)h_D(\xi) = p_1(\xi)$$

must be linear in ξ, and

$$p_3(\xi) = q(\xi)h_D(\xi)^3 - q(-\xi)h_D(-\xi)^3 = 2q(\xi)h_D(\xi)^3$$

must be a homogeneous polynomial of degree 3. Combining the last two equations we can write

$$p_3(\xi) = h_D(\xi)^2 p_1(\xi).$$

Since $h_D(\xi)$ is bounded, it follows that $p_3(\xi)$ must be divisible (in the polynomial ring) by $p_1(\xi)$, and hence $h_D(\xi)^2$ must be a quadratic polynomial, which completes the proof of Theorem 7.2 in this case.

The case when the domain D is not necessarily symmetric is somewhat more complicated. We then have to consider the condition

$$h_D(\xi)h_D(-\xi) \quad \text{is a polynomial}$$

instead of the condition that $h_D(\xi)^2$ is a polynomial. Note that support functions are (positively) homogeneous of degree 1, so if the function $h_D(\xi)h_D(-\xi)$ is a polynomial, it must be a homogeneous quadratic polynomial. Let $D_a = D + a$ with $a \in \mathbb{R}^n$ be the translated domain, and note that $h_{D_a}(\xi) = h_D(\xi) + \langle a, \xi \rangle$. If $D = -D$ and $h_D(\xi)^2$ is a polynomial, then $h_{D_a}(\xi)^2$ is in general not a polynomial, but

$$h_{D_a}(\xi) h_{D_a}(-\xi) = (h_D(\xi) + \langle \xi, a \rangle)(h_D(\xi) - \langle \xi, a \rangle) = h_D(\xi)^2 - \langle \xi, a \rangle^2$$

is a homogeneous quadratic polynomial for every $a \in \mathbb{R}^n$. This observation has the following important converse.

Proposition 7.3. *Assume that D is a convex, bounded domain for which the product $h_{D_a}(\xi) h_{D_a}(-\xi)$ is a homogeneous quadratic polynomial for every $a \in \mathbb{R}^n$. Then the boundary of D is an ellipsoid.*[1]

Proof. Take an arbitrary $a \in \mathbb{R}^n \setminus \{0\}$, for instance $a = (1, 0, \ldots, 0)$. Then $\langle a, \xi \rangle = \xi_1$. The assumption implies that

$$h_{D_a}(\xi) h_{D_a}(-\xi) - h_D(\xi) h_D(-\xi) + \xi_1^2$$
$$= (h_D(\xi) + \xi_1)(h_D(-\xi) - \xi_1) - h_D(\xi) h_D(-\xi) + \xi_1^2$$
$$= \xi_1(h_D(-\xi) - h_D(\xi))$$

is equal to a homogeneous quadratic polynomial $p_2(\xi)$. This implies that the polynomial $p_2(\xi)$ is divisible by the linear factor ξ_1 and hence the quotient must be another linear factor, so

$$h_D(\xi) - h_D(-\xi) = -2\langle b, \xi \rangle$$

for some $b \in \mathbb{R}^n$. But this implies

$$h_{D_b}(\xi) - h_{D_b}(-\xi) = h_D(\xi) + \langle b, \xi \rangle - (h_D(-\xi) - \langle b, \xi \rangle)$$
$$= h_D(\xi) - h_D(-\xi) + 2\langle b, \xi \rangle = -2\langle b, \xi \rangle + 2\langle b, \xi \rangle = 0,$$

which shows that $h_{D_b}(\xi)$ is even. Since D, and hence D_b, is convex, D_b is uniquely determined by its support function, and it follows that D_b is symmetric with respect to the origin. By the assumption $h_{D_b}(\xi) h_{D_b}(-\xi)$ is a homogeneous quadratic polynomial, so $h_{D_b}(\xi)^2$ has the same property, and by (7.3) this implies that the boundary of D_b is an ellipsoid, and hence so is the boundary of D. $\qquad\square$

We can now complete the proof of Theorem 7.2 for the case when $g(\xi, p)$ is a distribution of order zero. However, instead of writing the proof as short as possible we have chosen to present the calculations in a way that rather easily can be generalized to the case when g is a distribution of higher order. We have to show that the equations in (7.2) imply that the boundary of D is an ellipsoid.

---

1 Section 6 in [14] could have been omitted if we had known this fact at the time.

To shorten formulas we will write

$$h_D(\xi) = h, \quad h_D(-\xi) = \check{h} \quad \text{and} \quad q(\xi) = q, \quad q(-\xi) = \check{q}.$$

The infinite system (7.2) can then be written in matrix form:

$$\begin{pmatrix} 1 & 1 \\ h & -\check{h} \\ h^2 & \check{h}^2 \\ h^3 & -\check{h}^3 \\ \cdots & \cdots \end{pmatrix} \begin{pmatrix} q \\ \check{q} \end{pmatrix} = \begin{pmatrix} p_0 \\ p_1 \\ p_2 \\ p_3 \\ \cdots \end{pmatrix}. \tag{7.4}$$

Denote the sequence of 2×2 submatrices of the big matrix to the left by M_0, M_1, etc. Introduce the column vectors

$$Q = \begin{pmatrix} q \\ \check{q} \end{pmatrix}, \quad P_0 = \begin{pmatrix} p_0 \\ p_1 \end{pmatrix}, \quad P_1 = \begin{pmatrix} p_1 \\ p_2 \end{pmatrix}, \quad P_2 = \begin{pmatrix} p_2 \\ p_3 \end{pmatrix}, \quad \text{etc.} \tag{7.5}$$

Then we have the equations $M_0 Q = P_0$, $M_1 Q = P_1$, etc., and more generally

$$M_k Q = P_k \quad \text{for all } k \geq 0. \tag{7.6}$$

The matrices M_k form a geometric series in the sense that

$$M_{k+1} = S M_k = M_k T \quad \text{for all } k \geq 0 \tag{7.7}$$

with

$$S = \begin{pmatrix} 0 & 1 \\ h\check{h} & h - \check{h} \end{pmatrix} \quad \text{and} \quad T = \begin{pmatrix} h & 0 \\ 0 & -\check{h} \end{pmatrix}.$$

This makes it easy to eliminate Q from the system (7.4). In fact,

$$SP_0 = SM_0 Q = M_1 Q = P_1, \tag{7.8}$$

and similarly

$$SP_k = P_{k+1} \quad \text{for all } k \geq 0. \tag{7.9}$$

Viewing a row of columns as a matrix we can then form matrix identities by combining pairs of equations in (7.9) as follows:

$$S(P_0, P_1) = (P_1, P_2), \tag{7.10}$$

and more generally

$$S^k(P_0, P_1) = (P_k, P_{k+1}) \quad \text{for all } k \geq 0. \tag{7.11}$$

Using the product law for determinants in (7.10) we see that $h\check{h} = -\det S$ must be a rational function, provided

$$\det(P_0, P_1) = p_0 p_2 - p_1^2 \neq 0 \tag{7.12}$$

as a polynomial. Assuming (7.12) for a moment we prove that $h\check{h} = -\det S$ must be a polynomial by applying (7.11) for sufficiently large k instead of (7.10), which leads to a contradiction unless $h\check{h}$ is a polynomial. To prove (7.12) we first note that the following identity of matrices is valid:

$$(P_0, P_1) = (M_0 Q, M_1 Q) = (M_0 Q, M_0 TQ) = M_0(Q, TQ). \tag{7.13}$$

The matrix M_0 is non-singular for all ξ, since

$$\det M_0 = -(h + \check{h}),$$

$h(\xi)$ is strictly positive for all ξ if the origin is contained in D, and the expression $h(\xi) + h(-\xi)$ is invariant under translations of the coordinate system. Moreover,

$$\det(Q, TQ) = \det \begin{pmatrix} q & hq \\ \check{q} & -\check{h}\check{q} \end{pmatrix} = -q\check{q}(h + \check{h}). \tag{7.14}$$

It remains to show that $q\check{q}$ cannot be identically zero.

Lemma 7.4. *Assume that (7.2) holds and that $q(\xi) \neq 0$ for some ξ. Then $q(\xi)q(-\xi) \neq 0$ for some ξ.*

Proof. Solving q and $\check{q}$ from the first two equations in the system (7.4) gives

$$q = (p_0\check{h} + p_1)/(h + \check{h}), \quad \check{q} = (p_0 h - p_1)/(h + \check{h}). \tag{7.15}$$

Since q is assumed not to vanish identically, already the first equation of (7.15) shows that p_0 and p_1 cannot both vanish identically. Choosing the coordinate system so that the origin is in D, we have $h = h_D(\xi) > 0$ for all ξ and hence $h\check{h} > 0$ for all ξ. A translation of the coordinate system does not change $q(\xi)$. Recall that p_0 is constant and that p_1 is a homogeneous first-degree polynomial with real coefficients. Choose ξ^0 so that $p_1(\xi^0) = 0$. Then

$$q\check{q} = p_0^2 h\check{h}/(h + \check{h})^2 > 0 \quad \text{at } \xi^0$$

if $p_0 \neq 0$. On the other hand, if $p_0 = 0$, then by (7.15) we have $q\check{q} = -p_1^2/(h + \check{h})^2$, which cannot be identically zero. $\qquad\square$

By means of Lemma 7.4 together with (7.13) and (7.14) we can now conclude that the polynomial $\det(P_0, P_1)$ cannot be identically zero, and hence by (7.10) we conclude that $\det S = -h\check{h}$ must be a rational function. As already mentioned, using (7.11) for arbitrarily large k we now obtain a contradiction unless $h\check{h}$ is a polynomial. The same must be true if the domain D is replaced by the translate $D + a$ for arbitrary $a \in \mathbb{R}^n$, because

$$\int g(\xi, p)(p + \langle a, \xi \rangle)^k \, dp = \sum_{j=0}^{k} \binom{k}{j} \int g(\xi, p) p^j \, dp \, \langle a, \xi \rangle^{k-j}$$

$$= \sum_{j=0}^{k} \binom{k}{j} p_j(\xi) \langle a, \xi \rangle^{k-j},$$

and each term in the last expression is clearly a homogeneous polynomial of degree k. An application of Proposition 7.3 now finishes the proof of Theorem 7.2 for the case when $g(\xi, p)$ is a distribution of order zero.

We now sketch the proof of Theorem 7.2 for the case when $g(\xi, p)$ is a distribution of order at most 1. An arbitrary even distribution $g(\xi, p)$ on $S^{n-1} \times \mathbb{R}$ of order at most 1 that is supported on

$$\{(\xi, p);\ p = h_D(\xi)\} \cup \{(\xi, p);\ p = -h_D(-\xi)\}$$

can be written $g = g_0 + g_1$, where

$$g_0(\xi, p) = q_0(\xi)\delta(p - h_D(\xi)) + q_0(-\xi)\delta(p + h_D(-\xi)),$$
$$g_1(\xi, p) = q_1(\xi)\delta'(p - h_D(\xi)) - q_1(-\xi)\delta'(p + h_D(-\xi)).$$

The minus sign between the terms in the expression for g_1 is needed to make $g_1(\xi, p)$ even, because $\delta'(\cdot)$ is odd. The matrix form of the system analogous to (7.4) then becomes

$$\begin{pmatrix} 1 & 0 & 1 & 0 \\ h & -1 & -\check{h} & 1 \\ h^2 & -2h & \check{h}^2 & -2\check{h} \\ h^3 & -3h^2 & -\check{h}^3 & 3\check{h}^2 \\ h^4 & -4h^3 & \check{h}^4 & -4\check{h}^3 \\ \cdots & \cdots & \cdots & \cdots \end{pmatrix} \begin{pmatrix} q_0 \\ q_1 \\ \check{q}_0 \\ \check{q}_1 \end{pmatrix} = \begin{pmatrix} p_0 \\ p_1 \\ p_2 \\ p_3 \\ p_4 \\ \cdots \end{pmatrix}. \tag{7.16}$$

Denote the sequence of 4×4 submatrices of the big matrix in (7.16) by M_0, M_1, etc., and in analogy with (7.5) introduce the column vectors

$$Q = (q_0, q_1, \check{q}_0, \check{q}_1)^t \quad \text{and} \quad P_k = (p_k, p_{k+1}, p_{k+2}, p_{k+3})^t, \quad k = 0, 1, \ldots.$$

Then (7.6), (7.7), (7.8), and (7.9) are valid with

$$S = \begin{pmatrix} 0 & 1 & 0 & 0 \\ 0 & 0 & 2 & 0 \\ 0 & 0 & 0 & 3 \\ \sigma_4 & -\sigma_3 & \sigma_2 & -\sigma_1 \end{pmatrix}, \quad T = \begin{pmatrix} h & 1 & 0 & 0 \\ 0 & h & 0 & 0 \\ 0 & 0 & -\check{h} & 1 \\ 0 & 0 & 0 & -\check{h} \end{pmatrix},$$

where

$$\sigma_1 = 2(h - \check{h}), \quad \sigma_2 = -h^2 + 4h\check{h} - \check{h}^2, \quad \sigma_3 = -2h\check{h}(h - \check{h}), \quad \sigma_4 = -h^2\check{h}^2.$$

Note that σ_j is up to sign equal to the elementary symmetric polynomial in four variables of degree j, evaluated at $(h, h, -\check{h}, -\check{h})$. In this case

$$\det S = \det T = h^2\check{h}^2 = h(\xi)^2 h(-\xi)^2.$$

We can eliminate Q in the same way as before. Forming 4×4 matrix identities analogous to (7.10) and (7.11) we prove in the same way as before that $h^2\check{h}^2$ is a rational function and in fact a polynomial, provided the determinant

$$\det(P_0, P_1, P_2, P_3) \quad \text{is different from the 0-polynomial.} \tag{7.17}$$

It remains only to prove (7.17). A calculation [14, (4.17)] shows that

$$\det(P_0, P_1, P_2, P_3) = \det M_0 \det(Q, TQ, T^2Q, T^3Q) = c\, q_1\check{q}_1(h + \check{h})^4$$

with $c \neq 0$, so it is enough to prove the analog of Lemma 7.4 showing that $q_1\check{q}_1 = q_1(\xi)q_1(-\xi)$ cannot be identically zero. However, we have not found an easy and elementary proof of this fact for the case when g is a distribution of higher order.[2] One way is to use – the admittedly rather awkward – Lemma 5.2 from [14]. A more elegant, but less elementary way is to use the result from [15] showing that $q_1(\xi)$ must be real analytic [15, Theorem 2]. This implies that $q_1(\xi)q_1(-\xi)$ cannot be identically zero, since $q_1(\xi)$ is assumed not to be identically zero. This completes the sketch of the proof of Theorem 7.2 when f is a distribution of order at most 1.

7.3 Singularities of the boundary of the support and singularities of the distribution

From a different point of view one could say that the results of [13] and [14] inferred information about the regularity of the support of the distribution $g = \mathcal{R}f$ from regularity properties of the distribution itself. Indeed, the fact that a hypersurface Σ in the

---

2 Added in proof: a short proof will appear in a forthcoming article.

manifold $\mathcal{P}_n$ of hyperplanes in $\mathbb{R}^n$ is the set of tangent planes to an ellipsoid is an expression of very strong regularity of Σ. And the assumption that $\xi \mapsto \int g(\xi,p)p^k dp$ is a polynomial for every k implies a microlocal regularity property of g. In fact, already the weaker property that $\xi \mapsto \int g(\xi,p)\phi(p)dp$ is real analytic in a neighborhood of ξ^0 for every real analytic $\phi(p)$ is equivalent to every co-normal of the line $\gamma_{\xi^0} : p \mapsto (\xi^0,p)$ being absent in the analytic wave front set of g, in Hörmander's notation $WF_A(g) \cap N^*(\gamma_{\xi^0}) = \emptyset$. Here $WF_A(g)$ denotes the analytic wave front set of g and $N^*(\gamma_{\xi^0})$ denotes the set of co-normals in $T^*(\mathcal{P}_n)$ to the line γ_{ξ^0}. Geometrically γ_{ξ^0} is the set of all hyperplanes that are orthogonal to ξ^0, which is of course a hypersurface in $\mathcal{P}_n$. Our arguments in [13] and [14], as briefly sketched above, can be used to prove the following *local* statement. If a distribution g (which need not be a Radon transform) is assumed to be supported on a hypersurface Σ in a real analytic manifold and γ is a smooth curve that intersects Σ transversally, then $WF_A(g) \cap N^*(\gamma) = \emptyset$ implies that the surface Σ is real analytic in a neighborhood of the intersection point, and more. A theorem of this kind was presented in [15]. This result is closely related to a key step in Hörmander's famous proof of Holmgren's uniqueness theorem.

Bibliography

[1]   M. Agranovsky, *On polynomially integrable domains in Euclidean spaces*, in: Complex analysis and dynamical systems, Trends in mathematics, Springer, Cham, 2018.

[2]   M. Agranovsky, *On algebraically integrable domains*, Contemp. Math. **733** (2019), 33–44.

[3]   M. Agranovsky, *Locally polynomially integrable surfaces and finite stationary phase expansions*, J. Anal. Math. **141** (2020), 23–47.

[4]   M. Agranovsky, *Domains with algebraic X-ray transform*, Anal. Math. Phys. **12**(2) (2022), 60, 17 pp.

[5]   M. Agranovsky, A. Koldobsky, D. Ryabogin and V. Yaskin, *An analogue of polynomially integrable bodies in even-dimensional spaces*. arXiv:2211.12693.

[6]   Archimedes, *On conoids and spheroids*.

[7]   V. I. Arnold, A. N. Varchenko and S. M. Gusein-Zade, *Singularities of differentiable maps*, Monodromy and asymptotics of integrals, vol. 2, Birkhäuser, Basel, 1988.

[8]   V. I. Arnold and V. A. Vassiliev, *Newton's Principia read 300 years later*, Not. Am. Math. Soc. **36**(9) (1989), 1148–1154.

[9]   V. I. Arnold, *Huygens and Barrow, Newton and Hooke, pioneers in mathematical analysis and catastrophe theory from evolvents to quasicrystals*, Birkhäuser, Basel, 1990.

[10]   V. I. Arnold, *Arnold's problems*, 2nd ed., Springer, Berlin, 2004.

[11]   M. Atiyah and R. Bott, *The moment map and equivariant cohomology*, Topology **23** (1984), 1–28.

[12]   J. Bernhard, *Finite stationary phase expansions*, Asian J. Math. **9** (2005), 187–198.

[13]   J. Boman, *A hypersurface containing the support of a Radon transform must be an ellipsoid. I: The symmetric case*, J. Geom. Anal. **31** (2021), 2726–2741.

[14]   J. Boman, *A hypersurface containing the support of a Radon transform must be an ellipsoid. II: The general case*, J. Inverse Ill-Posed Probl. **29** (2021), 351–367.

[15]   J. Boman, *Regularity of a distribution and of the boundary of its support*, J. Geom. Anal. **32** (2022), 300. https://doi.org/10.1007/s12220-022-01021-y.

[16]   N. Bourbaki, *Groupes et algébres de Lie. Ch. IV, V, VI*, Hermann, Paris, 1968.

[17] R. J. Gardner, *Geometric tomography*, Cambridge Univ. Press, Cambridge, 1995.

[18] M. Goresky and R. MacPherson, *Stratified Morse theory*, Springer, 1988.

[19] S. Helgason, *Integral geometry and Radon transforms*, Springer, New York, 2011.

[20] S.-T. Hu, *Homotopy theory*, Academic Press, 1959.

[21] P. Jordan and J. von Neumann, *On inner products in linear metric spaces*, Ann. Math. **36** (1935), 719–723.

[22] A. Koldobsky, *Fourier analysis in convex geometry*, Amer. Math. Soc., Providence, RI, 2005.

[23] A. Koldobsky, A. Merkurjev and V. Yaskin, *On polynomially integrable convex bodies*, Adv. Math. **320** (2017), 876–886.

[24] F. Natterer, *The mathematics of computerized tomography*, B. G. Teubner, Stuttgart, 1986.

[25] I. Newton, *Philosophiae naturalis principia mathematica*, London, 1687.

[26] W. Rudin, *Functional analysis*, 2nd ed., McGraw-Hill, New York, 1991.

[27] D. Ryabogin and V. Yaskin, *Detecting symmetry in star bodies*, J. Math. Anal. Appl. **395** (2012), 509–514.

[28] V. A. Vassiliev, *Applied Picard–Lefschetz theory*, Amer. Math. Soc., Providence, RI, 2002.

[29] V. A. Vassiliev, *Newton's Lemma XXVIII on integrable ovals in higher dimensions and reflection groups*, Bull. Lond. Math. Soc. **47**(2) (2015), 290–300.

[30] V. A. Vassiliev, *Lacunas and local algebraicity of volume functions*, J. Singul. **18** (2018), 350–357.

[31] V. A. Vassiliev, *New examples of locally algebraically integrable bodies*, Math. Notes **106**(6) (2019), 894–898. arXiv:1902.07235.

[32] V. A. Vassiliev, *Algebroidally integrable bodies*, Arnold Math. J. **6**(2) (2020), 291–309. arXiv:2003.04665.

[33] V. A. Vassiliev, *Ramified volumes and reflection groups*, MCCME, Moscow, 2021, 104 pp. (in Russian), https://biblio.mccme.ru/node/74704/shop.

[34] *Newton's Theorem About Ovals*, Wikipedia.

Gabriele Bianchi

The covariogram problem

Abstract: The covariogram g_X of a measurable set X in $\mathbb{R}^n$ is the function which associates to each $x \in \mathbb{R}^n$ the measure of the intersection of X with $X + x$. We are interested in understanding what information about a set can be obtained from its covariogram. Matheron asked whether a convex body K is determined from the knowledge of g_K, and this is known as the covariogram problem.

The covariogram appears in very different contexts, and the covariogram problem can be rephrased in different terms. For instance, it is equivalent to determining the characteristic function 1_K of K from the modulus of its Fourier transform $\widehat{1_K}$ in $\mathbb{R}^n$, a particular instance of the phase retrieval problem. The covariogram problem has also a discrete counterpart.

We survey the known results and the methods.

Keywords: Autocorrelation, convex body, covariogram, discrete covariogram, geometric tomography, homometric sets, phase retrieval, set covariance

MSC 2020: Primary 42B10, 52A20, Secondary 32A60, 52B11, 60D05

1 Introduction

Let X be a Lebesgue-measurable set of finite measure in the Euclidean space $\mathbb{R}^n$. The *covariogram* g_X of X is the function on $\mathbb{R}^n$ defined by

$$g_X(x) = \mathcal{H}^n(X \cap (X + x)), \quad x \in \mathbb{R}^n, \tag{1.1}$$

where $\mathcal{H}^n$ stands for the n-dimensional Hausdorff measure. This function was introduced by Matheron in his book [52, Section 4.3] on random sets and it coincides with the *autocorrelation* of the characteristic function 1_X of X, that is,

$$g_X = 1_X * 1_{(-X)}. \tag{1.2}$$

The covariogram g_X is clearly unchanged with respect to translations and reflections of X, where, throughout the chapter, *reflection* means reflection in a point. In 1986 Matheron [51, p. 20] asked the following question and conjectured a positive answer for the case $n = 2$ (Matheron's conjecture).

Covariogram Problem. Does the covariogram determine a convex body in $\mathbb{R}^n$, among all convex bodies, up to translations and reflections?

Gabriele Bianchi, Dipartimento di Matematica e Informatica "U. Dini", Università di Firenze, Viale Morgagni 67/a, I-50134 Firenze, Italy, e-mail: gabriele.bianchi@unifi.it

https://doi.org/10.1515/9783110775389-002

We recall that a *convex body* in $\mathbb{R}^n$ is a compact convex set with non-empty interior, and we refer to the next section for all unexplained definitions. More generally, what information about a set, not necessarily convex, can be obtained from its covariogram? The covariogram appears in very different contexts, and the covariogram problem can be rephrased in different terms. Indeed, it is equivalent to any of the following problems.

Problem 1.1. Determine a convex body K from the knowledge, for each unit vector u in $\mathbb{R}^n$, of the distribution of the lengths of the chords of K parallel to u.

Problem 1.2. Determine a convex body K from the distribution of $W - Z$, where W and Z are independent random variables uniformly distributed over K.

Problem 1.3. Determine the characteristic function 1_K of a convex body K from the modulus of its Fourier transform $\widehat{1_K}$.

In Problem 1.1 a random chord parallel to u is obtained by taking the intersection of K with a random line L_u parallel to u, conditioned on $K \cap L_u \neq \emptyset$. Matheron [52, p. 86] explained the relation between Problem 1.1 and the covariogram of a set; see also Nagel [55]. Remark 2.3 in the next section explains this equivalence in detail. Blaschke [62, Section 4.2] asked whether the distribution of the lengths of *all* chords (that is, not separated direction by direction) of a planar convex body determines that body, up to isometries in $\mathbb{R}^2$. Mallows and Clark [50] constructed polygonal examples that show that the answer is negative in general. Gardner, Gronchi, and Zong [33] observed that the distribution of the lengths of the chords of K parallel to u coincides, up to a multiplicative factor, with the *rearrangement* of the X-ray of K in the direction u, and rephrased Problem 1.1 in these terms.

Problem 1.2 was asked by Adler and Pyke [1] in 1991. By (1.2) we see that the distribution of $W - Z$ coincides with $g_K / \mathcal{H}^n(K)^2$. Since $g_K(o) = \mathcal{H}^n(K)$, knowing the covariogram is equivalent to knowing the distribution of $W - Z$. In stochastic geometry, integrals of functions of $W - Z$ can be written in terms of the covariogram, as in the formula (called Borel's overlap formula in [21])

$$\int_{X \times X} f(x - y)\, dx\, dy = \int_{\mathbb{R}^n} f(z)\, g_X(z)\, dz,$$

valid, say, when $X \in \mathcal{L}^n$ is bounded and $f \in L^1_{\mathrm{loc}}(\mathbb{R}^n)$. Recent studies showing connections between the covariogram and stereology and stochastic geometry are [22, 37, 57, 58].

Problem 1.3 is a special case of the *phase retrieval problem*, where 1_K is replaced by a function with compact support. The equivalence of the problems comes by applying the Fourier transform to (1.2). One obtains, for $x \in \mathbb{R}^n$,

$$\widehat{g_K}(x) = \widehat{1_K}(x)\widehat{1_{-K}}(x) = \widehat{1_K}(x)\overline{\widehat{1_K}(x)} = |\widehat{1_K}(x)|^2. \tag{1.3}$$

The phase retrieval problem arises naturally and frequently in various areas of science, such as X-ray crystallography, electron microscopy, optics, astronomy, and remote sensing, in which only the magnitude of the Fourier transform can be measured and the phase is lost. The literature is vast; see the surveys [42, 44, 48, 54, 59], as well as the references given there. Phase retrieval is fundamentally underdetermined without additional constraints, and our problem is the phase retrieval problem with the constraint $f = 1_K$, the characteristic function of a convex body K in $\mathbb{R}^n$.

Baake and Grimm [8] have observed that the covariogram problem is relevant for the inverse problem of finding the atomic structure of a quasicrystal from its X-ray diffraction image. It turns out that quasicrystals can often be described by means of the so-called cut-and-project scheme. In this scheme a quasiperiodic discrete subset S of $\mathbb{R}^m$, $m \in \mathbb{N}$, which models the atomic structure of a quasicrystal is described as the orthogonal projection of $Z \cap (\mathbb{R}^m \times W)$ onto $\mathbb{R}^m$, where W is a subset of $\mathbb{R}^n$ and Z is a lattice in $\mathbb{R}^m \times \mathbb{R}^n$. For many quasicrystals, the lattice Z can be recovered from the diffraction image of S. Thus, in order to determine S, it is necessary to know W. The covariogram problem enters at this point, since g_W can be obtained from the diffraction image of S. Note that the set W is in many cases a convex body.

The first partial solution of Matheron's conjecture was given by Nagel [55] in 1993, who confirmed it for all convex polygons. A complete positive answer in the plane is contained in the combined works of Bianchi [14] and Averkov and Bianchi [4]. In higher dimensions, the first result was a positive one valid for a class of convex polytopes by Goodey, Schneider, and Weil [38]. This result implies that most (in the sense of Baire category) convex bodies are determined by their covariograms. It is easy to see that centrally symmetric convex bodies are determined. Bianchi [15] proved a positive answer for convex polytopes in $\mathbb{R}^3$, and in [14] found counterexamples in $\mathbb{R}^n$, for any $n \geq 4$, which may be chosen to be polytopes. Whether arbitrary convex bodies in $\mathbb{R}^3$ are determined is not yet known. Regarding strictly convex bodies $K \subset \mathbb{R}^n$, $n \geq 3$, of class C_+^m, Bianchi [18] proves a positive answer when m is higher than a threshold which depends on n.

Methods of Fourier analysis are relevant in attacking the covariogram problem. Studies of the asymptotic behavior of $\widehat{1_K}(\xi)$, where $\xi \in \mathbb{R}^n$ and $|\xi| \to \infty$, initiated by Haviland and Wintner [39], provide information about curvatures, for strictly convex and sufficiently smooth convex bodies, at points of the boundary with opposite outer normals (see Section 3.1 for the planar case). For the same class of convex bodies, studies of the zero set of $\widehat{1_K}(\zeta)$, seen as a function of $\zeta \in \mathbb{C}^n$, by Kobayashi [45] play a fundamental role in resolving some ambiguities in their determination (see Section 3.5). Some results on the phase retrieval problem applied to our situation show a connection between the determination of K and the irreducibility of $\widehat{1_K}$. In Section 7 we explain this and also why this property of $\widehat{1_K}$ is related to the Pompeiu problem in integral geometry.

Section 3 is devoted to the covariogram problem and is divided into subsections. Section 3.1 explains the ideas behind the proof of the positive result in the plane; Section 3.4 does the same for the result on polytopes in $\mathbb{R}^3$; Sections 3.2 and 3.3 treat the examples of non-determination in dimension $n \geq 4$ and the results for convex polytopes in dimension $n \geq 3$, respectively. Section 3.6 presents the associated problem of determination from the cross covariogram. The *cross covariogram* $g_{K,L}$ of two measurable sets

$K, L \subset \mathbb{R}^n$ is the function that associates to $x \in \mathbb{R}^n$ the volume $\mathcal{H}^n(K \cap (L + x))$. It appears naturally in our study, and it is also natural to ask whether $g_{K,L}$ determines both K and L, up to the inherent ambiguities. Surprisingly, in certain classes of sets it does. One such class is that of C_+^8-regular planar convex bodies. The case of convex polygons is also completely solved, with the understanding of exactly which pairs are determined and which pairs are not. Section 4 is devoted to algorithms for reconstruction. Section 5 presents examples of information that can be obtained from the covariogram of sets that are not necessarily convex. In particular, it deals with the possibility of recognizing whether a set is convex, whether it is centrally symmetric, and whether it is radial. Section 6 is devoted to the counterpart of the covariogram in the discrete case. We explain the relation between the continuous and the discrete covariogram and how this was used in [33] to construct a pair of non-congruent non-convex polygons with equal covariograms. Later a whole family of such pairs was found; each set in these pairs is a horizontally and vertically convex union of lattice squares. These examples show that the convexity assumption in the covariogram problem cannot be significantly weakened. Baake and Grimm [8] use the pair in [33] to construct two different quasicrystal model sets in $\mathbb{R}^2$ with equal diffraction images.

Some aspects of the covariogram problem have been neglected in this survey. We briefly mention two.

In many situations where we have a positive answer to the covariogram problem, knowledge of the covariogram on a proper subset of its domain still suffices. For instance, the results in Bianchi, Gardner, and Kiderlen [19] show that if a convex body is determined, up to translations and reflections, by its covariogram, then it is also so determined by its values at certain countable sets of points, even, almost surely, when these values are contaminated with noise. The recent paper by Engel and Laasch [25] proves that if $P, P' \subset \mathbb{R}^3$ are convex polytopes, $E \subset \mathbb{R}^3$ is the non-empty intersection of an open set with a sphere and $|\widehat{1_P}(\xi)| = |\widehat{1_{P'}}(\xi)|$, for each $\xi \in E$, then $P = P'$, up to translations and reflections. Averkov and Bianchi [3] also find some results where restricted information about the covariogram is sufficient for determination.

Substituting in the definition of the covariogram the volume with a different functional, like the surface area or other valuations, one defines different covariograms, and for each of them there is a corresponding covariogram problem. Averkov and Bianchi [5] study, for planar convex bodies, the problems associated to the perimeter covariogram and to the width covariogram.

2 Definitions, notations, and preliminaries

As usual, S^{n-1} denotes the unit sphere, B^n the unit ball, and o the origin in the Euclidean n-space $\mathbb{R}^n$. If $x, y \in \mathbb{R}^n$, then $\langle x, y \rangle$ is the scalar product of x and y, while $|x|$ is the norm of x. If $\zeta \in \mathbb{C}^n$ and $\zeta = x + iy$, with $x, y \in \mathbb{R}^n$, then $\mathrm{Re}\,\zeta$ and $\mathrm{Im}\,\zeta$ denote x and

y, respectively. Moreover, $|\zeta| = (|\operatorname{Re}\zeta|^2 + |\operatorname{Im}\zeta|^2)^{1/2}$ denotes the norm of ζ. For $\delta > 0$, $B(x,\delta)$ denotes $\{y \in \mathbb{R}^n : |y - x| < \delta\}$.

If X and Y are sets in $\mathbb{R}^n$, we denote by $\operatorname{lin}X$, $\operatorname{aff}X$, $\operatorname{conv}X$, $\operatorname{cl}X$, $\operatorname{int}X$, and 1_X the *linear hull*, *affine hull*, *convex hull*, *closure*, *interior*, and *characteristic function* of X, respectively. Also, $\operatorname{relint}X$ is the *relative interior* of X, that is, the interior of X relative to $\operatorname{aff}X$. The symbol $|X|$ denotes the *cardinality* of X. If $t \in \mathbb{R}$, then $tX = \{tx : x \in X\}$, $X + Y = \{x + y : x \in X, y \in Y\}$ denotes the *Minkowski sum* of X and Y, and

$$DX = X + (-X)$$

denotes the *difference set* of X. A set is *o-symmetric* if it is centrally symmetric, with center at the origin.

A *lattice set* is a finite subset of $\mathbb{Z}^n$ and a *lattice body* is a subset P of $\mathbb{R}^n$ which can be written as $P = A + [0,1]^n$, where A is a lattice set. We call P the lattice body associated to A and A the lattice set associated to P. A lattice set whose associated lattice body has connected interior is called a *polyomino*. Lattice bodies associated to polyominoes are called *lattice animals* (or polyominoes, by many authors). A lattice set A is *convex* if $A = (\operatorname{conv}A) \cap \mathbb{Z}^n$.

If $u \in S^{n-1}$, then $u^\perp = \{x \in \mathbb{R}^n : \langle x, u \rangle = 0\}$. If E is a linear subspace of $\mathbb{R}^n$, then $X|E$ is the orthogonal projection of X on E and $x|E$ is the projection of a vector $x \in \mathbb{R}^n$ on E. The symbol $\mathcal{R}_{\frac{\pi}{2}}$ denotes a counterclockwise rotation in $\mathbb{R}^2$ by $\pi/2$.

For $i \in \{0, 1, \ldots, n\}$, we write $\mathcal{H}^i$ for the i-dimensional Hausdorff measure in $\mathbb{R}^n$. We write ω_n for the surface area of the unit ball in $\mathbb{R}^n$. We denote by $\mathcal{C}^n$, $\mathcal{M}^n$, and $\mathcal{L}^n$ the class of non-empty compact sets, $\mathcal{H}^n$-measurable sets, and $\mathcal{H}^n$-measurable sets of finite measure, respectively, in $\mathbb{R}^n$. A compact set X is *regular* if $X = \operatorname{cl}\operatorname{int}X$. The *Hausdorff distance* $\delta(C,D)$ between two sets $C, D \in \mathcal{C}^n$ is defined as

$$\delta(C,D) = \min\{\varepsilon \geq 0 : C \subset D + \varepsilon B^n, D \subset C + \varepsilon B^n\}.$$

Let $\mathcal{K}^n$ be the class of non-empty compact convex subsets of $\mathbb{R}^n$ and let $\mathcal{K}^n_n$ be the class of *convex bodies*, i. e., members of $\mathcal{K}^n$ with interior points. The treatise of Schneider [66] is an excellent general reference for convex geometry. For $K \in \mathcal{K}^n$, the function

$$h_K(u) = \max\{\langle u, y \rangle : y \in K\},$$

for $u \in \mathbb{R}^n$, is the *support function* of K,

$$w_K(u) = h_K(u) + h_K(-u)$$

is its *width function*, and

$$b_K(u) = \mathcal{H}^{n-1}(K|u^\perp),$$

for $u \in S^{n-1}$, is its *brightness function*. Note that

$$w_{DK} = 2w_K.$$

Any $K \in \mathcal{K}^n$ is uniquely determined by its support function. Given $u \in S^{n-1}$, the *support set of K in direction u* is

$$K_u = \{x \in K : \langle x, u \rangle = h_K(u)\}.$$

The support sets are also called *exposed faces of K*. Note that [66, Theorem 1.7.5(c)]

$$(DK)_u = K_u + (-K)_u. \tag{2.1}$$

The *Blaschke body* ∇K of $K \in \mathcal{K}^n_n$ is the unique o-symmetric convex body such that

$$b_{\nabla K} = b_K.$$

We say that a convex body K is in the class C^m, for $m \in \mathbb{N}$, if ∂K is an m-differentiable manifold, and we write $K \in C^m_+$, for $m \geq 2$, if $K \in C^m$ and the Gauss curvature of ∂K is positive everywhere. We say that $K \in C^\infty_+$ if $K \in C^m_+$ for each $m \in \mathbb{N}$. When $K \in C^2_+$, $\nu_K : \partial K \to S^{n-1}$ denotes the *Gauss map* and $\tau_K(u)$ denotes the *Gauss curvature* of ∂K at the point $\nu_K^{-1}(u)$ in ∂K with outer normal $u \in S^{n-1}$.

Given a face F of a convex polytope $P \subset \mathbb{R}^n$, the *normal cone* of P at F is denoted by $N(P, F)$ and is the set of all outer normal vectors to P at x, where $x \in \operatorname{relint} F$, together with o. The *support cone of P at F* is the set

$$\operatorname{cone}(P, F) = \{a(y - x) : y \in P, \ a \geq 0\},$$

where $x \in \operatorname{relint} F$. Neither definition depends on the choice of x. Two faces F and G of P are *antipodal* if $\operatorname{relint} N(P, F) \cap (-\operatorname{relint} N(P, G)) \neq \emptyset$.

If $Y \in \mathcal{M}^n$, then

$$\Theta(Y, x) = \lim_{r \to 0^+} \frac{\mathcal{H}^n(Y \cap (rB^n + x))}{\mathcal{H}^n(rB^n + x)} \tag{2.2}$$

is the *density of Y at x*, provided the limit exists. For $t \in [0, 1]$, define $Y^t = \{x \in \mathbb{R}^n : \Theta(Y, x) = t\}$. The *essential boundary $\partial^e Y$ of Y* is $\partial^e Y = \mathbb{R}^n \setminus (Y^0 \cup Y^1)$. The *perimeter* Per Y of Y is Per $Y = \mathcal{H}^{n-1}(\partial^e Y)$, while the *directional variation $V_u(Y)$ of Y in the direction $u \in S^{n-1}$* is

$$V_u(Y) = \int_{u^\perp} \mathcal{H}^0(\partial^e Y \cap (l_u + x)) \, d\mathcal{H}^{n-1}(x),$$

where l_u is the line through o parallel to u. If $K \in \mathcal{K}^n_n$, then $\partial^e K = \partial K$, perimeter coincides with surface area, and $V_u(K) = 2b_K(u)$.

The *X-ray of $Y \in \mathcal{L}^n$ in the direction $u \in S^{n-1}$* is defined for $\mathcal{H}^{n-1}$-almost all $x \in u^\perp$ by

$$X_u Y(x) = \mathcal{H}^1(Y \cap (l_u + x)).$$

Given a function f defined on a subset of $\mathbb{R}^n$, $\operatorname{supp} f$, ∇f, and $D^2 f$ denote its support, its gradient, and its Hessian, respectively. We say that $f \in C_0^\infty(\mathbb{R}^n)$ if f is m-times differentiable for each $m \in \mathbb{N}$ and $\operatorname{supp} f$ is compact.

An *entire function* is a complex-valued function that is holomorphic over the whole $\mathbb{C}^n$. An entire function f is of *exponential type* if there exist $a, b \in \mathbb{R}$ and $m \in \mathbb{Z}$ such that $|f(\zeta)| \le a(1 + |\zeta|)^m e^{b|\operatorname{Im}\zeta|}$, for each $\zeta \in \mathbb{C}^n$.

The *Fourier transform* of a function $f \in L^2(\mathbb{R}^n)$ with compact support is defined for $\zeta \in \mathbb{C}^n$ as

$$\widehat{f}(\zeta) = \int_{\mathbb{R}^n} e^{i\langle x,\zeta\rangle} f(x)\, dx.$$

By the Paley–Wiener theorem, $\widehat{f}$ is an entire function of exponential type whose restriction to $\mathbb{R}^n$ belongs to L^2. The version of this theorem for distributions asserts that $\widehat{f}$ is an entire function of exponential type if and only if f is a distribution with compact support; see [60, Theorem 7.23]. Distributions will enter this paper only very marginally and we refer to Rudin [60] for their definition.

Taking Fourier transforms in (1.2) for $\zeta \in \mathbb{C}^n$ gives the relation

$$\widehat{g_K}(\zeta) = \widehat{1_K}(\zeta)\,\overline{\widehat{1_K}(\bar{\zeta})}. \tag{2.3}$$

2.1 Properties of the covariogram

Given $X, Y \in \mathcal{M}^n$, the *cross covariogram* of X and Y is the function

$$g_{X,Y}(x) = \mathcal{H}^n(X \cap (Y + x)),$$

where $x \in \mathbb{R}^n$ is such that $\mathcal{H}^n(X \cap (Y + x))$ is finite. Clearly, $g_{X,X} = g_X$.

The translation of X and Y by the same vector and the substitution of X with $-Y$ and of Y with $-X$ leave $g_{X,Y}$ unchanged. Let X' and Y' be in $\mathcal{M}^n$. We call (X, Y) and (X', Y') *trivial associates* when one pair is obtained from the other one via a combination of the two operations above, that is, when either $(X, Y) = (X' + x, Y' + x)$ or $(X, Y) = (-Y' + x, -X' + x)$, for some $x \in \mathbb{R}^n$. When dealing with the ordinary covariogram the previous definition simplifies to the following one: X and X' are called *trivial associates* if $X' = X + x$ or $X' = -X + x$ for some $x \in \mathbb{R}^n$.

The following propositions list some properties of the covariogram.

Proposition 2.1. *Let $X \in \mathcal{L}^n$.*
(a) *For all $x \in \mathbb{R}^n$, $0 \le g_X(x) \le g_X(o) = \mathcal{H}^n(X)$.*
(b) *The function g_X is even.*
(c) *We have $\int_{\mathbb{R}^n} g_X(x)\, dx = \mathcal{H}^n(X)^2$.*
(d) *The function g_X is uniformly continuous in $\mathbb{R}^n$ and $\lim_{|x|\to\infty} g_X(x) = 0$. Moreover, for all $x, y \in \mathbb{R}^n$,*

$$|g_X(x) - g_X(y)| \le g_X(o) - g_X(x - y).$$

(e) *The right directional derivative of g_X at o in direction $u \in S^{n-1}$ can be expressed as*

$$\frac{\partial^+ g_X}{\partial u}(o) = -\frac{1}{2} V_u(A).$$ (2.4)

(f) *The covariogram g_X is Lipschitz if and only if X has finite perimeter $\mathrm{Per}\, X$. Moreover, the Lipschitz constant of g_X equals $(1/2) \sup_{u \in S^{n-1}} V_u(X)$, and*

$$\mathrm{Per}\, X = -\frac{1}{\omega_{n-1}} \int_{S^{n-1}} \frac{\partial^+ g_X}{\partial u}(o)\, d\mathcal{H}^{n-1}(u).$$

These properties, in the generality of measurable sets, are proved in Galerne [29]. Some of them are immediate, like Items (a), (b), and (c), while others were already known in the case of convex bodies [52] and of full-dimensional compact $\mathcal{U}_{PR}$ sets [56] (a family of sets that consists of certain unions of sets of positive reach).

Proposition 2.2. *Let $K, L \in \mathcal{K}_n^n$ and let $C, D \in \mathcal{C}^n$ be regular.*
(a) *We have $\mathrm{supp}\, g_C = DC$ and $\mathrm{supp}\, g_{C,D} = C + (-D)$.*
(b) *The function $g_{K,L}^{1/n}$ is concave on its support. In particular, g_K is log-concave.*
(c) *The following inequalities hold:*

$$2^n g_C(o) \leq \mathcal{H}^n(\mathrm{supp}\, g_C),$$ (2.5)

$$\mathcal{H}^n(\mathrm{supp}\, g_K) \leq \binom{2n}{n} g_K(o).$$ (2.6)

Equality in (2.5) holds precisely when C is convex and centrally symmetric, while equality in (2.6) holds precisely when K is a simplex.
(d) *If $\mathcal{H}^{n-1}(\partial K \cap (\partial K + x)) = 0$, then $\nabla g_K(x)$ exists and*

$$\nabla g_K(x) = -\int_{\partial K \cap (K+x)} v(y)\, d\mathcal{H}^{n-1}(y),$$ (2.7)

where $v(y)$ denotes the unit outer normal vector to $y \in \partial K$, defined $\mathcal{H}^{n-1}$-almost everywhere. If $u \in S^{n-1}$, $r > 0$, and $ru \in \mathrm{int}\, \mathrm{supp}\, g_K$, then

$$\frac{\partial g_K}{\partial u}(ru) = -\mathcal{H}^{n-1}((K \cap (K + ru))|u^\perp).$$ (2.8)

Moreover,

$$\frac{\partial^+ g_K}{\partial u}(o) = -b_K(u).$$ (2.9)

For arbitrary sets $C, D \in \mathcal{L}^n$, Item (a) is not valid, even in the case of the ordinary covariogram. The property $x \notin DC$ is equivalent to $C \cap (C+x) = \emptyset$ and therefore the inclusion $\mathrm{supp}\, g_C \subset DC$ is still valid. However, the other inclusion may be false. For instance, if C is the Cantor ternary set in $[0,1]$, then $\mathrm{supp}\, g_C = \emptyset$, since $\mathcal{H}^1(C) = 0$, while $DC = [-1,1]$.

We give a proof of Item (a) below, since we could not find one in the literature valid in the class of regular sets in C^n.

Item (b) was first observed by Gardner and Zhang [35] and we give their proof below. The set

$$K(\delta) = \{x \in \mathbb{R}^n : g_K(x) \geq \delta\}$$

is called the *convolution body* of K. This notion is due to Kiener [43], who noted that $K(\delta)$ is convex, as Item (b) implies.

Formulas (2.5) and (2.6), together with their equality cases, are an immediate consequence of the general Brunn–Minkowski inequality [30, p. 362] and of the Rogers–Shephard inequality [66, p. 530], respectively, together with Item (a) and the fact that the value of the covariogram at o equals the volume of the set.

Formula (2.9) shows that g_K provides the brightness function b_K of K. Formula (2.8) and the interpretation of the right-hand side of (2.8) as

$$-\mathcal{H}^{n-1}(\{x \in u^\perp : X_u K(x) \geq r\}) \tag{2.10}$$

proves a connection between the covariogram and the X-rays of a convex body first observed by Gardner, Gronchi, and Zhong [33]: Knowing g_K is equivalent to knowing the rearrangement of $X_u K$ for each $u \in S^{n-1}$.

Formulas (2.7) and (2.9) are present in Matheron [51]. Regarding the existence of ∇g_K, [51] simply writes that this happens almost everywhere, due to the concavity property in Item (b). The existence of the derivatives in (2.7) and (2.8) is proved by Meyer, Reisner, and Schmuckenschläger [53], who also deal with the second-order derivatives of g_K.

Proof. a). The property $x \notin C + (-D)$ is equivalent to $C \cap (D + x) = \emptyset$ and therefore $\mathrm{supp}\, g_{C,D} \subset C + (-D)$. If $x \in C + (-D)$ and $x = c - d$ with $c \in C$ and $d \in D$, then $c = d + x \in C \cap (D + x)$. For any $\varepsilon > o$ there exist $c' \in B(c, \varepsilon) \cap \mathrm{int}\, C$ and $d' \in B(d, \varepsilon) \cap \mathrm{int}\, D$, due to C and D being regular. If $y = c' - (d' + x)$, then $|y| \leq 2\varepsilon$ and $C \cap (D + x + y)$ has non-empty interior, since it contains $c' = d' + x + y$. Therefore, $g_{C,D}(x + y) > 0$. Thus, $B(x, 2\varepsilon)$ contains points where $g_{C,D}$ is positive, which proves $C + (-D) \subset \mathrm{supp}\, g_{C,D}$.

b). For $x, y \in \mathbb{R}^n$ and $a \in [0,1]$, we have

$$K \cap (L + (1-a)x + ay) = K \cap ((1-a)(L+x) + a(L+y))$$
$$\supset (1-a)(K \cap (L+x)) + a(K \cap (L+y)).$$

Using the Brunn–Minkowski inequality we obtain

$$g_{K,L}((1-a)x + ay)^{1/n} \geq \mathcal{H}^n((1-a)(K \cap (L+x)) + a(K \cap (L+y)))^{1/n}$$
$$\geq (1-a)\mathcal{H}^n(K \cap (L+x))^{1/n} + a\mathcal{H}^n(K \cap (L+y))^{1/n}$$
$$= (1-a)g_{K,L}(x)^{1/n} + ag_{K,L}(y)^{1/n}. \qquad \square$$

Remark 2.3. Some formulas in Propositions 2.1 and 2.2 explain the equivalence between the covariogram problem and Problem 1.1. Let Z_u be the length of the chord $L_u \cap K$, where L_u is a random line parallel to $u \in S^{n-1}$ conditioned on $L_u \cap K \neq \emptyset$. Formula (2.8), with the right-hand side interpreted as in (2.10), shows that the probability of the event $\{Z_u \geq r\}$, for $r > 0$, is equal to $-(\partial g_K/\partial u)(ru)/b_K(u)$. Integrating the latter expression with respect to r we determine $f(ru) = g_K(ru)/b_K(u)$. Consequently, the distribution of Z_u for each $u \in S^{n-1}$ determines $f(ru)/f(0u) = g_K(ru)/\mathcal{H}^n(K)$, for every $r > 0$ and every $u \in S^{n-1}$. The latter is equivalent to the determination of $g_K(x)/\mathcal{H}^n(K)$ for every $x \in \mathbb{R}^n$. Integration of $g_K/\mathcal{H}^n(K)$ over $\mathbb{R}^n$ yields $\mathcal{H}^n(K)$, by Proposition 2.1(c), and we determine g_K. Conversely, g_K determines the distribution of Z_u by (2.9).

In the plane, ∇g_K has a simple geometric interpretation. Consider an arbitrary $x \in$ int supp g_K and assume that $\partial K \cap (\partial K + x)$ consists of two points. Then there exist points $p_i(x)$, $i \in \{1, \dots, 4\}$, in counterclockwise order on ∂K, such that $x = p_1(x) - p_2(x) = p_4(x) - p_3(x)$; see Figure 1. These points define a parallelogram

$$P(x) = \mathrm{conv}\{p_1(x), \dots, p_4(x)\} \tag{2.11}$$

inscribed in K, whose edges are translates of $[o, x]$ and $[o, D(x)]$, where

$$D(x) = p_1(x) - p_4(x).$$

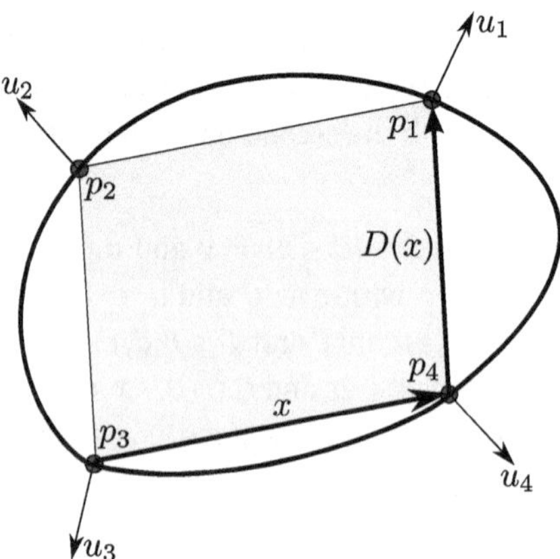

Figure 1: $P(x)$ and the vector $D(x)$, a rotation of $\nabla g_K(x)$.

Proposition 2.4. *Let $K, L \in \mathcal{K}_2^2$ and $x \in$ int supp $g_K \setminus \{o\}$ be such that $\partial K \cap (\partial K + x)$ consists of two points. Then $\nabla g_K(x) = \mathcal{R}_{\frac{\pi}{2}} D(x)$.*

If $g_L = g_K$, then any parallelogram inscribed in K has a translate inscribed in L.

The representation of $\nabla g_K(x)$ in terms of $D(x)$ was already observed in [51]. The second claim is proved in [14] and is related to the first one. This is easily understood when K and L are strictly convex. In this case any parallelogram P inscribed in K coincides with $P(x)$ when x is chosen so that $[o, x]$ is a translate of an edge of P. The parallelograms $P(K, x)$ and $P(L, x)$ are translates of each other, since both have two edges that are translates of $[o, x]$ and two edges that are translates of $D(K, x) = D(L, x)$, where these vectors coincide due to the first claim.

3 The covariogram problem

The focus on covariograms of *convex* bodies is natural. There exist non-congruent non-convex polygons, even (see Figure 7) horizontally and vertically convex ones, with the same covariogram, indicating that the convexity assumption cannot be significantly weakened. Sections 5 and 6 contain a discussion of this example and information on the determination of non-convex sets (including the case of discrete sets) by the covariogram (or by its discrete version).

Another preliminary observation is that objects which are centrally symmetric with respect to some point are easy to determine, up to translations, in the class of centrally symmetric objects. If K_1 and K_2 are convex bodies which are centrally symmetric with respect to a_1 and a_2, respectively, and $g_{K_1} = g_{K_2}$, this follows from the formula

$$2(K_1 - a_1) = D(K_1 - a_1) = \operatorname{supp} g_{K_1} = \operatorname{supp} g_{K_2} = D(K_2 - a_2) = 2(K_2 - a_2).$$

An analogous result holds in a much larger class. Cabo and Janssen [23] prove that if f and h are even functions in $L^1(\mathbb{R}^n)$ with compact support and with the same autocorrelation (i. e., $f(\cdot) * f(-\cdot) = h(\cdot) * h(-\cdot)$), then $f = \pm h$ almost everywhere. Note that the autocorrelation of the characteristic function of a set is its covariogram. This implies the following result, of which the elegant proof is taken from [33].

Theorem 3.1 (Cabo and Janssen [23]). *A centrally symmetric regular compact subset C of $\mathbb{R}^n$ is determined by g_C, up to translations, in the class of centrally symmetric regular compact sets.*

Proof. If D is a centrally symmetric regular compact set with $g_C = g_D$, then (1.3) implies $|\widehat{1_C}|^2 = |\widehat{1_D}|^2$. Up to translations, we may assume that C and D are o-symmetric. Fourier transforms of even functions are real-valued and the previous condition becomes $(\widehat{1_C})^2 = (\widehat{1_D})^2$. Therefore, $\widehat{1_C}(x) = \pm\widehat{1_D}(x)$, for each $x \in \mathbb{R}^n$. Since Fourier transforms of functions with compact supports are analytic and any analytic function is determined by its value on a set with a limit point, we conclude that $\widehat{1_C} = \pm\widehat{1_D}$. Fourier inversion yields $1_C = 1_D$ almost everywhere and since C and D are regular we have $C = D$. $\square$

We remark that here we are not asking whether a symmetric object is determined in the class of *all* objects. The answer to this last question is more subtle. It is known that a centrally symmetric convex body is determined in the class of all regular compact sets, but the same question is already open for the determination of a centrally symmetric regular compact set in the class of all regular compact sets. This will be explained in Section 5.2, when speaking of the possibility of recognizing the central symmetry of a set from its covariogram.

A great deal of effort has been spent on the determination of convex bodies from the combined information in their width and brightness functions. Since the covariogram of a convex body determines both functions, by Proposition 2.2(a) and (2.9), the question is directly related to the covariogram problem. The many known results do not add to what we know about the latter, but the interested reader can consult [31, Chapter 3 and Notes 3.3 and 3.6] and the update for [31] available on its author's website.

3.1 Complete answer in the plane

The first answer to the covariogram problem in the plane was a positive one for convex polygons proved by Nagel [55] in 1993. Bianchi [14] and Bianchi and Averkov [4] prove the following theorem, which confirms Matheron's conjecture.

Theorem 3.2. *Every planar convex body is determined among all planar convex bodies by its covariogram, up to translations and reflections.*

This is the combination of the following two results.

Theorem 3.3 (Bianchi [14]). *Let K be a planar convex body such that one of the following holds: K is not strictly convex; K is not in C^1. Then K is determined among all planar convex bodies by its covariogram, up to translations and reflections.*

Theorem 3.4 (Averkov and Bianchi [4]). *Let K and L be planar strictly convex bodies in C^1 with equal covariograms. Then $K = L$, up to translations and reflections.*

In the rest of this section K and L will denote planar convex bodies with $g_K = g_L$.

A unified proof of Theorem 3.2 would be very welcome but it is still missing. The proofs of Theorems 3.3 and 3.4 both rely on two ingredients. One is the following.

Proposition 3.5 (Bianchi [14]). *If K and L have a common non-degenerate boundary arc, then $K = L$, up to translations and reflections.*

The other ingredient is the proof that K and L do have a boundary arc in common, up to translations and reflections. The proof of Proposition 3.5 is different according to whether the common arc is strictly convex or not, even though the structures of the proofs for the two cases are similar. The proofs of the other step differ very much from Theorem 3.3 to Theorem 3.4, with smaller differences present even between the different cases handled by Theorem 3.3.

We first present a sketch of the proof of Proposition 3.5 assuming, for simplicity, that K and L are strictly convex and C^1-regular. Under this assumption the proof of Proposition 3.5 was substantially already present in Bianchi, Segala, and Volčič [20], and the version that we present is taken from there.

Sketch of the proof of Proposition 3.5

Let E be a maximal (with respect to inclusion) arc in $\partial K \cap \partial L$. The portion of ∂K antipodal to E is contained in ∂L too. By this we mean that if U is the subset of S^1 consisting of the vectors u such that

$$K_u = L_u \subset E,$$

then, for each $u \in U$,

$$K_{-u} = L_{-u}.$$

This comes from $DK = \operatorname{supp} g_K = \operatorname{supp} g_L = DL$ and (2.1). Thus, $\partial K \cap \partial L$ also contains the boundary arc $F = \{K_{-u} : u \in U\}$ antipodal to E. The arc F is maximal too.

The crucial point in the proof of Proposition 3.5 is the next lemma. We shall only give its proof, though further work is required to prove Proposition 3.5.

Lemma 3.6. *The arcs E and F are reflections of each other.*

Proof. Suppose, on the contrary, that E and F are not reflections of each other. We aim to obtain a contradiction by showing that E (and hence F) is not maximal, i. e., that $\partial K \cap \partial L$ contains an arc strictly larger than E. It follows from the discussion above that K and L have the same tangents at the endpoints of E and F and that these tangents are pairwise parallel. Moreover, U has length less than π and thus E can be represented as the graph of a convex function.

We need a definition. Suppose that X and Y are arcs of ∂K corresponding to opposite arcs V and $-V$ of S^1. Let z be one of the endpoints of Y. We denote by $\bar{Y}$ the convex curve formed by Y and the appropriate half of the tangent to Y at z. We say that *the point z can be captured by the arc X* if an appropriate translation of X intersects $\bar{Y}$ in two points determining an arc of $\bar{Y}$ containing z in its relative interior.

Claim 1. Let E and F be disjoint arcs in the boundary of a planar strictly convex body K corresponding to $U \subset S^1$ and $-U \subset S^1$, respectively, which are not reflections of each other. Then one arc has an endpoint which can be captured by the other.

To see this, let u denote an endpoint of U. The point K_u is an endpoint of E and K_{-u} is an endpoint of F. Changing, if necessary, the coordinate system, we may assume that $K_u = (0,0)$, that $u = (0,-1)$, and that locally the arc E is represented by the graph of a convex function defined in a right neighborhood of 0. Let $\tilde{F} = -(F + K_u - K_{-u})$. Then $\tilde{F}$ is tangent to E at K_u. Either $\tilde{F} \subset E$ or $E \subset \tilde{F}$ or there is a point (x,y) on one arc such that the other contains a point (x',y), with $x' > x$. The first two alternatives are impossible, since E and $\tilde{F}$ are strictly convex arcs with the same set of outer normals, so if $\tilde{F} \subset E$ or $E \subset \tilde{F}$, then $E = \tilde{F}$, contradicting our assumption.

We may assume that $(x, y) \in E$. Then the map $z \to -z + (x, y)$ takes $\tilde{F}$ to a translate of F with one endpoint at (x, y) and with a point (the image of (x', y)) on the negative x-axis. The origin is thus an endpoint of E which is captured by F. This proves Claim 1.

By Claim 1, we may assume that there is an endpoint z of E which can be captured by F via a translation by a vector p. As in the proof of Claim 1, we may assume that z is the origin and that the arc E is represented by the graph of a convex function g defined in a right neighborhood of 0, and we may also assume that $g(0) = 0$ and $g'(0)$ is finite. It is possible to extend the definition of g to a left neighborhood of 0 so that it represents a portion of ∂K adjacent to E. Let f be the concave function whose graph is $F + p$. The arc $F + p$ intersects E in a point (b, c) with $b > 0$ and moreover, possibly by changing the translation, we may assume that $F + p$ also intersects the graph of g in a point with a negative abscissa a.

If we show that the covariogram determines the boundary of $K \cap (K + p)$, we are done, since this means that the arc E is not maximal in $\partial K \cap \partial L$.

The covariogram gives the area of $K \cap (K + p - (0, t))$ for every $t > 0$. If we denote by $[a_t, b_t]$ the interval where $f(x) - t \geq g(x)$, then

$$g_K(p - (0, t)) = \int_{a_t}^{0} (f(x) - t - g(x))\, dx + \int_{0}^{b_t} (f(x) - t - g(x))\, dx.$$

The latter integral is known for any $t \in [0, f(0)]$, since by assumption, f and g are known on $[0, b]$. Therefore, we can deduce from the covariogram the value of

$$\int_{a_t}^{0} (f(x) - t - g(x))\, dx,$$

for any $t \in [0, f(0)]$. By assumption, f is known on $[a, 0]$. We now claim that this information is sufficient to determine g on $[a, 0]$.

Claim 2. Suppose that f is a continuous strictly increasing function on $[a, 0]$, with $f(0) > 0$. If g is continuous and strictly decreasing on $[a, 0]$ such that $g(a) > f(a)$ and $g(0) = 0$, then g is determined in a left neighborhood of 0 by the areas

$$A_t = \mathcal{H}^2(\{(x, y) : x \in [a, 0],\ g(x) \leq y \leq f(x) - t\}), \tag{3.1}$$

for $0 \leq t \leq f(0)$.

Indeed, let a_t be the point where $g(a_t) = f(a_t) - t$. Then $a_0 < 0$. The function $h(t) = a_t$ is continuous since h is the inverse of the increasing and continuous function $f - g$ restricted to $[a_0, 0]$. An elementary calculation shows that for every $\delta > 0$,

$$\delta a_{t+\delta} \leq A(t) - A(t + \delta) \leq \delta a_t.$$

It follows that $(A(t) - A(t + \delta))/\delta \to a_t = h(t)$, as $\delta \to 0$, because h is continuous.

We see from this that h is determined on its natural domain $[0, f(0)]$, and therefore so is its inverse $f - g$, defined on $[a_0, 0]$. But f is determined by assumption, so g is determined on $[a_0, 0]$. This proves Claim 2 and hence Lemma 3.6. $\qquad\square$

Determination of an arc of the boundary

We will describe this step in two cases, that of convex bodies in C_+^2, where it is particularly simple, and that of strictly convex C^1-regular bodies, historically the last case to be solved.

Determination of an arc of the boundary: convex bodies in C_+^2

We recall that, for $u \in S^1$, $\tau_K(u)$ denotes the Gauss curvature of ∂K at the point K_u with outer normal u.

Proposition 3.7 (Bianchi, Segala, and Volčič [20]). *If the planar convex body K is in C_+^2, then g_K determines the non-ordered pair $\{\tau_K(u), \tau_K(-u)\}$, for $u \in S^1$.*

The analogous result is valid also for C_+^2-convex bodies in $\mathbb{R}^n$ [18, Proposition 3.1].

In the planar case the information about the couple above is contained in the asymptotic behavior of g_K near the point $p = (\operatorname{supp} g_K)_u$. Since $\operatorname{supp} g_K = DK$, $p = K_u - K_{-u}$, by (2.1). Studying the behavior of g_K near p is equivalent to studying the behavior of the area of $K \cap (K + x)$ for x such that $K \cap (K + x)$ is contained in a small neighborhood of K_u. For these x, the boundary of $K \cap (K + x)$ consists of a portion of ∂K near K_u and a translation of a portion of ∂K near K_{-u}. The next formula expresses this area in terms of the curvatures and gives a proof of Proposition 3.7. Choose a reference system so that $u = (0, 1)$ and let $p = (p_1, p_2)$. For brevity, let $a = \tau_K(u)$ and $b = \tau_K(-u)$. Then, for (x_1, x_2) in a neighborhood of o such that $(p_1 + x_1, p_2 + x_2) \in \operatorname{supp} g_K$,

$$g_K(p_1 + x_1, p_2 + x_2) = \frac{2}{3} \frac{(-2(a+b)x_2 - abx_1^2)^{\frac{3}{2}}}{(a+b)^2}\left(1 + o(|(x_1, x_2)|)\right).$$

An alternative proof of Proposition 3.7, valid under the stronger assumption $K \in C_+^4$, derives from (1.3) and the study of the asymptotic behavior at infinity of $\widehat{1_K}$ by Haviland and Wintner [39]. The result in [39] yields the following asymptotic expansion, as $|t| \to \infty$, for $\widehat{g_K}$:

$$\widehat{g_K}(tu) = \frac{2\pi}{t^3}\left(\frac{1}{a} + \frac{1}{b} - \frac{2}{\sqrt{ab}}\sin(|t|w_K(u)) + o\left(\frac{1}{\sqrt{|t|}}\right)\right).$$

It remains to prove that Proposition 3.7 implies that K and L have a boundary arc in common, when both K and L are C_+^2-regular.

If K is centrally symmetric, the curvatures at antipodal points are equal and Proposition 3.7 implies $\tau_K(u) = \tau_L(u)$, for each u. The curvature determines the point $K_u = (x(u), y(u))$ of ∂K via the parametric representation (see for instance [28, p. 79])

$$x(u) = x(v) + \int_{\theta(v)}^{\theta(u)} \frac{-\sin t}{\tau_K(\cos t, \sin t)}\, dt, \quad y(u) = y(v) + \int_{\theta(v)}^{\theta(u)} \frac{\cos t}{\tau_K(\cos t, \sin t)}\, dt, \tag{3.2}$$

where $\theta(u)$ denotes the angular coordinate of $u \in S^1$. If $v \in S^1$ is fixed and L is translated so that $K_v = L_v$, then $K = L$ follows from (3.2).

If K is not centrally symmetric, the continuity of the curvature implies that given any component U of $\{u \in S^1 : \tau_K(u) \neq \tau_K(-u)\}$, we have, possibly after a reflection of L,

$$\tau_K(u) = \tau_L(u), \quad \text{for each } u \in U.$$

If $v \in U$ is fixed and L is translated so that $K_v = L_v$, then (3.2) implies that $K_u = L_u$, for $u \in U$.

Determination of an arc of the boundary: strictly convex C^1-regular bodies

We recall some notation introduced at the end of Section 2. For $x \in \operatorname{int} \operatorname{supp} g_K \setminus \{o\}$, let $p_i(x)$, $i \in \{1, \ldots, 4\}$, be points in counterclockwise order on ∂K such that $x = p_1(x) - p_2(x) = p_4(x) - p_3(x)$; see Figure 1. Let $u_i(x)$ be the unit outer normal vector to ∂K at $p_i(x)$ and let $P(x)$ be the parallelogram $\operatorname{conv}\{p_1(x), \ldots, p_4(x)\}$. The crucial point is that the outer normals of K are determined by g_K, up to the ambiguities arising from reflections of the body.

Proposition 3.8 (Averkov and Bianchi [4]). *Let K be a strictly convex C^1-regular body. Then, for every $x \in \operatorname{int} \operatorname{supp} g_K \setminus \{o\}$ with $\det G(x) \neq -1$, the set $\{u_1(x), -u_3(x)\}$ is determined by g_K.*

Here

$$G(x) = G(K, x) = \left(\frac{\partial^2 g_K}{\partial x_i \partial x_j}(x) \right)$$

is the Hessian matrix of g_K at x. The existence of the second-order derivatives at $x \in \operatorname{int} \operatorname{supp} g_K \setminus \{o\}$ is proved for strictly convex C^1-regular bodies in $\mathbb{R}^n$ in [53, Theorem 2.5], while the planar case was already treated in [52, pp. 12–18]. The Hessian $G(x)$ contains information about the vectors $u_1(x), \ldots, u_4(x)$, as expressed in the next proposition. For $x, y \in \mathbb{R}^2$, denote by $\det(x, y)$ the determinant of the matrix whose first column is x and the second is y.

The next goal is to outline the proof of Proposition 3.8, based on the following two lemmas.

Lemma 3.9. *Let K be a planar strictly convex C^1-regular body. Then $g_K(x)$ is twice continuously differentiable at every $x \in \operatorname{int} \operatorname{supp} g_K \setminus \{o\}$. Furthermore, for every $x \in \operatorname{int} \operatorname{supp} g_K \setminus \{o\}$, the following statements hold:*
(a) *The Hessian $G(x)$ is given by*

$$G(x) = \frac{u_2 u_1^\mathsf{T}}{\det(u_2, u_1)} - \frac{u_3 u_4^\mathsf{T}}{\det(u_3, u_4)} = \frac{u_1 u_2^\mathsf{T}}{\det(u_2, u_1)} - \frac{u_4 u_3^\mathsf{T}}{\det(u_3, u_4)}.$$

(b) *The determinant of $G(x)$ depends continuously on x and satisfies*

$$1 + \det G = \frac{\det(u_2, u_4) \det(u_1, u_3)}{\det(u_3, u_4) \det(u_1, u_2)}. \tag{3.3}$$

(c) *The vectors u_1 and u_3 and $G(x)$ are related by*

$$u_1^\mathsf{T} G(x)^{-1} u_3 = 0. \tag{3.4}$$

One can tell from $G(x)$ whether K is centrally symmetric. We say that a chord of K is an *affine diameter* if the normal vectors at ∂K at the endpoints of the chord are parallel.

Lemma 3.10. *Let K be a planar strictly convex C^1-regular body. The following are equivalent:*
(a) *K is centrally symmetric.*
(b) *At least one diagonal of each parallelogram inscribed in K is an affine diameter of K.*
(c) *The covariogram g_K is a solution of the* Monge–Ampère *differential equation*

$$\det G(x) = -1, \quad \textit{for } x \in \operatorname{int} \operatorname{supp} g_K \setminus \{o\}.$$

If one diagonal of the parallelogram $P(x)$ is an affine diameter, then u_1 is parallel to u_3 or u_2 is parallel to u_4 and (3.3) implies that $\det G(x) = -1$.

We are now ready to sketch the proof of Proposition 3.8. Due to the assumptions of Proposition 3.8 and (3.3) we have $u_1(x) \neq -u_3(x)$. We prove that there is a $y \neq x$ such that the parallelograms $P(x)$ and $P(y)$ satisfy $p_1(x) = p_1(y)$ and $p_3(x) = p_3(y)$. This clearly implies that $u_1(x) = u_1(y)$ and $u_3(x) = u_3(y)$. Thus, $u_1(x)$ and $u_3(x)$ satisfy the system given by the two equations obtained by evaluating (3.4) at both x and y. Lemma 5.3 in [4] expresses the vectors $u_1(x)$ and $u_3(x)$ in terms of the eigenvectors of $G(x)G(y)^{-1}$. In order to make this expression of $u_1(x)$ and $u_3(x)$ dependent only on the covariogram, it remains to prove that the property that $P(x)$ and $P(y)$ have a diagonal in common is shared by convex bodies with equal covariograms. The latter is done in [4, Proposition 5.4].

We now sketch how Proposition 3.8 is used to prove that if K and L are strictly convex C^1-regular bodies with $g_K = g_L$, then ∂K and ∂L have an arc in common, up to translations and reflections. Choose $x_0 \in \operatorname{int} \operatorname{supp} g_K \setminus \{o\}$ such that $\det G(x_0) \neq -1$. We claim that if x belongs to a suitable neighborhood U of x_0 and $p_3(K, x) = p_3(K, x_0)$, then $p_3(L, x) = p_3(L, x_0)$. Indeed, Proposition 3.8 together with a continuity argument allows us to prove that when x is close to x_0 and $u_3(K, x) = u_3(K, x_0)$, we have $u_3(L, x) = u_3(L, x_0)$. In view of the strict convexity of K and L, this proves the claim.

Now let $x(t)$, for $t \in [0, 1]$, be a parametrization of a curve contained in U with the property that, for each $t \in [0, 1]$, the parallelograms $P(K, x_0)$ and $P(K, x(t))$ are such that $p_3(K, x_0) = p_3(K, x(t))$. The previous claim implies that the arc of ∂K formed by the locus of $p_4(K, x(t))$, as t varies in $[0, 1]$, is a translate of the arc of ∂L formed by the locus of $p_4(L, x(t))$. Therefore, up to translations, ∂K and ∂L have an arc in common.

3.2 Counterexamples in $\mathbb{R}^n$ for any $n \geq 4$

We explain in Section 6 that the covariogram problem has a discrete counterpart which asks whether a finite set $A \subset \mathbb{R}^n$ is determined by its discrete covariogram (see (6.1) for the definition) and that there exists a construction that produces different sets (possibly multisets, sets with repetitions allowed) with equal discrete covariograms. If $B, C \subset \mathbb{R}^n$ are finite sets, then the multisets $B + C$ and $B + (-C)$ have the same discrete covariogram. When one tries to construct counterexamples to Matheron's conjecture via a similar procedure one immediately encounters two problems: the requirement that the corresponding Minkowski sums are sets, not multisets, and the requirement that they are convex. Choosing convex bodies in linear subspaces that intersect only in o solves both of these problems.

Theorem 3.11 (Bianchi [14]). *Let $\mathbb{R}^n = E \oplus F$ be the direct sum of the linear subspaces E and F and let $H \subset E$ and $K \subset F$ be convex bodies. Then the convex bodies $H + K$ and $H + (-K)$ in $\mathbb{R}^n$ have the same covariogram.*

If neither H nor K is centrally symmetric, then $H + K$ and $H + (-K)$ are not equal up to translations and reflections.

Proof. The property of having equal covariograms is invariant under linear maps, and the same is true for the property of being equal up to translations and reflections, since

$$g_{\mathcal{L}K}(x) = (\det \mathcal{L}) g_K(\mathcal{L}^{-1}x) \quad \text{and}$$

$$\mathcal{L}K = \mathcal{L}(\pm L) + y \quad \Longleftrightarrow \quad K = \pm L + \mathcal{L}^{-1}y,$$

for any invertible linear map $\mathcal{L}$, $x, y \in \mathbb{R}^n$ and $K, L \in \mathcal{K}^n$. We may therefore assume that E and F are orthogonal subspaces.

In this case, if we write $x = (x_1, x_2) \in E \oplus F$ and $\dim E = n_1$, $\dim F = n_2$, and $H \subset E$ and $K \subset F$ are convex bodies, then

$$\begin{aligned}
g_{H+K}(x_1, x_2) &= \mathcal{H}^n((H + K) \cap (H + K + (x_1, x_2))) \\
&= \mathcal{H}^n((H \cap (H + x_1)) + (K \cap (K + x_2))) \\
&= \mathcal{H}^{n_1}(H \cap (H + x_1))\mathcal{H}^{n_2}(K \cap (K + x_2)) \\
&= g_H(x_1) g_K(x_2).
\end{aligned} \tag{3.5}$$

A similar formula holds for $H + (-K)$ and the invariance of the covariogram with respect to reflections implies that $g_{H+K} = g_{H+(-K)}$.

To prove the second claim, suppose on the contrary that

$$H + K = H + (-K) + y \quad \text{or} \quad H + K = -(H + (-K)) + y, \tag{3.6}$$

for some $y \in \mathbb{R}^n$. We may assume that both H and K have centroid at the origin, since translations of H and K result in translations of $H + K$ and $H + (-K)$. Since the centroids of $H + K$ and $H + (-K)$ are at the origin, we have $y = 0$. Then by the cancelation law for Minkowski addition [66, p. 48], the first equality in (3.6) implies that $K = -K$, that is, K is centrally symmetric, a contradiction. Similarly, the second equality in (3.6) implies that H is centrally symmetric, again a contradiction. $\qquad\square$

Convex sets which are not centrally symmetric exist when the dimension of the ambient space is at least two, yielding the following corollary.

Corollary 3.12 (Bianchi [14]). *If $n \geq 4$ there exist convex bodies in $\mathcal{K}^n_n$ with the same covariogram which are not equal up to translations and reflections.*

This example is better understood if seen in the context of the decomposition of a convex body into direct summands. For $K \in \mathcal{K}^n_n$, we write

$$K = K_1 \oplus \cdots \oplus K_s \tag{3.7}$$

if $K = K_1 + \cdots + K_s$ for suitable convex bodies K_i lying in linear subspaces E_i of $\mathbb{R}^n$ such that $E_1 \oplus \cdots \oplus E_s = \mathbb{R}^n$. If a representation $K = L \oplus M$ is only possible when $\dim L = 0$ or $\dim M = 0$, then K is *directly indecomposable*. Each K with $\dim K \geq 1$ has a representation, unique up to the order of the summands, as in (3.7), with $\dim K_i \geq 1$ and K_i directly indecomposable.

If at least two of the summands of K, say K_1 and K_2, are not centrally symmetric, then $(-K_1) \oplus K_2 \oplus \cdots \oplus K_s$ has the same covariogram as K and is not equal to K up to translations and reflections. Two questions arise naturally:

(a) If $H \in \mathcal{K}^n_n$ and $g_H = g_K$, does H have a similar structure to K?

(b) If each directly indecomposable summand K_i is determined, up to translations and reflections, among the convex bodies in E_i by g_{K_i}, considered as a function defined in E_i, can the structure of H be understood?

The next theorem gives a positive answer to these questions.

Theorem 3.13 (Bianchi [15]). *Let $K \in \mathcal{K}^n_n$ and let E_i and K_i, $i = 1, \ldots, n$, be as in (3.7). If $H \in \mathcal{K}^n_n$ and $g_H = g_K$, then $H = H_1 \oplus \cdots \oplus H_s$, where, for each i, H_i is a directly indecomposable convex body contained in E_i and $g_{H_i} = g_{K_i}$.*

If in addition for each i, $g_{K_i} : E_i \to \mathbb{R}$ determines K_i among the convex bodies in E_i, up to translations and reflections, then H is a translation of $\sigma_1 K_1 \oplus \cdots \oplus \sigma_s K_s$, for suitable $\sigma_1, \ldots, \sigma_s \in \{-1, 1\}$.

In view of this, to understand the covariogram problem for general convex bodies it suffices to study it for indecomposable bodies. This last problem is however widely open, and the examples of non-determination described above are the only ones known.

For the proof of Theorem 3.13 we refer to [15]. Briefly, the first claim follows from the uniqueness of the decomposition into direct summands, the equality $DH = DK$, and a lemma that proves that a convex body is directly indecomposable if and only if its difference body is directly indecomposable. The second claim is a direct consequence of the first and the fact that the covariogram can be written in terms of the covariograms of the direct summands, with a formula similar to (3.5).

Before we conclude this section we prove, for later use, that if $E, F, H,$ and K are as in the statement of Theorem 3.11, then $H + K$ is not in the class C^1. This property is invariant under linear maps and we may therefore assume that E and F are orthogonal subspaces. For x in the boundary of H relative to E, let $N_E(H, x)$ (or $N(H, x)$) be the normal cone of H at x relative to E (or relative to $\mathbb{R}^n$, respectively). For y in the boundary of K relative to F, let $N_F(K, y)$ and $N(K, y)$ be defined in an analogous way. Schneider [66, (2.4) and Theorem 2.2.1(a)] proves that

$$N(H + K, x + y) = N(H, x) \cap N(K, y)$$
$$= \left(N_E(H, x) + F \right) \cap \left(E + N_F(K, y) \right) = N_E(H, x) + N_F(K, y).$$

This implies that the dimension of $N(H + K, x + y)$ is larger than 1. Thus, $\partial(H + K)$ is not C^1-regular at $x + y$.

3.3 Polytopes in $\mathbb{R}^n$, $n \geq 3$

In dimensions higher than 2 the covariogram problem has only partial results. The situation is better understood in the case of polytopes.

Theorem 3.14 (Goodey, Schneider, and Weil [38]). *If $P \in \mathcal{K}_n^n$, $n \geq 3$, is a polytope such that P and $-P$ are in general relative position and all its 2-dimensional faces are triangles, then P is determined by g_P, up to translations and reflections, in the class $\mathcal{K}^n$.*

The polytopes P and $-P$ are said to be in *general relative position* if for any two faces F and G of P lying in antipodal parallel supporting hyperplanes of P, $F \cap (G + x)$ contains at most one point, for any $x \in \mathbb{R}^n$. In $\mathbb{R}^3$, for instance, this means that P does not have pairs of parallel antipodal facets, or pairs of parallel antipodal edges, or an edge antipodal and parallel to a facet.

The proof of Theorem 3.14 is based on the Brunn–Minkowski inequality, together with its equality cases, and on a result about the decomposition of convex bodies in terms of sums of other convex bodies. Schneider [65] proves that the assumptions on P imply that every summand of DP is of the form $aP + (1 - a)(-P) + x$ with $a \in [0,1]$ and $x \in \mathbb{R}^n$. If $K \in \mathcal{K}^n$ satisfies $g_K = g_P$, then $\mathcal{H}^n(K) = \mathcal{H}^n(P)$ and $DK = DP$, by Propositions 2.1(a) and 2.2(a). The formula $K + (-K) = DK = DP$ says that K is a summand of DP and therefore

$$K = aP + (1 - a)(-P) + x.$$

If $a = 0$ or $a = 1$, then $K = -P + x$ or $K = P + x$, and we are done. If $a \in (0,1)$, then

$$\begin{aligned}
\mathcal{H}^n(K)^{1/n} &= \mathcal{H}^n\big(aP + (1 - a)(-P) + x\big)^{1/n} = \mathcal{H}^n\big(aP + (1 - a)(-P)\big)^{1/n} \\
&\geq a\mathcal{H}^n(P)^{1/n} + (1 - a)\mathcal{H}^n(-P)^{1/n} \\
&= \mathcal{H}^n(P)^{1/n} = \mathcal{H}^n(K)^{1/n}.
\end{aligned}$$

Thus, the Brunn–Minkowski inequality holds with equality and this implies that P and $-P$ are homothetic, i. e., P is centrally symmetric with respect to some point. Therefore, $K = aP + (1 - a)(-P) + x$ is a translate of P.

Theorem 3.15 (Bianchi [15]). *Let $P \in \mathcal{K}_3^3$ be a polytope. Then g_P determines P, in the class $\mathcal{K}^3$, up to translations and reflections.*

The proof of Theorem 3.15 is described in Section 3.4.

Theorem 3.11, when H and K are convex polytopes, has the following corollary.

Corollary 3.16. *For each $n \geq 4$, there exist pairs of polytopes in $\mathcal{K}_n^n$ with the same covariogram which are not equal up to translations and reflections.*

3.4 Some problems and ideas from the proof of Theorem 3.15

The proof of Theorem 3.15 is done in three steps and is contained in [15, 16, 17]. We first briefly describe all three steps and then each step in more detail. Let P and P' be convex polytopes in $\mathbb{R}^3$ with non-empty interior such that $g_P = g_{P'}$.

The *first step* consists in proving that ∂P and $\partial P'$ coincide locally up to translations and reflections. What this means is expressed by the next proposition.

Proposition 3.17. *Let P and P' be convex polytopes in $\mathbb{R}^3$ with non-empty interior such that $g_P = g_{P'}$. If $w \in S^2$, then there exists $\sigma = \sigma(w) \in \{-1, 1\}$ and $x = x(\sigma) \in \mathbb{R}^3$ such that*

$$\begin{aligned}
P_w &= (\sigma P')_w + x && \text{and} && \operatorname{cone}(P, P_w) = \operatorname{cone}(\sigma P', (\sigma P')_w), \\
P_{-w} &= (\sigma P')_{-w} + x && \text{and} && \operatorname{cone}(P, P_{-w}) = \operatorname{cone}(\sigma P', (\sigma P')_{-w}).
\end{aligned} \tag{3.8}$$

We recall that P_w is the face of P with outer normal w and that $\operatorname{cone}(P, P_w)$ is the support cone of P at P_w. Condition (3.8) implies that, for each proper face of P, be it a facet, an edge, or a vertex, after possibly a reflection and a translation, P and P' coincide in a neighborhood of that face and of the antipodal one.

One may wonder if the validity, for each w, of the local conditions (3.8) is sufficient to guarantee that $P' = P$, up to translations and reflections, but this is not the case, as Example 3.23 below shows. It can be proved that when (3.8) holds with $\sigma = 1$ (or with $\sigma = -1$) for each $w \in S^2$, then P' (or $-P'$, respectively) is a translate of P. A priori, however, (3.8) may hold with $\sigma = 1$ for some w, with $\sigma = -1$ for some w, and, possibly, both with $\sigma = 1$ and with $\sigma = -1$ for other w.

The set $\mathrm{int}(\partial P \cap \partial(\sigma(w)P' + x(\sigma)))$ (in this section the terms boundary, interior, and neighborhood of a subset of ∂P refer to the relative topology induced on ∂P by the Euclidean topology in $\mathbb{R}^3$) may have multiple components which depend on w. The *second step* of the proof consists in a study of these components and of their boundaries that leads to a choice of w such that the corresponding components satisfy certain convenient properties.

In the *third step* we use the setting prepared in the second step to conclude the proof by contradiction, by identifying some $y \in \mathbb{R}^3$ such that $g_P(y) \neq g_{P'}(y)$.

First step

In order to prove Proposition 3.17 we investigate two related problems. The first problem helps in proving the equalities in the first column in (3.8). In this chapter we explain only how to prove them when both P_w and P_{-w} are facets.

Assume that P_w and P_{-w} are facets and let $F = P_w|w^\perp$ and $G = P_{-w}|w^\perp$. We consider $P \cap (P + x)$ for x such that $P \cap (P + x) \neq \emptyset$ and the plane $\mathrm{aff}\, P_{-w} + x$ has a small distance, say ε, from the plane $\mathrm{aff}\, P_w$. In this situation, $P \cap (P + x)$ is approximately equal to a parallelepiped, with height ε and base a translate of $F \cap (G + y)$, where $y = x|w^\perp$. Thus,

$$\mathcal{H}^3(P \cap (P + x)) = \varepsilon \mathcal{H}^2(F \cap (G + y)) + o(\varepsilon) = \varepsilon g_{F,G}(y) + o(\varepsilon).$$

This formula proves that g_P determines the cross covariogram $g_{F,G}(y)$, for each $y \in w^\perp$. We thus encounter a first problem.

Problem 3.18 (Cross covariogram problem for polygons). Does the cross covariogram of the convex polygons $F, G \subset \mathbb{R}^2$ determine the pair (F, G), among all pairs of convex polygons, up to trivial associates?

A detailed description of its solution is presented in Section 3.6. Here we just anticipate that, for each choice of some real parameters, there exist four different pairs of parallelograms $(H_1, K_1), \ldots, (H_4, K_4)$ such that, for $i = 1, 3$, $g_{H_i, K_i} = g_{H_{i+1}, K_{i+1}}$, but (H_i, K_i) is not a trivial associate of (H_{i+1}, K_{i+1}), and that, up to affine transformations, the previous counterexamples are the only ones.

Thus, $g_{F,G}$ alone is not sufficient to determine F and G, and we have to get from g_P other information that eliminates the ambiguities due to the presence of these pairs of parallelograms. Rufibach [61, p. 14] was the first to observe the possibility of determining $g_F + g_G$ from g_P. His idea led to the next proposition.

Proposition 3.19. *Let* $P \subset \mathbb{R}^n$ *be a convex polytope with non-empty interior and let* $w \in S^{n-1}$, $F = P_w|w^\perp$, *and* $G = P_{-w}|w^\perp$. *The covariogram* g_P *determines both* $(g_F + g_G)(y)$ *and* $g_{F,G}(y)$, *for each* $y \in w^\perp$.

A proof of Proposition 3.19 has been presented [15] based on the expression of the second-order distributional derivative of g_P given in the next lemma.

Lemma 3.20. *Let* $P \subset \mathbb{R}^n$, $n \geq 2$, *be a convex polytope with non-empty interior. Let* $F_1, \ldots, F_m$ *be its facets, let* v_i *be the unit outer normal of P at* F_i, *for* $i = 1, \ldots, m$, *let* $w \in S^{n-1}$, *and let* $I_p = \{(i,j) : F_i \text{ is parallel to } F_j\}$ *and* $I_{np} = \{(i,j) : F_i \text{ is not parallel to } F_j\}$. *Then, for* $\phi \in C_0^\infty(\mathbb{R}^n)$, *we have*

$$-\frac{\partial^2 g_P}{\partial w^2}(\phi) = \sum_{(i,j)\in I_{np}} \frac{\langle w, v_i\rangle\langle w, v_j\rangle}{\sqrt{1 - \langle v_i, v_j\rangle^2}} \int_{\mathbb{R}^n} \mathcal{H}^{n-2}(F_i \cap (F_j + z))\, \phi(z)\, dz$$

$$+ \sum_{(i,j)\in I_p} \langle w, v_i\rangle\langle w, v_j\rangle \int_{F_i+(-F_j)} \mathcal{H}^{n-1}(F_i \cap (F_j + z))\, \phi(z)\, d\mathcal{H}^{n-1}(z). \qquad (3.9)$$

Both terms on the right-hand side of (3.9) *are determined by* g_P.

Proof of Proposition 3.19. The distribution defined by the second term on the right-hand side in (3.9) determines its support

$$S(P, w) = \bigcup\{F_i + (-F_j) : (i,j) \in I_p, \langle v_i, w\rangle \neq 0\}$$

and, for $\mathcal{H}^{n-1}$-almost each $x \in S(P, w)$, the expression

$$\sum_{(i,j)\in I_p} \langle w, v_i\rangle\langle w, v_j\rangle\mathcal{H}^{n-1}(F_i \cap (F_j + x)). \qquad (3.10)$$

The set $S(P, w)$ is contained in DP and consists of differences $F_i + (-F_j)$ of distinct parallel facets and of differences $F_i + (-F_i)$, with i such that $\langle v_i, w\rangle \neq 0$. The difference $F_i + (-F_j)$ is the facet $(DP)_{v_i}$ of DP. The difference $F_i + (-F_i)$ is contained in $v_i^\perp$.

Given $w \in S^2$, this information tells us whether P has zero, one, or two facets orthogonal to w. The polytope P has at least one facet orthogonal to w if and only if $(DP)_w$ is a facet of DP. It has two facets orthogonal to w if and only if $(DP)_w$ is a facet of DP and $(DP)_w \subset S(P, w)$. If it has two facets, then the part of the distribution supported in $w^\perp$ determines $g_F + g_G$, while the part supported in $(DP)_w$ determines $g_{F,G}$. If P has only one facet orthogonal to w and this facet is P_w, say, then the same holds, with the difference that now $g_F + g_G = g_F$ and $g_{F,G} = 0$. If P has no facet orthogonal to w, then $g_F + g_G = 0$ and $g_{F,G} = 0$. □

The next result decouples the information given by Proposition 3.19.

Lemma 3.21. *Let F, F', G, and G' be convex bodies in $\mathbb{R}^n$. If*

$$\begin{cases} g_F + g_G = g_{F'} + g_{G'}, \\ g_{F,G} = g_{F',G'}, \end{cases} \tag{3.11}$$

then either $g_F = g_{F'}$ and $g_G = g_{G'}$ or $g_F = g_{G'}$ and $g_G = g_{F'}$.

Proof. Applying the Fourier transform to the equalities in (3.11) we arrive, with the help of (1.3), at the system

$$\begin{cases} |\widehat{1_F}|^2 + |\widehat{1_G}|^2 = |\widehat{1_{F'}}|^2 + |\widehat{1_{G'}}|^2, \\ |\widehat{1_F}|^2|\widehat{1_G}|^2 = |\widehat{1_{F'}}|^2|\widehat{1_{G'}}|^2. \end{cases}$$

For each $\xi \in \mathbb{R}^n$, the previous system implies that either we have $|\widehat{1_F}(\xi)| = |\widehat{1_{F'}}(\xi)|$ and $|\widehat{1_G}(\xi)| = |\widehat{1_{G'}}(\xi)|$ or we have $|\widehat{1_F}(\xi)| = |\widehat{1_{G'}}(\xi)|$ and $|\widehat{1_G}(\xi)| = |\widehat{1_{F'}}(\xi)|$. A priori, the alternative may depend on ξ. The Fourier transform of a function with compact support is analytic and therefore the squared moduli of the previous transforms are analytic. Since any analytic function is determined by its values on a set with a limit point, we conclude that the previous alternative does not depend on ξ. Going back to covariograms via Fourier inversion, this means that either $g_F = g_{F'}$ and $g_G = g_{G'}$ or $g_F = g_{G'}$ and $g_G = g_{F'}$. $\qquad\square$

We are now ready to prove the equalities in the first column of (3.8) when both P_w and P_{-w} are facets. Let F and G be as above and let $F' = P'_w|w^\perp$ and $G' = P'_{-w}|w^\perp$. The faces P'_w and P'_{-w} are facets too, because otherwise $g_{F',G'} \equiv 0 \neq g_{F,G}$. If (F, G) and (F', G') are trivial associates, then, up to a reflection and/or translation of P', the equalities in the first column of (3.8) hold.

Now assume that (F, G) and (F', G') are not trivial associates. Theorem 3.30 states that (F, G) and (F', G') are, respectively, trivial associates of $(\mathcal{T}H_i, \mathcal{T}K_i)$ and $(\mathcal{T}H_j, \mathcal{T}K_j)$, for some affine transformation $\mathcal{T}$ and different indices i, j, with either $i, j \in \{1, 2\}$ or $i, j \in \{3, 4\}$.

Proposition 3.19 implies $g_{\mathcal{T}H_i} + g_{\mathcal{T}K_i} = g_{\mathcal{T}H_j} + g_{\mathcal{T}K_j}$. Lemma 3.21 and the positive answer to the covariogram problem in the plane imply that, up to translations and reflections, either $H_i = H_j$ and $K_i = K_j$ or $H_i = K_j$ and $K_i = H_j$. In view of the definition of these sets (see Figure 2) this is false.

It remains to prove the formulas in (3.8) regarding the support cones. For this purpose, P. Mani-Levitska, in a message to the author, suggested studying the following problem.

Problem 3.22 (Cross covariogram problem for polyhedral cones). Let A and B be convex polyhedral cones in $\mathbb{R}^n$, $n \geq 2$, with apex o and $A \cap B = \{o\}$. Does the cross covariogram of A and B determine the pair (A, B), among all pairs of convex polyhedral cones, up to trivial associates?

To see the relevance of this problem, suppose that P_w and P_{-w} are vertices and $x \in \mathbb{R}^3$ is chosen so that $P_{-w} + x$ is close to P_w. Then

$$P \cap (P + x) = A \cap (B + y),$$

where $A = \mathrm{cone}(P, P_w)$, $B = \mathrm{cone}(P, P_{-w})$, and $y = P_{-w} - P_w + x$. Thus, g_P determines $g_{A,B}(y)$ for each y in a neighborhood of o (and also for each $y \in \mathbb{R}^3$, since $g_{A,B}$ is 3-homogeneous). If we were able to determine A and B from $g_{A,B}$, we would be able to determine $\mathrm{cone}(P, P_w)$ and $\mathrm{cone}(P, P_{-w})$ from g_P, at least when P_w and P_{-w} are vertices.

A partial answer to Problem 3.22 in $\mathbb{R}^3$, which is sufficient for the purpose of proving Theorem 3.15, is given in Bianchi [18, Proposition 5.1]. Bianchi [16, Theorem 1.3] completely solves the problem in the plane, also describing some situations of non-determination.

Second step

The next example shows how to construct polytopes P and P' which satisfy (3.8) for every $w \in S^2$ such that $P \neq P'$, up to translations and reflections.

Example 3.23. Let $P \subset \mathbb{R}^3$ be a convex polytope such that $\Gamma \cup (-\Gamma) \subset \partial P$, where Γ is a simple closed curve such that $\Gamma \cap (-\Gamma) = \emptyset$. The union $\Gamma \cup (-\Gamma)$ disconnects ∂P into three components $\Sigma_j, j = 1, 2, 3$. Let $\partial \Sigma_1 = \Gamma, \partial \Sigma_2 = -\Gamma$, and $\partial \Sigma_3 = \Gamma \cup -\Gamma$. Choose P in such a way that $\Sigma_1 \neq -\Sigma_2, \Sigma_3 \neq -\Sigma_3$, and there exists a neighborhood W of Γ in ∂P which contains all faces of P intersecting Γ and $-W$ contains all faces of P intersecting $-\Gamma$.

Define P' as the polytope whose boundary is $\Sigma_1 \cup \Sigma_2 \cup (-\Sigma_3)$. The polytope P can be chosen so that $P' \neq P$, up to translations and reflections. We claim that (3.8) holds for each w. Indeed, if w is such that $P_w \cap \Gamma \neq \emptyset$, then $P_w \subset W$ and (3.8) holds both with $\sigma = -1$ and $x = o$ and with $\sigma = 1$ and $x = o$. If $P_w \cap (-\Gamma) \neq \emptyset$, then the same holds. If $P_w \subset (\mathrm{int}\,\Sigma_1) \cup (\mathrm{int}\,\Sigma_2)$, then (3.8) holds with $\sigma = 1$ and $x = o$. If $P_w \subset \mathrm{int}\,\Sigma_3$, then (3.8) holds with $\sigma = -1$ and $x = o$.

The construction above can be iterated and made more complex by considering other pairs of curves in ∂P which are reflections of each other, possibly with respect to a point different from o, not intersecting Γ and $-\Gamma$, and substituting one of the components of ∂P less all these curves with its reflection.

In the second step we study the components of

$$\mathrm{int}(\partial P \cap \partial(\sigma(w)P' + x(\sigma))) \tag{3.12}$$

and their boundaries when w varies in S^2. When P and P' are as in Example 3.23 and w is such that $P_w \subset \Sigma_1$, the set in (3.12) is $\mathrm{int}(\partial P \cap \partial P')$. This set has a component Σ_+ containing Σ_1 and a different "antipodal" component Σ_- containing Σ_2 (we assume here that P has been chosen so that $\Sigma_+ \neq \Sigma_-$). They satisfy

$$\Sigma_+ \neq -\Sigma_- \quad \text{and} \quad \partial\Sigma_+ = -\partial\Sigma_-.$$

The first formula holds because $\Sigma_1 \neq -\Sigma_2$, and the second one holds because both boundaries are contained in $\Sigma_3 \cap (-\Sigma_3)$, which is o-symmetric.

If we leave Example 3.23 and pass to the general case, there may exist w such that, if one defines Σ_+ and Σ_- as the components of the set in (3.12) containing P_w and P_{-w}, respectively, then $\partial\Sigma_+$ is not a reflection, with respect to some point, of $\partial\Sigma_-$. This can be seen if one modifies Example 3.23 as follows. Assume that ∂P contains, besides Γ and $-\Gamma$, a closed simple curve Λ and its reflection $-\Lambda+2z$ with respect to $z \neq o$, with $\Lambda \subset \operatorname{int}\Sigma_1$ and $-\Lambda + 2z \subset \operatorname{int}\Sigma_2$. In this case one can define P' starting from P and not only substituting Σ_3 with $-\Sigma_3$, but also exchanging the component Σ_4 of $\Sigma_1 \setminus \Lambda$ bounded by Λ with the reflection with respect to z of the component of $\Sigma_2 \setminus (-\Lambda + 2z)$ bounded by $-\Lambda + 2z$. If $P_w \subset \Sigma_1 \setminus \Sigma_4$, then it is not true that $\partial\Sigma_+$ is a reflection, with respect to some point, of $\partial\Sigma_-$. Indeed, in this case $\partial\Sigma_+$ has two components, and to obtain the two components of $\partial\Sigma_-$ one has to reflect one component of $\partial\Sigma_+$ with respect to o and the other component of $\partial\Sigma_+$ with respect to z.

The second step in the proof consists in proving in the general case that if $P \neq P'$, up to translations and reflections, then it is always possible to choose $w \in S^2$ so that there exist $z \in \mathbb{R}^3$ and two antipodal components $\Sigma_+ \supset P_w$ and Σ_- of the set in (3.12) such that

$$\Sigma_+ \neq -\Sigma_- + 2z \quad \text{and} \quad \partial\Sigma_+ = \partial(-\Sigma_- + 2z). \tag{3.13}$$

Third step

In this step we use the structure discovered in Step 2 to conclude and prove that, if $P \neq P'$, up to translations and reflections, then we can find $y \in \mathbb{R}^3$ such that $g_P(y) \neq g_{P'}(y)$. For the details we refer to [15, p. 1804].

3.5 Smooth convex bodies in $\mathbb{R}^n$, $n \geq 3$

In this section we deal with convex bodies that are at least C_+^2-regular. Every such body is directly indecomposable (we have proved at the end of Section 3.2 that the direct sum of two lower-dimensional convex bodies is not C^1-regular) and we do not have to worry about the examples in Section 3.2.

The covariogram problem for C_+^2-bodies is still open, even in $\mathbb{R}^3$, and the only results available are positive ones for bodies with higher regularity. To prove these results it has been useful to connect the covariogram problem to some studies regarding the zero set $\mathcal{Z}(K) = \{\zeta \in \mathbb{C}^n : \widehat{1_K}(\zeta) = 0\}$ of the Fourier transform $\widehat{1_K}$ seen as a function on $\mathbb{C}^n$. This set plays a role in attempts to solve the famous Pompeiu problem, a long-standing open problem in integral geometry (see, for instance, Garofalo and Segala [36] and Machado and Robins [49]), which we describe in detail in Section 7. Here we focus on the work of

Kobayashi [45, 46] regarding the geometric information about K contained in $\mathcal{Z}(K)$. In 1986 Kobayashi [45] posed the following problem.

Problem 3.24. Does the zero set $\mathcal{Z}(K) = \{\zeta \in \mathbb{C}^n : \widehat{1_K}(\zeta) = 0\}$ determine the convex body K, among all convex bodies, up to translations?

(Note that a translation of K leaves $\mathcal{Z}(K)$ unchanged.) In the class of C_+^∞-convex bodies, Problem 3.24 has been solved by Kobayashi [45] in the planar case, but it is still open for $n \geq 3$. In connection with Problem 3.24, Kobayashi studies the asymptotic behavior at infinity of $\mathcal{Z}(K)$, in any dimension but only in the case of C_+^∞-convex bodies. It turns out that this asymptotic behavior contains information about the width function of K and the ratio of the Gauss curvatures of ∂K at antipodal points (see Proposition 3.26 and Problem 3.27 below).

In Bianchi [18] Kobayashi's result regarding the asymptotics of $\mathcal{Z}(K)$ is proved under weaker regularity assumptions, replacing $K \in C_+^\infty$ by $K \in C_+^{r(n)}$, where $r(n)$ is as in Theorem 3.25. This is a key tool in the following positive answer to the covariogram problem for $C_+^{r(n)}$-convex bodies in $\mathbb{R}^n$, $n \geq 2$.

Theorem 3.25 (Bianchi [18]). *Let $n \geq 2$ and define $r(n) = 8$ when $n = 2, 4, 6$, $r(n) = 9$ when $n = 3, 5, 7$, and $r(n) = [(n - 1)/2] + 5$ when $n \geq 8$. Let H and K be convex bodies in $\mathbb{R}^n$ of class $C_+^{r(n)}$. Then $g_H = g_K$ implies $H = K$, up to translations and reflections.*

Note that Theorem 3.25 only proves that the covariogram determines a $C_+^{r(n)}$-body among $C_+^{r(n)}$-bodies, and it is not known whether the determination holds among all convex bodies.

We have explained in Section 3.1 that if K is C_+^2 regular, then for each $u \in S^{n-1}$, g_K provides the non-ordered pair $\{\tau_K(u), \tau_K(-u)\}$. Thus, if H is of class C_+^2 and $g_H = g_K$, the continuity of the curvature implies that given any component U of $\{u \in S^{n-1} : \tau_K(u) \neq \tau_K(-u)\}$, we have, possibly after a reflection of H,

$$\tau_H(u) = \tau_K(u), \quad \text{for each } u \in U. \tag{3.14}$$

If (3.14) were true for each $u \in S^{n-1}$, then H and K would coincide, up to a translation, by the uniqueness part of Minkowski's theorem [66, Theorem 7.2.1]. However, a priori the reflection that makes (3.14) valid may depend on the component U. The key ingredient in resolving this ambiguity, when the body is $C_+^{r(n)}$-regular, is the fact that the maps $F_{m,K}$ appearing in the statement of the next proposition are analytic.

Proposition 3.26 (Kobayashi [45], Bianchi [18]). *Let $S = \{\zeta \in \mathbb{C}^n : \zeta = zu, \text{ with } z \in \mathbb{C}, u \in S^{n-1}\}$, where we identify zu and $(-z)(-u)$, for each $z \in \mathbb{C}$ and $u \in S^{n-1}$. Let K be a convex body in $\mathbb{R}^n$ of class $C_+^{r(n)}$, where $r(n)$ is as in Theorem 3.25. Then there exists a positive integer $m(K)$ such that*

$$\mathcal{Z}(K) \cap S = \left(C(K) \cup \bigcup_{m=m(K)}^{\infty} \mathcal{Z}_m(K) \right),$$

where $C(K)$ is a bounded set and the union is disjoint. Moreover, for each integer $m \geq m(K)$, there exists an analytic map $F_{m,K} : S^{n-1} \to \mathbb{C}$ such that

$$\mathcal{Z}_m(K) = \{F_{m,K}(u)\, u \, : u \in S^{n-1}\}, \tag{3.15}$$

where

$$F_{m,K}(u) = \frac{\pi(4m + n - 1)}{2w_K(u)} + i\, \frac{\ln \tau_K(-u) - \ln \tau_K(u)}{2w_K(u)} + O\!\left(\frac{1}{m}\right) \tag{3.16}$$

and $O(1/m) \to 0$ as $m \to \infty$, uniformly in $u \in S^{n-1}$.

Indeed, formula (2.3) implies that

$$\{\zeta \in \mathbb{C}^n : \widehat{g_K}(\zeta) = 0\} = \mathcal{Z}(K) \cup \overline{\mathcal{Z}(K)}.$$

Thus, g_K gives the real part of $\mathcal{Z}(K)$ and hence, in view of (3.16), the width function of K. This is nothing new, since DK, the support of g_K, already determines w_K. But g_K also determines the imaginary part of $\mathcal{Z}(K)$, up to conjugation. In view of (3.16), it determines the imaginary part of $\mathcal{Z}(K)$, up to reflections of K. Thus, if H and K are as in Theorem 3.25 and, possibly, we have reflected H so that (3.14) holds for $u \in U$, then

$$F_{m,H}(u) = F_{m,K}(u) \quad \text{for } m \text{ large enough and } u \in U. \tag{3.17}$$

We can then use the analyticity of these maps to deduce that (3.17) holds for $u \in S^{n-1}$, which implies that (3.14) holds for $u \in S^{n-1}$. This concludes the sketch of the proof of Theorem 3.25.

We restate here [45, Problem 1.13] in the class $C_+^{r(n)}$.

Problem 3.27. If $H, K \in \mathcal{K}_n^n$ are in $C_+^{r(n)}$ and, for $u \in S^{n-1}$,

$$w_H(u) = w_K(u) \quad \text{and} \quad \frac{\tau_H(-u)}{\tau_H(u)} = \frac{\tau_K(-u)}{\tau_K(u)},$$

is $H = K$, up to translations?

A positive answer implies, due to Proposition 3.26, a positive answer to Problem 3.24 in $C_+^{r(n)}$. An answer is known only for $n = 2$, and is positive in that case [45, Corollary 2.3.10].

3.6 Determination from cross covariogram

We restate the cross covariogram problem in greater generality.

Problem 3.28 (Cross covariogram problem). Does $g_{H,K}$ determine the pair (H, K) of closed convex sets among all pairs of closed convex sets, up to trivial associates?

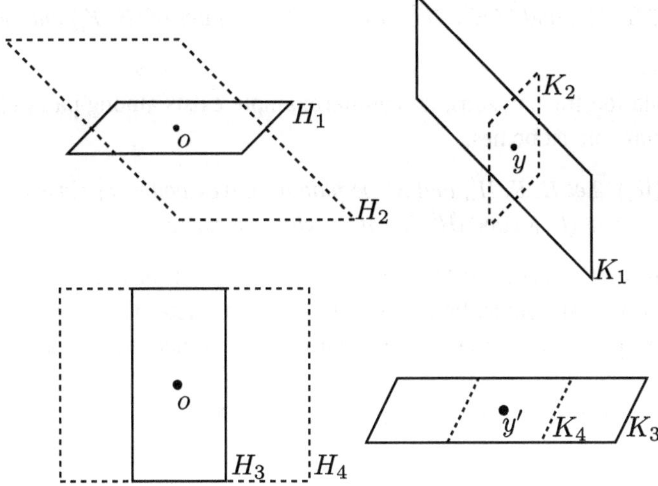

Figure 2: Here, $g_{H_1,K_1} = g_{H_2,K_2}$ and $g_{H_3,K_3} = g_{H_4,K_4}$. Moreover, up to affine transformations, these are the only pairs of planar convex polygons with equal cross covariograms.

When H and K are convex polygons, and also when they are planar convex cones, a complete answer is given in Bianchi [16]. When H and K are sufficiently regular planar convex bodies, the solution can be found in Bianchi [18]. In the case of polygons (and of planar cones as well) there are examples of non-determination.

Example 3.29. Let α, β, γ, δ, α', β', γ', and δ' be positive real numbers, $m \in \mathbb{R}$, $y, y' \in \mathbb{R}^2$, $I_1 = [(-1,0),(1,0)]$, $I_2 = 1/\sqrt{2}\,[(-1,-1),(1,1)]$, $I_3 = [(0,-1),(0,1)]$, $I_4 = 1/\sqrt{2}\,[(1,-1),(-1,1)]$, and $I_5 = (1/\sqrt{1+m^2})\,[(-m,-1),(m,1)]$. Assume that either $m = 0$, $\alpha' \neq \gamma'$, and $\beta' \neq \delta'$ or $m \neq 0$ and $\alpha' \neq \gamma'$. We define four pairs of parallelograms as follows (see Figure 2):

$$\begin{aligned}
H_1 &= \alpha I_1 + \beta I_2, & K_1 &= \gamma I_3 + \delta I_4 + y, \\
H_2 &= \alpha I_1 + \delta I_4, & K_2 &= \beta I_2 + \gamma I_3 + y, \\
H_3 &= \alpha' I_1 + \beta' I_3, & K_3 &= \gamma' I_1 + \delta' I_5 + y', \\
H_4 &= \gamma' I_1 + \beta' I_3, & K_4 &= \alpha' I_1 + \delta' I_5 + y'.
\end{aligned}$$

For $i = 1, 3$, we have $g_{H_i,K_i} = g_{H_{i+1},K_{i+1}}$ but (H_i, K_i) is not a trivial associate of (H_{i+1}, K_{i+1}).

The next theorem proves that, up to affine transformations, the previous counterexamples are the only ones.

Theorem 3.30 (Bianchi [16]). *Let H and K be convex polygons and let H' and K' be planar convex bodies with $g_{H,K} = g_{H',K'}$. If (H,K) is a not a trivial associate of (H',K'), then there are an affine transformation $\mathcal{T}$ and different indices i,j, with either $i,j \in \{1,2\}$ or*

$i, j \in \{3, 4\}$, *such that* $(\mathcal{T}H, \mathcal{T}K)$ *and* $(\mathcal{T}H', \mathcal{T}K')$ *are trivial associates of* (H_i, K_i) *and of* (H_j, K_j), *respectively.*

Contrary to the situation for polygons, no counterexample exists among pairs of sufficiently regular planar convex bodies.

Theorem 3.31 (Bianchi [18]). *Let* H, K, H', *and* K' *be planar convex bodies of class* C_+^8. *Then* $g_{H,K} = g_{H',K'}$ *implies that* (H, K) *and* (H', K') *are trivial associates.*

In summary, the information provided by the cross covariogram of convex polygons or of sufficiently smooth planar convex bodies is rich enough to determine not only one unknown body, as required by Matheron's conjecture, but two bodies, with a few exceptions.

Problem 3.27 is also relevant in trying to extend Theorem 3.31 to $\mathbb{R}^n$, $n > 2$.

4 Algorithms for reconstruction

None of the uniqueness proofs provide a method for actually reconstructing a convex body from its covariogram. For the phase retrieval problems, many algorithms have been developed, motivated by the diverse applications. We refer the interested reader to [10, 26, 27] for a description. We are aware of only three papers dealing specifically with the reconstruction from the covariogram. Schmitt [64] gives an explicit reconstruction procedure for a convex polygon when no pair of its edges are parallel, an assumption removed in an algorithm due to Benassi and D'Ercole [12]. Bianchi, Kiderlen, and Gardner [19] solve the following three problems. In each, K is a convex body in $\mathbb{R}^n$.

Problem 4.1 (Reconstruction from covariograms). Construct an approximation to K from a finite number of noisy (i. e., taken with error) measurements of g_K.

Problem 4.2 (Phase retrieval for characteristic functions of convex bodies: squared modulus). Construct an approximation to K (or, equivalently, to 1_K) from a finite number of noisy measurements of $|\widehat{1_K}|^2$.

Problem 4.3 (Phase retrieval for characteristic functions of convex bodies: modulus). Construct an approximation to K from a finite number of noisy measurements of $|\widehat{1_K}|$.

In both [64] and [12], all the exact values of the covariogram are supposed to be available. In contrast, the set of algorithms in [19] for Problem 4.1 take as input only a finite number of values of the covariogram of K. Moreover, these measurements are corrupted by errors, modeled by zero mean random variables with uniformly bounded p-th moments, where p is at most 6 and usually 4. It is assumed that K is determined by its covariogram, has its centroid at the origin, and is contained in a known bounded region of $\mathbb{R}^n$, which for convenience is taken to be the unit cube $[-1/2, 1/2]^n$. Two different methods have been provided [19] for reconstructing, for each suitable $k \in \mathbb{N}$, a convex polytope P_k that approximates K or its reflection $-K$. Each method involves two algorithms, an initial algorithm that produces suitable outer unit normals to the facets of P_k and a common main algorithm that goes on to actually construct P_k.

In the first method, the covariogram of K is measured, *multiple times*, at the origin and at vectors $(1/k)u_i$, $i = 1, \ldots, k$, where the u_i's are mutually non-parallel unit vectors that span $\mathbb{R}^n$. From these measurements, the initial algorithm NoisyCovBlaschke constructs an o-symmetric convex polytope Q_k that approximates ∇K, the Blaschke body of K. The crucial property of ∇K is that when K is a convex polytope, each of its facets is parallel to some facet of ∇K. It follows that the outer unit normals to the facets of P_k can be taken to be among those of Q_k. NoisyCovBlaschke utilizes (2.9), i. e., the fact that $-(\partial^+ g_K/\partial u)(o)$ equals the brightness function $b_K(u)$. This connection allows most of the work to be done by an algorithm, designed earlier by Gardner and Milanfar (see [34] and the references given there) for reconstructing an o-symmetric convex body from finitely many noisy measurements of its brightness function.

The second method achieves the same goal with a quite different approach. This time the covariogram of K is measured once at each point in a cubic array in $[-1,1]^n$ of side length $1/k$. From these measurements, the initial algorithm NoisyCovDiff(φ) constructs an o-symmetric convex polytope Q_k that approximates the difference body DK. The set DK has precisely the same property as ∇K, that when K is a convex polytope, each of its facets is parallel to some facet of DK. Furthermore, DK is just the support of g_K. The known property that $g_K^{1/n}$ is concave can therefore be combined with techniques from multiple regression. NoisyCovDiff(φ) employs a Gasser–Müller-type kernel estimator for g_K, with suitable kernel function φ, bandwidth, and threshold parameter.

The output Q_k of either initial algorithm forms part of the input to the main common algorithm NoisyCovLSQ. The covariogram of K is now measured *again*, once at each point in a cubic array in $[-1,1]^n$ of side length $1/k$. Using these measurements, Noisy-CovLSQ finds a convex polytope P_k, each of whose facets is parallel to some facet of Q_k, whose covariogram fits best the measurements in the least squares sense.

These algorithms are strongly consistent. Whenever K is determined among convex bodies, up to translations and reflections, by its covariogram, it has been shown [19] that, almost surely,

$$\min\{\delta(K, P_k), \delta(-K, P_k)\} \to 0$$

as $k \to \infty$. (If K is not so determined, the algorithms still construct a sequence (P_k) whose accumulation points exist and have the same covariogram as K.) From a theoretical point of view, this completely solves Problem 4.1.

The basic idea in [19] to solve Problem 4.2 is simple enough: Use (1.3) and the measurements of $|\widehat{1_K}|^2$ at points in a suitable cubic array to approximate g_K via its Fourier series and feed the resulting values into the algorithms for Problem 4.1. Two major technical obstacles arise. The new estimates of g_K are corrupted by noise that now involves

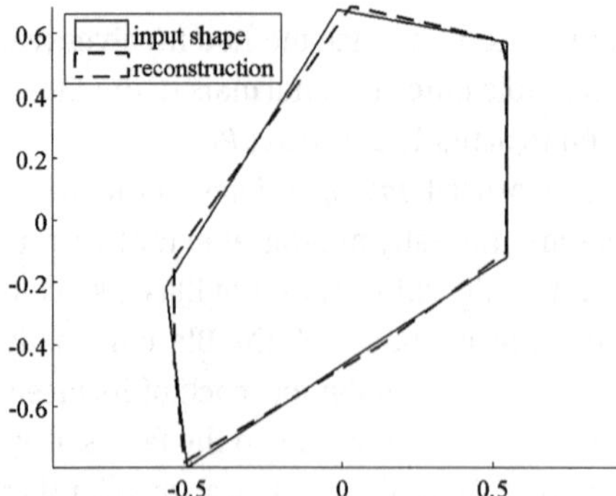

Figure 3: Pentagon, no noise.

dependent random variables, and a new deterministic error appears as well. A substitute for the strong law of large numbers must be proved, and the deterministic error must be controlled using Fourier analysis and the fortunate fact that g_K is Lipschitz. In the end the basic idea works, assuming that for suitable $1/2 < \gamma < 1$, measurements of $|\widehat{1_K}|^2$ are taken at the points in $(1/k^\gamma)\mathbb{Z}^n$ contained in the cubic window $[-k^{1-\gamma}, k^{1-\gamma}]^n$, whose size increases with k at a rate depending on the parameter γ. The three resulting algorithms, NoisyMod^{2}LSQ, NoisyMod^{2}Blaschke, and NoisyMod^{2}Diff(φ), are stated in detail and, with suitable restrictions on γ, proved to be strongly consistent under the same hypotheses as for Problem 4.1.

Three algorithms [19] have also been constructed for Problem 4.3. Again there is a basic simple idea, namely, to take two independent measurements at each of the points in the same cubic array as in the previous paragraph, multiply the two, and feed the resulting values into the algorithms for Problem 4.2. No serious extra technical difficulties arise, and it has been proved that the three new algorithms are strongly consistent under the same hypotheses as for Problem 4.2. This provides a complete theoretical solution to the phase retrieval problem for characteristic functions of convex bodies.

The study in [19] is a theoretical one. Convergence rates are given for NoisyCov-Diff(ϕ), and hence for the two related algorithms for phase retrieval, but are missing for the other algorithms. In particular, proving convergence rates for NoisyCovLSQ would need suitable stability versions of the uniqueness results for the covariogram problem, which are not available. Figures 3, 4, 5, and 6, taken from [19], present the reconstructions obtained in some experiments of a rudimentary implementation of Noisy-CovBlaschke and NoisyCovLSQ in the planar case. They are based on Gaussian $N(0, \sigma^2)$ noise and $k = 60$ equally spaced directions in NoisyCovBlaschke and $k = 8$ in Noisy-CovLSQ.

The website Geometric Tomography [32], a project of R. J. Gardner, offers a GUI to access a basic implementation of the algorithms for Problem 4.1 in the plane.

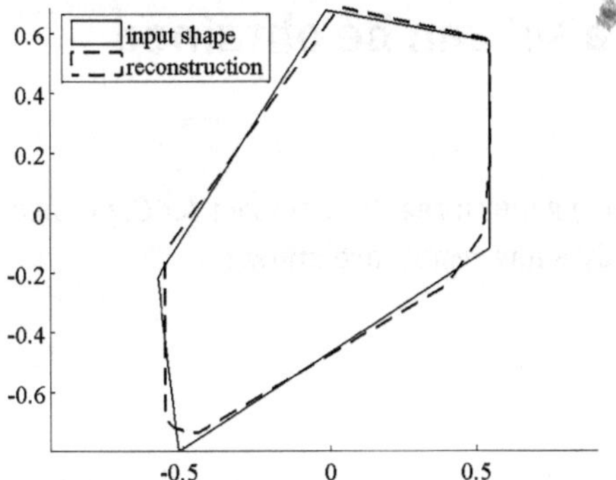

Figure 4: Pentagon, $\sigma = 0.01$.

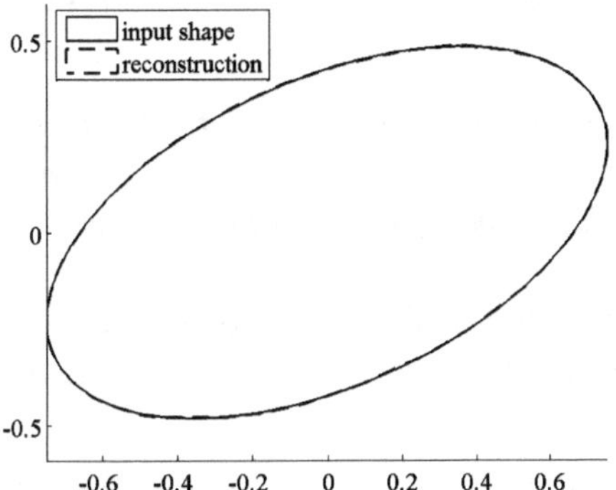

Figure 5: Ellipse, no noise.

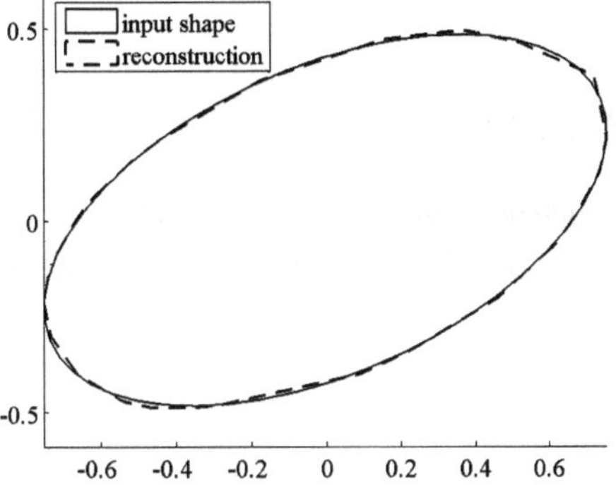

Figure 6: Ellipse, $\sigma = 0.01$.

5 What information about a set can be obtained from its covariogram?

A natural question is what information about a general regular compact set C, not necessarily convex, can be obtained from g_C. Only a few results are known.

5.1 Recognizing convexity

There are some properties of the covariogram of a convex set that may help in distinguishing it from a non-convex one, for instance, the concavity of $g_C^{1/n}$ on its support, the convexity of supp g_C, and inequality (2.6) between $g_C(o)$ and the volume of supp g_C coming from the Rogers–Shephard inequality. Benassi, Bianchi, and D'Ercole [11] give some other answers to this problem, mostly in the case of planar sets. Their main result is synthesized in the following theorem.

Theorem 5.1. *Let $\mathcal{D}$ be the class of regular compact sets in $\mathbb{R}^2$ whose interiors have at most two components and let $\mathcal{P}$ denote the class of regular compact sets in $\mathbb{R}^2$ whose boundaries consist of a finite number of closed disjoint polygonal curves, each with finitely many edges.*

If $C \in \mathcal{D}$, then the information provided by supp g_C and $(\partial^+ g_C/\partial u)(o)$, for all $u \in S^1$, determines whether C is convex. If $C \in \mathcal{P}$, then the information provided by supp g_C and the discontinuities of ∇g_C determines whether C is convex.

We explain the result regarding the class $\mathcal{D}$. If $C \subset \mathbb{R}^2$ is a convex body and $u \in S^1$, the formula

$$b_C(u) = w_C(\mathcal{R}_{\frac{\pi}{2}} u) = \frac{1}{2} w_{DC}(\mathcal{R}_{\frac{\pi}{2}} u)$$

and Proposition 2.2(a) allow (2.9) to be written as

$$-\frac{\partial^+ g_C}{\partial u}(o) = \frac{1}{2} w_{\operatorname{supp} g_C}(\mathcal{R}_{\frac{\pi}{2}} u). \tag{5.1}$$

If $C \in \mathcal{D}$ is not convex, then there exists $u \in S^1$ such that $C|u^\perp$ is an interval and $C \cap (l_u + x)$ has at least two components for a set of $x \in u^\perp$ of positive $\mathcal{H}^1$-measure. For such u one can prove that either $(\partial^+ g_C/\partial u)(o)$ does not exist or

$$-\frac{\partial^+ g_C}{\partial u}(o) > \frac{1}{2} w_{\operatorname{supp} g_C}(\mathcal{R}_{\frac{\pi}{2}} u), \tag{5.2}$$

violating (5.1). To see in a particular example why strict inequality holds in (5.2), imagine what happens when C is the union of two disjoint convex bodies C_1 and C_2 and u is as above. In this case, $-(\partial^+ g_C/\partial u)(o) = -(\partial^+ (g_{C_1} + g_{C_2})/\partial u)(o) = b_{C_1}(u) + b_{C_2}(u) = \mathcal{H}^1(C_1|u^\perp) + \mathcal{H}^1(C_2|u^\perp)$, while the term on the right-hand side in (5.2) is

$$\frac{1}{2}w_{DC}(\mathcal{R}_{\frac{\pi}{2}}u) = \frac{1}{2}w_{D(\operatorname{conv}C)}(\mathcal{R}_{\frac{\pi}{2}}u) = w_{\operatorname{conv}C}(\mathcal{R}_{\frac{\pi}{2}}u)$$

$$= \mathcal{H}^1(\operatorname{conv}(C_1 \cup C_2)|u^\perp) < \mathcal{H}^1(C_1|u^\perp) + \mathcal{H}^1(C_2|u^\perp). \qquad (5.3)$$

The first equality in (5.3) is due to the fact that $C|u^\perp$ is an interval, while the last inequality holds because $C_1|u^\perp$ and $C_2|u^\perp$ overlap, a consequence of the assumption about $C \cap (l_u + x)$.

When $C \in \mathcal{P}$, the result rests ultimately on Lemma 3.20, which remains valid for non-convex elements of $\mathcal{P}$. This lemma expresses the discontinuities of ∇g_C through the singular part of the distributional derivative $\partial^2 g_C/\partial w^2$, for $w \in S^1$, i. e., through the distribution defined by the second sum in (3.9). This distribution is supported in the set formed by the differences of any two parallel edges of C, including the differences of an edge with itself. This is a finite union of segments. Moreover, if $x \neq o$ belongs to this set, it provides the length of $\partial C \cap (\partial C + x)$. It has been proved [11] that this information and the knowledge of $\operatorname{supp} g_C$ distinguish between convex and non-convex sets in $\mathcal{P}$.

A consequence of Theorem 5.1 is a strengthening of Theorem 3.2.

Corollary 5.2. *Every planar convex body is determined within the class $\mathcal{D} \cup \mathcal{P}$ by its covariogram, up to translations and reflections.*

5.2 Recognizing symmetry properties

The covariogram g_C is an even function, independently of any symmetry property of C. When C is convex, recognizing from g_C whether C is centrally symmetric is possible.

Theorem 5.3. (a) *Let $C \in \mathcal{C}^n$ be regular. The set C is convex and centrally symmetric if and only if $2^n g_C(o) = \mathcal{H}^n(\operatorname{supp} g_C)$. If this equality holds, then $C = (1/2)\operatorname{supp} g_C$, up to translations.*

(b) *A centrally symmetric convex body is determined by its covariogram, up to translations, in the class of all regular compact sets.*

Proof. Item (a) is a consequence of the equality condition in inequality (2.5) in Proposition 2.2 and $\operatorname{supp} g_C = DC = 2C$. Item (b) follows from Item (a). $\qquad \square$

In contrast to this, we do not know of a way of recognizing from g_C the central symmetry of a non-convex set C.

It is possible to recognize the radial symmetry of the set from its covariogram. A result of Lawton [47, Corollary 1] yields the following theorem.

Theorem 5.4. *Let $n \geq 2$ and let $C \subset \mathbb{R}^n$ be a regular compact set such that g_C is radially symmetric. Then a translation of C is radially symmetric and C is determined by g_C, up to translations and reflections, in the class of regular compact sets.*

Lawton proves the corresponding result for real-valued $L^2(\mathbb{R}^n)$-functions with compact support using techniques from the theory of functions of several complex variables.

More precisely, the result is a consequence of a representation formula for entire functions of exponential type such that the modulus of their restriction to $\mathbb{R}^n$ is radially symmetric and in $L^2(\mathbb{R}^n)$.

6 The discrete covariogram

There is a counterpart to the covariogram in the discrete case. The *discrete covariogram* g_A of a finite subset A of $\mathbb{R}^n$ is defined by

$$g_A(x) = |A \cap (A + x)|, \tag{6.1}$$

for $x \in \mathbb{R}^n$. When no confusion can arise, we shall refer to the discrete covariogram of a finite set simply as its covariogram. As in the case of the ordinary (continuous) covariogram, it is unchanged by a translation or a reflection, and its support is DA. Note that

$$g_A(x) = |\{y \in A : y - x \in A\}|,$$

i. e., $g_A(x)$ is the number of "chords" of A that are translates of the line segment $[o, x]$. Thus, the covariogram can be identified with the *multiset* $A + (-A)$, that is, the set DA where each element is repeated with multiplicity. In particular, $g_A = g_B$ if and only if A and B have the same set of chords, each repeated with multiplicity, and this is true if and only if $A + (-A)$ and $B + (-B)$ are equal as multisets.

Finite sets with equal covariograms are sometimes called *homometric*. Also, multisets A and B such that the multisets $A+(-A)$ and $B+(-B)$ are equal are called *homometric*. When two sets are homometric and not equal up to translations and reflections, we say that they are *non-trivially homometric*. We refer to the survey paper of Senechal [67] for an introduction to homometric sets. Here we mention only a few facts.

If A and B are multisets, then the multisets

$$A + B \quad \text{and} \quad A + (-B) \tag{6.2}$$

are homometric. Indeed,

$$(A + B) + (-(A + B)) = A + (-A) + B + (-B) = (A + (-B)) + (-(A + (-B))).$$

If $|A| = 2$ or $|B| = 2$, then A or B is centrally symmetric and the two sets in (6.2) are equal up to translations and reflections. Thus, one cannot construct four-, six-, and eight-point non-trivially homometric pairs this way, but nine-point pairs abound. If A and B are sets, not multisets, and each point of $A + B$ (and of $A + (-B)$) can be written in a unique way as a sum of a point of A and a point of B (or of $-B$, respectively), then $A + B$ and $A + (-B)$ are sets with equal covariograms. Thus, for example, $\{0, 1, 3, 8, 9, 11, 12, 13, 15\}$

and $\{0, 1, 3, 4, 5, 7, 12, 13, 15\}$ in $\mathbb{R}$ have equal covariograms and arise from the above construction by taking $A = \{6, 7, 9\}$ and $B = \{-6, 2, 6\}$. Not every homometric set can be constructed by this procedure. For example, $\{0, 1, 2, 5, 7, 9, 12\}$ and $\{0, 1, 5, 7, 8, 10, 12\}$ have equal covariograms, but do not arise from the above construction. Indeed, if they did, we would have $|A||B| = |A + B| = 7$ and hence either $|A| = 1$ or $|B| = 1$, an impossibility.

Gardner, Gronchi, and Zong [33, Theorem 4.5] establish the following connection between the discrete and the continuous covariogram.

Theorem 6.1. *Let A and B be finite subsets of $\mathbb{R}^n$ with equal discrete covariograms. If X is a bounded Lebesgue-measurable set such that*

$$\mathcal{H}^n(A + X) = |A|\mathcal{H}^n(X) \quad and \quad \mathcal{H}^n(B + X) = |B|\mathcal{H}^n(X),$$

then $A + X$ and $B + X$ have equal continuous covariograms.

The assumption $\mathcal{H}^n(A + X) = |A|\mathcal{H}^n(X)$ says that there are no overlaps in the sum $A + X$, i. e., in the union $\bigcup_{a \in A}(X + a)$, except for sets of measure zero. A consequence of this theorem is that if A and B are lattice sets with equal discrete covariograms, then the associated lattice bodies $A + [0, 1]^n$ and $B + [0, 1]^n$ have equal continuous covariograms.

Gardner, Gronchi, and Zong [33] present a pair of non-congruent non-convex polygons with equal covariograms; see Figure 7. They are the lattice bodies associated with non-trivially homometric planar polyominoes which can be written as $A+B$ and $A+(-B)$, where A and B are the lattice sets in Figure 8.

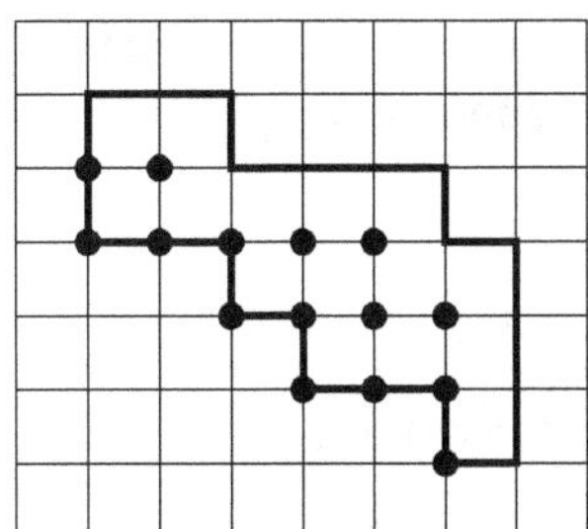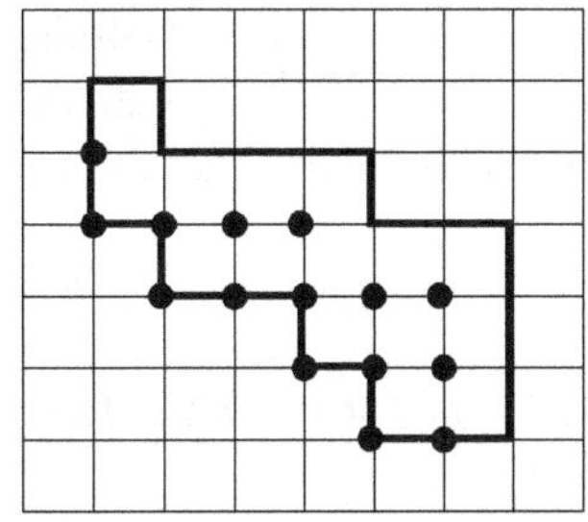

Figure 7: Two non-congruent non-convex polygons with equal covariograms. They arise as lattice bodies of two homometric convex polyominoes.

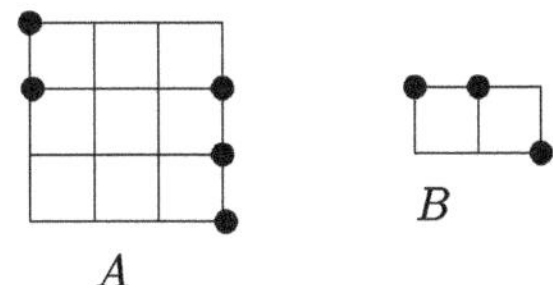

Figure 8: The polyominoes in Figure 7 are equal to $A + B$ and $A + (-B)$.

Another pair of planar lattice bodies with equal covariograms, made of nine squares and not equal up to translations and reflections, appeared in [24, Figure 1]. The polyominoes in the examples in [33] and in [24] are convex, and the associated lattice bodies are both horizontally and vertically convex. In the example in [24] one lattice body is the reflection with respect to a line of the other one.

Gardner, Gronchi, and Zhong [33] prove that a centrally symmetric finite set A is determined by g_A, up to translations, in the class of centrally symmetric finite sets, thus extending Theorem 3.1 to the discrete case. Averkov [2] considerably strengthens this result by proving that the determination holds in the class of all finite sets.

Theorem 6.2 (Averkov [2]). *A centrally symmetric finite subset A of $\mathbb{R}^n$ is determined by g_A, up to translations, in the class of all finite sets.*

Proof. Let B be a finite subset of $\mathbb{R}^n$ with $g_A = g_B$.

Suppose that $n = 1$; this case contains the heart of the proof. Let $a = \max \operatorname{supp} g_A$. Then

$$DA = \operatorname{supp} g_A \subset [-a, a] \quad \text{and} \quad a \in DA, \tag{6.3}$$

and analogous formulas hold for B. We may assume that $a > 0$, because otherwise A is a singleton; the same is true for B, and therefore B is a translate of A. The first formula in (6.3) implies that a translate of A is contained in $[0, a]$. We may thus assume, up to translations, that

$$A \subset [0, a] \quad \text{and} \quad B \subset [0, a].$$

The second formula in (6.3) implies $0, a \in A$. Similarly, $0, a \in B$. The set A is symmetric about $a/2$. Let $A \cap [a/2, a] = \{y_1, y_2, \ldots, y_m\}$, with $m \in \mathbb{N}$ and appropriate $y_1 < y_2 < \cdots < y_m = a$. We show by (reverse) induction that the sets

$$A_k = ([0, a - y_k] \cup [y_k, a]) \cap A \quad \text{and} \quad B_k = ([0, a - y_k] \cup [y_k, a]) \cap B$$

coincide, for every $k = 1, \ldots, m$. For $k = m$ this follows from $0, a \in A \cap B$.

Suppose that $A_{k+1} = B_{k+1}$.

We claim that $B \cap (y_k, y_{k+1}) = \emptyset$. Assume the contrary and let $x \in B \cap (y_k, y_{k+1})$. All pairs (x_0, x_1) with $x_0, x_1 \in A$ and $x_1 - x_0 = x$ satisfy $x_0, x_1 \in A_{k+1} = B_{k+1}$. Indeed, $x_1 = x_0 + x \geq x > y_k$ and $x_0 = x_1 - x \leq a - x < a - y_k$. Furthermore, B possesses at least one further pair (x_0, x_1) with $x_0, x_1 \in B$ and $x_1 - x_0 = x$, since we may set $x_0 = 0$ and $x_1 = x$. By the definition of the points y_k, we have $A \cap (y_k, y_{k+1}) = \emptyset$ and therefore $x \notin A$. Hence $g_B(x) > g_A(x)$, a contradiction. An analogous argument proves that $B \cap (a - y_{k+1}, a - y_k) = \emptyset$.

Next we show that $\{a - y_k, y_k\} \subset B$. We look at the pairs (x_0, x_1) with $x_0, x_1 \in A$ and $x_1 - x_0 = y_k$. Since $x_1 = x_0 + y_k \geq y_k$ and equality holds if and only if $x_0 = 0$, either $(x_0, x_1) = (0, y_k)$ or $x_1 \in A_{k+1}$. Analogously, either $(x_0, x_1) = (a - y_k, a)$ or $x_0 \in A_{k+1}$. Thus,

$$g_A(y_k) = \left|\{(x_0, x_1) : x_0, x_1 \in A, x_1 - x_0 = y_k\}\right|$$
$$= \left|\{(x_0, x_1) : x_0, x_1 \in A_{k+1}, x_1 - x_0 = y_k\}\right| + \left|\{(0, y_k), (a - y_k, a)\}\right|$$
$$= \left|\{(x_0, x_1) : x_0, x_1 \in B_{k+1}, x_1 - x_0 = y_k\}\right| + \left|\{(0, y_k), (a - y_k, a)\}\right|$$
$$\geq g_B(x)$$

and equality holds if and only if $(0, y_k), (a - y_k, a) \in B \times B$, i.e., $a - y_k, y_k \in B$. This concludes the proof that $A_k = B_k$ and, by induction, that $A_1 = B_1$. To prove that $A = B$ it remains to prove that $B \cap (a - y_1, y_1) = \emptyset$, and this can be done via analogous arguments.

Now assume that $n > 1$. We argue by induction on n and assume the claim is true for dimension $n - 1 \geq 1$. Without loss of generality, we assume that o is the centroid of both A and B. Let

$$U = S^{n-1} \setminus \{(x_1 - x_2)/|x_1 - x_2| : x_1, x_2 \in \operatorname{supp} g_A, x_1 \neq x_2\}.$$

If $u \in U$, then the orthogonal projection of $\mathbb{R}^n$ onto $u^\perp$ is injective on the sets A and B. It is not difficult to prove that this projection maps sets with equal covariograms to sets with equal covariograms. The inductive hypothesis, the central symmetry of $A|u^\perp$, and the assumption about the centroids of A and B prove that $A|u^\perp = B|u^\perp$. This is true for every $u \in U$ and hence for infinitely many $u \in S^{n-1}$. Heppes [40] proves that a finite set with cardinality k is determined by its orthogonal projections in $k + 1$ mutually non-parallel directions. This implies $A = B$. $\qquad\qquad\square$

This result applies also to lattice sets and, due to Theorem 6.1, it implies the following corollary.

Corollary 6.3 (Averkov [2]). *A centrally symmetric lattice body $A \subset \mathbb{R}^n$ is determined by g_A, up to translations, in the class of all lattice bodies.*

Averkov and Langfeld [6, 7] study the problem of determination from the covariogram in the class of *convex* lattice sets in $\mathbb{Z}^2$. The polyominoes associated to the examples of non-determination in [33, 24] discussed above are convex and therefore we cannot expect a global positive answer. In [6] it is shown that if a planar convex lattice set A samples $\operatorname{conv} A$ well enough (that is, if, in a certain sense, A is close enough to a convex polygon), then the determination from g_A is similar to the determination in the case of convex polygons.

We need some terminology. For $u \in \mathbb{R}^2 \setminus \{o\}$, let $A_u = \{x \in A : \langle x, u \rangle = h_A(u)\}$ denote the support set of A in the direction u. A support set A_u which contains more than one element will be called an *edge* of A with outer normal u. Let

$$U(A) = \{(u_1, u_2) \in \mathbb{Z}^2 \setminus \{o\} : (u_1, u_2) \text{ is an outer normal to an edge of } A$$
$$\text{and } u_1 \text{ and } u_2 \text{ are relatively prime}\}.$$

To measure the number of lattice points on the edges of A and the difference between the number of points on one edge and the number of points on the antipodal parallel edge, we introduce the following functions:

$$m'(A) = \min\{|A_u|, u \in U(A)\},$$
$$m''(A) = \min\{|A_u| - |A_{-u}| + 1 : u \in \mathbb{Z}^2 \setminus \{o\} \text{ and } |A_u| > |A_{-u}| > 1\},$$
$$m(A) = \min\{m'(A), m''(A)\},$$

where we use the convention that $\min \emptyset = \infty$.

Theorem 6.4 (Averkov and Langfeld [6]). *Let A be a convex lattice set in $\mathbb{Z}^2$. Then $m'(A)$, $m''(A)$, $m(A)$, and $U(A) \cup U(-A)$ are determined by g_A. Let $l \in \mathbb{N}$ be such that $U(A) \cup U(-A) \subset \{-l, \ldots, l\}^2$. If $m(A) > 4l^4 + 2l^2 + 1$, then A is determined by g_A in the class of convex lattice sets in $\mathbb{Z}^2$.*

Thus, for a given collection of prescribed edge normals, A is determined if all its edges have sufficiently large cardinality and the difference between cardinalities of parallel edges is either zero or sufficiently large.

Averkov and Langfeld [6, 7] also make substantial progress towards understanding the structure of non-trivially homometric pairs of convex lattice sets in $\mathbb{Z}^n$.

Example 6.5 ([6]). Let $k \in \mathbb{N} \setminus \{0\}$, $w_1 = (-k - 1, 1)$, $w_2 = (k, 1)$, and

$$L = \mathbb{Z}w_1 + \mathbb{Z}w_2.$$

Choose A to be any finite subset of L which is convex with respect to L (i. e., $A = (\operatorname{conv} A) \cap L$) such that each edge of the polygon conv A is parallel either to w_1 or to w_2 or to $w_2 - w_1 = (-1, 2)$. Let

$$B = (\{0, \ldots, k\} \times \{0\}) \cup (\{0, \ldots, k - 1\} \times \{1\}).$$

The lattice sets $A + B$ and $A + (-B)$ in $\mathbb{Z}^2$ are convex (with respect to $\mathbb{Z}^2$) and have the same covariogram. If A is not centrally symmetric, $A + B$ and $A + (-B)$ are not equal up to translations and reflections. See Figure 9.

Up to linear transformations of $\mathbb{Z}^2$ and up to translations of K and L, the non-trivially homometric pairs H and K from [24, 33] are members of the family presented in Example 6.5. The example in [24] is obtained by taking $k = 1$ and $A = \{(0, 0), (2, -1), (1, -2)\}$, and that in [33] is obtained by taking $k = 1$ and $A = \{(0, 0), (1, 1), (1, -2), (2, -1), (3, 0)\}$ and applying to H and K the linear transformation $(x, y) \to (x - y, y)$.

The lattice L in Example 6.5 is 2-dimensional and B is convex with respect to $\mathbb{Z}^2$. Moreover $\mathbb{Z}^2 = L + B$ and this is a *direct sum* (which, in this setting, means that each element of $\mathbb{Z}^2$ can be written in a unique way as a sum of elements of L and B). This implies that the translations of B by vectors in L tile $\mathbb{Z}^2$. It has been proved [7, Theorem 2.4]

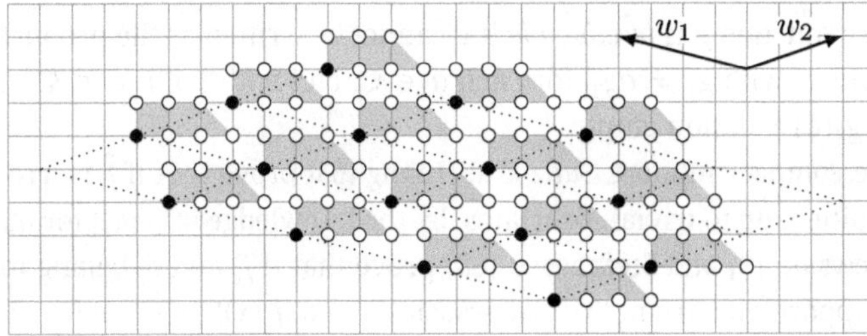

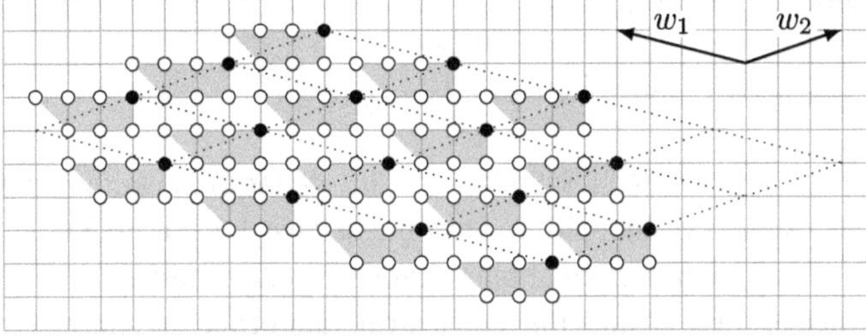

Figure 9: The sets $A + B$ (the union of the black and white points above) and $A + (-B)$ (below). The elements of A are drawn as black points and the convex hulls of the translates of B and $-B$ are indicated by gray polygons (from [6]).

that if B is a convex subset of $\mathbb{Z}^2$ and $L \subset \mathbb{Z}^2$ is a 2-dimensional lattice such that $\mathbb{Z}^2$ is the direct sum $L + B$, then the sets $A + B$ and $A + (-B)$ are non-trivially homometric if and only if, up to linear transformations in $\mathbb{Z}^2$ and translations, A, B, and L are those described in Example 6.5.

One may wonder whether there are non-trivially homometric pairs of *convex* lattice sets in $\mathbb{Z}^2$ that do not arise from the construction $A + B$ and $A + (-B)$. Averkov and Langfeld [6] write that they performed an exhaustive computer search of such pairs among lattice sets which are contained in $\{1, \dots, 6\} \times \{1, \dots, 5\}$ without finding any.

The same authors [7, Example 5.5] present the first examples of non-trivially homometric convex sets in $\mathbb{Z}^n$, for any $n \geq 2$, which are intrinsically n-dimensional, in the sense that they are not lifted from $\mathbb{Z}^2$ by taking Cartesian products.

7 Connections to Fourier analysis

In the previous sections we have already seen applications of results from Fourier analysis in studying the problem of determination from the covariogram. Here we present some other connections. They come from the literature on the phase retrieval problem and deal with the irreducibility of $\widehat{1_K}$. This connection also shows a link between the covariogram problem and the Pompeiu problem in integral geometry.

We say that an entire function g is *irreducible* if g cannot be written as the product of entire functions g_1 and g_2 with $g_1 \neq a g_2$, for each $a \in \mathbb{C}$, and with both $\{\zeta \in \mathbb{C}^n : g_1(\zeta) = 0\}$ and $\{\zeta \in \mathbb{C}^n : g_2(\zeta) = 0\}$ non-empty.

Let $f \in L^2(\mathbb{R}^n)$ have compact support. Sanz and Huang [63] prove that if $\hat{f}$ is irreducible, then f is determined, up to trivial associates, by the knowledge of $|\hat{f}(x)|$ for all $x \in \mathbb{R}^n$. Barakat and Newsam [9] and Stefanescu [68] prove that if f_1 and f_2 belong to $L^2(\mathbb{R}^2)$, have compact support, and are not trivial associates and $|\hat{f}_1(x)| = |\hat{f}_2(x)|$ for all $x \in \mathbb{R}^2$, then there exist entire functions g_1 and g_2 such that $\{\zeta \in \mathbb{C}^2 : g_1(\zeta) = 0\}$ and $\{\zeta \in \mathbb{C}^2 : g_2(\zeta) = 0\}$ are both non-empty and

$$\hat{f}_1(\zeta) = g_1(\zeta) g_2(\zeta) \quad \text{and} \quad \hat{f}_2(\zeta) = e^{i(c + \langle d, \zeta \rangle)} g_1(\zeta) \overline{g_2(\overline{\zeta})}, \tag{7.1}$$

for $\zeta \in \mathbb{C}^2$ and suitable $c \in \mathbb{R}$ and $d \in \mathbb{R}^2$. I. S. Stefanescu, in a letter to the author, has expressed the opinion that a similar result holds in any dimension $n \geq 2$. It is not known whether the property that $\hat{f}$ is not irreducible implies that f is not determined by $|\hat{f}|$.

What is the significance of these results for the covariogram problem? All the examples of non-determination presented in Section 3.2 arise from a factorization of $\widehat{1_K}$ as in (7.1). Indeed if E, F, H, and K are as in Theorem 3.11 and E and F are orthogonal subspaces, then

$$1_{H+K} = \delta_H * \delta_K \quad \text{and} \quad 1_{H+(-K)} = \delta_H * \delta_{-K},$$

where δ_H and δ_K are the distributions defined for $\phi \in C_0^\infty(\mathbb{R}^n)$ by

$$\delta_H(\phi) = \int_H \phi(x, 0)\, dx, \quad \delta_K(\phi) = \int_K \phi(0, y)\, dy$$

(here dx and dy indicate integration with respect to Lebesgue measure in E and in F, respectively), and δ_{-K} is defined similarly. By the Paley–Wiener theorem, $\widehat{\delta_H}$, $\widehat{\delta_K}$, and $\widehat{\delta_{-K}}$ are entire functions in $\mathbb{C}^n$ of exponential type. Clearly, $\widehat{\delta_{-K}}(\zeta) = \overline{\widehat{\delta_K}(\overline{\zeta})}$, and we have

$$\widehat{1_{H+K}}(\zeta) = \widehat{\delta_H}(\zeta)\widehat{\delta_K}(\zeta) \quad \text{and} \quad (\widehat{1_{H+(-K)}})(\zeta) = \widehat{\delta_H}(\zeta)\overline{\widehat{\delta_K}(\overline{\zeta})},$$

as in (7.1).

In view of these results it would be interesting to study the following problem.

Problem 7.1. Find explicit geometric conditions on a convex body K which guarantee that $\widehat{1_K}$ is irreducible.

To appreciate the difficulty in answering this question, consider the following subproblem.

Determine for which convex bodies K the function $\widehat{1_K}$ is the product of a non-trivial polynomial and an entire function.

We need some notation. Given a polynomial $p(\zeta) = \sum_{|l| \le m} c_l \zeta^l$, where $m \in \mathbb{N}$, $l = (l_1, \ldots, l_n)$ denotes a multi-index, $c_l \in \mathbb{C}$, $|l| = l_i + \cdots + l_n$, and $\zeta^l = \zeta_1^{l_1} \cdots \zeta_n^{l_n}$, let $p(D)$ denote the differential operator

$$p(D) = \sum_{|l| \le m} (i)^{-|l|} c_l (\partial^{l_1}/\partial x_1^{l_1}) \cdots (\partial^{l_n}/\partial x_n^{l_n}),$$

where $\partial^0/\partial x_i^0$ denotes the identity operator. It has been stated [60, Theorem 8.4] that

$$\widehat{1_K} = fp,$$

with f entire and p a polynomial, if and only if the equation

$$p(D)u = 1_K \tag{7.2}$$

has a solution u in the class of distributions with support contained in K. Here $\hat{u} = f$ and (7.2) has to be understood in the sense of distributions. The theorem of supports for convolutions [41, Theorem 4.3.3] and elementary considerations imply that if a solution u to (7.2) exists, then its support is K.

A particular instance of this problem has received much attention. When $p(\zeta) = \zeta_1^2 + \cdots + \zeta_n^2 - c$, for some $c > 0$, (7.2) becomes

$$\begin{cases} \Delta u + cu = -1 & \text{in } K, \\ u = \frac{\partial u}{\partial v} = 0 & \text{on } \partial K, \end{cases} \tag{7.3}$$

where v denotes the exterior normal to ∂K. Let $E \subset \mathbb{R}^n$ be a bounded simply connected Lipschitz domain. The Pompeiu problem asks whether there exists a non-zero continuous function $f : \mathbb{R}^n \to \mathbb{R}$ such that

$$\int_{\mathcal{T}(E)} f \, dx = 0 \quad \text{for all rigid motions } \mathcal{T} \text{ in } \mathbb{R}^n$$

only when E is a ball. It is known that the Pompeiu problem is equivalent to asking whether a solution to (7.3) (with K replaced by E) exists for some $c > 0$ only if E is a ball (see Berenstein [13]). As far as we know, these problems are still open.

The example of a ball implies that the irreducibility condition is not necessary for determination by covariogram. Indeed, when K is a ball a solution to (7.3) exists and $\widehat{1_K}$ factors. On the other hand, in any dimension a ball K is uniquely determined by g_K, as Theorem 5.4 implies.

Bibliography

[1]   R. J. Adler and R. Pyke, *Problem 91–3*, Inst. Math. Stat. Bull. **20** (1991), 409.

[2]   G. Averkov, *Detecting and reconstructing centrally symmetric sets from the autocorrelation: two discrete cases*, Appl. Math. Lett. **22** (2009), 1476–1478.

[3]   G. Averkov and G. Bianchi, *Retrieving convex bodies from restricted covariogram functions*, Adv. Appl. Probab. **39** (2007), 613–629.

[4]   G. Averkov and G. Bianchi, *Confirmation of Matheron's conjecture on the covariogram of a planar convex body*, J. Eur. Math. Soc. **11** (2009), 1187–1202.

[5]   G. Averkov and G. Bianchi, *Covariograms generated by valuations*, Int. Math. Res. Not. **2015**(19) (2015), 9277–9329.

[6]   G. Averkov and B. Langfeld, *On the reconstruction of planar lattice-convex sets from the covariogram*, Discrete Comput. Geom. **48** (2012), 216–238.

[7]   G. Averkov and B. Langfeld, *Homometry and direct-sum decompositions of lattice-convex sets*, Discrete Comput. Geom. **56** (2016), 216–249.

[8]   M. Baake and U. Grimm, *Homometric model sets and window covariograms*, Z. Kristallogr. **222** (2007), 54–58.

[9]   R. Barakat and G. Newsam, *Necessary conditions for a unique solution to two-dimensional phase recovery*, J. Math. Phys. **25** (1984), 3190–3193.

[10] H. H. Bauschke, P. L. Combettes and D. R. Luke, *Phase retrieval, error reduction algorithm, and Fienup variants: a view from convex optimization*, J. Opt. Soc. Am. A **19** (2002), 1334–1345.

[11] C. Benassi, G. Bianchi and G. D'Ercole, *Covariogram of non-convex sets*, Mathematika **56** (2010), 267–284.

[12] C. Benassi and G. D'Ercole, *An algorithm for reconstructing a convex polygon from its covariogram*, Rend. Ist. Mat. Univ. Trieste **39** (2007), 457–476.

[13] C. A. Berenstein, *An inverse spectral theorem and its relation to the Pompeiu problem*, J. Anal. Math. **37** (1980), 128–144.

[14] G. Bianchi, *Matheron's conjecture for the covariogram problem*, J. Lond. Math. Soc. (2) **71** (2005), 203–220.

[15] G. Bianchi, *The covariogram determines three-dimensional convex polytopes*, Adv. Math. **220** (2009), 1771–1808.

[16] G. Bianchi, *The cross covariogram of a pair of polygons determines both polygons, with a few exceptions*, Adv. Appl. Math. **42** (2009), 519–544.

[17] G. Bianchi, *Geometric tomography of convex cones*, Discrete Comput. Geom. **41** (2009), 61–76.

[18] G. Bianchi, *The covariogram and Fourier–Laplace transform in* $\mathbb{C}^n$, Proc. Lond. Math. Soc. (3) **113** (2016), 1–23.

[19] G. Bianchi, R. J. Gardner and M. Kiderlen, *Phase retrieval for characteristic functions of convex bodies and reconstruction from covariograms*, J. Am. Math. Soc. **24** (2011), 293–343.

[20] G. Bianchi, F. Segala and A. Volčič, *The solution of the covariogram problem for plane* C^2_+ *convex bodies*, J. Differ. Geom. **60** (2002), 177–198.

[21] A. J. Cabo and A. J. Baddeley, *Line transects, covariance functions and set convergence*, Adv. Appl. Probab. **27** (1995), 585–605.

[22] A. J. Cabo and A. J. Baddeley, *Estimation of mean particle volume using the set covariance function*, Adv. Appl. Probab. **35** (2003), 27–46.

[23] A. J. Cabo and R. H. P. Janssen, *Cross-covariance functions characterise bounded regular closed sets*, CWI Technical Report, Amsterdam, 1994.

[24] A. Daurat, Y. Gérard and M. Nivat, *Some necessary clarifications about the chords' problem and the partial digest problem*, Theor. Comput. Sci. **347** (2005), 432–436.

[25] K. Engel and B. Laasch, *The modulus of the Fourier transform on a sphere determines 3-dimensional convex polytopes*, J. Inverse Ill-Posed Probl. **30** (2022), 475–483.

[26] J. R. Fienup, *Phase retrieval algorithms: a comparison*, Appl. Opt. **21** (1982), 2758–2769.

[27] J. R. Fienup, T. R. Crimmins and W. Holsztyński, *Reconstruction of the support of an object from the support of its autocorrelation*, J. Opt. Soc. Am. **72** (1982), 610–624.

[28] M. Gage and R. S. Hamilton, *The heat equation shrinking convex plane curves*, J. Differ. Geom. **23** (1986), 69–96.

[29] B. Galerne, *Computation of the perimeter of measurable sets via their covariogram. Applications to random sets*, Image Anal. Stereol. **30** (2011), 39–51.

[30] R. J. Gardner, *The Brunn–Minkowski inequality*, Bull. Am. Math. Soc. (N. S.) **39** (2002), 355–405.

[31] R. J. Gardner, *Geometric tomography*, 2nd edn, Encyclopedia of mathematics and its applications, vol. 58, Cambridge University Press, New York, 2006.

[32] R. J. Gardner, *Geometric tomography*. http://www.geometrictomography.com, 2013. Accessed on February 8, 2023.

[33] R. J. Gardner, P. Gronchi and C. Zong, *Sums, projections, and sections of lattice sets, and the discrete covariogram*, Discrete Comput. Geom. **34** (2005), 391–409.

[34] R. J. Gardner, M. Kiderlen and P. Milanfar, *Convergence of algorithms for reconstructing convex bodies and directional measures*, Ann. Stat. **34** (2006), 1331–1374.

[35] R. J. Gardner and G. Zhang, *Affine inequalities and radial mean bodies*, Am. J. Math. **120** (1998), 505–528.

[36] N. Garofalo and F. Segala, *Asymptotic expansions for a class of Fourier integrals and applications to the Pompeiu problem*, J. Anal. Math. **56** (1991), 1–28.

[37] A. Gasparyan and V. K. Oganyan, *Orientation-dependent distribution of the length of a random segment and covariogram*, Izv. Nats. Akad. Nauk Armenii Mat. **50** (2015), 3–12.

[38] P. Goodey, R. Schneider and W. Weil, *On the determination of convex bodies by projection functions*, Bull. Lond. Math. Soc. **29** (1997), 82–88.

[39] E. K. Haviland and A. Wintner, *On the Fourier transforms of distribution on convex curves*, Duke Math. J. **2** (1936), 712–721.

[40] A. Heppes, *On the determination of probability distributions of more dimensions by their projections*, Acta Math. Acad. Sci. Hung. **7** (1956), 403–410.

[41] L. Hörmander, *The analysis of linear partial differential operators. I. Distribution theory and Fourier analysis*, Classics in mathematics, Springer, Berlin, 2003. Reprint of the second (1990) ed.

[42] N. E. Hurt, *Phase retrieval and zero crossings*, Mathematical methods in image reconstruction, vol. 52, Kluwer Academic Publishers Group, Dordrecht, 1989.

[43] K. Kiener, *Extremalität von Ellipsoiden und die Faltungsungleichung von Sobolev*, Arch. Math. (Basel) **46** (1986), 162–168.

[44] M. V. Klibanov, P. E. Sacks and A. V. Tikhonravov, *The phase retrieval problem*, Inverse Probl. **11** (1995), 1–28.

[45] T. Kobayashi, *Asymptotic behaviour of the null variety for a convex domain in a non-positively curved space form*, J. Fac. Sci., Univ. Tokyo, Sect. 1A, Math. **36** (1989), 389–478.

[46] T. Kobayashi, *Convex domains and the Fourier transform on spaces of constant curvature*, in: Lecture notes of the Unesco-Cimpa School on "Invariant differential operators on Lie groups and homogeneous spaces", P. Torasso, ed., WuHan University, China, 1991.

[47] W. Lawton, *Uniqueness results for the phase-retrieval problem for radial functions*, J. Opt. Soc. Am. **71** (1981), 1519–1522.

[48] D. R. Luke, J. V. Burke and R. G. Lyon, *Optical wavefront reconstruction: theory and numerical methods*, SIAM Rev. **44** (2002), 169–224.

[49] F. C. Machado and S. Robins, *The null set of a polytope, and the Pompeiu property for polytopes*, 2021, arXiv:2104.01957.

[50] C. L. Mallows and J. M. C. Clark, *Linear-intercept distributions do not characterize plane sets*, J. Appl. Probab. **7** (1970), 240–244.

[51] G. Matheron, *Le covariogramme géometrique des compacts convexes des R^2*, Technical report N-2/86/G, Centre de Géostatistique, Ecole Nationale Supérieure des Mines de Paris, 1986.

[52] G. Matheron, *Random sets and integral geometry*, John Wiley & Sons, New York–London–Sydney, 1975.

[53] M. Meyer, S. Reisner and M. Schmuckenschläger, *The volume of the intersection of a convex body with its translates*, Mathematika **40** (1993), 278–289.

[54] R. P. Millane, *Recent advances in phase retrieval*, in: Image reconstruction from incomplete data IV (P. J. Bones, M. A. Fiddy and R. P. Millane, eds.), Proc. SPIE, vol. 6316, 2006.

[55] W. Nagel, *Orientation-dependent chord length distributions characterize convex polygons*, J. Appl. Probab. **30** (1993), 730–736.

[56] J. Rataj, *On set covariance and three-point test sets*, Czechoslov. Math. J. **54**(129) (2004), 205–214.

[57] M. Reitzner, M. Schulte and C. Thäle, *Limit theory for the Gilbert graph*, Adv. Appl. Math. **88** (2017), 26–61.

[58] M. Reitzner, E. Spodarev and D. Zaporozhets, *Set reconstruction by Voronoi cells*, Adv. Appl. Probab. **44** (2012), 938–953.

[59] J. Rosenblatt, *Phase retrieval*, Commun. Math. Phys. **95** (1984), 317–343.

[60] W. Rudin, *Functional analysis*, 2nd edn., International series in pure and applied mathematics, McGraw-Hill, Inc., New York, 1991.

[61] K. Rufibach, *Phasenrekonstruktion von Gleichverteilungen in R^d*, Diploma Thesis, University of Bern, Switzerland, 2001.

[62] L. A. Santaló, *Integral geometry and geometric probability*, 2nd edn., Cambridge University Press, Cambridge, 2004.

[63] J. L. C. Sanz and T. S. Huang, *Phase reconstruction from magnitude of band-limited multidimensional signals*, J. Math. Anal. Appl. **104** (1984), 302–308.

[64] M. Schmitt, *On two inverse problems in mathematical morphology*, in: Mathematical morphology in image processing, Opt. engrg., vol. 34, Dekker, New York, 1993, pp. 151–169.

[65] R. Schneider, *Polytopes and Brunn–Minkowski theory*, in: *Polytopes: abstract, convex and computational*, Scarborough, ON, 1993, NATO adv. sci. inst. ser. C: Math. phys. sci., vol. 440, Kluwer Acad. Publ., Dordrecht, 1994, pp. 273–299.

[66] R. Schneider, *Convex bodies: the Brunn–Minkowski theory*, 2nd expanded edn., Encycl. math. appl., vol. 151, Cambridge University Press, Cambridge, 2014.

[67] M. Senechal, *A point set puzzle revisited*, Eur. J. Comb. **29** (2008), 1933–1944.

[68] I. S. Stefanescu, *On the phase retrieval problem in two dimensions*, J. Math. Phys. **26** (1985), 2141–2160.

Károly J. Böröczky

The logarithmic Minkowski conjecture and the L_p-Minkowski problem

Abstract: The current state of the art concerning the L_p-Minkowski problem as a Monge–Ampère equation on the sphere and Lutwak's logarithmic Minkowski conjecture about the uniqueness of even solutions in the $p = 0$ case are surveyed and connections to many related problems are discussed.

Keywords: Minkowski problem, L_p Brunn–Minkowski theory, Monge–Ampeère equations

MSC 2020: Primary 35J96, Secondary 52A40

1 Introduction

The Minkowski problem forms the core of various areas in fully non-linear PDEs and convex geometry (see Trudinger and Wang [207] and Schneider [203]), which was extended to the L_p-Minkowski theory by Lutwak [168, 169, 170] where $p = 1$ corresponds to the classical case. The classical Minkowski existence theorem due to Minkowski and Aleksandrov characterizes the surface area measure S_K of a convex body K in $\mathbb{R}^n$; more precisely, it solves the Monge–Ampère equation

$$\det(\nabla^2 h + h\,\mathrm{Id}) = f$$

on the sphere S^{n-1} where a convex body K with C_+^2-boundary provides a solution if $h = h_K|_{S^{n-1}}$ for the support function h_K of K, and in this case, $1/f(u)$ is the Gaussian curvature at the $x \in \partial K$ where u is an exterior normal for $u \in S^{n-1}$. The so-called logarithmic Minkowski (log-Minkowski) or L_0-Minkowski problem

$$h\,\det(\nabla^2 h + h\,\mathrm{Id}) = f \tag{1.1}$$

was posed by Firey in his seminal paper [98]. It seeks to characterize the cone volume measure $dV_K = \frac{1}{n} h_K dS_K$ of a convex body K containing the origin o and to determine

Acknowledgement: I would like to thank Emanuel Milman, Erwin Lutwak, Gaoyong Zhang, Shibing Chen, Qi-Rui Li, Christos Saroglou, and Martin Henk for illuminating discussions during the preparation of the manuscript. I am also grateful for the hospitality of ETH Zürich during the completion of the survey. Author was supported by NKFIH grant K 132002.

Károly J. Böröczky, Alfréd Rényi Institute of Mathematics, Realtanoda u. 13-15, H-1053 Budapest, Hungary, e-mail: Boroczky.karoly.j@renyi.hu

https://doi.org/10.1515/9783110775389-003

whether the even solution is unique if f is even. The latter problem is the so-called log-Minkowski conjecture by Lutwak. However, the log-Minkowski problem has received due attention only after finding its place as part of Lutwak's L_p-Minkowski problem

$$h^{1-p} \det(\nabla^2 h + h\,\mathrm{Id}) = f$$

in the 1990s, where the cases $p = 1$ and $p = 0$ are the classical and the log-Minkowski problem, respectively. For $p \geq 0$, the L_p-Minkowski problem is intimately related to the L_p-version of the Brunn–Minkowski inequality/conjecture

$$
\begin{aligned}
V\big((1-\lambda)K +_p \lambda C\big)^{\frac{p}{n}} &\geq (1-\lambda)\,V(K)^{\frac{p}{n}} + \lambda\,V(C)^{\frac{p}{n}} \quad \text{if } p > 0, \\
V\big((1-\lambda)K +_p \lambda C\big) &\geq V(K)^{1-\lambda} V(C)^{\lambda} \quad \text{if } p = 0
\end{aligned}
\tag{1.2}
$$

for $\lambda \in (0,1)$ and convex bodies K, C containing the origin. Here (1.2) is the classical Brunn–Minkowski inequality if $p = 1$, a theorem of Firey [97] if $p > 1$, and assuming that K and C are origin-symmetric, a conjecture being the central theme of this chapter if $p \in [0,1)$. Actually, the conjecture has been recently verified if $p \in (0,1)$ is close to 1; more precisely, combining Kolesnikov and Milman [148] and Chen, Huang, Li, and Liu [63] proves that the L_p-Brunn–Minkowski conjecture holds if $p \in [p_n, 1)$ and K, C are origin-symmetric convex bodies for an explicit $p_n \in (0,1)$.

The main goal of this survey is to inspire the resolution of the log-Brunn–Minkowski conjecture (cf. (1.2) when $p = 0$) or Lutwak's essentially equivalent log-Minkowski conjecture (the Monge–Ampère equation (1.1) on S^{n-1} has a unique even solution if f is even, positive, and C^∞).

Its versatility is an intriguing aspect of the log-Minkowski conjecture; namely, uniqueness of the even solutions of a Monge–Ampère equation on the sphere is equivalent to some strengthening of the Brunn–Minkowski inequality for origin-symmetric convex bodies, to an inequality for the Gaussian density, and to some spectral gap estimates for certain self-adjoint operators, and in turn to displacement convexity of certain functionals of probability measures on the sphere in optimal transportation.

In this survey, we review some related aspects of the classical Brunn–Minkowski theory in Section 2, the state of the art concerning the log-Minkowski problem and the log-Minkowski conjecture in Section 3, Lutwak's L_p-Minkowski problem and the L_p-Minkowski conjecture in Section 4, and some variants of the L_p-Minkowski problem in Section 5.

2 Classical Brunn–Minkowski theory

This section not only serves as an introduction into the relevant aspects of Brunn–Minkowski theory, but also introduces the basic ideas and tools used in the upcoming sections. For a thorough discussion on the subject and related problems from various perspectives, see Artstein-Avidan, Giannopoulos, and Milman [11, 12], Ball [14], Gardner [100], Leichtweiss [153], and Schneider [203].

We call a compact convex set in $\mathbb{R}^n$ with non-empty interior a convex body. The family of convex bodies in $\mathbb{R}^n$ is denoted by $\mathcal{K}^n$, we write $\mathcal{K}^n_o$ ($\mathcal{K}^n_{(o)}$) to denote the subfamily of $K \in \mathcal{K}^n$ with $o \in K$ ($o \in \operatorname{int} K$), and we write $\mathcal{K}^n_e$ to denote the family of origin-symmetric convex bodies in $\mathbb{R}^n$. The support function of a compact convex set K is $h_K(u) = \max_{x \in K} \langle u, x \rangle$ for $u \in \mathbb{R}^n$, and hence h_K is convex and homogeneous (the latter property says that $h_K(\lambda u) = \lambda h_K(u)$ for $\lambda \geq 0$). In turn, for any convex and homogeneous function h on $\mathbb{R}^n$, there exists a unique compact convex set K such that $h = h_K$. We note that differences of support functions are dense among continuous functions on the sphere; more precisely, functions of the form $(h_K - h_C)|_{S^{n-1}}$ for convex bodies K and C with C^∞_+-boundary in $\mathbb{R}^n$ are dense in $C(S^{n-1})$ with respect to the L_∞-metric.

We say that convex bodies K and C are homothetic if $K = \gamma C + z$ for $\gamma > 0$ and $z \in \mathbb{R}^n$. We write $V(X)$ to denote the Lebesgue measure of a measurable subset X of $\mathbb{R}^n$ (with $V(\emptyset) = 0$) and $\mathcal{H}$ to denote the $(n-1)$-Hausdorff measure normalized in a way such that it coincides with the $(n-1)$-dimensional Lebesgue measure on $(n-1)$-dimensional affine subspaces. For $X, Y \subset \mathbb{R}^n$ and $\alpha, \beta \in \mathbb{R}$, the Minkowski linear combination is $\alpha X + \beta Y = \{\alpha x + \beta y : x \in X, \ y \in Y\}$, which is convex compact if X and Y are convex compact. We write B^n to denote the unit Euclidean ball centered at the origin o, and we equip the space of compact convex sets of $\mathbb{R}^n$ with topology induced by the Hausdorff metric (sometimes called Hausdorff distance); namely, if K and C are compact convex sets, then their Hausdorff distance is

$$\delta_H(K, C) = \min\{r \geq 0 : K \subset C + rB^n \text{ and } C \subset K + rB^n\}.$$

The Brunn–Minkowski inequality says that if $\alpha, \beta > 0$ and K, C are convex bodies in $\mathbb{R}^n$, then

$$V(\alpha K + \beta C)^{\frac{1}{n}} \geq \alpha V(K)^{\frac{1}{n}} + \beta V(C)^{\frac{1}{n}}, \tag{2.1}$$

with equality if and only if K and C are homothetic. We note that the Brunn–Minkowski inequality (2.1) also holds if K and C are bounded Borel subsets of $\mathbb{R}^n$ (note that a Minkowski linear combination of measurable subsets may not be measurable; therefore, the outer measure is used in that case). The Brunn–Minkowski inequality famously yields the isoperimetric inequality; namely, the surface area of a bounded Borel set X of given volume is minimized by balls. Naturally, one needs a suitable notion of surface area. It is the Hausdorff measure $\mathcal{H}(\partial X)$ if X is a convex body, or more generally, ∂X is the finite union of the images of Lipschitz functions defined on bounded subsets of $\mathbb{R}^{n-1}$ (see Schneider [203] or Ambrosio, Colesanti, and Villa [6]), but the right notion is finite perimeter (see Maggi [175]). Fusco, Maggi, and Pratelli [99] proved an optimal stability version of the isoperimetric inequality in terms of the symmetric difference metric (whose result was extended to the Brunn–Minkowski inequality by Figalli, Maggi, and Pratelli [91, 92]; see Theorem 2.1 below).

Because of the homogeneity of the Lebesgue measure, an equivalent form of the Brunn–Minkowski inequality (2.1) is that if K, C are convex bodies in $\mathbb{R}^n$ and $\lambda \in (0,1)$, then

$$V((1-\lambda)K + \lambda C) \geq V(K)^{1-\lambda}V(C)^{\lambda}, \tag{2.2}$$

with equality if and only if K and C are translates. A big advantage of this product form of the Brunn–Minkowski inequality is that it is dimension invariant.

The first stability forms of the Brunn–Minkowski inequality were due to Minkowski himself (see Groemer [108]). If the distance of the convex bodies K and C is measured in terms of the so-called Hausdorff distance, then Diskant [79] and Groemer [107] provided close to optimal stability versions (see Groemer [108]). However, the natural distance is in terms of the volume of the symmetric difference, and the optimal result is due to Figalli, Maggi, and Pratelli [91, 92]. To define the "homothetic distance" $A(K, C)$ of convex bodies K and C, let $\alpha = |K|^{\frac{-1}{n}}$ and $\beta = V(C)^{\frac{-1}{n}}$, and let

$$A(K, C) = \min\{V(\alpha K \Delta(x + \beta C)) : x \in \mathbb{R}^n\}.$$

In addition, let

$$\sigma(K, C) = \max\left\{\frac{V(C)}{V(K)}, \frac{V(K)}{V(C)}\right\}.$$

Theorem 2.1 (Figalli, Maggi, Pratelli). *For $\gamma^*(n) > 0$ depending on n and any convex bodies K and C in $\mathbb{R}^n$,*

$$V(K + C)^{\frac{1}{n}} \geq \left(V(K)^{\frac{1}{n}} + V(C)^{\frac{1}{n}}\right)\left[1 + \frac{\gamma^*(n)}{\sigma(K, C)^{\frac{1}{n}}} \cdot A(K, C)^2\right].$$

Here the exponent 2 of $A(K, C)^2$ is optimal; see Figalli, Maggi, and Pratelli [92]. We note that prior to [92], the only known error term in the Brunn–Minkowski inequality was of order $A(K, C)^{\eta}$ with $\eta \geq n$, coming from the estimates of Diskant [79] and Groemer [107] in terms of the Hausdorff distance.

Figalli, Maggi, and Pratelli [92] proved a factor of the form $\gamma^*(n) = cn^{-14}$ for some absolute constant $c > 0$, which was first improved to cn^{-7} by Segal [202] and subsequently to $cn^{-5.5}$ by Kolesnikov and Milman [148, Theorem 12.12]. The currently best-known bound for $\gamma^*(n)$ is $cn^{-5}(\log n)^{-10}$, which follows by combining the general estimate of Kolesnikov and Milman [148, Theorem 12.2] with the polylogarithmic bound of Klartag and Lehec [143] on the Cheeger constant of a convex body in isotropic position, improving on Yuansi Chen"s work [69] on the Kannan–Lovász–Simonovits conjecture. Harutyunyan [120] conjectured that $\gamma^*(n) = cn^{-2}$ is the optimal order of the constant and showed that it cannot be of smaller order. Actually, Segal [202] observed that Dar's conjecture in [78] would imply that we may choose $\gamma^*(n) = cn^{-2}$ for some absolute constant $c > 0$. Here Dar's conjectured strengthening of the Brunn–Minkowski inequality states in [78] that if K and C are convex bodies in $\mathbb{R}^n$ and $M = \max_{x \in \mathbb{R}^n} V(K \cap (x + C))$, then

$$V(K + C)^{\frac{1}{n}} \geq M^{\frac{1}{n}} + \left(\frac{V(K)V(C)}{M} \right)^{\frac{1}{n}}. \tag{2.3}$$

Dar's conjecture is only known to hold in the plane (see Xi and Leng [209]) and in some very specific cases in higher dimensions (see Dar [78]).

Eldan and Klartag [83] discuss "isomorphic" stability versions of the Brunn–Minkowski inequality under conditions of the type $|\frac{1}{2} K + \frac{1}{2} C| \leq 5\sqrt{|K| \cdot |C|}$ and consider, for example, the L^2-Wasserstein distance of the uniform measures on suitable affine images of K and C.

We note that stability versions of the Brunn–Minkowski inequality have been verified even if K or C is not convex. The essentially optimal estimate Theorem 2.1 (with much worse factor $\gamma^*(n)$) is verified if K is bounded measurable and C is a convex body by Barchiesi and Julin [18] (improving on the estimate in Carlen and Maggi [60]), if $n \geq 2$ and $K = C$ is a bounded Borel set by Hintum, Spink, and Tiba [125], and if $n = 2$ and K and C are bounded Borel sets by Hintum, Spink, and Tiba [126]. If $n \geq 3$ and K and C are bounded Borel sets, then only a much weaker estimate in terms of $A(K, C)$ is known, proved by Figalli and Jerison [89, 90]. On the other hand, a better error term of order $A(X, Y)$ holds if $n = 1$ according to Freiman and Christ (see Christ [71]).

It was proved by Minkowski that if K and C are convex bodies and $\alpha, \beta \geq 0$, then

$$V(\alpha K + \beta C) = \sum_{i=0}^{n} \binom{n}{i} V(K, C; i)\alpha^{n-i}\beta^i, \tag{2.4}$$

where $V(K, C; i)$ are the so-called mixed volumes. For fixed i, $V(K, C; i)$ is positive and continuous in both variables and satisfies $V(\alpha K, \beta C; i) = \alpha^{n-i}\beta^i V(K, C; i)$ for $\alpha, \beta > 0$ and $V(K, C; i) = V(C, K; n-i)$ and $V(\Phi K + x, \Phi C + y; i) = V(K, C; i)$ for $x, y \in \mathbb{R}^n$ and $\Phi \in SL(n)$. Many mixed volumes have geometric meaning; for example, $V(K, K; i) = V(K, C; 0) = V(K)$ and $\frac{1}{n} V(K, B^n; 1) = \mathcal{H}(\partial K)$ is the surface area of K. In addition, if $i = 1, \ldots, n-1$, then $V(K, B^n; n-i)$ is proportional to the mean i-dimensional projection of K according to the Kubota formula (see Leichtweiss [153] and Schneider [203]).

It follows from the Brunn–Minkowski inequality (2.1) that the function $f(\lambda) = V((1-\lambda)K + \lambda C)^{\frac{1}{n}}$ is concave on $[0, 1]$. Combining $f'(0) \geq f(1) - f(0)$ and (2.4) leads to the famous Minkowski inequality

$$V(K, C; 1)^n \geq V(K)^{n-1} V(C), \tag{2.5}$$

with equality if and only if K and C are homothetic. The Minkowski inequality is actually equivalent to the Brunn–Minkowski inequality because it implies that the function $f(\lambda) = V((1 - \lambda)K + \lambda C)^{\frac{1}{n}}$ is concave on $[0,1]$, which in turn yields (2.1). Considering the second derivative $f''(\lambda)$ leads to Minkowski's second inequality

$$V(K, C; 1)^2 \geq V(K)V(K, C; 2), \qquad (2.6)$$

which in turn is also equivalent to the Brunn–Minkowski inequality (2.1). The rather involved equality case of (2.6) has been only recently clarified by van Handel and Shenfeld [118].

Actually, Minkowski defined the mixed volume $V(C_1, \ldots, C_n)$ of n convex bodies via the identity

$$V(\lambda_1 K_1 + \cdots + \lambda_m K_m) = \sum_{i_1, \ldots, i_n = 1}^{m} V(K_{i_1}, \ldots, K_{i_n}) \cdot \lambda_{i_1} \cdots \cdots \lambda_{i_n}, \qquad (2.7)$$

for $K_1, \ldots, K_m \in \mathcal{K}^n$ and $\lambda_1, \ldots, \lambda_m \geq 0$, where $V(C_1, \ldots, C_n) \geq 0$ is symmetric and continuous in its variables (see [203]) and $V(K, C; i)$ means i copies of C and $n - i$ copies of K. A far reaching generalization of Minkowski's first and second inequalities is the Alexandrov–Fenchel inequality

$$V(K_1, K_2, K_3, \ldots, K_n)^2 \geq V(K_1, K_1, K_3, \ldots, K_n)V(K_2, K_2, K_3, \ldots, K_n)$$

(see Alexandrov [1, 5] and Schneider [203]). Equality in the Alexander–Fenchel inequality is much better understood now due to van Handel and Shenfeld [118, 119], where the case of polytopes was clarified in [119].

Let us summarize some equivalent formulations of the Brunn–Minkowski inequality that hold for all convex bodies K and C in $\mathbb{R}^n$:
- $V(\alpha K + \beta C)^{\frac{1}{n}} \geq \alpha V(K)^{\frac{1}{n}} + \beta V(C)^{\frac{1}{n}}$ (cf. (2.1));
- $V((1 - \lambda)K + \lambda C) \geq V(K)^{1-\lambda}V(C)^{\lambda}$ (cf. (2.2));
- $f(\lambda) = V((1 - \lambda)K + \lambda C)^{\frac{1}{n}}$ is concave on $[0,1]$;
- Minkowski's inequality $V(K, C; 1)^n \geq V(K)^{n-1}V(C)$ (cf. (2.5));
- Minkowski's second inequality $V(K, C; 1)^2 \geq V(K)V(K, C; 2)$ (cf. (2.6)).

The classical Minkowski problem is concerned with the characterization of the so-called surface area measure S_K of a convex body K. Let $\partial' K$ denote the subset of the boundary of K such that there exists a unique exterior unit normal vector $\nu_K(x)$ at any point $x \in \partial' K$. It is well known that $\mathcal{H}(\partial K \setminus \partial' K) = 0$ and $\partial' K$ is a Borel set (see Schneider [203]). The function $\nu_K : \partial' K \to S^{n-1}$ is the spherical Gauss map that is continuous on $\partial' K$. The surface area measure S_K of K is a Borel measure on S^{n-1} satisfying that $S_K(\eta) = \mathcal{H}(\nu_K^{-1}(\eta))$ for any Borel set $\eta \subset S^{n-1}$. The surface area measure is the first variation of the volume; namely, if C is any convex body in $\mathbb{R}^n$, then

$$nV(K,C;1) = \lim_{\varepsilon \to 0^+} \frac{V(K + \varepsilon C) - V(K)}{\varepsilon} = \int_{S^{n-1}} h_C \, dS_K. \tag{2.8}$$

It also follows that the Minkowski inequality (2.5) can be written in the following form: If $V(K) = V(C)$ holds for $K, C \in \mathcal{K}^n$, then

$$\int_{S^{n-1}} h_C \, dS_K \geq \int_{S^{n-1}} h_K \, dS_K, \tag{2.9}$$

with equality if and only if K and C are translates.

To consider some examples for the surface area measure, if P is a polytope with facets $F_1, \ldots, F_k$ and exterior unit normals $u_1, \ldots, u_k$, then S_P is concentrated onto $\{u_1, \ldots, u_k\}$ and $S_P(u_i) = \mathcal{H}(F_i)$ for $i = 1, \ldots, k$. On the other hand, if ∂K is C_+^2, that is, C^2 with positive Gaussian curvature, and the Gaussian curvature at the point of ∂K with exterior unit normal $u \in S^{n-1}$ is $\kappa(u) = \kappa(K, u)$, then

$$dS_K = \kappa^{-1} \, d\mathcal{H} = \det(\nabla^2 h + h \, \mathrm{Id}) \, d\mathcal{H} \tag{2.10}$$

on S^{n-1}, where $h = h_K|_{S^{n-1}}$ and ∇h and $\nabla^2 h$ are the gradient and the Hessian of h with respect to a moving orthonormal frame. In particular, S_K is absolute continuous in this case.

We note that if ∂K is C_+^2 for $K \in \mathcal{K}^n$ and $h = h_K|_{S^{n-1}}$, then for any $u \in S^{n-1}$, the differential operator

$$D^2 h(u) = \nabla^2 h(u) + h(u) \, \mathrm{Id} \tag{2.11}$$

is the restriction of the Hessian of h_K (in $\mathbb{R}^n$) at λu to an operator $u^\perp \mapsto u^\perp$ for any $\lambda > 0$ and the eigenvalues of $D^2 h(u)$ are the radii of curvature at $x \in \partial K$ where u the exterior unit normal is. In turn, for any given $h \in C^m(S^{n-1})$ with $m \geq 2$, $h = h_K|_{S^{n-1}}$ for $K \in \mathcal{K}^n$ with C^m (C_+^m)-boundary if and only if $D^2 h(u)$ is positive semidefinite (positive definite) for $u \in S^{n-1}$.

Now the Minkowski problem asks for necessary and sufficient conditions for a Borel measure μ on S^{n-1} such that

$$\mu = S_K \tag{2.12}$$

for a convex body K. The solution together with its uniqueness was provided by Minkowski [181, 182] if the measure μ is discrete (and hence the convex body is a polytope) or absolutely continuous. Minkowski's solution was extended to any general measure μ by Alexandrov [2, 3, 5]; namely, there exists a convex body K with $\mu = S_K$ if and only if

$$\mu(L \cap S^{n-1}) < \mu(S^{n-1}) \quad \text{for any linear } (n-1)\text{-subspace } L \subset \mathbb{R}^n, \tag{2.13}$$

$$\int_{S^{n-1}} u \, dS_K(u) = o, \tag{2.14}$$

and moreover, $S_K = S_C$ holds for convex bodies K and C if and only if K and C are translates. Essentially complete solutions were published also by Fenchel and Jensen [86] and Lewy [155] about the same time. In particular, the Monge–Ampère equation on the sphere S^{n-1} corresponding to the Minkowski problem is

$$\det(\nabla^2 h + h \, \mathrm{Id}) = f, \tag{2.15}$$

where f is a given non-negative function with positive integral. If the given Borel measure μ on S^{n-1} is not absolutely continuous with respect to the Lebesgue measure, then $h = h_K|_{S^{n-1}}$ is a solution of (2.15) in the Alexandrov sense if (2.12) holds. We note that the surface area measure S_K is actually the Monge–Ampère measure corresponding to h (see Trudinger and Wang [207] and Böröczky and Fodor [38, Section 7]).

The regularity of the solution of the Minkowski problem (2.15) is well investigated by Nirenberg [188], Cheng and Yau [68], and Pogorelov [190], showing eventually that if f is positive and C^k for $k \geq 1$, then h is C^{k+2}. Finally, Caffarelli [56, 57] proves that if f is positive and C^α for $\alpha \in (0,1)$ (namely, $|f(x) - f(y)| \leq C\|x - y\|^\alpha$ for $x, y \in S^{n-1}$ and constant $C > 0$), then the solution h is $C^{2,\alpha}$ (see also Böröczky and Fodor [38, Section 7] on how to connect results on Monge–Ampère equations on $\mathbb{R}^n$ to Monge–Ampère equations on S^{n-1} and Chen, Liu, and Wang [67] for an extension of [56, 57]).

Turning to proofs, one of the elegant arguments proving the Brunn–Minkowski inequality (2.1) is due to Hilbert. It is based on a spectral gap estimate for a differential operator (see (4.10) and Bonnesen and Fenchel [31]). This approach was further developed by Alexandrov [1, 5], leading to the Alexandrov–Fenchel inequality, by van Handel and Shenfeld [118, 119] to characterize equality in the Alexandrov–Fenchel inequality in certain cases, and by Milman and Kolesnikov [148], leading to the L_p-Minkowski inequality Theorem 4.4, improving the Brunn–Minkowski inequality for origin-symmetric convex bodies (see the end of Section 4). Another fundamental approach proving the Brunn–Minkowski inequality is initiated by Gromov's influential appendix to Milman and Schechtman [180] using ideas by Knothe [150], providing a proof of the isoperimetric inequality using optimal (mass) transport, and the argument can be readily extended to the Brunn–Minkowski inequality (2.1) and the Prékopa–Leindler inequality (2.16) below. This approach even led to the stability version of Theorem 2.1 by Figalli, Maggi, and Pratelli [91, 92]. We note that the original argument of Brunn and Minkowski for (2.1) (see Bonnesen and Fenchel [31]) can also be considered as a version of the mass transportation approach.

For the Minkowski problem (2.12), the variational method seeks the minimum of $\int_{S^{n-1}} h_C \, d\mu$ over all convex bodies C with $V(C) = 1$, where μ satisfies (2.13) and (2.14). It follows from (2.14) that the integral is invariant under translation of C, and hence the existence of a minimizer C_0 can be established. The fact that S_{C_0} is proportional to μ

follows via Alexandrov's lemma, Lemma 2.2, extending (2.8) (see Schneider [203, Theorem 7.5.3]).

Lemma 2.2 (Alexandrov). *Given a convex body K in $\mathbb{R}^n$ and continuous functions $h_t, g :$ $S^{n-1} \to \mathbb{R}$, let us assume that the Wulff shape $K_t = \{x \in \mathbb{R}^n : \langle x, u \rangle \leq h_t(u) \,\forall u \in S^{n-1}\}$ is a convex body and $\lim_{t \to 0} \frac{h_t(u) - h_K(u)}{t} = g(u)$ uniformly in $u \in S^{n-1}$. Then*

$$\lim_{t \to 0} \frac{V(K_t) - V(K)}{t} = \int_{S^{n-1}} g \, dS_K.$$

The classical functional form of the Brunn–Minkowski inequality is the Prékopa–Leindler inequality due to Prékopa [191] and Leindler [154] in the 1-dimensional case. It was generalized in Prékopa [192, 193] and Borell [32] (cf. also Marsiglietti [177], Bueno and Pivovarov [55], Brascamp and Lieb [50], Kolesnikov and Werner [149], and Bobkov, Colesanti, and Fragalà [28]). Various applications are provided and surveyed in Ball [13], Barthe [19, 20], Fradelizi and Meyer [95], and Gardner [100]. The following multiplicative version from [13] is often more useful and is more convenient for geometric applications.

Theorem 2.3 (Prékopa–Leindler). *If $\lambda \in (0, 1)$ and h, f, g are non-negative integrable functions on $\mathbb{R}^n$ satisfying $h((1 - \lambda)x + \lambda y) \geq f(x)^{1-\lambda} g(y)^{\lambda}$ for $x, y \in \mathbb{R}^n$, then*

$$\int_{\mathbb{R}^n} h \geq \left(\int_{\mathbb{R}^n} f \right)^{1-\lambda} \cdot \left(\int_{\mathbb{R}^n} g \right)^{\lambda}. \tag{2.16}$$

For a convex function $W : \mathbb{R}^n \to (-\infty, \infty]$, we say that the function $\varphi = e^{-W}$ is log-concave, where $e^{-\infty} = 0$ (in other words, $\log \varphi$ is concave for $\varphi : \mathbb{R}^n \to [0, \infty)$). According to Dubuc [81], if equality holds in (2.3) assuming $\int_{\mathbb{R}^n} h > 0$, then h is log-concave, and there exist $a > 0$ and $z \in \mathbb{R}^n$ such that $f(x) = a^{\lambda} h(x - \lambda z)$ and $g(x) = a^{-(1-\lambda)} h(x + (1 - \lambda)z)$ for almost all $x \in \mathbb{R}^n$. Stability versions of the Prékopa–Leindler inequality in terms of the L_1-distance have been established by Böröczky and De [34] in the log-concave case and by Böröczky, Figalli, and Ramos [37] for any functions where the case of log-concave functions in one variable has been dealt with earlier by Ball and Böröczky [15]. A stability version of the Prékopa–Leindler inequality of a somewhat different nature is due to Bucur and Fragalà [54].

An "isomorphic" stability result for the Prékopa–Leindler inequality in terms of the transportation distance is obtained in Eldan [82, Lemma 5.2]. By rather standard considerations, one can show that non-isomorphic stability results in terms of transportation distance imply stability in terms of L_1-distance (e. g., such implication is attained by combining Proposition 2.9 in Bubeck, Eldan, and Lehec [53] and Proposition 10 in Eldan and Klartag [83]). However, the current result in [82], due to its isomorphic nature, falls short of being able to obtain a meaningful bound in terms of the L_1-distance.

Brascamp and Lieb [50] proved a local version of the Prékopa–Leindler inequality for log-concave functions [50, Theorem 4.2], which is equivalent to a Poincare-type so-called Brascamp–Lieb inequality [50, Theorem 4.1]. The paper by Livshyts [164] provides a stability version of this Brascamp–Lieb inequality, and Bolley, Cordero-Erausquin, Fujita, Gentil, and Guillin [29] prove a more general inequality.

Our final topic in this section is the Blaschke–Sataló inequality (2.17). For $K \in \mathcal{K}^n_{(o)}$ and $u \in S^{n-1}$, the radial function $\varrho_K(u) > 0$ satisfies $\varrho_K(u)u \in \partial K$, and the polar (dual) $K^* \in \mathcal{K}^n_{(o)}$ of K is defined by $\varrho_{K^*}(u) = h_K(u)$ for $u \in S^{n-1}$. Next, the centroid of a convex body K in $\mathbb{R}^n$ is $\sigma_K = \frac{1}{V(K)} \int_K x \, dx$, which is invariant under affine transformations. For $K \in \mathcal{K}^n$, we call it centered if $\sigma_K = o$, and Kannan, Lovász, and Simonovits [141] prove that there exists a centered ellipsoid E such that $E \subset K \subset nE$ in this case.

According to the Blaschke–Santaló inequality (see Santaló [198], Luwak [168], or Schneider [203]), if $K \in \mathcal{K}^n$ is centered, then $V(K^*)V(K) \leq V(B^n)^2$, or equivalently,

$$\int_{S^{n-1}} h_K^{-n} \, d\mathcal{H} \leq \frac{nV(B^n)^2}{V(K)}, \tag{2.17}$$

with equality if and only if K is a centered ellipsoid.

The Blaschke–Santaló inequality can be proved for example via the Brunn–Minkowski inequality (see Ball [13] in the origin-symmetric case and Meyer and Pajor [176] in the general case). Various equivalent formulations are discussed in the beautiful survey by Lutwak [168] (see also Böröczky [33] and Schneider [203]). Stability versions of the Blaschke–Santaló inequality are verified in Böröczky [33] and Ball and Böröczky [16]. Following Ball [13], functional versions of the Blaschke–Santaló inequality have been obtained by Artstein-Avidan, Klartag, and Milman [10], Fradelizi and Meyer [95], Lehec [151, 152], Kolesnikov and Werner [149], and Kalantzopoulos and Saroglou [140].

Recently, various breakthrough stability results about geometric functional inequalities have been obtained. Stronger versions of the functional Blaschke–Santaló inequality are provided by Barthe, Böröczky, and Fradelizi [21], of the Borell–Brascamp–Lieb inequality by Ghilli and Salani [105], Rossi and Salani [195, 196], and Balogh and Kristály [17] (later even on Riemannian manifolds), of the Sobolev inequality by Figalli and Zhang [94] (extending Bianchi and Egnell [26] and Figalli and Neumayer [93]), Nguyen [187], and Wang [213], of the log-Sobolev inequality by Gozlan [106], and of some related inequalities by Caglar and Werner [58], Cordero-Erausquin [74], and Kolesnikov and Kosov [145]. Another functional version of the Brunn–Minkowski inequality is provided by Artstein-Avidan, Florentin, and Segal [9].

3 Cone volume measure, log-Minkowski problem, log-Brunn–Minkowski conjecture

Given a convex body K containing the origin, the cone volume measure is defined as $dV_K = \frac{1}{n} h_K \, dS_K$, and hence the total measure is $V_K(S^{n-1}) = V(K)$. The name originates from the fact that if P is a polytope with facets $F_1, \ldots, F_k$ and exterior unit normals $u_1, \ldots, u_k$, then V_P is concentrated onto $\{u_1, \ldots, u_k\}$ and $V_P(u_i) = \frac{h_P(u_i)}{n} \cdot \mathcal{H}(F_i)$ is the volume of the cone $\mathrm{conv}\{o, F_i\}$ for $i = 1, \ldots, k$. We note that the Monge–Ampère equation on the sphere S^{n-1} corresponding to the log-Minkowski problem is

$$h \det(\nabla^2 h + h\,\mathrm{Id}) = nf \tag{3.1}$$

for a non-negative measurable function f on S^{n-1} with $0 < \int_{S^{n-1}} f\, d\mathcal{H} < \infty$. It follows via Caffarelli [56, 57] that if f is positive and C^α for $\alpha \in (0,1)$, then the solution h is $C^{2,\alpha}$, and if f is positive and C^k for integer $k \geq 1$, then the solution h is C^{k+2}. For a finite non-trivial Borel measure μ on S^{n-1}, a non-negative function h on S^{n-1} that is the restriction of the support function h_K for a convex body K is the solution of (3.1) in the Alexandrov sense if

$$d\mu = dV_K = \frac{1}{n} h_K\, dS_K. \tag{3.2}$$

A characteristic feature of the cone volume measure is that it intertwines with linear transformations; more precisely, $V_{\Phi K} = |\det \Phi| \cdot (\Phi^{-t})_* V_K$ for any $K \in \mathcal{K}_o^n$ and $\Phi \in \mathrm{GL}(n, \mathbb{R})$. We note that if $u \in S^{n-1}$ is an exterior normal at an $x \in \partial K$, then $\Phi^{-t} u$ is an exterior normal at $\Phi x \in \partial(\Phi K)$, and the push forward measure $\Psi_* \mu$ on S^{n-1} for a Borel measure μ on S^{n-1} and $\Psi \in \mathrm{GL}(n)$ is defined (with a slight abuse of notation) in a way such that if $\omega \subset S^{n-1}$ is Borel, then

$$\Psi_* \mu(\omega) = \mu\left(\left\{ \frac{\Psi^{-1}(u)}{\|\Psi^{-1}(u)\|} : u \in \omega \right\} \right).$$

The cone volume measure was introduced by Firey [98], and it has been a widely used tool since the paper by Gromov and Milman [109]; see for example Barthe, Guédon, Mendelson, and Naor [22], Naor [184], and Paouris and Werner [189]. The still open log-Minkowski problem (3.2) or (3.1) was posed by Firey [98] in 1974, who showed that if f is a positive constant function, then (3.1) has a unique even solution coming from the suitable centered ball. For a positive constant function f, the general uniqueness result without the evenness condition is due to Andrews [7] if $n = 2, 3$ and due to Brendle, Choi, and Daskalopoulos [51] if $n \geq 4$. It is known that uniqueness of the solution may not hold if f is not a constant function (see, for example, Chen, Li, and Zhu [66]). However, the celebrated "log-Minkowski conjecture" by Lutwak [169] from 1993 states that (3.1) has a unique even solution if f is even and positive (Conjecture 3.1 is a more restricted version).

Conjecture 3.1 (log-Minkowski conjecture #1). *If f is a positive even C^∞-function in (3.1), then* (3.1) *has a unique even solution.*

As we explain below, the following logarithmic analog of Minkowski's inequality (2.9) is an intimately related form of the log-Minkowski conjecture (see Böröczky,

Lutwak, Yang, and Zhang [44] for origin-symmetric bodies and Böröczky and Kalantzopoulos [43] for centered convex bodies).

Conjecture 3.2 (log-Minkowski conjecture #2). *If K and C are convex bodies in $\mathbb{R}^n$ whose centroid is the origin, then*

$$\int_{S^{n-1}} \log \frac{h_C}{h_K}\, dV_K \geq \frac{V(K)}{n} \log \frac{V(C)}{V(K)}, \tag{3.3}$$

with equality if and only if $K = K_1 + \cdots + K_m$ and $C = C_1 + \cdots + C_m$ for compact convex sets $K_1, \ldots, K_m, C_1, \ldots, C_m$ of dimension at least one, where $\sum_{i=1}^{m} \dim K_i = n$ and K_i and C_i are dilates, $i = 1, \ldots, m$.

In particular, the more precise form of the log-Minkowski conjecture, Conjecture 3.1, is that if $K, C \in \mathcal{K}^n$ are centered, then $V_K = V_C$ implies the equality conditions in Conjecture 3.2.

We note that the choice of the right translates of K and C is important in Conjecture 3.2 according to the examples by Nayar and Tkocz [185] and that Conjecture 3.2 is invariant under application of the same non-singular linear transformation to C and K. An equivalent form of Conjecture 3.2 is that if $V(C) = V(K)$ for centered C and K, then

$$\int_{S^{n-1}} \log h_C\, dV_K \geq \int_{S^{n-1}} \log h_K\, dV_K, \tag{3.4}$$

where the case of equality is like in Conjecture 3.2.

Let us explain why Conjecture 3.1 is equivalent to the case of Conjecture 3.2 when $K \in \mathcal{K}_e^n$ has C_+^∞-boundary. Since V_K satisfies the strict subspace concentration condition (see below), Böröczky, Lutwak, Yang, and Zhang [45] prove that the function $C \mapsto \int_{S^{n-1}} \log h_C\, dV_K$ of origin-symmetric convex bodies C with $V(C) = V(K)$ attains its minimum; moreover, whenever it attains its minimum at some $C = \tilde{C}$, then $V_{\tilde{C}} = V_K$. In turn, the stated equivalence follows. In addition, it follows by approximation that Conjecture 3.1 yields inequality (3.3) for any pair of origin-symmetric convex bodies K and C without the case of equality.

In $\mathbb{R}^2$, Conjecture 3.2 is verified in Böröczky, Lutwak, Yang, and Zhang [44] for origin-symmetric convex bodies, but it is still open in the general case, even in the plane. In higher dimensions, Conjecture 3.2 is proved for complex bodies (cf. Rotem [197]) and if there exist n independent linear reflections that are common symmetries of K and C (cf. Böröczky and Kalantzopoulos [43], and even a stability version is verified by Böröczky and De [36]). The latter type of bodies include unconditional convex bodies; this case was handled earlier by Saroglou [199]. In addition, Conjecture 3.2 is verified if C is origin-symmetric and K is a zonoid by van Handel [116] (with equality only clarified when K has C_+^2-boundary) or if C is a centered convex body and K is a centered ellipsoid by Guan and Ni [110]. The latter case directly follows from the Jensen inequality and the Blaschke–Santaló inequality (2.17), as assuming that $K = B^n$ and $V(C) = V(B^n)$, we have

$$\exp\left(\int_{S^{n-1}} \log h_C \cdot \frac{1}{V(B^n)} dV_K \right) = \exp\left(\int_{S^{n-1}} \log h_C \cdot \frac{1}{nV(B^n)} d\mathcal{H} \right)$$

$$\geq \left(\int_{S^{n-1}} h_C^{-n} \cdot \frac{1}{nV(B^n)} d\mathcal{H} \right)^{\frac{-1}{n}} \geq 1.$$

For origin-symmetric K and C, Conjecture 3.2 is proved when K is close to be an ellipsoid (with equality only clarified when K has C_+^2-boundary) by a combination of the local estimates by Kolesnikov and Milman [148] and the use of the continuity method for PDEs by Chen, Huang, Li, and Liu [63]. Here closeness to an ellipsoid means that there exist some $c_n > 0$ depending only on n and an origin-symmetric ellipsoid E such that $E \subset K \subset (1+c_n)E$. Another even more recent proof of this result is due to Putterman [194]. We note that an analogous result holds for linear images of Hausdorff neighborhoods of l_q balls for $q > 2$ if the dimension n is high enough according to [148] and the method of [63]. Actually, Milman [179] provides rather generous explicit curvature pinching bounds for ∂K in order for Conjecture 3.2 to hold, and proves that for any origin-symmetric convex body M there exists an origin-symmetric convex body K with C_+^∞-boundary and $M \subset K \subset 8M$ such that Conjecture 3.2 holds for any origin-symmetric convex body C. Additional local versions of Conjecture 3.2 are due to Colesanti, Livshyts, and Marsiglietti [72], Kolesnikov and Livshyts [147], and Hosle, Kolesnikov, and Livshyts [127]. We review Kolesnikov and Milman's approach in [148] based on the Hilbert–Brunn–Minkowski operator at the end of Section 4.

Xi and Leng [209] consider a version of Conjecture 3.2 where the convex bodies K and C in $\mathbb{R}^n$ are translated by vectors depending on both K and C. We set $r(K,C) = \max\{t > 0 : \exists x, x + tC \subset K\}$ and $R(K,C) = \min\{t > 0 : \exists x, K \subset x + tC\}$, and say that K and C are in dilated position if $o \in K \cap C$ and

$$r(K,C)\, C \subset K \subset R(K,C)\, C. \tag{3.5}$$

We observe that $r(C,K)\, K \subset C \subset R(C,K)\, K$ in this case. Now for any convex bodies K and C there exist $z \in K$ and $w \in C$ such that $K - z$ and $C - w$ are in dilated position. If $n = 2$ and K and C are in dilated position, then Xi and Leng [209] proved (3.4) including the characterization of equality. Actually, Dar's conjecture (2.3) for convex planar bodies in dilated position was even verified in [209] (no need for translation in this case).

Let us discuss the existence of the solution of the log-Minkowski problem (3.1) or (3.2). Following partial and related results by Andrews [7], Chou and Wang [70], He, Leng, and Li [121], Henk, Schürman, and Wills [124], Stancu [204], and Xiong [212], Böröczky, Lutwak, Yang, and Zhang [45] characterized even cone volume measures by the so-called subspace concentration conditions (i) and (ii) in Theorem 3.3.

Theorem 3.3. *There exists an origin-symmetric convex body $K \in \mathcal{K}_e^n$ with $\mu = V_K$ for a non-trivial finite even Borel measure μ on S^{n-1} if and only if:*

(i) $\mu(L \cap S^{n-1}) \leq \frac{\dim L}{n} \cdot \mu(S^{n-1})$ *for any proper linear subspace $L \subset \mathbb{R}^n$;*

(ii) $\mu(L \cap S^{n-1}) = \frac{\dim L}{n} \cdot \mu(S^{n-1})$ *in (i) is equivalent to the existence of a complementary linear subspace $L' \subset \mathbb{R}^n$ with $\operatorname{supp}\mu \subset L \cup L'$.*

We observe that V_K satisfies (ii) if and only if $K = C + C'$, where $C \subset L^{\perp}$ and $C' \subset L'^{\perp}$ are compact convex sets. A finite Borel measure μ on S^{n-1} satisfies the strict subspace concentration condition if $\mu(L \cap S^{n-1}) < \frac{\dim L}{n} \cdot \mu(S^{n-1})$ for any proper linear subspace $L \subset \mathbb{R}^n$.

Given a non-trivial finite even Borel measure μ on S^{n-1} that is invariant under n reflections $\Phi_1, \ldots, \Phi_n$ through n independent linear hyperplanes, Böröczky and Kalantzopoulos [43] proved that $\mu = V_K$ for a convex body K in $\mathbb{R}^n$ invariant under $\Phi_1, \ldots, \Phi_n$ if and only if μ satisfies the subspace concentration conditions (i) and (ii) for any proper linear subspace $L \subset \mathbb{R}^n$ invariant under $\Phi_1, \ldots, \Phi_n$. Actually, the statement also holds if $\Phi_1, \ldots, \Phi_n$ are linear reflections (see [43] for details). For a centered convex body $K \in \mathcal{K}^n$, Böröczky and Henk [41] (see Henk and Linke [122] for the case of polytopes) verified that V_K satisfies the subspace concentration conditions (i) and (ii), but V_K satisfies some additional conditions as well. On the other hand, if $V_K(L \cap S^{n-1}) \geq (1 - \varepsilon) \cdot \frac{\dim L}{n} \cdot V(K)$ holds for $K \in \mathcal{K}^n$, a proper linear subspace $L \subset \mathbb{R}^n$, and a small $\varepsilon > 0$, then K is close to the sum of two complementary lower-dimensional compact convex sets according to Böröczky and Henk [42]. We note that Freyer, Henk, and Kipp [96] even verified certain so-called affine subspace concentration conditions for the cone volume measure of centered polytopes.

Much less is known, not even a conjecture about the characteristic properties of a cone volume measure on S^{n-1}, not even in the plane. Chen, Li, and Zhu [66] proved that if a non-trivial finite Borel measure μ on S^{n-1} satisfies the subspace concentration conditions (i) and (ii), then μ is a cone volume measure. On the other hand, Böröczky and Hegedűs [39] characterized the restriction of a cone volume measure to a pair of antipodal points.

As Lutwak, Yang, and Zhang [171] conjectured, a Borel probability measure μ on S^{n-1} satisfies the subspace concentration conditions (i) and (ii) if and only if there exists an isotropic linear image $\Phi_* \mu$ for a $\Phi \in \operatorname{GL}(n)$ according to Böröczky, Lutwak, Yang, and Zhang [46] (extending the work by Carlen and Cordero-Erausquin [59] in the discrete case and Klartag [142] in the strict subspace concentration condition case). Here the probability measure μ on S^{n-1} is isotropic if $n \int_{S^{n-1}} u \otimes u \, d\mu(u) = \operatorname{Id}_n$, or in other words, $\|x\| = n \int_{S^{n-1}} \langle x, u \rangle^2 \, d\mu(u)$ for any $x \in \mathbb{R}^n$.

Next we turn to the logarithmic Brunn–Minkowski conjecture/inequality. For $\lambda \in (0, 1)$ and $K, C \in \mathcal{K}_o^n$, we define their logarithmic or L_0-linear combination by the formula

$$(1 - \lambda)K +_0 \lambda C = \{x \in \mathbb{R}^n : \langle x, u \rangle \leq h_K(u)^{1-\lambda} h_C(u)^{\lambda} \ \forall u \in S^{n-1}\}.$$

The L_0-linear combination is linear invariant; namely, if $\Phi \in GL(n)$, then $\Phi((1-\lambda)K +_0 \lambda C) = (1-\lambda)\Phi(K) +_0 \lambda\Phi(C)$. Moreover, the L_0-linear combination is a convex body if $\{h_K = 0\} = \{h_C = 0\}$ (for example, when $K, C \in \mathcal{K}^n_{(o)}$). We note that $(1-\lambda)(aK) +_0 \lambda(\beta C) = \alpha^{1-\lambda}\beta^\lambda((1-\lambda)K +_0 \lambda C)$ for $\alpha, \beta > 0$. The L_0-linear combination of polytopes is always a polytope, but the boundary of the L_0-linear combination of convex bodies with C^2_+-boundaries may contain segments, and hence may not be C^2_+. A functional analog of the L_0-addition is presented by Crasta and Fragalà [77].

We observe that $(1-\lambda)K +_0 \lambda C \subset (1-\lambda)K + \lambda C$ for any convex bodies K and C containing the origin interior, but $(1-\lambda)K +_0 \lambda C$ might be much smaller than $(1-\lambda)K + \lambda C$. For example, if $a > 0$ is large, $n = 2$, $K = [\frac{-1}{a}, \frac{1}{a}] \times [-a, a]$, and $C = [-a, a] \times [\frac{-1}{a}, \frac{1}{a}]$, then

$$\frac{1}{2}K +_0 \frac{1}{2}C = [-1, 1]^2,$$

$$\frac{1}{2}K + \frac{1}{2}C = \left[-\frac{1}{2}\left(a + \frac{1}{a}\right), \frac{1}{2}\left(a + \frac{1}{a}\right)\right]^2. \tag{3.6}$$

Böröczky, Lutwak, Yang, and Zhang [44] conjectured the following for origin-symmetric convex bodies, and Martin Henk proposed the version with centered convex bodies (see also [43]).

Conjecture 3.4 (log-Brunn–Minkowski conjecture). *If $\lambda \in (0, 1)$ and K and C are centered convex bodies in $\mathbb{R}^n$, then*

$$V((1-\lambda)K +_0 \lambda C) \geq V(K)^{1-\lambda}V(C)^\lambda, \tag{3.7}$$

with equality if and only if $K = K_1 + \cdots + K_m$ and $C = C_1 + \cdots + C_m$ for compact convex sets $K_1, \ldots, K_m, C_1, \ldots, C_m$ of dimension at least one, where $\sum_{i=1}^m \dim K_i = n$ and K_i and C_i are dilates, $i = 1, \ldots, m$.

The log-Brunn–Minkowski conjecture, Conjecture 3.4, is a significant strengthening of the Brunn–Minkowski inequality for centered convex bodies (see (3.6)). Given $K, C \in \mathcal{K}^n_o$, if (3.7) holds for all $\lambda \in (0, 1)$, then the log-Minkowski inequality (3.3) follows by considering $\frac{d}{d\lambda}V((1-\lambda)K +_0 \lambda C)|_{\lambda=0^+}$ and using Alexandrov's lemma, Lemma 2.2, according to [44]. On the other hand, the argument in [44] shows that if $\mathcal{F}$ is any family of convex bodies closed under L_0-linear combination, then the log-Minkowski inequality (3.3) for all $K, C \in \mathcal{F}$ is equivalent to the logarithmic Brunn–Minkowski inequality (3.7) for all $K, C \in \mathcal{F}$ and $\lambda \in (0, 1)$. In particular, the equivalence holds for the family of origin-symmetric convex bodies. According to Kolesnikov and Milman [148] and Putterman [194], taking the second derivative of $\lambda \mapsto V((1-\lambda)K +_0 \lambda C)$ for origin-symmetric convex bodies K and C in $\mathbb{R}^n$ leads to the conjectured inequality

$$\frac{V(K, C; 1)^2}{V(K)} \geq \frac{n-1}{n}V(K, C; 2) + \frac{1}{n}\int_{S^{n-1}} \frac{h_C^2}{h_K^2}\, dV_K, \tag{3.8}$$

which is a strengthened from of Minkowski's second inequality (2.6) and is equivalent to the log-Brunn–Minkowski conjecture without the characterization of equality. More precisely, it has been proved [148] that for a fixed $K \in \mathcal{K}_e^n$ with C_+^2-boundary, (3.8) for all smooth $C \in \mathcal{K}_e^n$ is equivalent to a local form of the log-Brunn–Minkowski around K, and the global statement has been verified [194]. Therefore, we have the following three equivalent forms of the log-Brunn–Minkowski conjecture for origin-symmetric convex bodies K and C in $\mathbb{R}^n$ (without the characterization of equality in the case of the third formulation):

- $V((1-\lambda)K +_0 \lambda C) \geq V(K)^{1-\lambda}V(C)^\lambda$ as in (3.7);
- $\int_{S^{n-1}} \log \frac{h_C}{h_K}\, dV_K \geq \frac{V(K)}{n} \log \frac{V(C)}{V(K)}$ as in (3.3);
- $\frac{V(K,C;1)^2}{V(K)} \geq \frac{n-1}{n}\, V(K,C;2) + \frac{1}{n}\int_{S^{n-1}} \frac{h_C^2}{h_K^2}\, dV_K$ as in (3.8).

Another equivalent formulation using the Hilbert–Brunn–Minkowski operator (4.10) is due to Kolesnikov and Milman [148], and is discussed at the end of Section 4 (see (4.12)). In addition, Saroglou [199] verified that the log-Brunn–Minkowski inequality for any origin-symmetric convex bodies is equivalent to the so-called B-property: For any origin-symmetric convex body K in $\mathbb{R}^n$ and $n \times n$ positive definite diagonal matrix Φ, the function $s \mapsto V([-1,1]^n \cap \Phi^s K)$ of $s \in \mathbb{R}$ is log-concave. Yet another equivalent formulation of the log-Brunn–Minkowski conjecture for origin-symmetric convex bodies in $\mathbb{R}^n$ using the "strong B-property" is due to Nayar and Tkocz [186]: For any $N > n$ and n-dimensional linear subspace L of $\mathbb{R}^N$, the n-volume of $L \cap \prod_{i=1}^N [-e^{t_i}, e^{t_i}]$ is a log-concave function of $(t_1, \ldots, t_N) \in \mathbb{R}^N$. Actually, an analogous property of the cross-polytopes has been proved [186].

Saroglou [200] proved that if the log-Brunn–Minkowski conjecture (3.7) holds for any origin-symmetric convex bodies K and C and $\lambda \in (0,1)$, then it holds for any even log-concave measure μ on $\mathbb{R}^n$, namely,

$$\mu((1-\lambda)K +_0 \lambda C) \geq \mu(K)^{1-\lambda}\mu(C)^\lambda. \tag{3.9}$$

In turn, the argument in [200] shows that (3.9) for the Gaussian measure $\mu = \gamma_n$ implies the log-Brunn–Minkowski conjecture (3.7) for origin-symmetric convex bodies. Finally, Kolesnikov [144] provides another equivalent formulation of the log-Brunn–Minkowski conjecture for origin-symmetric convex bodies in terms of displacement convexity of certain functionals of probability measures on the sphere in optimal transportation.

The log-Brunn–Minkowski conjecture, Conjecture 3.4, is still open but has been verified in various cases. In $\mathbb{R}^2$, Conjecture 3.4 is verified by Böröczky, Lutwak, Yang, and Zhang [44] for origin-symmetric convex bodies, but it is still open for general centered planar convex bodies. For unconditional convex bodies, the L_0-linear combination contains the so-called coordinatewise product (see Saroglou [199]); therefore, the corresponding inequality for the coordinatewise product by Uhrin [206], Bollobás and Leader

[30], and Cordero-Erausquin, Fradelizi, and Maurey [75] following from the Prékopa–Leindler inequality (Theorem 2.3) yields Conjecture 3.4. The equality case of the log-Brunn–Minkowski inequality for unconditional convex bodies was clarified by Saroglou [199] (see also [43]). The log-Brunn–Minkowski conjecture, Conjecture 3.4, for convex bodies invariant under reflections through n independent linear hyperplanes is due to Böröczky and Kalantzopoulos [43]. In addition, Conjecture 3.4 is proved for complex bodies by Rotem [197].

Conjecture 3.4 holds for origin-symmetric convex bodies in a neighborhood of a fixed centered ellipsoid E, or more precisely, for origin-symmetric K and C provided $E \subset K, C \subset (1 + c_n)E$, where $c_n > 0$ depends only on n. In this form, the statement is due to Chen, Huang, Li, and Liu [63], extending the local estimate by Kolesnikov and Milman [148] (an analogous result holds for linear images of l_q balls for $q > 2$ if the dimension n is high enough according to [148] and the method of [63]). If $\mathbb{R}^3$, some additional partial results are obtained by Chen, Feng, and Liu [62]. We note that the case when K and C are in a C^2-neighborhood of E was handled earlier by Colesanti, Livshyts, and Marsiglietti [72].

In some cases when uniqueness of the solution of the log-Minkowski problem is known, even the stability of the solution has been established. For example, Böröczky and De [35] established this among convex bodies invariant under n given reflections through linear hyperplanes. Concerning Firey's classical result that the only origin-symmetric solution of the log-Minkowski problem (3.1) with constant f is the centered ball, Ivaki [136] verified a stability version. Next Chen, Feng, and Liu [62] proved the uniqueness results if $n = 3$ and a possibly non-even f is C^α-close to a constant function. It is an intriguing open problem to verify the uniqueness of the solution for a possibly non-even f close to a constant function or to have a stability version of the uniqueness result (see Andrews [7] and Brendle, Choi, and Daskalopoulos [51]) if $n \geq 4$.

If $n = 2$ and K and C are in dilated position (3.5), then Xi and Leng [209] proved (3.7) for any $\lambda \in (0, 1)$ including the characterization of equality using the same method as in the case of the planar Dar conjecture. It is an intriguing question what the relation between Dar's conjecture (2.3) and the log-Brunn–Minkowski conjecture, Conjecture 3.4, is, with respect to whether one of them implies the other for origin-symmetric convex bodies.

We note that there exist $\eta_2 > \eta_1 > 0$ depending on n such that if $\lambda \in (0, 1)$ and K and C are centered convex bodies in $\mathbb{R}^n$, then

$$\eta_1 V(K)^{1-\lambda}V(C)^\lambda \leq V((1 - \lambda)K +_0 \lambda C) \leq \eta_2 V(K)^{1-\lambda}V(C)^\lambda, \tag{3.10}$$

which estimates indicate why proving the log-Brunn–Minkowski conjecture, Conjecture 3.4, is so notoriously difficult. Conjecture 3.4 states that $\eta_1 = 1$, but here we only verify that $\eta_1 = n^{-n}$ and $\eta_2 = n^{3n/2}$ work. According to Kannan, Lovász, and Simonovits [141], there exist centered ellipsoids $E' \subset K$ and $E \subset C$ such that $K \subset nE'$ and $C \subset nE$. After a linear transform, we may assume that $E' = B_2^n$ and E is unconditional. Since

Conjecture 3.4 holds for the unconditional convex bodies B_2^n and E, we deduce that $\eta_1 = n^{-n}$ works in (3.10). For the upper bound, let $a_1, \ldots, a_n$ be the half-axes of E, and hence $C \subset \widetilde{C} = \prod_{i=1}^{n}[-na_i, na_i]$ and $K \subset \widetilde{K} = [-n, n]^n$ with $V(\widetilde{C}) \leq n^{3n/2}V(C)$ and $V(\widetilde{K}) \leq n^{3n/2}V(K)$. Since $V((1-\lambda)\widetilde{K} +_0 \lambda\widetilde{C}) = V(\widetilde{K})^{1-\lambda}V(\widetilde{C})^{\lambda}$, it follows that $\eta_2 = n^{3n/2}$ works in (3.10).

The validity of the log-Minkowski (or log-Brunn–Minkowski) conjecture is also supported by the fact that various consequences of it have been verified. For example, the L_p-Minkowski conjecture has been proved when $p \in (0, 1)$ is close to 1 (see Theorem 4.4). Next we turn to results about the canonical Gaussian probability measure γ_n on $\mathbb{R}^n$. One possible consequence of the log-Brunn–Minkowski conjecture, Conjecture 3.4, is the earlier celebrated "B-inequality" by Cordero-Erausquin, Fradelizi, and Maurey [75], stating that $\gamma_n(e^t K)$ is a log-concave function of $t \in \mathbb{R}$ for any origin-symmetric $K \in \mathcal{K}^n$. The Gardner–Zvavitch conjecture in [103] states that if K and C are origin-symmetric convex bodies in $\mathbb{R}^n$, then

$$\gamma_n((1-\lambda)K + \lambda C)^{\frac{1}{n}} \geq (1-\lambda)\gamma_n(K)^{\frac{1}{n}} + \lambda\gamma_n(C)^{\frac{1}{n}}. \tag{3.11}$$

It was proved by Livshyts, Marsiglietti, Nayar, and Zvavitch [73] that the log-Brunn–Minkowski conjecture would imply the Gardner–Zvavitch conjecture. After various attempts, the conjecture was finally verified by Eskenazis and Moschidis [84]. Kolesnikov and Livshyts [146] verified that if the exponents $\frac{1}{n}$ in (3.11) are changed into $\frac{1}{2n}$, then this modified Gardner–Zvavitch conjecture holds for any pair of centered convex bodies K and C.

We note that independently of the log-Brunn–Minkowski conjecture, various Brunn–Minkowski-type inequalities have been proved and conjectured for the Gaussian measure, the most famous ones being the Ehrhardt inequality and the Gaussian isoperimetric inequality (see Livshyts [164]).

Colesanti, Livshyts, and Marsiglietti [72] conjectured the following generalization of the Gardner–Zvavitch conjecture: If μ is an even log-concave measure on $\mathbb{R}^n$, then

$$\mu((1-\lambda)K + \lambda C)^{\frac{1}{n}} \geq (1-\lambda)\mu(K)^{\frac{1}{n}} + \lambda\mu(C)^{\frac{1}{n}} \tag{3.12}$$

holds for any origin-symmetric convex bodies K and C. According to Livshyts, Marsiglietti, Nayar, and Zvavitch [73], the log-Brunn–Minkowski conjecture, Conjecture 3.4, would imply the conjecture (3.12). Cordero-Erausquin and Rotem [76] proved (3.12) if μ is a rotationally symmetric log-concave measure. In addition, Livshyts [165] verified that (3.12) holds for any even log-concave measure on $\mathbb{R}^n$ and origin-symmetric convex bodies K and C if the exponents $\frac{1}{n}$ in (3.12) are changed into $n^{-4-o(1)}$.

4 Lutwak's L_p-Minkowski theory

The rapidly developing new L_p-Brunn–Minkowski theory (where $p = 1$ is the classical case and $p = 0$ corresponds to the cone volume measure) initiated by Lutwak [168, 169, 170] has become a main research area in modern convex geometry and geometric analysis. For $p \in \mathbb{R}$ and $K \in \mathcal{K}_o^n$, the L_p-surface area measure $S_{K,p}$ on S^{n-1} is defined by

$$dS_{K,p} = h_K^{1-p}\, dS_K, \tag{4.1}$$

where if $p > 1$ and $o \in \partial K$, then we assume that $S_K(\{h_K = 0\}) = 0$. In particular, $S_{K,1} = S_K$ and $S_{K,0} = nV_K$. For $p \in \mathbb{R}$, the Monge–Ampère equation on S^{n-1} corresponding to the L_p-Minkowski problem is

$$
\begin{aligned}
\det(\nabla^2 h + h\,\mathrm{Id}) &= h^{p-1}f \quad \text{if } p > 1,\\
h^{1-p}\det(\nabla^2 h + h\,\mathrm{Id}) &= f \quad \text{if } p \le 1,
\end{aligned}
\tag{4.2}
$$

where $f \in L_1(S^{n-1})$ is non-negative with $\int_{S^{n-1}} f\,d\mathcal{H} > 0$, and for a finite non-trivial Borel measure μ on S^{n-1}, a convex body $K \in \mathcal{K}_o^n$ is an Alexandrov solution of the L_p-Minkowski problem if

$$
\begin{aligned}
dS_K &= h_K^{p-1}\, d\mu \quad \text{if } p > 1,\\
h_K^{1-p}\, dS_K &= d\mu \quad \text{if } p \le 1.
\end{aligned}
\tag{4.3}
$$

If $p > 1$ and $p \ne n$, then Hug, Lutwak, Yang, and Zhang [134] (improving on Chou and Wang [70]) prove that (4.3) has an Alexandrov solution if and only if the μ is not concentrated onto any closed hemisphere, and the solution is unique. If in addition $p > n$, then the unique solution of (4.3) satisfies $o \in \mathrm{int}\,K$, and hence $S_{K,p} = \mu$. However, examples in [134] show that if $1 < p < n$, then it may happen that the density function f is positive continuous in (4.2) and $o \in \partial K$ holds for the unique Alexandrov solution. If $p = n$, then $S_{K,n} = S_{\lambda K,n}$ holds for $\lambda > 0$; therefore, all what is known (see [134]) is that for any measure μ not concentrated onto any closed hemisphere, there exists a convex body $K \in \mathcal{K}_o^n$ and $c > 0$ such that $\mu = c \cdot S_{K,n}$.

The case $p = 1$ is the classical Minkowski problem (see Section 2), and the case $p = 0$ is the log-Minkowski problem (see Section 3).

If $p \in (0, 1)$ and the measure μ is not concentrated onto any great subsphere, then Chen, Li, and Zhu [65] prove that there exists an Alexandrov solution $K \in \mathcal{K}_o^n$ of (4.3) with $S_{K,p} = \mu$. For $p \in (0, 1)$, complete characterization of L_p-surface area measures is only reported if $n = 2$ by Böröczky and Trinh [49]; namely, a finite non-trivial Borel measure μ on S^1 is an L_p-surface area measure if and only if $\mathrm{supp}\,\mu$ does not consist of a pair of antipodal points. Finally, let $p \in (0, 1)$ and $n \ge 3$, and let us assume that $1 \le \dim L \le n - 1$, where L is the linear hull of $\mathrm{supp}\,\mu$ in $\mathbb{R}^n$. If $\mathrm{supp}\,\mu$ is contained in a closed hemisphere centered at a point of $L \cap S^{n-1}$, then μ is an L_p-surface area measure according to Bianchi,

Böröczky, Colesanti, and Yang [24]. On the other hand, Saroglou [201] proved that if $\mu(\omega)$ is the Lebesgue measure of $\omega \cap L$ for any Borel $\omega \subset S^{n-1}$, then μ is not an L_p-surface area measure.

If $-n < p < 0$ and $f \in L_{\frac{n}{n+p}}(S^{n-1})$ in (4.2), then (4.2) has a solution according to Bianchi, Böröczky, Colesanti, and Yang [24]. If $p < 0$ and the μ in (4.3) is discrete satisfying that μ is not concentrated on any closed hemisphere and any n unit vectors in the support of μ are independent, then the L_p-Minkowski problem can be solved [217].

The $p = -n$ case of the L_p-Minkowski problem is the critical case because of its link with the SL(n)-invariant centro-affine curvature. If $K \in \mathcal{K}_{(o)}^n$ has C_+^2-boundary, then its centro-affine curvature at $u \in S^{n-1}$ is $\kappa_0(K,u) = \frac{\kappa(K,u)}{h_K(u)^{n+1}}$, where $\kappa(K,u)$ is the Gaussian curvature at the point with exterior normal u. It is well known to be SL(n)-invariant in the sense that $\kappa_0(\Phi K, u) = \kappa_0(K, \frac{\Phi^t u}{\|\Phi^t u\|})$ for $\Phi \in$ SL(n) (see Hug [135] or Ludwig [166]). It follows from (2.10) that if $K \in \mathcal{K}_{(o)}^n$ has C_+^2-boundary, then $dS_{K,-n}(u) = \kappa_0(K,u)^{-1} d\mathcal{H}(u)$; therefore, solving the L_p-Minkowski problem (4.2) for $p = -n$ and positive C^α-function f is equivalent to reconstructing a convex body $K \in \mathcal{K}_{(o)}^n$ from its centro-affine curvature function.

All in all, the centro-affine (L_{-n}-)Minkowski problem is wide open. If $p = -n$ and the f in (4.2) is unconditional and satisfies certain additional technical conditions, then Jian, Lu, and Zhu [139] verify the existence of a solution of (4.2). Moreover, Li, Guang, and Wang [114] solve a variant of the centro-affine Minkowski problem. On the other hand, Chou and Wang [70] prove an implicit condition on possible functions f in (4.2) such that f^{-1} is a centro-affine curvature (see also [24]), and Du [80] constructs an explicit example of a positive C^α-f such that (4.2) has no solution when $p = -n$.

In the supercritical case $p < -n$, Li, Guang, and Wang [112] have recently achieved a breakthrough by proving that for any positive C^2-f, there exists a C^4-solution of (4.2). In addition, the authors of [112] verify that if $p < -n$ and $1/c < f < c$ for a constant $c > 1$ in (4.2), then there exists a $C^{1,\alpha}$-Alexandrov solution $h_K|_{S^{n-1}}$ satisfying (4.3), where $d\mu = f\,d\mathcal{H}$. In their paper, Guang, Li, and Wang [112] combine a flow argument with homology calculations. On the other hand, Du [80] constructs a non-negative C^α-function f that is positive everywhere but a fixed pair of antipodal points and (4.2) has no solution, not even in the Alexandrov sense. It is not surprising that the flow argument works in the supercritical case, as Milman [178] points out the limitations of the variational argument in this case. For a discrete measure μ satisfying that μ is not concentrated on any closed hemisphere and any n unit vectors in the support of μ are independent, Zhu [217] solves the L_p-Minkowski problem (4.3) for $p < 0$.

If $p > -n$, then while flow arguments are also known (see, e. g., Bryan, Ivaki, and Scheuer [52]), the most common argument to find a solution of (4.2) is based on the variational method; namely, one considers the infimum of $\int_{S^{n-1}} h_C^p f\,d\mathcal{H}$ for a suitable family of convex bodies $C \in \mathcal{K}_{(o)}^n$ with $V(C) = 1$ when f is positive and continuous (see [24] or Chou and Wang [70]). The existence of some minimizer C_0 follows via the Blaschke–Santaló inequality (2.17) as $p > -n$, and the fact that $dS_{C_0,p} = \lambda f\,d\mathcal{H}$ for some constant factor $\lambda > 0$ follows via the Alexandrov lemma, Lemma 2.2. The case of more general measures than the ones with positive continuous density functions follows by approximation. For the variational approach, it is also common to use discrete measures on S^{n-1} (corresponding to polytopes; see [134, 216, 40, 217]).

Concerning the smoothness of the solution of the L_p-Minkowski problem (4.2), if f is positive and C^α and h is positive (equivalently, $o \in \operatorname{int} K$ for the corresponding convex body K), then h is $C^{2,\alpha}$ by Cafarelli [56, 57] (see [38, 25]). Assuming that f is positive and continuous, it is known that $o \in \operatorname{int} K$ if $p \geq n$ (see [134]) or $p \leq 2 - n$ (see [25]). On the other hand, if $2 - n < p < n$, $p \neq 1$, then there exists a positive C^α-function f on S^{n-1} such that $o \in \partial K$ holds for the Alexandrov solution of (4.3) with $d\mu = f\, d\mathcal{H}$; see [134] if $1 < p < n$ and [25] if $2 - n < p < 1$. Additional results about the smoothness of the solution are provided by [25] in the case $2 - n < p < 1$.

Now we discuss the uniqueness of the solution of the L_p-Minkowski problem (4.3). As we have seen, if $p > 1$ and $p \neq n$, then Hug, Lutwak, Yang, and Zhang [134] proved that the Alexandrov solution of the L_p-Minkowski problem (4.3) is unique. However, if $p < 1$, then the solution of the L_p-Minkowski problem (4.2) may not be unique even if f is positive and continuous. Examples are provided by Chen, Li, and Zhu [65, 66] if $p \in [0, 1)$, and Milman [178] shows that for any $C \in \mathcal{K}_{(0)}$, one finds $q \in (-n, 1)$ such that if $p < q$, then there exist multiple solutions to the L_p-Minkowski problem (4.3) with $\mu = S_{C,p}$; in other words, there exists $K \in \mathcal{K}_{(0)}$ with $K \neq C$ and $S_{K,p} = S_{C,p}$. In addition, Jian, Lu, and Wang [138] and Li, Liu, and Lu [158] prove that for any $p < 0$, there exists a positive even C^∞-function f with rotational symmetry such that the L_p-Minkowski problem (4.2) has multiple positive even C^∞-solutions. We note that in the case of the centro-affine Minkowski problem $p = -n$, Li [157] even verified the possibility of existence of infinitely many solutions without affine equivalence, and Stancu [205] proved that if an origin-symmetric convex body K with C_+^∞-boundary is a unique solution to the L_p-Minkowski problem (4.3) up to linear equivalence for $p = -n$ with $\mu = S_{K,-n}$, then it is a unique solution for $p = 0$ with $\mu = S_{K,0}$.

The case when f is a constant function in the L_p-Minkowski problem (4.2) has received special attention since Firey [98]. Through the work of Lutwak [169], Andrews [7], Andrews, Guan, and Ni [8], and Brendle, Choi, and Daskalopoulos [51], it has been clarified that the only solutions are centered balls if $p > -n$ and centered ellipsoids if $p = -n$, and there are several solutions if $p < -n$. Stability versions of these results have been obtained by Ivaki [136], but still no stability version is known in the case $p \in [0, 1)$ if we allow any solutions of (4.2), not only even ones.

In particular, concerning uniqueness, the only significant question left open is the uniqueness of even solutions of the L_p-Minkowski problem (4.2) when f is a positive even C^∞-function and $p \in [0, 1)$. In the case of $p = 0$, this is Lutwak's log-Minkowski conjecture, Conjecture 3.1. If $p \in (0, 1)$, it is also conjectured that the L_p-Minkowski problem (4.2) has a unique even solution for any positive, C^∞, and even f. More generally, we have the following conjecture (see Böröczky, Lutwak, Yang, and Zhang [44] for origin-symmetric bodies).

Conjecture 4.1 (L_p-Minkowski conjecture #1). *If $p \in (0,1)$ and K and C are centered convex bodies in $\mathbb{R}^n$ with $S_{K,p} = S_{C,p}$, then $K = C$.*

Before presenting what is known about the L_p-Minkowski conjecture, let us discuss its relation to the L_p-Brunn–Minkowski theory for $p \geq 0$. More precisely, the cases $p = 0$ and $p = 1$ have been discussed in Sections 2 and 3.

For $p > 0$, $\alpha, \beta > 0$, and $K, C \in \mathcal{K}_o^n$, we define the L_p-linear combination by the formula

$$(1 - \lambda)K +_p \lambda C = \{x \in \mathbb{R}^n : \langle x, u \rangle^p \leq \alpha\, h_K(u)^p + \beta\, h_C(u)^p \ \forall u \in S^{n-1}\}.$$

The L_p-linear combination is linear invariant; namely, if $\Phi \in GL(n)$, then $\Phi(\alpha K +_p \beta C) = \alpha\, \Phi(K) +_p \beta\, \Phi(C)$. If $p \in (0,1)$, then the L_p-linear combination of polytopes is always a polytope, but the boundary of the L_p-linear combination of convex bodies with C_+^2-boundaries may contain segments, and hence may not be C_+^2. On the other hand, if $p > 1$, then for any $\alpha, \beta > 0$, Minkowski's inequality yields $h_{\alpha K +_p \beta C}^p = \alpha\, h_K^p + \beta\, h_C^p$, as the L_p-linear combination was defined by Firey [97] in this case. According to Firey [97], if $p > 1$ and $K, C \in \mathcal{K}_o^n$, then the Brunn–Minkowski inequality yields the L_p-Brunn–Minkowski inequality

$$V(\alpha K +_p \beta C)^{\frac{p}{n}} \geq \alpha\, V(K)^{\frac{p}{n}} + \beta\, V(C)^{\frac{p}{n}} \tag{4.4}$$

for any $\alpha, \beta > 0$ with equality if and only if K and C are dilates, or equivalently,

$$V((1 - \lambda)K +_p \lambda C) \geq V(K)^{1-\lambda} V(C)^{\lambda} \tag{4.5}$$

for $\lambda \in (0,1)$ with equality if and only if $K = C$.

For $p > 0$ and $K, C \in \mathcal{K}_{(o)}^n$, analogously to the classical mixed volumes, Lutwak [168] introduced the L_p-mixed volume

$$V_p(K, C) = \frac{p}{n} \lim_{t \to 0^+} \frac{V(K +_p t C) - V(K)}{t} = \frac{1}{n} \int_{S^{n-1}} h_C^p\, dS_{K,p} = \int_{S^{n-1}} \frac{h_C^p}{h_K^p}\, dV_K,$$

and hence $V_1(K, C) = V(K, C; 1)$. Considering the first derivative of $\lambda \mapsto V((1-\lambda)K +_p \lambda C)^{\frac{p}{n}}$ yields the L_p-Minkowski inequality

$$V_p(K, C) \geq V(K)^{\frac{n-p}{n}} V(C)^{\frac{p}{n}} \tag{4.6}$$

for $p > 1$ and $K, C \in \mathcal{K}_{(o)}^n$ with equality if and only if K and C are dilates. An equivalent form is that

$$\int_{S^{n-1}} h_C^p\, dS_{K,p} \geq \int_{S^{n-1}} h_K^p\, dS_{K,p} \tag{4.7}$$

if $p > 1$, $K, C \in \mathcal{K}_{(o)}^n$, and $V(K) = V(C)$ with equality if and only if $K = C$.

We recall that the Brunn–Minkowski inequality (2.1) holds for bounded Borel subsets K and C of $\mathbb{R}^n$ as well. When $p > 1$, the L_p-Brunn–Minkowski inequality has also been extended to certain families of non-convex sets by Zhang [214], Ludwig, Xiao, and Zhang [167], and Lutwak, Yang, and Zhang [172].

If $p \in (0,1)$, then translating a cube shows that neither the L_p-Brunn–Minkowski inequality nor the L_p-Minkowski inequality holds for general $K, C \in \mathcal{K}^n_{(o)}$. However, Böröczky, Lutwak, Yang, and Zhang [44] conjecture that they hold for at least origin-symmetric convex bodies (see Böröczky and Kalantzopoulos [43] for centered convex bodies).

Conjecture 4.2 (L_p-Minkowski conjecture #2). *If $p \in (0,1)$, then (4.6) or equivalently (4.7) holds for centered $K, C \in \mathcal{K}^n$.*

Conjecture 4.3 (L_p-Brunn–Minkowski conjecture). *If $p \in (0,1)$, then (4.4) or equivalently (4.5) holds for centered $K, C \in \mathcal{K}^n$.*

The fact that the forms Conjectures 4.1 and 4.2 (including the characterization of equality) of the L_p-Minkowski conjecture are equivalent follows from (4.7) and the variational method as described above.

According to the Jensen inequality, $(1 - \lambda)K +_q \lambda \subset (1 - \lambda)K +_p \lambda C$ for $p > q \geq 0$. It follows for example via (4.5) (or via (4.11)) that if $0 \leq q < p < 1$, then the L_q-Brunn–Minkowski conjecture (or equivalently the L_q-Minkowski conjecture) yields the L_p-Brunn–Minkowski conjecture (or equivalently the L_q-Minkowski conjecture). In particular, the L_p-Brunn–Minkowski conjecture, Conjecture 4.3, for $p \in (0,1)$ is a strengthening of the Brunn–Minkowski inequality for centered convex bodies on the one hand, and follows from the log-Brunn–Minkowski conjecture, Conjecture 3.4, on the other hand. In addition, Kolesnikov and Milman [148] prove that knowing the L_p-Minkowski inequality (4.6) for some $p \in (0,1)$ yields even the characterization of the equality case for the L_q-Minkowski inequality when $q \in (p,1)$.

Let $p \in (0,1)$. Given $K, C \in \mathcal{K}^n_o$, if (4.4) holds for all $\alpha, \beta > 0$, then the L_p-Minkowski inequality (4.6) follows by considering $\frac{d}{d\lambda} V((1 - \lambda)K +_0 \lambda C)|_{\lambda=0^+}$ and using Alexandrov's lemma, Lemma 2.2, according to [44]. On the other hand, the argument in [44] shows that if $\mathcal{F}$ is any family of convex bodies closed under L_p-linear combination, then the L_p-Minkowski inequality (4.6) for all $K, C \in \mathcal{F}$ is equivalent to the L_p-Brunn–Minkowski inequality (4.4) for all $K, C \in \mathcal{F}$ and $\alpha, \beta > 0$. In particular, the equivalence holds for the family of origin-symmetric convex bodies. According to Kolesnikov and Milman [148] and Putterman [194], taking the second derivative of $\lambda \mapsto V((1 - \lambda)K +_p \lambda C)^{\frac{p}{n}}$ for origin-symmetric convex bodies K and C in $\mathbb{R}^n$ leads to the conjectured inequality

$$\frac{V(K,C;1)^2}{V(K)} \geq \frac{n-1}{n-p} V(K,C;2) + \frac{1-p}{n-p} \int_{S^{n-1}} \frac{h_C^2}{h_K^2} \, dV_K, \tag{4.8}$$

which is again a strengthened form of Minkowski's second inequality (2.6), and is equivalent to the L_p-Brunn–Minkowski conjecture without the characterization of equality. More precisely, the authors of [148] prove that for a fixed $K \in \mathcal{K}_e^n$ with C_+^2-boundary, (3.8) for all smooth $C \in \mathcal{K}_e^n$ is equivalent to a local form of the L_p-Brunn–Minkowski around K, and the global statement is verified in [194]. We note that van Handel [116] presents an approach relating the equality case of (4.8) to the equality case of (4.6) for a fixed $K \in \mathcal{K}_e^n$ with C_+^2-boundary.

In summary, we have the following three equivalent forms of the L_p-Brunn–Minkowski conjecture for $p \in (0,1)$ and origin-symmetric convex bodies K and C in $\mathbb{R}^n$ (without the characterization of equality in the case of the third formulation):

- $V((1 - \lambda)K +_p \lambda C) \geq V(K)^{1-\lambda} V(C)^\lambda$ for $\lambda \in (0,1)$;

- $V_p(K, C) \geq V(K)^{\frac{n-p}{n}} V(C)^{\frac{p}{n}}$;

- $\dfrac{V(K,C;1)^2}{V(K)} \geq \dfrac{n-1}{n-p} V(K, C; 2) + \dfrac{1-p}{n-p} \int_{S^{n-1}} \dfrac{h_C^2}{h_K^2}\, dV_K$.

Let us discuss the cases when Conjectures 4.1, 4.2, and 4.3 have been verified. They have been verified in the planar $n = 2$ case by Böröczky, Lutwak, Yang, and Zhang [44]. The most spectacular result due to the combination of the local result by Kolesnikov and Milman [148] and the local to global approach based on Schrauder estimates in PDEs by Chen, Huang, Li, and Liu [63] (see Puttermann [194] for an Alexandrov-type argument for the local to global approach) is that the L_p-Minkowski and L_p-Brunn–Minkowski conjectures hold for origin-symmetric convex bodies if $p \in (0,1)$ is close to 1.

Theorem 4.4. *If $n \geq 3$ and $p \in (p_n, 1)$, where $0 < p_n < 1 - \frac{c}{n(\log n)^{10}}$ for an absolute constant $c > 0$, then the L_p-Brunn–Minkowski and L_p-Minkowski conjectures (4.4), (4.5), (4.6), and (4.7) hold for $K, C \in \mathcal{K}_e^n$, including the characterization of the equality cases.*

The paper by Kolesnikov and Milman [148] provides an explicit estimate for p_n, depending on the Cheeger or Poincaré constants subject to the celebrated Kannan–Lovász–Simonovits conjecture [141]. Our estimate for p_n comes from the upper bound $c(\log n)^5$ by Klartag and Lehec [143] for the Poincaré constant, where $c > 0$ is an absolute constant.

Otherwise, the known cases of the L_p-Minkowski and L_p-Brunn–Minkowski conjectures for origin-symmetric bodies follow from the known cases of the log-Minkowski and log-Brunn–Minkowski conjectures. Let $p \in (0,1)$ and $n \geq 3$. Then (4.4), (4.5), (4.6), and (4.7) hold if K and C are invariant under reflections through fixed n independent linear hyperplanes (cf. [43]) and K and C are origin-symmetric complex bodies (cf. [197]). In addition, the L_p-Minkowski conjecture (4.6) and (4.7) hold for $K, C \in \mathcal{K}_e^n$ (together with characterization of equality if ∂K is C_+^3) if either K is a zonoid according to [116] or there exists a centered ellipsoid E with $E \subset K \subset (1 + c_n)E$ where $c_n > 0$ depends only on n according to [63] (an analogous result holds for linear images of l_q balls for $q > 2$ if the dimension n is high enough according to [148]).

Concerning the L_p-Brunn–Minkowski conjecture, Hosle, Kolesnikov, and Livshyts [127] and Kolesnikov and Livshyts [147] present certain natural generalizations and approaches.

In the final part of Section 4, we discuss how David Hilbert's elegant operator theoretic proof of the Brunn–Minkowski inequality has led to recent new approaches initiated by Kolesnikov and Milman [148] towards the L_p-Minkowski conjecture (see also Putterman [194] and van Handel [116]). Here we present Kolesnikov and Milman's version of the Hilbert–Brunn–Minkowski operator based on [148] because this modified operator $\mathcal{L}_K$ intertwines with linear transformations (cf. [148, Theorem 5.8]).

The mixed discriminant $D_\ell(B_1, \ldots, B_\ell)$ of ℓ positive definite $\ell \times \ell$ matrices can be defined via the identity

$$\det{}_\ell(\lambda_1 A_1 + \cdots + \lambda_m A_m) = \sum_{i_1,\ldots,i_\ell=1}^{m} D_\ell(A_{i_1}, \ldots, A_{i_\ell}) \cdot \lambda_{i_1} \cdots \lambda_{i_\ell} \tag{4.9}$$

for $\lambda_1, \ldots, \lambda_m \in \mathbb{R}$ and positive definite $\ell \times \ell$ matrices $A_1, \ldots, A_m$, where $D_\ell(A_{i_1}, \ldots, A_{i_\ell}) > 0$ is symmetric in its variables and $D_\ell(A, \ldots, A) = \det A$ (see van Handel and Shenfeld [117, 119] or Kolesnikov and Milman [148]). The similarity between (2.7) and (4.9) is not a coincidence as

$$V(K_1, \ldots, K_n) = \frac{1}{n} \int_{S^{n-1}} h_{K_n} D_{n-1}(D^2 h_{K_1}, \ldots, D^2 h_{K_{n-1}}) \, d\mathcal{H}$$

for $K_1, \ldots, K_n \in \mathcal{K}^n$ with C_+^2-boundary.

For $K \in \mathcal{K}_{(o)}^n$ with C_+^2-boundary, Kolesnikov and Milman [148] define the Hilbert–Brunn–Minkowski operator $\mathcal{L}_K : C^2(S^{n-1}) \to C^2(S^{n-1})$ by the formula

$$\mathcal{L}_K f = \frac{D_{n-1}(D^2(f h_K), D^2 h_K, \ldots, D^2 h_K)}{D_{n-1}(D^2 h_K, \ldots, D^2 h_K)} - f. \tag{4.10}$$

Following Hilbert's footsteps, it was verified in [148] that the operator $\mathcal{L}_K$ is elliptic, and hence admits a unique self-adjoint extension in $L^2(dV_K)$, and has discrete spectrum. The operator $-\mathcal{L}_K$ is positive semidefinite, its smallest eigenvalue is $\lambda_0(\mathcal{L}_K) = 0$, whose eigenspace consists of the constant functions. As Hilbert (see also van Handel and Shenfeld [117, 119] or Kolesnikov and Milman [148]) proved, the next eigenvalue is $\lambda_1(-\mathcal{L}_K) = 1$, corresponding to the n-dimensional eigenspace spanned by the linear functions; moreover, this fact is equivalent to the Brunn–Minkowski inequality for any convex bodies.

If K is origin-symmetric, then $-\mathcal{L}_K$ can be restricted to the space of even functions in $C^2(S^{n-1})$, and $\lambda_{1,e}(-\mathcal{L}_K) > 1$ holds for the smallest positive eigenvalue of this restricted operator because linear functions are odd. Here the linear invariance yields that $\lambda_{1,e}(-\mathcal{L}_K) = \lambda_{1,e}(-\mathcal{L}_{\Phi K})$ for $\Phi \in \mathrm{GL}(n)$. A key result by Kolesnikov and Milman [148] improves the estimate $\lambda_{1,e}(-\mathcal{L}_K) > 1$ uniformly; more precisely,

$$\lambda_{1,e}(-\mathcal{L}_K) \geq \frac{n - p_n}{n - 1}$$

for any $K \in \mathcal{K}_e^n$ with C_+^2-boundary where the explicit $p_n \in (0, 1 - \frac{c}{n(\log n)^{10}})$ is the same as in Theorem 4.4. The connection to the L_p-Minkowski conjecture for fixed $p \in [0,1)$ is another key result by Kolesnikov and Milman [148], as developed further by Putterman [194], namely,

$$\lambda_{1,e}(-\mathcal{L}_K) \geq \frac{n - p}{n - 1} \tag{4.11}$$

is equivalent to saying that (4.8) holds for any $C \in \mathcal{K}_e^n$. In particular, given $p \in [0,1)$, the L_p-Minkowski conjecture follows if (4.11) holds for all $K \in \mathcal{K}_e^n$ with C_+^∞-boundary, and the log-Minkowski conjecture and (3.7) are equivalent to saying that

$$\lambda_{1,e}(-\mathcal{L}_K) \geq \frac{n}{n - 1} \tag{4.12}$$

for all $K \in \mathcal{K}_e^n$ with C_+^∞-boundary. If $K_m \in \mathcal{K}_e^n$ with C_+^2-boundary tends to a cube, then $\lambda_{1,e}(-\mathcal{L}_{K_m})$ tends to $\frac{n}{n-1}$ according to [148]; therefore, the log-Minkowski conjecture states that cubes "minimize" $\lambda_{1,e}(-\mathcal{L}_K)$. On the other hand, it has been calculated [148] that $\lambda_{1,e}(-\mathcal{L}_K) = \frac{2n}{n-1}$ if K is a centered Euclidean ball, and Milman [178] verifies that centered ellipsoids maximize $\lambda_{1,e}(-\mathcal{L}_K)$ among all $K \in \mathcal{K}_e^n$ with C_+^2-boundary.

5 Some variants of the L_p-Minkowski problem

We note that Livshyts [163] considers a version of the Minkowski problem with a given measure on $\mathbb{R}^n$ acting as a weight on the surface of the convex body.

Considering the variation of the i-th intrinsic volume of a convex body K (or equivalently, variation of $V(B^n, K; i)$) for $i = 2, \ldots, n-1$ instead of the volume of K leads to the so-called Christoffel–Minkowski problem, which asks to determine a convex body when its $(i - 1)$-th area measure on S^{n-1} is prescribed (see Guan and Ma [110] and Guan and Xia [111]). We note that for $K \in \mathcal{K}^n$ with C_+^2-boundary, S_K is then the $(n - 1)$-th surface area measure, and the j-th area measure is defined using the j-th symmetric function of the principal radii of curvatures instead of the reciprocal of the Gaussian curvature. The L_p-Christoffel–Minkowski problem is discussed by Guan and Xia [111], Hu, Ma, and Shen [128], and Bryan, Ivaki, and Scheuer [52] in the case $p > 1$ and by Bianchini, Colesanti, Pagnini, and Roncoroni [27] in the case $p \in [0,1)$, where again $p = 1$ corresponds to the classical case.

The Minkowski problem on the sphere is solved by Guang, Li, and Wang [113] (see [113] for related references as well), and in the hyperbolic space, partial results, also about the hyperbolic Christoffel–Minkowski problem, are obtained by Gerhardt [104].

The Gaussian surface area measure of a $K \in \mathcal{K}^n$ is defined by Huang, Xi, and Zhao [132], and significant results about the even Gaussian Minkowski problem were obtained. These results are extended to the not necessarily even case by Feng, Liu, and Xu [88], and the L_p-Gaussian Minkowski problem is considered by Liu [161].

Next we discuss the L_p-dual Minkowski problem, which is a common generalization of the L_p-Minkowski problem and the Alexandrov problem. In order to define the dual curvature measures, let $K \in \mathcal{K}^n_{(o)}$. Recall that the radial function $\varrho_K(u) > 0$ satisfies $\varrho_K(u)u \in \partial K$ for any $u \in S^{n-1}$. For a Borel set $\omega \subset S^{n-1}$, its $\mathcal{H}$ measurable inverse radial Gauss image $\alpha^*(\omega)$ is the set of $u \in S^{n-1}$ such that there exists $v \in \omega$ that is an exterior normal at $\varrho_K(u)u$ (see Huang, Lutwak, Yang, and Zhang [130]). Now for any $q \in \mathbb{R}$, in [130] the q-th dual curvature measure of the Borel set $\omega \subset S^{n-1}$ is defined by

$$\widetilde{C}_{K,q}(\omega) = \frac{1}{n} \int\limits_{\alpha^*(\omega)} \varrho_K^n \, d\mathcal{H}.$$

In particular, $\widetilde{C}_{K,n} = V_K$ is the cone volume measure (discussed in Section 3) and $n\widetilde{C}_{K,0}$ is the Alexandrov integral curvature measure of the polar K^*. The Monge–Ampère equation corresponding to the q-th dual Minkowski problem is

$$\left(\|\nabla h\|^2 + h^2 \right)^{\frac{q-n}{2}} \cdot h \det(\nabla^2 h + h \operatorname{Id}) = f. \tag{5.1}$$

The Alexandrov problem, namely, the characterization of $\widetilde{C}_{K,0}$, has been solved by Aleksandrov [4, 5] (see also Böröczky, Lutwak, Yang, Zhang, and Zhao [48]), and for the L_p-version of the Alexandrov problem posed by Huang, Lutwak, Yang, and Zhang [131], see Mui [183] and Wu, Wu, and Xiang [208]). If $q \neq 0, n$, then the following results are known:

-   If $q < 0$, then any Borel measure on S^{n-1} not concentrated on a closed hemisphere is a q-th dual Minkowski curvature measure according to Zhao [215] and Li, Sheng, and Wang [160].
-   If $0 < q < n$, then an even Borel measure on S^{n-1} is a q-th dual Minkowski curvature measure if and only if

$$\mu(L \cap S^{n-1}) < \frac{\dim L}{q} \cdot \mu(S^{n-1})$$

    for any proper linear subspace L of $\mathbb{R}^n$ according to Böröczky, Lutwak, Yang, Zhang, and Zhao [47] where one needs to add that μ is not concentrated onto a great subsphere if $q < 1$.
-   If $q \geq n + 1$ and $K \in \mathcal{K}^n$ is origin-symmetric, then Henk and Pollehn [123] prove

$$\widetilde{C}_{K,q}(L \cap S^{n-1}) < \frac{q - n + \dim L}{q} \cdot \widetilde{C}_{K,q}(S^{n-1}).$$

-   If $q > 0$ and $n = 2$, then (5.1) has a solution for any measurable f provided $\frac{1}{c} < f < c$ for a $c > 1$ according to Chen and Li [64].

In particular, it is an intriguing open problem to characterize an even q-th dual Minkowski curvature measure on S^{n-1} if $q > n$.

For $p, q \in \mathbb{R}$, Lutwak, Yang, and Zhang [173] define the q-th L_p-dual Minkowski curvature measure on S^{n-1} by $d\widetilde{C}_{K,p,q} = h_K^{-p} d\widetilde{C}_{K,q}$, and hence $\widetilde{C}_{K,0,q} = \widetilde{C}_{K,q}$ and $\widetilde{C}_{K,p,n} = \frac{1}{n} S_{K,p}$. Given a Borel measure μ on S^{n-1}, the simplest version of the q-th L_p-dual Minkowski problem asks for a $K \in \mathcal{K}_{(o)}^n$ with $\mu = \widetilde{C}_{K,p,q}$, and the corresponding Monge–Ampère equation is

$$h^{1-p} \det(\nabla^2 h + h \operatorname{Id}) = \left(\|\nabla h\|^2 + h^2 \right)^{\frac{n-q}{2}} \cdot f. \tag{5.2}$$

Improving on [38, 133], Chen and Li [61] prove that if $p > 0$ and $q \neq p$, then any Borel measure not concentrated on a closed hemisphere is a q-th L_p-dual Minkowski curvature measure (more precisely, if $p \leq q$, then some modification of the Monge–Ampère equation might be needed).

Uniqueness of the solution of the q-th L_p-dual Minkowski problem (5.2) is thoroughly investigated by Li, Liu, and Lu [158]. The case when $n = 2$ and f is a constant function has been completely clarified by Li and Wan [156].

Some other important related variants of the Minkowski problem currently considered are the so-called chord measures (cf. Lutwak, Yang, Xi, and Zhang [174]) and the L_p-Minkowski problem for log-concave functions (cf. Fang, Xing, and Ye [85]).

Starting with Haberl, Lutwak, Yang, and Zhang [115], Orlicz versions of the L_p-Minkowski problem have been intensively investigated, where the function $t \mapsto t^{1-p}$ in (4.2) is replaced by certain $\varphi : (0, \infty) \to (0, \infty)$, and hence (4.2) is replaced by

$$\varphi(h) \det(\nabla^2 h + h \operatorname{Id}) = f,$$

where f is a given non-negative function on S^{n-1}. Typically, the solution is only up to a constant factor, that is, there exists some $c > 0$ such that $\varphi(h) \det(\nabla^2 h + h \operatorname{Id}) = c \cdot f$. The known existence results about the L_p-Minkowski problem have been generalized to the Orlicz L_p-Minkowski problem where φ replaces t^p by Huang and He [129] if $p > 1$ (see also Xie [210]), by Jian and Lu [137] if $p \in (0, 1)$, and by Bianchi, Böröczky, and Colesanti [23] if $p \in (-n, 0)$.

Orlicz versions of the Alexandrov problem are considered by Li, Sheng, Ye, and Yi [159] and Feng, Hu, and Liu [87], and versions of the L_p-dual Minkowski problem in general are considered by Gardner, Hug, Weil, Xing, and Ye [101, 102], Xing, Ye, and Zhu [211], and Liu and Lu [162].

Bibliography

[1]   A. D. Aleksandrov, *Zur Theorie der gemischten Volumina von konvexen Körpern II*, Mat. Sb. N. S. **2** (1937), 1205–1238.

[2] A. D. Aleksandrov, *On the theory of mixed volumes. I. Extension of certain concepts in the theory of convex bodies*, Mat. Sb. N. S. **2** (1937), 947–972 (Russian; German summary).

[3] A. D. Aleksandrov, *On the theory of mixed volumes. III. Extension of two theorems of Minkowski on convex polyhedra to arbitrary convex bodies*, Mat. Sb. N. S. **3** (1938), 27–46 (Russian; German summary).

[4] A. D. Aleksandrov, *Existence and uniqueness of a convex surface with a given integral curvature*, C. R. (Dokl.) Acad. Sci. URSS **35** (1942), 131–134.

[5] A. D. Aleksandrov, *Selected works. Part I*, Gordon and Breach Publishers, Amsterdam, 1996.

[6] L. Ambrosio, A. Colesanti and E. Villa, *Outer Minkowski content for some classes of closed sets*, Math. Ann. **342** (2008), 727–748.

[7] B. Andrews, *Gauss curvature flow: the fate of rolling stone*, Invent. Math. **138** (1999), 151–161.

[8] B. Andrews, P. Guan and L. Ni, *Flow by the power of the Gauss curvature*, Adv. Math. **299** (2016), 174–201.

[9] S. Artstein-Avidan, D. I. Florentin and A. Segal, *Functional Brunn–Minkowski inequalities induced by polarity*, Adv. Math. **364** (2020), 107006, 19 pp.

[10] S. Artstein-Avidan, B. Klartag and V. D. Milman, *On the Santaló point of a function and a functional Santaló inequality*, Mathematika **54** (2004), 33–48.

[11] S. Artstein-Avidan, A. Giannopoulos and V. D. Milman, *Asymptotic geometric analysis. Part I*, Mathematical surveys and monographs, vol. 202, American Mathematical Society, Providence, RI, 2015.

[12] S. Artstein-Avidan, A. Giannopoulos and V. D. Milman, *Asymptotic geometric analysis. Part II*, Mathematical surveys and monographs, vol. 261, American Mathematical Society, Providence, RI, 2021.

[13] K. M. Ball, *Isoperimetric problems in ℓ_p and sections of convex sets*, PhD thesis, University of Cambridge, 1986.

[14] K. M. Ball, *An elementary introduction to modern convex geometry*, in: Flavors of geometry, S. Levy, ed., Cambridge University Press, 1997, pp. 1–58.

[15] K. M. Ball and K. J. Böröczky, *Stability of the Prékopa–Leindler inequality*, Mathematika **56** (2010), 339–356.

[16] K. M. Ball and K. J. Böröczky, *Stability of some versions of the Prékopa–Leindler inequality*, Monatshefte Math. **163** (2011), 1–14.

[17] Z. M. Balogh and A. Kristály, *Equality in Borell–Brascamp–Lieb inequalities on curved spaces*, Adv. Math. **339** (2018), 453–494.

[18] M. Barchiesi and V. Julin, *Robustness of the Gaussian concentration inequality and the Brunn–Minkowski inequality*, Calc. Var. Partial Differ. Equ. **56** (2017), 80, 12 pp.

[19] F. Barthe, *Autour de l'inégalité de Brunn–Minkowski*, Mémoire d'Habilitation, 2008.

[20] F. Barthe, *On a reverse form of the Brascamp–Lieb inequality*, Invent. Math. **134** (1998), 335–361.

[21] F. Barthe, K. J. Böröczky and M. Fradelizi, *Stability of the functional forms of the Blaschke–Santaló inequality*, Monatshefte Math. **173**(2) (2014), 135–159.

[22] F. Barthe, O. Guédon, S. Mendelson and A. Naor, *A probabilistic approach to the geometry of the l_p^n-ball*, Ann. Probab. **33** (2005), 480–513.

[23] G. Bianchi, K. J. Böröczky and A. Colesanti, *The Orlicz version of the L_p Minkowski problem for $-n < p < 0$*, Adv. Appl. Math. **111** (2019), 101937, 29 pp.

[24] G. Bianchi, K. J. Böröczky, A. Colesanti and D. Yang, *The L_p-Minkowski problem for $-n < p < 1$ according to Chou–Wang*, Adv. Math. **341** (2019), 493–535.

[25] G. Bianchi, K. J. Böröczky and A. Colesanti, *Smoothness in the L_p Minkowski problem for $p < 1$*, J. Geom. Anal. **30** (2020), 680–705.

[26] G. Bianchi and H. Egnell, *A note on the Sobolev inequality*, J. Funct. Anal. **100** (1991), 18–24.

[27] C. Bianchini, A. Colesanti, D. Pagnini and A. Roncoroni, *On p-Brunn–Minkowski inequalities for intrinsic volumes, with $0 \le p < 1$*, Math. Ann. (2022), 10.1007/s00208-022-02454-0.

[28] S. G. Bobkov, A. Colesanti and I. Fragalà, *Quermassintegrals of quasi-concave functions and generalized Prékopa–Leindler inequalities*, Manuscr. Math. **143** (2014), 131–169.

[29] F. Bolley, D. Cordero-Erausquin, Y. Fujita, I. Gentil and A. Guillin, *New sharp Gagliardo–Nirenberg–Sobolev inequalities and an improved Borell–Brascamp–Lieb inequality*, Int. Math. Res. Not. **2020** (2020), 3042–3083.

[30] B. Bollobás and I. Leader, *Products of unconditional bodies*, in: Geometric aspects of functional analysis, Israel, 1992–1994, Oper. theory adv. appl., vol. 77, Birkhauser, Basel, 1995, pp. 13–24.

[31] T. Bonnesen and W. Fenchel, *Theory of convex bodies*, BCS Associates, Moscow, ID, 1987.

[32] C. Borell, *Convex set functions in d-space*, Period. Math. Hung. **6** (1975), 111–136.

[33] K. J. Böröczky, *Stability of the Blaschke–Santaló and the affine isoperimetric inequality*, Adv. Math. **225** (2010), 1914–1928.

[34] K. J. Böröczky and A. De, *Stability of the Prékopa–Leindler inequality for log-concave functions*, Adv. Math. **386** (2021), 107810.

[35] K. J. Böröczky and A. De, *Stable solution of the log-Minkowski problem in the case of hyperplane symmetries*, J. Differ. Equ. **298** (2021), 298–322.

[36] K. J. Böröczky and A. De, *Stability of the log-Brunn–Minkowski inequality in the case of many hyperplane symmetries*, arXiv:2101.02549.

[37] K. J. Böröczky, A. Figalli and J. P. G. Ramos, *A quantitative stability result for the Prékopa–Leindler inequality for arbitrary measurable functions*, submitted, arXiv:2201.11564.

[38] K. J. Böröczky and F. Fodor, *The L_p dual Minkowski problem for $p > 1$ and $q > 0$*, J. Differ. Equ. **266** (2019), 7980–8033.

[39] K. J. Böröczky and P. Hegedűs, *The cone volume measure of antipodal points*, Acta Math. Hung. **146** (2015), 449–465.

[40] K. J. Böröczky, P. Hegedűs and G. Zhu, *On the discrete logarithmic Minkowski problem*, Int. Math. Res. Not. **2016** (2016), 1807–1838.

[41] K. J. Böröczky and M. Henk, *Cone-volume measure of general centered convex bodies*, Adv. Math. **286** (2016), 703–721.

[42] K. J. Böröczky and M. Henk, *Cone-volume measure and stability*, Adv. Math. **306** (2017), 24–50.

[43] K. J. Böröczky and P. Kalantzopoulos, *Log-Brunn–Minkowski inequality under symmetry*, Trans. Am. Math. Soc. **375** (2022), 5987–6013.

[44] K. J. Böröczky, E. Lutwak, D. Yang and G. Zhang, *The log-Brunn–Minkowski-inequality*, Adv. Math. **231** (2012), 1974–1997.

[45] K. J. Böröczky, E. Lutwak, D. Yang and G. Zhang, *The logarithmic Minkowski problem*, J. Am. Math. Soc. **26** (2013), 831–852.

[46] K. J. Böröczky, E. Lutwak, D. Yang and G. Zhang, *Affine images of isotropic measures*, J. Differ. Geom. **99** (2015), 407–442.

[47] K. J. Böröczky, E. Lutwak, D. Yang, G. Zhang and Y. Zhao, *The dual Minkowski problem for symmetric convex bodies*, Adv. Math. **356** (2019), 106805.

[48] K. J. Böröczky, E. Lutwak, D. Yang, G. Zhang and Y. Zhao, *The Gauss image problem*, Commun. Pure Appl. Math. **73** (2020), 1406–1452.

[49] K. J. Böröczky and H. T. Trinh, *The planar L_p-Minkowski problem for $0 < p < 1$*, Adv. Appl. Math. **87** (2017), 58–81.

[50] H. J. Brascamp and E. H. Lieb, *On extensions of the Brunn–Minkowski and Prékopa–Leindler theorems, including inequalities for log concave functions, and with an application to the diffusion equation*, J. Funct. Anal. **22** (1976), 366–389.

[51] S. Brendle, K. Choi and P. Daskalopoulos, *Asymptotic behavior of flows by powers of the Gaussian curvature*, Acta Math. **219** (2017), 1–16.

[52] P. Bryan, M. Ivaki and J. Scheuer, *A unified flow approach to smooth, even Lp-Minkowski problems*, Anal. PDE **12** (2019), 259–280.

[53] S. Bubeck, R. Eldan and J. Lehec, *Sampling from a log-concave distribution with projected Langevin Monte Carlo*, Discrete Comput. Geom. **59** (2018), 757–783.

[54] D. Bucur and I. Fragalà, *Lower bounds for the Prékopa–Leindler deficit by some distances modulo translations*, J. Convex Anal. **21** (2014), 289–305.

[55] J. R. Bueno and P. Pivarov, *A stochastic Prékopa–Leindler inequality for log-concave functions*, Commun. Contemp. Math. **23**(02) (2021), 2050019, arXiv:1912.06904. 10.1142/S0219199720500194.

[56] L. A. Caffarelli, *Interior $W^{2,p}$-estimates for solutions of the Monge–Ampère equation*, Ann. Math. (2) **131** (1990), 135–150.

[57] L. A. Caffarelli, *A localization property of viscosity solutions to the Monge–Ampère equation and their strict convexity*, Ann. Math. (2) **131** (1990), 129–134.

[58] U. Caglar and E. M. Werner, *Stability results for some geometric inequalities and their functional versions*, in: Convexity and concentration, IMA vol. math. appl., vol. 161, Springer, New York, 2017, pp. 541–564.

[59] E. A. Carlen and D. Cordero-Erausquin, *Subadditivity of the entropy and its relation to Brascamp–Lieb type inequalities*, Geom. Funct. Anal. **19** (2009), 373–405.

[60] E. A. Carlen and F. Maggi, *Stability for the Brunn–Minkowski and Riesz rearrangement inequalities, with applications to Gaussian concentration and finite range non-local isoperimetry*, Can. J. Math. **69** (2017), 1036–1063.

[61] H. Chen and Q-R. Li, *The L_p dual Minkowski problem and related parabolic flows*, J. Funct. Anal. **281** (2021), 109139, 65 pp.

[62] S. Chen, Y. Feng and W. Liu, *Uniqueness of solutions to the logarithmic Minkowski problem in $\mathbb{R}^3$*, Adv. Math. **411** (2022), 108782, 18 pp.

[63] S. Chen, Y. Huang, Q.-R. Li and J. Liu, *The L_p-Brunn–Minkowski inequality for $p < 1$*, Adv. Math. **368** (2020), 107166.

[64] S. Chen and Q.-R. Li, *On the planar dual Minkowski problem*, Adv. Math. **333** (2018), 87–117.

[65] S. Chen, Q.-R. Li and G. Zhu, *On the L_p Monge–Ampère equation*, J. Differ. Equ. **263** (2017), 4997–5011.

[66] S. Chen, Q.-R. Li and G. Zhu, *The Logarithmic Minkowski Problem for non-symmetric measures*, Trans. Am. Math. Soc. **371** (2019), 2623–2641.

[67] S. Chen, J. Liu and X.-J. Wang, *Global regularity for the Monge–Ampère equation with natural boundary condition*, Ann. Math. (2) **194** (2021), 745–793.

[68] S.-Y. Cheng and S.-T. Yau, *On the regularity of the solution of the n-dimensional Minkowski problem*, Commun. Pure Appl. Math. **29** (1976), 495–561.

[69] Y. Chen, *An almost constant lower bound of the isoperimetric coefficient in the KLS conjecture*, Geom. Funct. Anal. **31** (2021), 34–61.

[70] K. S. Chou and X. J. Wang, *The L_p-Minkowski problem and the Minkowski problem in centroaffine geometry*, Adv. Math. **205** (2006), 33–83.

[71] M. Christ, *An approximate inverse Riesz–Sobolev inequality*, 2012, arXiv:1112.3715.

[72] A. Colesanti, G. Livshyts and A. Marsiglietti, *On the stability of Brunn–Minkowski type inequalities*, J. Funct. Anal. **273** (2017), 1120–1139.

[73] A. Colesanti and G. Livshyts, *A note on the quantitative local version of the log-Brunn–Minkowski inequality*, in: The mathematical legacy of Victor Lomonosov – operator theory, Adv. anal. geom., vol. 2, De Gruyter, Berlin, 2020, pp. 85–98.

[74] D. Cordero-Erausquin, *Transport inequalities for log-concave measures, quantitative forms, and applications*, Can. J. Math. **69** (2017), 481–501, arXiv:1504.06147.

[75] D. Cordero-Erausquin, M. Fradelizi and B. Maurey, *The (B) conjecture for the Gaussian measure of dilates of symmetric convex sets and related problems*, J. Funct. Anal. **214** (2004), 410–427.

[76] D. Cordero-Erausquin and L. Rotem, *Improved log-concavity for rotationally invariant measures of symmetric convex sets*, Ann. Probab., accepted, arXiv:2111.05110.

[77] G. Crasta and I. Fragalà, *On a geometric combination of functions related to Prékopa–Leindler inequality*, arXiv:2204.11521.

[78] S. Dar, *A Brunn–Minkowski-type inequality*, Geom. Dedic. **77** (1999), 1–9.

[79] V. I. Diskant, *Stability of the solution of a Minkowski equation*, Sib. Mat. Zh. **14** (1973), 669–673 (Russian); Eng. transl.: Siberian Math. J. **14** (1974), 466–473.

[80] S.-Z. Du, *On the planar L_p-Minkowski problem*, J. Differ. Equ. **287** (2021), 37–77.

[81] S. Dubuc, *Critères de convexité et inégalités integralés*, Ann. Inst. Fourier Grenoble **27**(1) (1977), 135–165.

[82] R. Eldan, *Thin shell implies spectral gap up to polylog via a stochastic localization scheme*, Geom. Funct. Anal. **23** (2013), 532–569.

[83] R. Eldan and B. Klartag, *Pointwise estimates for marginals of convex bodies*, J. Funct. Anal. **254** (2008), 2275–2293.

[84] A. Eskenazis and G. Moschidis, *The dimensional Brunn–Minkowski inequality in Gauss space*, J. Funct. Anal. **280** (2021), 108914, 19 pp.

[85] N. Fang, S. Xing and D. and, *Geometry of log-concave functions: the L_p Asplund sum and the L_p Minkowski problem*, Calc. Var. Partial Differ. Equ. **61** (2022), 45, 37 pp.

[86] W. Fenchel and B. Jessen, *Mengenfunktionen und konvexe Körper*, Mat.-Fys. Medd. Danske Vid. Selsk. **16** (1938), 1–31.

[87] Y. Feng, S. Hu and W. Liu, *Existence and uniqueness of solutions to the Orlicz Aleksandrov problem*, Calc. Var. Partial Differ. Equ. **61** (2022), 148, 23 pp.

[88] Y. Feng, W. Liu and L. Xu, *Existence of non-symmetric solutions to the Gaussian Minkowski problem*, J. Geom. Anal. **33** (2023), 89, 39 pp.

[89] A. Figalli and D. Jerison, *Quantitative stability for sumsets in $\mathbb{R}^n$*, J. Eur. Math. Soc. **17** (2015), 1079–1106.

[90] A. Figalli and D. Jerison, *Quantitative stability for the Brunn–Minkowski inequality*, Adv. Math. **314** (2017), 1–47.

[91] A. Figalli, F. Maggi and A. Pratelli, *A refined Brunn–Minkowski inequality for convex sets*, Ann. IHP **26** (2009), 2511–2519.

[92] A. Figalli, F. Maggi and A. Pratelli, *A mass transportation approach to quantitative isoperimetric inequalities*, Invent. Math. **182** (2010), 167–211.

[93] A. Figalli and R. Neumayer, *Gradient stability for the Sobolev inequality: the case $p \geq 2$*, J. Eur. Math. Soc. **21** (2019), 319–354.

[94] A. Figalli, Yi Ru-Ya Zhang, *Sharp gradient stability for the Sobolev inequality*, arXiv:2003.04037.

[95] M. Fradelizi and M. Meyer, *Some functional forms of Blaschke-Santaló inequality*, Math. Z. **256** (2007), 379–395.

[96] A. Freyer, M. Henk and C. Kipp, *Affine subspace concentration conditions for centered polytopes*, Mathematika **69** (2023), 458–472.

[97] W. J. Firey, *p-means of convex bodies*, Math. Scand. **10** (1962), 17–24.

[98] W. J. Firey, *Shapes of worn stones*, Mathematika **21** (1974), 1–11.

[99] N. Fusco, F. Maggi and A. Pratelli, *The sharp quantitative isoperimetric inequality*, Ann. Math. **168**(3) (2008), 941–980.

[100] R. J. Gardner, *The Brunn–Minkowski inequality*, Bull. Am. Meteorol. Soc. **29** (2002), 335–405.

[101] R. Gardner, D. Hug, W. Weil, S. Xing and D. Ye, *General volumes in the Orlicz–Brunn–Minkowski theory and a related Minkowski problem I*, Calc. Var. Partial Differ. Equ. **58** (2019), 12, 35 pp.

[102] R. Gardner, D. Hug, S. Xing and D. Ye, *General volumes in the Orlicz–Brunn–Minkowski theory and a related Minkowski problem II*, Calc. Var. Partial Differ. Equ. **59** (2020), 15, 33 pp.

[103] R. Gardner and A. Zvavitch, *Gaussian Brunn–Minkowski-type inequalities*, Trans. Am. Math. Soc. **362** (2010), 5333–5353.

[104] C. Gerhardt, *Minkowski type problems for convex hypersurfaces in hyperbolic space*, arXiv:math/0602597.

[105] D. Ghilli and P. Salani, *Quantitative Borell-Brascamp-Lieb inequalities for power concave functions*, J. Convex Anal. **24** (2017), 857–888.

[106] N. Gozlan, *The deficit in the Gaussian log-Sobolev inequality and inverse Santalo inequalities*, Int. Math. Res. Not. **2022**(17) (2022), 12940–12983, arXiv:2007.05255.

[107] H. Groemer, *On the Brunn–Minkowski theorem*, Geom. Dedic. **27** (1988), 357–371.

[108] H. Groemer, *Stability of geometric inequalities*, in: Handbook of convex geometry, P. M. Gruber and J. M. Wills, eds., North-Holland, Amsterdam, 1993, pp. 125–150.

[109] M. Gromov and V. D. Milman, *Generalization of the spherical isoperimetric inequality for uniformly convex Banach Spaces*, Compos. Math. **62** (1987), 263–282.

[110] P. Guan and X. Ma, *The Christoffel–Minkowski problem. I. Convexity of solutions of a Hessian equation*, Invent. Math. **151**(3) (2003), 553–577.

[111] P. Guan and C. Xia, *L^p Christoffel–Minkowski problem: the case* $1 < p < k + 1$, Calc. Var. Partial Differ. Equ. **57** (2018), 69, 23 pp.

[112] Q. Guang, Q-R. Li and X.-J. Wang, *The L_p-Minkowski problem with super-critical exponents*, arXiv:2203.05099.

[113] Q. Guang, Q-R. Li and X.-J. Wang, *The Minkowski problem in the sphere*, https://person.zju.edu.cn/person/attachments/2021-02/01-1612235356-841400.pdf.

[114] Q. Guang, Q-R. Li and X.-J. Wang, *Existence of convex hypersurfaces with prescribed centroaffine curvature*, https://person.zju.edu.cn/person/attachments/2022-02/01-1645171178-851572.pdf.

[115] C. Haberl, E. Lutwak, D. Yang and G. Zhang, *The even Orlicz Minkowski problem*, Adv. Math. **224**(6) (2010), 2485–2510.

[116] R. van Handel, *The local logarithmic Brunn–Minkowski inequality for zonoids*, arXiv:2202.09429.

[117] R. van Handel and Y. Shenfeld, *Mixed volumes and the Bochner method*, Proc. Am. Math. Soc. **147**(12) (2019), 5385–5402.

[118] R. van Handel and Y. Shenfeld, *The extremals of Minkowski's quadratic inequality*, Duke Math. J. **171** (2022), 957–1027, arXiv:1902.10029.

[119] R. van Handel and Y. Shenfeld, The extremals of the Alexandrov–Fenchel inequality for convex polytopes, *Acta Math.*, accepted, arXiv:2011.04059.

[120] D. Harutyunyan, *Quantitative anisotropic isoperimetric and Brunn–Minkowski inequalities for convex sets with improved defect estimates*, ESAIM Control Optim. Calc. Var. **24** (2018), 479–494.

[121] B. He, G. Leng and K. Li, *Projection problems for symmetric polytopes*, Adv. Math. **207** (2006), 73–90.

[122] M. Henk and E. Linke, *Cone-volume measures of polytopes*, Adv. Math. **253** (2014), 50–62.

[123] M. Henk and H. Pollehn, *Necessary subspace concentration conditions for the even dual Minkowski problem*, Adv. Math. **323** (2018), 114–141.

[124] M. Henk, A. Schürman and J. M. Wills, *Ehrhart polynomials and successive minima*, Mathematika **52** (2006), 1–16.

[125] P. van Hintum, H. Spink and M. Tiba, *Sharp stability of Brunn–Minkowski for homothetic regions*, J. Eur. Math. Soc. **24** (2022), 4207–4223.

[126] P. van Hintum, H. Spink and M. Tiba, *Sharp quantitative stability of the planar Brunn–Minkowski inequality*, arXiv:1911.11945.

[127] J. Hosle, A. V. Kolesnikov and G. V. Livshyts, *On the L_p-Brunn–Minkowski and dimensional Brunn–Minkowski conjectures for log-concave measures*, J. Geom. Anal. **21** (2021), 5799–5836.

[128] C. Hu, X. Ma and C. Shen, *On the Christoffel–Minkowski problem of Firey's p-sum*, Calc. Var. Partial Differ. Equ. **21**(2) (2004), 137–155.

[129] Q. Huang and B. He, *On the Orlicz Minkowski problem for polytopes*, Discrete Comput. Geom. **48** (2012), 281–297.

[130] Y. Huang, E. Lutwak, D. Yang and G. Zhang, *Geometric measures in the dual Brunn–Minkowski theory and their associated Minkowski problems*, Acta Math. **216** (2016), 325–388.

[131] Y. Huang, E. Lutwak, D. Yang and G. Zhang, *The L_p-Aleksandrov problem for L_p-integral curvature*, J. Differ. Geom. **110** (2018), 1–29.

[132] Y. Huang, D. Xi and Y. Zhao, *The Minkowski problem in Gaussian probability space*, Adv. Math. **385** (2021), 107769.

[133] Y. Huang and Y. Zhao, *On the L_p dual Minkowski problem*, Adv. Math. **332** (2018), 57–84.

[134] D. Hug, E. Lutwak, D. Yang and G. Zhang, *On the L_p Minkowski problem for polytopes*, Discrete Comput. Geom. **33** (2005), 699–715.

[135] D. Hug, *Contributions to affine surface area*, Manuscr. Math. **91** (1996), 283–301.

[136] M. Ivaki, *On the stability of the L_p-curvature*, J. Funct. Anal. **283** (2022), 109684.

[137] H. Jian and J. Lu, *Existence of the solution to the Orlicz–Minkowski problem*, Adv. Math. **344** (2019), 262–288.

[138] H. Jian, J. Lu and X-J. Wang, *Nonuniqueness of solutions to the L_p-Minkowski problem*, Adv. Math. **281** (2015), 845–856.

[139] H. Jian, J. Lu and G. Zhu, *Mirror symmetry solutions to the centro-affine Minkowski problem*, Calc. Var. Partial Differ. Equ. **55** (2016), 41, 22 pp.

[140] P. Kalantzopoulos and C. Saroglou, *On a j-Santaló Conjecture*, arXiv:2203.14815.

[141] R. Kannan, L. Lovász and M. Simonovits, *Isoperimetric problems for convex bodies and a localization lemma*, Discrete Comput. Geom. **13** (1995), 541–559.

[142] B. Klartag, *On nearly radial marginals of high-dimensional probability measures*, J. Eur. Math. Soc. **12** (2010), 723–754.

[143] B. Klartag and J. Lehec, *Bourgain's slicing problem and KLS isoperimetry up to polylog*, arXiv:2203.15551.

[144] A. V. Kolesnikov, *Mass transportation functionals on the sphere with applications to the logarithmic Minkowski problem*, Mosc. Math. J. **20** (2020), 67–91.

[145] A. V. Kolesnikov and E. D. Kosov, *Moment measures and stability for Gaussian inequalities*, Theory Stoch. Process. **22** (2017), 47–61.

[146] A. V. Kolesnikov and G. V. Livshyts, *On the Gardner–Zvavitch conjecture: symmetry in inequalities of Brunn–Minkowski type*, Adv. Math. **384** (2021), 107689, 23 pp.

[147] A. V. Kolesnikov and G. V. Livshyts, *On the Local version of the Log-Brunn–Minkowski conjecture and some new related geometric inequalities*, Int. Math. Res. Not. **18** (2022), 14427–14453.

[148] A. V. Kolesnikov and E. Milman, *Local L_p-Brunn–Minkowski inequalities for $p < 1$*, Mem. Am. Math. Soc. **277** (2022), 1360.

[149] A. V. Kolesnikov and E. M. Werner, *Blaschke–Santalo inequality for many functions and geodesic barycenters of measures*, Adv. Math. **396** (2022), 108110, 44 pp.

[150] H. Knothe, *Contributions to the theory of convex bodies*, Mich. Math. J. **4** (1957), 39–52.

[151] J. Lehec, *A direct proof of the functional Santaló inequality*, C. R. Math. Acad. Sci. Paris **347**(1–2) (2009), 55–58.

[152] J. Lehec, *Partitions and functional Santaló inequalities*, Arch. Math. (Basel) **92**(1) (2009), 89–94.

[153] K. Leichtweiß, *Affine geometry of convex bodies*, Johann Ambrosius Barth Verlag, Heidelberg, 1998.

[154] L. Leindler, *On a certain converse of Hölder's inequality. II*, Acta Sci. Math. **33** (1972), 217–223.

[155] H. Lewy, *On differential geometry in the large. I. Minkowski problem*, Trans. Am. Math. Soc. **43** (1938), 258–270.

[156] H. Li and Y. Wan, *Classification of solutions for the planar isotropic L_p dual Minkowski problem*, arXiv:2209.14630.

[157] Q.-R. Li, *Infinitely many solutions for centro-affine Minkowski problem*, Int. Math. Res. Not. **2019** (2019), 5577–5596.

[158] Q.-R. Li, J. Liu and J. Lu, *Non-uniqueness of solutions to the dual L_p-Minkowski problem*, Int. Math. Res. Not. **2022** (2022), 9114–9150.

[159] Q-R. Li, W. Sheng, D. Ye and C. Yi, *A flow approach to the Musielak–Orlicz–Gauss image problem*, Adv. Math. **403** (2022), 108379, 40 pp.

[160] Q.-R. Li, W. Sheng and X.-J. Wang, *Flow by Gauss curvature to the Aleksandrov and dual Minkowski problems*, J. Eur. Math. Soc. **22** (2020), 893–923.

[161] J. Liu, *The L_p-Gaussian Minkowski problem*, Calc. Var. Partial Differ. Equ. **61** (2022), 28, 23 pp.

[162] Y. Liu and J. Lu, *A flow method for the dual Orlicz–Minkowski problem*, Trans. Am. Math. Soc. **373**(8) (2020), 5833–5853.

[163] G. V. Livshyts, *An extension of Minkowski's theorem and its applications to questions about projections for measures*, Adv. Math. **356** (2019), 106803, 40 pp.

[164] G. V. Livshyts, *On a conjectural symmetric version of Ehrhard's inequality*, submitted, arXiv:2103.11433.

[165] G. V. Livshyts, *A universal bound in the dimensional Brunn–Minkowski inequality for log-concave measures*, submitted, arXiv:2107.00095.

[166] M. Ludwig, *General affine surface areas*, Adv. Math. **224** (2010), 2346–2360.

[167] M. Ludwig, J. Xiao and G. Zhang, *Sharp convex Lorentz–Sobolev inequalities*, Math. Ann. **350** (2011), 169–197.

[168] E. Lutwak, *Selected affine isoperimetric inequalities*, in: Handbook of convex geometry, North-Holland, Amsterdam, 1993, pp. 151–176.

[169] E. Lutwak, *The Brunn–Minkowski-Firey theory. I. Mixed volumes and the Minkowski problem*, J. Differ. Geom. **38** (1993), 131–150.

[170] E. Lutwak, *The Brunn–Minkowski-Firey theory. II. Affine and geominimal surface areas*, Adv. Math. **118** (1996), 244–294.

[171] E. Lutwak, D. Yang and G. Zhang, *L_p John ellipsoids*, Proc. Lond. Math. Soc., (3) **90** (2005), 497–520.

[172] E. Lutwak, D. Yang and G. Zhang, *The Brunn–Minkowski–Firey inequality for nonconvex sets*, Adv. Appl. Math. **48** (2012), 407–413.

[173] E. Lutwak, D. Yang and G. Zhang, *L_p dual curvature measures*, Adv. Math. **329** (2018), 85–132.

[174] E. Lutwak, D. Yang, D. Xi and G. Zhang, *Chord measures in integral geometry and their Minkowski problems*, submitted.

[175] F. Maggi, *Sets of finite perimeter and geometric variational problems. An introduction to geometric measure theory*, Cambridge University Press, Cambridge, 2012.

[176] M. Meyer and A. Pajor, *On the Blaschke–Santaló inequality*, Arch. Math. (Basel) **55** (1990), 82–93.

[177] A. Marsiglietti, *Borell's generalized Prékopa–Leindler inequality: a simple proof*, J. Convex Anal. **24** (2017), 807–817.

[178] E. Milman, *A sharp centro-affine isospectral inequality of Szegő–Weinberger type and the L_p-Minkowski problem*, J. Differ. Geom., accepted, arXiv:2103.02994.

[179] E. Milman, Centro-affine differential geometry and the log-Minkowski problem, arXiv:2104.12408.

[180] V. D. Milman and G. Schechtman, *Asymptotic theory of finite-dimensional normed spaces*, Lecture notes in mathematics, vol. 1200, Springer, Berlin, 1986. With an appendix by M. Gromov.

[181] H. Minkowski, *Allgemeine Lehrsätze über die konvexen Polyeder*, Nachr. Ges. Wiss. Göttingen (1897), 198–219.

[182] H. Minkowski, *Volumen und Oberfäche*, Math. Ann. **57** (1903), 447–495.

[183] S. Mui, *On the L_p Aleksandrov problem for negative p*, Adv. Math. **408** (2022), 108573, 26 pp.

[184] A. Naor, *The surface measure and cone measure on the sphere of l_p^n*, Trans. Am. Math. Soc. **359** (2007), 1045–1079.

[185] P. Nayar and T. Tkocz, *A note on a Brunn–Minkowski inequality for the Gaussian measure*, Proc. Am. Math. Soc. **141** (2013), 4027–4030.

[186] P. Nayar and T. Tkocz, *On a convexity property of sections of the cross-polytope*, Proc. Am. Math. Soc. **148** (2020), 1271–1278.

[187] V. H. Nguyen, *New approach to the affine Polya–Szego principle and the stability version of the affine Sobolev inequality*, Adv. Math. **302** (2016), 1080–1110.

[188] L. Nirenberg, *The Weyl and Minkowski problems in differential geometry in the large*, Commun. Pure Appl. Math. **6** (1953), 337–394.

[189] G. Paouris and E. Werner, *Relative entropy of cone measures and L_p centroid bodies*, Proc. Lond. Math. Soc. **104** (2012), 253–286.

[190] A. V. Pogorelov, *The Minkowski multidimensional problem*, V. H. Winston & Sons, Washington, D. C., 1978.

[191] A. Prékopa, *Logarithmic concave measures with application to stochastic programming*, Acta Sci. Math. **32** (1971), 301–316.

[192] A. Prékopa, *On logarithmic concave measures and functions*, Acta Sci. Math. **34** (1973), 335–343.

[193] A. Prékopa, *New proof for the basic theorem of logconcave measures*, Alkalmaz. Mat. Lapok **1** (1975), 385–389 (Hungarian).

[194] E. Putterman, *Equivalence of the local and global versions of the L_p-Brunn–Minkowski inequality*, J. Funct. Anal. **280** (2021), 108956.

[195] A. Rossi and P. Salani, *Stability for Borell–Brascamp–Lieb inequalities*, in: Geometric aspects of functional analysis, Lecture notes in math., vol. 2169, Springer, Cham, 2017, pp. 339–363.

[196] A. Rossi and P. Salani, *Stability for a strengthened Borell–Brascamp–Lieb inequality*, Appl. Anal. **98** (2019), 1773–1784.

[197] L. Rotem, *A letter: The log-Brunn–Minkowski inequality for complex bodies*, arXiv:1412.5321.

[198] L. A. Santaló, *An affine invariant for convex bodies of n-dimensional space*, Port. Math. **8** (1949), 155–161 (Spanish).

[199] C. Saroglou, *Remarks on the conjectured log-Brunn–Minkowski inequality*, Geom. Dedic. **177** (2015), 353–365.

[200] C. Saroglou, *More on logarithmic sums of convex bodies*, Mathematika **62** (2016), 818–841.

[201] C. Saroglou, *A non-existence result for the L_p-Minkowski problem*, arXiv:2109.06545.

[202] A. Segal, *Remark on stability of Brunn–Minkowski and isoperimetric inequalities for convex bodies*, in: Geometric aspects of functional analysis, Lecture notes in math., vol. 2050, Springer, Heidelberg, 2012, pp. 381–391.

[203] R. Schneider, *Convex bodies: the Brunn–Minkowski theory*, Cambridge University Press, 2014.

[204] A. Stancu, *The discrete planar L_0-Minkowski problem*, Adv. Math. **167** (2002), 160–174.

[205] A. Stancu, *Prescribing centro-affine curvature from one convex body to another*, Int. Math. Res. Not. **2022** (2022), 1016–1044.

[206] B. Uhrin, *Curvilinear extensions of the Brunn–Minkowski–Lusternik inequality*, Adv. Math. **109**(2) (1994), 288–312.

[207] N. S. Trudinger and X.-J. Wang, *The Monge–Ampere equation and its geometric applications*, in: Handbook of geometric analysis, Adv. lect. math., vol. 7, Int. Press, Somerville, MA, 2008, pp. 467–524.

[208] C. Wu, D. Wu and N. Xiang, *The L_p Gauss image problem*, Geom. Dedic. **216**(6) (2022), 62.

[209] D. Xi and G. Leng, *Dar's conjecture and the log-Brunn–Minkowski inequality*, J. Differ. Geom. **103** (2016), 145–189.

[210] F. Xie, *The Orlicz Minkowski Problem for general measures*, Proc. Am. Math. Soc. **150** (2022), 4433–4445.

[211] S. Xing, D. Ye and B. Zhu, *The general dual-polar Orlicz–Minkowski problem*, J. Geom. Anal. **32** (2022), 91, 40 pp.

[212] G. Xiong, *Extremum problems for the cone-volume functional of convex polytopes*, Adv. Math. **225** (2010), 3214–3228.

[213] T. Wang, *The affine Polya–Szego principle: equality cases and stability*, J. Funct. Anal. **265** (2013), 1728–1748.

[214] G. Zhang, *The affine Sobolev inequality*, J. Differ. Geom. **53** (1999), 183–202.

[215] Y. Zhao, *The dual Minkowski problem for negative indices*, Calc. Var. Partial Differ. Equ. **56**(2) (2017), 18, 16 pp.

[216] G. Zhu, *The L_p Minkowski problem for polytopes for $0 < p < 1$*, J. Funct. Anal. **269** (2015), 1070–1094.

[217] G. Zhu, *The L_p Minkowski problem for polytopes for $p < 0$*, Indiana Univ. Math. J. **66** (2017), 1333–1350.

Komla Domelevo and Stefanie Petermichl

Bellman functions and continuous time

Abstract: The purpose of this text is both instructive and historic. We give a review of the classical Bellman technique mainly in the "weak" (dualized) form for dyadic martingales. From here, we approach techniques and novelties required to pass to their use for continuous time martingales with jumps. The historic part shows the development in dyadic analysis of a Bellman function for a specific problem. We then study this Bellman function and show it has some additional properties, useful for the analogous question in the continuous case.

Keywords: Weight, differential subordination, Bellman function

MSC 2020: 60G44, 60G46

1 Introduction

The term continuous time refers to the index of filtration of a stochastic process. Martingales arising from the Haar basis have discrete time, where smaller interval sizes relate to a later time. Martingales driven by Brownian motion are an example of stochastic processes with continuous time.

The purpose of this text is both instructive and historic. We give a review of the classical Bellman technique mainly in the "weak" (dualized) form for dyadic martingales. From here, we approach techniques and novelties required to pass to their use for continuous time martingales with jumps. The historic part shows the development in dyadic analysis of a Bellman function for a specific problem. We then study this Bellman function and show it has some additional properties, useful for the analogous question in the continuous case. We also attempt to ease the transition into the much more abstract language of probability, so the reader will likely gain a first intuition for this subject. Some of the arguments can look highly technical – and they are – continuous processes with jumps are a danger zone for mistakes and mishaps. But being guided in its limitations through something we know, namely dyadic analysis, one masters these technical aspects with much more ease.

Acknowledgement: The research of S. P. is supported by the ERC project CHRiSHarMa no. DLV-682402 and by the Alexander von Humboldt Stiftung.

Komla Domelevo, Stefanie Petermichl, Institute of Mathematics, University of Würzburg, Würzburg, Germany, e-mails: komla.domelevo@uni-wuerzburg.de, stefanie.petermichl@uni-wuerzburg.de

https://doi.org/10.1515/9783110775389-004

Both the continuous time and the presence of jumps force the understanding of new ideas from the reader. Our model of study is the Bellman function for the weighted problem in L^2. It has – in terms of Bellman functions – a long history, going back to the paper by NAZAROV–TREIL–VOLBERG [15], which has set the groundwork for the early advances in modern weighted theory in harmonic analysis and probability. It has also contributed to set the beginnings of the systematic use of Bellman functions in the weak form via extremal problems. In their paper, the authors show necessary and sufficient conditions for a dyadic martingale transform to be bounded in the L^2-two-weight setting. The methodology of their proof could be extended and used to get the first sharp result in the real-valued one-weight setting, for the dyadic martingale transform of WITTWER [24]. Sharpness in this setting means best control on growth with the necessary dyadic A_2 condition

$$Q_2[w] = \sup_I \left(\frac{1}{|I|} \int w \right) \left(\frac{1}{|I|} \int w^{-1} \right),$$

where the supremum runs over all dyadic intervals.

During the early days of weighted theory in harmonic analysis, before optimal weighted estimates were within reach, say, for the maximal operator or the Hilbert transform [11], similar questions were asked in probability theory, concerning stochastic processes with continuous-in-time filtrations [4, 12]. The difficulty that arises in the non-homogeneous setting, typically seen when these processes have jumps, were already observed back then and this restriction was made in one form or another in these papers. Certain basic facts about weights do not hold true for jump processes, such as the classical self-improvement (in its index) of the A_2 characteristic of the weight [4]. Another obstacle typical for working with weights is the non-convexity of the set inspired by the A_2 characteristic: $\{r, s \in \mathbb{R}_+ : 1 \leqslant rs \leqslant Q\}$ with $Q > 1$. Such continuity-in-space assumptions still appear regularly for these or other reasons when addressing weights; see [3, 16]. In this text we make no additional regularity assumptions.

WITTWER's proof subtly uses the homogeneity that arises from the dyadic filtration, where the underlying measure is Lebesgue in a crucial way. This homogeneity assumption has only recently been removed in the papers [21] and [13]. These authors work with discrete-in-time general filtrations with arbitrary underlying measure, where one martingale is a predictable multiplier of the other. A direct passage using the results for discrete-in-time filtrations to the continuous-in-time filtration case where one uses Burkholder's definition (below) is only possible in very special cases, such as predictable multipliers of stochastic integrals – this passage is explained in one of Burkholder's early works on L^p-estimates for pairs of differentially subordinate martingales [5]. Indeed, the literature in this direction is vast, with many delicate results by BANUELOS, BAUDOIN, or OSEKOWSKI [1, 2, 3]. These texts typically give estimates in the strong form, often making use of differential subordination.

In the dyadic case, Y is differentially subordinate to X if and only if $|\Delta Y| \leqslant |\Delta X|$ and $|Y_0| \leqslant |X_0|$. This is trivially satisfied in the case of a Haar multiplier. In the continuous case, we must pass through the quadratic variation process (defined later) and require $[X,X]_t - [Y,Y]_t$ to be non-negative and non-decreasing.

In this chapter, we tackle the sharp weighted estimate in full generality, using the notion of differential subordination of Burkholder and the martingale A_2 characteristic,

$$Q_2^{\mathcal{F}}[w] = \sup_\tau \text{ess} \cdot \sup_\omega (w)_\tau (w^{-1})_\tau,$$

with τ stopping times (defined later).

We start with the derivation of a complete Bellman function that solves again the estimate of WITTWER, meaning the dyadic, scalar-valued Haar multiplier's weighted bound,

$$\|T_\sigma f\|_{L^2(w)} \lesssim Q_2[w]\|f\|_{L^2(w)}.$$

Eventually, we prove that for L^2-integrable Hilbert-space-valued martingales Y, X with Y differentially subordinate to X, we have

$$\|Y\|_{L^2(w)} \lesssim Q_2^{\mathcal{F}}[w]\|X\|_{L^2(w)},$$

where the implied constant is numeric and does not depend upon the dimension, the pair of martingales, or the weight. It is well known that the linear growth in the quantity $Q_2[w]$ is sharp, already in the dyadic case.

This Bellman function proof does not give the strongest result to date in this direction nor is it the best or shortest proof. Indeed, in LACEY [13] the beautifully impressive so-called dyadic sparse operators are used and made their appearance in the top-down probabilistic logic, following a dyadic martingale's path and providing a pointwise domination, giving an extremely short and concise argument and providing a stronger result than WITTWER [24] or THIELE–TREIL–VOLBERG [21] in the discrete time setting. There is in [8] a stronger result from the authors, obtained via this new approach using sparse operators, but in a continuous time setting, also avoiding Bellman functions almost entirely.

However, the weighted problem is near perfect to explain the use of Bellman functions for continuous time martingales: It presents a collection of discoveries made during the last 20 years in the dyadic case, along with new observations and interpretations. In addition, it shows how hard one needs to work to overcome non-homogeneity and it instructs on the delicacy of continuous-in-time martingales. The Bellman function itself has also seen uses, for example in [7], for which, again, the explicit expression was of crucial importance. A sparse approach for the main result in that paper is not known. One might argue that the Bellman technique in the weak form is especially well adapted to semigroup problems while sparse domination has a strong martingale as opposed to semigroup flavor.

The construction of the Bellman function in [9] is kept to a minimum with very few explanations on how to come up with this expression. There have been a number of readers approaching us about this lack of intuition given in our paper. Other readers have found the probabilistic part of our paper too brutal and not sufficiently explained. In a way, we had written a text for very few readers, which should have been written so as to bring different groups together. In this survey, we will illustrate the history more clearly, illuminating the origins of the parts of the function and how it has been assembled. Indeed, the reader will once again see that the development of this function spans about 20 years. The scope of this chapter is not a full explanation of the Bellman technique. There are texts where this is much better accomplished. For a quick introduction, see [14], and for an in depth explanation, see the book [23]. The paper [9] is also very brisk in its use of probability theory. While there are experts in the Bellman function technique who master this background on this level it does present an obstacle to many. Again, the read of an excellent book in probability is the best measure here, but this chapter attempts to give the reader a jump start through the use of analogy: We will restrict ourselves to the definitions that are absolutely necessary, but we will attempt to show how they are put to use, guided by the dyadic case. We list a few facts on probability that are important here and taken from textbooks, explaining their meaning rather than repeating their proofs (for which we give the references). We hope this will ease the transition to a more structured study of stochastic processes for some readers and will motivate them to gain this expertise.

In this book chapter, we derive an explicit Bellman function of four variables adapted to the problem, starting with the much simpler and familiar dyadic case. The function has the usual conditions on its range and convexity. In the dyadic case as well as some simple continuous cases, one is free to split the estimate into pieces and succeeds with smaller Bellman functions for parts of the problem. This is not the case in the generality we treat here. A continuous subconvexity as well as a so-called discrete one-leg convexity, such as seen in [21] for two smaller Bellman functions (their functions make up a part of ours), are needed to treat non-homogeneity. We heavily use the explicit form of our Bellman function and its regularity properties in several parts in our proof to handle the delicacy of the irregular continuous-in-time processes with values in Hilbert space. The resulting function is in the "dualized" or "weak form," which is in contrast to the "strong form" of a Burkholder-type functional often seen when using the strong subordination condition. Indeed, the form of the strong differential subordination condition is adapted to work well for Burkholder-type functionals. The passage to its use in the weak form is accomplished through the use of the so-called ellipse lemma and requires a Bellman function solving the entire problem at once, as opposed to splitting the problem into pieces.

2 Dyadic martingales

Analysts often have a preference for orthonormal systems over martingales. But the dyadic Haar functions naturally give rise to both and will serve as an entry point for analysts with little background in probability. A trained probabilist might find the reverse direction interesting.

The dyadic intervals on the unit interval $I = [0, 1)$ are

$$\mathcal{D} = \{2^{-k}[0,1) + 2^{-k}n : k, n \in \mathbb{N}, n < 2^k\} = \bigcup_{k=0}^{\infty} \mathcal{A}_k.$$

Here, the collections of atoms of size 2^{-k} are

$$\mathcal{A}_k = \{2^{-k}[0,1) + 2^{-k}n : n \in \mathbb{N}, n < 2^k\}. \tag{2.1}$$

Adapted to the intervals $J \in \mathcal{D}$ is a simple wavelet system $\{h_J : J \in \mathcal{D}\}$, where

$$h_J = (-\chi_{J_-} + \chi_{J_+})|J|^{1/2}.$$

The collection forms an orthonormal basis for L^2. That is, a function belonging to L^2 can be written as

$$f(x) = \langle f \rangle_I \chi_I(x) + \sum_{J \in \mathcal{D}} \langle f, h_J \rangle h_J(x),$$

where $\langle f, h_J \rangle = \int_{\mathbb{R}} f(t) h_J(t) dt$ denotes the inner product in the Hilbert space L^2 and by $\langle f \rangle_I = |I|^{-1} \int_I f(t) dt$ we mean the average of a function f over the interval I; χ_I denotes the characteristic function of the interval I. We observe that $\langle \chi_K, h_J \rangle = 0$ unless $K \subsetneqq J$ and in this case $\langle \langle f \rangle_K \chi_K, h_J \rangle = \langle f, h_J \rangle$. We so obtain the multi-resolution analysis (MRA) property

$$\langle f \rangle_K \chi_K(x) = \chi_K(x) \sum_{J \in \mathcal{D}: K \subsetneqq J} \langle f, h_J \rangle h_J(x).$$

We can use this to express differences of averages. Let $J \subseteq I$ with, say, $J^{(n)} = I$ (the n-th ancestor). We have

$$\chi_J(\langle f \rangle_J - \langle f \rangle_I) = \chi_J \left(\sum_{K \in \mathcal{D}: J \subsetneqq K} \langle f, h_K \rangle h_K - \sum_{K \in \mathcal{D}: I \subsetneqq K} \langle f, h_K \rangle h_K \right).$$

Together this becomes $\chi_J(\sum_{k=1}^n \langle f, h_{J^{(k)}} \rangle h_{J^{(k)}})$. Observe that this is a telescoping sum. We can also express this telescoping series using a dyadic stochastic integral: Using naive martingale differences instead of the MRA property, we trivially write

$$\chi_J(\langle f\rangle_J - \langle f\rangle_I) = \chi_J(\langle f\rangle_J - \langle f\rangle_{J^{(1)}} + \langle f\rangle_{J^{(1)}} \cdots - \langle f\rangle_I).$$

Recall that on J this writes as $\chi_J \sum_{k=1}^n \Delta_{J^{(k)}} f$ with $\Delta_{J^{(k)}} f = \langle f\rangle_{J^{(k-1)}} - \langle f\rangle_{J^{(k)}}$. This notion of counting ancestors is very typical in analysis, but it can be somewhat confusing as in probability time goes the other way.

Let us more formally pass from the orthonormal system to the notion of filtration by thinking of $[0,1)$ as a probability space with Lebesgue measure. Denote by $\mathcal{F}_k$ the σ-algebra generated by $\mathcal{A}_k$ as in (2.1) and denote by the collection $\mathcal{D}_k$ the dyadic intervals contained in $\mathcal{F}_k$. In this sense, we obtain an increasing sequence of σ-algebras $\mathcal{F}_0 \subset \mathcal{F}_1 \subset \mathcal{F}_2 \subset \cdots$, where the sequence is called the dyadic filtration $\mathcal{F}$.

Before we continue, let us discuss conditional expectation in this simple, dyadic setting. Given the probability space $[0,1)$ endowed with Lebesgue measure, we write

$$\mathbb{E}(f) = \int_{[0,1)} f(x)dx,$$

meaning the expectation of f. Take now any $\mathcal{F}_n$. The conditional expectation of f given $\mathcal{F}_n$ will produce a certain function related to f, which is measurable in $\mathcal{F}_n$. It is denoted by $\mathbb{E}(f|\mathcal{F}_n)$ and has the property that it is $\mathcal{F}_n$-measurable and that for each $A \in \mathcal{F}_n$, we have

$$\int_A \mathbb{E}(f|\mathcal{F}_n)(x)dx = \int_A f(x)dx.$$

In the dyadic case, the measurable functions in $\mathcal{F}_n$ are precisely the ones constant on atoms in $\mathcal{A}_n$. This implies that $\mathbb{E}(f|\mathcal{F}_n)$ must be constant on atoms in $\mathcal{A}_n$. The integral condition becomes most restrictive for $I \in \mathcal{A}_n$ and we obtain

$$\mathbb{E}(f|\mathcal{F}_n) = \sum_{I \in \mathcal{A}_n} \langle f\rangle_I \chi_I.$$

Notice that the integral condition is equivalent to saying that $f - \mathbb{E}(f|\mathcal{F}_n)$ is orthogonal in the Hilbert space L^2 to indicator functions that are measurable in $\mathcal{F}_n$.

A sequence $F = (F_n)_{0 \leqslant n < \infty}$ of functions that are measurable in $\mathcal{F}_n$ for all n is called a sequence adapted to the filtration $\mathcal{F}$. Such an adapted sequence is called a martingale with respect to $\mathcal{F}$ if $F_n \in L^1(dx)$ and if $m \leqslant n$, then $\mathbb{E}(F_n|\mathcal{F}_m) = F_m$. Notice that it is customary to use capital letters for martingales or stochastic processes. We will not distinguish strictly between F arising from the function f.

The following convergence theorem is fundamental and intuitive in the dyadic case.

Theorem 1. *Let $\mathcal{F}$ be a filtration and let $\mathcal{F}_\infty = \bigcup_{n \geqslant 0} \mathcal{F}_n$. Let $1 \leqslant p < \infty$ and let $F \in L^p$. Then the adapted process $F_n = \mathbb{E}(F|\mathcal{F}_n)$ is a martingale such that $F_n \to \mathbb{E}(F|\mathcal{F}_\infty)$ in L^p when $n \to \infty$.*

To prove this, it is easy to see that we can assume that F is $\mathcal{F}_\infty$-measurable. Then observe that $\bigcup_{n\geqslant 0} L^p(\mathcal{F}_n)$ is dense in $L^p(\mathcal{F}_\infty)$.

The so-called martingale difference notation for terms arising in dyadic sums is

$$\Delta_I f = \chi_{I_\pm}(\langle f\rangle_{I_\pm} - \langle f\rangle_I).$$

Observe that these arise naturally as martingale differences, for instance on I_n,

$$F_n - F_{n-1} = \Delta_{I_n} f.$$

This difference notation generalizes more easily, as we will see below. Using this notation, we observe a simple case of the Itô formula by writing for a sequence of intervals $I_n \subset \cdots \subset I_1 \subset I$ such that $I_k \in \mathcal{A}_k$

$$\langle f\rangle_{I_n} = \langle f\rangle_I + (\langle f\rangle_{I_1} - \langle f\rangle_I) + \cdots + (\langle f\rangle_{I_n} - \langle f\rangle_{I_{n-1}})$$
$$= \langle f\rangle_I + \sum_{k=1}^{n} \Delta_{I_k} f,$$

valid on the set I_n.

Here is a different view on the dyadic system. It can also be regarded as living on the space $\Omega = \{-1, 1\}^{\mathbb{N}}$ equipped with the cylindrical measure. Denote by ε_n for $n \in \mathbb{N}$ the n-th coordinate of $\varepsilon \in \Omega$. The sign tosses ε_n are independent and take the values ± 1 with equal probability. Set $\mathcal{F}_0 = \{\varnothing, \Omega\}$ and $\mathcal{F}_n = \sigma(\varepsilon_1, \ldots \varepsilon_n)$, the σ-algebra generated by the random variables $\varepsilon_1, \ldots, \varepsilon_n$. A sequence d_n is adapted in this filtration iff d_0 is constant and iff d_n depends only on $\varepsilon_1, \ldots, \varepsilon_n$ so that $d_n = d_n(\varepsilon_1, \ldots, \varepsilon_n)$. See below in the continuous context the actual definition of the term adapted.

Notice that with this interpretation, the dyadic intervals of length 2^{-n} can be regarded as outcomes of n sign tosses $\varepsilon_1, \ldots, \varepsilon_n$. Via a trivial observation the dyadic sum

$$\sum_{J \in \mathcal{D}_n} \langle f, h_J\rangle h_J(x)$$

corresponds to the stochastic integral

$$\sum_{k=1}^{n} d_k(\varepsilon_1^x, \ldots, \varepsilon_k^x)\varepsilon_{k+1}^x$$

following the trajectory of sign tosses so that $J \in \mathcal{D}_n : x \in J$ is reached by the specific sequence of tosses $\varepsilon_1^x, \ldots, \varepsilon_n^x$.

3 The non-homogeneous case

Consider now more general discrete-in-time martingales. There is little change to the above if the time remains discrete – however, many mathematical problems become substantially more difficult. For example, non-homogeneity arises if we maintain a dyadic filtration, but reconsider the underlying measure. For instance, if the underlying measure is non-homogeneous, for example it has a density which is the characteristic function of a set, we have a case of a non-homogeneous setting: It can be interpreted that jump probabilities to the right or to the left have very different probabilities as opposed to 1/2 each in the classical dyadic case. In a more general situation, we may have a non-homogeneous underlying measure (or Lebesgue measure) coupled with a filtration where intervals have varying numbers of children. For that, let $(\Omega, \mathcal{F}_\infty, \mathbb{P})$ be a probability space with a non-decreasing sequence $\mathcal{F} = (\mathcal{F}_n)_{n \geqslant 0}$ of sub-σ-fields of $\mathcal{F}_\infty$ such that $\mathcal{F}_0$ contains all $\mathcal{F}_\infty$-null-sets. If $f = \{f_n\}_{n \in \mathbb{N}}$ is a martingale adapted to $\mathcal{F}$, we note $f_n = \sum_{k=0}^n \mathrm{d}f_k$, with the convention $\mathrm{d}f_0 := f_0$, and $\mathrm{d}f_k := f_k - f_{k-1}$, for $k \geqslant 1$. Similarly, if g is another adapted martingale, we note $g_n = \sum_{k=0}^n \mathrm{d}g_k$ with the same conventions. One says that g is differentially subordinate to f if one has for almost all $\omega \in \Omega$ and all $k \geqslant 0$ $|\mathrm{d}g_k| \leqslant |\mathrm{d}f_k|$.

4 Continuity in time

While we will review some of the definitions, a complete and comprehensive introduction goes beyond the scope of this text. The authors can recommend the beginning chapters of PROTTER [19]. The text is extremely concise and very general – most texts do not bother with jump processes or are too lengthy and written assuming no background in real analysis. With the dyadic case above as a model and basic mastery of measure theory, the text by PROTTER becomes easy to read and is well worth the reader's time. Some concepts are only seen when the time is continuous, but others can be illustrated using the dyadic case to ease into the material. Whenever we feel it is helpful, we will point it out.

In this light, let again $(\Omega, \mathcal{F}_\infty, \mathbb{P})$ be a probability space with a family $\mathcal{F} = (\mathcal{F}_t)_{t \geqslant 0}$ of sub-σ-fields of $\mathcal{F}_\infty$ such that $\mathcal{F}_0$ contains all $\mathcal{F}_\infty$-null-sets. It is assumed non-decreasing, meaning that $\mathcal{F}_s \subseteq \mathcal{F}_t$ if $s \leqslant t$, and right continuous, meaning that $\mathcal{F}_t = \cap_{s>t} \mathcal{F}_s$. To define the condition on the weights and in various locations in the proof, we will require the notion of stopping time.

Definition 1. A random variable $T : \Omega \to [0, \infty]$ is a stopping time if the event $\{T \leqslant t\}$ is measurable in $\mathcal{F}_t$.

In dyadic harmonic analysis a stopping time often refers to a collection of disjoint dyadic intervals. This, indeed, is a stopping time N in the dyadic filtration. If Ω is the space of the dyadic trajectories in $[0,1]$, then we see that a disjoint dyadic collection I_k gives rise to a stopping time. If ω passes through an I_k, then set for the stopping time $N(\omega) = \log_2 |I_k|$; otherwise set $N(\omega) = \infty$. It suffices to think of integer-valued stopping times. For this random variable, $\{\omega : N(\omega) \leqslant m\}$ consists of those trajectories that are passing through any I_k with $|I_k| \geqslant 2^{-m}$. This is a measurable set in $\mathcal{F}_m$. In the reverse direction, the measurability condition ensures that the intervals corresponding to $\{N = k\}$ for $0 \leqslant k \leqslant \infty$ form a disjoint collection of dyadic intervals.

The right continuity has no analog in the dyadic setting, but it has the following simple but important implication: $\{T < t\} \in \mathcal{F}_t, 0 \leqslant t \leqslant \infty$, if and only if T is a stopping time.

A stochastic process X is a collection of random variables $(X_t)_{0 \leqslant t < \infty}$. The process is said adapted if X_t is $\mathcal{F}_t$-measurable (for which one often abbreviates $X_t \in \mathcal{F}_t$) for all t. In part of our setting, these processes have values in Hilbert spaces.

We say that two stochastic processes X, Y are modifications of one another if for each t there holds $X_t = Y_t$ almost surely. They are called indistinguishable if almost surely for all t we have $X_t = Y_t$.

Notice that the null-sets are assumed to be in $\mathcal{F}_0$ and that in the case of modifications, the null-sets describing the differences between the processes depend upon each t. Since the range of time is not countable, their unions could have any measure. There is however a universal null-set N that works for all t in case the processes are indistinguishable. This means that the sample paths $t \mapsto X_t(\omega)$ and $t \mapsto Y_t(\omega)$ are the same for $\omega \notin N$.

Definition 2. A stochastic process X is called càdlàg if it almost surely has sample paths which are right continuous with left limits.

The "word" càdlàg stands for "continue à droite et limite à gauche." With this assumption on stochastic processes, something nice happens: If X and Y are càdlàg stochastic processes that are modifications of one another, then they are indistinguishable. Here it is of course sufficient to test one of these processes for càdlàg. Processes that are càdlàg provide a rich class for stopping times. So-called hitting times of open sets give rise to stopping times.

Definition 3. Let X be càdlàg and let Λ be a Borel set in the real line. Define $T(\omega) = \inf\{t > 0 : X_t \in \Lambda\}$. Then T is called a hitting time of Λ.

Theorem 2. *Let X be càdlàg and let Λ be an open set. The hitting time of Λ is a stopping time.*

It is more difficult to stop a process with jumps through a hitting time if the set is closed. We have to think what we want to do if the process jumps over the obstacle instead of approaching it. In the presence of jumps, the following is important.

Theorem 3. *Let X be an adapted càdlàg stochastic process and let Λ be a closed set. Then $T(\omega) = \inf\{t > 0 : X_t \in \Lambda \vee X_{t_-}(\omega) \in \Lambda\}$ is a stopping time.*

Here, $X_{t_-}(\omega) = \lim_{s \to t, s < t} X_s(\omega)$. The following concept of a stopping time σ-algebra is important. It describes knowledge at a certain time.

Definition 4. Let T be a stopping time. The stopping time σ-algebra $\mathcal{F}_T$ is

$$\{\Lambda \in \mathcal{F} : \Lambda \cap \{T \leqslant t\} \in \mathcal{F}_t, \forall t \geqslant 0\}.$$

This definition is not immediately intuitive, but let us think what this means in the dyadic case. We have seen that dyadic stopping times N give rise to a disjoint collection of dyadic intervals. The above definition of $\mathcal{F}_N$ is then the σ-algebra generated by these disjoint dyadic intervals. The stopping intervals become the atoms. Notice that a dyadic stochastic process sampled at a stopping time X_N is constant on the stopping intervals (rather, takes the same values for all trajectories passing through each stopping interval).

Let us also remark that if $\Lambda \in \mathcal{F}_T$, then the stochastic process $X_t = \chi_\Lambda \chi_{\{t \geqslant T\}}$ is a càdlàg process with $X_T = \chi_\Lambda$. We have what we call in analysis a testing condition fact for finite stopping times. It is not hard to prove that indicators of sets are representative.

Theorem 4. *Let T be a finite stopping time. Then*

$$\mathcal{F}_T = \sigma\{X_T : X \text{ all adapted càdlàg processes}\}.$$

So, $\mathcal{F}_T$ is the smallest σ-algebra containing all càdlàg adapted processes sampled at T. The stopping σ-algebra thus represents knowledge up to time T.

We give now a very brief introduction to martingales.

Definition 5 (Martingale). An adapted process $X = (X_t)_{0 \leqslant t < \infty}$ is called a martingale with respect to the filtration $\mathcal{F}$ if:
i. $X_t \in L^1$, that is, $\mathbb{E}(|X_t|) < \infty$,
ii. if $s \leqslant t$, then $\mathbb{E}(X_t | \mathcal{F}_s) = X_s$ almost surely.

If we change the equality in *ii.* to $\geqslant$, we get a submartingale, and if we change it to $\leqslant$, we get a supermartingale.

A martingale is only defined for finite times, yet sometimes we want to speak of its value at time ∞ and care is required.

Definition 6. A martingale X is said to be closed by a random variable Y if $\mathbb{E}(|Y|) < \infty$ and $X_t = \mathbb{E}(Y | \mathcal{F}_t)$ for finite t.

Notice that a random variable closing a martingale is not necessarily unique, nor does it always exist. The right concept is uniform integrability and the reader is encouraged to read about it in PROTTER [19]. Let us here just mention that all martingales have a càdlàg modification (so that they are indistinguishable). The following covers our bases.

Theorem 5. *Let X be a right continuous martingale such that $\sup_{0\leqslant t<\infty} \mathbb{E}(|X_t|) < \infty$. Then $Y = \lim_{t\to\infty} X_t$ exists almost surely and $\mathbb{E}(|Y|) < \infty$. Moreover, if X is a martingale closed by a random variable Z, then Y also closes X and $Y = \mathbb{E}(Z|\bigvee_{0\leqslant t<\infty} \mathcal{F}_t)$.*

Definition 7 (Uniform integrability). $(X_t)_{0\leqslant t<\infty}$ is uniformly integrable if

$$\lim_{n\to\infty} \sup_t \int_{\{|X_t|\geqslant n\}} |X_t|\,d\mathbb{P} = 0.$$

Theorem 6. *Let X be a right continuous martingale. Then $(X_t)_{0\leqslant t<\infty}$ is closed iff $(X_t)_{0\leqslant t<\infty}$ uniformly integrable iff $Y = \lim_{t\to\infty} X_t$ exists almost surely, $\mathbb{E}(|Y|) < \infty$, and $(X_t)_{0\leqslant t\leqslant\infty}$ is a martingale with $X_\infty = Y$.*

The notion of *ucp* convergence is explained below.

Definition 8 (ucp convergence). A sequence of processes $(H_n)_{n\geqslant 1}$ converges to a process H uniformly on compacts in probability (ucp) if, for each $t > 0$, the quantity $\sup_{0\leqslant s\leqslant t} |H^n_s - H_s|$ converges to 0 in probability.

We are interested in $\mathbb{H}$-valued càdlàg martingales, where $\mathbb{H}$ is a separable Hilbert space. In order to clearly define differential subordination in this setting, we make use of the square bracket or quadratic variation process.

Recall that the quadratic variation process of a martingale X is the process denoted by $[X,X] := ([X,X]_t)_{t\geqslant 0}$ and defined as (see, e. g., PROTTER [19])

$$[X,X]_t = X_t^2 - 2\int_0^t X_{s-}\,dX_s,$$

where we have set $X_{0-} = 0$. Similarly, the quadratic covariation of two semimartingales X and Y is the following process, also known as the bracket process:

$$[X,Y]_t := X_t Y_t - \int_0^t X_{s-}\,dY_s - \int Y_{s-}\,dX_s.$$

Most probabilists are used to the sharp bracket process, noted $\langle X,X\rangle_t$, which is a certain average of the square bracket (and we do not need it here). This object is not the same in the presence of jumps. Indeed, subordination with respect to the square bracket is a much more restrictive requirement.

Here are some elementary properties.

Theorem 7. *The quadratic variation process of X is càdlàg, increasing, and adapted. Moreover:*
i. $[X,X]_0 = X_0^2$ *and* $\Delta[X,X] = (\Delta X)^2$.

ii. *Let $(\sigma_n)_{n \geqslant 0}$ be a sequence of subdivisions, $\sigma_n := (T_0^n, \ldots, T_{k_n}^n)$, with $0 = T_0^n \leqslant T_1^n \leqslant \cdots \leqslant T_i^n \leqslant \cdots \leqslant T_{k_n}^n$, where T_i^n are stopping times. Assume further that $|\sigma_n| := \max_i |T_{i+1} - T_i|$ goes to 0 and that $T_{k_n}^n \to \infty$ as $n \to \infty$. Then*

$$X_0^2 + \sum_i \left(X^{T_{i+1}^n} - X^{T_i^n}\right)^2 \to [X, X]$$

with convergence in ucp.

Here, X^T denotes the process stopped at T: $X_t^T = X_{t \wedge T}$.

Definition (Differential subordination). Let X and Y be two adapted càdlàg semimartingales taking values in a separable Hilbert space. We say that Y is differentially subordinate by quadratic variation to X iff $[X, X]_t - [Y, Y]_t$ is a non-decreasing and non-negative function of $t \geqslant 0$.

Let us denote by X^c the unique continuous part of X (its construction is not completely obvious) with

$$[X, X]_t = |X_0|^2 + [X^c, X^c]_t + \sum_{0 < s \leqslant t} |\Delta X_s|^2.$$

We have $[X, X]_t^c = [X^c, X^c]_t$ and $\Delta[X, X]_t = |\Delta X_t|^2$, where $\Delta X_t := X_t - X_{t_-}$. We have the following obvious characterization distinguishing the continuous and jump parts.

Lemma. *If X and Y are semimartingales, then Y is differentially subordinate to X if and only if (i) $[X, X]_t^c - [Y, Y]_t^c$ is a non-negative and non-decreasing function of t, (ii) the inequality $|\Delta Y_t| \leqslant |\Delta X_t|$ holds for all $t > 0$, and (iii) $|Y_0| \leqslant |X_0|$.*

Now, we explain what a weight is and what the Muckenhoupt condition is, and we give a clarification of the statement in this setting.

Let again $(\Omega, \mathcal{F}_\infty, \mathbb{P})$ be a probability space with a non-decreasing right continuous family $\mathcal{F} := (\mathcal{F}_t)_{t \geqslant 0}$ of sub-σ-fields of $\mathcal{F}_\infty$ such that $\mathcal{F}_0$ contains all $\mathcal{F}_\infty$-null-sets. The measure $d\mathbb{P}$ is arbitrary (up to the obvious normalization). If X and Y are adapted càdlàg square integrable $\mathbb{H}$-valued martingales and Y is differentially subordinate to X, then it is obvious that

$$\|Y\|_2 \leqslant \|X\|_2. \tag{4.1}$$

Recall here that $\|X\|_2 := \sup_t \|X_t\|_2$, where

$$\|X_t\|_2^2 := \mathbb{E}|X_t|^2 = \int_\Omega \left|X_t(\omega)\right|^2 d\mathbb{P}(\omega). \tag{4.2}$$

Assume again that Y is differentially subordinate to X. We might insist on the underlying probability space at hand by saying in short that X and Y are $\mathbb{P}$-martingales and that Y is $\mathbb{P}$-differentially subordinate to X. The main concern of this chapter is to obtain sharp inequalities similar to (4.1) under a change of law in the definition of the L^2-norm according to [6]. Let w be a positive, uniformly integrable martingale (which we often identify with its closure w_∞) that we call a weight. Let $d\mathbb{Q} := d\mathbb{P}^w := w d\mathbb{P}$ and let $(\Omega, \mathcal{F}_\infty, \mathbb{Q})$ be a probability space with the same assumptions as $(\Omega, \mathcal{F}_\infty, \mathbb{P})$ but with a change of the probability law.

Let us describe a general change of measure. In our case the measures $\mathbb{P}$ and $\mathbb{Q}$ will of course be related via a density function, called a weight. Let $\mathbb{P}$ and $\mathbb{Q}$ be such that $(\Omega, \mathcal{F}_\infty, \mathbb{P})$ and $(\Omega, \mathcal{F}_\infty, \mathbb{Q})$ are two filtered probability spaces as described above. Does there exist a constant $C_{\mathbb{P},\mathbb{Q}} > 0$ depending only on $(\mathbb{P}, \mathbb{Q})$ – thus, the weight – such that if X and Y are uniformly integrable $\mathbb{P}$-martingales adapted to $\mathcal{F}$ and Y is $\mathbb{P}$-differentially subordinate to X, then

$$\|Y\|_{2,\mathbb{Q}} \leqslant C_{\mathbb{P},\mathbb{Q}} \|X\|_{2,\mathbb{Q}}?$$

We look for $C_{\mathbb{P},\mathbb{Q}} := C_w$ allowing to compare $\|Y\|_{2,\mathbb{Q}} := \|Y\|_{2,w}$ and $\|X\|_{2,\mathbb{Q}} := \|X\|_{2,w}$. We will also need the inverse weight $u = w^{-1}$ and we assume u is uniformly integrable. We will finally note that $d\mathbb{P}^u := u d\mathbb{P}$. It follows that $\mathbb{P}^w$ and $\mathbb{P}^u$ are probability measures on Ω up to the obvious normalizations. The necessary condition on the weight is the following.

Definition (A_2 class). Let $(\Omega, \mathcal{F}, (\mathcal{F}_t)_{t \geqslant 0}, \mathbb{P})$ be a filtered probability space. We say that the weight $w > 0$ is in the A_2 class iff the A_2 characteristic of the weight w, noted $Q_2^{\mathcal{F}}[w]$ and defined as

$$Q_2^{\mathcal{F}}[w] := \sup_\tau \mathrm{ess} \cdot \sup_\omega (w)_\tau (w^{-1})_\tau$$

with the first supremum running over all adapted stopping times is finite.

We often write $Q_2^{\mathcal{F}}[w] := \sup_\tau \mathrm{ess} \cdot \sup_\omega w_\tau u_\tau$, where $u := w^{-1}$ is the inverse weight.

5 Statement

Theorem 8 (Dyadic estimate). *Let $I = [0,1)$ and let h_J be the Haar system with $h_I = \chi_I \frac{1}{\sqrt{|I|}}$. Let $(\sigma_J)_{J \in \mathcal{D}}$ be a sequence taking values in $\{1, -1\}$. Let $T_\sigma : h_J \mapsto \sigma_J h_J$. Then*

$$\|T_\sigma\|_{L^2(w)} \leqslant Q_2[w] \|f\|_{L^2(w)}$$

uniformly in σ.

So, in Wittwer's result [24], $\mathcal{F}$ is the dyadic filtration and Y is a Haar multiplier of X.

Theorem 9 (Differential subordination under change of law in the general case). *Let* X *and* Y *be two adapted uniformly integrable càdlàg* $\mathbb{H}$*-valued martingales such that* Y *is differentially subordinate to X. Then*

$$\|Y\|_{L^2(w)} \leq Q_2^{\mathcal{F}}[w]\|X\|_{L^2(w)}.$$

In the result [21] the filtration is discrete but not homogeneous and again Y is a Haar multiplier of X. In the above, the paths are continuous and Y need not be a multiplier of X.

These results will be a consequence of the following bilinear estimates.

Proposition 1 (Dyadic bilinear estimate). *The following bilinear estimate holds:*

$$\sum_{J \subseteq I} |(f, h_J)(g, h_J)| \leq Q_2[w]\|f\|_{L^2(w)}\|g\|_{L^2(w)}.$$

It is elementary to see why this estimate gives rise to the norm estimate for T_σ. Just dualize and choose σ_I so that $\sigma_I(f, h_I)(g, h_I) = |(f, h_I)(g, h_I)|$; see below for details.

Proposition 2 (Bilinear estimate). *Let* X *and* Y *be two adapted uniformly integrable càdlàg* $\mathbb{H}$*-valued martingales such that* Y *is differentially subordinate to X. Let w be an admissible weight in the $\mathbf{A}_2$ class. Then*

$$\mathbb{E} \int_0^\infty |d[Y, Z]_t| \leq Q_2^{\mathcal{F}}[w]\|X\|_w\|Z\|_u.$$

We will explain how to find a Bellman function for these bilinear estimates.
In the dyadic real-valued case, we use

$$V := (x, y, r, s) \in \mathbb{R} \times \mathbb{R} \times \mathbb{R}_+^* \times \mathbb{R}_+^* =: \mathbb{S}$$

so that $1 \leq rs \leq Q$ (which we call our domain).

A Bellman function with fewer properties is sufficient to solve this case. Here is what is needed.

Lemma 1 (Existence and properties of the Bellman function for the dyadic problem). *There exists a function B on the domain $\mathcal{D}_Q := \{V \in \mathbb{S} : 1 \leq rs \leq Q\}$ so that*

$$0 \leq B(V) \leq \frac{|x|^2}{r} + \frac{|y|^2}{s}$$

and

$$-B(V) + \frac{1}{2}B(V_+) + \frac{1}{2}B(V_-) \geq \frac{1}{Q}|\Delta x||\Delta y|.$$

We will give an explicit expression and its historic motivation. In the continuous case, more is needed. Here, too, we have an explicit expression of the function described below. Let us define the quadruplet V as

$$V := (x, y, r, s) \in \mathbb{H} \times \mathbb{H} \times \mathbb{R}_+^* \times \mathbb{R}_+^* =: \mathbb{S}.$$

The variables (x, y) will be associated to $\mathbb{H}$-valued martingales, whereas the variables (r, s) will be associated to $\mathbb{R}$-valued martingales for the weights. We introduce the domain $\mathcal{D}_Q$:

$$\mathcal{D}_Q := \{V \in \mathbb{S} : 1 \leqslant rs \leqslant Q\}.$$

We will often restrict our attention to truncated weights, that is, given $0 < \varepsilon < 1$, variables (r, s) are bounded below and above:

$$\mathcal{D}_Q^\varepsilon := \{V \in \mathcal{D}_Q : \varepsilon \leqslant r \leqslant \varepsilon^{-1}, \varepsilon \leqslant s \leqslant \varepsilon^{-1}\}.$$

Lemma 2 (Existence and properties of the Bellman function for the continuous problem).
There exists a function $B(V) = B_Q$ that is C^1 on $\mathcal{D}_Q^\varepsilon$ and piecewise C^2 with the estimate

$$B(V) \leqslant \frac{|x|^2}{r} + \frac{|y|^2}{s},$$

and on each subdomain where it is C^2 we have

$$d^2 B \geqslant \frac{2}{Q} |dx||dy|.$$

Whenever V and V_0 are in the domain, the function has the property

$$B(V) - B(V_0) - dB(V_0)(V - V_0) \geqslant \frac{2}{Q} |x - x_0||y - y_0|.$$

Moreover, we have the estimates

$$|(\partial_x^2 B dx, dx)| \leqslant \varepsilon^{-1}|dx|^2, \quad |(\partial_y^2 B dy, dy)| \leqslant \varepsilon^{-1}|dy|^2$$

with the implied constants independent of V and (dx, dy).

6 Existence and properties of the Bellman function

6.1 The homogeneous, real-valued dyadic case

This is the case considered in the 1990s. Given is the so-called Haar multiplier mentioned above, that is, a sequence (σ_j) indexed by dyadic intervals, taking values in ± 1. A Haar expansion for f is mapped by extension as follows:

$$h_J \mapsto \sigma_J h_J,$$

so that $X = \langle f \rangle_I \chi_I + \sum_{J \subseteq I} (f, h_J) h_J$ and $Y = \sigma_I \langle f \rangle_I \chi_I + \sum_{J \subseteq I} \sigma_J (f, h_J) h_J$.

In the martingale language, this is a predictable multiplier of a dyadic stochastic integral, acting on the random increments. Predictable means that σ_J is known when the next sign toss is happening, namely, the one determining whether we go to J_- or J_+. Bluntly, this means that in the product $\sigma_J h_J$ the sign toss σ_J is constant on J while h_J is not. The differential subordination here is trivial:

$$|\sigma_J \Delta_J f| \leqslant |\Delta_J f|.$$

Proof of dyadic estimate knowing existence of B. We sketch the proof. Note that the most classical of Bellman arguments used concavity instead of convexity, using two more linear variables representing $X = \langle f^2 w \rangle_I$ and $Y = \langle g^2 w^{-1} \rangle_I$. The concave function would be $\tilde{B}(X, Y, V) = X + Y - B(V)$. The reader can write the concavity proof as an exercise. We prefer to use convexity here to be closer to our proof for the continuous case.

By duality and after standard localizing to I, we must estimate

$$\left| \left\langle \frac{1}{|I|} \sum_{J \subseteq I} \sigma_J (f, h_J) h_J, \frac{1}{|I|} \sum_{K \subseteq I} (g, h_K) h_K \right\rangle_{L^2} \right| \leqslant Q \langle f^2 w \rangle_I \langle g^2 w^{-1} \rangle_I.$$

Taking the worst signs on the left and using orthogonality, this is equivalent to

$$\sum_{J \subseteq I} |(f, h_J)(g, h_J)| \leqslant Q \langle f^2 w \rangle_I \langle g^2 w^{-1} \rangle_I. \tag{6.1}$$

When multiplying f by λ and g by λ^{-1}, the left-hand side remains unchanged. Taking infima in $\lambda > 0$, we see it suffices to prove

$$\sum_{J \subseteq I} |(f, h_J)(g, h_J)| \leqslant Q(\langle f^2 w \rangle_I + \langle g^2 w^{-1} \rangle_I).$$

Iterating, we obtain

$$\sum_{J \subseteq I, |J| = 2^{-n}|I|} |J| B(V_J) \geqslant \frac{1}{Q} \sum_{J \subseteq I, |J| < 2^{-n}|I|} |J| |\Delta_J f| |\Delta_J g|.$$

As $n \to \infty$, the upper bounds of the Bellman function of the left-hand side approximate the Riemann integrals $\langle f^2 w \rangle_I$ and $\langle g^2 w^{-1} \rangle_I$. $\qquad \square$

A few historic remarks are in order. The first proofs of this kind are known as the classical four-sum proofs. When trying to estimate the dualized form in (6.1), first write

$$|(f, h_J)(g, h_J)| = |(fw, h_J)_{w^{-1}} (gw^{-1}, h_J)_w| \tag{6.2}$$

and follow the strategy to replace the Haar basis (h_J) by weighted Haar bases that are orthonormal in $L^2(w)$ and $L^2(w^{-1})$, respectively. With the Ansatz

$$h_J = a_J^w \chi_J + \beta_J^w h_J^w, \quad h_J^w \perp_w 1, \quad \|h_J^w\|_w = 1,$$

one easily calculates that with $\Delta_J w = \langle w \rangle_{J_+} - \langle w \rangle_{J_-}$

$$a_J^w = \frac{(w, h_J)}{\langle w \rangle_J |J|} = \frac{\Delta_J w}{2\langle w \rangle_J |J|^{1/2}}, \quad \beta_J^w = \left(\langle w \rangle_J - \frac{(w, h_J)^2}{\langle w \rangle_J |J|} \right)^{1/2} \leqslant \langle w \rangle_J^{1/2}.$$

Indeed, $\langle w \rangle_{J_\pm} \leqslant 2\langle w \rangle_J$ thanks to homogeneity, so $\frac{(w, h_J)^2}{\langle w \rangle_J |J|} \leqslant \langle w \rangle_J$. Inserting these into (6.2) reduces the estimate to four sums. By Cauchy–Schwarz and $\beta_J^w \beta_J^w \leqslant Q$, we easily estimate the sum with two Haar functions

$$\sum_{J \subseteq I} \left| \beta_J^{w^{-1}} \beta_J^w (fw, h_J^{w^{-1}})_{w^{-1}} (gw^{-1}, h_J^w)_w \right| \leqslant Q\|f\|_w \|g\|_{w^{-1}}.$$

A mixed sum becomes

$$\sum_{J \subseteq I} \left| \beta_J^w a_J^{w^{-1}} (f, \chi_J)(gw^{-1}, h_J^w)_w \right| \leqslant \left(\sum_{J \subseteq I} \frac{\langle w \rangle_J |\Delta_J w^{-1}|^2}{4\langle w^{-1} \rangle_J^2} \langle f \rangle_J^2 |J| \right)^{1/2} \|g\|_{w^{-1}}$$

with the other mixed sum symmetric to it. When two characteristic functions are present, we get a bilinear term

$$\sum_{J \subseteq I} \left| a_J^{w^{-1}} a_J^w (f, \chi_J)(g, \chi_J) \right| \leqslant \sum_{J \subseteq I} \frac{|\Delta_J w^{-1}||\Delta_J w|}{4\langle w^{-1} \rangle_J \langle w \rangle_J} \langle f \rangle_J \langle g \rangle_J |J|.$$

So the sum with only cancellative terms has a trivial estimate. The two mixed sums that were split by Cauchy–Schwarz respond to a trivial estimate and a Carleson estimate. Indeed, the Carleson embedding theorem with measure w can be formulated as

$$\frac{1}{|K|} \sum_{J \subseteq K} \mu_J \leqslant C\langle v \rangle_K \quad \forall K \quad \Leftrightarrow \quad \frac{1}{|I|} \sum_{J \subseteq I} \frac{\mu_J}{\langle v \rangle_J^2} \langle \varphi v \rangle_J^2 \leqslant 4C\langle \varphi^2 v \rangle_I.$$

Using this for $f = \varphi v$, $v = w^{-1}$, $\mu_J = \frac{1}{4}\langle w \rangle_J |\Delta_J w^{-1}|^2$ yields an estimate

$$\frac{1}{|I|} \sum_{J \subseteq I} \frac{\langle w \rangle_J |\Delta_J w^{-1}|^2}{4\langle w^{-1} \rangle_J^2} \langle f \rangle_J^2 |J| \leqslant CQ^2 \langle f^2 w \rangle_I,$$

provided

$$\frac{1}{|K|} \sum_{J \subseteq K} \langle w \rangle_J |\Delta_J w^{-1}|^2 \leqslant CQ^2 \langle w^{-1} \rangle_K \forall K.$$

Last, there is a bilinear term, which requires more care and a more complex Bellman function of a so-called bilinear embedding. Provided

$$\exists C : \forall K : \frac{1}{|K|} \sum_{J \subseteq K} \mu_J \leqslant C,$$

there exists C' so that we have

$$\frac{1}{|I|} \sum_{J \subseteq I} \mu_J \langle f \rangle_J \langle g \rangle_J \leqslant C' \left(\langle f^2 w \rangle_I \langle g^2 w^{-1} \rangle_I \right)^{1/2}.$$

Using this for $\mu_J = \frac{|\Delta_J w^{-1}||\Delta_J w|}{4 \langle w^{-1} \rangle_J \langle w \rangle_J} |J|$ provides the desired conclusion. Notice that other than in the classical Carleson lemma, the condition is not necessary for the conclusion. The original versions appear in [18, 17] with another two conditions that later turned out redundant (via the function M below). Morally, the core Bellman function pieces and many of the beautiful ideas go back to the text on the two-weight problem [15]. The assembly of the function in the one-weight case could be shortened and streamlined with the above nice and concise statement – after all, the goal was a linear weighted estimate and not sharp conditions for the two-weight case. Given that the embedding condition is not necessary, it had no application in two-weight questions.

With today's knowledge, there is a much easier proof for the last bilinear terms, but it does not provide a Bellman function, at least not explicitly. Similarly, the trivial estimates that just use orthogonality of the weighted Haar systems will require a Bellman proof to build the entire function. This is what we will sketch now in the dyadic case. One minor but important difference is the estimate we aim for on the right-hand side. Bellman functions are much more easily derived if the right-hand estimate is $\langle f^2 w \rangle_I + \langle g^2 w^{-1} \rangle_I$ instead of $\langle f^2 w \rangle_I^{1/2} \langle g^2 w^{-1} \rangle_I^{1/2}$. It is not hard to see (and explained below) why these estimates are equivalent. The Bellman functions we expose for the two Carleson terms are also not the original ones, but they will be more useful and accessible for the non-homogeneous cases.

Proof. (Dyadic Bellman function.) First, we write $V_{\pm} = V \pm \Delta V$ and calculate explicitly for the function

$$B_1(V) = \frac{x^2}{r} + \frac{y^2}{s}.$$

The determinant of its Hessian is 0, so while it is simple, it is not of the superconvex type that we often see in Bellman proofs. Rather, if we calculate the discrete Laplacian directly, we will obtain

$$- B_1(V) + \frac{1}{2} B_1(V_+) + \frac{1}{2} B_1(V_-)$$

$$\begin{aligned}
&= \frac{x^2\Delta r^2 + r^2\Delta x^2 - 2r\Delta rx\Delta x}{r(r^2 - \Delta r^2)} + \frac{y^2\Delta s^2 + s^2\Delta y^2 - 2s\Delta sy\Delta y}{s(s^2 - \Delta s^2)} \\
&= \frac{(x\Delta r - r\Delta x)^2}{r(r^2 - \Delta r^2)} + \frac{(y\Delta s - s\Delta y)^2}{s(s^2 - \Delta s^2)} \\
&\geqslant \frac{(x\Delta r - r\Delta x)^2}{4r^3} + \frac{(y\Delta s - s\Delta y)^2}{4s^3} \\
&\geqslant \frac{|x\Delta r - r\Delta x||y\Delta s - s\Delta y|}{2r^{3/2}s^{3/2}} \\
&\geqslant \frac{1}{Q}\frac{|x\Delta r - r\Delta x||y\Delta s - s\Delta y|}{2rs},
\end{aligned}$$

with the last inequality somewhat wasteful. From line 3 to 4 we used homogeneity: Averages of non-negative functions can at most double when passing from a dyadic interval to its children.

Next, our functions B_2, B_3, B_4 will contain elements for which it is a bit harder to directly estimate finite differences, and we will for the most part be satisfied with a rougher estimate, so we will use the easy fact that midpoint concavity is the same as concavity as per the use of the Hessian. For convex domains

$$\langle -d^2B(V)dV, dV\rangle \geqslant \langle A(V)dV, dV\rangle \quad \Longleftrightarrow \quad B(V) - \frac{1}{2}B(V_+) - \frac{1}{2}B(V_-) \geqslant \langle A(V)\Delta V, \Delta V\rangle.$$

To see this is not difficult. One uses tools such as the Taylor formula and integration against a tent paired with integration by parts. But be careful: The non-convexity of our domain containing $1 \leqslant rs \leqslant Q$ becomes a small obstacle in this line of reasoning, when integrating along the path between V_+ and V_-, observing that this path is sometimes not in the domain, even though V, V_+, V_- are. While in the homogeneous case, there is an easy fix at a moderate cost of slightly worse constants. Enlarging the domain somewhat by increasing Q by a factor fixes the problem. The detailed argument can be found in [22]. Let us now motivate our function

$$B_2(V) = \frac{x^2}{2r - \frac{1}{s(N+1)}} + \frac{y^2}{s}, \quad N(r,s) = \frac{\sqrt{rs}}{\sqrt{Q}}\left(1 - \frac{(rs)^2}{128Q^2}\right).$$

This function is indeed assembled backwards using two embedding theorems. Indeed,

$$B_2(V) = \frac{x^2}{r + M(r,s)} + \frac{y^2}{s}, \quad M(r,s) = r - \frac{1}{s(N(r,s) + 1)}.$$

Readers familiar with the Bellman proof of the Carleson lemma will immediately recognize the (very old and beautifully simple) Bellman function for the problem. The function M is the so-called self-improvement of the Carleson condition, meaning testing a condition resembling N will provide the embedding condition resembling M. In more detail, when $1 \leqslant rs \leqslant Q$, then

$$N \leqslant \left(1 - \frac{1}{128Q^2}\right)\frac{\sqrt{rs}}{\sqrt{Q}} < \frac{\sqrt{rs}}{\sqrt{Q}}$$

and $-\mathrm{d}^2 N \geqslant \frac{1}{Q^2}s^2(\mathrm{d}r)^2$ provides an estimate of the form

$$\frac{1}{|I|}\sum_{J\subseteq I}\langle w\rangle_J^2 |\Delta_I w^{-1}|^2 \leqslant Q^2.$$

The Bellman function

$$M(r,N) = r - \frac{1}{s(N+1)}$$

has its range $0 \leqslant M \leqslant r$ in the domain $0 \leqslant N \leqslant 1$ and $1 \leqslant rs$. It also has concavity as well as the derivative estimate $-\partial_N M \geqslant \frac{1}{4s}$. A standard Bellman argument provides the implication

$$\frac{1}{|J|}\sum_{K\subseteq J}\alpha_K \langle w\rangle_K \leqslant 1 \quad \Rightarrow \quad \frac{1}{|J|}\sum_{K\subseteq J}\alpha_K \leqslant \langle w^{-1}\rangle_J.$$

In the four-sum proof, this gives the required embedding condition we need uniformly for all J:

$$\frac{1}{|J|}\sum_{K\subseteq J}\langle w\rangle_K |\Delta_K w^{-1}|^2 \leqslant Q^2 \langle w^{-1}\rangle_J.$$

Assembling these elements backwards and using convexity of $\frac{x^2}{r+M}$ as well as its derivative estimate from below of $\frac{x^2}{4r}$, we obtain the dissipation estimate for B_2 as follows. For simplicity, we only treat

$$B_{2,1}(x,r,M(r,s)) = \frac{x^2}{r+M(r,s)}.$$

We start with an estimate from below with the increment of N. Due to the super-concavity of N, as long as $r_+ + r_- = 2r$ and $s_+ + s_- = 2s$ with all pairs (r,s) and $(r_\pm, s_\pm)$ in the domain, we obtain, when writing $(\Delta r)^2 = (r_+ - r_-)^2$,

$$N(r,s) - \frac{1}{2}N(r_+, s_+) - \frac{1}{2}N(r_-, s_-) \geqslant \frac{1}{Q^2}s^2(\Delta r)^2.$$

Now, we write

$$\Delta N = \Delta N(r_+, s_+, r_-, s_-) = N(r,s) - \frac{1}{2}N(r_+, s_+) - \frac{1}{2}N(r_-, s_-)$$

and estimate via a derivative estimate

$$M(r,s,N(r,s)) - M(r,s,N(r,s) - \Delta N) \geqslant \Delta N \frac{1}{4s}$$

and via concavity

$$M(r,s,N(r,s) - \Delta N) - \frac{1}{2}M(r_+,s_+,N(r_+,s_+)) - \frac{1}{2}M(r_-,s_-,N(r_-,s_-)) \geqslant 0,$$

and thus

$$M(r,s,N(r,s)) - \frac{1}{2}M(r_+,s_+,N(r_+,s_+)) - \frac{1}{2}M(r_-,s_-,N(r_-,s_-)) \geqslant \Delta N \frac{1}{4s}.$$

Let

$$\Delta M = \Delta M(r_+,s_+,r_-,s_-)$$
$$= M(r,s,N(r,s)) - \frac{1}{2}M(r_+,s_+,N(r_+,s_+)) - \frac{1}{2}M(r_-,s_-,N(r_-,s_-)).$$

Then

$$- B_{2,1}(x,r,M) + B_{2,1}(x,r,M - \Delta M)$$
$$\geqslant \Delta M \frac{x^2}{4r^2} \geqslant \Delta N \frac{1}{4s}\frac{x^2}{4r^2} \geqslant \frac{1}{Q^2}s^2\frac{1}{4s}\frac{x^2}{4r^2}(\Delta r)^2 = \frac{1}{4Q^2}\frac{sx^2}{r^2}(\Delta r)^2.$$

Using convexity, we get, when $x_+ + x_- = 2x$, $r_+ + r_- = 2r$, $M_+ + M_- = 2M - 2\Delta M$,

$$-B_{2,1}(x,r,M) + \frac{1}{2}B_{2,1}(x_+,r_+,M_+) - \frac{1}{2}B_{2,1}(x_-,r_-,M_-) \geqslant \frac{1}{Q^2}\frac{sx^2}{r^2}(\Delta r)^2.$$

Together with $B_{2,2}(y,s) = \frac{y^2}{s}$ and

$$-B_{2,2}(y,s) + \frac{1}{2}B_{2,2}(y_+,s_+) - \frac{1}{2}B_{2,2}(y_-,s_-) \geqslant \frac{(y\Delta s - s\Delta y)^2}{4s^3},$$

we get

$$-B_2(V) + \frac{1}{2}B_2(V_+) - \frac{1}{2}B_2(V_-) \geqslant \frac{1}{Qsr}|x||\Delta r||y\Delta s - s\Delta y|,$$

and by symmetry also

$$-B_3(V) + \frac{1}{2}B_3(V_+) - \frac{1}{2}B_3(V_-) \geqslant \frac{1}{Qsr}|y||\Delta s||x\Delta r - r\Delta x|.$$

Last, we consider three more functions, which share the responsibility of our last term, depending upon the constellation of the variables to each other. We have

$$B_4(V) = \sup_{a>0} \frac{x^2}{r + aK(r,s)} + \frac{y^2}{r + a^{-1}K(r,s)}$$

with

$$K(r,s) = \frac{\sqrt{rs}}{\sqrt{Q}}\left(1 - \frac{\sqrt{rs}}{8\sqrt{Q}}\right).$$

We have

$$0 \leqslant K \leqslant \left(1 - \frac{1}{8\sqrt{Q}}\right)\frac{\sqrt{rs}}{\sqrt{Q}} < \frac{\sqrt{rs}}{\sqrt{Q}} \leqslant 1,$$

and by direct calculation

$$-\langle \mathrm{d}^2 K \mathrm{d}W, \mathrm{d}W \rangle \geqslant \frac{1}{8Q}|dr||ds|.$$

Notice that B_4 is as a supremum of convex functions. It is possible to exhibit the extremal values and the α where they are attained, but this is not needed in the dyadic case. We use some tricks found in [15].

Let us assume $2|x|K < |y|r$ and $2|y|K < |x|s$. We note

$$L_4^\alpha(V,K) = \frac{x^2}{r + \alpha K} + \frac{y^2}{r + \alpha^{-1}K}, \quad L_4(V,K) = \sup_\alpha L_4^\alpha(V,K)$$

and denote by α' the value where the supremum at (V,K) is (almost) attained, so in particular,

$$L_4(V_\pm, K_\pm) \geqslant L_4^{\alpha'}(V_\pm, K_\pm).$$

This gives the first inequality below and the second one follows as before, using the mean value theorem, a derivative estimate, and convexity:

$$\begin{aligned}
&\frac{1}{2}L_4(V_+, K_+) + \frac{1}{2}L_4(V_-, K_-) - L_4(V,K) \\
&\geqslant \frac{1}{2}L_4^{\alpha'}(V_+, K_+) + \frac{1}{2}L_4^{\alpha'}(V_-, K_-) - L_4^{\alpha'}(V,K) \\
&\geqslant \frac{\partial L_4^{\alpha'}}{\partial K}(V,K)(\Delta K) \\
&\geqslant \frac{1}{Q}\left(\frac{x^2\alpha'}{(r + \alpha'K)^2} + \frac{y^2\alpha'^{-1}}{(s + \alpha'^{-1}K)^2}\right)|\Delta r||\Delta s| \\
&\geqslant \frac{1}{Q}\left(\frac{2|x||y|}{(r + \alpha'K)(s + \alpha'^{-1}K)}\right)|\Delta r||\Delta s|.
\end{aligned}$$

We want to estimate this further from below. To do so, notice that

$$L_4^\alpha(V,K) \geqslant \frac{x^2}{r} + \frac{y^2}{s} - \alpha K \frac{x^2}{r^2} - \alpha^{-1} K \frac{y^2}{s^2},$$

but because we assumed $2|x|K < |y|r$ and $2|y|K < |x|s$, we get

$$\inf_{\alpha}\left(\alpha K\frac{x^2}{r^2} + \alpha^{-1}K\frac{y^2}{s^2}\right) = K\frac{|x||y|}{rs} < \min\left(\frac{x^2}{2r}, \frac{y^2}{2s}\right),$$

so we deduce

$$L_4(V,K) = \frac{x^2}{r + \alpha'K} + \frac{y^2}{s + \alpha'^{-1}K} \geq \max\left(\frac{x^2}{r} + \frac{y^2}{2s}, \frac{x^2}{2r} + \frac{y^2}{s}\right). \tag{6.3}$$

Had we $r + \alpha'K > 2r$ or $s + \alpha'^{-1}K > 2s$ combined with the trivial estimates $s + \alpha'^{-1}K > s$ and $r + \alpha'K > r$, respectively, we could in each constellation contradict the above inequality (6.3). We deduce that $r + \alpha'K \leq 2r$ and $s + \alpha'^{-1}K \leq 2s$, so we deduce under the assumptions $2|x|K < |y|r$ and $2|y|K < |x|s$ the inequality

$$\frac{1}{2}B_4(V_+) + \frac{1}{2}B_4(V_-) - B_4(V) \geq \frac{1}{Q}\frac{|x||y|}{rs}|\Delta r||\Delta s|.$$

When $2|x|K \geq |y|r$ or $2|y|K \geq |x|s$, then we use respectively

$$B_5 = \frac{x^2}{2r - \frac{1}{s(K(r,s)+1)}} + \frac{y^2}{s}, \quad B_6 = \frac{x^2}{r} + \frac{y^2}{2s - \frac{1}{r(K(r,s)+1)}}$$

to get the same dissipation as for B_4, following similar steps from above when we treated B_2 and in addition using the assumptions on the size of K. Since B_4, B_5, and B_6 are convex everywhere, we deduce the global superconvexity for $B_7 = B_4 + B_5 + B_6$:

$$\frac{1}{2}B_7(V_+) + \frac{1}{2}B_7(V_-) - B_7(V) \geq \frac{1}{Qrs}|x||y||\Delta r||\Delta s|.$$

Now, we use the lower triangle inequality and choose constants C_1, C_2, C_3, and C_7 (which can be chosen independent of Q) such that

$$C_1\left(\frac{1}{2}B_1(V_+) + \frac{1}{2}B_1(V_-) - B_1(V)\right) \geq \frac{1}{Qrs}(|x\Delta r| - |r\Delta x|)(+|y\Delta s| - |s\Delta y|),$$

$$C_2\left(\frac{1}{2}B_2(V_+) + \frac{1}{2}B_2(V_-) - B_2(V)\right) \geq \frac{1}{Qrs}|x\Delta r|(-|y\Delta s| + |s\Delta y|),$$

$$C_3\left(\frac{1}{2}B_3(V_+) + \frac{1}{2}B_3(V_-) - B_3(V)\right) \geq \frac{1}{Qrs}(-|x\Delta r| + |r\Delta x|)|y\Delta s|,$$

$$C_7\left(\frac{1}{2}B_7(V_+) + \frac{1}{2}B_7(V_-) - B_7(V)\right) \geq \frac{1}{Qrs}|x\Delta r||y\Delta s|$$

gives us for $B(V) = C_1B_1(V) + C_2B_2(V) + C_3B_3(V) + C_7B_7(V)$ that

$$\frac{1}{2}B(V_+) + \frac{1}{2}B(V_-) - B(V) \geq \frac{1}{Q}|\Delta x||\Delta y|.$$

$\square$

6.2 The non-homogeneous dyadic case

Even though the function above can in principle be reused, several steps require a lot more care. In particular, the jump probabilities are no longer 1/2, so many of the short-cuts where we used Δx, Δy, Δr, and Δs are no longer possible. Indeed, it is required that the functions K and N have the so-called one-leg superconcavities, which lead to one-sided dissipation estimates. These were carefully calculated in [21]. It is not surprising that no difference is required for any of the other participating Bellman functions (the ones we plug K and N into), seen that Carleson lemmas, much like maximal function estimates, are not bothered by non-homogeneity. In the text [21] these well-known embedding theorems were replaced by the so-called outer measure. We did not see a use or possibility for this in the passage to the continuous case.

6.3 The continuous, general case

This case requires an array of technical changes. Some, but very few of these, are due to the fact that we are using Hilbert-space-valued martingales. The main differences arise due to the continuity of the process, the requirements in the Itô formula, and the presence of jumps. These two combinations make the closed expression of B a necessity and not a cute luxury.

Proof of Lemma 2. We give an explicit expression for the continuous problem. Let again $V = (x, y, r, s)$ and $W = (r, s)$. We first consider again for $r, s > 0$ and $1 \leqslant rs \leqslant Q$ the function

$$B_1(x, y, r, s) = \frac{\langle x, x \rangle}{r} + \frac{\langle y, y \rangle}{s}.$$

Then $0 \leqslant B_1 \leqslant \frac{\langle x,x \rangle}{r} + \frac{\langle y,y \rangle}{s}$ and

$$\langle d^2 B_1 dV, dV \rangle$$

$$= \frac{2}{r} \langle dx, dx \rangle + \frac{2\langle x, x \rangle}{r^3} (dr)^2 - 4 \frac{\langle x, dx \rangle}{r^2} dr + \frac{2}{s} \langle dy, dy \rangle + \frac{2\langle y, y \rangle}{s^3} (ds)^2 - 4 \frac{\langle y, dy \rangle}{s^2} ds$$

$$= \frac{2}{r} \left\langle dx - \frac{x}{r} dr, dx - \frac{x}{r} dr \right\rangle + \frac{2}{s} \left\langle dy - \frac{y}{s} ds, dy - \frac{y}{s} ds \right\rangle$$

$$\geqslant 0.$$

Letting $V_0 = (x_0, y_0, r_0, s_0)$ and $V = (x, y, r, s)$ we also calculate the one-sided convexity estimate

$$- (B_1(V_0) - B_1(V) + dB_1(V_0)(V - V_0))$$

$$= -\left(\frac{x_0^2}{r_0} - \frac{x^2}{r} + \frac{2x_0}{r_0}(x - x_0) - \frac{x_0^2}{r_0^2}(r - r_0) \right)$$

$$-\left(\frac{y_0^2}{s_0} - \frac{y^2}{s} + \frac{2y_0}{s_0}(y - y_0) - \frac{y_0^2}{s_0^2}(s - s_0)\right)$$

$$= r\left\langle \frac{x}{r} - \frac{x_0}{r_0}, \frac{x}{r} - \frac{x_0}{r_0} \right\rangle + s\left\langle \frac{y}{s} - \frac{y_0}{s_0}, \frac{y}{s} - \frac{y_0}{s_0} \right\rangle.$$

We now consider again the two functions from [21]

$$K(r,s) = \frac{\sqrt{rs}}{\sqrt{Q}}\left(1 - \frac{\sqrt{rs}}{8\sqrt{Q}}\right) \quad \text{and} \quad N(r,s) = \frac{\sqrt{rs}}{\sqrt{Q}}\left(1 - \frac{(rs)^2}{128Q^2}\right)$$

in the domain $1 \leqslant rs \leqslant Q$. We recall that

$$0 \leqslant K \leqslant 1,$$
$$0 \leqslant N \leqslant 1.$$

So in particular, we have $rs \geqslant rs - K^2 > rs(1 - \frac{1}{Q})$. One calculates that

$$-\langle \mathrm{d}^2 K \, \mathrm{d}W, \mathrm{d}W \rangle \geqslant \frac{1}{8Q}|\mathrm{d}r||\mathrm{d}s|,$$

$$-\langle \mathrm{d}^2 N \, \mathrm{d}W, \mathrm{d}W \rangle \geqslant \frac{1}{Q^2}s^2(\mathrm{d}r)^2, \quad \text{and}$$

$$-(\mathrm{d}^2 N \, \mathrm{d}W, \mathrm{d}W) \geqslant \frac{1}{Q^2}r^2(\mathrm{d}s)^2.$$

Also, whenever W, W_0 are in the domain, then

$$K(W_0) - K(W) + \mathrm{d}K(W_0)(W - W_0) \geqslant \frac{1}{Q}|r - r_0||s - s_0|,$$

$$N(W_0) - N(W) + \mathrm{d}N(W_0)(W - W_0) \geqslant \frac{1}{Q^2}s_0 s|r - r_0|^2,$$

$$N(W_0) - N(W) + \mathrm{d}N(W_0)(W - W_0) \geqslant \frac{1}{Q^2}r_0 r|s - s_0|^2.$$

These one-leg concavities are crucial and very delicate and were proved in [21].

Let now

$$B_2 = \frac{\langle x, x \rangle}{2r - \frac{1}{s(N(r,s)+1)}} + \frac{\langle y, y \rangle}{s}.$$

One checks easily by calculation of their Hessians that

$$F(x,r,M) = \frac{\langle x, x \rangle}{r + M}, \quad G(r,s,N) = \frac{1}{s(N + 1)}$$

are convex everywhere. In order to estimate the Hessian of B_2 from below, one merely requires estimates of derivatives of these functions, similar to the dyadic case. We have

$$-\partial_M F = \frac{\langle x, x \rangle}{(r + M)^2} \geqslant \frac{\langle x, x \rangle}{4r^2} \quad \text{and} \quad -\partial_N G = \frac{1}{s(N + 1)^2} \geqslant \frac{1}{4s}.$$

Since $0 \leq r - \frac{1}{s(N(r,s)+1)} \leq r$, we know that $0 \leq B_2 \leq \frac{|x|^2}{r} + \frac{|y|^2}{s}$. Now the Hessian estimate becomes

$$\langle \mathrm{d}^2 B_2 \mathrm{d}V, \mathrm{d}V \rangle$$

$$\geq \frac{\langle x, x \rangle}{4r^2} \frac{1}{s(N+1)^2} \frac{1}{Q^2} |\mathrm{d}r|^2 s^2 + \frac{2}{s} \left\langle \mathrm{d}y - \frac{y}{s}\mathrm{d}s, \mathrm{d}y - \frac{y}{s}\mathrm{d}s \right\rangle$$

$$\geq \frac{|x|^2 s}{Q^2 r^2} |\mathrm{d}r|^2 + \frac{2}{s} \left\langle \mathrm{d}y - \frac{y}{s}\mathrm{d}s, \mathrm{d}y - \frac{y}{s}\mathrm{d}s \right\rangle$$

$$\geq \frac{|x|}{Q} |\mathrm{d}r| \left| \mathrm{d}y - \frac{y}{s}\mathrm{d}s \right|.$$

If we had no jumps, this would be sufficient. But this function has the additional important property

$$-(B_2(V_0) - B_2(V) + \mathrm{d}B_2(V_0)(V - V_0)) \geq \frac{\langle x_0, x_0 \rangle}{Q^2 r_0^2} s(r - r_0)^2 + s\left\langle \frac{y}{s} - \frac{y_0}{s_0}, \frac{y}{s} - \frac{y_0}{s_0} \right\rangle.$$

To see this, write

$$\frac{\langle x, x \rangle}{2r - \frac{1}{s(N(r,s)+1)}} = H(x, r, s, N(r,s)), \quad H(x, r, s, N) = \frac{\langle x, x \rangle}{2r - \frac{1}{s(N+1)}},$$

where H is convex and

$$-\partial_N H \geq \frac{\langle x, x \rangle}{Q^2 r^2 s}.$$

Now since H is convex, with $P_0 = (x_0, r_0, s_0, N_0)$ and $P = (x, r, s, N)$ we have $H(P) \geq H(P_0) + \mathrm{d}H(P - P_0)$. So

$$H(P) - H(P_0) - \partial_x H(P_0)(x - x_0) - \partial_r H(P_0)(r - r_0) - \partial_s H(P_0)(s - s_0)$$
$$\geq -\partial_N H(P_0)(N_0 - N),$$

with

$$N(r_0, s_0) - N(r, s) + \partial_r N(r_0, s_0)(r - r_0) + \partial_s N(r_0, s_0)(s - s_0) \geq \frac{1}{Q^2} r_0 r |s - s_0|^2.$$

With $N_0 = N(r_0, s_0)$ and $N = N(r, s)$ the above becomes

$$B_2(V) - B_2(V_0) - \mathrm{d}B_2(V_0)(V - V_0) \geq \frac{\langle x_0, x_0 \rangle}{Q^2 r_0^2} s|r - r_0|^2 + s\left\langle \frac{y}{s} - \frac{y_0}{s_0}, \frac{y}{s} - \frac{y_0}{s_0} \right\rangle,$$

where we used the lower derivative estimate and the chain rule. Analogously,

$$B_3 = \frac{\langle x, x \rangle}{r} + \frac{\langle y, y \rangle}{2s - \frac{1}{r(N(r,s)+1)}}$$

has the same size estimates as

$$\langle \mathrm{d}^2 B_3 \mathrm{d}V, \mathrm{d}V \rangle \geq \frac{|y|}{Q} |\mathrm{d}s| \left| \mathrm{d}x - \frac{x}{r} \mathrm{d}r \right|$$

and one-leg convexity

$$B_3(V) - B_3(V_0) - \mathrm{d}B_3(V_0)(V - V_0) \geq \frac{\langle y_0, y_0 \rangle}{Q^2 s_0^2} r |s - s_0|^2 + r \left\langle \frac{x}{r} - \frac{x_0}{r_0}, \frac{x}{r} - \frac{x_0}{r_0} \right\rangle.$$

Let us now work on the last part. We consider again the function

$$H_4(V, K) = \sup_{0 < \alpha} \beta_4(\alpha, V, K) = \sup_{0 < \alpha} \left(\frac{\langle x, x \rangle}{r + \alpha K} + \frac{\langle y, y \rangle}{s + \alpha^{-1} K} \right).$$

But what we did in the dyadic case will no longer suffice here. We will calculate the supremum directly and study the smoothness properties of this function.

Testing for critical points,

$$\partial_\alpha \beta_4 = -\frac{\langle x, x \rangle K}{(r + \alpha K)^2} + \frac{\langle y, y \rangle K}{(\alpha s + K)^2}.$$

So $\partial_\alpha \beta_4 = 0$ if and only if

$$\alpha = \alpha' = \frac{|y| r - |x| K}{|x| s - |y| K}.$$

Since only $\alpha > 0$ are admissible, we require that $|y|r - |x|K$ and $|x|s - |y|K$ have the same sign. To determine the sign change of $\partial_\alpha \beta_4$ at α', consider

$$-\frac{|x|}{r + \alpha K} + \frac{|y|}{\alpha s + K} = \frac{(|y|r - |x|K) - \alpha(|x|s - |y|K)}{(r + \alpha K)(\alpha s + K)}.$$

If the signs are negative, then the sign change is from negative to positive; otherwise it is from positive to negative. For a maximum to be attained at $\alpha' > 0$ we require that both numerator and denominator be positive. Then, if K is relatively small, meaning that $|y|r - |x|K$ and $|x|s - |y|K$ are positive, we have

$$H_4(x, y, r, s, K)$$
$$= \beta_4(\alpha', x, y, r, s)$$
$$= \frac{\langle x, x \rangle(|x|s - |y|K)}{r(|x|s - |y|K) + (|y|r - |x|K)K} + \frac{\langle y, y \rangle(|y|r - |x|K)}{s(|y|r - |x|K) + (|x|s - |y|K)K}$$
$$= \frac{\langle x, x \rangle s - 2|x||y|K + \langle y, y \rangle r}{rs - K^2}.$$

Observe that by the above considerations on K, the denominator is never 0. The case $|x| = 0$ or $|y| = 0$ corresponds to other parts of the domain, so when K is small in the sense above, this function is in $\mathcal{C}^2$.

When $|y|r - |x|K \leqslant 0$ or $|x|s - |y|K \leqslant 0$, the supremum is attained at the boundary and $H_4 = \frac{\langle y,y \rangle}{s}$ or $H_4 = \frac{\langle x,x \rangle}{r}$. Thanks to the size restrictions on K, we never have both $|x|s - |y|K \leqslant 0$ and $|y|r - |x|K \leqslant 0$ unless $x, y = 0$. Indeed,

$$|x|(|x|s - |y|K) + |y|(|y|r - |x|K)$$

$$= \frac{|x|^2}{r} - 2\frac{|x||y|}{rs}K + \frac{|y|^2}{s}$$

$$= \left(\frac{|x|}{\sqrt{r}} - \frac{|y|}{\sqrt{s}} \right)^2 + \frac{2|x||y|}{\sqrt{rs}}\left(1 - \frac{K}{\sqrt{rs}} \right).$$

With $1 - \frac{K}{\sqrt{rs}} > 0$ we see that the above is never negative and the quantity vanishing implies $x = y = 0$. If $|x|s - |y|K \leqslant 0$ and $|y|r - |x|K > 0$, then $\frac{\langle x,x \rangle}{r} < \frac{\langle y,y \rangle}{s}$ and $H_4 = \frac{\langle y,y \rangle}{s}$; if $|y|r - |x|K \leqslant 0$ and $|x|s - |y|K > 0$, then $H_4 = \frac{\langle x,x \rangle}{r}$.

Notice that when $\frac{\langle x,x \rangle}{r} = \frac{\langle y,y \rangle}{s}$ and $x, y \neq 0$, then $|y|r - |x|K > 0$ and $|x|s - |y|K > 0$. Indeed, we have seen that $\frac{|x|^2}{r} - 2\frac{|x||y|}{rs}K + \frac{|y|^2}{s} > 0$. Thus, $\frac{\langle x,x \rangle}{r} = \frac{\langle y,y \rangle}{s} > \frac{|x||y|}{rs}K$ and $|y|r - |x|K > 0$ and $|x|s - |y|K > 0$.

Thus, $H_4 \in \mathcal{C}^2$ for these parts of the domain. We also see from these considerations that in order to see $H_4 \in \mathcal{C}^1$ we only need to check the cuts $|x|s - |y|K = 0$ and $|y|r - |x|K \geqslant 0$ as well as $|y|r - |x|K = 0$ and $|x|s - |y|K \geqslant 0$.

When $|y|r - |x|K > 0$ and $|x|s - |y|K > 0$ (we call this part of the domain R_1),

$$(\partial_x H_4, dx) = 2\frac{\langle dx, x \rangle}{|x|}\frac{|x|s - |y|K}{rs - K^2},$$

$$\partial_r H_4 = -\frac{(|x|s - |y|K)^2}{(rs - K^2)^2},$$

$$\partial_K H_4 = -2\frac{(|x|s - |y|K)(|y|r - |x|K)}{(rs - K^2)^2}.$$

We first prove that $\partial_x H_4$ is continuous throughout. Recall that we have to treat three regions: R_1 and R_2, where $|y|r - |x|K > 0$ and $|x|s - |y|K \leqslant 0$, and R_3, where $|x|s - |y|K \leqslant 0$ and $|y|r - |x|K > 0$. Inside R_2 we have $H_4 = \frac{\langle y,y \rangle}{s}$ and thus $\partial_x H_4 = 0$. Inside R_3 we have $H_4 = \frac{\langle x,x \rangle}{r}$ and thus $\partial_x H_4 = \frac{2\langle x,dx \rangle}{r}$. Inside R_1 we have

$$\partial_x H_4 = 2\frac{\langle dx, x \rangle}{|x|}\frac{|x|s - |y|K}{rs - K^2} = 2\langle x, dx \rangle\frac{|x|s - |y|K}{r(|x|s - |y|K) + (|y|r - |x|K)K}.$$

We have three cases. First, let us approach a boundary point of R_1 from within R_1 so that $|y|r - |x|K > 0$ and $|x|s - |y|K = 0$. Assume therefore that $|y|r - |x|K \sim a > 0$ and $0 < |x|s - |y|K < \varepsilon$. We have $|\langle \partial_x H_4, dx \rangle| \leqslant 2|dx|\frac{\varepsilon}{rs-K^2} \leqslant \varepsilon|dx|$ since $rs - K^2$ is bounded below. Letting $\varepsilon \to 0$ shows continuity in this point. Second, let us approach a boundary point $|x|s - |y|K > 0$ and $|y|r - |x|K = 0$ from within R_1. Assume therefore that $|x|s - |y|K \sim a > 0$ and $0 < |y|r - |x|K < \varepsilon$. We show that we have $(\partial_x H_4, dx) \le \frac{\varepsilon}{a}|dx|$. Since

$$\frac{1}{r} - (|y|r - |x|K)K \frac{|x|s - |y|K}{r^2(|x|s - |y|K)^2}$$
$$\le \frac{(|x|s - |y|K)}{r(|x|s - |y|K) + (|y|r - |x|K)K}$$
$$\le \frac{1}{r},$$

we have

$$\left| \frac{2\langle x, dx\rangle(|x|s - |y|K)}{r(|x|s - |y|K) + (|y|r - |x|K)K} - \frac{2\langle x, dx\rangle}{r} \right|$$
$$\le 2|\langle x, dx\rangle| \frac{(|y|r - |x|K)K}{r^2(|x|s - |y|K)}$$
$$\le |x||dx|\frac{\varepsilon}{a}.$$

Since $0 < |y|r - |x|K < \varepsilon$ and s, r, K are controlled, one can deduce from $|x|s - |y|K \sim a$ that $|x| \sim a$. Last, let us approach $|y|r - |x|K = 0$ and $|x|s - |y|K = 0$. To this end, one can see that if $0 < |y|r - |x|K < \varepsilon$ and $0 < |x|s - |y|K < \varepsilon$, then $|x|, |y| \le \varepsilon$, establishing continuity in the third case.

The $\partial_r H_4$ derivative is similar since the term $\frac{|x|s - |y|K}{rs - K^2}$ reappears as a square and in R_3 notice that $H_4 = \frac{\langle x, x\rangle}{r}$, so $\partial_r H_4 = -\frac{\langle x, x\rangle}{r^2}$. It is easy to see that the derivative $\partial_K H_4$ is zero in R_2 and R_3 as well as when approaching the boundary of R_1.

These derivatives are representative by symmetry and the function is therefore in C^1. As a consequence,

$$B_4(x, y, r, s) = H_4(x, y, r, s, K(r, s)) \in C^1.$$

Function B_4 is a supremum of convex functions. We show directly that $-\partial_K B_4 \ge 0$ everywhere and that in $R_1' \subset R_1$ where $|y|r - 2|x|K > 0$ and $|x|s - 2|y|K > 0$ we have $-\partial_K B_4 \ge \frac{|x||y|}{rs}$. Recall that

$$-\partial_K B_4 = 2\frac{(|x|s - |y|K)(|y|r - |x|K)}{(rs - K^2)^2}$$
$$= 2\frac{(|x|s - |y|K)(|y|r - |x|K)|x||y|}{(r(|x|s - |y|K) + K(|y|r - |x|K))(s(|y|r - |x|K) + K(|x|s - |y|K))},$$

so $-\partial_K B_4 \ge c\frac{|x||y|}{rs}$ if

$$\frac{rs}{c} \ge K^2 + rs + \frac{Kr(|x|s - |y|K)}{|y|r - |x|K} + \frac{Ks(|y|r - |x|K)}{|x|s - |y|K}.$$

Now $K^2 \leqslant 1 \leqslant rs$, and when $|y|r - 2|x|K \geqslant 0$, then $|y|r - |x|K \geqslant |x|K$. Similarly, $|x|s - |y|K \geqslant |y|K$. So the last two terms are bounded by $\frac{Kr|x|s}{|x|K} + \frac{Ks|y|r}{|y|K} = 2rs$. Therefore, the choice $c = 1/4$ works. In R_1',

$$(\mathrm{d}^2 B_4 \mathrm{d}V, \mathrm{d}V) \geqslant 4\frac{|x||y|}{8rsQ}|\mathrm{d}r||\mathrm{d}s| = \frac{|x||y|}{2rsQ}|\mathrm{d}r||\mathrm{d}s|.$$

We now need to add more functions with the good concavity for other K. The reader is encouraged to compare this with the dyadic case above. Let

$$B_5 = \frac{\langle x,x \rangle}{2r - \frac{1}{s(K(r,s)+1)}} + \frac{\langle y,y \rangle}{s}.$$

Since $0 \leqslant r - \frac{1}{s(K(r,s)+1)} \leqslant r$, we know that $0 \leqslant B_5 \leqslant \frac{|x|^2}{r} + \frac{|y|^2}{s}$. Now the Hessian estimate becomes

$$(\mathrm{d}^2 B_5 \mathrm{d}V, \mathrm{d}V) \geqslant \frac{\langle x,x \rangle}{4r^2}\frac{1}{s(K+1)^2}\frac{1}{8Q}|\mathrm{d}r||\mathrm{d}s| \geqslant \frac{|x|^2}{128Qsr^2}|\mathrm{d}r||\mathrm{d}s|.$$

B_5 is convex, and when $2|x|K \geqslant |y|r$, then

$$(\mathrm{d}^2 B_5 \mathrm{d}V, \mathrm{d}V) \geqslant \frac{|x||y|}{256KQsr}|\mathrm{d}r||\mathrm{d}s| \geqslant \frac{|x||y|}{256Qsr}|\mathrm{d}r||\mathrm{d}s|.$$

With

$$B_6 = \frac{\langle x,x \rangle}{r} + \frac{\langle y,y \rangle}{2s - \frac{1}{r(K(r,s)+1)}},$$

we have $0 \leqslant B_6 \leqslant \frac{|x|^2}{r} + \frac{|y|^2}{s}$ convex and when $2|y|K \geqslant |x|s$, then

$$(\mathrm{d}^2 B_6 \mathrm{d}V, \mathrm{d}V) \geqslant \frac{|x||y|}{256Qsr}|\mathrm{d}r||\mathrm{d}s|.$$

Together, for $B_7 = B_4 + B_5 + B_6$ we have

$$(\mathrm{d}^2 B_7 \mathrm{d}V, \mathrm{d}V) \geqslant \frac{|x||y|}{Qsr}|\mathrm{d}r||\mathrm{d}s|.$$

Through similar considerations as above, we have discrete one-leg convexity,

$$B_7(V) - B_7(V_0) - \mathrm{d}B_7(V_0)(V - V_0) \geqslant \frac{|x_0||y_0|}{Qs_0 r_0}|r - r_0||s - s_0|.$$

Letting for appropriate fixed c_i

$$B = c_1 B_1 + c_2 B_2 + c_3 B_3 + c_7 B_7, \tag{6.4}$$

we obtain $0 \leqslant B \leqslant \frac{|x|^2}{r} + \frac{|y|^2}{s}$ and $d^2 B \geqslant \frac{2}{Q}|dx||dy|$ in the regions where $B \in C^2$. Indeed,

$$(d^2 B_1 dV, dV) \geqslant \frac{4}{Q}|dx||dy| + \frac{4|x||y|}{Qrs}|dr||ds| - \frac{4|y|}{Qs}|dx||ds| - \frac{4|x|}{Qr}|dy||dr|,$$

$$(d^2 B_2 dV, dV) \geqslant \frac{\sqrt{3}|x|}{2Qr}|dy||dr| - \frac{\sqrt{3}|y|}{2Qrs}|dr||ds|,$$

$$(d^2 B_3 dV, dV) \geqslant \frac{\sqrt{3}|y|}{2Qs}|dx||ds| - \frac{\sqrt{3}|x|}{2Qrs}|dr||ds|,$$

$$(d^2 B_7 dV, dV) \geqslant \frac{|x||y|}{256Qrs}|dr||ds|,$$

where the last inequality holds in the regions where the function $B_4 \in C^2$. The weighted sum of these inequalities according to (6.4) yields the desired inequality on convexity:

$$B_1(V) - B_1(V_0) - dB_1(V_0)(V - V_0)$$
$$\geqslant \frac{rs}{Q}\left|\frac{x}{r} - \frac{x_0}{r_0}\right|\left|\frac{y}{s} - \frac{y_0}{s_0}\right|$$
$$\geqslant \frac{rs}{Q}\left|\left\langle \frac{x}{r} - \frac{x_0}{r_0}, \frac{y}{s} - \frac{y_0}{s_0}\right\rangle\right|,$$

$$B_2(V) - B_2(V_0) - dB_2(V_0)(V - V_0)$$
$$\geqslant \frac{s}{Q}\frac{|x_0|}{r_0}|r - r_0|\left|\frac{y}{s} - \frac{y_0}{s_0}\right|$$
$$\geqslant \frac{s}{Q}\left|\left\langle \frac{x_0}{r_0}, \frac{y}{s} - \frac{y_0}{s_0}\right\rangle\right||r - r_0|,$$

$$B_3(V) - B_3(V_0) - dB_3(V_0)(V - V_0)$$
$$\geqslant \frac{r}{Q}\frac{|y_0|}{s_0}|s - s_0|\left|\frac{x}{r} - \frac{x_0}{r_0}\right|$$
$$\geqslant \frac{r}{Q}\left|\left\langle \frac{x}{r} - \frac{x_0}{r_0}, \frac{y_0}{s_0}\right\rangle\right||s - s_0|,$$

$$B_7(V) - B_7(V_0) - dB_7(V_0)(V - V_0)$$
$$\geqslant \frac{1}{Q}|x_0||y_0||r - r_0||s - s_0|$$
$$\geqslant \frac{1}{Q}|\langle x_0, y_0\rangle||r - r_0||s - s_0|.$$

Notice that the last inequalities also remain true when we replace x by Θx and x_0 by Θx_0, where the rotation Θ is chosen so that $\Theta(x - x_0)$ and $y - y_0$ have the same direction and thus we may assume that $\langle x - x_0, y - y_0\rangle = |x - x_0||y - y_0|$.

Summing the above inequalities gives

$$Q(B(V) - B(V_0) - dB(V_0)(V - V_0))$$

$$\geq \left\langle \left(\frac{x}{r} - \frac{x_0}{r_0}\right)r, \left(\frac{y}{s} - \frac{y_0}{s_0}\right)s + \frac{y_0}{s_0}(s - s_0) \right\rangle$$

$$+ \left\langle \frac{x_0}{r_0}(r - r_0), \left(\frac{y}{s} - \frac{y_0}{s_0}\right)s + \frac{y_0}{s_0}(s - s_0) \right\rangle$$

$$= \left\langle \left(\frac{x}{r} - \frac{x_0}{r_0}\right)r, y - y_0 \right\rangle + \left\langle \frac{x_0}{r_0}(r - r_0), y - y_0 \right\rangle$$

$$= \langle x - x_0, y - y_0 \rangle = |x - x_0||y - y_0|$$

and we have proved the one-leg convexity. It remains to bound the second derivatives in x and y. Let ε be the cutoff of the weights so that $\varepsilon \leq r, s \leq \varepsilon^{-1}$. We calculate

$$\left\langle \partial_x^2 \frac{\langle x, x \rangle}{r} dx, dx \right\rangle = \frac{2\langle dx, dx \rangle}{r} \leq \varepsilon^{-1}\langle dx, dx \rangle,$$

$$\left\langle \partial_x^2 \frac{\langle x, x \rangle}{r + M(r, s)} dx, dx \right\rangle = \frac{2\langle dx, dx \rangle}{r + M(r, s)} \leq \frac{2\langle dx, dx \rangle}{r} \leq \varepsilon^{-1}\langle dx, dx \rangle,$$

$$\left\langle \partial_x^2 \frac{\langle x, x \rangle s - 2|x||y|K + \langle y, y \rangle r}{rs - K^2} dx, dx \right\rangle$$

$$= \frac{2\langle dx, dx \rangle s}{rs - K^2} - \frac{2|y|K}{rs - K^2}\left\langle \frac{\langle dx, dx \rangle}{|x|} - \frac{\langle x, dx \rangle^2}{|x|^3} \right\rangle \leq \varepsilon^{-1}\langle dx, dx \rangle,$$

where the last implied constant uses the lower bound for

$$rs - K^2 > rs\left(1 - \frac{1}{Q}\right) \geq 1 - \frac{1}{Q}.$$

We used the fact that $\langle x, dx \rangle^2 \leq \langle x, x \rangle \langle dx, dx \rangle \Rightarrow \frac{\langle dx, dx \rangle}{|x|} - \frac{\langle x, dx \rangle^2}{|x|^3} \geq 0$. These imply that for $V \in \mathcal{D}_{Q, \varepsilon}$

$$\langle \partial_x^2 B(V) dx, dx \rangle \leq \varepsilon^{-1}|dx|^2. \tag{6.5}$$

This concludes the proof of Lemma 2. $\qquad\square$

Convexities of the form $d^2 B(V) \geq 2|dx||dy|$ can be self-improved using the following interesting lemma.

Lemma 3 (Ellipse lemma, DRAGICEVIC–TREIL–VOLBERG [10]). *Let* $\mathbb{H}$ *be a Hilbert space with* A, B *two positive definite operators on* $\mathbb{H}$. *Let* T *be a self-adjoint operator on* $\mathbb{H}$ *such that*

$$\langle Th, h \rangle \geq 2\langle Ah, h \rangle^{1/2}\langle Bh, h \rangle^{1/2}$$

for all $h \in \mathbb{H}$. *Then there exists* $\tau > 0$ *satisfying*

$$\langle Th, h \rangle \geq \tau\langle Ah, h \rangle + \tau^{-1}\langle Bh, h \rangle$$

for all $h \in \mathbb{H}$.

For our specific Bellman function, we will need a quantitative version.

Lemma 4 (Quantitative ellipse lemma for B). *Let $V \in \mathcal{D}_Q^\varepsilon$. Assume moreover that B is $\mathcal{C}^2$ at V. Then there exists $\tau(V) > 0$ such that*

$$Q\mathrm{d}_V^2 B(V) \geq \tau(V)|\mathrm{d}x|^2 + (\tau(V))^{-1}|\mathrm{d}y|^2.$$

Moreover, we have the bound

$$Q^{-1}\varepsilon \leq \tau(V) \leq Q\varepsilon^{-1}.$$

Proof of Lemma 4. (Quantitative ellipse lemma for B.) Let $V \in \mathcal{D}_Q^\varepsilon$. We have already seen in Lemma 2 that

$$\mathrm{d}_V^2 B(V) \geq \frac{2}{Q}|\mathrm{d}x||\mathrm{d}y|.$$

The ellipse lemma [10] implies the existence of $\tau(V)$ such that for all vectors $\mathrm{d}x$ and $\mathrm{d}y$ we have

$$Q\mathrm{d}_V^2 B(V) \geq \tau(V)|\mathrm{d}x|^2 + (\tau(V))^{-1}|\mathrm{d}y|^2.$$

We can estimate $\tau(V)$ by testing the Hessian on any $\mathrm{d}V$ of the form $\mathrm{d}V = (\mathrm{d}x, 0, 0, 0)$,

$$\tau(V)|\mathrm{d}x|^2 \leq Q\langle \mathrm{d}_V^2 B(V)\mathrm{d}V, \mathrm{d}V\rangle = Q\langle \partial_x^2 B(V)\mathrm{d}x, \mathrm{d}x\rangle \leq Q\varepsilon^{-1}|\mathrm{d}x|^2,$$

where the last inequality follows from (6.5). Hence $\tau(V) \leq Q\varepsilon^{-1}$ as claimed. The same bound holds for $(\tau(V)^{-1})$ by testing against $\mathrm{d}V = (0, \mathrm{d}y, 0, 0)$. Finally, we have proved that for all $V \in \mathcal{D}_Q^\varepsilon$,

$$Q^{-1}\varepsilon \leq \tau(V) \leq Q\varepsilon^{-1}. \qquad \square$$

We now address the lack of smoothness of B. All functions aside from H_4 that appear are at least in $\mathcal{C}^2$. We apply a standard mollifying procedure via convolution with φ_ℓ directly on $H_4(x, y, r, s, K)$, now only taking *real* variables with x, y positive, $1 < rs < Q$, and $0 < K < 1$. Here φ denotes a standard mollifying kernel in the five real variables $(x, y, r, s, K) \in \mathbb{R}^5$ with support in the corresponding unit ball, whereas $\varphi_\ell(\cdot) := \ell^{-5}\varphi(\cdot/\ell)$ denotes its scaled version with support of size ℓ. By slightly changing the constructions, the upper and lower estimates on the product rs can be modified at the cost of a multiplicative constant in the final estimate of the Bellman function. Also take into account that the weights are cut, and therefore bounded above and below. Further, we will assume that the positive variables x and y are bounded below. These considerations give us enough room to smooth the function H_4. It is important that H_4 is at least in $\mathcal{C}^1$ and its second-order partial derivatives exist almost everywhere. So we have $\mathrm{d}^2(H_4 * \varphi_\ell) = (\mathrm{d}^2 H_4) * \varphi_\ell$. Last, we are observing that as long as the norms of vectors $|x|$ and $|y|$ are bounded away from 0, our function $H_4 * \varphi_\ell$ mollified in $\mathbb{R}^5$ remains smooth when taking vector variables (observe that the final Bellman function only depends upon $|x|$ and $|y|$). It is important that the smoothing happens before the function is composed with K; we therefore preserve fine convexity properties, in particular also the much needed one-leg convexity. Size estimates change slightly, but are recovered when the mollifying parameter goes to 0. These details either are standard and appear in numerous articles on Bellman functions or are an easy consequence of reading the construction of the Bellman function above.

Lemma 5 (Regularized Bellman function and its properties). *Let $\varepsilon > 0$ be given. Let $0 < \ell \leqslant \varepsilon/2$. There exists a function $B_\ell(x, y, r, s)$ defined with domain*

$$\mathcal{D}_Q^{\varepsilon,\ell} := \{V \in \mathcal{D}_Q^\varepsilon;\ |x| \geqslant \ell, |y| \geqslant \ell\} \subset \mathcal{D}_Q^\varepsilon$$

such that for all $V_0, V \in \mathcal{D}_Q^{\varepsilon,\ell}$, we have

$$B_\ell \leqslant (1 + \ell)\left(\frac{|x|^2}{r} + \frac{|y|^2}{s}\right),$$

$$\mathrm{d}_V^2 B_\ell(V) \geqslant \frac{2}{Q}|\mathrm{d}x||\mathrm{d}y|, \tag{6.6}$$

$$B_\ell(V) - B_\ell(V_0) - \mathrm{d}_V B_\ell(V_0)(V - V_0) \geqslant \frac{1}{Q}|\Delta x||\Delta y| = \frac{1}{Q}|x - x_0||y - y_0|, \tag{6.7}$$

and moreover the quantitative ellipse lemma now holds in the form

$$Q\mathrm{d}_V^2 B_\ell(V) \geqslant \tau_\ell(V)|\mathrm{d}x|^2 + \left(\tau_\ell(V)\right)^{-1}|\mathrm{d}y|^2,$$

where $\tau_\ell := \tau_\ell(V)$ is a continuous function of its arguments and

$$Q^{-1}\varepsilon \leqslant \tau_\ell(V) \leqslant Q\varepsilon^{-1}.$$

7 Dissipation estimates

Let $V := (X, Z, u, w)$ be a càdlàg adapted martingale with values in $\mathcal{D}_Q^\varepsilon$. In order to bound away from the $\mathbb{H}$-valued martingale $X := (X^1, X^2, \ldots)$, it is classical to introduce the $\mathbb{R} \times \mathbb{H}$-valued martingales $X^a := (a, X^1, X^2, \ldots)$, where $a > 0$. It follows that $\|X^a\|^2 = \|X\|^2 + a^2$ and $\|X^a\| \geqslant a$, and the same construction holds for Z. We note $V^a := (X^a, Z^a, u, w)$. Given a smoothing parameter $\ell > 0$, take $a \geqslant \ell$. Then it follows that

$$V \in \mathcal{D}_Q^\varepsilon \quad \Rightarrow \quad V^a \in \mathcal{D}_Q^{\varepsilon,\ell}.$$

The main result of this section is the following dissipation estimate.

Proposition 3 (Dissipation estimate). *Let $\varepsilon > 0$, $\ell > 0$ as defined above. Let V be a càdlàg adapted martingale with $V \in \mathcal{D}_Q^\varepsilon$. Let $F_t := \mathbb{E}(|X_\infty|^2 w_\infty^\varepsilon | \mathcal{F}_t)$ and $G_t := \mathbb{E}(|Z_\infty|^2 u_\infty^\varepsilon | \mathcal{F}_t)$. Let finally $a \geq \ell$. We have*

$$Q(1+\ell)\big(\mathbb{E}F_t + \mathbb{E}G_t + 2a^2 \varepsilon^{-1}\big)$$

$$\geq \frac{1}{2}\mathbb{E}\int_0^t \tau_\ell(V_{s-})\mathrm{d}[X,X]_s^c + \big(\tau_\ell(V_{s-})\big)^{-1}\mathrm{d}[Z,Z]_s^c + \mathbb{E}\sum_{0 < s \leq t}|\Delta X_s||\Delta Z_s|.$$

We need the following preliminary lemma.

Lemma 6 (Comparison of quadratic forms in stochastic integrals). *Let $\mathcal{Q}$ denote the set of quadratic forms from $\mathbb{R}^m \times \mathbb{R}^m \to \mathbb{R}$. Let $A := (A_{\alpha\beta})_{1 \leq \alpha,\beta \leq m}$ and $B := (B_{\alpha\beta})_{1 \leq \alpha,\beta \leq m}$ be two $\mathcal{Q}$-valued càdlàg processes. Assume for all $t \geq 0$ and a. s. that $A(t) \geq B(t)$ (resp. $A(t) \geq |B(t)|$), in the sense*

$$\forall \mathrm{d}V \in \mathbb{R}^m, \quad \langle A\mathrm{d}V, \mathrm{d}V \rangle \geq \langle B\mathrm{d}V, \mathrm{d}V \rangle \quad (\text{resp. } \langle A\mathrm{d}V, \mathrm{d}V \rangle \geq |\langle B\mathrm{d}V, \mathrm{d}V \rangle|).$$

Abbreviating $A_{s-} : \mathrm{d}[V,V]_s = \sum_{\alpha,\beta}(A_{\alpha\beta})_{s-}\mathrm{d}[V_\alpha, V_\beta]_s$, for all $t \geq 0$ we have

$$\mathbb{E}\int_0^t A_{s-} : \mathrm{d}[V,V]_s \geq \mathbb{E}\int_0^t B_{s-} : \mathrm{d}[V,V]_s$$

$$\left(\text{resp. } \mathbb{E}\int_0^t A_{s-} : \mathrm{d}[V,V]_s \geq \mathbb{E}\int_0^t |B_{s-} : \mathrm{d}[V,V]_s|\right).$$

Proof of Lemma 6. (Comparison of quadratic forms in stochastic integrals.) With the hypotheses above, let us consider the case $A(t) \geq B(t)$, the case $A(t) \geq |B(t)|$ being treated in the same manner. Given $t \geq 0$, assume that

$$\int_0^t A_{s-} : \mathrm{d}[V,V]_s = \sum_{\alpha,\beta}\int_0^t (A_{\alpha\beta})_{s-}\mathrm{d}[V_\alpha, V_\beta]_s < \infty$$

for otherwise the claim is proved. Given the process V, let $\sigma_n := (0 \leq T_0^n \leq T_1^n \leq \cdots \leq T_i^n \leq \cdots \leq T_{k_n}^n \leq t)$ denote a random partition of stopping times tending to the identity as n tends to infinity. Given α and β, $A_{\alpha\beta}$ is an $\mathbb{R}$-valued càdlàg process. It follows (see, e. g., [19]) that the stochastic integral

$$\int_0^t A_{\alpha\beta}(s-)\mathrm{d}[V_\alpha, V_\beta]_s \tag{7.1}$$

is the limit in *ucp* as n tends to infinity of sums

$$S_{\alpha\beta}^{A} := \sum_{i=0}^{k_n-1} A_{\alpha\beta}(T_i^n)(V_\alpha^{T_{i+1}^n} - V_\alpha^{T_i^n})(V_\beta^{T_{i+1}^n} - V_\beta^{T_i^n})$$

involving the stopping times defined above. Since $A \geqslant B$, summing w. r. t. α, β yields, for any $s \in [0, t]$,

$$\left(\sum_{\alpha,\beta} S_{\alpha\beta}^{A}\right)(s)$$

$$:= \sum_{\alpha,\beta} \sum_{i=0}^{k_n-1} A_{\alpha\beta}(T_i^n)(V_{\alpha,s}^{T_{i+1}^n} - V_{\alpha,s}^{T_i^n})(V_{\beta,s}^{T_{i+1}^n} - V_{\beta,s}^{T_i^n})$$

$$= \sum_{i=0}^{k_n-1} \sum_{\alpha,\beta} A_{\alpha\beta}(T_i^n)(V_{\alpha,s}^{T_{i+1}^n} - V_{\alpha,s}^{T_i^n})(V_{\beta,s}^{T_{i+1}^n} - V_{\beta,s}^{T_i^n})$$

$$\geqslant \sum_{i=0}^{k_n-1} \sum_{\alpha,\beta} B_{\alpha\beta}(T_i^n)(V_{\alpha,s}^{T_{i+1}^n} - V_{\alpha,s}^{T_i^n})(V_{\beta,s}^{T_{i+1}^n} - V_{\beta,s}^{T_i^n})$$

$$\geqslant \left(\sum_{\alpha,\beta} S_{\alpha\beta}^{B}\right)(s),$$

with an obvious definition for $S_{\alpha\beta}^{B}$. Passing to the limit in the sums $\sum_{\alpha,\beta}$ gives the result. $\qquad\square$

Proof of Proposition 3. (Dissipation estimates.)

Step 1

We first consider the finite-dimensional case. Let V be a càdlàg adapted martingale with $V \in \mathcal{D}_Q^\varepsilon$. Then $V^a \in \mathcal{D}_Q^{\varepsilon,\ell}$. We denote by $X^{a,m}$ the projection of $X^a \in \mathbb{R}\times\mathbb{H}$ onto $\mathbb{R}\times\mathbb{R}^m$, and we introduce accordingly $Z^{a,m}$ and $V^{a,m}$. Notice that $[X^a, X^a] = a^2 + [X, X]$ and similarly $[X^{a,m}, X^{a,m}] = a^2 + [X^m, X^m]$. Since $V^{a,m} \in \mathcal{D}_Q^{\varepsilon,\ell}$, where B_ℓ is C^2, we can apply Itô's formula and obtain, for all $t > 0$, almost sure paths,

$$B_\ell(V_t^{a,m}) = B_\ell(V_0^{a,m})$$

$$+ \int_{0+}^{t} \mathrm{d}_V B(V_{s-}^{a,m}) \mathrm{d}V_s^m + \frac{1}{2} \int_{0+}^{t} \mathrm{d}_V^2 B_\ell(V_{s-}^{a,m}) : \mathrm{d}[V^m, V^m]_s^c$$

$$+ \sum_{0<s\leqslant t} \{B_\ell(V_s^{a,m}) - B_\ell(V_{s-}^{a,m}) - \mathrm{d}_V B_\ell(V_{s-}^{a,m})\Delta V_s^m\}.$$

Thanks to Lemma 4 and Lemma 6, the concavity properties (6.6) of B_ℓ imply for the continuous part

$$\frac{1}{2} \int_{0+}^{t} \mathrm{d}_V^2 B_\ell(V_{s-}^{a,m}) : \mathrm{d}[V^m, V^m]_s^c \geqslant \frac{1}{2Q} \int_{0+}^{t} \tau_\ell(V_{s-}^{a,m})\mathrm{d}[X^m, X^m]_s^c + (\tau_\ell(V_{s-}^{a,m}))^{-1}\mathrm{d}[Z^m, Z^m]_s^c.$$

Also, the concavity properties (6.7) of B_ℓ imply for the jump part

$$B_\ell(V_s^{a,m}) - B_\ell(V_{s-}^{a,m}) - \mathrm{d}_V B_\ell(V_{s-}^{a,m})\Delta V_s^m \geq \frac{1}{Q}|\Delta X_s^m||\Delta Z_s^m|.$$

Plugging the continuous and jump dissipation estimates into Itô's formula yields for all times, almost sure paths,

$$B_\ell(V_t^{a,m})$$

$$\geq B_\ell(V_0^{a,m}) + \int_{0+}^{t} \mathrm{d}_V B_\ell(V_{s-}^{a,m})\mathrm{d}V_s^m$$

$$+ \frac{1}{2Q}\int_0^t \tau_\ell(V_{s-}^{a,m})\mathrm{d}[X^m,X^m]_s^c + (\tau_\ell(V_{s-}^m))^{-1}\mathrm{d}[Z^m,Z^m]_s^c + \frac{1}{Q}\sum_{0<s\leq t}|\Delta X_s^m||\Delta Z_s^m|.$$

Step 2

For technical reasons in the proof, we want to work with bounded martingales that we obtain through a usual stopping procedure. Recall that V is a càdlàg adapted martingale with $V \in \mathcal{D}_Q^\varepsilon$ and $V^a \in \mathcal{D}_Q^{\varepsilon,\ell}$. For all $M \in \mathbb{N}$, define the stopping time T_M as $T_M := \inf\{t > 0;$ $|V^a|_t^2 + [V^a,V^a]_t > M^2\}$, so that T_M is a stopping time that tends to infinity as M goes to infinity. It follows that V^{a,T_M} is a local martingale and that V^{a,T_M-} and $[V^a,V^a]^{T_M-}$ are bounded semimartingales. Let $m \in \mathbb{N}^*$ and let $V^{a,m}$ be the projection of V^a onto $\mathbb{R}^m \subset \mathbb{H}$. For each M, there exists a sequence $\{T_{M,k}\}_{k\geq 1}$ of stopping times such that $T_{M,k} \uparrow T_M$ as $k \uparrow \infty$ and such that $(V^{a,m})^{T_{M,k}}$ is a martingale. Since $|V^{a,m}| \leq |V^a|$, it follows that $(V^{a,m})^{T_{M,k}-}$ is a bounded semimartingale, to which we can apply the dissipation estimate of Step 1 above and obtain

$$B_\ell(V_{t\wedge T_{M,k}-}^{a,m})$$

$$\geq B_\ell(V_0^{a,m}) + \int_{0+}^{t\wedge T_{M,k}-} \mathrm{d}_V B_\ell(V_{s-}^{a,m})\mathrm{d}V_s^m$$

$$+ \frac{1}{2Q}\int_0^{t\wedge T_{M,k}-} \tau_\ell(V_{s-}^{a,m})\mathrm{d}[X^m,X^m]_s^c + (\tau_\ell(V_{s-}^{a,m}))^{-1}\mathrm{d}[Z^m,Z^m]_s^c + \frac{1}{Q}\sum_{0<s<t\wedge T_{M,k}}|\Delta X_s^m||\Delta Z_s^m|$$

$$= B_\ell(V_0^{a,m}) + \int_{0+}^{t\wedge T_{M,k}} \mathrm{d}_V B_\ell(V_{s-}^{a,m})\mathrm{d}V_s^m$$

$$+ \frac{1}{2Q}\int_0^{t\wedge T_{M,k}-} \tau_\ell(V_{s-}^{a,m})\mathrm{d}[X^m,X^m]_s^c + (\tau_\ell(V_{s-}^{a,m}))^{-1}\mathrm{d}[Z^m,Z^m]_s^c$$

$$+ \frac{1}{Q}\sum_{0<s<t\wedge T_{M,k}}|\Delta X_s^m||\Delta Z_s^m| - \mathrm{d}_V B_\ell(V_{t\wedge T_{M,k}-}^{a,m})\Delta V_{t\wedge T_{M,k}}^m.$$

Taking expectation and then letting $k \to \infty$, the dominated convergence theorem yields

$$\mathbb{E}B_\ell(V^{a,m}_{t\wedge T_M-}) \geq \mathbb{E}B_\ell(V^{a,m}_0)$$

$$+ \frac{1}{2Q}\mathbb{E}\int_0^{t\wedge T_M} \tau_\ell(V^{a,m}_{s-})\mathrm{d}[X^m,X^m]^c_s + (\tau_\ell(V^{a,m}_{s-}))^{-1}\mathrm{d}[Z^m,Z^m]^c_s$$

$$+ \frac{1}{Q}\mathbb{E}\sum_{0<s<t\wedge T_M}|\Delta X^m_s||\Delta Z^m_s| - \mathbb{E}\{\mathrm{d}_V B_\ell(V^{a,m}_{t\wedge T_M-})\Delta V^m_{t\wedge T_M}\}. \tag{7.2}$$

Observe that we used size properties of B_ℓ, the definition of the stopping time $T_{M,k}$, the estimate of the τ_ℓ provided by Lemma 5, and the size control of the weights.

Step 3

Now, we wish to return to the infinite-dimensional case, which does not add too much hassle. First recall that

$$0 \leq B_\ell(V) \leq (1+\ell)(X^2/u + Y^2/w).$$

Let

$$F_t := \mathbb{E}(|X_\infty|^2 w_\infty|\mathcal{F}_t), \quad G_t := \mathbb{E}(|Z_\infty|^2 u_\infty|\mathcal{F}_t),$$
$$F^a_t := \mathbb{E}(|X^a_\infty|^2 w_\infty|\mathcal{F}_t), \quad G^a_t := \mathbb{E}(|Z^a_\infty|^2 u_\infty|\mathcal{F}_t)$$

and notice that

$$F^a_t = \mathbb{E}((|X_\infty|^2 + a^2)w_\infty|\mathcal{F}_t) \leq F_t + \mathbb{E}(a^2 w^\varepsilon_\infty|\mathcal{F}_t) \leq F_t + a^2\varepsilon^{-1}.$$

It follows, thanks to the Jensen inequality, that

$$B_\ell(V^a_t) \leq C_0(1+\ell)(F^a_t + G^a_t) \leq (1+\ell)(F_t + G_t + 2a^2\varepsilon^{-1}).$$

A similar inequality holds for $V^{a,m}$ and in particular

$$B_\ell(V^{a,m}_{t\wedge T_M-}) \leq (1+\ell)(F_{t\wedge T_M-} + G_{t\wedge T_M-} + 2a^2\varepsilon^{-1}).$$

Hence, the dominated convergence theorem implies that $\mathbb{E}B_\ell(V^{a,m}_{t\wedge T_M-})$ converges when m goes to infinity towards $\mathbb{E}B_\ell(V^a_{t\wedge T_M-})$.

Let us consider the first term in the last integral of Step 2 (inequality (7.2)). We write

$$\mathbb{E}\int_0^{t\wedge T_M} \tau_\ell(V^{a,m}_{s-})\mathrm{d}[X^m,X^m]^c_s$$

$$= \mathbb{E} \int_0^{t \wedge T_M} \tau_\ell(V_{s-}^a) \mathrm{d}[X,X]_s^c$$

$$+ \mathbb{E} \int_0^{t \wedge T_M} (\tau_\ell(V_{s-}^{a,m}) - \tau_\ell(V_{s-}^a)) \mathrm{d}[X,X]_s^c$$

$$+ \mathbb{E} \int_0^{t \wedge T_M} \tau_\ell(V_{s-}^{a,m}) \mathrm{d}([X^m,X^m]^c - [X,X]^c)_s.$$

The uniform boundedness and continuity of τ_ℓ, the square integrability of X, and the dominated convergence theorem imply that the second term of the right-hand side converges to zero. The last term can be bounded above using the estimates for τ_ℓ. We have

$$\left| \mathbb{E} \int_0^{t \wedge T_M} \tau_\ell(V_{s-}^{a,m}) \mathrm{d}([X^m,X^m]^c - [X,X]^c)_s \right|$$

$$\leq \frac{Q}{\varepsilon} \mathbb{E} \int_0^{t \wedge T_M} |\mathrm{d}([X^m,X^m]^c - [X,X]^c)_s|$$

$$\leq \frac{Q}{\varepsilon} \mathbb{E} \int_0^{t \wedge T_M} \mathrm{d}([X,X]^c - [X^m,X^m]^c)_s$$

$$\leq \frac{Q}{\varepsilon} (\mathbb{E}[X,X]_{t \wedge T_M}^c - \mathbb{E}[X^m,X^m]_{t \wedge T_M}^c),$$

where we used the fact that for m fixed, $[X,X]^c - [X^m,X^m]^c$ is a non-negative non-decreasing process. The last expression in the last line tends to zero when $m \to \infty$ by the monotone convergence theorem. We prove in a similar manner the convergence

$$\mathbb{E} \sum_{0 < s < t \wedge T_M} |\Delta X_s^m||\Delta Z_s^m| \xrightarrow[m \to \infty]{} \mathbb{E} \sum_{0 < s < t \wedge T_M} |\Delta X_s||\Delta Z_s|.$$

Finally, since $|V_{t \wedge T_M-}^{a,m}| \leq |V_{t \wedge T_M-}^a|$ for all m, $\mathrm{d}_V B_\ell$ is continuous and bounded on compact sets, $|\Delta V_{t \wedge T_M}^m|^2 \leq |\Delta V_{t \wedge T_M}|^2 \leq [V,V]_t$, and $\mathbb{E}[V,V]_t = \mathbb{E}|V_t|^2 < \infty$, the dominated convergence theorem ensures that

$$-\mathbb{E}\{\mathrm{d}_V B_\ell(V_{t \wedge T_M-}^{a,m})\Delta V_{t \wedge T_M}^m\} \to -\mathbb{E}\{\mathrm{d}_V B_\ell(V_{t \wedge T_M-}^a)\Delta V_{t \wedge T_M}\}.$$

Collecting all terms,

$$\mathbb{E}B_\ell(V_{t \wedge T_M-}^a)$$

$$\geq \frac{1}{2Q} \mathbb{E} \int_0^{t \wedge T_M} \tau_\ell(V_{s-}^a)\mathrm{d}[X,X]_s^c + (\tau_\ell(V_{s-}^a))^{-1}\mathrm{d}[Z,Z]_s^c$$

$$+ \frac{1}{Q} \mathbb{E} \sum_{0 < s < t \wedge T_M} |\Delta X_s||\Delta Z_s| - \mathbb{E}\{d_V B_\ell(V^a_{t \wedge T_M -})\Delta V_{t \wedge T_M}\}.$$

Step 4

Now we add the contribution of the possible jumps occurring at T_M. We have seen in Step 1 the dissipation estimate along one jump

$$B_\ell(V^a_{t \wedge T_M}) - B_\ell(V^a_{t \wedge T_M -}) - d_V B_\ell(V^a_{t \wedge T_M -})\Delta V_{t \wedge T_M} \geq \frac{1}{Q}|\Delta X_{t \wedge T_M}||\Delta Z_{t \wedge T_M}|.$$

Taking expectation and adding the contribution of Step 3 yields

$$\mathbb{E}B_\ell(V^a_{t \wedge T_M})$$
$$\geq \frac{1}{2Q} \mathbb{E} \int_0^{t \wedge T_M} \tau_\ell(V^a_{s-})d[X,X]^c_s + (\tau_\ell(V^a_{s-}))^{-1}d[Z,Z]^c_s + \frac{1}{Q}\mathbb{E} \sum_{0 < s \leq t \wedge T_M} |\Delta X_s||\Delta Z_s|. \tag{7.3}$$

Step 5

We will pass to the limit $M \to \infty$. Recall again that $0 \leq B_\ell(V) \leq (1 + \ell)(X^2/u + Y^2/w)$. Using Doob's inequality for square integrable martingales, we have for all M

$$\mathbb{E}B_\ell(V^a_{t \wedge T_M}) \leq \varepsilon^{-1}(1 + \ell)(\mathbb{E}X^2 + \mathbb{E}Y^2) < \infty.$$

So, by the dominated convergence theorem, $\mathbb{E}B_\ell(V^a_{t \wedge T_M}) \to \mathbb{E}B_\ell(V^a_t)$ as $M \to \infty$. The monotone convergence theorem for the integral in the right-hand side of inequality (7.3) therefore yields in the limit $M \to \infty$

$$(1 + \ell)(\mathbb{E}F_t + \mathbb{E}G_t + 2a^2\varepsilon^{-1})$$
$$\geq \mathbb{E}B_\ell(V^a_t)$$
$$\geq \frac{1}{2Q}\mathbb{E} \int_0^t \tau_\ell(V^a_{s-})d[X,X]^c_s + (\tau_\ell(V^a_{s-}))^{-1}d[Z,Z]^c_s + \frac{1}{Q}\mathbb{E} \sum_{0 < s \leq t} |\Delta X_s||\Delta Z_s|.$$

This concludes the proof of Proposition 3. $\qquad\square$

8 Truncation of the weights

Due to several technicalities in the proof, we have used weights bounded from above and away from 0. In order to pass to the general case, we cut a possibly unbounded weight above and below and show that this operation does not increase the characteristic of the weight. This is convenient and has been used in several places; here we extend

[20] to the martingale setting. This line of reasoning is particularly nice, since it does not increase the characteristic at all, not even by a constant. We need the following preliminary lemmas.

Lemma 7 (Truncation from above). *For $a > 0$ let $M = \{w \leqslant a\}$ and $H = \{w > a\}$. Now take $w_{\bar{a}} = w\chi_M + a\chi_H$. Then $Q_2^{\mathcal{F}}[w_{\bar{a}}] \leqslant Q_2^{\mathcal{F}}[w]$.*

Proof of Lemma 7. Let τ be a stopping time and let us decompose

$$
\begin{aligned}
\mathbb{E}(w|\mathcal{F}_\tau) &\\
&= \mathbb{E}(w\chi_M|\mathcal{F}_\tau) + \mathbb{E}(w\chi_H|\mathcal{F}_\tau) \\
&= \mathbb{E}(\chi_M|\mathcal{F}_\tau)\mathbb{E}_M(w|\mathcal{F}_\tau) + \mathbb{E}(\chi_H|\mathcal{F}_\tau)\mathbb{E}_H(w|\mathcal{F}_\tau),
\end{aligned}
$$

where for example $\mathbb{E}_M(w|\mathcal{F}_\tau)$ means expectation is taken with respect to the measure $\chi_M d\mathbb{P}$. Write as usual $\mathbb{E}(\chi_M|\mathcal{F}_\tau) = (\chi_M)_\tau$. We have

$$
\begin{aligned}
&\mathbb{E}(w|\mathcal{F}_\tau)\mathbb{E}(w^{-1}|\mathcal{F}_\tau) - \mathbb{E}(w_{\bar{a}}|\mathcal{F}_\tau)\mathbb{E}(w_{\bar{a}}^{-1}|\mathcal{F}_\tau) \\
&\quad = \big((\chi_M)_\tau \mathbb{E}_L(w|\mathcal{F}_\tau) + (\chi_H)_\tau \mathbb{E}_H(w|\mathcal{F}_\tau)\big) \\
&\qquad \times \big((\chi_M)_\tau \mathbb{E}_L(w^{-1}|\mathcal{F}_\tau) + (\chi_H)_\tau \mathbb{E}_H(w^{-1}|\mathcal{F}_\tau)\big) \\
&\qquad - \big((\chi_M)_\tau \mathbb{E}_L(w|\mathcal{F}_\tau) + (\chi_H)_\tau a\big)\big((\chi_M)_\tau \mathbb{E}_L(w^{-1}|\mathcal{F}_\tau) + (\chi_H)_\tau a^{-1}\big) \\
&\quad = (\chi_M)_\tau (\chi_H)_\tau \big(\mathbb{E}_M(w|\mathcal{F}_\tau)\mathbb{E}_H(w^{-1}|\mathcal{F}_\tau) + \mathbb{E}_M(w^{-1}|\mathcal{F}_\tau)\mathbb{E}_H(w|\mathcal{F}_\tau) \\
&\qquad - \mathbb{E}_M(w|\mathcal{F}_\tau)a^{-1} - \mathbb{E}_M(w^{-1}|\mathcal{F}_\tau)a\big) \\
&\qquad + (\chi_H)_\tau^2\big(\mathbb{E}_H(w|\mathcal{F}_\tau)\mathbb{E}_H(w^{-1}|\mathcal{F}_\tau) - 1\big).
\end{aligned}
$$

The last term is positive thanks to the Jensen inequality. Let us observe that also

$$
\begin{aligned}
&\mathbb{E}_M(w|\mathcal{F}_\tau)\mathbb{E}_H(w^{-1}|\mathcal{F}_\tau) + \mathbb{E}_M(w^{-1}|\mathcal{F}_\tau)\mathbb{E}_H(w|\mathcal{F}_\tau) - \mathbb{E}_M(w|\mathcal{F}_\tau)a^{-1} - \mathbb{E}_M(w^{-1}|\mathcal{F}_\tau)a \\
&\quad = \mathbb{E}_M(w|\mathcal{F}_\tau)\mathbb{E}_H(w^{-1} - a^{-1}|\mathcal{F}_\tau) + \mathbb{E}_M(w^{-1}|\mathcal{F}_\tau)\mathbb{E}_H(w - a|\mathcal{F}_\tau) \\
&\quad = \mathbb{E}_H\left(\frac{w - a}{wa}\big(wa\mathbb{E}_M(w^{-1}|\mathcal{F}_\tau) - \mathbb{E}_M(w|\mathcal{F}_\tau)\big)\Big|\mathcal{F}_\tau\right) \\
&\quad \geqslant 0.
\end{aligned}
$$

Here the last inequality uses $\mathbb{E}_M(w^{-1}|\mathcal{F}_\tau) \geqslant a^{-1}$ and $\mathbb{E}_M(w|\mathcal{F}_\tau) \leqslant a$ and also $w - a \geqslant 0$ on H. This proves the lemma. $\qquad\square$

Lemma 8 (Two-sided truncation). *For $a > 0$ let $M = \{a^{-1} \leqslant w \leqslant a\}$, $L = \{w < a^{-1}\}$, and $H = \{w > a\}$. Then with $w_a = a^{-1}\chi_L + w\chi_M + a\chi_H$ we have $Q_2^{\mathcal{F}}[w_a] \leqslant Q_2^{\mathcal{F}}[w]$.*

Proof of Lemma 8. Let $w_{\bar{a}}$ be the weight obtained in the previous lemma. Apply now the previous lemma to $w_{\bar{a}}^{-1}$, truncating above by the same a. $\qquad\square$

9 Proof of the main results

Proof of Proposition 2. (Bilinear estimate.) Let $\lambda > 0$. Let Y be differentially subordinate to X. Then λY is differentially subordinate to λX. Let w be a weight in the A_2 class. Let w^ε be the ε-truncation of w. Use Proposition 3 with $V^{\varepsilon,\lambda} := (\lambda X, \lambda^{-1} Z, u^\varepsilon, w^\varepsilon)$ and $Q = Q_2^{\mathcal{F}}[w]$. Notice that since $Q_2^{\mathcal{F}}[w^\varepsilon] \leqslant Q_2^{\mathcal{F}}[w]$, $V^{\varepsilon,\lambda} \in \mathcal{D}_Q^{\varepsilon,\ell}$ using the differential subordination of λY w. r. t. λX, we have, for all $t > 0$,

$$Q_2^{\mathcal{F}}[w](1+\ell)(\mathbb{E}\lambda^2 F_t + \mathbb{E}\lambda^{-2} G_t + 2a^2\varepsilon^{-1})$$

$$\geqslant \frac{1}{2}\mathbb{E}\int_0^t \tau_\ell(V_{s-}^a)\,\mathrm{d}[\lambda X, \lambda X]_s^c + \left(\tau_\ell(V_{s-}^a)\right)^{-1}\mathrm{d}[\lambda^{-1}Z, \lambda^{-1}Z]_s^c + \mathbb{E}\sum_{0<s\leqslant t} |\lambda\Delta X_s||\lambda^{-1}\Delta Z_s|$$

$$\geqslant \frac{1}{2}\mathbb{E}\int_0^t \tau_\ell(V_{s-}^a)\,\mathrm{d}[\lambda Y, \lambda Y]_s^c + \left(\tau_\ell(V_{s-})\right)^{-1}\mathrm{d}[\lambda^{-1}Z, \lambda^{-1}Z]_s^c + \mathbb{E}\sum_{0<s\leqslant t} |\Delta Y_s||\Delta Z_s|.$$

Since for any $0 < \kappa < \infty$ and any $x \in \mathbb{H}, y \in \mathbb{H}$, we have $\kappa x^2 + \kappa^{-1} y^2 \geqslant 2|\langle x,y\rangle|$, it follows easily that

$$\frac{1}{2}\mathbb{E}\int_0^t \tau_\ell(V_{s-})\,\mathrm{d}[\lambda Y, \lambda Y]_s^c + \left(\tau_\ell(V_{s-})\right)^{-1}\mathrm{d}[\lambda^{-1}Z, \lambda^{-1}Z]_s^c + \mathbb{E}\sum_{0<s\leqslant t} |\Delta Y_s||\Delta Z_s|$$

$$\geqslant \mathbb{E}\int_0^t |\mathrm{d}[\lambda Y, \lambda^{-1}Z]_s^c| + \mathbb{E}\sum_{0<s\leqslant t} |\Delta Y_s||\Delta Z_s|$$

$$\geqslant \mathbb{E}\int_0^t |\mathrm{d}[Y, Z]_s^c| + \mathbb{E}\sum_{0<s\leqslant t} |\Delta Y_s||\Delta Z_s|$$

$$\geqslant \mathbb{E}\int_0^t |\mathrm{d}[Y, Z]_s|,$$

where all integrals and sums converge. Hence for all $\lambda > 0$

$$Q_2^{\mathcal{F}}[w](1+\ell)(\lambda^2\mathbb{E}F_t + \lambda^{-2}\mathbb{E}G_t + 2a^2\varepsilon^{-1}) \geqslant \mathbb{E}\int_0^t |\mathrm{d}[Y, Z]_s|.$$

We let now successively $\ell \to 0$. Then $a \to 0$. Choosing $\lambda^2 = (\mathbb{E}G_t)^{1/2}(\mathbb{E}F_t)^{-1/2}$, we can assume $\lambda > 0$ (otherwise the claim is trivial). We have

$$\mathbb{E}\int_0^t |\mathrm{d}[Y, Z]_s| \leqslant Q_2^{\mathcal{F}}[w](\mathbb{E}F_t)^{1/2}(\mathbb{E}G_t)^{1/2} \leqslant Q_2^{\mathcal{F}}[w]\|X\|_{2,w^\varepsilon}\|Z\|_{2,u^\varepsilon}.$$

The inequality above remains valid in the limit $t \to \infty$. Since the left-hand side does not depend on the truncation of the weight, it remains to observe that

$$\lim_{\varepsilon \to 0} \|X\|_{2,w^{\varepsilon}} = \|X\|_{2,w} \quad \text{and} \quad \lim_{\varepsilon \to 0} \|Z\|_{2,u^{\varepsilon}} = \|Z\|_{2,u}.$$

Indeed, since $X \in L^2(\Omega; d\mathbb{P}) \cap L^2(\Omega; d\mathbb{P}^w)$, we have, for all $0 < \varepsilon < 1$, a. s. $X_{\infty}^2 w_{\infty}^{\varepsilon} \leqslant X_{\infty}^2 + X_{\infty}^2 w_{\infty}$ and the limits above are a consequence of the dominated convergence theorem. The same reasoning applied to Z completes the proof of the bilinear embedding. $\square$

Proof of Theorem 9. (Differential subordination under change of law.) The proof of the main result is now straightforward since the proposition above allows us to estimate, for any test function $Z_{\infty} \in L^2(\Omega, d\mathbb{P}) \cap L^2(\Omega, d\mathbb{P}^u)$,

$$|(Y_{\infty}, Z_{\infty})| = \left| \int_0^{\infty} d[Y,Z]_s \right| \leqslant \int_0^{\infty} |d[Y,Z]_s| \leqslant Q_2^{\mathcal{F}}[w] \|X\|_{2,w} \|Z\|_{2,u},$$

which is exactly

$$\|Y\|_{2,w} \leqslant Q_2^{\mathcal{F}}[w] \|X\|_{2,w}.$$

This concludes the proof of Theorem 9. $\square$

Bibliography

[1]   R. Bañuelos and F. Baudoin, *Martingale transforms and their projection operators on manifolds*, Potential Anal. **38**(4) (2013), 1071–1089.

[2]   R. Bañuelos and A. Osekowski, *On the Bellman function of Nazarov, Treil and Volberg*, Math. Z. **278** (2014), 385–399.

[3]   R. Bañuelos and A. Osekowski, *Sharp weighted L^2 inequalities for square functions*, 2016, arXiv:1603.07618.

[4]   A. Bonami and D. Lepingle, *Fonction maximale et variation quadratique des martingales en présence d'un poids*, in: Séminaire de probabilités XIII, Univ. Strasbourg 1977/78, Lect. notes math., vol. 721, 1979, pp. 294–306.

[5]   D. L. Burkholder, *Boundary value problems and sharp inequalities for martingale transforms*, Ann. Probab. **12**(3) (1984), 647–702.

[6]   C. Dellacherie and P.-A. Meyer, *Probabilities and potential B. Theory of martingales*, North-Holland mathematics studies, vol. 72, North-Holland Publishing Co., Amsterdam, 1982.

[7]   K. Domelevo, C. Kriegler and S. Petermichl, *H^{∞} calculus for submarkovian semigroups on weighted L^2 spaces*, Math. Ann. **381**(3–4) (2021), 1137–1195.

[8]   K. Domelevo and S. Petermichl, *Continuous-time sparse domination*, 2016, arXiv:1607.06319.

[9]   K. Domelevo and S. Petermichl, *Differential subordination under change of law*, Ann. Probab. **47**(2) (2019), 896–925.

[10] O. Dragičević, S. Treil and A. Volberg, *A theorem about three quadratic forms*, Int. Math. Res. Not. **2008** (2008), rnn072.

[11] R. Hunt, B. Muckenhoupt and R. Wheeden, *Weighted norm inequalities for the conjugate function and Hilbert transform*, Trans. Am. Math. Soc. **176** (1973), 227–251.

[12] M. Izumisawa and N. Kazamaki, *Weighted norm inequalities for martingales*, Tôhoku Math. J. (2) **29**(1) (1977), 115–124.

[13] M. T. Lacey, *An elementary proof of the A_2 bound*, Isr. J. Math. **217**(1) (2017), 181–195.

[14] F. L. Nazarov and S. R. Treil, *The hunt for a Bellman function: applications to estimates for singular integral operators and to other classical problems of harmonic analysis*, Algebra Anal. **8**(5) (1996), 32–162.

[15] F. Nazarov, S. Treil and A. Volberg, *The Bellman functions and two-weight inequalities for Haar multipliers*, J. Am. Math. Soc. **12**(4) (1999), 909–928.

[16] A. Osekowski, *Sharp L^p-bounds for the martingale maximal function*, Tôhoku Math. J. (2) **70**(1) (2018), 121–138.

[17] S. Petermichl, *The sharp bound for the Hilbert transform on weighted Lebesgue spaces in terms of the classical A_p characteristic*, Am. J. Math. **129**(5) (2007), 1355–1375.

[18] S. Petermichl and J. Wittwer, *A sharp estimate for the weighted Hilbert transform via Bellman functions*, Mich. Math. J. **50**(1) (2002), 71–87.

[19] P. E. Protter, *Stochastic integration and differential equations*, Stochastic modelling and applied probability, vol. 21, Springer-Verlag, Berlin, 2005.

[20] A. Reznikov, V. Vasyunin and A. Volberg, *An observation: cut-off of the weight w does not increase the A_{p_1,p_2}-"norm" of w*, 2010, arXiv:1008.3635.

[21] C. Thiele, S. Treil and A. Volberg, *Weighted martingale multipliers in the non-homogeneous setting and outer measure spaces*, Adv. Math. **285** (2015), 1155–1188.

[22] S. Treil, *Sharp A_2 estimates of Haar shifts via Bellman function*, in: Recent trends in analysis. Proceedings of the conference in honor of Nikolai Nikolski on the occasion of his 70th birthday, Bordeaux, France, August 31–September 2, The Theta Foundation, Bordeaux, France, 2011, pp. 187–208.

[23] V. Vasyunin and A. Volberg, *The Bellman function technique in harmonic analysis*, Cambridge studies in advanced mathematics, vol. 186, Cambridge University Press, Cambridge, 2020.

[24] J. Wittwer, *A sharp estimate on the norm of the martingale transform*, Math. Res. Lett. **7**(1) (2000), 1–12.

Matthieu Fradelizi, Mathieu Meyer, and Artem Zvavitch

Volume product

Abstract: Our purpose here is to give an overview of known results and open questions concerning the volume product $\mathcal{P}(K) = \min_{z \in K} \mathrm{vol}(K)\,\mathrm{vol}((K - z)^*)$ of a convex body K in $\mathbb{R}^n$. We present a number of upper and lower bounds for $\mathcal{P}(K)$; in particular, we discuss the Mahler conjecture on the lower bound of $\mathcal{P}(K)$, which is still open. We also show connections of $\mathcal{P}(K)$ with different parts of modern mathematics, including geometric number theory, convex geometry, analysis, and harmonic analysis, as well as systolic and symplectic geometries and probability.

Keywords: Convex bodies, polar bodies, volume product, Mahler's conjecture, Blaschke–Santaló inequality, equipartitions

MSC 2020: 52A20, 52A40, 53A15, 52B10

1 Introduction

More or less attached to the name of Kurt Mahler (1903–1988) are at least two celebrated unsolved problems:

- Lehmer's problem on algebraic numbers.
- The lower bound for the volume product of convex bodies.

The celebrity of these two problems comes from the fact that they are both very natural and easy to state, but still unsolved and that for almost one century, a lot of mathematicians gave partial results, equivalent statements, or many generalizations. There are still a lot of interesting attempts to resolve those problems which appear every now and then and produce connections of those questions to many areas of modern mathematics.

Although we shall be interested here in the second one, for the curiosity of the reader we summarize the first one. Let α be an algebraic integer and let P be the minimal

Acknowledgement: We are grateful to Richard Gardner, Dmitry Faifman, Dylan Langharst, Erwin Lutwak, and Shlomo Reisner for many corrections, valuable discussions, and useful suggestions.

A. Z. is supported in part by the U. S. National Science Foundation Grant DMS-1101636, CNRS, and U. S. National Science Foundation under Grant No. DMS-1929284, while A. Z. was in residence at the Institute for Computational and Experimental Research in Mathematics in Providence, RI, during the Harmonic Analysis and Convexity semester program.

Matthieu Fradelizi, Mathieu Meyer, Univ Gustave Eiffel, Univ Paris Est Creteil, CNRS, LAMA UMR8050, F-77447, Marne-la-Vallée, France, e-mails: matthieu.fradelizi@univ-eiffel.fr, mathieu.meyer@univ-eiffel.fr

Artem Zvavitch, Department of Mathematical Sciences, Kent State University, Kent, OH 44242, USA, e-mail: zvavitch@math.kent.edu

https://doi.org/10.1515/9783110775389-005

polynomial of α, that is, the polynomial with integer coefficients with the smallest degree d such that the coefficient of x^d is 1 and $\mathcal{P}(\alpha) = 0$. Let $\alpha_1 = \alpha$ and let $\alpha_2, \ldots, \alpha_d \in \mathbb{C}$ be the other roots of P. The *Mahler measure* of α is the number $M(\alpha) := \prod_{k=1}^{d} \max(|\alpha_k|, 1)$. By a classical result of Kronecker, if $M(\alpha) = 1$, then α is a root of unity. But how near can α be from 1 when it is not a root of unity? Is there a constant $c > 1$, independent of the degree of α, such that $M(\alpha) > 1$ implies that $M(\alpha) > c$? Derrick Henry Lehmer conjectured in 1933 [122] that the answer to this question is positive (we note that for c depending on the degree of α a lot of estimates were given) and Mahler contributed to it, at least, by defining the measure to which his name was given [187, 195].

We shall be mainly concerned here with a second problem: Let K be a convex symmetric body in $\mathbb{R}^n$, which is the unit ball of an n-dimensional normed space E. Let K^* be the polar body of K, which is the unit ball of E^*, the dual of E. What are the bounds for the *volume product* $\mathcal{P}(E) = \mathcal{P}(K) := \mathrm{vol}(K)\,\mathrm{vol}(K^*)$? It appears that the best upper bounds are known for a long time (Blaschke in 1917 for $n = 2, 3$ [29], Santaló in 1949 for $n \geq 4$ [182]), but to find the exact lower bounds is still an open conjecture, although the asymptotic behavior of $\min\{\mathcal{P}(E); E\ n\text{-dimensional normed space}\}$ was discovered almost 40 year ago by Bourgain and V. Milman [38]. This problem has a lot of generalizations and specializations. One can ask a series of very natural questions, including: What happens if K is no longer centrally symmetric? What happens for special classes of convex bodies? Is there a functional version of the volume product? What are the possible applications and connections inside and outside convex geometry? We must also note that a lot of new methods were used, in particular from functional analysis, harmonic analysis, topology, differential geometry, and probability, to prove properties of the volume product and to attack different cases of this question. We shall try here to explain just some of them and summarize the others.

The chapter is structured in the following way. We introduce the volume product and prove its basic properties in Section 1.1. In Section 1.2, we describe the methods of shadow systems which turn out to be essential in the study of the bounds for the volume product. In Section 2, we discuss the upper bound for the volume product – the Blaschke–Santaló inequality. We present different proofs, including a proof via Steiner symmetrizations and a harmonic analysis approach; we also discuss a number of extensions of this inequality. In Section 3, we discuss the conjectured lower bound – the Mahler conjecture. We present a solution in a number of partial cases, including the case of zonoids, of unconditional bodies and of dimension 2 and a very recent solution for symmetric 3-dimensional bodies. We also present here an approach to stability results to both upper and lower bounds. Section 4 is dedicated to the asymptotic lower estimates for the volume product, in particular to the Bourgain–Milman inequality. Here, we extend our presentation to the harmonic analytic and complex analytic approach to the volume product. Section 5 is dedicated to the functional inequalities related to the volume product with a special connection to transport inequalities. In Section 6, we discuss the generalization of the volume product to the case of many functions and bodies. Finally, in Section 7, we present a sample of connections of the bounds for the volume product to other inequalities, including the slicing conjecture, Viterbo's conjecture, applications to the geometry of numbers, and isosystolic inequalities.

We refer the reader to [7, 8, 65, 66, 79, 115, 170, 184, 194] for many additional information on convex bodies, volume and mixed volume, duality, and other core objects in analysis, geometry, and convexity used in this survey.

1.1 Notations and results before 1980

A convex body K in $\mathbb{R}^n$ is a convex compact subset of $\mathbb{R}^n$ with non-empty interior denoted int(K). We say that $L \subset \mathbb{R}^n$ is centrally symmetric if $L = -L$. Let K be a convex body in $\mathbb{R}^n$ such that $0 \in$ int(K); for $x \in \mathbb{R}^n$, we define

$$\|x\|_K = \inf\{t > 0;\ tx \in K\}$$

to be the *gauge* of K; in particular, when K is a convex symmetric body, $x \mapsto \|x\|_K$ is the norm on $\mathbb{R}^n$ for which K is the closed unit ball. We endow $\mathbb{R}^n$ with its natural scalar product, denoted $\langle\ ,\ \rangle$, the associated Euclidean norm denoted $|\cdot|$; the Euclidean ball of radius one is denoted B_2^n. The canonical Lebesgue measure of a Borel set $A \subset \mathbb{R}^n$ is denoted by vol(A).

Let K be a convex body. If $0 \in$ int(K), the *polar body* K^* is defined by

$$K^* = \{y \in \mathbb{R}^n; \langle x,y \rangle \le 1 \text{ for all } x \in K\}. \tag{5.1}$$

Then K^* is also a convex body such that $0 \in$ int(K^*) and if $T : \mathbb{R}^n \to \mathbb{R}^n$ is a linear isomorphism, one has

$$\left(T(K)\right)^* = \left(T^*\right)^{-1}(K^*),$$

where T^* is the adjoint of T. More generally, for a convex body K and $z \in$ int(K), one defines *the polar body K^z of K with respect to z* by

$$K^z = (K - z)^* + z.$$

The celebrated *bipolar theorem* asserts that if $0 \in$ int(K), then

$$(K^*)^* = K \quad \text{and consequently} \quad (K^z)^z = K$$

for any convex body K and any $z \in$ int(K). Let $h_K(y) = \max_{x \in K}\langle x,y \rangle$ be the *support function* of K. Note that $K^* = \{h_K \le 1\}$, i. e., $h_K(x) = \|x\|_{K^*}$, when $0 \in$ int(K). Moreover, if $z \in$ int(K),

$$K^z = z + \{y \in \mathbb{R}^n; h_K(y) - \langle z,y \rangle \le 1\}$$

and thus

$$\operatorname{vol}(K^z) = \int_{K^*} \frac{1}{(1 - \langle z, y \rangle)^{n+1}} \, dy.$$

It follows that the map $z \mapsto \operatorname{vol}(K^z)$ is a strictly convex positive C^∞-function on $\operatorname{int}(K)$.

A small effort is enough to prove that $\operatorname{vol}(K^z) \to +\infty$ when z approaches the boundary of K. Consider $\theta \in S^{n-1}$; by Brunn's theorem, the function $f_\theta : [-h_{K^*}(-\theta), h_{K^*}(\theta)] \to [0, \infty)$ defined by $f_\theta(t) := \operatorname{vol}_{n-1}(\{y \in K^*; \langle y, \theta \rangle = t\})$ satisfies that $f_\theta^{1/(n-1)}$ is concave. Hence, one has $f_\theta(t) \ge f_\theta(0)(1 - h_{K^*}^{-1}(\theta)t)^{n-1}$ for $0 \le t \le h_{K^*}(\theta)$. Let $r_K(\theta) = \min\{a \ge 0 : \theta \in aK\}$ be the radial function of K. Then $r_K(\theta) = h_{K^*}^{-1}(\theta)$ and letting $z = s\theta$ for $0 \le s < r_K(\theta)$, we get

$$\operatorname{vol}(K^z) = \int_{K^*} \frac{1}{(1 - \langle z, y \rangle)^{n+1}} \, dy = \int_{-h_{K^*}(-\theta)}^{h_{K^*}(\theta)} \frac{f_\theta(t)}{(1 - st)^{n+1}} \, dt$$

$$\ge f_\theta(0) \int_0^{h_{K^*}(\theta)} \frac{(1 - th_{K^*}^{-1}(\theta))^{n-1}}{(1 - st)^{n+1}} \, dt = \frac{f_\theta(0)}{n(r_K(\theta) - s)} \ge \frac{\min_{\theta \in S^{n-1}} f_\theta(0)}{n(r_K(\theta) - s)} \to +\infty,$$

when $s \to r_K(\theta)$, that is, $z \to \partial K$.

Consequently, the function $z \mapsto \operatorname{vol}(K^z)$ reaches its minimum on $\operatorname{int}(K)$ at a unique point $s(K)$, called the *Santaló point* of K. Computing its differential, we see that $s(K)$ is characterized by the fact that the centroid (center of mass) of $K^{s(K)}$ is $s(K)$ (see [154]). For $t > 0$ big enough, the sets $\{z \in \operatorname{int}(K); \operatorname{vol}(K^z) \le t\}$, called *Santaló regions of K*, were studied in [155] (see also [156]).

Definition 1. The Santaló point of a convex body K, denoted $s(K)$, is the unique point in $\operatorname{int}(K)$ such that

$$\operatorname{vol}(K^{s(K)}) = \min_{z \in K} \operatorname{vol}(K^z).$$

The *volume product* of K is

$$\mathcal{P}(K) := \operatorname{vol}(K) \operatorname{vol}(K^{s(K)}).$$

We mention the following facts:

- If K is centrally symmetric, then so is K^*, and one has $s(K) = 0 = s(K^*)$ and $\mathcal{P}(K) = \mathcal{P}(K^*)$. One has always $\mathcal{P}(K^{s(K)}) \le \mathcal{P}(K)$ and when K is not centrally symmetric, it may happen that $\mathcal{P}(K^{s(K)}) < \mathcal{P}(K)$.
- It is easy to see that $s(K)$ is the unique point of $\operatorname{int}(K)$ such that 0 is the center of mass of $(K - s(K))^*$ or $s(K)$ is the center of mass of $K^{s(K)}$.

-   The map $K \mapsto \mathcal{P}(K)$ is affine invariant, that is, if $A : \mathbb{R}^n \to \mathbb{R}^n$ is a one-to-one affine transform, then $\mathcal{P}(AK) = \mathcal{P}(K)$. This indicates that if E if an n-dimensional normed space with a closed unit ball B_E and $\phi : E \to \mathbb{R}^n$ is a one-to-one linear mapping, then $\mathcal{P}(E) := \mathcal{P}(\phi(B_E))$ does not depend on ϕ. In particular, this property makes $\mathcal{P}(E)$ to be an important tool in the local theory of normed space (see [170, 127, 194]).
-   Let K be a convex body such that $0 \in \text{int}(K)$ and let E be a linear subspace of $\mathbb{R}^n$. Then $K \cap E$ is a convex body in E, endowed with the scalar product inherited from the Euclidean structure of $\mathbb{R}^n$, and $(K \cap E)^*$ (with polarity inside E) can be identified with $P_E(K^*)$, where P_E is the orthogonal projection from $\mathbb{R}^n$ onto E. Consequently, when K is centrally symmetric, $\mathcal{P}(K \cap E) = \text{vol}_E(K \cap E)\,\text{vol}_E(P_E(K^*))$, where vol_E denotes the Lebesgue measure on E.

In view of these facts, a natural question is to compute, for fixed n, an upper and a lower bound of $\mathcal{P}(K)$ for all convex bodies K in $\mathbb{R}^n$. The existence of these bounds follows from the affine invariance which allows to consider the bounds of $K \mapsto \mathcal{P}(K)$ on a compact subset of the set of all convex bodies endowed with the Hausdorff metric. Indeed, if B_2^n is the Euclidean ball, by John's theorem, one may reduce to study $\mathcal{P}(K)$ for $B_2^n \subset K \subset nB_2^n$ in the general case or $B_2^n \subset K \subset \sqrt{n}B_2^n$ when K is centrally symmetric, which gives already rough but concrete estimates for these bounds.

It seems that the first one who dealt with this problem was Wilhelm Blaschke (1885–1962), who proved that for $n = 2$ and $n = 3$, $\mathcal{P}(K) \le \mathcal{P}(\mathcal{E})$, where $\mathcal{E}$ is any ellipsoid [29, 30]. Then, Mahler gave exact lower bounds for $\mathcal{P}(K)$ for $n = 2$ both in the general case and in the centrally symmetric case [137, 138]. In 1947, Luis Santaló (1911–2001) [182] extended the results of Blaschke to all n with the same tools as him. The case of equality for the upper bound was first proved much later in 1978 by Petty [169] (the argument given in [182] was not quite valid).

Theorem 1 (Blaschke–Santaló–Petty). *If $K \subset \mathbb{R}^n$ is a convex body, then*

$$\mathcal{P}(K) \le \mathcal{P}(B_2^n), \tag{5.2}$$

with equality if and only if K is an ellipsoid.

Bambah [20] gave rough lower bounds for $\mathcal{P}(K)$. Guggenheimer [81, 82] believed at some moment that he had a complete proof of the exact lower bound $\mathcal{P}(K) \ge \mathcal{P}([-1,1]^n) = \frac{4^n}{n!}$ for K centrally symmetric, but it appeared that his proof was incorrect. This was the situation in the 1980s, when new insights into the problem were provided by Saint-Raymond [179], Reisner [172, 173], Gordon and Reisner [74], and Bourgain and Milman [38].

We conclude this section by introducing an important tool in this theory.

1.2 Shadow systems

Definition 2. A *shadow system* along a direction $\theta \in S^{n-1}$ is a family of convex sets $K_t \subset \mathbb{R}^n$ which are defined by $K_t = \mathrm{conv}(\{x + ta(x)\theta; x \in M\})$, where M is a bounded subset in $\mathbb{R}^n$, $a : M \to \mathbb{R}$ is a bounded function, and $t \in I$, an interval of $\mathbb{R}$ (and where $\mathrm{conv}(A)$ denotes the closed convex hull of a set $A \subset \mathbb{R}^n$).

It may be observed that the classical Steiner symmetrization can be seen as a shadow system such that the volume of K_t remains constant (see Remark 1 below). The notion of shadow system was introduced by Rogers and Shephard [177, 186] and can be explained via an idea of Shephard in [186], who pointed out that a shadow system of convex bodies in $\mathbb{R}^n$ can be seen as a family of projections of an $(n + 1)$-dimensional convex set on some n-dimensional subspace of $\mathbb{R}^{n+1}$. Namely, let $e_1, e_2, \ldots, e_{n+1}$ be an orthonormal basis of $\mathbb{R}^{n+1}$. Let M be as before a bounded subset of $\mathbb{R}^n$ (i. e., M is contained in the linear span of $e_1, e_2, \ldots, e_n$), let $a : M \to \mathbb{R}$ be a bounded function, let $\theta \in S^{n-1}$, and let I be an interval of $\mathbb{R}$. We define an $(n + 1)$-dimensional convex set $\tilde{K} \subset \mathbb{R}^{n+1}$ by

$$\tilde{K} = \mathrm{conv}\{x + a(x)e_{n+1} : x \in M\}.$$

For each $t \in I$, let $P_t : \mathbb{R}^{n+1} \to \mathbb{R}^n$ be the projection from $\mathbb{R}^{n+1}$ onto $\mathbb{R}^n$ along the direction $e_{n+1} - t\theta$. Then

$$P_t(\tilde{K}) = \mathrm{conv}(\{x + ta(x)\theta; x \in M\}) = K_t.$$

This interpretation permits to see that $t \mapsto \mathrm{vol}(K_t)$ is a convex function [177]. This result is a powerful tool for obtaining geometric inequalities of isoperimetric type.

The following theorem connects shadow systems with the volume product. It was proved by Campi and Gronchi [43] when the bodies K_t are centrally symmetric and by Meyer and Reisner in the general case [151] (see also [63]).

Theorem 2. *Let* $(K_t)_{t\in(a,b)}$ *be a shadow system of convex bodies in* $\mathbb{R}^n$*. Then the function* $t \mapsto \mathrm{vol}((K_t)^{s(K_t)})^{-1}$ *is convex on* (a, b)*.*

With the previous notations, if K_t are centrally symmetric, then $s(K_t) = 0$ and thus

$$(K_t)^{s(K_t)} = K_t^* = \left(P_t(\tilde{K})\right)^*.$$

As discussed above, the polar of the orthogonal projection of a convex body on a subspace E is the section of its polar by E. But here P_t is not an orthogonal projection, and we get

$$\left(P_t(\tilde{K})\right)^* = P_{e_{n+1}^\perp}\left(\tilde{K}^* \cap (e_{n+1} - t\theta)^\perp\right),$$

where $P_{e_{n+1}^\perp}$ is the orthogonal projection from $\mathbb{R}^{n+1}$ onto $\mathbb{R}^n = e_{n+1}^\perp$. In that context, Campi–Gronchi's theorem appears as another formulation, with a new proof, of Busemann's theorem [41] about the central hyperplane sections of a centrally symmetric body (see also [153]). This point of view was put forward in [48], where such properties were also generalized to more general measures than the Lebesgue measure.

An important component related to Theorem 2, which is proved in [151, Proposition 7] (see also [152]), is the case when both $\mathrm{vol}(K_t)$ and $\mathrm{vol}((K_t)^{s(K_t)})^{-1}$ are affine functions of $t \in (a, b)$. In this case, all the bodies K_t are affine images of each other by special affine transformations. This has been useful in identifying the cases of equality in inequalities involving volume products, as well as in the proof of the results of [152, 63]. The proof of this component, which involves some ordinary differential equations (ODEs), was extended in [159, Proposition 6.1] to generalize this result.

2 The Blaschke–Santaló inequality

The original proofs of the Blaschke–Santaló inequality [29, 181, 182, 123] were based on the affine isoperimetric inequality. We show now how this inequality implies the Blaschke–Santaló inequality and how conversely the Blaschke–Santaló inequality implies it. We refer to [184, Section 10] and to [185, 124] for details about the tools used in this section. If M is a smooth convex body with positive curvature everywhere, its affine surface area $\mathcal{A}(M)$ is defined by

$$\mathcal{A}(M) = \int_{S^{n-1}} f_M(u)^{\frac{n}{n+1}} du,$$

where $f_M : S^{n-1} \to \mathbb{R}_+$ is the curvature function, i. e., the density of the measure σ_M on S^{n-1} with respect to the Haar measure on S^{n-1}, where for a Borel subset A of S^{n-1}, $\sigma_M(A)$ is the $(n-1)$-Hausdorff measure of the set of all points in ∂M such that their unit normal to M is in A. The *affine isoperimetric inequality* says that at a fixed volume, ellipsoids have the largest affine surface area among all convex bodies with positive continuous curvature. In other words,

$$\mathcal{A}(M)^{n+1} \le n^{n+1} v_n^2 \, \mathrm{vol}(M)^{n-1}, \tag{5.3}$$

where $v_n = \mathrm{vol}(B_2^n)$. Let L be another convex body. By Hölder's inequality, one has

$$\mathcal{A}(M)^{n+1} \le \left(\int_{S^{n-1}} h_L(u) f_M(u) du \right)^n \int_{S^{n-1}} h_L^{-n}(u) du$$
$$= n^{n+1} V(M[n-1], L)^n \, \mathrm{vol}(L^*), \tag{5.4}$$

where $V(M[n-1], L) = V(M[n-1], L[1]) = \frac{1}{n} \int_{S_{n-1}} h_L(u) f_M(u) du$ is a mixed volume of M and L, which can also be defined by the formula, for $t \ge 0$,

$$\mathrm{vol}(M + tL) = \sum_{k=1}^{n} \binom{n}{k} V(M[n-k], L[k]) t^k.$$

We refer to [184] for precise definitions and properties of mixed volumes. Using inequality (5.3) and Minkowski's first inequality

$$V(M[n-1], L)^n \geq \mathrm{vol}(M)^{n-1} \mathrm{vol}(L),$$

one gets

$$\mathcal{A}(M)^{n+1} \leq n^{n+1} v_n^2 \, \mathrm{vol}(M)^{n-1} \leq n^{n+1} v_n^2 V(M[n-1], L)^n \, \mathrm{vol}(L)^{-1}. \tag{5.5}$$

Now given a convex body K, let $s = s(K)$ be its Santaló point; then 0 is the centroid of $K - s$ so that

$$\int_{S^{n-1}} u h_{K-s}(u)^{-(n+1)} du = 0,$$

and thus by Minkowski's existence theorem (see [184, Section 8.2]), there exists a convex body M such that $f_M = h_{K-s}^{-(n+1)}$. Set $L = K - s$. Then there is equality in (5.4) so that by (5.5)

$$n^{n+1} V(M[n-1], K-s)^n \, \mathrm{vol}((K-s)^*) = \mathcal{A}(M)^{n+1}$$
$$\leq n^{n+1} v_n^2 V(M[n-1], K)^n \, \mathrm{vol}(K-s)^{-1},$$

which gives the Blaschke–Santaló inequality (5.2).

Conversely, let M be a convex body with positive curvature and suppose that 0 is the Santaló point of M and that the Blaschke–Santaló inequality holds for M. By (5.4) with $L = M$, we get

$$\mathcal{A}(M)^{n+1} \leq n^{n+1} V(M[n-1], M)^n \, \mathrm{vol}(M^*) \leq n^{n+1} v_n^2 \, \mathrm{vol}(M)^{n-1},$$

which is the affine isoperimetric inequality.

For examples of other results of this type and relations between affine surface area and the volume product, see Petty [168, 169], Lutwak [130, 133], Li, Schütt, and Werner [126], Naszódi, Nazarov, and Ryabogin [160], and Hug [94], who gave a proof of the affine isoperimetric inequality using Steiner symmetrization and studied the cases of equality.

2.1 A proof of the Blaschke–Santaló inequality in the centrally symmetric case

In [146, 147], Meyer and Pajor used the classical Steiner symmetrization for giving a proof which we sketch in this section. Various symmetrizations of sections appeared also in [179, 17].

Step 1: We prove first that if H is a linear hyperplane of $\mathbb{R}^n$ and $S_H K$ is the *Steiner symmetral of K with respect to H,* as will be defined below, then

$$\mathrm{vol}(K^*) \le \mathrm{vol}((S_H K)^*).$$

To simplify notations, suppose that $H = \mathbb{R}^{n-1} \times \{0\}$; as before, let $P_H : \mathbb{R}^n \mapsto H$ be the orthogonal projection onto H. Then K may be described as follows:

$$K = \{x + se_n; x \in P_H K, s \in I(x)\},$$

where for $x \in P_H K, I(x) = \{s \in \mathbb{R}; x+se_n \in K\}$ is a closed interval. The Steiner symmetral of K with respect to H, defined as

$$S_H K = \left\{ x + se_n; x \in PK, s \in \frac{I(x) - I(x)}{2} \right\},$$

is a convex body symmetric with respect to H, such that

$$\mathrm{vol}(K) = \mathrm{vol}(S_H K).$$

For $t \in \mathbb{R}$, let $K^*(t) := \{y \in H; y + te_n \in K^*\}$ be the section of K^* by the hyperplane $\{x_n = t\}$ and $J = \{t \in \mathbb{R}; K^*(t) \ne \emptyset\}$. Then

$$K^* = \{y + te_n; t \in J, y \in K^*(t)\}.$$

By the symmetry of K^*, one has $K(-x) = -K(x)$ for every $x \in P_H K$, so that for every $t \in J$, $y_1 \in K^*(t), y_2 \in K^*(-t)$, and $s_1, s_2 \in K(x)$, one has

$$\langle x, y_1 \rangle + s_1 t \le 1 \quad \text{and} \quad \langle x, y_2 \rangle - s_2 t \le 1.$$

Adding these two inequalities, we get that for every $x \in P_H K$, $s = \frac{s_1 - s_2}{2} \in \frac{1}{2}(I(X) - I(X))$, $t \in J$, and $y = \frac{y_1 + y_2}{2} \in \frac{1}{2}(K^*(t) + K^*(-t))$, one has

$$\langle x, y \rangle + st \le 1.$$

Thus, for every $t \in J$,

$$\frac{1}{2}(K^*(t) + K^*(-t)) \subset (S_H K)^*(t). \tag{5.6}$$

By the symmetry of K, one has $K^*(-t) = -K^*(t)$. It follows from Brunn–Minkowski's theorem that $\mathrm{vol}_{n-1}((S_H K)^*(t)) \ge \mathrm{vol}_{n-1}(K^*(t))$, and integrating in t we get

$$\mathrm{vol}((S_H K)^*) = \int \mathrm{vol}_{n-1}((S_H K)^*(t))dt \ge \int \mathrm{vol}_{n-1}(K^*(t))dt = \mathrm{vol}(K^*).$$

One thus gets $\mathcal{P}(S_H K) \ge \mathcal{P}(K)$.

Step 2: It is well known that there exists a sequence of hyperplanes $(H_m)_m$ such that if $K_0 = K$ and for $m \geq 1$, $K_m := S_{H_m} K_{m-1}$, then the sequence $(K_m)_m$ converges to the Euclidean ball $R_K B_2^n$ with the same volume as K (see for example Section 10.3 in [184]). Thus,

$$\mathcal{P}(K) \leq \mathcal{P}(K_{n-1}) \leq \mathcal{P}(K_n) \leq \mathcal{P}(R_K B_2^n) = \mathcal{P}(B_2^n).$$

The case of equality in the Blaschke–Santaló inequality was first proved by Petty [169], using sharp differential geometry arguments. When K is centrally symmetric, a new elementary proof was given by Saint-Raymond [179], using Minkowski symmetrization of the hyperplane sections (see also [17]). These ideas were generalized by Meyer and Pajor [147] to give an elementary proof for the general case and a somewhat stronger result, also based on Steiner's symmetrizations.

Theorem 3 (Meyer–Pajor [147]). *For every convex body K and every hyperplane H separating $\mathbb{R}^n$ into two closed half-spaces H_+ and H_-, denoting $\lambda = \frac{\mathrm{vol}(H_+ \cap K)}{\mathrm{vol}(K)}$, there exists $z \in H$ such that $\mathrm{vol}(K)\,\mathrm{vol}(K^z) \leq \frac{\mathcal{P}(\mathcal{E})}{4\lambda(1-\lambda)}$.*

Remark 1. Notice that the Steiner symmetral of a convex body K with respect to a direction $u \in S^{n-1}$ can be written as a part of a shadow system in the following way. For $y \in P_{u^\perp} K$, let $I(y) = \{s \in \mathbb{R}; y + su \in K\}$. For $t \in [-1,1]$, let

$$K_t = \left\{ y + su; s \in \frac{1+t}{2} I(y) - \frac{1-t}{2} I(y) \right\}.$$

Then $K_1 = K$, $K_0 = S_{u^\perp} K$, and for every $t \in [-1,1]$, K_{-t} is the reflection of K_t with respect to the hyperplane $u^\perp$. This implies that the function $t \mapsto \mathcal{P}(K_t)$ is even. Moreover, since $\mathrm{vol}(K_t) = \mathrm{vol}(K)$ is constant for $t \in [-1,1]$, using Theorem 2, the function $t \mapsto \mathcal{P}(K_t)^{-1}$ is convex. It follows that $\mathcal{P}(K_t)$ is maximized at 0, which proves that the volume product is non-decreasing by Steiner symmetrization for any convex body, recovering Meyer–Pajor's result [147]. Using again an appropriate sequence of Steiner symmetrizations, this gives the general Blaschke–Santaló inequality. The preceding observation is due to [151].

2.2 A harmonic analysis proof of the Blaschke–Santaló inequality

We follow the work of Bianchi and Kelli [28]. Harmonic analysis plays an essential role in the study of duality and the volume product. The main idea was discovered by Nazarov [161], who used it to provide a proof of the Bourgain–Milman inequality [38], and it was adopted by Bianchi and Kelly to give a very elegant proof of the Blaschke–Santaló inequality. We refer to [189, 91] for basic facts in harmonic analysis.

Let K be a convex symmetric body in $\mathbb{R}^n$. Let $F \in L_2(\mathbb{R}^n)$ such that its Fourier transform $\widehat{F} = 0$ a. e. on $\mathbb{R}^n \setminus K$. Then F is the restriction to $\mathbb{R}^n$ of the entire function still denoted F defined by $F(z) = \int_K e^{2\pi i \langle z,\xi \rangle} \widehat{F}(\xi) d\xi$ for $z \in \mathbb{C}^n$, which satisfies the following inequality, giving a first hint on why the theory is so useful:

$$|F(iy)| = \left| \int_K e^{-2\pi \langle y,\xi \rangle} \widehat{F}(\xi) d\xi \right| \le \int_K e^{2\pi \sup_{\xi \in K} |\langle \xi, y \rangle|} |\widehat{F}(\xi)| d\xi = e^{2\pi \|y\|_{K^*}} \int_K |\widehat{F}(\xi)| d\xi.$$

Thus, for some $C > 0$, one has $|F(iy)| \le Ce^{2\pi \|y\|_{K^*}}$ for all $y \in \mathbb{R}^n$, i.e., F is *of exponential type K^**. This fact is the elementary part of the following classical theorem.

Theorem 4 (Paley–Wiener). *Let $F \in L_2(\mathbb{R}^n)$ and let K be a convex symmetric body. Then the following are equivalent:*
- *F is the restriction to $\mathbb{R}^n$ of an entire function of exponential type K^*.*
- *The support of $\widehat{F}$ is contained in K.*

Now we are ready to present the following proof.

Proof of the Blaschke–Santaló inequality adapted from Bianchi and Kelly [28]. Let $F = \frac{1}{\mathrm{vol}(K)} \widehat{\mathbf{1}_K}$, where $\mathbf{1}_K$ is the characteristic function of K:

$$\mathbf{1}_K(x) = \begin{cases} 1, & \text{for } x \in K, \\ 0, & \text{for } x \notin K. \end{cases}$$

Then $F(0) = 1$, $F \in L_2(\mathbb{R}^n)$, and F is continuous and even and can be extended as an entire function on $\mathbb{C}^n$, still denoted F, as

$$F(z_1, \ldots, z_n) = \frac{1}{\mathrm{vol}(K)} \int_K e^{2i\pi(\sum_{k=1}^n z_i x_i)} dx_1 \cdots dx_n.$$

For $\theta \in S^{n-1}$ and $z \in \mathbb{C}$, let $F_\theta(z) = F(z\theta)$. Then, by the easy part of the Paley–Wiener theorem, F_θ is an even entire function of exponential type $[-\|\theta\|_{K^*}^{-1}, \|\theta\|_{K^*}^{-1}]$. Moreover, since F is entire and even, there exists an entire function $H_\theta : \mathbb{C} \to \mathbb{C}$ such that $H_\theta(z^2) = F_\theta(z) = F(z\theta)$. Finally, we define $R_\theta : \mathbb{C}^n \to \mathbb{C}$ as a radial extension of F_θ by

$$R_\theta(z_1, \ldots, z_n) = H_\theta(z_1^2 + \cdots + z_n^2).$$

Note that $z \mapsto R_\theta(z)$ is an entire function which satisfies $R_\theta(0) = F(0) = 1$. Moreover, since F_θ is of exponential type $[-\|\theta\|_{K^*}^{-1}, \|\theta\|_{K^*}^{-1}]$, R_θ is of exponential type $\|\theta\|_{K^*}^{-1} B_2^n$. It follows from the Paley–Wiener theorem that the support of the restriction to $\mathbb{R}^n$ of $\widehat{R_\theta}$ is contained in $(\|\theta\|_{K^*}^{-1} B_2^n)^* = \|\theta\|_{K^*} B_2^n$. Since $R_\theta \in L_2(\mathbb{R}^n)$, one may write, using the Plancherel equality and the Cauchy–Schwarz inequality with $v_n = \mathrm{vol}(B_2^n)$,

$$\int_{\mathbb{R}^n} |R_\theta(x)|^2 dx = \int_{\|\theta\|_{K^*} B_2^n} |\widehat{R_\theta}(x)|^2 dx \ge \frac{|\int_{\|\theta\|_{K^*} B_2^n} \widehat{R_\theta}(x) dx|^2}{v_n \|\theta\|_{K^*}^n} \tag{5.7}$$

$$= \frac{\left|\int_{\mathbb{R}^n} \widehat{R_\theta}(x)\,dx\right|^2}{v_n \|\theta\|_{K^*}^n} = \frac{R_\theta(0)}{v_n \|\theta\|_{K^*}^n} = \frac{1}{v_n \|\theta\|_{K^*}^n}.$$

If $|x| = r = |re_1|$, one has

$$R_\theta(x) = F(|x|\theta) = F(r\theta) = R_\theta(re_1), \tag{5.8}$$

so that R_θ is rotation invariant on $\mathbb{R}^n$. Thus,

$$\frac{1}{\mathrm{vol}(K)} = \int_{\mathbb{R}^n} |\widehat{F}(x)|^2 dx = \int_{\mathbb{R}^n} |F(x)|^2 dx = \int_{S^{n-1}} \int_0^{+\infty} |F(r\theta)|^2 r^{n-1} dr d\theta$$

$$= \int_{S^{n-1}} \int_0^{+\infty} |R_\theta(re_1)|^2 r^{n-1} dr d\theta = \frac{1}{nv_n} \int_{\mathbb{R}^n} |R_\theta(x)|^2 dx.$$

It follows that

$$\frac{1}{\mathrm{vol}(K)} = \frac{1}{nv_n} \int_{S^{n-1}} \int_{\mathbb{R}^n} |R_\theta(x)|^2 dx d\theta \geq \frac{1}{nv_n^2} \int_{S^{n-1}} \|\theta\|_{K^*}^{-n} d\theta = \frac{\mathrm{vol}(K^*)}{\mathrm{vol}(B_2^n)^2},$$

which is the Blaschke–Santaló inequality. $\qquad\square$

Bianchi and Kelly [28] also provided a proof of the equality case. Indeed, assume that we have equality in the Blaschke–Santaló inequality. Then we must have equality in (5.7). Thus, for every $\theta \in S^{n-1}$ and some $c_\theta \in \mathbb{R}$, one has $\widehat{R_\theta} = c_\theta \mathbf{1}_{\|\theta\|_{K^*} B_2^n}$ on $\mathbb{R}^n$ and

$$R_\theta(x) = \int_{\mathbb{R}^n} e^{-2i\pi\langle x,y\rangle} \widehat{R_\theta}(y)\,dy = c_\theta \int_{\|\theta\|_{K^*} B_2^n} e^{-2i\pi\langle x,y\rangle}\,dy.$$

Moreover, since $R_\theta(0) = 1$, one gets $c_\theta \,\mathrm{vol}(\|\theta\|_{K^*} B_2^n) = 1$. Next, by (5.8), we get

$$\frac{1}{\mathrm{vol}(K)} \int_K e^{-2i\pi r\langle\theta,y\rangle}\,dy = F(r\theta) = \frac{1}{\mathrm{vol}(\|\theta\|_{K^*} B_2^n)} \int_{\|\theta\|_{K^*} B_2^n} e^{-2i\pi r\langle\theta,y\rangle}\,dy. \tag{5.9}$$

If M is a convex body, $\theta \in S^{n-1}$, and $t \in \mathbb{R}$, let $A_{M,\theta}(t) = \mathrm{vol}_{n-1}(M \cap (\theta^\perp + t\theta))$. One has

$$\widehat{A_{K,\theta}}(r) = \int_{\mathbb{R}} e^{-2i\pi rt} A_{K,\theta}(t)\,dt = \int_K e^{-2i\pi r\langle\theta,y\rangle}\,dy.$$

Inverting the Fourier transform, it follows from (5.9) that for all $t \in \mathbb{R}$ and $\theta \in S^{n-1}$,

$$\frac{1}{\mathrm{vol}(K)} A_{K,\theta}(t) = \frac{A_{\|\theta\|_{K^*} B_2^n,\theta}(t)}{\mathrm{vol}(\|\theta\|_{K^*} B_2^n)} = \frac{A_{B_2^n,\theta}(t\|\theta\|_{K^*}^{-1})}{\|\theta\|_{K^*}\,\mathrm{vol}(B_2^n)}. \tag{5.10}$$

Now, for $\theta \in S^{n-1}$, one has by (5.10)

$$\int_K \langle x, \theta \rangle^2 dx = \int_{\mathbb{R}} t^2 A_{K,\theta}(t)dt = \frac{\mathrm{vol}(K)}{\|\theta\|_{K^*} \mathrm{vol}(B_2^n)} \int_{\mathbb{R}} t^2 A_{B_2^n,\theta}(t\|\theta\|_{K^*}^{-1})dt$$

$$= \frac{\mathrm{vol}(K)}{\mathrm{vol}(B_2^n)} \|\theta\|_{K^*}^2 \int_{\mathbb{R}} u^2 A_{B_2^n,\theta}(u)du,$$

and since by the rotation invariance $A_{B_2^n,\theta}(u)$ does not depend on $\theta \in S^{n-1}$, one gets for some $c > 0$ and all $\theta \in S^{n-1}$

$$\|\theta\|_{K^*} = c \left(\int_K \langle x, \theta \rangle^2 dx \right)^{1/2},$$

which proves that K^*, and thus K, is an ellipsoid (the last arguments are inspired by [148]).

2.3 Further results and generalizations

Let us present a few results which may be considered as offspring of the Blaschke–Santaló inequality.

2.3.1 Stability

K. Börözcky [35] established a stability version of the Blaschke–Santaló inequality, later improved by K. Ball and K. Börözcky in [19]. Let $d_{BM}(K, L)$ be the Banach–Mazur distance between two convex bodies K and L in $\mathbb{R}^n$:

$$d_{BM}(K, L) = \inf\{d > 0 : K - x \subseteq T(L - y) \subseteq d(K - x), \text{ for some } T \in GL(n) \text{ and } x, y \in \mathbb{R}^n\}.$$

The following stability theorem was proved in [19].

Theorem 5. *If K is a convex body in $\mathbb{R}^n$, $n \geq 3$, such that for some $\varepsilon > 0$ one has*

$$(1 + \varepsilon)\mathcal{P}(K) \geq \mathcal{P}(B_2^n),$$

then

$$\log(d_{BM}(K, B_2^n)) \leq c_n \varepsilon^{\frac{1}{3(n+1)}} |\log \varepsilon|^{\frac{2}{3(n+1)}},$$

where c_n is an absolute constant depending on n only.

If in the above theorem we assume that K is symmetric, then the exponent of ε can be improved to $2/3(n + 1)$.

2.3.2 Local and restricted maxima

After having proved that convex bodies with maximal volume product are ellipsoids, one may ask about the local maxima of the volume product in the sense of Hausdorff distance. Using Theorem 2, it was proved in [152] that any local maximum is an ellipsoid, which gives another proof of Blaschke–Santaló's inequality.

One may also investigate maxima among certain classes of bodies not containing ellipsoids. For instance, in $\mathbb{R}^2$, among polygons with $m \geq 4$ vertices, the maxima are the affine images of regular polygons with m vertices [151]. For $n \geq 3$, the much more complicated situation was investigated by [3] using shadow systems. In particular, it was proved in [3] that a polytope with maximal volume product among polytopes with at most m vertices is simplicial (all its facets are simplices) and has exactly m vertices. It was also proved that, among polytopes with at most $n+2$ vertices, the volume product is maximized by $\mathrm{conv}(\Delta_{\lceil \frac{n}{2} \rceil}, \Delta_{\lfloor \frac{n}{2} \rfloor})$, where $\Delta_{\lceil \frac{n}{2} \rceil}$ and $\Delta_{\lfloor \frac{n}{2} \rfloor}$ are simplices living in complementary subspaces of dimensions $\lceil \frac{n}{2} \rceil$ and $\lfloor \frac{n}{2} \rfloor$, respectively (by definition, for $\alpha \notin \mathbb{Z}$, $\lfloor \alpha \rfloor$ is the integer part of α and $\lceil \alpha \rceil = \lfloor \alpha \rfloor + 1$, for $\alpha \in \mathbb{Z}$, $\lceil \alpha \rceil = \lfloor \alpha \rfloor = \alpha$). It is conjectured in [3] that, for $1 \leq k \leq n$, among polytopes with at most $n + k$ vertices, the convex hull of k simplices living in complementary subspaces of dimensions $\lceil \frac{n}{k} \rceil$ or $\lfloor \frac{n}{k} \rfloor$ has maximal volume product.

Among unit balls of finite-dimensional Lipschitz-free spaces, which are polytopes with at most $(n + 1)^2$ extreme points, some preliminary results were established in [2] and it was shown that the maximizers of the volume product are simplicial polytopes.

2.3.3 L_p-centroid inequalities

In a series of works by Lutwak, Yang, and Zhang [136, 134, 135], Blaschke–Santaló's inequality appears as a special case of a family of isoperimetric inequalities involving the so-called L_p-centroid bodies and L_p-projection bodies. More precisely, consider a compact star-shaped body K in $\mathbb{R}^n$ and $p \in [1, \infty]$. The polar L_p-centroid body $\Gamma_p^* K$ is defined via its norm:

$$\|x\|_{\Gamma_p^* K}^p = \frac{1}{c_{n,p} \, \mathrm{vol}(K)} \int_K |\langle x, y \rangle|^p dy.$$

Here the normalization constant $c_{n,p}$ is chosen so that $\Gamma_p^* B_2^n = B_2^n$. It was proved in [136] that for all $p \in [1, \infty]$

$$\mathrm{vol}(K) \, \mathrm{vol}(\Gamma_p^* K) \leq \mathrm{vol}(B_2^n)^2, \tag{5.11}$$

with equality if and only if K is an ellipsoid centered at the origin. It turns out that if K is a centrally symmetric convex body, then $\Gamma_\infty^* K = K^*$, and thus the symmetric case of the Blaschke–Santaló inequality follows from (5.11) when $p = \infty$. A stronger version of (5.11) was proved in [134]:

$$\operatorname{vol}(\Gamma_p K) \geq \operatorname{vol}(K),$$

for any star body in $\mathbb{R}^n$ and $p \in [1, \infty]$. This inequality, for $p = 1$, links the theory to the Busemann–Petty centroid inequality [166]; see also [65, 184]. If K and L are compact subsets of $\mathbb{R}^n$, then for $p \geq 1$, it was proved in [135, Corollary 6.3] that for some $c(p, n) > 0$, one has

$$\int_{K \times L} |\langle x, y \rangle|^p \, dx dy \geq c(p, n)(\operatorname{vol}(K) \operatorname{vol}(L))^{\frac{n+p}{n}}$$

with equality if and only if K and L are, up to sets of measure 0, dilates of polar-reciprocal, origin-centered ellipsoids. When $p \to +\infty$, one gets the following version of the symmetric Blaschke–Santaló inequality in [135]: If K, L are compact subsets of $\mathbb{R}^n$, then

$$\operatorname{vol}(B_2^n)^2 \max_{x \in K, y \in L} |\langle x, y \rangle|^n \geq \operatorname{vol}(K) \operatorname{vol}(L).$$

2.3.4 Connection to affine quermassintegrals

Affine quermassintegrals were defined by Lutwak [129]. For $1 \leq k \leq n$, the k-th affine quermassintegral of a convex body K is

$$\Phi_k(K) = \frac{v_n}{v_k} \left(\int_{Gr(k,n)} \operatorname{vol}_k(P_F K)^{-n} \sigma_{n,k}(dF) \right)^{-1/n},$$

where $Gr(k, n)$ is the Grassmann manifold of k-dimensional linear subspaces F of $\mathbb{R}^n$, $\sigma_{n,k}$ is the Haar probability measure on $Gr(k, n)$, and P_F is the orthogonal projection onto F. It was proved by Grinberg [77] that $\Phi_k(K)$ is invariant under volume preserving affine transformations. Let $R_K > 0$ satisfy $\operatorname{vol}(R_K B_2^n) = \operatorname{vol}(K)$. Lutwak [131] conjectured that for any convex body K in $\mathbb{R}^n$ and any $k = 1, \ldots, n - 1$, one has

$$\Phi_k(K) \geq \Phi_k(R_K B_2^n) \tag{5.12}$$

with equality if and only if K is an ellipsoid. This conjecture was open for quite a long time. Lutwak proved that, for $k = 1$, it follows directly from the Blaschke–Santaló inequality (and the fact that the case $k = n - 1$ is connected to an inequality of Petty [167, 168]). Recently, E. Milman and Yehudayoff [159] proved that this conjecture is true. As one of the steps in the proof, they showed that $\Phi_k(K) \geq \Phi_k(S_H K)$, generalizing the previous result of [146]. In addition, a simplified proof of the Petty projection inequality was presented in [159]. Those interesting results suggest that (5.12) can be viewed as a generalization of the Blaschke–Santaló inequality.

2.3.5 A conjecture of K. Ball

Keith Ball [15] conjectured that if K is a convex symmetric body in $\mathbb{R}^n$, then

$$\int_K \int_{K^*} \langle x, y \rangle^2 dx dy \leq \int_{B_2^n} \int_{B_2^n} \langle x, y \rangle^2 dx dy = \frac{n}{(n+2)^2} \operatorname{vol}(B_2^n)^2. \tag{5.13}$$

He proved a kind of reverse inequality:

$$\frac{n(\operatorname{vol}(K)\operatorname{vol}(K^*))^{\frac{n+2}{n}}}{(n+2)^2 \operatorname{vol}(B_2^n)^{\frac{4}{n}}} \leq \int_K \int_{K^*} \langle x, y \rangle^2 dx dy,$$

which shows that inequality (5.13) is stronger than the Blaschke–Santaló inequality. In [15, 17], (5.13) was proved for unconditional bodies. Generalizations are considered in [102, 51] (see Section 7.3 for the latter).

2.3.6 Stochastic and log-concave measure extensions

Following the ideas initiated in [164], the authors of [48] pursued a probabilistic approach of the Blaschke–Santaló inequality for symmetric bodies and established the following result.

Theorem 6. *For $N, n \geq 1$, let $(\Omega, \mathcal{B}, P)$ be a probability space and:*
- *Let $X_1, \ldots, X_N : \Omega \to \mathbb{R}^n$ be independent random vectors, whose laws have densities with respect to the Lebesgue measure which are bounded by 1.*
- *Let $Z_1, \ldots, Z_N : \Omega \to \mathbb{R}^n$ be independent random vectors uniformly distributed in rB_2^n with $\operatorname{vol}(rB_2^n) = 1$.*
- *Let μ be the rotation invariant measure on $\mathbb{R}^n$ with density $e^{\varphi(|x|)}$, $x \in \mathbb{R}^n$, with respect to the Lebesgue measure, where $\varphi : \mathbb{R}_+ \to \mathbb{R}_+$ is a non-increasing function.*
- *Let $C_{X,N}(\omega) = \operatorname{conv}(\pm X_1(\omega), \ldots, \pm X_N(\omega))$ and $C_{Z,N}(\omega) = \operatorname{conv}(\pm Z_1(\omega), \ldots, \pm Z_N(\omega))$ for $\omega \in \Omega$.*

Then for all $t \geq 0$, one has $P(\{\omega \in \Omega; \mu(C_{X,N}(\omega)^*) \geq t\}) \leq P(\{\omega \in \Omega; \mu(C_{Z,N}(\omega)^*) \geq t\})$.

It follows of course that the same comparison holds in expectation. The tools used there are shadow systems as in the work of Campi and Gronchi [43], together with the rearrangement inequalities of Rogers [176] and Brascamp–Lieb–Luttinger [39]. Applying Theorem 6 to $X_1, \ldots, X_N$ uniformly distributed on a convex body K and using the fact that when $N \to +\infty$, the sequence of random polytopes $P_{K,N} := \operatorname{conv}(\pm X_1, \ldots, \pm X_N)$ converges almost surely to K in the Hausdorff metric, we deduce that for measures μ, as in Theorem 6, one has

$$\mu(K^*) \le \mu((R_K B_2^n)^*) = \mu\left(\frac{B_2^n}{R_K}\right), \quad \text{where } R_K = \left(\frac{\text{vol}(K)}{\text{vol}(B_2^n)}\right)^{\frac{1}{n}}.$$

Since clearly $\mu(K) \le \mu(R_K B_2^n)$, we deduce that $\mu(K)\mu(K^*) \le \mu(R_K B_2^n)\mu(B_2^n/R_K)$. If, moreover, $t \mapsto \varphi(e^t)$ is concave, then $t \mapsto \mu(e^t B_2^n)$ is also log-concave (see [47]). Thus, it follows that for such measures μ and for any symmetric convex body K, one has

$$\mu(K)\mu(K^*) \le \mu(B_2^n)^2. \tag{5.14}$$

It was proved in [49] that under those hypotheses, $t \mapsto \mu(e^t K)$ is log-concave (extending the same property for Gaussian measures established in [47]). It was asked in [46] whether (5.14) holds for all symmetric log-concave measures μ.

We shall prove (5.14) when moreover μ has an unconditional density f with respect to the Lebesgue measure (a function $f : \mathbb{R}^n \to \mathbb{R}$ is called *unconditional* if for some basis $e_1, \ldots, e_n$ of $\mathbb{R}^n$, one has, for all $(\varepsilon_1, \ldots, \varepsilon_n) \in \{-1; 1\}^n$ and $(x_1, \ldots, x_n) \in \mathbb{R}^n$, $f(\sum_{i=1}^n x_i e_i) = f(\sum_{i=1}^n \varepsilon_i x_i e_i)$).

Theorem 7. *If μ is a measure on $\mathbb{R}^n$ with an unconditional and log-concave density with respect to the Lebesgue measure and K is a symmetric convex body in $\mathbb{R}^n$, then $\mu(K)\mu(K^*) \le \mu(B_2^n)^2$.*

Proof. We apply first a linear transform making the density of μ unconditional with respect to the canonical basis of $\mathbb{R}^n$. Let H be a coordinate hyperplane and let $S_H K$ be the Steiner symmetral of K with respect to H. Using (5.6) as in the proof of Meyer and Pajor [146] (see Section 2.1 above), we get $\mu(K^*) \le \mu((S_H K)^*)$. Moreover, it is easy to see that $\mu(K) \le \mu(S_H K)$. Thus, denoting by L the convex body obtained from K after n successive Steiner symmetrizations with respect to the coordinate hyperplanes, we get $\mu(K)\mu(K^*) \le \mu(L)\mu(L^*)$. We are now reduced to the case when μ and K are unconditional. Using the classical Prékopa–Leindler inequality (see for example [170, page 3]), it was shown in [58] that then $\mu(L)\mu(L^*) \le \mu(B_2^n)^2$. $\square$

2.3.7 Blaschke–Santaló-type inequality on the sphere

Another inequality of Blaschke–Santaló type was established by Gao, Hug, and Schneider [64] on the sphere. We define the polar of $A \subset S^{n-1}$ by

$$A^\circ := \{y \in S^{n-1}; \langle x, y \rangle \le 0, \text{ for all } x \in A\}.$$

If $\text{pos}(A) := \{tx; x \in A, t \ge 0\}$, then $A^\circ = (\text{pos}(A))^* \cap S^{n-1}$. Let σ be the Haar probability measure on S^{n-1}. A *spherical cap* is the non-empty intersection of S^{n-1} with a half-space. This work was further generalized by Hu and Li [93], who proved a number of Blaschke–Santaló-type inequalities in the sphere and hyperbolic space.

Theorem 8 ([64]). *Let A be a non-empty measurable subset of S^{n-1} and let C be a spherical cap such that $\sigma(A) = \sigma(C)$. Then $\sigma(A^\circ) \leq \sigma(C^\circ)$. If moreover A is closed and $\sigma(A) < 1/2$, there is equality if and only if A is a spherical cap.*

Two proofs were given in [64]. One of them uses a special type of symmetrization called two-point symmetrization and for the equality case the results of [14]. Hack and Pivovarov [83] gave a stochastic extension of Theorem 7 in the spirit of Theorem 6.

3 Mahler conjecture. Special cases

The problem of the lower bound of $\mathcal{P}(K)$ is not yet solved, although significant progress has been made in recent years. The first results are due to Mahler for $n = 2$, who proved that $\mathcal{P}(K) \geq \mathcal{P}(\Delta_2) = \frac{27}{4}$, where Δ_2 is a triangle, and that in the centrally symmetric case $\mathcal{P}(K) \geq \mathcal{P}([-1,1]^2) = \frac{8}{3}$ (see also [193]). For the proofs, Mahler used polygons, and thus he could not give the case of equality. Observe that Mahler continued to be interested in this problem [139, 140]. The case of equality in dimension 2 was obtained by Meyer [145] for general bodies and by Reisner [172] (see also [179, 144, 193]) for centrally symmetric bodies. What happens in dimension $n \geq 3$? There are two conjectures; the first one, but not the second one, was formulated explicitly by Mahler [137].

Conjecture 1. *For every convex body K in $\mathbb{R}^n$, one has*

$$\mathcal{P}(K) \geq \mathcal{P}(\Delta_n) = \frac{(n+1)^{n+1}}{(n!)^2},$$

where Δ_n is a simplex in $\mathbb{R}^n$, with equality if and only if $K = \Delta_n$.

Conjecture 2. *For every centrally symmetric convex body K in dimension n, one has*

$$\mathcal{P}(K) \geq \mathcal{P}(B_\infty^n) = \frac{4^n}{n!},$$

where $B_\infty^n = [-1,1]^n$ is a cube, with equality if and only if K is a Hanner polytope (see Definition 4 below).

3.1 The conjectured minimum in the symmetric case is not unique

To understand Conjecture 2 and different phenomena related to it, we define Hanner polytopes [86], as well as the ℓ_1-sum $E \oplus_1 F$ and the ℓ_∞-sum $E \oplus_\infty F$ of two normed spaces E and F.

Definition 3. Let $(E, \|\cdot\|_E)$ and $(F, \|\cdot\|_F)$ be two normed spaces. Then on $E \times F$, we define two norms, the norm of the ℓ_∞-sum $E \oplus_\infty F$ of E and F and the norm of their ℓ_1-sum $E \oplus_1 F$, as follows:

-   $\|(x,y)\|_\infty = \max(\|x\|_E, \|y\|_F)$.
-   $\|(x,y)\|_1 = \|x\|_E + \|y\|_F$.

We note that if E and F are normed spaces, then the unit ball of their ℓ_∞-sum is the Minkowski sum of the unit balls of E and F in $E \times F$ and the unit ball of their ℓ_1-sum is their convex hull. Analogously, if we consider two convex bodies $K \subset \mathbb{R}^{n_1}$ and $L \subset \mathbb{R}^{n_2}$, we define two convex bodies in $\mathbb{R}^{n_1+n_2}$:

-   $K \oplus_\infty L = K \times \{0\} + \{0\} \times L = \{x_1 + x_2 : x_1 \in K, x_2 \in L\}$, their ℓ_∞-sum.
-   $K \oplus_1 L = \mathrm{conv}(K \times \{0\}, \{0\} \times L)$, their ℓ_1-sum.

One major property of ℓ_1- and ℓ_∞-sums is that

$$(K \oplus_\infty L)^* = K^* \oplus_1 L^*. \tag{5.15}$$

Now we are ready to define Hanner polytopes.

Definition 4. In dimension 1, Hanner polytopes are symmetric segments. Suppose that Hanner polytopes are defined in all dimensions $m \leq n - 1$. A Hanner polytope in dimension n is the unit ball of an n-dimensional normed space H such that for some k-dimensional subspace E, $1 \leq k \leq n$, and $(n - k)$-dimensional subspace F of H, whose unit balls are Hanner polytopes, $1 \leq k \leq n - 1$, H is the ℓ_∞-sum or the ℓ_1-sum of E and F.

Let us now discuss the basic properties of Hanner polytopes:

-   In $\mathbb{R}^2$, there is a unique (up to isomorphism) Hanner polytope, which is the square.
-   In $\mathbb{R}^3$, there are exactly two (up to isomorphism) Hanner polytopes, which are the cube and the centrally symmetric octahedron.
-   In $\mathbb{R}^4$, there are, up two isomorphisms, four different classes of Hanner polytopes, including two which are not isomorphic to the cube or the cross-polytope. And in $\mathbb{R}^n$, their number increases quickly with n.
-   The normed spaces whose unit balls K are Hanner polytopes are up to isometry exactly those which satisfy the $(3 - 2)$-intersection property: For any three vectors u_1, u_2, and u_3, if $(K + u_i) \cap (K + u_j) \neq \emptyset$, for all $1 \leq i < j \leq 3$, then the intersection of all three balls is not empty [87].
-   A Hanner polytope is unconditional (see Definition 5 below).
-   If K is a Hanner polytope, then so is K^*. This follows from (5.15).
-   If $K \subset \mathbb{R}^{n_1}$ and $L \subset \mathbb{R}^{n_2}$ are two convex bodies, then

$$\mathcal{P}(K \oplus_\infty L) = \mathcal{P}(K \oplus_1 L) = \frac{n_1! n_2!}{(n_1 + n_2)!} \mathcal{P}(K)\mathcal{P}(L).$$

-   Using induction, it follows that the volume product of a Hanner polytope in $\mathbb{R}^n$ is $\frac{4^n}{n!}$.

In some sense, Conjecture 1 seems easier than Conjecture 2 because up to an isomorphism, there is only one proposed minimum. But polarity is established with respect to the Santaló point of a convex body K, which is not always well located, so that one has to prove that for every $z \in \text{int}(K)$, $\text{vol}(K)\,\text{vol}(K^z) \geq \mathcal{P}(\Delta_n)$. Observe however that if K has minimal volume product among all other convex bodies, then its Santaló point is also its center of gravity.

3.2 The planar case

First, note that the conjecture holds with the case of equality for $n = 2$ (see Mahler [137] and Meyer [145] for another proof and the case of equality). Let us sketch a proof of the planar case and use this opportunity to give an example of how the method of shadow systems as well as Theorem 2 can be used; note that the method in this case can be traced back to the original proof from [137] and is almost identical for the general and the symmetric case. We concentrate on the general case.

Proof. (Lower bound in $\mathbb{R}^2$.) It is enough to show that $\mathcal{P}(T) \geq \mathcal{P}(\Delta_2)$ for all convex polygons $T \subset \mathbb{R}^2$. The main idea is to remove vertices of T. We use induction on the number k of vertices. Let T be a polygon with $k \geq 4$ vertices. Suppose that $T = \text{conv}(v_1, v_2, v_3, \ldots, v_k)$, with $v_1, v_2, v_3, \ldots, v_k$, written in clockwise order. We shall prove that $\mathcal{P}(T) \geq \mathcal{P}(Q)$, for a polygon Q with only $k-1$ vertices. For $i \neq j$, let $\ell_{i,j}$ be a line through v_i and v_j. Let $\theta \in S^1$ be parallel to the line $\ell_{1,k-1}$ and define $T_t = \text{conv}(v_1, v_2, \ldots, v_{k-1}, v_k + t\theta)$ (i. e., we move v_k on a line parallel to $\ell_{1,k-1}$). The line $\{v_k + t\theta; t \in \mathbb{R}\}$ meets $\ell_{k-1,k}$ at v_k' when $t = a$ and $\ell_{1,2}$ at v_1' when $t = b$. Since $T_0 = T$, one may assume that $a < 0 < b$. It is easy to see that, for $t \in [a,b]$, $t \mapsto T_t$ is a shadow system with $\text{vol}(T_t) = \text{vol}(T)$. By Theorem 2, $t \mapsto \mathcal{P}(T_t)^{-1}$ is convex on the interval $[a,b]$ and thus is maximal at its end points. Thus, $\mathcal{P}(T) \geq \min(\mathcal{P}(T_a), \mathcal{P}(T_b))$, where $T_a = \text{conv}(v_1, \ldots, v_{k-2}, v_k')$ and $T_b = \text{conv}(v_1', v_2, \ldots, v_{k-1})$ are polygons with only $k - 1$ vertices. $\square$

Remark 2. The above method was used to prove a number of partial cases of Mahler's conjectures (see [150, 63, 3, 2, 183]). Unfortunately, there seems to be no way to generalize this approach to dimension 3 and higher. One of the reasons is that a vertex v of a polytope P may be a vertex of a lot of non-simplicial faces. And how can v be "moved" without breaking the combinatorial structure of P? When the combinatorial structure of P is broken, it is difficult to compute volumes.

Remark 3. In [171], Rebollo Bueno established also stochastic versions of the planar case of Mahler's conjectures. With the notations of Section 2.3.6, he proved that for any centrally symmetric convex body K in the plane and any $r \geq 1$,

$$\mathbb{E}\left(\text{vol}(P_{K,N}^*)^{-r}\right) \leq \mathbb{E}\left(\text{vol}(P_{Q,N}^*)^{-r}\right),$$

where Q is a square with $\text{vol}(Q) = \text{vol}(K)$. For $r = 1$ and $N \to +\infty$, this gives back the planar case of Mahler's conjecture. The same type of result is also established in [171] for general convex bodies in the plane.

3.3 The case of zonoids

The conjecture holds for zonoids and polars of zonoids, with equality for cubes (see Reisner [172, 173] and Gordon, Meyer, and Reisner [73] for a second proof). We recall that a *zonoid* in $\mathbb{R}^n$ is a Hausdorff limit of *zonotopes*, that is, of finite sums of segments. Since a segment is symmetric with respect to its midpoint, any zonotope, and thus any zonoid, is centrally symmetric. From now, when speaking of a zonoid Z, we shall suppose that $Z = -Z$. Also, the polar bodies of zonoids can be seen as the unit balls of finite-dimensional subspaces of $L_1([0,1], dx)$. Observe that every convex centrally symmetric body in $\mathbb{R}^2$ is a zonoid. We refer to [33, 70, 184] for basic properties of zonoids.

Proof. (The lower bound of volume product for zonoids [73].) For a zonoid $Z \subset \mathbb{R}^n$, there exists a measure μ on S^{n-1} such that $h_Z(x) = \frac{1}{2} \int_{S^{n-1}} |\langle x, u\rangle| d\mu(u)$ for all $x \in \mathbb{R}^n$. Since $\mathrm{vol}(Z) = \frac{1}{n} \int_{S^{n-1}} \mathrm{vol}_{n-1}(P_{u^\perp}Z) d\mu(u)$, one has

$$\mathrm{vol}(Z^*) \int_{S^{n-1}} \mathrm{vol}_{n-1}(P_{u^\perp}Z) d\mu(u) = n\, \mathrm{vol}(Z)\,\mathrm{vol}(Z^*) = (n+1)\,\mathrm{vol}(Z) \int_{Z^*} h_Z(x) dx$$

$$= \frac{n+1}{2}\,\mathrm{vol}(Z) \int_{S^{n-1}} \left(\int_{Z^*} |\langle x, u\rangle| dx \right) d\mu(u).$$

It follows that for some $u \in S^{n-1}$, one has

$$\mathrm{vol}(Z^*)\,\mathrm{vol}_{n-1}(P_{u^\perp}Z) \leq \frac{n+1}{2}\,\mathrm{vol}(Z) \int_{Z^*} |\langle x, u\rangle| dx.$$

Now $\int_{Z^*} |\langle x, u\rangle| dx = 2 \int_0^\infty t f(t) dt$, where $f(t) = \mathrm{vol}(Z^* \cap (u^\perp + tu))$ is the $(n-1)$ dimensional volume of the sections of Z^* by hyperplanes parallel to $u^\perp$. Note that $f(0) = \mathrm{vol}(Z^* \cap u^\perp)$ and $2 \int_0^\infty f(t) dt = \mathrm{vol}(Z^*)$. By the Brunn–Minkowski theorem, the function $f^{\frac{1}{n-1}}$ is concave on its support. By a classical estimate for $1/(n-1)$ concave functions (see for instance [158]),

$$\int_0^\infty t f(t) dt \leq \frac{n}{n+1} \frac{\left(\int_0^\infty f(t) dt\right)^2}{f(0)},$$

with equality if and only if $f(t) = f(0)(1 - ct)_+^{n-1}$, for some $c > 0$ and all $t \geq 0$. This gives

$$\int_{Z^*} |\langle x, u\rangle| dx \leq 2 \frac{n}{n+1} \frac{4^{-1}\,\mathrm{vol}(Z^*)^2}{\mathrm{vol}_{n-1}(Z^* \cap u^\perp)} = \frac{n}{2(n+1)} \frac{\mathrm{vol}(Z^*)^2}{\mathrm{vol}_{n-1}(Z^* \cap u^\perp)},$$

and thus

$$\text{vol}(Z^*)\,\text{vol}_{n-1}(P_{u^\perp}Z) \le \frac{n+1}{2}\,\text{vol}(Z)\frac{n}{2(n+1)}\frac{\text{vol}(Z^*)^2}{\text{vol}_{n-1}(Z^* \cap u^\perp)},$$

so that

$$\text{vol}(Z)\,\text{vol}(Z^*) \ge \frac{4}{n}\,\text{vol}_{n-1}(P_{u^\perp}Z)\,\text{vol}_{n-1}(Z^* \cap u^\perp),$$

which allows to conclude by induction, with the case of equality, since $P_{u^\perp}Z$ is a zonoid in dimension $n-1$ and $(P_{u^\perp}Z)^* = Z^* \cap u^\perp$. $\qquad\square$

Remark 4. Campi and Gronchi [44] presented a very interesting inequality on the volume of L_p-zonotopes, which gives inequality, in particular, another proof of the above result. It is interesting to note that the proof in [44] is based on the shadow systems technique. Another proof using shadow systems was presented by Saroglou in [183].

Remark 5. Marc Meckes [143] gave another proof of Mahler's conjecture for zonoids, based on the notion of *magnitude* introduced by Leinster [125], which is a numerical isometric invariant for metric spaces. He studied the magnitude of a convex body in hypermetric normed spaces (which include $\ell_p^n, p \in [1,2]$) and proved a new upper bound for magnitude on such spaces using the Holmes–Thompson intrinsic volumes of their unit balls.

3.4 The case of unconditional bodies

Definition 5. Let K in $\mathbb{R}^n$ be a convex body. We say that K is *unconditional* if for some basis $e_1, \ldots, e_n$ of $\mathbb{R}^n$ one has $x_1 e_1 + \cdots + x_n e_n \in K$ if and only if $|x_1|e_1 + \cdots + |x_n|e_n \in K$. We say that K is *almost unconditional* if for some basis $e_1, \ldots, e_n$ of $\mathbb{R}^n$ for every $1 \le i \le n$, one has $P_i K = K \cap H_i$, where H_i is the linear span of $\{e_j, j \ne i\}$ and P_i is the linear projection from $\mathbb{R}^n$ onto H_i parallel to e_i.

If K is unconditional, after a linear transformation which does not change $\mathcal{P}(K)$, we may suppose that $(e_1, \ldots, e_n)$ is the canonical basis of $\mathbb{R}^n$. Unconditional bodies are almost unconditional and centrally symmetric. Observe also that if K is unconditional (resp. almost unconditional) with respect to some basis, then K^* is also unconditional (resp. almost unconditional) with respect to the dual basis.

We follow the proof of [144] of the inequality $\mathcal{P}(K) \ge \mathcal{P}(B_\infty^n)$ (the first proof was given in [179]). We do not prove the case of equality (Hanner polytopes), which is more involved.

Proof. We use induction on n. It is trivial for $n = 1$. We suppose that $e_1, \ldots, e_n$ is the canonical basis of $\mathbb{R}^n$. Let $K_+ = K \cap \mathbb{R}_+^n$, $K^*_{\ +} = K^* \cap \mathbb{R}_+^n$. Then $\mathcal{P}(K) = 4^n\,\text{vol}(K_+)\,\text{vol}(K^*_+)$. For $x \in \mathbb{R}_+^n$, one has

$$x \in K_+ \quad \text{if and only if } \langle x, y \rangle \le 1 \text{ for any } y \in K_+^*,$$
$$y \in K_+^* \quad \text{if and only if } \langle x, y \rangle \le 1 \text{ for any } x \in K_+.$$

For $1 \le i \le n$, $K_i := K \cap \{x_i = 0\}$ is an unconditional body in $\mathbb{R}^{n-1}$ and $(K_i)^* = (K^*)_i$. Let $(K_i)_+ = K_i \cap (\mathbb{R}^+)^n$. For $x = (x_1, \ldots, x_n) \in K_+$, let $C_i(x)$ be the convex hull of $\{x\}$ with $(K_i)_+$. Since $C_i(x)$ is a cone with apex x and basis $(K_i)_+$, one has

$$\mathrm{vol}(C_i(x)) = \frac{x_i}{n} \, \mathrm{vol}_{n-1}((K_i)_+).$$

Thus,

$$\mathrm{vol}(K_+) \ge \mathrm{vol}\left(\bigcup_{i=1}^{n} C_i(x) \right) = \sum_{i=1}^{n} \mathrm{vol}(C_i(x)) = \frac{1}{n} \sum_{i=1}^{n} x_i \, \mathrm{vol}_{n-1}((K_i)_+). \tag{5.16}$$

Let $a := \frac{1}{n \, \mathrm{vol}(K_+)} (\mathrm{vol}_{n-1}((K_1)_+), \ldots, \mathrm{vol}_{n-1}((K_n)_+))$ in $\mathbb{R}^n$. By (5.16) one has $\langle a, x \rangle \le 1$ for all $x \in K_+$, that is, $a \in K_+^*$. Also, $a^* := \frac{1}{n \, \mathrm{vol}(K_+^*)} (\mathrm{vol}_{n-1}((K_1^*)_+), \ldots, \mathrm{vol}_{n-1}((K_n^*)_+)) \in K_+$. Thus, $\langle a, a^* \rangle \le 1$, that is,

$$\frac{\sum_{i=1}^{n} \mathrm{vol}_{n-1}((K_i)_+) \, \mathrm{vol}_{n-1}((K_i^*)_+)}{n^2 \, \mathrm{vol}(K_+) \, \mathrm{vol}(K_+^*)} \le 1,$$

so that

$$\mathcal{P}(K) = 4^n \, \mathrm{vol}(K_+) \, \mathrm{vol}(K_+^*) \ge \frac{4^n}{n^2} \sum_{i=1}^{n} \mathrm{vol}_{n-1}((K_i)_+) \, \mathrm{vol}_{n-1}((K_i^*)_+).$$

For $1 \le i \le n$, one has $\mathrm{vol}_{n-1}(K_i) = 2^{n-1} \, \mathrm{vol}_{n-1}((K_i)_+)$ and $\mathrm{vol}_{n-1}(K_i^*) = 2^{n-1} \, \mathrm{vol}_{n-1}((K_i^*)_+)$. Since K_i are also unconditional, the induction hypothesis gives $\mathcal{P}(K_i) \ge \frac{4^{n-1}}{(n-1)!}$, $1 \le i \le n$. Thus,

$$\mathcal{P}(K) \ge \frac{4}{n^2} \sum_{i=1}^{n} \mathrm{vol}_{n-1}(K_i) \, \mathrm{vol}_{n-1}(K_i^*) \ge \frac{4}{n^2} \cdot n \cdot \frac{4^{n-1}}{(n-1)!} = \frac{4^n}{n!}. \qquad \square$$

Remark 6. A small modification of this proof allows to treat the case of almost unconditional centrally symmetric bodies. Note that every centrally symmetric body in $\mathbb{R}^2$ is almost unconditional.

3.5 The 3-dimensional symmetric case

The symmetric case in $\mathbb{R}^3$ was solved by Irieh and Shibota [95] in 2017 with a quite involved proof of about 60 pages. We would like here to highlight the main ideas and to connect it with the unconditional case presented above. We will use the shorter proof given in [57].

A symmetric body $K \subset \mathbb{R}^n, n \geq 3$, is not generally almost unconditional, and thus not unconditional. However, every planar convex body has an almost unconditional basis. For $n = 3$, the goal is to show that a 3-dimensional convex symmetric body K may still have core properties of an unconditional body. This is done with the help of the following equipartition result.

Theorem 9. *Let $K \subset \mathbb{R}^3$ be a symmetric convex body. Then there exist three planes H_1, H_2, and H_3 passing through the origin such that:*
- *they split K into eight pieces of equal volume, and*
- *for each $i = 1, 2, 3$, the section $K \cap H_i$ is split into four parts of equal area by the other two planes.*

Note that Theorem 9 belongs to the very rich theory of equipartitions. For example, a celebrated result of Hadwiger [85], answering a question of Grünbaum [80], shows that for any absolutely continuous finite measure in $\mathbb{R}^3$, there exist three planes for which any octant has 1/8 of the total mass. To prove Theorem 9, one can use a result of Klartag [109, Theorem 2.1]; we refer to [57] for details.

Our goal is to create an analog of formula (5.16). Consider a sufficiently regular oriented hypersurface $A \subset \mathbb{R}^n$ and define the vector

$$\vec{V}(A) = \int_A \vec{n_A}(x)dx,$$

where $\vec{n_A}(x)$ is the unit normal to A at x defined by its orientation. Next, for a convex body $K \subset \mathbb{R}^n$ with $0 \in \mathrm{int}(K)$, the orientation of a subset $A \subset \partial K$ is given by the outer normal $\vec{n_K}$ to K. If $C(A) := \{rx;\ 0 \leq r \leq 1, x \in A\}$, then

$$\mathrm{vol}(C(A)) = \frac{1}{n} \int_A \langle x, \vec{n_K}(x)\rangle dx.$$

The following is a key proposition for our proof.

Proposition 1. *Let $K \subset \mathbb{R}^n$ be a convex body with $0 \in \mathrm{int}(K)$ and let A be a Borel subset of ∂K with $\mathrm{vol}(C(A)) \neq 0$. Then for all $x \in K$,*

$$\frac{1}{n}\langle x, \vec{V}(A)\rangle \leq \mathrm{vol}(C(A)) \quad \text{and thus} \quad \frac{\vec{V}(A)}{n\,\mathrm{vol}(C(A))} \in K^*.$$

Proof. For all $x \in K$, we have $\langle x, \vec{n_K}(z)\rangle \leq \langle z, \vec{n_K}(z)\rangle$ for every $z \in \partial K$. Thus, for all $x \in K$,

$$\langle x, \vec{V}(A)\rangle = \int_A \langle x, \vec{n_K}(z)\rangle dz \leq \int_A \langle z, \vec{n_K}(z)\rangle dz = n\,\mathrm{vol}(C(A)).$$

$\square$

Corollary 1. *Let K be a convex body in $\mathbb{R}^n$ with $0 \in \mathrm{int}(K)$. If $A \subset \partial K$ and $B \subset \partial K^*$ are Borel subsets such that $\mathrm{vol}(C(A)) > 0$ and $\mathrm{vol}(C(B)) > 0$, then*

$$\langle \vec{V}(A), \vec{V}(B) \rangle \le n^2 \, \mathrm{vol}(C(A)) \, \mathrm{vol}(C(B)).$$

Proof. We use Proposition 1 to get $\dfrac{\vec{V}(A)}{n \, \mathrm{vol}(C(A))} \in K^*$ and $\dfrac{\vec{V}(B)}{n \, \mathrm{vol}(C(B))} \in K$. $\square$

Proof of Conjecture 2 for $n = 3$. Since the volume product is continuous, it is enough to prove the conjecture for a centrally symmetric, smooth, strictly convex body K (see [184, Section 3.4]). From the linear invariance of the volume product, we may assume that the equipartition property obtained in Theorem 9 is satisfied by the coordinate planes given by the canonical orthonormal basis (e_1, e_2, e_3). As in the unconditional case, we divide $\mathbb{R}^3$ and the body K into the octants defined by this basis, which define cones as in Corollary 1. The main issue is that in sharp contrast to the unconditional case, the dual cone to the cone defined as an intersection of K with an octant is not the intersection of K^* with this octant. We will need a bit of combinatorics to work around this issue.

For $\varepsilon \in \{-1; 1\}^3$, let the ε-octant be $\{x \in \mathbb{R}^3; \varepsilon_i x_i \ge 0 \text{ for } i = 1, 2, 3\}$ and for $L \subset \mathbb{R}^3$, let L_ε be the intersection of L with the ε-octant: $L_\varepsilon = \{x \in L; \varepsilon_i x_i \ge 0; \ i = 1, 2, 3\}$. Let $N(\varepsilon) := \{\varepsilon' \in \{-1, 1\}^3 : \sum_{i=1}^{3} |\varepsilon_i - \varepsilon_i'| = 2\}$. Then $\varepsilon' \in N(\varepsilon)$ iff $[\varepsilon, \varepsilon']$ is an edge $[-1, 1]^3$.

If $K_\varepsilon \cap K_{\varepsilon'}$ is a hypersurface, we define $K_\varepsilon \vec{\cap} K_{\varepsilon'}$ to be oriented according to the outer normals of ∂K_ε. Using the Stokes theorem, we obtain

$$\vec{V}(\partial K_\varepsilon) = \int_{\partial K_\varepsilon} \overrightarrow{n_{\partial K_\varepsilon}}(x)\,dx - \sum_{\varepsilon' \in N(\varepsilon)} \vec{V}(K_\varepsilon \vec{\cap} K_{\varepsilon'}).$$

Using the equipartition of the areas of $K \cap e_i^\perp$, we get

$$\vec{V}(\partial K_\varepsilon) = - \sum_{\varepsilon' \in N(\varepsilon)} \vec{V}(K_\varepsilon \vec{\cap} K_{\varepsilon'}) = \sum_{i=1}^{3} \frac{\mathrm{vol}(K \cap e_i^\perp)}{4} \varepsilon_i \vec{e_i}.$$

Let us look at the dual. Since K is strictly convex and smooth, there exists a diffeomorphism $\varphi : \partial K \to \partial K^*$ such that $\langle \varphi(x), x \rangle = 1$ for all $x \in \partial K$. We extend φ to $\mathbb{R}^3$ by homogeneity of degree 1: $\varphi(\lambda x) = \lambda \varphi(x)$ for $\lambda \ge 0$. Then

$$K^* = \bigcup_\varepsilon \varphi(K_\varepsilon) \quad \text{and} \quad \mathrm{vol}(K^*) = \sum_\varepsilon \mathrm{vol}(\varphi(K_\varepsilon)).$$

From the equipartition of volumes, one has

$$\mathrm{vol}(K) \, \mathrm{vol}(K^*) = \sum_\varepsilon \mathrm{vol}(K) \, \mathrm{vol}(\varphi(K_\varepsilon)) = 8 \sum_\varepsilon \mathrm{vol}(K_\varepsilon) \, \mathrm{vol}(\varphi(K_\varepsilon)).$$

From Corollary 1, we deduce that for $\varepsilon \in \{-1, 1\}^3$

$$\mathrm{vol}(K_\varepsilon) \, \mathrm{vol}(\varphi(K_\varepsilon)) \ge \frac{1}{9} \langle \vec{V}(\partial K_\varepsilon), \vec{V}(\varphi(\partial K_\varepsilon)) \rangle.$$

Thus,

$$\operatorname{vol}(K)\operatorname{vol}(K^*) \geq \frac{8}{9}\sum_{\varepsilon}\langle \vec{V}(\partial K_\varepsilon), \vec{V}(\varphi(\partial K_\varepsilon))\rangle$$

$$= \frac{8}{9}\sum_{\varepsilon}\left\langle \sum_{i=1}^{3}\frac{\operatorname{vol}(K\cap e_i^{\perp})}{4}\varepsilon_i\vec{e_i}, \vec{V}(\varphi(\partial K_\varepsilon))\right\rangle$$

$$= \frac{8}{9}\sum_{i=1}^{3}\frac{\operatorname{vol}(K\cap e_i^{\perp})}{4}\left\langle \vec{e_i}, \sum_{\varepsilon}\varepsilon_i\vec{V}(\varphi(\partial K_\varepsilon))\right\rangle.$$

Now we use the Stokes theorem for $\varphi(\partial K)$ to get

$$\vec{V}(\varphi((\partial K_\varepsilon))) = -\sum_{\varepsilon'\in N(\varepsilon)}\vec{V}(\varphi(K_\varepsilon\vec{\cap}K'_{\varepsilon'})).$$

The next step requires a careful computation of the sums following orientation of all surfaces, which gives many cancelations. Next one combines the correct parts of K and $\varphi(K)$ to get

$$\operatorname{vol}(K)\operatorname{vol}(K^*) \geq \frac{4}{9}\sum_{i=1}^{3}\operatorname{vol}_{n-1}(K\cap e_i^{\perp})\langle \vec{e_i}, V(\varphi(K\cap \vec{e_i}^{\perp}))\rangle$$

(see [57] for the precise computations). Let P_i be the orthogonal projection onto $e_i^{\perp}$. Then $P_i : \varphi(K\cap e_i^{\perp}) \to P_i(K^*)$ is a bijection. Using Cauchy's formula for the volume of projections, we get

$$\langle \vec{e_i}, V(\varphi(K\cap e_i^{\perp}))\rangle = \int_{\varphi(K\cap e_i^{\perp})} \langle \overrightarrow{n_{\varphi(K\cap e_i^{\perp})}}(x), \vec{e_i}\rangle dx$$

$$= \operatorname{vol}_{n-1}(P_i(\varphi(K\cap e_i^{\perp}))) = \operatorname{vol}_{n-1}(P_i(K^*)),$$

and if $\varepsilon = (\varepsilon_1,\ldots,\varepsilon_n)$,

$$\operatorname{vol}(K)\operatorname{vol}(K^*) \geq \frac{8}{9}\sum_{i=1}^{3}\frac{\operatorname{vol}_{n-1}(K\cap e_i^{\perp})}{4}\left\langle \vec{e_i}, \sum_{\varepsilon}\varepsilon_i\vec{V}(\varphi(\partial K_\varepsilon))\right\rangle.$$

Finally,

$$\operatorname{vol}(K)\operatorname{vol}(K^*) \geq \frac{4}{9}\sum_{i=1}^{3}\operatorname{vol}_{n-1}(K\cap e_i^{\perp})\operatorname{vol}_{n-1}(P_i(K^*))$$

$$= \frac{4}{9}\sum_{i=1}^{3}\operatorname{vol}_{n-1}(K\cap e_i^{\perp})\operatorname{vol}_{n-1}((K\cap e_i^{\perp})^*)$$

$$\geq \frac{4}{9}\times 3\times \frac{4^2}{2!} = \frac{4^3}{3!}.$$

$\square$

3.6 Further special cases where the conjectures hold

Let us list here a number of other special cases in which the conjectured inequality was proved:

- Symmetric polytopes in $\mathbb{R}^n$ with $2n + 2$ vertices for $n \leq 9$ (Lopez and Reisner [128]) and for any n (Karasev [104]).
- For $p \geq 1$, hyperplane sections through 0 of $B_p^n = \{(x_1,\ldots,x_n) \in \mathbb{R}^n; \sum_{i=1}^n |x_i|^p \leq 1\}$ (Karasev [104]). Karasev's proof of those results is, so far, one of the few concrete applications of symplectic geometry, through the billiards approach, to prove special cases of Mahler's conjecture.
- Bodies of revolution [149].
- Some bodies with many symmetries: Barthe and Fradelizi [22] established that a convex body K which is symmetric with respect to a family of hyperplanes whose intersection is reduced to one point satisfies Conjecture 1. More generally, it is proved in [22] that if K is invariant under the reflections fixing $P_1 \times \cdots \times P_k$, where for $1 \leq i \leq k$, P_i are regular polytopes or a Euclidean ball in a subspace E_i and $\mathbb{R}^n = E_1 \oplus \cdots \oplus E_k$, then $\mathcal{P}(K) \geq \mathcal{P}(P_1 \times \cdots \times P_k)$.
- Iriyeh and Shibata established similar results in [96, 97]. They determined the exact lower bound of the volume product of convex bodies invariant by some group of symmetries (many classical symmetry groups in dimension 3 [96] and for the special orthogonal group of the simplex and of the cube [97]).
- Polytopes in $\mathbb{R}^n$ with not more than $n + 3$ vertices [150].
- Almost unconditional symmetric bodies (Saint-Raymond [179]) with equality for Hanner polytopes (Meyer [144] and Reisner [174]). Also, in [179] a result is proved for unconditional sums of convex bodies: For $1 \leq i \leq m$, let $K_i \subset \mathbb{R}^{d_i}$ be convex symmetric bodies and let $L \subset \mathbb{R}^m$ be an unconditional body with respect to the canonical basis $e_1,\ldots,e_m$. We define *the unconditional sum of $K_1,\ldots,K_m$ with respect to L* by

$$K_1 \oplus_L \cdots \oplus_L K_m = \{(x_1,\ldots,x_m) \in \mathbb{R}^{d_1} \times \cdots \times \mathbb{R}^{d_m}; \|x_1\|_{K_1} e_1 + \cdots + \|x_m\|_{K_m} e_m \in L\}.$$

Clearly, $K_1 \oplus_L \cdots \oplus_L K_m$ is a symmetric convex body in $\mathbb{R}^{d_1+\cdots+d_m}$. Moreover it is easy to see that $(K_1 \oplus_L \cdots \oplus_L K_m)^* = K_1^* \oplus_{L^*} \cdots \oplus_{L^*} K_m^*$, and denoting $L_+ = L \cap \mathbb{R}_+^m$ and $^*L_+ = L^* \cap \mathbb{R}_+^m$, one has

$$\mathcal{P}(K_1 \oplus_L \cdots \oplus_L K_m)$$
$$= \left(\int_{(t_1,\ldots,t_m)\in L_+} \prod_{i=1}^m t_i^{d_i-1} dt_1 \cdots dt_m \right) \left(\int_{(t_1,\ldots,t_m)\in L_+^*} \prod_{i=1}^m t_i^{d_i-1} dt_1 \cdots dt_m \right) \prod_{i=1}^m \mathcal{P}(K_i)$$

and

$$\left(\int_{(t_1,\ldots,t_m)\in L_+} \prod_{i=1}^m t_i^{d_i-1} dt_1 \cdots dt_m \right) \left(\int_{(t_1,\ldots,t_m)\in L_+^*} \prod_{i=1}^m t_i^{d_i-1} dt_1 \cdots dt_m \right) \geq \frac{d_1! \times \cdots \times d_m!}{(d_1 + \cdots + d_m)!}.$$

Observe that it follows from [144] or [174] that there is equality in the last inequality if and only if L is a Hanner polytope. Finally, if $\mathcal{P}(K_i) \geq 4^i/i!, 1 \leq i \leq m$, then

$$\mathcal{P}(K_1 \oplus_L \cdots \oplus_L K_m) \geq \frac{4^{d_1 + \cdots + d_m}}{(d_1 + \cdots + d_m)!}.$$

- Although their volumes have been computed (see [180]), it is not known whether the unit ball of classical ideals of operators satisfies Conjecture 2.
- An interpretation of Conjecture 2 in terms of wavelets was given in [18].
- Connections of Mahler's conjecture and the Blaschke–Santaló inequality to the maximal and minimal of $\lambda_1(K)\lambda_1(K^*)$, where K is a convex body and $\lambda_1(K)$ is the first eigenvalue of the Laplacian on the relative interior of K with Dirichlet condition $u = 0$ on ∂K, was given in [40].

3.7 Local minimizers and stability results

One may investigate the properties of the local minimizers for $\mathcal{P}(K)$. A natural open question is whether such a minimizer must be a polytope. A number of results in this direction were proved by studying convex bodies with positive curvature. Stancu [188] proved that if K is a convex body, which is smooth enough and has a strictly positive Gauss curvature everywhere, then the volume product of K cannot be a local minimum. She showed it as a consequence of the fact that, for some $\delta(K) > 0$, one has

$$\mathrm{vol}(K_\delta)\,\mathrm{vol}\big((K_\delta)^*\big) \geq \mathrm{vol}(K)\,\mathrm{vol}(K^*) \geq \mathrm{vol}(K^\delta)\,\mathrm{vol}\big((K^\delta)^*\big),$$

for any $\delta \in (0, \delta(K))$, where K_δ and K^δ stand for the convex floating body and the illumination body associated with K with parameter δ, respectively. A stronger result for local minimizers was proved in [175]: If K is a convex body which is a local minimizer of the volume product, then K has no positive curvature at any point of its boundary. The study of local minimizers was continued in [88], where the authors computed the first and the second derivative of the volume product in terms of the support function. Those results may be seen as a hint toward the conjecture that a minimizer must be a polytope. We also note that [72] extended it to the functional case (see Section 5 below).

It is known that the conjectured global minimizers, that is, Hanner polytopes in the centrally symmetric case and simplices in the general case, are actually local minimizers. This question originates from the blog of Tao [191, 192], where a number of ideas that may lead to a better understanding of the volume product were discussed. Nazarov, Petrov, Ryabogin, and Zvavitch [162] were able to show that the cube and the cross-polytope are local minimizers. Kim and Reisner [107] generalized this result to the case of non-symmetric bodies, proving that the simplex is a local minimizer.

The most general result in the symmetric case was obtained by Kim [106], who considered the case of Hanner polytopes. More precisely, let

$$d_{BM}(K,L) = \inf\{d : d > 0, \text{ there exists } T \in GL(n) \text{ such that } K \subseteq TL \subseteq dK\}$$

be the Banach–Mazur multiplicative distance between two symmetric convex bodies $K, L \subset \mathbb{R}^n$. Then we have the following result.

Theorem 10. *There exist constants $\delta(n), c(n) > 0$ depending only on n such that if K is a symmetric convex body in $\mathbb{R}^n$ with*

$$\min\{d_{BM}(K,H) : H \text{ is a Hanner polytope in } \mathbb{R}^n\} = 1 + \delta,$$

for some $0 < \delta \le \delta(n)$, then

$$\mathcal{P}(K) \ge (1 + c(n)\delta) \cdot \mathcal{P}(B_\infty^n).$$

The above theorem was used in [108] to show the stability of the volume product around the class of unconditional bodies. The question of stability for minima and maxima was also treated in various cases [34, 36, 108, 35, 57]. A general approach to global stability of the volume product was considered in [57], where the following natural lemma was proved.

Lemma 1. *Let $(\mathcal{A}_1, d_1)$ be a compact metric space, let $(\mathcal{A}_2, d_2)$ be a metric space, let $f : \mathcal{A}_1 \to \mathcal{A}_2$ be a continuous function, and let D be a closed subset of $\mathcal{A}_2$. Then:*
(1) For any $\beta > 0$, there exists $\alpha > 0$ such that $d_1(x, f^{-1}(D)) \ge \beta$ implies $d_2(f(x), D) \ge \alpha$.
(2) If for some $c_1, c_2 > 0$, $d_1(x, f^{-1}(D)) < c_1$ implies $d_2(f(x), D) \ge c_2 d_1(x, f^{-1}(D))$, then for some $C > 0$, one has $d_1(x, f^{-1}(D)) \le c d_2(f(x), D)$ for every $x \in \mathcal{A}_1$.

Together with a local minima result (for example Theorem 10), Lemma 1 gives almost immediately a stability result for known bounds of the volume product. Let us illustrate this technique in the case of symmetric convex bodies in $\mathbb{R}^3$.

Theorem 11. *There exists an absolute constant $C > 0$ such that for every symmetric convex body $K \subset \mathbb{R}^3$ and $\delta > 0$ satisfying $\mathcal{P}(K) \le (1 + \delta)\mathcal{P}(B_\infty^3)$, one has*

$$\min\{d_{BM}(K, B_\infty^3), d_{BM}(K, B_1^3)\} \le 1 + C\delta.$$

Proof. Using the linear invariance of the volume product and John's theorem, we reduce to the case $B_2^3 \subseteq K \subseteq \sqrt{3}B_2^3$. Our metric space $\mathcal{A}_1$ will be the set of such bodies with the Hausdorff metric d_H. Let $\mathcal{A}_2 = \mathbb{R}$. Then $f : \mathcal{A}_1 \to \mathcal{A}_2$, defined by $f(K) = \mathcal{P}(K)$, is continuous on $\mathcal{A}_1$ (see for example [63]). Finally, let $D = \mathcal{P}(B_\infty^3)$. From the description of the equality cases (i. e., that K or K^* must be a parallelepiped) proved in [95, 57] we get

$$f^{-1}(D) = \{K \in \mathcal{A}_1; \mathcal{P}(K) = \mathcal{P}(B_\infty^3)\}$$
$$= \{K \in \mathcal{A}_1; K = SB_\infty^3 \text{ or } K = \sqrt{3}SB_1^3, \text{ for some } S \in SO(3)\}.$$

Note that B_∞^3 is in John position (see for example [7]) and thus if $B_2^3 \subset TB_\infty^3 \subset \sqrt{3}B_2^3$ for some $T \in GL(3)$, then $T \in SO(3)$.

Next, we show that the assumptions in the second part of Lemma 1 are satisfied. Since $d_{BM}(K^*, L^*) = d_{BM}(K, L)$, we may restate the $\mathbb{R}^3$ version of Theorem 10 in the following form: There are absolute constants $c_1, c_2 > 0$ such that for every symmetric convex body K in $\mathbb{R}^3$ satisfying $\min\{d_{BM}(K, B_\infty^3), d_{BM}(K, B_1^3)\} := 1 + d \le 1 + c_1$, one has $\mathcal{P}(K) \ge \mathcal{P}(B_\infty^3) + c_2 d$. To finish checking the assumption, note that for all K, L convex bodies such that $B_2^3 \subseteq K, L \subseteq \sqrt{3}B_2^3$, one has

$$d_{BM}(K, L) - 1 \le \min_{T \in GL(3)} d_H(TK, L) \le \sqrt{3}(d_{BM}(K, L) - 1). \tag{5.17}$$

Applying Lemma 1, we deduce that there exists $c > 0$ such that if $B_2^3 \subseteq K \subseteq \sqrt{3}B_2^3$, then

$$\min_{S \in SO(3)} \min(d_H(K, SB_\infty^3), d_H(K, S\sqrt{3}B_1^3)) \le c|\mathcal{P}(K) - \mathcal{P}(B_\infty^3)|.$$

Using (5.17) we conclude the proof. $\qquad\square$

4 Asymptotic estimates and Bourgain–Milman's theorem

If Conjecture 2 holds true for centrally symmetric bodies K, then one has

$$\frac{4}{n!^{\frac{1}{n}}} \le \mathcal{P}(K)^{\frac{1}{n}} \le \frac{\pi}{\Gamma(1 + \frac{n}{2})^{\frac{2}{n}}},$$

so that

$$\frac{4e + o(1)}{n} \le \mathcal{P}(K)^{\frac{1}{n}} \le \frac{2e\pi + o(1)}{n}. \tag{5.18}$$

Similarly, the truth of Conjecture 1 would imply that for any convex body K, one has

$$\mathcal{P}(K)^{\frac{1}{n}} \ge \mathcal{P}(\Delta_n)^{\frac{1}{n}} \ge \frac{e^2 + o(1)}{n},$$

so that the function $K \mapsto n\mathcal{P}(K)^{\frac{1}{n}}$ would vary between two positive constants. This last fact was actually proved by Bourgain and Milman [38] in 1986. Indeed, the upper bound is ensured by the Blaschke–Santaló inequality. For the lower bound, the first important step was made by Gordon and Reisner [74], who proved that

$$\mathcal{P}(K)^{\frac{1}{n}} \ge \frac{c}{n\log(n)}.$$

Then, Bourgain and Milman [38] proved that

$$P(K)^{\frac{1}{n}} \geq \frac{c}{n}. \tag{5.19}$$

For the original proof of (5.19) and other proofs of the same type, see [38, 127, 170]. The constant c obtained in those proofs was not at all explicit, and even if so, it was quite small. After having given a low-technology proof of Gordon–Reisner's result [117], G. Kuperberg [118] gave another proof of (5.19) based on differential geometry, and got the explicit constant $c = \pi e$ in (5.19) in the symmetric case, which is not far from the best possible bound $4e$ and is the best constant known for now. The best constant in the general (i. e., not necessarily symmetric) case may be obtained using the Rogers–Shephard inequality; see the end of this section. Using Fourier transform techniques, other proofs were given by Nazarov [161] (see also Blocki [31, 32], Berndtsson [26, 27], and Mastroianis and Rubinstein [141]). Giannopoulos, Paouris, and Vritsiou gave also a proof using classical techniques of the local theory of Banach spaces [69].

The isomorphic version of the lower bound in (5.18) is "the best possible step" one can make, before actually proving (or disproving) the Mahler conjecture. Indeed, assume we can achieve an asymptotic behavior better than $P(K) \geq c^n P(B_\infty^n)$, $0 < c < 1$, i. e., we have

$$a(n)P(B_\infty^n) \leq P(K) \quad \text{and} \quad \lim_{n \to \infty} a(n)/c^n = \infty, \tag{5.20}$$

but there is a dimension, say l, such that the Mahler conjecture is false in $\mathbb{R}^l$, i. e., there exists a convex symmetric body $K \subset \mathbb{R}^l$ such that $P(K) < P(B_\infty^l)$ or

$$P(K) \leq c_2 P(B_\infty^l), \quad \text{for some } 0 < c_2 < 1.$$

Let K' be the m-th direct sum of copies of K, $K' = K \oplus \cdots \oplus K \subset \mathbb{R}^n$, $n = ml$. Using the direct sum formula and inequality (5.20), we get

$$a(lm)P(B_\infty^{lm}) \leq P(K') = P(K \oplus \cdots \oplus K) \leq c_2^m P(B_\infty^{lm}) = \left(c_2^{1/l}\right)^{lm} P(B_\infty^{lm}).$$

This yields $a(n) \leq c^n$ for $n = ml$ and $c = c_2^{1/l}$, and we get a contradiction for m big enough with $\lim_{n \to \infty} a(n)/c^n = \infty$.

We note that (5.19) for general convex bodies follows (with a constant divided by two) from the symmetric case. Indeed, let L be a convex body in $\mathbb{R}^n$ and let $z \in \text{int}(L)$. Let $K = L - z$. Then by the Rogers–Shephard inequality [177], $\text{vol}(\frac{K-K}{2}) \leq 2^{-n}\binom{2n}{n}\text{vol}(K) \leq 2^n \text{vol}(K)$ and

$$\text{vol}\left(\left(\frac{K-K}{2}\right)^*\right) = \frac{1}{n}\int_{S^{n-1}}\left(\frac{h_K(u) + h_{-K}(u)}{2}\right)^{-n}d\sigma(u) \leq \frac{1}{n}\int_{S^{n-1}} h_K(u)^{-n}d\sigma(u) = \text{vol}(K^*).$$

It follows that

$$\mathrm{vol}(K)\,\mathrm{vol}(K^*) \geq 2^{-n}\mathcal{P}\left(\frac{K-K}{2}\right).$$

Since this holds for every $z \in \mathrm{int}(L)$, it follows that $\mathcal{P}(L) \geq 2^{-n}\mathcal{P}(\frac{L-L}{2})$. From this relation and Kuperberg's best bound $c = \pi e$ in (5.19) for symmetric bodies, it follows that for general convex bodies, (5.19) holds with $c = \pi e/2$.

4.1 Approach via Milman's quotient of subspace theorem

The next lemma is a consequence of the Rogers–Shephard inequality [177].

Lemma 2. *Let K be a convex symmetric body in $\mathbb{R}^n$, let E be an m-dimensional subspace of E, and let $E^\perp$ be its orthogonal subspace. Then*

$$\binom{n}{m}^{-2} \mathcal{P}(K \cap E)\mathcal{P}(K \cap E^\perp) \leq \mathcal{P}(K) \leq \mathcal{P}(K \cap E)\mathcal{P}(K \cap E^\perp).$$

The following result is the *quotient of subspace theorem* of V. Milman ([157], see [71] for a simple proof).

Theorem 12. *Let K be a convex symmetric body in $\mathbb{R}^n$, with n a multiple of 4. Then there exist a constant $c > 0$, independent of n, an $\frac{n}{2}$-dimensional subspace E of $\mathbb{R}^n$, an $\frac{n}{4}$-dimensional subspace F of E, and an ellipsoid $\mathcal{E} \subset F$ such that*

$$\mathcal{E} \subset P_F(E \cap K) \subset c\mathcal{E},$$

where, as before, P_F is the orthogonal projection onto F.

The proof of Bourgain–Milman's theorem by Pisier [170]. For a convex symmetric body $K \subset \mathbb{R}^n$, with n a multiple of 4, let $a_n(K) = n\mathcal{P}(K)^{\frac{1}{n}}$. Let E and F be the subspaces of $\mathbb{R}^n$ chosen in Theorem 12. By Lemma 2, for some constant $d > 0$ independent of n, one has

$$d\sqrt{a_{\frac{n}{2}}(K \cap E)a_{\frac{n}{2}}(K \cap E^\perp)} \leq a_n(K) \leq \sqrt{a_{\frac{n}{2}}(K \cap E)a_{\frac{n}{2}}(K \cap E^\perp)}$$

and

$$d\sqrt{a_{\frac{n}{4}}(P_F(K \cap E))a_{\frac{n}{4}}(K \cap E \cap F^\perp)} \leq a_{\frac{n}{2}}(K \cap E) \leq \sqrt{a_{\frac{n}{4}}(P_F(K \cap E))a_{\frac{n}{4}}(K \cap E \cap F^\perp)}.$$

Next, from Theorem 12, for some absolute constants $c', d' > 0$, one has

$$c' \leq a_{\frac{n}{4}}(P_F(K \cap E)) \leq d'.$$

It follows that for some universal constant $c > 0$, one has

$$a_n(K) \geq c(a_{\frac{n}{2}}(K \cap E^\perp))^{\frac{1}{2}}(a_{\frac{n}{4}}(K \cap E \cap F^\perp))^{\frac{1}{4}}. \tag{5.21}$$

Define now for every $n \geq 1$

$$a_n = \min\{a_m(L); 1 \leq m \leq n,\ L \text{ convex symmetric body in } \mathbb{R}^m\}.$$

Observing that $a_n > 0$, from (5.21) one obtains

$$a_n \geq c(a_n)^{\frac{1}{2}}(a_n)^{\frac{1}{4}}. \tag{5.22}$$

Thus, $a_n \geq c^4$. $\qquad\qquad\qquad\qquad\qquad\qquad\qquad\qquad\qquad\qquad\qquad\qquad\qquad\square$

4.2 Complex analysis approach

Let us very briefly discuss an approach via complex and harmonic analysis which was initiated by Nazarov [161]. We will follow here a work of Berndtsson [27], which is based on functional inequalities (central for the next section).

We will consider a special case of the Bergman spaces. Let $\psi : \mathbb{C}^n \to \mathbb{R} \cup \{+\infty\}$ be a convex function and let $\Omega = \{(x,y) \in \mathbb{R}^{2n} : \psi(x + iy) < \infty\}$. The *Bergman space* $A^2(e^{-\psi})$ is the Hilbert space of holomorphic functions f on Ω such that

$$\|f\|^2 = \int_{\Omega} |f(x + iy)|^2 e^{-\psi(x+iy)} \, dx dy < \infty.$$

The (diagonal) Bergman kernel B for $A^2(e^{-\psi})$ is defined as

$$B(z) = \sup_{f \in A^2(e^{-\psi})} \frac{|f(z)|^2}{\|f\|^2}.$$

Next, consider an even convex function $\phi : \mathbb{R}^n \to \mathbb{R} \cup \{+\infty\}$ such that $e^{-\phi(x)}$ is integrable over $\mathbb{R}^n$. For $\alpha \in \mathbb{C}$, consider the Bergman kernel $B_\alpha(z)$ corresponding to the function $\psi(z) = \phi(\mathrm{Re}(z)) + \phi(\mathrm{Re}(\alpha z))$. The main theorem in [27] is the claim that

$$B_i(0) \leq c^n B_1(0), \tag{5.23}$$

where c is an absolute constant (precisely computed in [27]). We note that B_1 is the Bergman kernel for $\psi(x + iy) = 2\phi(x)$, i.e., independent of $\mathrm{Im}(z)$, and that B_i is the Bergman kernel for $\psi(x+iy) = \phi(x)+\phi(y)$. It is essential to understand that the Bergman spaces corresponding to those densities are different. Thus, the connection is not immediate. For example, the function $f = 1$ belongs to the second space but does not belong to the space corresponding to $\psi(x + iy) = 2\phi(x)$. Using $f = 1$ we get

$$B_i(0) \geq \frac{1}{\int_{\mathbb{R}^n} e^{-\phi(x)} dx \int_{\mathbb{R}^n} e^{-\phi(y)} dy}.$$

Together with (5.23) this gives

$$B_1(0) \geq \frac{c^{-n}}{(\int_{\mathbb{R}^n} e^{-\phi(x)} dx)^2}, \tag{5.24}$$

which is an essential estimate for proving the Bourgain–Milman inequality. The proof of (5.23) in [27] is based on a very nice and tricky approach of "linking" B_i and B_1 via B_α. Indeed, it turns out that $b(\alpha) := \log B_\alpha(0)$ is subharmonic in $\mathbb{C}$ (see [25, 27]), and moreover $b(\alpha) \leq C + n \log |\alpha|^2$, which can be seen from the change of variables

$$\|f\|_\alpha^2 = \int_{\mathbb{C}^n} |f(z)|^2 e^{-(\phi(\mathrm{Re}(z))+\phi(\mathrm{Re}(\alpha z)))} dz$$

$$= |\alpha|^{-2n} \int_{\mathbb{C}^n} |f(z/\alpha)|^2 e^{-(\phi(\mathrm{Re}(z/\alpha))+\phi(\mathrm{Re}(z)))} dz.$$

Thus, $B_\alpha(0) = |\alpha|^{2n} B_{1/\alpha}(0)$. Moreover, $B_{1/\alpha}(0)$ is bounded as $\alpha \to \infty$. Thus, one can apply the Poisson representation formula in the upper half-plane to the function $b(\alpha) - n \log |\alpha|^2$ to get

$$\log B_i(0) = b(i) \leq \frac{1}{\pi} \int_{-\infty}^{\infty} \frac{b(s) - n \log(s^2)}{1 + s^2} ds = \frac{2}{\pi} \int_0^{\infty} \frac{b(s) - n \log(s^2)}{1 + s^2} ds.$$

Using the fact that $s \mapsto \phi(sx)$ is increasing on $(0, 1]$, one has, for $s \in (0, 1]$,

$$\|f\|_s^2 = \int_{\mathbb{C}^n} |f(z)|^2 e^{-(\phi(\mathrm{Re}(z))+\phi(\mathrm{Re}(sz)))} dz \geq \|f\|_1^2,$$

and hence $b(s) \leq b(1)$. If $s \geq 1$,

$$\|f\|_s^2 = |s|^{-2n} \int_{\mathbb{C}^n} |f(z/s)|^2 e^{-(\phi(\mathrm{Re}(z/\alpha))+\phi(\mathrm{Re}(z)))} dz \geq s^{-2n} \|f\|_1^2,$$

and thus $b(s) \leq b(1) + n \log s^2$. Putting those estimates together completes the proof of (5.23).

The next step is to adapt the Paley–Wiener space associated to a convex body (discussed in Theorem 4) to the case of a convex function. For a convex function $\varphi : \mathbb{R}^n \to \mathbb{R} \cup \{+\infty\}$, we denote by $PW(e^\varphi)$ the space of holomorphic functions f of the form

$$f(z) = \int_{\mathbb{R}^n} e^{\langle z, \xi \rangle} \tilde{f}(\xi) d\xi, \quad \text{where } z \in \mathbb{C}^n,$$

for which

$$\|f\|_{PW}^2 = \int_{\mathbb{R}^n} |\tilde{f}|^2 e^{\varphi} dt < \infty,$$

for some function $\tilde{f}$, so that the two formulas above make sense. The classical Paley–Wiener space discussed in Theorem 4 then corresponds to the case when $\varphi(x) = 0$ for $x \in K$ and $\varphi(x) = +\infty$ for $x \notin K$. For a convex function ψ on $\mathbb{R}^n$, let us consider its logarithmic Laplace transform given by

$$\Lambda\psi(\xi) = \log \int_{\mathbb{R}^n} e^{2\langle x,\xi\rangle} e^{-\psi} dx.$$

The second key ingredient in Berndtsson's proof is the fact that the spaces $PW(e^{\Lambda\psi})$ and $A^2(e^{-\psi})$ coincide and that

$$\|f\|_{A^2}^2 = (2\pi)^{2n} \|f\|_{PW(e^{\Lambda\psi})}^2. \tag{5.25}$$

This fact originates from the observation that any $f \in PW(e^{\Lambda\psi})$ is the Fourier–Laplace transform of $\tilde{f}$ and $e^{\langle x,t\rangle}\tilde{f}(t)$ belongs to $L_2(\mathbb{R}^n)$ for all x such that $\psi(x) < \infty$. Then, we apply Parseval's formula to get

$$\int_{\mathbb{R}^n} |f(x+iy)|^2 dy = (2\pi)^n \int_{\mathbb{R}^n} e^{2\langle x,t\rangle} |\tilde{f}(t)|^2 dt.$$

Multiplying the above equality by $e^{-\psi(x)}$ and integrating with respect to x, we get

$$\int_{\mathbb{R}^n}\int_{\mathbb{R}^n} |f(x+iy)|^2 e^{-\psi(x)} dxdy = (2\pi)^n \int_{\mathbb{R}^n} |\tilde{f}(t)|^2 e^{\Lambda\psi(t)} dt.$$

Thus, $f \in A^2(e^{-\psi})$ and the A^2-norm coincides with a multiple of the norm in $PW(e^{\Lambda\psi})$. This confirms that the Paley–Wiener space is isometrically embedded into the corresponding Bergman space and the rest follows from the observation that it is dense.

One can compute the Bergman kernel for $PW(e^{\Lambda\psi})$ and use (5.25) to show that the Bergman kernel for $A^2(e^{-\psi})$ is equal to

$$(2\pi)^{-n} \int_{\mathbb{R}^n} e^{2\langle x,t\rangle - \Lambda\psi(t)} dt. \tag{5.26}$$

We will use (5.26) to give an estimate from above of the value of the Bergman kernel at zero. The Legendre transform $\mathcal{L}\psi$ of a function $\psi : \mathbb{R}^n \to \mathbb{R} \cup \{+\infty\}$ is defined by

$$\mathcal{L}\psi(y) = \sup_{x\in\mathbb{R}^n}(\langle x,y\rangle - \psi(x)), \quad \text{for } y \in \mathbb{R}^n. \tag{5.27}$$

Consider the Bergman space $A^2(e^{-2\phi(x)})$, where $\phi : \mathbb{R}^n \to \mathbb{R} \cup \{\infty\}$ is convex and even (as in (5.24)). Then

$$B(0) \leq \pi^{-n} \frac{\int_{\mathbb{R}^n} e^{-\mathcal{L}\phi(y)} \, dy}{\int_{\mathbb{R}^n} e^{-\phi(x)} \, dx}. \tag{5.28}$$

Indeed, using (5.26) we get

$$B(0) \leq (2\pi)^{-n} \int_{\mathbb{R}^n} e^{-\Lambda(2\phi(t))} \, dt. \tag{5.29}$$

Note that, for any $y \in \mathbb{R}^n$, one has

$$e^{\Lambda(2\phi(t))} = 2^{-n} \int_{\mathbb{R}^n} e^{\langle t,u \rangle - 2\phi(u/2)} \, du = 2^{-n} e^{\langle t,y \rangle} \int_{\mathbb{R}^n} e^{\langle t,v \rangle - 2\phi(v/2+y/2)} \, dv$$
$$\geq 2^{-n} e^{\langle t,y \rangle - \phi(y)} \int_{\mathbb{R}^n} e^{\langle t,v \rangle - \phi(v)} \, dv,$$

where in the last inequality we used the convexity of ϕ. Using the fact that ϕ is even, we obtain

$$\int_{\mathbb{R}^n} e^{\langle t,v \rangle - \phi(v)} \, dv \geq \int_{\mathbb{R}^n} e^{-\phi(v)} \, dv$$

and

$$e^{\Lambda(2\phi(t))} \geq 2^{-n} e^{\langle t,y \rangle - \phi(y)} \int_{\mathbb{R}^n} e^{-\phi(v)} \, dv.$$

Taking the supremum over all $y \in \mathbb{R}^n$, we get

$$e^{\Lambda(2\phi(t))} \geq 2^{-n} e^{\mathcal{L}\phi(t)} \int_{\mathbb{R}^n} e^{-\phi(v)} \, dv.$$

Together with (5.29), this gives (5.28). Combining (5.28) with (5.24), we get the following theorem.

Theorem 13 (Functional version of the Bourgain–Milman inequality). *Let $\phi : \mathbb{R}^n \to \mathbb{R} \cup \{+\infty\}$ be even and convex; then for some $c > 0$ independent of n, one has*

$$\int_{\mathbb{R}^n} e^{-\phi(x)} \, dx \int_{\mathbb{R}^n} e^{-\mathcal{L}\phi(x)} \, dx \geq c^n.$$

Remark 7. Theorem 13 was first proved via the Bourgain–Milman inequality for symmetric convex bodies in [10] and then generalized to non-even functions in [60]. It implies the classical Bourgain–Milman inequality for convex bodies, as we shall see in the next section (see Remark 11 below).

5 Functional inequalities and a link with transport inequalities

We dedicate this section to the study of functional inequalities related to the volume product.

5.1 Upper bounds

The following general form of the functional Blaschke–Santaló inequality was proved by Ball [15] for f even, by Fradelizi and Meyer [61] for f log-concave, and by Lehec [120] in the general case.

Theorem 14. *Let $f : \mathbb{R}^n \to \mathbb{R}_+$ be Lebesgue integrable. There exists $z \in \mathbb{R}^n$ such that for any $\rho : \mathbb{R}_+ \to \mathbb{R}_+$ and any $g : \mathbb{R}^n \to \mathbb{R}_+$ measurable satisfying*

$$f(x + z)g(y) \le \rho(\langle x,y \rangle)^2 \quad \text{for all } x,y \in \mathbb{R}^n \text{ satisfying } \langle x,y \rangle > 0,$$

one has

$$\int f(x)\,dx \int g(y)\,dy \le \left(\int \rho(|x|^2)\,dx \right)^2.$$

If f is even, one can take $z = 0$.

Applying this result to $\rho = \mathbf{1}_{[0,1]}$ and $f = \mathbf{1}_K$, one recovers the Blaschke–Santaló inequality for convex sets. Applying it to $\rho(t) = e^{-t/2}$, it gives a proof of the following functional Blaschke–Santaló inequality for the Legendre transform due to Artstein, Klartag, and Milman [10] (see [121] for another proof).

Theorem 15. *Let $\varphi : \mathbb{R}^n \to \mathbb{R} \cup \{+\infty\}$ satisfy $0 < \int e^{-\varphi} < +\infty$. If for $x,y \in \mathbb{R}^n$, $\varphi_y(x) := \varphi(x + y)$, there exists $z \in \mathbb{R}^n$ such that*

$$\int_{\mathbb{R}^n} e^{-\varphi(x)}\,dx \int_{\mathbb{R}^n} e^{-\mathcal{L}(\varphi_z)(y)}\,dy \le \left(\int_{\mathbb{R}^n} e^{-\frac{|x|^2}{2}}\,dx \right)^2 = (2\pi)^n,$$

with equality if and only if $\varphi_z(x) = |Ax|^2$ for some invertible linear map A and some $z \in \mathbb{R}^n$.

Remark 8. In [121], Lehec deduced from Theorem 15 that if the "barycenter" $b(\varphi) :=$ $\int x e^{-\varphi(x)} dx / \int e^{-\varphi}$ satisfies $b(\varphi) = 0$, then

$$\int e^{-\varphi} \int e^{-\mathcal{L}\varphi} \le (2\pi)^n.$$

Indeed, for any z, one has $\mathcal{L}(\varphi_z)(y) = \mathcal{L}\varphi(y) - \langle y, z \rangle$. It follows that $\mathcal{L}((\mathcal{L}\varphi)_z)(y) = \mathcal{L}\mathcal{L}\varphi(y) - \langle y, z \rangle \le \varphi(y) - \langle y, z \rangle$. Using Jensen's inequality and $b(\varphi) = 0$, we get

$$\int e^{-\mathcal{L}((\mathcal{L}\varphi)_z)} \ge \int e^{-\varphi(y) + \langle y, z \rangle} dy \ge e^{\langle b(\varphi), z \rangle} \int e^{-\varphi} = \int e^{-\varphi}.$$

Applying Theorem 15 to $\mathcal{L}\varphi$, there exists thus a z such that

$$\int e^{-\varphi} \int e^{-\mathcal{L}\varphi} \le \int e^{-\mathcal{L}((\mathcal{L}\varphi)_z)} \int e^{-\mathcal{L}\varphi} \le (2\pi)^n.$$

As Lehec observed also, this gives a new proof of the result of Lutwak [132].

Proposition 2. *For a star-shaped body $K \subset \mathbb{R}^n$ (for all $(x, t) \in K \times [0, 1]$, one has $tx \in K$) with barycenter at 0, one has*

$$\operatorname{vol}(K) \operatorname{vol}(K^*) \le \operatorname{vol}(B_2^n)^2.$$

Proof. Let $\varphi(x) = \frac{\|x\|_K^2}{2}$. Then since

$$\int_{\mathbb{R}^n} x e^{-\frac{\|x\|_K^2}{2}} dx = \int_{\mathbb{R}^n} x \int_{\|x\|_K}^{+\infty} t e^{-\frac{t^2}{2}} dt\, dx = \int_0^{+\infty} t^{n+1} e^{-\frac{t^2}{2}} dt \int_K x\, dx = 0,$$

one has $b(\varphi) = 0$. Moreover, for any $y \in \mathbb{R}^n$, one has $\mathcal{L}\varphi(y) = \sup_x \langle x, y \rangle - \frac{\|x\|_K^2}{2} = \frac{\|y\|_{K^*}^2}{2}$ and $\int_{\mathbb{R}^n} e^{-\frac{\|x\|_K^2}{2}} dx = 2^{\frac{n}{2}} \Gamma(\frac{n}{2} + 1) \operatorname{vol}(K)$. $\square$

Before giving sketches of various proofs of Theorems 14 and 15, we need a lemma.

Lemma 3. *Let $\alpha, \beta, \gamma : \mathbb{R}_+ \to \mathbb{R}_+$ be measurable functions such that for every $s, t > 0$ one has $\alpha(s)\beta(t) \le \gamma(\sqrt{st})^2$. Then $\int_{\mathbb{R}_+} \alpha(t) dt \int_{\mathbb{R}_+} \beta(t) dt \le (\int_{\mathbb{R}_+} \gamma(t) dt)^2$.*

Proof. Define $f, g, h : \mathbb{R} \to \mathbb{R}$ by $f(x) = \alpha(e^x)e^x$, $g(x) = \beta(e^x)e^x$, and $h(x) = \gamma(e^x)e^x$. Then $f(x)g(y) \le h(\frac{x+y}{2})$ for all $x, y \in \mathbb{R}$. By the Prékopa–Leindler inequality (see [170, p. 3]) we get $\int_{\mathbb{R}} f(x) dx \int_{\mathbb{R}} g(x) dx \le (\int_{\mathbb{R}} h(x) dx)^2$. We conclude with a change of variables. $\square$

Proofs of Theorem 14. (1) In the case where f is even and ρ is decreasing, this proof is due to Ball [15]. For $s, t \in \mathbb{R}_+$, let $K_s = \{f \ge s\}$ and $L_t = \{g \ge t\}$. The hypothesis on f and g implies that $L_t \subset \rho^{-1}(\sqrt{st})K_s^*$. Since f is even, K_s is symmetric. We deduce from the Blaschke–Santaló inequality that for every $s, t \in \mathbb{R}_+$, if $\alpha(s) = \operatorname{vol}(K_s)$ and $\beta(t) = \operatorname{vol}(L_t)$, one has

$$\alpha(s)\beta(t) = \text{vol}(K_s)\,\text{vol}(L_t) \le \left(\rho^{-1}(\sqrt{st})\right)^n \text{vol}(K_s)\,\text{vol}(K_s^*) \le \left(\rho^{-1}(\sqrt{st})\right)^n \text{vol}(B_2^n)^2.$$

Denoting $\gamma(t) = (\rho^{-1}(t))^{n/2}\,\text{vol}(B_2^n)$, we apply Lemma 3 and use the fact that $\int_{\mathbb{R}^n} f(x)dx = \int_0^{+\infty} \alpha(s)ds$ to conclude the proof in this case.

(2) In the case where f is not supposed to be even, but is log-concave, the proof of Theorem 14 given in [61] uses the so-called Ball's body $K_f(z)$ associated to a log-concave function f, which is defined by

$$K_f(z) = \left\{ x \in \mathbb{R}^n;\; \int_0^{+\infty} r^{n-1} f(z+rx)dr \ge 1 \right\}.$$

It follows from Ball's results [16] that $K_f(z)$ is convex and that its radial function is $r_{K_f(z)}(x) = \left(\int_0^{+\infty} r^{n-1}f(z+rx)dr\right)^{\frac{1}{n}}$ for $x \in \mathbb{R}^n \setminus \{0\}$. If $x,y \in \mathbb{R}^n$ satisfy $\langle x,y \rangle > 0$, define, for $r \ge 0$, $\alpha(r) = r^{n-1}f(z+rx)$, $\beta(r) = r^{n-1}g(rx)$, and $\gamma(r) = r^{n-1}\rho(r^2\langle x,y\rangle)$. It follows from Lemma 3 that

$$\int_0^{+\infty} r^{n-1}f(z+rx)dr \int_0^{+\infty} r^{n-1}g(rx)dr \le \left(\int_0^{+\infty} r^{n-1}\rho(r^2\langle x,y\rangle)dr \right)^2.$$

This means that

$$\langle x,y \rangle \le \frac{c_n(\rho)}{r_{K_f(z)}(x)\,r_{K_g(0)}(y)}, \quad \text{where } c_n(\rho) := \left(\int_0^{+\infty} r^{n-1}\rho(r^2)dr \right)^{2/n},$$

or in other words, $K_g(0) \subset c_n(\rho)K_f(z)^*$. Moreover, one has

$$\int_{\mathbb{R}^n} f(x)dx = n\,\text{vol}(K_f(z)) \quad \text{for every } z \in \text{supp}(f).$$

Using Brouwer's fixed point theorem, it was proved in [61] that for some $z \in \mathbb{R}^n$, the center of mass of $K_f(z)$ is at the origin. The result follows then from the Blaschke–Santaló inequality.

This method was also used in [21] to prove stability versions of the functional forms of the Blaschke–Santaló inequality. $\qquad\qquad\square$

Proofs of Theorem 15. (1) The proof given in [10] attaches to $\varphi : \mathbb{R}^n \to \mathbb{R} \cup \{+\infty\}$, *supposed here to be even*, the functions $f_m(x) = (1 - \frac{\varphi(x)}{m})_+^m$, for $m \ge 1$, and the convex bodies

$$K_m(f_m) := \{(x,y) \in \mathbb{R}^{n+m};\, |y| \le f_m(\sqrt{m}x)^{1/m}\}.$$

When $m \to +\infty$, $f_m \to e^{-\varphi}$ and

$$m^{\frac{n}{2}} \frac{\mathrm{vol}(K_m(f_m))}{\mathrm{vol}(B_2^m)} = \int_{\mathbb{R}^n} f_m(x)dx \to \int_{\mathbb{R}^n} e^{-\varphi(x)} dx.$$

Moreover, $K_m(f_m)^* = K_m(\mathcal{L}_m f_m)$, where $\mathcal{L}_m(f_m)(y) = \inf_x \frac{(1-\frac{\langle x,y\rangle}{m})_+^m}{f_m(x)}$. Also, when $m \to +\infty$,

$$\mathcal{L}_m(f_m)(y) \to e^{-\mathcal{L}\varphi(y)} \quad \text{and} \quad m^{\frac{n}{2}} \frac{\mathrm{vol}(K_m(\mathcal{L}_m\varphi))}{\mathrm{vol}(B_2^m)} = \int_{\mathbb{R}^n} \mathcal{L}_m f_m(x)dx \to \int_{\mathbb{R}^n} e^{-\mathcal{L}\varphi(x)} dx.$$

One then applies the Blaschke–Santaló inequality to the bodies $K_m(f_m)$.

(2) Lehec's proof [121] of Theorem 15 uses induction on the dimension. For $n = 1$, choose $z \in \mathbb{R}$ such that $\int_z^{+\infty} e^{-\varphi(t)}dt = \int_{\mathbb{R}} e^{-\varphi(t)}dt/2$. For all $s, t \geq 0$, one has $\varphi_z(s) + \mathcal{L}(\varphi_z)(t) \geq st$. Thus, the functions $\alpha(s) = e^{-\varphi_z(s)}$, $\beta(t) = e^{-\mathcal{L}(\varphi_z)(t)}$, and $\gamma(u) = e^{-u^2/2}$ satisfy $\alpha(s)\beta(t) \leq \gamma(\sqrt{st})^2$, for every $s, t \geq 0$. It follows from Lemma 3 that

$$\int_0^{+\infty} e^{-\varphi_z(t)}dt \int_0^{+\infty} e^{-\mathcal{L}(\varphi_z(t))}dt = \int_{\mathbb{R}_+} \alpha(t)dt \int_{\mathbb{R}_+} \beta(t)dt \leq \left(\int_{\mathbb{R}_+} \gamma(u)du\right)^2 = \frac{\pi}{2}. \tag{5.30}$$

This inequality also holds on $\mathbb{R}_-$; adding the two inequalities, we get the result.

Now suppose that the result holds for n and let us perform the induction step. Let $\varphi : \mathbb{R}^{n+1} \to \mathbb{R} \cup \{+\infty\}$. If $X \in \mathbb{R}^{n+1}$, we denote $X = (x, s) \in \mathbb{R}^n \times \mathbb{R}$. Let

$$\mathcal{P}(\varphi) := \min_z \int_{\mathbb{R}^{n+1}} e^{-\varphi(X)}dX \int_{\mathbb{R}^{n+1}} e^{-\mathcal{L}(\varphi_z)(X)}dX.$$

For any invertible affine map A, one has $\mathcal{P}(\varphi \circ A) = \mathcal{P}(\varphi)$. Translating φ in the e_{n+1} direction, we may assume that

$$\int_{s>0} \int e^{-\varphi(x,s)}dxds = \int_{s<0} \int e^{-\varphi(x,s)}dxds.$$

Define $b_+(\varphi)$ and $b_-(\varphi)$ in $\mathbb{R}^{n+1}$ by

$$b_+(\varphi) = \frac{\int_{s>0} \int (x,s)e^{-\varphi(x,s)}dxds}{\int_{s>0} \int e^{-\varphi}dxds} \quad \text{and} \quad b_-(\varphi) = \frac{\int_{s<0} \int (x,s)e^{-\varphi(x,s)}dxds}{\int_{s<0} \int e^{-\varphi}dxds}.$$

Since $\langle b_+(\varphi), e_{n+1}\rangle > 0$ and $\langle b_-(\varphi), e_{n+1}\rangle < 0$, the point $\{z\} := [b_-(\varphi), b_+(\varphi)] \cap e_{n+1}^\perp$ is well-defined. By translating φ in the remaining directions, we may assume that $z = 0$. Let A be the linear invertible map defined by $Ax = x$ for $x \in \mathbb{R}^n$ and $Ae_{n+1} = b_+(\varphi)$. Then $b_+(\varphi \circ A) = A^{-1}b_+(\varphi) = e_{n+1}$. Changing φ into $\varphi \circ A$, we may assume that $b_+(\varphi) = e_{n+1}$. We define $\Phi, \Psi : \mathbb{R}^n \to \mathbb{R} \cup \{+\infty\}$ by

$$e^{-\Phi(x)} = \int_0^{+\infty} e^{-\varphi(x,t)}\,dt \quad \text{and} \quad e^{-\Psi(x)} = \int_0^{+\infty} e^{-\mathcal{L}\varphi(x,t)}\,dt.$$

Since $b_+(\varphi) = e_{n+1}$, we get $\int_{\mathbb{R}^n} x e^{-\Phi(x)}\,dx = \int_{t>0}\int_{\mathbb{R}^n} x e^{-\varphi(x,t)}\,dx\,dt = 0$. Hence, $b(\Phi) = 0$. From the induction hypothesis and the remark after Theorem 15, it follows that

$$\int_{\mathbb{R}^n} e^{-\Phi(x)}\,dx \int_{\mathbb{R}^n} e^{-\mathcal{L}\Phi(y)}\,dy \le (2\pi)^n. \tag{5.31}$$

For every $x,y \in \mathbb{R}^n$ and $s,t \in \mathbb{R}$, let $\varphi^x(s) = \varphi(x,s)$ and $(\mathcal{L}\varphi)^y(t) = \mathcal{L}\varphi(y,t)$. Applying again Lemma 3 as in (5.30), we get

$$\int_0^{+\infty} e^{-\varphi^x(s)}\,ds \int_0^{+\infty} e^{-\mathcal{L}(\varphi^x)(t)}\,dt \le \frac{\pi}{2}.$$

Since $\varphi^x(s)+(\mathcal{L}\varphi)^y(t) \ge \langle x,y\rangle + st$, one has $(\mathcal{L}\varphi)^y(t) - \langle x,y\rangle \ge \mathcal{L}(\varphi^x)(t)$. Thus, for $x,y \in \mathbb{R}^n$,

$$e^{-\Phi(x)-\Psi(y)} = \int_0^{+\infty} e^{-\varphi^x(s)}\,ds \int_0^{+\infty} e^{-(\mathcal{L}\varphi)^y(t)}\,dt \le \frac{\pi}{2} e^{-\langle x,y\rangle}.$$

This implies that $e^{-\Psi(y)} \le \frac{\pi}{2} e^{-\mathcal{L}\Phi(y)}$. Using (5.31), we get

$$\int_{\mathbb{R}^n} e^{-\Phi(x)}\,dx \int_{\mathbb{R}^n} e^{-\Psi(y)}\,dy \le \frac{\pi}{2}(2\pi)^n,$$

that is

$$\int_0^{+\infty}\int_{\mathbb{R}^n} e^{-\varphi(x,s)}\,dx\,ds \int_0^{+\infty}\int_{\mathbb{R}^n} e^{-\mathcal{L}\varphi(y,t)}\,dy\,dt \le \frac{\pi}{2}(2\pi)^n.$$

Adding this to the analogous bound for $s < 0$ and using $\int_{s>0}\int e^{-\varphi(x,s)}\,dx\,ds = \int_{s<0}\int e^{-\varphi(x,s)}\,dx\,ds$, we conclude. $\square$

Remark 9. Various L_p-versions of the functional Blaschke–Santaló inequalities have been given (see for instance [84]). Also, Blaschke–Santaló-type inequalities were established in the study of extremal general affine surface areas [68, 198, 92]. A consequence of the Blaschke–Santaló inequality was recently given in [197].

5.2 Lower bounds of the volume product of log-concave functions

Let $\varphi : \mathbb{R}^n \to \mathbb{R} \cup \{+\infty\}$ be convex. The *domain of φ* is $\mathrm{dom}(\varphi) := \{x \in \mathbb{R}^n; \varphi(x) < +\infty\}$. If $0 < \int e^{-\varphi} < +\infty$, we define the *functional volume product of φ* as

$$\mathcal{P}(\varphi) = \min_z \int_{\mathbb{R}^n} e^{-\varphi(x)}\,dx \int_{\mathbb{R}^n} e^{-\mathcal{L}(\varphi_z)(y)}\,dy.$$

If φ is even, this minimum is reached at 0. The following conjectures were proposed in [60].

Conjecture 3. *If $n \geq 1$ and $\varphi : \mathbb{R}^n \to \mathbb{R} \cup \{+\infty\}$ is a convex function such that $0 < \int e^{-\varphi} < +\infty$, then*

$$\int_{\mathbb{R}^n} e^{-\varphi(x)}\,dx \int_{\mathbb{R}^n} e^{-\mathcal{L}\varphi(y)}\,dy \geq e^n,$$

with equality if and only if there are a constant $c > 0$ and an invertible linear map T such that

$$e^{-\varphi(Tx)} = c \prod_{i=1}^{n} e^{-x_i} \mathbf{1}_{[-1,+\infty)}(x_i).$$

Conjecture 4. *If $n \geq 1$ and $\varphi : \mathbb{R}^n \to \mathbb{R} \cup \{+\infty\}$ is an even convex function such that $0 < \int e^{-\varphi} < +\infty$, then*

$$\int_{\mathbb{R}^n} e^{-\varphi(x)}\,dx \int_{\mathbb{R}^n} e^{-\mathcal{L}\varphi(y)}\,dy \geq 4^n,$$

with equality if and only if there exist a constant $c > 0$, two complementary subspaces F_1 and F_2, and two Hanner polytopes $K_1 \subset F_1$ and $K_2 \subset F_2$ such that for all $(x_1, x_2) \in F_1 \times F_2$,

$$e^{-\varphi(x_1+x_2)} = ce^{-\|x_1\|_{K_1}} \mathbf{1}_{K_2}(x_2).$$

Remark 10. With a different duality for a convex function φ, another Blaschke–Santaló and inverse Santaló inequality were obtained in [13, 54]. Another extension of the Blaschke–Santaló inequality and of its functional form was considered in [90], where duality is combined with the study of inequalities related to monotone non-trivial Minkowski endomorphisms.

Partial results toward the proofs of these conjectures are gathered in the following theorem.

Theorem 16. *Let $n \geq 1$ and let $\varphi : \mathbb{R}^n \to \mathbb{R} \cup \{+\infty\}$ be a convex function such that $0 < \int e^{-\varphi} < +\infty$. Then:*
(1) *Conjecture 3 holds for $n = 1$. It holds also for all $n \geq 1$ if there exists an invertible affine map T such that $\mathrm{dom}(\varphi \circ T) = \mathbb{R}^n_+$ and $\varphi \circ T$ is non-decreasing on P, in the sense that if $x_i \leq y_i$ for all $1 \leq i \leq n$, $(\varphi \circ T)(x_1, \ldots, x_n) \leq (\varphi \circ T)(y_1, \ldots, y_n)$.*
(2) *Conjecture 4 holds if $n = 1$ or $n = 2$. It holds also for all $n \geq 1$ if φ is unconditional, in the sense that there exists an invertible linear map T such that $(\varphi \circ T)(x_1, \ldots, x_n) = (\varphi \circ T)(|x_1|, \ldots, |x_n|)$ for all $(x_1, \ldots, x_n) \in \mathbb{R}^n$.*

(1) For $n = 1$, Conjecture 3 was proved in two different ways in [58, 60]. The case of non-decreasing convex functions on the positive octant was also proved in [60].

(2) For unconditional convex functions on $\mathbb{R}^n$, Conjecture 4 was established in two different ways in [59, 60], with the case of equality in [55]. In particular, this settles the general case $n = 1$. For $n = 2$, it was proved in [62].

Remark 11. There is a strong link between Conjectures 1 and 2 for convex bodies and their functional counterparts Conjectures 3 and 4. Indeed, as observed in [60], given a symmetric convex body K in $\mathbb{R}^n$, if $\varphi_K(x) = \|x\|_K$, we get $e^{-\mathcal{L}\varphi_K} = \mathbf{1}_{K^*}$, and integrating on level sets, $\mathcal{P}(\varphi_K) = n!\mathcal{P}(K)$. Therefore, if Conjecture 4 holds for φ_K, then Conjecture 2 holds for K. Reciprocally, if Conjecture 2 holds in $\mathbb{R}^n$ for every dimension n, then, given an even, convex function $\varphi : \mathbb{R}^n \to \mathbb{R} \cup \{+\infty\}$, we can apply it in dimension $n + m$ to the convex sets

$$K_m(\varphi) = \left\{(x,y) \in \mathbb{R}^n \times \mathbb{R}^m; \|y\|_\infty \le \left(1 - \frac{\varphi(mx)}{m}\right)_+\right\}.$$

Using

$$\mathrm{vol}_{n+m}(K_m(\varphi)) = \frac{2^m}{m^n} \int_{\mathbb{R}^n} \left(1 - \frac{\varphi(x)}{m}\right)_+^m dx$$

and

$$K_m(\varphi)^* = \left\{(x,y) \in \mathbb{R}^n \times \mathbb{R}^m; \|y\|_1 \le \inf_{\varphi(x') \le m} \frac{(1 - \langle x, x'\rangle)_+}{1 - \frac{\varphi(x')}{m}}\right\},$$

it is proved in [60] that when $m \to +\infty$, the inequality $\mathcal{P}(K_m(\varphi)) \ge \frac{4^{n+m}}{(n+m)!}$ gives $\mathcal{P}(\varphi) \ge 4^n$. In a similar way, if Conjecture 3 holds in dimension $n + 1$, given a convex body K in $\mathbb{R}^n$ with Santaló point at the origin, we apply it to $\varphi : \mathbb{R}^n \times \mathbb{R} \to \mathbb{R} \cup \{+\infty\}$ defined by

$$e^{-\varphi(x,t)} = \mathbf{1}_{[-n-1,+\infty)}(t)\mathbf{1}_{(t+n+1)K}(x)e^{-t}.$$

Then the Legendre transform of φ is

$$e^{-\mathcal{L}\varphi(y,s)} = \mathbf{1}_{(-\infty,1]}(s)\mathbf{1}_{(1-s)K^*}(y)e^{(n+1)(s-1)},$$

and

$$\mathcal{P}(\varphi) = \frac{(n!)^2 e^{n+1}}{(n+1)^{n+1}}|K||K^*|.$$

This proves that if Conjecture 3 holds for φ, then

$$\mathcal{P}(K) \ge \mathcal{P}(\Delta_n) = \frac{(n+1)^{n+1}}{(n!)^2},$$

which is Conjecture 1 for K. Lastly, as shown in [60], one can adapt the arguments for even functions to prove that, given a convex function $\varphi : \mathbb{R}^n \to \mathbb{R} \cup \{+\infty\}$, Conjecture 2 applied to a well-chosen sequence of bodies $\Delta_m(\varphi)$ in dimension $n+m$ gives Conjecture 3 for φ when $m \to +\infty$.

It was also proved in [72] that if $\mathcal{P}(\varphi)$ is minimal, then φ has no positive Hessian at any point. Asymptotic estimates hold too: in the even case, it was proved in [113] that for some constant $c > 0$, one has, for all even convex functions φ and all $n \geq 1$, $\mathcal{P}(\varphi) \geq c^n$. This was generalized to all convex functions in [60].

5.3 Volume product and transport inequalities

Maurey [142] introduced the following property (τ): Let μ be a measure on $\mathbb{R}^n$ and let $c : \mathbb{R}^n \times \mathbb{R}^n \to \mathbb{R}_+$ be a lower semicontinuous function (called a *cost function*); we say that the couple (μ, c) satisfies *property* (τ) if for any continuous and bounded function $f : \mathbb{R}^n \to \mathbb{R}$, setting

$$Q_c f(y) = \inf_x (f(x) + c(x,y)) \quad \text{for } y \in \mathbb{R}^n,$$

one has

$$\int_{\mathbb{R}^n} e^{-f(x)} d\mu(x) \int_{\mathbb{R}^n} e^{Q_c f(y)} d\mu(y) \leq 1.$$

Maurey [142] showed that if γ_n is the standard Gaussian probability measure on $\mathbb{R}^n$, with density $(2\pi)^{-n/2} e^{-|x|^2/2}$ and $c_2(x,y) = \frac{1}{2}|x - y|^2$, then as a consequence of the Prékopa–Leindler inequality, $(\gamma_n, \frac{c_2}{2})$ satisfies property (τ).

In [10], it was pointed out that the functional form of the Blaschke–Santaló inequality for the Legendre transform (Theorem 15) is equivalent to an improved property (τ) for even functions: We say that the pair (γ_n, c_2) satisfies the *even property* (τ) if for any even function f, one has

$$\int_{\mathbb{R}^n} e^{-f(x)} d\gamma_n(x) \int_{\mathbb{R}^n} e^{Q_{c_2} f(y)} d\gamma_n(y) \leq 1. \tag{5.32}$$

This equivalence follows from the change of functions $\varphi(x) = f(x) + \frac{|x|^2}{2}$ and the fact that

$$-\mathcal{L}\varphi(y) = \inf_x \left(f(x) + \frac{|x|^2}{2} - \langle x, y \rangle \right) = Q_{c_2} f(y) + \frac{|y|^2}{2}.$$

A direct proof of (5.32) was then given by Lehec in [119]. It follows from Remark 8 above, due to Lehec [121], that (5.32) also holds as soon as $\int_{\mathbb{R}^n} x e^{-f(x)} d\gamma_n(x) = 0$.

Moreover, as shown for example in [76, Proposition 8.2], there is a general equivalence between property (τ) and symmetrized forms of transport-entropy inequalities. These transport-entropy inequalities were introduced by Talagrand [190], who showed that, for every probability measure v on $\mathbb{R}^n$, one has

$$W_2^2(v, \gamma_n) \le 2H(v|\gamma_n), \tag{5.33}$$

where W_2 is the *Kantorovich–Wasserstein distance* defined by

$$W_2^2(v, \gamma_n) = \inf\left\{ \int_{\mathbb{R}^n \times \mathbb{R}^n} |x - y|^2 d\pi(x, y); \pi \in \Pi(v, \gamma_n) \right\},$$

where $\Pi(v, \gamma_n)$ is the set of probability measures on $\mathbb{R}^n \times \mathbb{R}^n$ whose first marginal is v and whose second marginal is γ_n; H is the *relative entropy* defined for $dv = f d\gamma_n$ by

$$H(v|\gamma_n) = - \int_{\mathbb{R}^n} f \log f d\gamma_n.$$

Using this type of equivalence between property (τ) and transport-entropy inequalities, Fathi [53] proved the following symmetrized form of Talagrand's transport-entropy inequality: If v_1 (or v_2) is centered, in the sense that $\int x dv_1(x) = 0$, then

$$W_2^2(v_1, v_2) \le 2(H(v_1|\gamma_n) + H(v_2|\gamma_n)). \tag{5.34}$$

He showed actually that (5.34) is equivalent to the functional form of Blaschke–Santaló's inequality (Theorem 15). Applying (5.34) to $v_1 = \gamma_n$, one recovers Talagrand's inequality (5.33). In his proof, Fathi used a reverse logarithmic Sobolev inequality for log-concave functions established in [11] under some regularity assumptions, removed later with a simplified proof in [42].

In a similar way, Gozlan [75] gave equivalent transport-entropy forms of Conjectures 3 and 4 and of Bourgain–Milman's asymptotic inequality. This work was continued in [56], where new proofs of the 1-dimensional case of Conjectures 3 and 4 are also provided.

6 Generalization to many functions and bodies

The following intriguing conjecture was proposed by Kolesnikov and Werner [116].

Conjecture 5. *Let $\rho : \mathbb{R} \to \mathbb{R}^+$ be increasing, and for $m \ge 2$, let $f_i : \mathbb{R}^n \to \mathbb{R}$, $i = 1, \ldots, m$, be even Lebesgue integrable functions satisfying*

$$\prod_{i=1}^{m} f_i(x_i) \leq \rho\left(\sum_{1 \leq i < j \leq m} \langle x_i, x_j \rangle \right) \quad \text{for all } x_1, \ldots, x_m \in \mathbb{R}^n.$$

Then

$$\prod_{i=1}^{m} \int_{\mathbb{R}^n} f_i(x_i) dx_i \leq \left(\int_{\mathbb{R}^n} \rho^{\frac{1}{m}} \left(\frac{m(m-1)}{2} |u|^2 \right) du \right)^{m}.$$

Conjecture 5 was proved by Kolesnikov and Werner when the functions f_i are unconditional. Observe that Conjecture 5 is a functional form of a new conjectured Blaschke–Santaló inequality involving more than two convex bodies. Indeed, for $1 \leq i \leq m$, let K_i be star bodies, let $f_i(x) = v_n(2\pi)^{-n/2} e^{-\|x\|_{K_i}^2/2}$, and let $\rho(t) = e^{-t/(m-1)}$. Since $\text{vol}(K_i) = v_n(2\pi)^{-n/2} \int_{\mathbb{R}^n} e^{-\|x\|_{K_i}^2/2} dx$, we get from Conjecture 5 the following.

Conjecture 6. *Let $m \geq 2$ and let $K_1, \ldots, K_m$ be symmetric convex bodies in $\mathbb{R}^n$ such that*

$$\sum_{1 \leq i < j \leq m} \langle x_i, x_j \rangle \leq \frac{m-1}{2} \sum_{i=1}^{m} \|x_i\|_{K_i}^2, \quad \text{for all } x_1, \ldots, x_m \in \mathbb{R}^n. \tag{5.35}$$

Then $\prod_{i=1}^{m} \text{vol}(K_i) \leq \text{vol}(B_2^n)^m$.

Conjecture 5 has been confirmed in [116] for unconditional bodies; for $m \geq 3$, it was shown then that there is equality if and only if $K_i = B_2^n$, for $i = 1, \ldots, m$.

This direction was further developed by Kalantzopoulos and Saroglou [102], who generalized the polarity condition (5.35). For $2 \leq p \leq m$ and $x_1, \ldots, x_m \in \mathbb{R}^n$, let

$$S_p(x_1, \ldots, x_m) = \binom{m}{p}^{-1} \sum_{l=1}^{n} s_p(x_1(l), \ldots, x_m(l)),$$

where $x_i = \sum_{l=1}^{n} x_i(l) e_l$ and s_p is the elementary symmetric polynomial in m variables of degree p. The case $p = 2$ corresponds to the sum of scalar products, i. e.,

$$S_2(x_1, \ldots, x_m) = \frac{2}{m(m-1)} \sum_{1 \leq i < j \leq m} \langle x_i, x_j \rangle.$$

In [102] the following *p-Santaló conjecture* was proposed.

Conjecture 7. *Let $2 \leq p \leq m$ be two integers. If $K_1, \ldots, K_m$ are symmetric convex bodies in $\mathbb{R}^n$, such that*

$$S_p(x_1, \ldots, x_m) \leq 1, \quad \text{for all } x_i \in K_i,$$

then $\prod_{i=1}^{m} \text{vol}(K_i) \leq \text{vol}(B_p^n)^m$, where B_p^n is the unit ball of the ℓ_p^n-norm.

Kalantzopoulos and Saroglou [102] were able to confirm Conjecture 7 when $p = m$ and in the case of unconditional convex bodies for all $p = 2, \ldots, m$. Moreover, when $p = 2$, it is enough to assume that only $K_3, \ldots, K_m$ are unconditional. In all of those known cases, the conjectured inequality is actually sharp for $K_1 = \cdots = K_m = B_p^n$. A functional analog of Conjecture 7 was also proposed in [102].

7 Links to other inequalities

In this section we present just a sample of connections of the volume product to other inequalities in convex geometry. As before, we refer to the books [7, 8, 65, 66, 79, 115, 170, 184, 194] and especially to the amazing diagram of connections of different open problems in convex geometry constructed by Richard Gardner in [66, Figure 1].

7.1 Slicing conjecture

Klartag [110] found a connection between a sharp version of Bourgain's slicing conjecture and Mahler's conjecture for general convex bodies (Conjecture 1). The covariance matrix of a convex body K in $\mathbb{R}^n$ is the $n \times n$ matrix $\mathrm{Cov}(K)$ defined by

$$\mathrm{Cov}(K)_{i,j} = \frac{\int_K x_i x_j dx}{\mathrm{vol}(K)} - \frac{\int_K x_i dx}{\mathrm{vol}(K)} \frac{\int_K x_j dx}{\mathrm{vol}(K)}.$$

The isotropic constant L_K is defined as $L_K^{2n} = \det(\mathrm{Cov}(K)) \, \mathrm{vol}(K)^{-2}$. It is well known that L_K is bounded from below by an absolute positive constant which is reached for ellipsoids. Bourgain's slicing problem asks whether for some universal constant $C > 0$, one has $L_K \leq C$ for every convex body K. The name *slicing conjecture* comes from the following very interesting equivalent reformulation: Is it true that for some universal $c > 0$, every convex body of volume 1 in $\mathbb{R}^n$ has an hyperplane section with $(n-1)$-dimensional volume greater than c? (See [158] for equivalent statements.) The boundedness of L_K by an absolute constant is still an open question. Bourgain [37] proved that $L_K \leq Cn^{1/4}$ up to a logarithmic factor, which was removed by Klartag [111]. Chen [45] proved that $L_K \leq C_\varepsilon n^\varepsilon$ for every $\varepsilon > 0$. Then, Klartag and Lehec [112] established a polylogarithmic bound $L_K \leq C \log^5 n$, which was then further improved to $L_K \leq C \log^{2.2} n$ by Jambulapati, Lee, and Vempala [100] and very recently Klartag [114] proved that $L_K \leq C \sqrt{\log n}$. A *strong version of the slicing conjecture* asks the following.

Conjecture 8. *For any convex body K in $\mathbb{R}^n$ one has*

$$L_K \leq L_{\Delta_n} = \frac{(n!)^{1/n}}{(n+1)^{\frac{n+1}{2n}} \sqrt{n+2}}. \tag{5.36}$$

Let K be a local minimizer of the volume product among the set of all convex bodies in $\mathbb{R}^n$ endowed with the Hausdorff distance. Then Klartag [110] was able to prove that

$$\mathrm{Cov}(K^*) \geq (n+2)^{-2}\mathrm{Cov}(K)^{-1}.$$

Taking the determinant and raising to the power $1/n$, one gets

$$\frac{1}{n+2} \leq L_K L_{K^*} \mathcal{P}(K)^{1/n}. \tag{5.37}$$

Thus, combining (5.37) and (5.36),

$$\frac{1}{n+2} \leq L_K L_{K^*} \mathcal{P}(K)^{1/n} \leq \frac{(n!)^{2/n}}{(n+1)^{\frac{n+1}{n}}(n+2)} \mathcal{P}(K)^{1/n}.$$

Thus, we proved the following theorem.

Theorem 17 (Klartag). *The strong version of Bourgain's slicing conjecture given in Conjecture* 8 *implies Conjecture* 1 *(Mahler's conjecture) for general convex bodies.*

In connection with his proof of the Bourgain–Milman inequality, Kuperberg asked in [118] whether the quantity

$$\frac{1}{\mathrm{vol}(K)\,\mathrm{vol}(K^*)} \int\limits_K \int\limits_{K^*} \langle x,y \rangle^2 dx dy$$

is maximized for ellipsoids in the class of convex symmetric bodies $K \subset \mathbb{R}^n$. Alonso-Gutiérrez [4] proved that this conjecture implies both the Blaschke–Santaló inequality and the hyperplane conjecture and that it holds true for B_p^n, the unit ball of ℓ_p^n, for $p \geq 1$. The connection to the hyperplane conjecture was also studied in [67]. Kuperberg had not much hope for his conjecture and Klartag [110] showed that it is false in high dimensions, even in the case of unconditional bodies.

7.2 Symplectic geometry and Viterbo's conjecture

Artstein-Avidan, Karasev, and Ostrover in [9] discovered an amazing connection between the volume product and symplectic geometry. Let (X, ω) be a symplectic manifold, where X is a smooth manifold with a closed non-degenerate two-form ω, for instance, $(\mathbb{R}^{2n}, \omega_{st})$, where $\mathbb{R}^{2n} = \mathbb{R}_p^n \times \mathbb{R}_q^n$ and $\omega_{st} = \sum dp_i \wedge dq_i$. A core fact in symplectic geometry states that symplectic manifolds have no local invariants (except the dimension). This, clearly, makes the structure very different from that of Riemannian manifolds. The first examples of global symplectic invariants were introduced by Gromov [78] and are known as Gromov's non-squeezing theorem. Gromov's work inspired the introduction of global symplectic invariants – symplectic capacities – which may be seen as a way to measure the symplectic size of sets in $\mathbb{R}^{2n}$. More precisely, a *symplectic capacity* c on $(\mathbb{R}^{2n}, \omega_{st})$ is a mapping $c : \mathcal{S}(\mathbb{R}^{2n}) \to \mathbb{R}_+$, where $\mathcal{S}(\mathbb{R}^{2n})$ is the set of all subsets of $\mathbb{R}^{2n}$, which satisfies the following conditions:

- Monotonicity: $c(U) \le c(V)$, for all $U \subset V$.
- Conformality: $c(\phi(U)) = |a|c(U)$, for all diffeomorphism ϕ such that $\phi^* \omega_{st} = a\omega_{st}$.
- Normalization: $c(B_2^{2n}) = c(B_2^2 \times \mathbb{R}^{2(n-1)}) = \pi$.

The following is the conjecture of Viterbo [196] for symplectic capacities of convex bodies.

Conjecture 9. *For any symplectic capacity c and any convex body Σ in $\mathbb{R}^{2n}$, one has*

$$\frac{c(\Sigma)}{c(B_2^{2n})} \le \left(\frac{\mathrm{vol}_{2n}(\Sigma)}{\mathrm{vol}_{2n}(B_2^{2n})} \right)^{\frac{1}{n}}.$$

Conjecture 9 is of isoperimetric type: Indeed, it claims that among all convex bodies in $\mathbb{R}^{2n}$ of a given fixed volume, the Euclidean ball of the same volume has the maximal symplectic capacity. It is open even for $n = 2$, but it holds for certain classes of convex bodies, including ellipsoids [89], and up to a universal multiplicative constant [12]. The following was proved in [9].

Theorem 18. *Conjecture* 9 *implies Conjecture* 2.

More precisely, it was proved in [9] that for any convex symmetric body $K \subset \mathbb{R}^n$, $c_{HZ}(K \times K^*) = 4$, where c_{HZ} denotes the Hofer–Zehnder capacity, which is one of the important symplectic capacities. This fact together with Conjecture 9 and the normalization property of c_{HZ} immediately gives an affirmative answer to Conjecture 2:

$$\frac{4^n}{\pi^n} = \left(\frac{c_{HZ}(K \times K^*)}{c_{HZ}(B_2^{2n})} \right)^n \le \frac{\mathrm{vol}_{2n}(K \times K^*)}{\mathrm{vol}_{2n}(B_2^{2n})} = \frac{n!\,\mathrm{vol}_{2n}(K \times K^*)}{\pi^n}.$$

We refer to [9, 163] for more details on these connections. The connections of Conjecture 2 with symplectic geometry were further studied in [1, 24, 104, 105]. In [178], Viterbo's conjecture was connected with Minkowski versions of worm problems, inspired by the well-known Moser worm problem from geometry. For the special case of Lagrangian products, this relation provides further links to systolic Minkowski billiard inequalities and Mahler's conjecture.

7.3 Funk geometry

A very interesting connection of the volume product with Funk geometry was recently discovered by Faifman [51]. We refer to [165] for a detailed introduction to Finsler manifolds and Funk geometry. We will recall a few of the most basic ideas.

A non-reversible Finsler manifold (M, F) is a smooth manifold M equipped with a smooth function F on the tangent bundle of M which, when restricted on any tangent space, is the gauge of some convex body. The crucial difference with Riemannian geometry is the lack of inner product. The tangent unit ball at a point $x \in M$ is denoted by $B_x M$ and consists of all vectors v in the tangent space $T_x M$ such that $F(x, v) \le 1$.

For a convex body K in a fixed affine space, the Funk metric on the interior of K is given by $B_x K = K$, i. e., at any point x in the interior of K, the body K with origin at x is the unit ball. We continue in the following way. Consider $x, y \in \mathrm{int}(K)$ and let $R(x, y)$ be the ray starting at x passing through y. Let $a(x, y) = R(x, y) \cap \partial K$. Then the Funk metric, defined for $x \ne y \in \mathrm{int}(K)$, is

$$d_K^F(x, y) = \log \frac{|x - a(x, y)|}{|y - a(x, y)|},$$

and $d_K^F(x, x) = 0$. The Funk metric is projective, i. e., straight segments are geodesics. The outward ball of radius $r > 0$ and center $z \in \mathrm{int}(K)$ is

$$B_K^F(z, r) = \{x \in \mathrm{int}(K) : d_K^F(z, x) \le r\} = (1 - e^{-r})(K - z) + z.$$

The Holmes–Thompson volume of $A \subset \mathrm{int}(K)$ is defined as

$$\mathrm{vol}_K^F(A) = \frac{1}{v_n} \int_A \mathrm{vol}(K^x) dx.$$

Asymptotically as $r \to 0$, the volume of $B_K^F(z, r)$ behaves as $v_n^{-1} \mathrm{vol}_{2n}(K \times K^z) r^n$. It was also shown in [23] that for a strictly convex and smooth body K, when $r \to +\infty$, the volume of $B_K^F(z, r)$ behaves as $c_n e^{\frac{n-1}{2} r} \mathcal{A}(K, z)$, where $c_n > 0$ depends only on n and $\mathcal{A}(K, z)$ is the centro-affine surface area of K defined by

$$\mathcal{A}(K, z) = \int_{\partial K} \frac{\kappa_K^{1/2}(x)}{\langle x - z, n_K(x) \rangle^{(n-1)/2}} dx,$$

where $\kappa_k(x)$ is the Gauss curvature of ∂K at point x and $n_K(x)$ is an outer normal vector. Note that $\mathcal{A}(K, 0) = \mathcal{A}(K)$. The following duality relation for vol_K^F, for centrally symmetric K, is proved in [51]:

$$\mathrm{vol}_K^F(B_K^F(0, r)) = \mathrm{vol}_{K^*}^F(B_{K^*}^F(0, r)).$$

The existence of an analog of the Santaló point $s(K)$ of a convex body K in the Funk geometry was proved in [52]: For any $r > 0$, there is a unique point $s_r(K) \in \mathrm{int}(K)$ that minimizes the Funk volume of $B_K^F(q, r)$. One has $s_r(K) = 0$ for symmetric K and $s_r(K) \to s(K)$ as $r \to 0$. Let

$$M_r(K) = v_n \, \mathrm{vol}_K^F(B_K^F(s_r(K), r)).$$

The following conjecture was proposed in [51].

Conjecture 10. *For all $r > 0$, $M_r(K)$ is maximal when K is an ellipsoid.*

The limiting cases of Conjecture 10 are the Blaschke–Santaló inequality as $r \to 0$ and the centro-affine isoperimetric inequality as $r \to \infty$. Faifman [51] was able to show that Conjecture 10 holds for unconditional bodies K. The idea of the proof includes the generalization of the conjecture of K. Ball (see inequality (5.13)), namely

$$\int_K \int_{K^*} \langle x, y \rangle^{2j} dx dy \leq \int_{B_2^n} \int_{B_2^n} \langle x, y \rangle^{2j} dx dy, \tag{5.38}$$

for all $j \in \mathbb{N}$, which Faifman was able to confirm for K unconditional. A lower bound for the quantity $M_r(K)$ was proposed in [52].

Conjecture 11. *For $r > 0$, $M_r(K)$ is minimized by simplices in general and by Hanner polytopes for symmetric bodies K.*

The limiting case as $r \to 0$ in Conjecture 11 for symmetric K is Conjecture 2 and the limiting case as $r \to +\infty$ is a conjecture of Kalai [101] on the minimization of the flag number of K. Conjecture 11 is proved in [52] for unconditional bodies and follows from an interesting new inequality discovered in [52] and proved for unconditional bodies:

$$\int_H \int_{H^*} \langle x, y \rangle^{2j} dx dy \leq \int_K \int_{K^*} \langle x, y \rangle^{2j} dx dy, \tag{5.39}$$

where H is a Hanner polytope in $\mathbb{R}^n$ and $j \in \mathbb{N}$. The proof of (5.39) in [52] is based on the functional inverse Santaló inequality [59].

7.4 Geometry of numbers and isosystolic inequalities

The volume product is a standard tool in the geometry of numbers. The connection goes back to the theorem of Mahler [138] (see [38], [79, Chapter 3], or [50]) on the bound of the successive minima of a convex body and its dual.

Let us here present yet another connection of the volume product with the geometry of numbers and the systolic geometry discovered by Álvarez Paiva, Balacheff, and Tzanev [6].

Minkowski's first theorem in the geometry of numbers states that if K is a symmetric convex body in $\mathbb{R}^n$ with $\mathrm{vol}(K) \geq 2^n$, then K contains at least one non-zero integer point (in $\mathbb{Z}^n$). The symmetry assumption is needed as there are convex bodies K of large volume containing the origin and no other integer point. We know that such bodies must be "flat" [103], and Álvarez Paiva and Balacheff [5] conjectured that the volume of their polars K^* is not too small.

Conjecture 12. *Let $K \subset \mathbb{R}^n$ be a convex body such that $\mathrm{int}(K) \cap \mathbb{Z}^n = \{0\}$. Then $\mathrm{vol}(K^*) \geq (n+1)/n!$, with equality if and only if K is a simplex with vertices in $\mathbb{Z}^n$ and no other integer points than its vertices and 0.*

In [6], Conjecture 12 was proved in $\mathbb{R}^2$ and an isomorphic bound for $\mathrm{vol}(K^*)$ was given in all dimensions. Namely, for some absolute constant $c > 0$, one has $\mathrm{vol}(K^*) \geq c^n(n + 1)/n!$ for any convex body K in $\mathbb{R}^n$ such that $\mathrm{int}(K)$ contains no integer point other than the origin. The proof of this fact in [6] uses the Bourgain–Milman inequality, and it is shown that this isomorphic version of Conjecture 12 is actually equivalent to it.

Conjecture 12 can be further generalized to a conjecture in systolic geometry. We refer to [6] for exact statements and definitions. We mention here a version of the conjecture in the language of Finsler geometry (see Section 7.3). The Holmes–Thompson volume of a Finsler manifold (M, F) is defined as

$$\mathrm{vol}_{HT}(M, F) = \frac{1}{v_n} \int_M \mathrm{vol}((B_x M)^*) dx.$$

Conjecture 13. *For any Finsler metric F on $\mathbb{RP}^n$, there exists a closed non-contractible geodesic with length bounded by $\frac{(n! v_n)^{1/n}}{2} \mathrm{vol}_{HT}(\mathbb{RP}^n, F)^{1/n}$.*

We recall that a set which can be reduced to one of its points by a continuous deformation is said to be contractible. For $n = 2$, Conjecture 13 follows from the works of Ivanov [98, 99]. The following theorem was proved in [6].

Theorem 19. *Conjecture 13 implies Conjecture 2 for centrally symmetric bodies.*

The proof of Theorem 19 uses the Finsler metric on a convex symmetric body K which coincides at each point with the norm corresponding to K. By identifying the points x and $-x$ in ∂K, we obtain a length space (a space in which the intrinsic metric coincides with the original metric) on $\mathbb{RP}^n$. We denote this Finsler space by $(\mathbb{RP}^n, d_K)$. It turns out that one has

$$\mathrm{vol}(\mathbb{RP}^n, d_K) = \frac{1}{v_n} \mathcal{P}(K)$$

and that the length of the systoles (the shortest non-contractible geodesics) in $(\mathbb{RP}^n, d_K)$ is equal to 2. Combining those commutations and assuming that Conjecture 13 holds, we get from Conjecture 13 a proof of Conjecture 2 for symmetric convex bodies in $\mathbb{R}^n$.

Bibliography

[1]   A. Akopyan, A. Balitskiy, R. Karasev and A. Sharipova, *Elementary approach to closed billiard trajectories in asymmetric normed spaces*, Proc. Am. Math. Soc. **144**(10) (2016), 4501–4513.

[2]   M. Alexander, M. Fradelizi, L. García-Lirola and A. Zvavitch, *Geometry and volume product of finite dimensional Lipschitz-free spaces*, J. Funct. Anal. **280**(4) (2021), 108849, 38 pp.

[3]   M. Alexander, M. Fradelizi and A. Zvavitch, *Polytopes of maximal volume product*, Discrete Comput. Geom. **62**(3) (2019), 583–600.

[4]   D. Alonso-Gutiérrez, *On an extension of the Blaschke–Santaló inequality and the hyperplane conjecture*, J. Math. Anal. Appl. **344**(1) (2008), 292–300.

[5]   J. C. Álvarez Paiva and F. Balacheff, *Contact geometry and isosystolic inequalities*, Geom. Funct. Anal. **24**(2) (2014), 648–669.

[6]   J. C. Álvarez Paiva, F. Balacheff and K. Tzanev, *Isosystolic inequalities for optical hypersurfaces*, Adv. Math. **301** (2016), 934–972.

[7]   S. Artstein-Avidan, A. Giannopoulos and V. D. Milman, *Asymptotic geometric analysis. Part I*, Mathematical surveys and monographs, vol. 202, American Mathematical Society, Providence, RI, 2015.

[8]   S. Artstein-Avidan, A. Giannopoulos and V. D. Milman, *Asymptotic geometric analysis. Part II*, Mathematical surveys and monographs, vol. Number 262, American Mathematical Society, Providence, RI, 2021.

[9]   S. Artstein-Avidan, R. Karasev and Y. Ostrover, *From symplectic measurements to the Mahler conjecture*, Duke Math. J. **163**(11) (2014), 2003–2022.

[10]   S. Artstein-Avidan, B. Klartag and V. D. Milman, *On the Santaló point of a function and a functional Santaló inequality*, Mathematika **54** (2004), 33–48.

[11]   S. Artstein-Avidan, B. Klartag, C. Schütt and E. Werner, *Functional affine-isoperimetry and an inverse logarithmic Sobolev inequality*, J. Funct. Anal. **262**(9) (2012), 4181–4204.

[12]   S. Artstein-Avidan, V. D. Milman and Y. Ostrover, *The M-ellipsoid, symplectic capacities and volume*, Comment. Math. Helv. **83** (2008), 359–369.

[13]   S. Artstein-Avidan and B. A. Slomka, *A note on Santaló inequality for the polarity transform and its reverse*, Proc. Am. Math. Soc. **143**(4) (2015), 1693–1704.

[14]   G. Aubrun and M. Fradelizi, *Two-point symmetrization and convexity*, Arch. Math. (Basel) **82**(3) (2004), 282–288.

[15]   K. Ball, *Isometric problems in ℓ_p and sections of convex sets*, PhD thesis, Cambridge, 1986.

[16]   K. Ball, *Logarithmically concave functions and sections of convex sets in $\mathbb{R}^n$*, Stud. Math. **88**(1) (1988), 69–84.

[17]   K. Ball, *Some remarks on the geometry of convex sets*, in: Geometric aspects of functional analysis, Lecture notes in mathematics, vol. 317, Springer, 1988, pp. 224–231.

[18]   K. Ball, *Mahler's conjecture and wavelets*, Discrete Comput. Geom. **13** (1995), 271–277.

[19]   K. Ball and K. J. Böröczky, *Stability of some versions of the Prékopa–Leindler inequality*, Monatshefte Math. **163**(1) (2011), 1–14.

[20]   R. P. Bambah, *Polar reciprocal convex bodies*, Proc. Camb. Philos. Soc. **51** (1955), 377–378.

[21]   F. Barthe, K.-J. Böröczky and M. Fradelizi, *Stability of the functional forms of the Blaschke–Santaló inequality*, Monatshefte Math. **173**(2) (2014), 135–159.

[22]   F. Barthe and M. Fradelizi, *The volume product of convex bodies with many hyperplane symmetries*, Am. J. Math. **135** (2013), 1–37.

[23]   G. Berck, A. Bernig and C. Vernicos, *Volume entropy of Hilbert geometries*, Pac. J. Math. **245**(2) (2010), 201–225.

[24]   M. Berezovik and R. Karasev, *Symplectic polarity and Mahler's conjecture*, arXiv:2211.14630.

[25]   B. Berndtsson, *Subharmonicity properties of the Bergman kernel and some other functions associated to pseudoconvex domains*, Ann. Inst. Fourier (Grenoble) **56**(6) (2006), 1633–1662.

[26]   B. Berndtsson, *Complex integrals and Kuperberg's proof of the Bourgain–Milman theorem*, Adv. Math. **388** (2021), 107927, 10 pp.

[27]   B. Berndtsson, *Bergman kernels for Paley–Wiener spaces and Nazarov's proof of the Bourgain–Milman theorem*, Pure Appl. Math. Q. **18**(2) (2022), 395–409.

[28] G. Bianchi and M. Kelli, *A Fourier analytic proof of the Blaschke–Santaló inequality*, Proc. Am. Math. Soc. **143**(11) (2015), 4901–4912.

[29] W. Blaschke, *Über affine Geometrie VII: Neue Extremeigenschaften von Ellipse und Ellipsoid*, Leipz. Ber. **69** (1917), 306–318 and 436–453.

[30] W. Blaschke, *Vorlesungen über Differentialgeometrie II: Affine Differentialgeometrie*, Springer, Berlin, 1923.

[31] Z. Blocki, *A lower bound for the Bergman kernel and the Bourgain–Milman inequality*, in: Geometric aspects of functional analysis, B. Klartag and E. Milman, eds., Springer, 2014, pp. 53–63.

[32] Z. Blocki, *On Nazarov's complex analytic approach to the Mahler conjecture and the Bourgain–Milman inequality*, in: Complex Analysis and Geometry, F. Bracci et al., eds., Springer, 2015, pp. 89–98.

[33] E. D. Bolker, *A class of convex bodies*, Trans. Am. Math. Soc. **145** (1969), 323–345.

[34] K. J. Böröczky, E. Makai, M. Meyer and S. Reisner, *Volume product in the plane – lower estimates with stability*, Studia Sci. Math. Hung. **50** (2013), 159–198.

[35] K. J. Böröczky, *Stability of the Blaschke–Santaló and the affine isoperimetric inequality*, Adv. Math. **225**(4) (2010), 1914–1928.

[36] K. J. Böröczky and D. Hug, *Stability of the inverse Blaschke–Santaló inequality for zonoids*, Adv. Appl. Math. **44** (2010), 309–328.

[37] J. Bourgain, *On the distribution of polynomials on high-dimensional convex sets*, Lect. Notes Math. **1469** (1991), 127–137.

[38] J. Bourgain and V. D. Milman, *New volume ratio properties for convex symmetric bodies in $\mathbb{R}^n$*, Invent. Math. **88** (1987), 319–340.

[39] H. J. Brascamp, E. H. Lieb and J. M. Luttinger, *A general rearrangement inequality for multiple integrals*, J. Funct. Anal. **17** (1974), 227–237.

[40] D. Bucur and I. Fragala, *Blaschke–Santaló and Mahler inequalities for the first eigenvalue of the Dirichlet Laplacian*, Proc. Lond. Math. Soc. **113**(3) (2016), 387–417.

[41] H. Busemann, *A theorem on convex bodies of the Brunn–Minkowski type*, Proc. Natl. Acad. Sci. USA **35** (1949), 27–31.

[42] U. Caglar, M. Fradelizi, O. Guédon Olivier, J. Lehec, C. Schütt and E. Werner, *Functional versions of L_p-affine surface area and entropy inequalities*, Int. Math. Res. Not. **4** (2016), 1223–1250.

[43] S. Campi and P. Gronchi, *On volume product inequalities for convex sets*, Proc. Am. Math. Soc. **134**(8) (2006), 2393–2402.

[44] S. Campi and P. Gronchi, *Volume inequalities for L_p-zonotopes*, Mathematika **53**(1) (2006), 71–80.

[45] Y. Chen, *An almost constant lower bound of the isoperimetric coefficient in the KLS conjecture*.

[46] D. Cordero-Erausquin, *Santaló's inequality on $\mathbb{C}^n$ by complex interpolation*, C. R. Acad. Sci. Paris, Ser. I **334** (2002), 767–772.

[47] D. Cordero-Erausquin, M. Fradelizi and B. Maurey, *The (B)-conjecture for the Gaussian measure of dilates of symmetric convex sets and related problems*, J. Funct. Anal. **214** (2004), 410–427.

[48] D. Cordero-Erausquin, M. Fradelizi, G. Paouris and P. Pivovarov, *Volume of the polar of random sets and shadow systems*, Math. Ann. **362**(3–4) (2015), 1305–1325.

[49] D. Cordero-Erausquin and L. Rotem, *Several results regarding the (B)-conjecture*, in: Geometric aspects of functional analysis. Vol. I, Lecture notes in math., vol. 2256, Springer, Cham, 2020, pp. 247–262.

[50] J.-H. Evertse, *Mahler's work on the geometry of numbers*, in: Documenta mathematica. Extra volume Mahler selecta, 2019, pp. 29–43.

[51] D. Faifman, A Funk perspective on billiards, projective geometry and Mahler volume, J. Differ. Geom., to appear, arXiv:2012.12159.

[52] D. Faifman, C. Vernicos and C. Walsh, Forthcoming paper.

[53] M. Fathi, *A sharp symmetrized form of Talagrand's transport-entropy inequality for the Gaussian measure*, Electron. Commun. Probab. **23** (2018), 81.

[54] I. D. Florentin and A. Segal, *A Santaló-type inequality for $\mathcal{J}$ transform*, Commun. Contemp. Math. **23**(1) (2021), 1950090.

[55]   M. Fradelizi, Y. Gordon, M. Meyer and S. Reisner, *The case of equality for an inverse Santaló functional inequality*, Adv. Geom. **10**(4) (2010), 621–630.

[56]   M. Fradelizi, N. Gozlan and S. Zugmeyer, *The Transport proofs of some functional inverse Santaló inequalities*, arXiv:2109.00871.

[57]   M. Fradelizi, A. Hubard, M. Meyer, E. Roldán-Pensado and A. Zvavitch, *Equipartitions and Mahler volumes of symmetric convex bodies*, Am. J. Math. **144**(5) (2022), 1201–1219.

[58]   M. Fradelizi and M. Meyer, *Functional inequalities related to Mahler's conjecture*, Monatshefte Math. **159**(1–2) (2010), 13–25.

[59]   M. Fradelizi and M. Meyer, *Increasing functions and inverse Santaló inequality for unconditional functions*, Positivity **12**(3) (2008), 407–420.

[60]   M. Fradelizi and M. Meyer, *Some functional inverse Santaló inequalities*, Adv. Math. **218**(5) (2008), 1430–1452.

[61]   M. Fradelizi and M. Meyer, *Some functional forms of Blaschke–Santaló inequality*, Math. Z. **256**(2) (2007), 379–395.

[62]   M. Fradelizi and E. Nakhle, *The functional form of Mahler's conjecture for even log-concave functions in dimension 2*, Int. Math. Res. Not. **2022** (2022), rnac120.

[63]   M. Fradelizi, M. Meyer and A. Zvavitch, *An application of shadow systems to Mahler's conjecture*, Discrete Comput. Geom. **48**(3) (2012), 721–734.

[64]   F. Gao, D. Hug and R. Schneider, *Intrinsic volumes and polar sets in spherical space*, Math. Notae **41**(2001/02) (2003), 159–176; Homage to Luis Santaló. Vol. 1 (Spanish).

[65]   R. Gardner, *Geometric tomography*, 2nd edn., Encyclopedia of mathematics and its applications, vol. 58, Cambridge University Press, New York, 2006.

[66]   R. Gardner, *Geometric tomography: update to the second edition*, https://www.wwu.edu/faculty/ gardner/research.html.

[67]   A. Giannopoulos, *Problems on convex bodies* (in Greek), Ph. D. thesis written under the supervision of Prof. Papadopoulou, University of Crete, Greece, 1993.

[68]   O. Giladi, H. Huang, C. Schütt and E. M. Werner, *Constrained convex bodies with extremal affine surface areas*, J. Funct. Anal. **279** (2020), 3, 23 pp.

[69]   A. Giannopoulos, G. Paouris and B. Vritsiou, *The isotropic position and the reverse Santaló inequality*, Isr. J. Math. **203**(1) (2015), 1–22.

[70]   P. Goodey and W. Weil, *Zonoids and generalisations*, in: Handbook of convex geometry, vols. A, B, North-Holland, Amsterdam, 1993, pp. 1297–1326.

[71]   Y. Gordon, *On Milman's inequality and random subspaces which escape through a mesh in $\mathbb{R}^n$*, in: Geometric aspects of functional analysis, Lecture notes in math., vol. 1317, Springer, Berlin, 1988, pp. 84–106.

[72]   Y. Gordon and M. Meyer, *On the minima of the functional Mahler product*, Houst. J. Math. **40**(2) (2014), 385–393.

[73]   Y. Gordon, M. Meyer and S. Reisner, *Zonoids with minimal volume product – a new proof*, Proc. Am. Math. Soc. **104** (1988), 273–276.

[74]   Y. Gordon and S. Reisner, *Some aspects of volume estimates to various parameters in Banach spaces*, in: Proceedings of research workshop on Banach space theory, Iowa City, Iowa, 1981, Univ. Iowa, Iowa City, IA, 1982, pp. 23–53.

[75]   N. Gozlan, *The deficit is the Gaussian log-Sobolev inequality and inverse Santaló inequalities*, Int. Math. Res. Not. **17** (2022), 13396–13446.

[76]   N. Gozlan and C. Léonard, *Transport inequalities. A survey*, Markov Process. Relat. Fields **16**(4) (2010), 635–736.

[77]   E. L. Grinberg, *Isoperimetric inequalities and identities for k-dimensional cross-sections of convex bodies*, Math. Ann. **291** (1991), 75–86.

[78]   M. Gromov, *Pseudoholomorphic curves in symplectic manifolds*, Invent. Math. **82** (1985), 307–347.

[79]   P. M. Gruber, *Convex and discrete geometry*, Grundlehren der mathematischen Wissenschaften [Fundamental principles of mathematical sciences], vol. 336, Springer, Berlin, 2007, xiv+578 pp.

[80]   B. Grünbaum, *Partitions of mass-distributions and of convex bodies by hyperplanes*, Pac. J. Math. **10** (1960), 1257–1261.

[81]   H. Guggenheimer, *Polar reciprocal convex bodies*, Isr. J. Math. **14** (1973), 309–316.

[82]   H. Guggenheimer, *Correction to: Polar reciprocal convex bodies, Israel J. Math. 14 (1973), no. 3, 309–316*, Isr. J. Math. **14**(3) (1973), 309–316.

[83]   T. Hack and P. Pivovarov, *Randomized Urysohn-type inequalities*, Mathematika **67**(1) (2021), 100–115.

[84]   J. Haddad, C. H. Jiménez and M. A. Montenegro, *Asymmetric Blaschke–Santaló functional inequalities*, J. Funct. Anal. **278**(2) (2020), 108319.

[85]   H. Hadwiger, *Simultane Vierteilung zweier Körper*, Arch. Math. (Basel) **17** (1966), 274–278.

[86]   O. Hanner, *Intersections of translates of convex bodies*, Math. Scand. **4** (1956), 65–87.

[87]   A. B. Hansen and A. Lima, *The structure of finite-dimensional Banach spaces with the 3.2. intersection property*, Acta Math. **146**(1–2) (1981), 1–23.

[88]   E. M. Harrell, A. Henrot and J. Lamboley, *On the local minimizers of the Mahler volume*, J. Convex Anal. **22**(3) (2015), 809–825.

[89]   D. Hermann, *Non-equivalence of symplectic capacities for open sets with restricted contact type boundary*, 1998, preprint.

[90]   G. C. Hofstatter and F. E. Schuster, *Blaschke Santaló inequalities for Minkowski and Asplund endomorphisms*, Int. Math. Res. Not. **2023**(2) (2023), 1378–1419.

[91]   L. Hörmander, *The analysis of linear partial differential operators I*, Springer-Verlag, 1983.

[92]   S. Hoehner, *Extremal general affine surface areas*, J. Math. Anal. Appl. **505** (2022), 2, 21 pp.

[93]   Y. Hu and H. Li, *Blaschke–Santaló type inequalities and quermassintegral inequalities in space forms*, Adv. Math. **413** (2023), 108826.

[94]   D. Hug, *Contributions to affine surface area*, Manuscr. Math. **91**(3) (1996), 283–301.

[95]   H. Iriyeh and M. Shibata, *Symmetric Mahler's conjecture for the volume product in the three dimensional case*, Duke Math. J. **169**(6) (2020), 1077–1134.

[96]   H. Iriyeh and M. Shibata, *Minimal volume product of three dimensional convex bodies with various discrete symmetries*, Discrete Comput. Geom. **68** (2022), 738–773.

[97]   H. Iriyeh and M. Shibata, *Minimal volume product of convex bodies with certain discrete symmetries and its applications*, preprint, arXiv:2203.13990.

[98]   S. Ivanov, *On two-dimensional minimal fillings*, St. Petersburg Math. J. **13** (2002), 17–25.

[99]   S. Ivanov, *Filling minimality of Finslerian 2-discs*, Tr. Mat. Inst. Steklova **273** (2011), 192–206, Sovremennye Problemy Matematiki.

[100]   A. Jambulapati, Y. T. Lee and S. S. Vempala, *A slightly improved bound for the KLS constant*, arXiv:2208.11644.

[101]   G. Kalai, *The number of faces of centrally-symmetric polytopes*, Graphs Comb. **5**(1) (1989), 389–391.

[102]   P. Kalantzopoulos and C. Saroglou, *On a j-Santaló Conjecture*, arXiv:2203.14815.

[103]   R. Kannan and L. Lovász, *Covering minima and lattice-point-free convex bodies*, Ann. Math. (2) **128**(3) (1988), 577–602.

[104]   R. Karasev, *Mahler's conjecture for some hyperplane sections*, Isr. J. Math. **241** (2021), 795–815.

[105]   R. Karasev and A. Sharipova, *Viterbo's conjecture for certain Hamiltonians in classical mechanics*, Arnold Math. J. **5**(4) (2019), 483–500.

[106]   J. Kim, *Minimal volume product near Hanner polytopes*, J. Funct. Anal. **266**(4) (2014), 2360–2402.

[107]   J. Kim and S. Reisner, *Local minimality of the volume-product at the simplex*, Mathematika **57** (2011), 121–134.

[108]   J. Kim and A. Zvavitch, *Stability of the reverse Blaschke–Santaló inequality for unconditional convex bodies*, Proc. Am. Math. Soc. **143**(4) (2015), 1705–1717.

[109]   B. Klartag, *Convex geometry and waist inequalities*, Geom. Funct. Anal. **27**(1) (2017), 130–164.

[110]   B. Klartag, *Isotropic constants and Mahler volumes*, Adv. Math. **330** (2018), 74–108.

[111] B. Klartag, *On convex perturbations with a bounded isotropic constant*, Geom. Funct. Anal. **16** (2006), 1274–1290.

[112] B. Klartag and J. Lehec, *Bourgain's slicing problem and KLS isoperimetry up to polylog*, arXiv:2203.15551.

[113] B. Klartag and V. D. Milman, *Geometry of log-concave functions and measures*, Geom. Dedic. **112** (2005), 169–182.

[114] B. Klartag, *Logarithmic bounds for isoperimetry and slices of convex sets*, arXiv:2303.14938.

[115] A. Koldobsky, *Fourier analysis in convex geometry*, Mathematical surveys and monographs, vol. 116, American Mathematical Society, Providence, RI, 2005, vi+170 pp.

[116] A. Kolesnikov and E. W. Werner, *Blaschke–Santaló inequality for many functions and geodesic barycenters of measures*, Adv. Math. **396** (2022), 108110.

[117] G. Kuperberg, *A low-technology estimate in convex geometry*, Int. Math. Res. Not. **9** (1992), 181–183.

[118] G. Kuperberg, *From the Mahler conjecture to Gauss linking integrals*, Geom. Funct. Anal. **3** (2008), 870–892.

[119] J. Lehec, *The symmetric property (τ) for the Gaussian measure*, Ann. Fac. Sci. Toulouse Math. (6) **17**(2) (2008), 357–370.

[120] J. Lehec, *Partitions and functional Santaló inequalities*, Arch. Math. **92**(1) (2009), 89–94.

[121] J. Lehec, *A direct proof of the functional Santaló inequality*, C. R. Acad. Sci. Paris, Ser. I **347** (2009), 55–58.

[122] D. H. Lehmer, *Factorization of certain cyclotomic functions*, Ann. Math. **34** (1933), 461–469.

[123] K. Leichtweiss, *Geometric convexity and differential geometry*, in: Convexity and its applications, Birkhäuser, Basel, 1983, pp. 163–169.

[124] K. Leichtweiss, *Affine geometry of convex bodies*, Johann Ambrosius Barth Verlag, Heidelberg, 1998.

[125] T. Leinster, *The magnitude of metric spaces*, Doc. Math. **18** (2013), 857–905.

[126] B. Li, C. Schütt and E. W. Werner, *Affine invariant maps for log-concave functions*, J. Geom. Anal. **32**(4) (2022), 123, 50 pp.

[127] I. Lindenstrauss and V. Milman, *Local theory of normed spaces and convexity*, in: Handbook of convex geometry, vol. 2, North Holland, Amsterdam, 1993, pp. 1151–1220.

[128] M. A. Lopez and S. Reisner, *A special case of Mahler's conjecture*, Discrete Comput. Geom. **20** (1998), 151–176.

[129] E. Lutwak, *A general isepiphanic inequality*, Proc. Am. Math. Soc. **90**(3) (1984), 415–421.

[130] E. Lutwak, *On the Blaschke–Santaló inequality*, in: Discrete geometry and convexity, New York, 1982, Ann. New York Acad. Sci., vol. 440, New York Acad. Sci., New York, 1985, pp. 106–112.

[131] E. Lutwak, *Inequalities for Hadwiger's harmonic Quermassintegrals*, Math. Ann. **280**(1) (1988), 165–175.

[132] E. Lutwak, *Extended affine surface area*, Adv. Math. **85**(1) (1991), 39–68.

[133] E. Lutwak, *Selected affine isoperimetric inequalities*, in: Handbook of convex geometry, vol. 1, North Holland, Amsterdam, 1993, pp. 1263–1277.

[134] E. Lutwak, D. Yang and G. Zhang, L^p *affine isoperimetric inequalities*, J. Differ. Geom. **56**(1) (2000), 111–132.

[135] E. Lutwak, D. Yang and G. Zhang, *Moment-entropy inequalities*, Ann. Probab. **32** (2004), 757–774.

[136] E. Lutwak and G. Zhang, *Blaschke–Santaló inequalities*, J. Differ. Geom. **47**(1) (1997), 1–16.

[137] K. Mahler, *Ein Minimalproblem für konvexe Polygone*, Mathematica (Zutphen) **B7** (1938), 118–127.

[138] K. Mahler, *Ein übertragungsprinzip für konvexe Körper*, Čas. Pěst. Math. Fys. **68**(3) (1939), 93–102.

[139] K. Mahler, *Polar analogues of two theorems by Minkowski*, Bull. Aust. Math. Soc. **11** (1974), 121–129.

[140] K. Mahler, *An addition to a note of mine: polar analogues of two theorems by Minkowski (Bull. Australian. Math. Soc. 11 (1974) 121–129)*, Bull. Aust. Math. Soc. **14**(3) (1976), 397–398.

[141] V. Mastrantonis and Y. A. Rubinstein, *The Nazarov proof of the non-symmetric Bourgain–Milman inequality*, arXiv:2206.06188.

[142] B. Maurey, *Some deviation inequalities*, Geom. Funct. Anal. **1**(2) (1991), 188–197.

[143] M. Meckes, *Magnitude and Holmes–Thompson intrinsic volumes of convex bodies*, Can. Math. Bull., to appear, arXiv:2206.02600.

[144] M. Meyer, *Une caractérisation volumique de certains espaces normés*, Isr. J. Math. **55** (1986), 317–326.

[145] M. Meyer, *Convex bodies with minimal volume product in* $\mathbb{R}^2$, Monatshefte Math. **12** (1991), 297–301.

[146] M. Meyer and A. Pajor, *On Santaló's inequality*, Lecture notes in mathematics, vol. 1376, Springer, 1989, pp. 261–263.

[147] M. Meyer and A. Pajor, *On the Blaschke–Santaló inequality*, Arch. Math. (Basel) **55**(1) (1990), 82–93.

[148] M. Meyer and S. Reisner, *Characterizations of ellipsoids by section-centroid location*, Geom. Dedic. **31**(3) (1989), 345–355.

[149] M. Meyer and S. Reisner, *Inequalities involving integrals of polar-conjugate concave functions*, Monatshefte Math. **125** (1998), 219–227.

[150] M. Meyer and S. Reisner, *Shadow systems and volumes of polar convex bodies*, Mathematika **53**(1) (2006), 129–148.

[151] M. Meyer and S. Reisner, *On the volume product of polygons*, Abh. Math. Semin. Univ. Hamb. **81** (2011), 93–100.

[152] M. Meyer and S. Reisner, *Ellipsoids are the only local maximizers of the volume product*, Mathematika **65**(3) (2019), 500–504.

[153] M. Meyer and S. Reisner, *The convex intersection body of a convex body*, Glasg. Math. J. **53**(3) (2011), 523–534.

[154] M. Meyer, C. Schütt and E. Werner, *New affine measures of symmetry for convex bodies*, Adv. Math. **228**(5) (2011), 2920–2942.

[155] M. Meyer and E. Werner, *The Santaló-regions of a convex body*, Trans. Am. Math. Soc. **350**(11) (1998), 45694591.

[156] M. Meyer and E. Werner, *On the p-affine surface area*, Adv. Math. **152**(2) (2000), 288313.

[157] V. D. Milman, *Almost Euclidean quotient spaces of subspaces of a finite-dimensional normed space*, Proc. Am. Math. Soc. **94**(3) (1985), 445–449.

[158] V. Milman and A. Pajor A, *Isotropic position and inertia ellipsoids and zonoids of the unit ball of a normed n-dimensional space*, in: Geometric aspects of functional analysis (1987–88), Lecture notes in math., vol. 1376, Springer, Berlin, 1989, pp. 64–104.

[159] E. Milman and A. Yehudayoff, *Sharp isoperimetric inequalities for affine quermassintegrals*, J. Am. Math. Soc., to appear, arXiv:2005.04769.

[160] M. Naszódi, F. Nazarov and D. Ryabogin, *Fine approximation of convex bodies by polytopes*, Am. J. Math. **142**(3) (2020), 809–820.

[161] F. Nazarov, *The Hörmander proof of the Bourgain–Milman Theorem*, in: GAFA seminar notes, Lecture notes in mathematics, vol. 2050, 2012, pp. 335–343.

[162] F. Nazarov, F. Petrov, D. Ryabogin and A. Zvavitch, *A remark on the Mahler conjecture: local minimality of the unit cube*, Duke Math. J. **154** (2010), 419–430.

[163] Y. Ostrover, *A tale of two conjectures: from Mahler to Viterbo*, Tel Aviv University, 2018, https://www.ias.edu/sites/default/files/video/IAS-2018.pdf.

[164] G. Paouris and P. Pivovarov, *A probabilistic take on isoperimetric-type inequalities*, Adv. Math. **230** (2012), 1402–1422.

[165] A. Papadopoulos and M. Troyanov (eds.), *Handbook of Hilbert geometry*, IRMA lectures in mathematics and theoretical physics, vol. 22, European Mathematical Society (EMS), Zürich, 2014.

[166] C. M. Petty, *Centroid surfaces*, Pac. J. Math. **11** (1961), 1535–1547.

[167] C. M. Petty, *Isoperimetric problems*, in: Proceedings of the Conference on Convexity and Combinatorial Geometry, Univ. Oklahoma, Norman, Okla., 1971, 1971, pp. 26–41.

[168] C. M. Petty, *Geominimal surface area*, Geom. Dedic. **3** (1974), 77–97.

[169] C. M. Petty, *Affine isoperimetric problems*, Ann. N.Y. Acad. Sci. **440** (1985), 113–127.

[170] G. Pisier, *The volume of convex bodies and Banach space geometry*, Tracts in math., Cambridge University Press, Cambridge, 1994.

[171] J. Rebollo Bueno, *Stochastic reverse isoperimetric inequalities in the plane*, Geom. Dedic. **215** (2021), 401–413.

[172] S. Reisner, *Random polytopes and the volume–product of symmetric convex bodies*, Math. Scand. **57**(2) (1985), 386–392.

[173] S. Reisner, *Zonoids with minimal volume-product*, Math. Z. **192** (1986), 339–346.

[174] S. Reisner, *Minimal volume product in Banach spaces with a 1-unconditional basis*, J. Lond. Math. Soc. **36** (1987), 126–136.

[175] S. Reisner, C. Schütt and E. M. Werner, *Mahler's conjecture and curvature*, Int. Math. Res. Not. **2012**(1) (2012), 1–16.

[176] C. A. Rogers, *A single integral inequality*, J. Lond. Math. Soc. **32** (1957), 102–108.

[177] C. A. Rogers and G. C. Shephard, *Some extremal problems for convex bodies*, Mathematika **5** (1958), 93–102.

[178] D. Rudolf, *Viterbo's conjecture as a worm problem*, arXiv:2203.02043.

[179] J. Saint-Raymond, *Sur le volume des corps convexes symétriques*, in: Séminaire d'Initiation à l'Analyse, 1980-81, Université Paris VI, 1981.

[180] J. Saint-Raymond, *Le volume des idéaux d'opérateurs classiques*, Stud. Math. **80**(1) (1984), 63–75.

[181] L. A. Santaló, *Un Invariante afin para las curvas convexas del plano*, Math. Notae **8** (1949), 103–111.

[182] L. A. Santaló, *Un invariante afin para los cuerpos convexos del espacio de n dimensiones*, Port. Math. **8**(4) (1949), 155–161.

[183] C. Saroglou, *Shadow systems: remarks and extensions*, Arch. Math. (Basel) **100** (2013), 389–399.

[184] R. Schneider, *Convex bodies: the Brunn–Minkowski theory*, second expanded edn., Encyclopedia of mathematics and its applications, vol. 151, Cambridge University Press, Cambridge, 2014.

[185] C. Schütt and E. W. Werner, *Affine surface area*, arXiv:2204.01926.

[186] G. C. Shephard, *Shadow systems of convex bodies*, Isr. J. Math. **2** (1964), 229–236.

[187] C. Smyth, *The Mahler measure of algebraic numbers: a survey*, in: Number theory and polynomials, J. McKe and C. Smyth, eds., London mathematical society lecture note series, vol. 352, Cambridge University Press, 2008, pp. 322–349.

[188] A. Stancu, *Two volume product inequalities and their applications*, Can. Math. Bull. **52** (2009), 464–472.

[189] E. M. Stein and G. Weiss, *Introduction to Fourier analysis on Euclidean spaces*, Princeton mathematical series, vol. 32, Princeton University Press, Princeton, N. J., 1971.

[190] M. Talagrand, *Transportation cost for Gaussian and other product measures*, Geom. Funct. Anal. **6**(3) (1996), 587–600.

[191] T. Tao, *Santalo's inequality*, http://terrytao.wordpress.com/2007/03/08/open-problem-the-mahler-conjecture-on-convex-bodies/.

[192] T. Tao, *Structure and randomness: pages from year one of a mathematical blog*, American Mathematical Society, Providence, RI, 2008.

[193] M. C. H. Tointon, *The Mahler conjecture in two dimensions via the probabilistic method*, Am. Math. Mon. **25**(9) (2018), 820–828.

[194] N. Tomczak-Jaegermann, *Banach–Mazur distances and finite-dimensional operator ideals*, Pitman monographs and surveys in pure and applied mathematics, vol. 38, Longman Scientific & Technical, Harlow, 1989, copublished in the United States with John Wiley & Sons, Inc., New York, 1989, xii+395 pp.

[195] L. Verger-Gaugry, *A proof of the Conjecture of Lehmer*, 2021. https://doi.org/10.48550/arXiv.1911.10590.

[196] C. Viterbo, *Metric and isoperimetric problems in symplectic geometry*, J. Am. Math. Soc. **13** (2000), 411–431.

[197] C. Vernicos and D. Yang, *A centro-projective inequality*, C. R. Math. Acad. Sci. Paris **357**(8) (2019), 681–685.

[198] D. Ye, L_p *geominimal surface areas and their inequalities*, Int. Math. Res. Not. **9** (2015), 2465–2498.

Apostolos Giannopoulos, Alexander Koldobsky, and Artem Zvavitch

Inequalities for sections and projections of convex bodies

Abstract: This chapter belongs to the area of geometric tomography, which is the study of geometric properties of solids based on data about their sections and projections. We describe a new direction in geometric tomography where different volumetric results are considered in a more general setting, with volume replaced by an arbitrary measure. Surprisingly, such a general approach works for a number of volumetric results. In particular, we discuss the Busemann–Petty problem on sections of convex bodies for arbitrary measures and the slicing problem for arbitrary measures. We present generalizations of these questions to the case of functions. A number of generalizations of questions related to projections, such as the problem of Shephard, are also discussed, as well as some questions in discrete tomography.

Keywords: Convex bodies, sections, Radon transform, intersection body

MSC 2020: 52A20, 53A15, 52B10

1 Introduction

The Busemann–Petty problem asks whether origin-symmetric convex bodies in $\mathbb{R}^n$ with uniformly smaller $(n-1)$-dimensional volume of central hyperplane sections necessarily have smaller n-dimensional volume. The slicing problem of Bourgain asks whether

Acknowledgement: We are grateful to Dylan Langharst and Michael Roysdon for many corrections, valuable discussions, and useful suggestions.

The first named author is supported by the Hellenic Foundation for Research and Innovation (H. F. R. I.) under the "First Call for H. F. R. I. Research Projects to support Faculty members and Researchers and the procurement of high-cost research equipment grant" (Project Number: 1849). The second named author was supported in part by the U. S. National Science Foundation Grant DMS-2054068. The third named author was supported in part by the U. S. National Science Foundation Grant DMS-2000304 and the United States - Israel Binational Science Foundation (BSF). Both the second and the third named author were supported in part by the U. S. National Science Foundation under Grant No. DMS-1929284 while in residence at the Institute for Computational and Experimental Research in Mathematics in Providence, RI, during the Harmonic Analysis and Convexity semester program.

Apostolos Giannopoulos, Department of Mathematics, National and Kapodistrian University of Athens, Panepistimiopolis 157-84, Athens, Greece, e-mail: apgiannop@math.uoa.gr

Alexander Koldobsky, Department of Mathematics, University of Missouri, Columbia, MO 65211, USA, e-mail: koldobskiya@missouri.edu

Artem Zvavitch, Department of Mathematical Sciences, Kent State University, Kent, OH, USA, e-mail: zvavitch@math.kent.edu

https://doi.org/10.1515/9783110775389-006

every symmetric convex body of volume 1 in $\mathbb{R}^n$ has a central hyperplane section whose $(n-1)$-dimensional volume is greater than an absolute constant. We look at these and other results and problems of convex geometry from a more general point of view, replacing volume by an arbitrary measure. Though common sense suggests that the setting of arbitrary measures is too general to produce significant results, we present several situations where such generalizations are very much possible. In particular, it was shown in [127] that the solution of the Busemann–Petty problem (affirmative if $n \leq 4$ and negative if $n \geq 5$) is exactly the same for an arbitrary measure with positive density in place of volume. A version of the slicing problem for arbitrary measures was proved in [85], namely, for any probability density f on an origin-symmetric convex body K of volume 1, there exists a hyperplane H in $\mathbb{R}^n$ so that the integral of f over $K \cap H$ is greater than $\frac{1}{2\sqrt{n}}$.

The chapter is organized as follows. In Section 2 we briefly introduce the most essential basic notation and facts required for our exposition. In Section 3, first we present volume estimates from orthogonal projections and sections and then we continue with generalizations and variants of these inequalities. In Section 4 we discuss the Busemann–Petty problem and its generalizations with emphasis on the general setting with measures in place of volume. Section 5 covers results related to projections of convex bodies. These include the Shephard problem, which is the projection analog of the Busemann–Petty problem, Milman's problem, which can be considered as a mixed Busemann–Petty–Shephard problem, and slicing-type inequalities for the surface area of projections. Section 6 deals with comparison and slicing inequalities for the surface area of convex bodies. In Section 7 we discuss volume difference inequalities which allow to estimate the error in tomographic calculations. Finally, in Section 8 we present what is known about discrete analogs of the slicing problem.

2 Notation and definitions

In this section we will introduce a few basic notations and definitions needed for this chapter. We refer the reader to [3, 4, 27, 28, 44, 45, 78, 95, 117] for a wealth of additional information on objects and tools from convex geometry, geometric tomography, and Fourier analysis used in this survey.

We work in $\mathbb{R}^n$, which is equipped with the standard inner product $\langle \cdot, \cdot \rangle$. We denote by B_2^n and S^{n-1} the Euclidean unit ball and sphere, respectively. We write $|\cdot|$ for volume in the appropriate dimension, ω_n for the volume of B_2^n, and σ for the rotationally invariant probability measure on S^{n-1}. The Grassmann manifold $G_{n,k}$ of all k-dimensional subspaces of $\mathbb{R}^n$ is equipped with the Haar probability measure $\nu_{n,k}$. For every $1 \leq k \leq n-1$ and $H \in G_{n,k}$ we denote by P_H the orthogonal projection from $\mathbb{R}^n$ onto H. The letters c, c', c_1, c_2, etc., denote absolute positive constants which may change from line to line. Whenever we write $a \approx b$, we mean that there exist absolute constants $c_1, c_2 > 0$ such that $c_1 a \leq b \leq c_2 a$.

A convex body in $\mathbb{R}^n$ is a compact convex subset K of $\mathbb{R}^n$ with non-empty interior. We say that K is origin-symmetric if $-K = K$ and we say that K is centered if its barycenter $\frac{1}{|K|}\int_K x\,dx$ is at the origin. The support function of a convex body K is defined by $h_K(y) = \max\{\langle x,y\rangle : x \in K\}$, and the mean width of K is

$$w(K) = \int_{S^{n-1}} h_K(\xi)\,d\sigma(\xi).$$

A closed bounded set K in $\mathbb{R}^n$ is called a star body if every straight line passing through the origin crosses the boundary of K at exactly two points different from the origin, the origin is an interior point of K, and the Minkowski functional of K defined by

$$\|x\|_K = \min\{a \geq 0 : x \in aK\}$$

is a continuous function on $\mathbb{R}^n$. We use the polar formula for the volume $|K|$ of a star body K:

$$|K| = \frac{1}{n}\int_{S^{n-1}} \|\xi\|_K^{-n}\,d\xi. \tag{2.1}$$

If f is an integrable function on K, then

$$\int_K f = \int_{S^{n-1}}\left(\int_0^{\|\xi\|_K^{-1}} r^{n-1}f(r\xi)\,dr\right)d\xi. \tag{2.2}$$

For $1 \leq k \leq n-1$, the $(n-k)$-dimensional spherical Radon transform $\mathcal{R}_{n-k} : C(S^{n-1}) \to C(G_{n,n-k})$ is a linear operator defined by

$$\mathcal{R}_{n-k}g(H) = \int_{S^{n-1}\cap H} g(x)\,dx \quad \text{for all } H \in G_{n,n-k}$$

for every function $g \in C(S^{n-1})$.

For every $H \in G_{n,n-k}$, the $(n-k)$-dimensional volume of the section of a star body K by H can be written as

$$|K \cap H| = \frac{1}{n-k}\mathcal{R}_{n-k}(\|\cdot\|_K^{-n+k})(H). \tag{2.3}$$

More generally, for an integrable function f and any $H \in G_{n,n-k}$,

$$\int_{K\cap H} f = \mathcal{R}_{n-k}\left(\int_0^{\|\cdot\|_K^{-1}} r^{n-k-1}f(r\,\cdot)\,dr\right)(H). \tag{2.4}$$

The class of intersection bodies $\mathcal{I}_n$ was introduced by Lutwak [105]. We consider a generalization of this concept due to Zhang [125]. We say that an origin-symmetric star body D in $\mathbb{R}^n$ is a generalized k-intersection body, and we write $D \in \mathcal{BP}_k^n$ if there exists a finite Borel non-negative measure ν_D on $G_{n,n-k}$ so that for every $g \in C(S^{n-1})$

$$\int\limits_{S^{n-1}} \|x\|_D^{-k} g(x)\,dx = \int\limits_{G_{n,n-k}} R_{n-k}g(H)\,d\nu_D(H). \tag{2.5}$$

When $k = 1$ we get the original Lutwak class of intersection bodies $\mathcal{BP}_1^n = \mathcal{I}_n$.

Let $\mathcal{A}$ be a class of star bodies in $\mathbb{R}^n$ which is invariant with respect to invertible linear transformations. We denote by

$$d_{BM}(K, \mathcal{A}) = \inf\{a > 0 : \exists D \in \mathcal{A} \text{ such that } K \subset D \subset aK\}$$

the Banach–Mazur distance from K to $\mathcal{A}$. We also define the smaller volume ratio distance

$$d_{\mathrm{vr}}(K, \mathcal{A}) = \inf\{(|K|/|D|)^{1/n} : D \subset K,\ D \in \mathcal{A}\}$$

and the outer volume ratio distance

$$d_{\mathrm{ovr}}(K, \mathcal{A}) = \inf\{(|D|/|K|)^{1/n} : K \subset D,\ D \in \mathcal{A}\}$$

from K to $\mathcal{A}$.

Minkowski's fundamental theorem states that if $K_1, \dots, K_m$ are non-empty, compact convex subsets of $\mathbb{R}^n$, then the volume of $t_1 K_1 + \cdots + t_m K_m$ is a homogeneous polynomial of degree n in $t_i > 0$, that is,

$$|t_1 K_1 + \cdots + t_m K_m| = \sum_{1 \leq i_1, \dots, i_n \leq m} V(K_{i_1}, \dots, K_{i_n}) t_{i_1} \cdots t_{i_n},$$

where the coefficients $V(K_{i_1}, \dots, K_{i_n})$ are chosen to be invariant under permutations of their arguments. The coefficient $V(K_1, \dots, K_n)$ is the mixed volume of $K_1, \dots, K_n$; we refer to [117] for a detailed exposition of the definition and main properties of mixed volumes. In particular, if K and D are two convex bodies in $\mathbb{R}^n$, then the function $|K + tD|$ is a polynomial in $t \in [0, \infty)$:

$$|K + tD| = \sum_{j=0}^{n} \binom{n}{j} V_{n-j}(K, D)\, t^j,$$

where $V_{n-j}(K, D) = V((K, n-j), (D, j))$ is the j-th mixed volume of K and D (we use the notation (D, j) for $D, \dots, D$ j times). If $D = B_2^n$, then we set $W_j(K) := V_{n-j}(K, B_2^n) = V((K, n-j), (B_2^n, j))$; this is the j-th quermassintegral of K. The mixed volume $V_{n-1}(K, D)$ can be expressed as

$$V_{n-1}(K,D) = \frac{1}{n} \int_{S^{n-1}} h_D(\xi)\,d\sigma_K(\xi), \tag{2.6}$$

where σ_K is the surface area measure of K; this is the Borel measure on S^{n-1} defined by

$$\sigma_K(A) = \lambda(\{x \in \mathrm{bd}(K) : \text{the outer normal to } K \text{ at } x \text{ belongs to } A\}),$$

where λ is the Hausdorff measure on $\mathrm{bd}(K)$. In particular, if σ_K is absolutely continuous with respect to λ, then the density of σ_K is called the curvature function and is usually denoted as f_K. The surface area $S(K) := \sigma_K(S^{n-1})$ of K satisfies

$$S(K) = nW_1(K).$$

Volume and mixed volumes in general satisfy a number of very useful inequalities. The first one is the Brunn–Minkowski inequality $|K + L|^{1/n} \geq |K|^{1/n} + |L|^{1/n}$, whenever K, L, and $K + L$ are measurable and non-empty.

Direct consequences of the Brunn–Minkowski inequality are Minkowski's first inequality

$$V_{n-1}(K,L) \geq |K|^{(n-1)/n}|L|^{1/n} \tag{2.7}$$

and Minkowski's second inequality

$$V(K,L)^2 \geq |K|V((L,2),(K,n-2)), \tag{2.8}$$

for two convex, compact subsets K and L of $\mathbb{R}^n$.

A zonoid is the limit of Minkowski sums of line segments in the Hausdorff metric. Equivalently, an origin-symmetric convex body Z is a zonoid if and only if its polar body Z° is the unit ball of an n-dimensional subspace of an L_1-space, i. e., if there exists a positive measure μ (the supporting measure of Z) on S^{n-1} such that

$$h_Z(x) = \|x\|_{Z^\circ} = \frac{1}{2} \int_{S^{n-1}} |\langle x,y\rangle|\,d\mu(y).$$

The class of origin-symmetric zonoids with non-empty interior coincides with the class of projection bodies. Recall that the projection body ΠK of a convex body K is the symmetric convex body whose support function is defined by

$$h_{\Pi K}(\xi) = |P_{\xi^\perp}(K)| = \frac{1}{2} \int_{S^{n-1}} |\langle \xi,y\rangle|\,d\sigma_K(y), \quad \xi \in S^{n-1},$$

where $\xi^\perp = \{x \in \mathbb{R}^n : \langle x,\xi\rangle = 0\}$ is the central hyperplane perpendicular to ξ and $P_{\xi^\perp}(K)$ denotes the orthogonal projection of $K \subset \mathbb{R}^n$ onto $\xi^\perp$.

3 Volume estimates from orthogonal projections and sections

Estimating the volume of a convex body from the volumes of its orthogonal projections or sections is a classical question in convex geometry. A well-known such estimate is the famous Loomis–Whitney inequality, which asserts that for any convex body (actually, any compact set) K

$$|K|^{n-1} \leq \prod_{i=1}^{n} |P_{e_i^\perp}(K)|, \tag{3.1}$$

where $\{e_1, \ldots, e_n\}$ is an orthonormal basis of $\mathbb{R}^n$ (see [104]). Equality holds in (3.1) if and only if K is an orthogonal parallelepiped such that $\pm e_i$ are the normal vectors of its facets. A dual inequality, in which the volume of K is estimated by the volumes of the sections $K \cap e_i^\perp$, was proved by Meyer in [106]: for every convex body K in $\mathbb{R}^n$ one has

$$|K|^{n-1} \geq \frac{n!}{n^n} \prod_{i=1}^{n} |K \cap e_i^\perp|, \tag{3.2}$$

with equality if and only if $K = \mathrm{conv}\{\pm\lambda_1 e_1, \ldots, \pm\lambda_n e_n\}$ for some $\lambda_i > 0$.

Both inequalities admit various generalizations. In order to state one of them, let $s > 0$ and say that the subspaces $F_1, \ldots, F_r$ form an s-uniform cover of $\mathbb{R}^n$ with weights $c_1, \ldots, c_r > 0$ if

$$s I_n = \sum_{i=1}^{r} c_i P_{F_i}, \tag{3.3}$$

where I_n is the identity operator on $\mathbb{R}^n$. Then, as an application of the multi-dimensional geometric Brascamp–Lieb inequality, one may show that, for every compact subset K of $\mathbb{R}^n$, we have

$$|K|^s \leq \prod_{i=1}^{r} |P_{F_i}(K)|^{c_i}. \tag{3.4}$$

On the other hand, using Barthe's geometric reverse Brascamp–Lieb inequality, it was proved in [102] that if K is a convex body in $\mathbb{R}^n$ with $0 \in \mathrm{int}(K)$ and $F_1, \ldots, F_r$ are subspaces as above, then

$$|K|^s \geq \frac{1}{(n!)^s} \prod_{i=1}^{r} (d_i! |K \cap F_i|)^{c_i}, \tag{3.5}$$

where $d_i = \dim(F_i)$. A special case of these inequalities occurs when $u_1, \ldots, u_m$ are unit vectors in $\mathbb{R}^n$ and $c_1, \ldots, c_m$ are positive real numbers that satisfy John's condition $I_n = \sum_{i=j}^{m} c_j u_j \otimes u_j$. Then if K is a convex body in $\mathbb{R}^n$ with $0 \in \mathrm{int}(K)$, we have

$$\frac{n!}{n^n} \prod_{j=1}^{m} |K \cap u_j^\perp|^{c_j} \le |K|^{n-1} \le \prod_{j=1}^{m} |P_{u_j^\perp}(K)|^{c_j}. \tag{3.6}$$

To see this, observe that if $P_j = P_{u_j^\perp}$, then $u_j \otimes u_j = I_n - P_j$; hence John's condition may be written as $I_n = \sum_{j=1}^{m} c_j(I_n - P_j)$, which implies that

$$(n-1)I_n = \sum_{j=1}^{m} c_j P_j, \tag{3.7}$$

because $\sum_{j=1}^{m} c_j = n$. The assumption that the interior of K contains the origin is needed only for the left-hand side inequality. The right-hand side inequality in (3.6) was proved by Ball in [7], while the left-hand side inequality was obtained by Li and Huang in [100].

Another extension of the Loomis–Whitney inequality, which can be put in the same framework, has been established in [19]. For every non-empty $\tau \subset [n]$, where $[n] = \{1, \ldots, n\}$, we set $F_\tau = \mathrm{span}\{e_j : j \in \tau\}$ and $E_\tau = F_\tau^\perp$. Given an integer $s \ge 1$, we say that the (not necessarily distinct) sets $\sigma_1, \ldots, \sigma_r \subseteq [n]$ form an s-uniform cover of $[n]$ if every $j \in [n]$ belongs to exactly s of the sets σ_i. The uniform cover inequality of Bollobás and Thomason states that, for every compact subset K of $\mathbb{R}^n$ which is the closure of its interior, we have

$$|K|^s \le \prod_{i=1}^{r} |P_{F_{\sigma_i}}(K)|. \tag{3.8}$$

Note that if $(\sigma_1, \ldots, \sigma_r)$ is an s-uniform cover of $[n]$, then setting $F_i = F_{\sigma_i} = \mathrm{span}(\{e_j : j \in \sigma_i\})$, $i \in [r]$, we have $s I_n = \sum_{i=1}^{r} P_{F_i}$. Thus, (3.8) is an immediate consequence of (3.4). Also, (3.5) implies that if K is a convex body in $\mathbb{R}^n$ with $0 \in \mathrm{int}(K)$, then

$$|K|^s \ge \frac{1}{(n!)^s} \prod_{i=1}^{r} |\sigma_i|! \, |K \cap F_i|. \tag{3.9}$$

To recover Meyer's inequality from (3.5), we use the particular case $F_i = e_i^\perp$, $i \in [n]$, so that we have $(n-1)I_n = \sum_{i=1}^{n} P_{e_i^\perp}$. Applying (3.9) with $s = n-1$ and $|\sigma_i| = \dim(F_i) = n-1$ we get

$$|K|^{n-1} \ge \frac{n!}{n^n} \prod_{i=1}^{n} |K \cap e_i^\perp|$$

for any convex body K in $\mathbb{R}^n$ with $0 \in \mathrm{int}(K)$. It is also not hard to see that the left-hand side inequality in (3.6) is a consequence of (3.5).

Local Loomis–Whitney-type inequalities were studied in many works, including [55, 37], where it was proved that for any convex body K in $\mathbb{R}^n$ and a pair of orthogonal vectors $u, v \in S^{n-1}$, one has

$$|K|\big|P_{[u,v]^{\perp}}(K)\big| \leq \frac{2(n-1)}{n}\big|P_{u^{\perp}}(K)\big|\big|P_{v^{\perp}}(K)\big|. \tag{3.10}$$

We refer to [39] for a simple proof of this inequality and a number of equivalent restatements.

Many restricted variants of the Loomis–Whitney inequality and of the uniform cover inequality, estimating the volume of a convex body from the volumes of a smaller set of sections or projections, were obtained in [26]. See also [120, 2, 39] for some sharp results in this direction.

4 Comparison and slicing inequalities for functions

4.1 The comparison problem for functions

In 1956, Busemann and Petty [29] posed the problem if, for any origin-symmetric convex bodies K, L in $\mathbb{R}^n$, the inequalities

$$|K \cap \xi^{\perp}| \leq |L \cap \xi^{\perp}| \quad \text{for all } \xi \in S^{n-1} \tag{4.1}$$

imply $|K| \leq |L|$. The problem was solved at the end of the 1990s in a sequence of papers [99, 5, 51, 23, 105, 111, 42, 43, 123, 76, 77, 124, 50]. The answer is affirmative if $n \leq 4$, and it is negative if $n \geq 5$. We refer the reader to [78, p. 3] or [44, p. 343] for the history of the solution.

The isomorphic Busemann–Petty problem, posed in [108], asks whether the inequalities in (4.1) imply $|K| \leq C|L|$, where C is an absolute constant. This question is still open and is equivalent to the slicing problem of Bourgain; see below. A very recent result of Klartag [70] shows that the constant can be estimated by $C\sqrt{\log n}$, where $C > 0$ is an absolute constant.

An extension of the Busemann–Petty problem to arbitrary measures in place of volume was considered in [127]. Let K, L be origin-symmetric convex bodies in $\mathbb{R}^n$, and let f be a locally integrable non-negative function on $\mathbb{R}^n$. Suppose that for every $\xi \in S^{n-1}$

$$\int_{K \cap \xi^{\perp}} f(x)\, dx \leq \int_{L \cap \xi^{\perp}} f(x)\, dx, \tag{4.2}$$

where integration is with respect to Lebesgue measure on $\xi^{\perp}$. Does it necessarily follow that

$$\int_{K} f(x)\, dx \leq \int_{L} f(x)\, dx?$$

It was proved in [127] that, for any strictly positive function f, the solution is the same as in the case of volume (where $f \equiv 1$): affirmative if $n \leq 4$ and negative if $n \geq 5$.

In view of this result, it is natural to ask the isomorphic question again. Do the inequalities in (4.2) imply that

$$\int_K f(x)\,dx \le s_n \int_L f(x)\,dx,$$

where the constant s_n does not depend on f, K, and L? It was proved in [96] that the answer is affirmative, namely, $s_n \le \sqrt{n}$.

The argument in [96] is based on a more general estimate. It was proved in [96] that the inequalities in (4.2) imply

$$\int_K f(x)\,dx \le d_{BM}(K,\mathcal{I}_n) \int_L f(x)\,dx. \tag{4.3}$$

By John's theorem [64] and the fact that the class $\mathcal{I}_n$ contains ellipsoids, we have $d_{BM}(K,\mathcal{I}_n) \le \sqrt{n}$, which proves the $\sqrt{n}$ estimate in the isomorphic Busemann–Petty problem for functions. It is not known whether the $\sqrt{n}$ estimate is optimal. Another open question is whether the Banach–Mazur distance in (4.3) can be replaced by the smaller outer volume ratio distance $d_{ovr}(K,\mathcal{I}_n)$.

A slightly different estimate with the outer volume ratio distance instead of the Banach–Mazur distance was proved in [92]. Namely, if K,L are star bodies in $\mathbb{R}^n$ and f and g are non-negative locally integrable functions on $\mathbb{R}^n$ with $\|g\|_\infty = g(0) = 1$, then the inequalities

$$\int_{K\cap\xi^\perp} f(x)\,dx \le \int_{L\cap\xi^\perp} g(x)\,dx \quad \text{for all } \xi \in S^{n-1}$$

imply

$$\int_K f(x)\,dx \le d_{ovr}(K,\mathcal{I}_n)\frac{n}{n-1}|K|^{\frac{1}{n}}\left(\int_L g(x)\,dx\right)^{\frac{n-1}{n}}. \tag{4.4}$$

4.2 The slicing problem for functions

The slicing problem of Bourgain [20, 21] asks whether there exists a constant C so that, for any $n \in \mathbb{N}$ and any origin-symmetric convex body K in $\mathbb{R}^n$,

$$|K|^{\frac{n-1}{n}} \le C \max_{\xi \in S^{n-1}} |K \cap \xi^\perp|. \tag{4.5}$$

In other words, is it true that every origin-symmetric convex body K in $\mathbb{R}^n$ of volume 1 has a hyperplane section whose $(n-1)$-dimensional volume is greater than an absolute constant?

The problem remains open. Bourgain [22] proved that $C \leq O(n^{1/4})$ up to a logarithmic factor which was removed by Klartag [69]. Chen [32] proved that $C \leq O(n^\epsilon)$ for every $\epsilon > 0$, and Klartag and Lehec [73] established a polylogarithmic bound $C \leq O(\log^4 n)$. The method of [73] was slightly refined in [63], where it was shown that $C \leq O(\log^{2.2226} n)$. Very recently, Klartag [70] established the bound $C \leq O(\sqrt{\log n})$. The answer is known to be affirmative for some special classes of convex bodies. For unconditional convex bodies this was observed by Bourgain; see also [108, 66, 17], for unit balls of subspaces of L_p it was proved in [8, 65, 107], for intersection bodies in [44, Th. 9.4.11], for zonoids, duals of bodies with bounded volume ratio in [108], for the Schatten classes in [97], and for k-intersection bodies in [90, 86]. Other partial results on the problem include [6, 24, 33, 35, 58, 68, 72, 109, 36, 10]; see the book [27] and the surveys [75, 45] for details.

A generalization of the slicing problem to arbitrary functions was suggested in [80]. Does there exist a constant T_n depending only on the dimension so that, for every origin-symmetric convex body K in $\mathbb{R}^n$ and every non-negative integrable function f on K,

$$\int_K f(x)\,dx \leq T_n\,|K|^{1/n} \max_{\xi \in S^{n-1}} \int_{K \cap \xi^\perp} f(x)\,dx? \tag{4.6}$$

In other words, is it true that the sup-norm of the Radon transform of any probability density on a convex body of volume 1 is bounded from below by a constant depending only on the dimension? The case where $f \equiv 1$ corresponds to the slicing problem of Bourgain.

It was proved in [82, 83] that the answer to this question is affirmative with $T_n \leq O(\sqrt{n})$. A different proof, based on the Blaschke–Petkantschin formula (see [118]) was given in [31]. Inequality (4.6) holds true with an absolute constant in place of T_n for intersection bodies, unconditional convex bodies, and duals of convex bodies with bounded volume ratio [85] and for the unit balls of n-dimensional subspaces of L_p, $p > 2$, with $C = O(\sqrt{p})$ [89] (note that the unit balls of subspaces of L_p with $0 < p \leq 2$ are intersection bodies).

These results follow from a more general inequality proved in [85] for any origin-symmetric star body K in $\mathbb{R}^n$ and any integrable non-negative function f on K,

$$\int_K f(x)\,dx \leq 2\,d_{\mathrm{ovr}}(K, \mathcal{I}_n)|K|^{\frac{1}{n}} \max_{\xi \in S^{n-1}} \int_{K \cap \xi^\perp} f(x)\,dx. \tag{4.7}$$

Now assuming that K is an origin-symmetric convex body, by John's theorem [64] we get $d_{\mathrm{ovr}}(K, \mathcal{I}_n) \leq \sqrt{n}$. Also, the distance is bounded by an absolute constant for unconditional convex bodies [85] and for the unit balls of subspaces of L_p, $p > 2$ [107, 89]. Clearly, if K is an intersection body, the distance is 1. The proof of inequality (4.7) in [85] is based on a stability result for sections of star bodies. However, in Section 4.3 we present the proof of a more general result which implies (4.7).

The estimate $T_n \leq O(\sqrt{n})$ is optimal. Klartag and the second named author showed in [71] that there exists an origin-symmetric convex body M in $\mathbb{R}^n$ and a probability density f on M so that

$$\int_{M \cap H} f \leq c \frac{\sqrt{\log \log n}}{\sqrt{n}} |M|^{-1/n},$$

for every affine hyperplane H in $\mathbb{R}^n$, where c is an absolute constant. The convex body M which provides the example is a Gluskin-type random polytope generated by properly scaled random vectors θ_j uniformly distributed on the sphere S^{n-1}, while the density f is the density of an appropriate convolution of the standard Gaussian measure on $\mathbb{R}^n$ with the sum of the Dirac masses of the θ_j's, restricted to M. The logarithmic term was later removed by Klartag and Livshyts [74], who added a "random rounding" technique to the previous construction. So, finally $T_n \geq c \sqrt{n}$.

Another estimate of this kind, involving also non-central sections, was proved in [18]. Namely, there exists an absolute constant C so that

$$\int_K f(x)\, dx \leq C \sqrt{p}\, d_{\mathrm{ovr}}(K, L_p^n) |K|^{1/n} \sup_H \int_{K \cap H} f(x)\, dx, \tag{4.8}$$

for any $p \geq 1$, $n \in \mathbb{N}$, and any origin-symmetric convex body K in $\mathbb{R}^n$, where $d_{\mathrm{ovr}}(K, L_p^n)$ is the outer volume ratio distance from K to the class L_p^n of the unit balls of n-dimensional subspaces of $L_p([0,1])$, and the supremum is taken over all affine hyperplanes H in $\mathbb{R}^n$.

4.3 A quotient inequality for sections of functions

A general inequality which implies both (4.4) and (4.7) was proved in [57]. Let K and L be star bodies in $\mathbb{R}^n$ and let f, g be non-negative continuous functions on K and L, respectively, with $\|g\|_\infty = g(0) = 1$. Then

$$\frac{\int_K f}{\left(\int_L g\right)^{\frac{n-1}{n}} |K|^{\frac{1}{n}}} \leq \frac{n}{n-1} d_{\mathrm{ovr}}(K, \mathcal{I}_n) \max_{\xi \in S^{n-1}} \frac{\int_{K \cap \xi^\perp} f}{\int_{L \cap \xi^\perp} g}.$$

In fact, for any integer $0 < k < n$ we have

$$\frac{\int_K f}{\left(\int_L g\right)^{\frac{n-k}{n}} |K|^{\frac{k}{n}}} \leq \frac{n}{n-k} \left(d_{\mathrm{ovr}}(K, \mathcal{BP}_k^n)\right)^k \max_{H \in G_{n,n-k}} \frac{\int_{K \cap H} f}{\int_{L \cap H} g}. \tag{4.9}$$

For the proof of (4.9) fix $\delta > 0$ and let $D \in \mathcal{BP}_k^n$ be a body such that $K \subset D$ and

$$|D|^{\frac{1}{n}} \leq (1 + \delta)\, d_{\mathrm{ovr}}(K, \mathcal{BP}_k^n) |K|^{\frac{1}{n}}. \tag{4.10}$$

Write v_D for the measure on $G_{n,n-k}$ corresponding to L according to the definition (2.5). Let ε be such that

$$\int_{K\cap H} f \le \varepsilon \int_{L\cap H} g, \quad \text{for all } H \in G_{n,n-k}.$$

By (2.4), we have

$$\mathcal{R}_{n-k}\left(\int_0^{\|\cdot\|_K^{-1}} r^{n-k-1} f(r\cdot)\, dr\right)(H) \le \varepsilon\, \mathcal{R}_{n-k}\left(\int_0^{\|\cdot\|_L^{-1}} r^{n-k-1} g(r\cdot)\, dr\right)(H)$$

for every $H \in G_{n,n-k}$. Integrating both sides of the latter inequality with respect to v_D and using the definition (2.5), we get

$$\int_{S^{n-1}} \|x\|_D^{-k}\left(\int_0^{\|x\|_K^{-1}} r^{n-k-1} f(rx)\, dr\right) dx \le \varepsilon \int_{S^{n-1}} \|x\|_D^{-k}\left(\int_0^{\|x\|_L^{-1}} r^{n-k-1} g(rx)\, dr\right) dx, \qquad (4.11)$$

which is equivalent to

$$\int_K \|x\|_D^{-k} f(x)\, dx \le \varepsilon \int_L \|x\|_D^{-k} g(x)\, dx. \qquad (4.12)$$

Since $K \subset D$, we have $1 \ge \|x\|_K \ge \|x\|_D$ for every $x \in K$. Therefore,

$$\int_K \|x\|_D^{-k} f(x)\, dx \ge \int_K \|x\|_K^{-k} f(x)\, dx \ge \int_K f.$$

On the other hand, we may apply [108, Lemma 2.1]. Indeed, recall that $g(0) = \|g\|_\infty = 1$ and the same is true for the function $\chi_L(x) g(x)$; moreover, the proof of [108, Lemma 2.1], also works under the assumption that the body is star-shaped. Thus, we get

$$\left(\frac{\int_L \|x\|_D^{-k} g(x)\, dx}{\int_D \|x\|_D^{-k}\, dx}\right)^{1/(n-k)} \le \left(\frac{\int_L g(x)\, dx}{\int_D dx}\right)^{1/n}.$$

Since $\int_D \|x\|_D^{-k}\, dx = \frac{n}{n-k}|D|$, we can estimate the right-hand side of (4.12) by

$$\int_L \|x\|_D^{-k} g(x)\, dx \le \varepsilon \frac{n}{n-k}\left(\int_L g\right)^{\frac{n-k}{n}} |D|^{\frac{k}{n}}.$$

Applying (4.10) and sending δ to zero, we see that the latter inequality in conjunction with (4.12) implies

$$\int_K f \le \varepsilon \, \frac{n}{n-k} \left(d_{\mathrm{ovr}}(K, \mathcal{B}\mathcal{P}_k^n)\right)^k \left(\int_L g\right)^{\frac{n-k}{n}} |K|^{\frac{k}{n}}.$$

Now put $\varepsilon = \max_{H \in G_{n,n-k}} \frac{\int_{K \cap H} f}{\int_{L \cap H} g}$ and the result follows.

Note the following immediate consequences of (4.9). If we add the assumption that

$$\int_{K \cap H} f \le \int_{L \cap H} g$$

for all $H \in G_{n,n-k}$, then we get the generalization of (4.4):

$$\int_K f \le \frac{n}{n-k} \left(d_{\mathrm{ovr}}(K, BP_k^n)\right)^k |K|^{\frac{k}{n}} \left(\int_L g\right)^{\frac{n-k}{n}}.$$

If we choose $L = B_2^n$ and $g \equiv 1$, then we obtain the slicing inequality

$$\int_K f \le \frac{n}{n-k} \left(d_{\mathrm{ovr}}(K, BP_k^n)\right)^k |K|^{\frac{k}{n}} \max_H \int_{K \cap H} f,$$

which generalizes (4.7). Choosing $K = L$ and $g \equiv 1$ we obtain another variant of the slicing inequality for functions:

$$\frac{\int_K f}{|K|} \le \frac{n}{n-k} \left(d_{\mathrm{ovr}}(K, BP_k^n)\right)^k \max_H \frac{\int_{K \cap H} f}{|K \cap H|}.$$

5 Projections of convex bodies

5.1 The Shephard problem

Shephard's problem [119] is "dual" to the Busemann–Petty problem. Let K and L be two origin-symmetric convex bodies in $\mathbb{R}^n$ and suppose that

$$\left|P_{\xi^\perp}(K)\right| \le \left|P_{\xi^\perp}(L)\right| \tag{5.1}$$

for every $\xi \in S^{n-1}$. Does it follow that $|K| \le |L|$?

The answer is affirmative if $n = 2$, but shortly after it was posed, Shephard's question was answered in the negative for all $n \ge 3$. This was done independently by Petty in [112], who gave an explicit counterexample in $\mathbb{R}^3$, and by Schneider in [116] for all $n \ge 3$. In particular, Schneider in [116] proved that the answer is affirmative if the body L having projections of larger volume is a projection body; we refer to [93] for harmonic analysis proofs of these facts. After these counterexamples, one might try to relax the question, asking for the smallest constant C_n (or the order of growth of this constant C_n as $n \to \infty$) for which if K, L are origin-symmetric convex bodies in $\mathbb{R}^n$ and $|P_{\xi^\perp}(K)| \le |P_{\xi^\perp}(L)|$ for all $\xi \in S^{n-1}$, then $|K| \le C_n |L|$. Such a constant C_n does exist, and a simple argument, based on John's theorem, shows that $C_n \le c\sqrt{n}$, where $c > 0$ is an absolute constant. On the other hand, K. Ball has proved in [7] that this simple estimate is optimal: One has $C_n \approx \sqrt{n}$.

The lower-dimensional Shephard problem is the following question. Let $1 \le k \le n-1$ and let $S_{n,k}$ be the smallest constant $S > 0$ with the following property: For every pair of convex bodies K and L in $\mathbb{R}^n$ that satisfy $|P_F(K)| \le |P_F(L)|$ for all $F \in G_{n,n-k}$, one has $|K|^{\frac{1}{n}} \le S|L|^{\frac{1}{n}}$. Is it true that there exists an absolute constant $C > 0$ such that $S_{n,k} \le C$ for all n and k?

Goodey and Zhang [59] proved that $S_{n,k} > 1$ if $n - k > 1$. General estimates for $S_{n,k}$ are provided in [52]: If K and L are two convex bodies in $\mathbb{R}^n$ such that $|P_F(K)| \le |P_F(L)|$, for every $F \in G_{n,n-k}$, then

$$|K|^{\frac{1}{n}} \le c_1 \sqrt{\frac{n}{n-k}} \log\left(\frac{en}{n-k}\right)|L|^{\frac{1}{n}},$$

where $c_1 > 0$ is an absolute constant. It follows that $S_{n,k}$ is bounded by an absolute constant if $\frac{n}{n-k}$ is bounded. Also, under the same assumptions and using results from [110] one can prove a general estimate which is logarithmic in n and valid for all k:

$$|K|^{\frac{1}{n}} \le \frac{c_1 \min w(\tilde{L})}{\sqrt{n}}|L|^{\frac{1}{n}} \le c_2(\log n)|L|^{\frac{1}{n}},$$

where $c_1, c_2 > 0$ are absolute constants and the minimum is over all linear images $\tilde{L}$ of L that have volume 1. The second inequality follows from the fact that if $\tilde{L}$ is a convex body of volume 1 in $\mathbb{R}^n$ which is in the minimal mean width position (i. e., $w(\tilde{L}) \le w(T(\tilde{L}))$ for all $T \in SL(n)$), then $w(\tilde{L}) \le c\sqrt{n}(\log n)$ for some absolute constant $c > 0$. This is a consequence of well-known results of Lewis, Figiel and Tomczak-Jaegermann, and Pisier (see [3, Chapter 6] for a complete discussion).

An extension of Shephard's problem to the case of more general measures was first considered by Livshyts [101], who studied the case of p-concave and $1/p$-homogeneous measures. Kryvonos and Langharst [98] further extended the results from [101] and studied the isomorphic case of the question.

5.2 A variant of the Busemann–Petty and Shephard problems

A variant of the Busemann–Petty and Shephard problems was proposed by V. Milman: Assume that K and L are origin-symmetric convex bodies in $\mathbb{R}^n$ and satisfy

$$|P_{\xi^\perp}(K)| \le |L \cap \xi^\perp| \tag{5.2}$$

for all $\xi \in S^{n-1}$. Does it follow that $|K| \leq |L|$? An affirmative answer to this question was given by the first two authors in [52]. Also the lower-dimensional analog of the problem has an affirmative answer, and moreover, one can drop the symmetry assumptions and even the assumption of convexity for L. More precisely, if K is a convex body in $\mathbb{R}^n$ and L is a compact subset of $\mathbb{R}^n$ such that, for some $1 \leq k \leq n-1$,

$$|P_F(K)| \leq |L \cap F|$$

for all $F \in G_{n,n-k}$, then

$$|K| \leq |L|.$$

The proof exploits the Busemann–Straus/Grinberg inequality (see [30, 60])

$$\int\limits_{G_{n,k}} |K \cap E|^n dv_{n,k}(E) \leq \frac{\omega_k^n}{\omega_n^k} |K|^k, \tag{5.3}$$

which is true for every bounded Borel set K in $\mathbb{R}^n$ and $1 \leq k \leq n-1$, and the classical Alexandrov's inequalities in the following form: If K is a convex body in $\mathbb{R}^n$, then the sequence

$$Q_k(K) = \left(\frac{1}{\omega_k} \int\limits_{G_{n,k}} |P_F(K)| \, dv_{n,k}(F) \right)^{1/k} \tag{5.4}$$

is decreasing in k. This is a consequence of the Alexandrov–Fenchel inequality (see the books of Burago and Zalgaller [28] and Schneider [117]). In particular, for every $1 \leq k \leq n-1$, we have

$$\left(\frac{|K|}{\omega_n} \right)^{\frac{1}{n}} \leq \left(\frac{1}{\omega_{n-k}} \int\limits_{G_{n,n-k}} |P_F(K)| \, dv_{n,n-k}(F) \right)^{\frac{1}{n-k}} \leq w(K). \tag{5.5}$$

With these tools one proceeds as follows. Let K be a convex body in $\mathbb{R}^n$ and let L be a compact subset of $\mathbb{R}^n$. Assume that for some $1 \leq k \leq n-1$ we have $|P_F(K)| \leq |L \cap F|$ for all $F \in G_{n,n-k}$. From (5.5) we get

$$\left(\frac{|K|}{\omega_n} \right)^{\frac{n-k}{n}} \leq \frac{1}{\omega_{n-k}} \int\limits_{G_{n,n-k}} |P_F(K)| \, dv_{n,n-k}(F).$$

Our assumption, Hölder's inequality, and the Busemann–Straus/Grinberg inequality give

$$\frac{1}{\omega_{n-k}} \int\limits_{G_{n,n-k}} |P_F(K)| \, dv_{n,n-k}(F) \leq \frac{1}{\omega_{n-k}} \int\limits_{G_{n,n-k}} |L \cap F| \, dv_{n,n-k}(F)$$

$$\leq \frac{1}{\omega_{n-k}} \left(\int_{G_{n,n-k}} |L \cap F|^n \, dv_{n,n-k}(F) \right)^{1/n}$$

$$\leq \frac{1}{\omega_{n-k}} \frac{\omega_{n-k}}{\omega_n^{\frac{n-k}{n}}} |L|^{\frac{n-k}{n}} = \left(\frac{|L|}{\omega_n} \right)^{\frac{n-k}{n}}. \tag{5.6}$$

Therefore, $|K| \leq |L|$.

Hosle [61] proved that if the condition (5.2) is reversed, then $|L| \leq \sqrt{n}|K|$. He also suggests a way to generalize the solution of Milman's problem to log-concave measures, p-concave measures, and measures with $1/p$-homogeneous densities in place of volume.

5.3 Surface area of projections

The second named author obtained a number of "slicing-type" inequalities about the surface area of hyperplane projections of projection bodies. In [81] he proved that if Z is a projection body in $\mathbb{R}^n$, then

$$|Z|^{\frac{1}{n}} \min_{\xi \in S^{n-1}} S(P_{\xi^{\perp}}(Z)) \leq b_n S(Z), \tag{5.7}$$

where $S(A)$ denotes the surface area of A and

$$b_n = (n-1)\omega_{n-1}/(n\omega_n^{\frac{n-1}{n}}) \approx 1.$$

This inequality is sharp; there is equality if $Z = B_2^n$. Conversely, in [84] he proved that if Z is a projection body in $\mathbb{R}^n$ which is a dilate of a body in isotropic position (see, for example, [27]), then

$$|Z|^{\frac{1}{n}} \max_{\xi \in S^{n-1}} S(P_{\xi^{\perp}}(Z)) \geq c(\log n)^{-2} S(Z), \tag{5.8}$$

where $c > 0$ is an absolute constant.

Similar inequalities for the surface area of hyperplane projections of an arbitrary convex body K in $\mathbb{R}^n$ were studied in [56]. In what follows, we denote by ∂_K the minimal surface area parameter of K, which is the quantity

$$\partial_K := \min\{S(T(K))/|T(K)|^{\frac{n-1}{n}} : T \in \mathrm{Aff}(n)\}, \tag{5.9}$$

where $\mathrm{Aff}(n)$ denotes the class of invertible affine transformations of $\mathbb{R}^n$. From the isoperimetric inequality and K. Ball's reverse isoperimetric inequality [9] it is known that $c_1 \sqrt{n} \leq \partial_K \leq c_2 n$ for every convex body K in $\mathbb{R}^n$, where $c_1, c_2 > 0$ are absolute constants. It was proved in [56] that there exists an absolute constant $c_1 > 0$ such that, for every convex body K in $\mathbb{R}^n$,

$$|K|^{\frac{1}{n}} \min_{\xi \in S^{n-1}} S(P_{\xi^{\perp}}(K)) \le \frac{2b_n \partial_K}{n \omega_n^{\frac{1}{n}}} S(K) \le \frac{c_1 \partial_K}{\sqrt{n}} S(K). \tag{5.10}$$

This inequality, which generalizes (5.7), is sharp, e. g., for the Euclidean unit ball. Since $c_1 \partial_K / \sqrt{n} \le c \sqrt{n}$ for every convex body K in $\mathbb{R}^n$, we get the general upper bound

$$|K|^{\frac{1}{n}} \min_{\xi \in S^{n-1}} S(P_{\xi^{\perp}}(K)) \le c \sqrt{n} S(K). \tag{5.11}$$

The proof is based on the next result from [55], which is a consequence of (3.10). If K is a convex body in $\mathbb{R}^n$, then

$$\frac{S(P_{\xi^{\perp}}(K))}{|P_{\xi^{\perp}}(K)|} \le \frac{2(n-1)}{n} \frac{S(K)}{|K|} \tag{5.12}$$

for every $\xi \in S^{n-1}$. It follows that

$$|K| \min_{\xi \in S^{n-1}} S(P_{\xi^{\perp}}(K)) \le \frac{2(n-1)}{n} S(K) \min_{\xi \in S^{n-1}} |P_{\xi^{\perp}}(K)|.$$

Next, we observe that

$$\min_{\xi \in S^{n-1}} |P_{\xi^{\perp}}(K)| = \min_{\xi \in S^{n-1}} h_{\Pi K}(\xi) = r(\Pi K),$$

where $r(A)$ is the inradius of A, i. e., the largest $r > 0$ such that $rB_2^n \subseteq A$. We write

$$r(\Pi K) \le \left(\frac{|\Pi K|}{\omega_n} \right)^{\frac{1}{n}} \le \frac{\omega_{n-1} \partial_K}{n \omega_n} |K|^{\frac{n-1}{n}},$$

where the upper estimate for $|\Pi K|$ is observed in [54]. Combining the above we get

$$|K| \min_{\xi \in S^{n-1}} S(P_{\xi^{\perp}}(K)) \le \frac{2(n-1)\omega_{n-1} \partial_K}{n^2 \omega_n} S(K)|K|^{\frac{n-1}{n}}.$$

Inequality (5.8) can also be generalized, starting with the next fact. If K is a convex body in $\mathbb{R}^n$, then

$$\int_{S^{n-1}} S(P_{\xi^{\perp}}(K)) \, d\sigma(\xi) \ge c \, S(K)^{\frac{n-2}{n-1}}.$$

This inequality implies that if K is in some of the "classical positions" (we refer to [3, 108] for discussion on those "classical positions," including, minimal surface area, minimal mean width, and isotropic, John, or Löwner position), then

$$|K|^{\frac{1}{n}} \int_{S^{n-1}} S(P_{\xi^{\perp}}(K)) \, d\sigma(\xi) \ge c \, S(K), \tag{5.13}$$

where $c > 0$ is an absolute constant. In particular, if K is a convex body in $\mathbb{R}^n$, which is in any of the "classical positions," then

$$|K|^{\frac{1}{n}} \max_{\xi \in S^{n-1}} S(P_{\xi^{\perp}}(K)) \geq c\, S(K).$$

Recall that a $(\log n)^2$-term appeared in (5.8). The estimate in (5.13) is stronger and, for bounds of this type, there is no need to assume that K is a projection body. In fact, the estimate continues to hold as long as

$$S(K)^{\frac{1}{n-1}} \leq c|K|^{\frac{1}{n}}$$

for an absolute constant $c > 0$. This is a mild condition which is satisfied not only by the classical positions but also by all *reasonable* positions of K.

The same questions may be studied for the quermassintegrals $V_{n-k}(K) = V((K, n - k), (B_2^n, k))$ of a convex body K and the corresponding quermassintegrals of its hyperplane projections. The proofs employ the same tools as in the surface area case. The main additional ingredient is a generalization of (5.12) (proved in [37]) to subspaces of arbitrary dimension and quermassintegrals of any order: If K is a convex body in $\mathbb{R}^n$ and $0 \leq p \leq k \leq n$, then, for every $F \in G_{n,k}$,

$$\frac{V_{n-p}(K)}{|K|} \geq \frac{1}{\binom{n-k+p}{n-k}} \frac{V_{k-p}(P_F(K))}{|P_F(K)|}. \tag{5.14}$$

This inequality allows one to obtain further generalizations; one can compare the surface area of a convex body K to the minimal, average, or maximal surface area of its lower-dimensional projections $P_F(K)$, $F \in G_{n,k}$, for any given $1 \leq k \leq n - 1$. We refer to [39, 38] to further study of inequalities related to (5.14).

6 Surface area

The question whether it is possible to have a version of the slicing inequality for the surface area instead of volume has been formulated as follows: Is it true that there exists a constant α_n depending (or not) on the dimension n so that

$$S(K) \leq \alpha_n |K|^{\frac{1}{n}} \max_{\xi \in S^{n-1}} S(K \cap \xi^{\perp}) \tag{6.1}$$

for every origin-symmetric convex body K in $\mathbb{R}^n$? A lower-dimensional slicing problem may also be formulated; for any $2 \leq k \leq n - 1$ one may ask for a constant $\alpha_{n,k}$ such that

$$S(K) \leq \alpha_{n,k}^k |K|^{\frac{k}{n}} \max_{H \in G_{n,n-k}} S(K \cap H) \tag{6.2}$$

for every origin-symmetric convex body K in $\mathbb{R}^n$. Moreover, one may replace surface area by any other quermassintegral and pose the corresponding question.

A negative answer was given in [25]. For any $n \geq 2$ and any $\alpha > 0$ one may find an origin-symmetric convex body K in $\mathbb{R}^n$ such that

$$S(K) > \alpha |K|^{\frac{1}{n}} \max_{\xi \in S^{n-1}} S(P_{\xi^\perp}(K)) \geq \alpha |K|^{\frac{1}{n}} \max_{\xi \in S^{n-1}} S(K \cap \xi^\perp).$$

In fact, it is shown that one may construct an origin-symmetric ellipsoid $\mathcal{E}$ such that

$$S(\mathcal{E}) > \alpha |\mathcal{E}|^{\frac{1}{n}} \max_{\xi \in S^{n-1}} S(P_{\xi^\perp}(\mathcal{E})).$$

In order to do this, for a given ellipsoid $\mathcal{E}$ in $\mathbb{R}^n$ one needs to know the $(n-1)$-dimensional section of $\mathcal{E}$ that has the largest surface area. This is a natural question of independent interest. It is shown in [25] that if $\mathcal{E}$ is an origin-symmetric ellipsoid in $\mathbb{R}^n$ and $a_1 \leq a_2 \leq \cdots \leq a_n$ are the lengths and $e_1, e_2, \ldots, e_n$ are the corresponding directions of its semiaxes, then

$$S(\mathcal{E} \cap \xi^\perp) \leq S(P_{\xi^\perp}(\mathcal{E})) \leq S(\mathcal{E} \cap e_1^\perp) \tag{6.3}$$

for every $\xi \in S^{n-1}$. This information is then combined with a formula of Rivin [113] for the surface area of an ellipsoid: If $\mathcal{E}$ is an ellipsoid as above, then

$$S(\mathcal{E}) = n|\mathcal{E}| \int_{S^{n-1}} \left(\sum_{i=1}^{n} \frac{\xi_i^2}{a_i^2} \right)^{1/2} d\sigma(\xi). \tag{6.4}$$

Assume that there exists a constant $a_n > 0$ such that we have the following inequality for ellipsoids:

$$S(\mathcal{E}) \leq a_n |\mathcal{E}|^{1/n} \max_{\xi \in S^{n-1}} S(\mathcal{E} \cap \xi^\perp). \tag{6.5}$$

Then we have

$$\max_{\xi \in S^{n-1}} S(\mathcal{E} \cap \xi^\perp) = S(\mathcal{E} \cap e_1^\perp) = (n-1)|\mathcal{E} \cap e_1^\perp| \int_{S^{n-2}} \left(\sum_{i=2}^{n} \frac{\xi_i^2}{a_i^2} \right)^{1/2} d\sigma(\xi),$$

and assuming, as we may, that $\prod_{i=1}^{n} a_i = 1$, we can rewrite (6.5) as

$$n\omega_n \cdot \frac{1}{d_n} \mathbb{E}\left[\left(\sum_{i=1}^{n} \frac{g_i^2}{a_i^2} \right)^{1/2} \right] \leq a_n \omega_n^{1/n} \cdot (n-1)\omega_{n-1} \frac{1}{a_1} \cdot \frac{1}{d_{n-1}} \mathbb{E}\left[\left(\sum_{i=2}^{n} \frac{g_i^2}{a_i^2} \right)^{1/2} \right],$$

where $d_n \sim \sqrt{n}$. After some calculations we see that

$$a_n \geq c\left(\frac{1 + \sum_{i=2}^{n} \frac{a_1^2}{a_i^2}}{\sum_{i=2}^{n} \frac{1}{a_i^2}}\right)^{1/2}$$

for some absolute constant $c > 0$. Choosing $a_2 = \cdots = a_n = r$ and $a_1 = r^{-(n-1)}$ we see that

$$\left(\frac{1 + \sum_{i=2}^{n} \frac{a_1^2}{a_i^2}}{\sum_{i=2}^{n} \frac{1}{a_i^2}}\right)^{1/2} = \left(\frac{1 + \frac{n-1}{r^{2n}}}{\frac{n-1}{r^2}}\right)^{1/2} = \left(\frac{1}{r^{2n-2}} + \frac{r^2}{n-1}\right)^{1/2} \to \infty$$

as $r \to \infty$. So, we arrive at a contradiction.

In fact, one can prove a more general analog of (6.3); for any k-dimensional subspace H and any $0 \leq j \leq k - 1$ we have

$$W_j(\mathcal{E} \cap F_k) \leq W_j(\mathcal{E} \cap H) \leq W_j(\mathcal{E} \cap E_k)$$

and

$$W_j(P_{F_k}(\mathcal{E})) \leq W_j(P_H(\mathcal{E})) \leq W_j(P_{E_k}(\mathcal{E})),$$

where $F_k = \mathrm{span}\{e_1, \ldots, e_k\}, E_k = \mathrm{span}\{e_{n-k+1}, \ldots, e_n\}$, and $W_j(A) = V((A, k-j), (B_2^k, k-j))$ is the j-th quermassintegral of a k-dimensional convex body A. These results generalize a known fact for the maximal and minimal volume of k-dimensional sections and projections of ellipsoids. As a consequence one can obtain a more general negative result about all the quermassintegrals of sections and projections of convex bodies.

In [25] some positive results are stated for variants of this question, which were strengthened in [103]. An example is the next inequality: If K is an origin-symmetric convex body in $\mathbb{R}^n$, then for any $0 \leq j \leq n - k - 1 \leq n - 1$ we have

$$\alpha_{n,k,j}\binom{n}{k}^{-1}\frac{W_j(K)}{|K|} \leq \int_{G_{n,n-k}} \frac{W_j(K \cap F)}{|K \cap F|} dv_{n,n-k}(F) \leq \alpha_{n,k,j}\binom{n-j}{k}\frac{W_j(K)}{|K|},$$

where $\alpha_{n,k,j}$ is a constant depending only on n, k, j.

The analog of the Busemann–Petty problem for surface area was studied by Koldobsky and König in [87]: If K and L are two convex bodies in $\mathbb{R}^n$ such that $S(K \cap \xi^\perp) \leq S(L \cap \xi^\perp)$ for all $\xi \in S^{n-1}$, does it then follow that $S(K) \leq S(L)$? Answering a question of Pelczynski, they prove that the central $(n - 1)$-dimensional section of the cube $B_\infty^n = [-1, 1]^n$ that has maximal surface area is the one that corresponds to the unit vector $\xi_0 = \frac{1}{\sqrt{2}}(1, 1, 0, \ldots, 0)$ (exactly as in the case of volume), i. e.,

$$\max_{\xi \in S^{n-1}} S(B_\infty^n \cap \xi^\perp) = S(B_\infty^n \cap \xi_0^\perp) = 2((n - 2)\sqrt{2} + 1).$$

Comparing with a ball of suitable radius one gets that the answer to the Busemann–Petty problem for surface area is negative in dimensions $n \geq 14$.

It is natural to ask whether an isomorphic version of the problem has an affirmative answer. Assuming that there is a constant γ_n such that if K and L are origin-symmetric convex bodies in $\mathbb{R}^n$ that satisfy

$$S(K \cap \xi^\perp) \leq S(L \cap \xi^\perp)$$

for all $\xi \in S^{n-1}$, then $S(K) \leq \gamma_n S(L)$, one can see that there is some constant $c(n)$ such that

$$S(K) \leq c(n)S(K)^{\frac{1}{n-1}} \max_{\xi \in S^{n-1}} S(K \cap \xi^\perp) \tag{6.6}$$

for every convex body K in $\mathbb{R}^n$. It was proved in [25] that an inequality of this type holds true in general. If K is a convex body in $\mathbb{R}^n$, then

$$S(K) \leq A_n S(K)^{\frac{1}{n-1}} \max_{\xi \in S^{n-1}} S(K \cap \xi^\perp),$$

where $A_n > 0$ is a constant depending only on n. The result is first proved for an arbitrary ellipsoid and then it is extended to any convex body, using John's theorem. In [103] a direct proof of a more general result is given, showing that an inequality as (6.6) holds for any k and j, where k is the codimension of the subspaces and j is the order of the quermassintegral that we consider: Let K be an origin-symmetric convex body in $\mathbb{R}^n$. For every $0 \leq j \leq n - k - 1 \leq n - 1$ we have

$$W_j(K)^{n-k-j} \leq a_{n,k,j} \max_{F \in G_{n,n-k}} W_j(K \cap F)^{n-j},$$

where $a_{n,k,j} > 0$ is a constant depending only on n, k, and j. The proof of this inequality exploits the Blaschke–Petkantschin formula and some integral-geometric results of Dann, Paouris, and Pivovarov from [34].

7 Volume difference inequalities

Volume difference inequalities estimate the error in computations of volume of a body out of the areas of its sections and projections. Starting with the case of sections, let $\gamma_{n,k}$ be the smallest constant $\gamma > 0$ with the property that

$$|K|^{\frac{n-k}{n}} - |L|^{\frac{n-k}{n}} \leq \gamma^k \max_{F \in G_{n,n-k}} (|K \cap F| - |L \cap F|) \tag{7.1}$$

for all $1 \leq k < n$ and all origin-symmetric convex bodies K and L in $\mathbb{R}^n$ such that $L \subset K$. The question is whether there exists an absolute constant C so that $\sup_{n,k} \gamma_{n,k} \leq C$. Note that without extra assumptions on K and L, inequality (7.1) cannot hold with any $\gamma > 0$, because of the counterexamples to the Busemann–Petty problem. Note also that if we apply (7.1) with $L = \beta B_2^n$ as $\beta \to 0$, we get the slicing inequality.

It was proved in [79] for $k = 1$ and in [88] for $1 < k < n$ that if $K \in \mathcal{BP}_k^n$ and L is any origin-symmetric star body in $\mathbb{R}^n$, then (7.1) is true in the form

$$|K|^{\frac{n-k}{n}} - |L|^{\frac{n-k}{n}} \leq c_{n,k}^k \max_{F \in G_{n,n-k}} \left(|K \cap F| - |L \cap F|\right), \tag{7.2}$$

where $c_{n,k}^k = \omega_n^{\frac{n-k}{n}}/\omega_{n-k}$. One can check that $c_{n,k} \in (\frac{1}{\sqrt{e}}, 1)$ for all n, k.

In [53], inequality (7.2) was extended to arbitrary origin-symmetric star bodies. Let $1 \leq k < n$ and let K and L be origin-symmetric star bodies in $\mathbb{R}^n$ such that $L \subset K$. Then

$$|K|^{\frac{n-k}{n}} - |L|^{\frac{n-k}{n}} \leq c_{n,k}^k d_{\mathrm{ovr}}^k(K, \mathcal{BP}_k^n) \max_{F \in G_{n,n-k}} \left(|K \cap F| - |L \cap F|\right). \tag{7.3}$$

The outer volume ratio distance was estimated in [91]. If K is an origin-symmetric convex body in $\mathbb{R}^n$, then

$$d_{\mathrm{ovr}}(K, \mathcal{BP}_k^n) \leq c \sqrt{n/k}(\log(en/k))^{\frac{3}{2}}, \tag{7.4}$$

where $c > 0$ is an absolute constant. Therefore, (7.3) provides an affirmative answer to the question for sections of proportional dimension.

The volume difference inequality (7.3) can be extended to arbitrary measures in place of volume as follows. Let f be a bounded non-negative measurable function on $\mathbb{R}^n$. Let μ be the measure with density f so that $\mu(B) = \int_B f$ for every Borel set B in $\mathbb{R}^n$. Also, for every $F \in G_{n,n-k}$ we write $\mu(B \cap F) = \int_{B \cap F} f$, where we integrate the restriction of f to F against the Lebesgue measure on F. For any $1 \leq k < n$, any pair of origin-symmetric star bodies K and L in $\mathbb{R}^n$ such that $L \subset K$, and any measure μ with even non-negative continuous density, we have

$$\mu(K) - \mu(L) \leq \frac{n}{n-k} c_{n,k}^k |K|^{\frac{k}{n}} d_{\mathrm{ovr}}^k(K, \mathcal{BP}_k^n) \max_{F \in G_{n,n-k}} \left(\mu(K \cap F) - \mu(L \cap F)\right). \tag{7.5}$$

In the opposite direction, for any measure in $\mathbb{R}^n$ with bounded density g,

$$(\mu(K) - \mu(L))^{\frac{n-k}{n}} \geq \frac{c_{n,k}^k}{\|g\|_\infty^{\frac{k}{n}}} \left(\int_{G_{n,n-k}} (\mu(K \cap F) - \mu(L \cap F))^{\frac{n}{n-k}} \, dv_{n,n-k}(F) \right)^{\frac{n-k}{n}}. \tag{7.6}$$

In particular,

$$(\mu(K) - \mu(L))^{\frac{n-k}{n}} \geq c_{n,k}^k \frac{1}{\|g\|_\infty^{\frac{k}{n}}} \min_{F \in G_{n,n-k}} \left(\mu(K \cap F) - \mu(L \cap F)\right). \tag{7.7}$$

There is also a result for projections. Let β_n be the smallest constant $\beta > 0$ satisfying

$$\beta\left(|L|^{\frac{n-1}{n}} - |K|^{\frac{n-1}{n}}\right) \geq \min_{\xi \in S^{n-1}} \left(\left|P_{\xi^{\perp}}(L)\right| - \left|P_{\xi^{\perp}}(K)\right|\right) \tag{7.8}$$

for all origin-symmetric convex bodies K, L in $\mathbb{R}^n$ whose curvature functions f_K and f_L exist and satisfy $f_K(\xi) \leq f_L(\xi)$ for all $\xi \in S^{n-1}$. Then, $\beta_n \simeq \sqrt{n}$, i. e., there exist absolute constants $a, b > 0$ such that for all $n \in \mathbb{N}$,

$$a\sqrt{n} \leq \beta_n \leq b\sqrt{n}.$$

Note that without an extra condition on K and L, (7.8) cannot hold in general with any $\beta > 0$, because of the counterexamples to the Shephard problem.

It was proved in [79, 81] that if L is a projection body and K is an origin-symmetric convex body, then

$$|L|^{\frac{n-1}{n}} - |K|^{\frac{n-1}{n}} \geq c_{n,1} \min_{\xi \in S^{n-1}} \left(\left|P_{\xi^{\perp}}(L)\right| - \left|P_{\xi^{\perp}}(K)\right|\right). \tag{7.9}$$

The condition $f_K \leq f_L$ is not needed for (7.9) because we assume that L is a projection body. This inequality is extended in [53] to arbitrary origin-symmetric convex bodies as follows. Suppose that K and L are origin-symmetric convex bodies in $\mathbb{R}^n$ and their curvature functions exist and satisfy $f_K(\xi) \leq f_L(\xi)$ for all $\xi \in S^{n-1}$. Then

$$d_{\mathrm{vr}}(L, \Pi)\left(|L|^{\frac{n-1}{n}} - |K|^{\frac{n-1}{n}}\right) \geq c_{n,1} \min_{\xi \in S^{n-1}} \left(\left|P_{\xi^{\perp}}(L)\right| - \left|P_{\xi^{\perp}}(K)\right|\right). \tag{7.10}$$

Again by K. Ball's volume ratio estimate, for any convex body K in $\mathbb{R}^n$, $d_{\mathrm{vr}}(K, \Pi) \leq \sqrt{n}$. In fact, this distance can be of the order $\sqrt{n}$, up to an absolute constant.

For more on volume difference inequalities, see [94].

8 Discrete versions

Let $\mathbb{Z}^n$ be the standard integer lattice in $\mathbb{R}^n$. Given an origin-symmetric convex body K, define $\#K = \mathrm{card}(K \cap \mathbb{Z}^n)$, the number of points of $\mathbb{Z}^n$ in K. The original proof of the Loomis–Whitney inequality (3.1) is based on a discretization technique, i. e., to consider a set K which is decomposed into a union of equal disjoint cubes and restate the question in combinatorial language. In particular, this combinatorial version implies (and is actually equivalent to) the following discrete variant of the Loomis–Whitney inequality:

$$\#K \leq \left(\prod_{i=1}^{n} \#(P_{e_i^{\perp}}(K))\right)^{\frac{1}{n-1}}. \tag{8.1}$$

There are a number of very interesting tomographic questions related to the number of integer points in the projections of a convex body; we refer to [46, 115, 126]. We also note that in recent years there have been several attempts to translate questions and facts from classical convexity to more general settings including discrete geometry. Here we will concentrate on inequalities concerning sections of a convex body. The properties of sections of convex bodies with respect to the integer lattice were extensively studied in discrete tomography; see [46, 47, 48, 49, 62, 115], where many interesting new properties were proved and a series of exciting open questions were proposed. It is interesting to note that after translation many questions become quite non-trivial and counterintuitive, and the answer may be quite different from the one in the continuous case. In addition, finding the relation between the geometry of a convex set and the number of integer points contained in the set is always a non-trivial task. One can see this, for example, from the history of Khinchin's flatness theorem (see [11, 12, 13, 67]).

Around 2013 the second named author asked if it is possible to provide a discrete analog of inequality (4.6).

Question 1. Does there exist a constant d_n such that

$$\#K \leq d_n \max_{\xi \in S^{n-1}} (\#(K \cap \xi^{\perp})) |K|^{\frac{1}{n}},$$

for all origin-symmetric convex bodies $K \subset \mathbb{R}^n$ containing n linearly independent lattice points?

We note here that we require that K contains n linearly independent lattice points in order to eliminate the degenerate case of a body (for example, a box of the form $[-\delta, \delta]^{n-1} \times [-20, 20]$) whose maximal section contains all lattice points in the body, but whose volume may be arbitrarily close to 0 by considering sufficiently small $\delta > 0$.

Thus, the methods applied to attack this question are quite different from the methods described in the previous sections and seem to require use of tools from the geometry of numbers. Let us start with the simplest case and show that the constant d_2 exists, i. e., it is independent from the origin-symmetric planar convex body K. We will use two classical statements (see, for example, [122]).

Minkowski's first theorem. *Let $K \subset \mathbb{R}^n$ be an origin-symmetric convex body such that $|K| \geq 2^n$. Then K contains at least one non-zero element of $\mathbb{Z}^n$.*

Pick's theorem. *Let P be an integral 2-dimensional convex polygon. Then $|P| = I + \frac{1}{2}B - 1$, where I is the number of lattice points in the interior of P and B is the number of lattice points on its boundary. Here, a polygon is called integral if it can be described as the convex hull of its lattice points.*

Following [1], consider an origin-symmetric planar convex body K and let $s = \sqrt{|K|/4}$. By Minkowski's first theorem, there exists a non-zero vector $u \in \mathbb{Z}^2 \cap \frac{1}{s}K$. Then $su \in K$ and

$$\#(L_u \cap K) \geq 2\lfloor s \rfloor + 1,$$

where $\lfloor s \rfloor$ is the integer part of s and L_u is the line containing u and the origin. Next, consider the convex hull P of the lattice points inside K. Since P is an integral 2-dimensional convex polygon, by Pick's theorem we obtain

$$|P| = I + \frac{1}{2}B - 1 \geq \frac{I+B}{2} - \frac{1}{2},$$

using the fact that $I \geq 1$. Thus,

$$\#P = I + B \leq 2|P| + 1 \leq \frac{5}{2}|P|,$$

since the area of an origin-symmetric integral convex polygon is at least 2. It follows that

$$\#K = \#P \leq \frac{5}{2}|P| \leq \frac{5}{2}|K| = \frac{5}{2}(2s)|K|^{\frac{1}{2}} < 4(2\lfloor s \rfloor + 1)|K|^{\frac{1}{2}}$$

$$\leq 4 \max_{\xi \in S^1} \#(K \cap \xi^{\perp})|K|^{\frac{1}{2}}.$$

Unfortunately, there seems to be no straightforward generalization of the above approach to higher dimensions. This is partially due to the fact that the hyperplane sections of dimension higher than 1 are much harder to study, but is also due to the lack of a direct analog of Pick's formula. It is also essential to note that, in general, the constant d_n is dependent on n. Indeed, for the cross-polytope $B_1^n = \{x \in \mathbb{R}^n : \|x\|_1 \leq 1\}$ we have $\#B_1^n = 2n + 1$, $\max\{\#(B_1^n \cap \xi^{\perp}) : \xi \in S^{n-1}\} = \#(B_1^n \cap e_1^{\perp}) = 2n - 1$, and $|B_1^n|^{1/n} \sim n^{-1}$. Thus, d_n must be greater than cn. Using the special structure of unconditional bodies, it was proved in [1] that d_n is of order n for this class of convex bodies.

Another example which illustrates this situation is the classical Brunn principle (see for example [44]), which tells us that for an origin-symmetric convex body K one has

$$|K \cap \xi^{\perp}| \geq |K \cap (t\xi + \xi^{\perp})|, \quad \text{for all } \xi \in S^{n-1}, \, t \in \mathbb{R}.$$

The above statement is not true if volume is replaced by the number of integer points. Indeed, let $Q = [0,1]^{n-1} \subset \mathbb{R}^{n-1}$ and let K be the convex hull of $Q + e_n$ and $-Q - e_n$. Then $\#(K \cap e_n^{\perp}) = 1$, but $\#(K \cap (e_n + e_n^{\perp})) = 2^{n-1}$. The following analog of Brunn's concavity principle was proved in [1, 40]:

$$\#(K \cap \xi^{\perp}) \geq 2^{1-n}\#(K \cap (t\xi + \xi^{\perp})), \quad \text{for all } \xi \in S^{n-1}, \, t \in \mathbb{R}. \tag{8.2}$$

To prove that the constant d_n is bounded for a general origin-symmetric convex body, one may use the discrete analog of John's theorem [14, 121, 122, 16], which gives an approximation of an origin-symmetric convex body by a symmetric generalized arithmetic progression. This approach was used in [1] to prove that $d_n \leq O(n)^{7n/2}$. The latter estimate can be slightly improved: Using [16] one can show that $d_n \leq O(n)^n$, which is still far from optimal. Finally, the following fact was proved in [1]:

$$\#K \leq O(1)^n n^{n-m} \max(\#(K \cap H))|K|^{\frac{n-m}{n}},$$

where the maximum is taken over all m-dimensional linear subspaces $H \subset \mathbb{R}^n$. In particular, $d_n \leq C^n$ for some large $C > 0$. The proof of this fact is based on Minkowski's second theorem, the discrete Brunn principle (8.2), and the Bourgain–Milman inequality. We refer to [1] for more details. Here we would like to present a probabilistic-harmonic analysis approach to Question 1 which is due to Oded Regev [114]. We will show that there is a distribution of directions ξ for which $\#(K \cap \xi^{\perp})$ is large enough, with a positive probability. For the convenience of the reader we need to provide a few standard definitions and prove some technical estimates. A lattice $\Lambda \subset \mathbb{R}^n$ is the set of all integer linear combinations of n linearly independent vectors in $\mathbb{R}^n$. We notice that $\Lambda = T\mathbb{Z}^n$, where $T \in GL(n)$, $\det(T) \neq 0$, and we denote $\det(\Lambda) = \det(T)$. We will also consider the dual lattice $\Lambda^* = \{x \in \mathbb{R}^n : \langle x, y \rangle \in \mathbb{Z}, \text{ for all } y \in \Lambda\}$. Finally for any $s > 0$ consider the function $\rho_s(x) : \mathbb{R}^n \to (0,1]$, defined by $\rho_s(x) = e^{-\pi|x|^2/s^2}$, where $|x|$ is the Euclidean norm of x, and for a countable set $A \subset \mathbb{R}^n$ let

$$\rho_s(A) = \sum_{x \in A} e^{-\pi|x|^2/s^2}.$$

We will use the Poisson summation formula (see for example [15, Lemma 17.2]). Consider a function $f : \mathbb{R}^n \to \mathbb{C}$ and let

$$\widehat{f}(y) = \int_{\mathbb{R}^n} e^{-2\pi i \langle x, y \rangle} f(x)\, dx \quad (y \in \mathbb{R}^n)$$

be the Fourier transform of f. Assume that f and $\widehat{f}$ decay sufficiently fast, i. e., there exist positive constants C, δ such that $|f(x)|, |\widehat{f}(x)| \leq \frac{C}{1+|x|^{n+\delta}}$ for all $x \in \mathbb{R}^n$ (notice that this condition it trivially satisfied by ρ_s). Then

$$\sum_{x \in \Lambda} f(x) = \frac{1}{\det(\Lambda)} \sum_{y \in \Lambda^*} \widehat{f}(y). \tag{8.3}$$

Using the fact that $\widehat{\rho}_s(y) = s^n \rho_{1/s}(y)$ together with (8.3) we get

$$\rho_s(\Lambda) = \sum_{x \in \Lambda} \rho_s(x) = \frac{1}{\det(\Lambda)} \sum_{y \in \Lambda^*} \widehat{\rho}_s(y) = \frac{s^n}{\det(\Lambda)} \rho_{1/s}(\Lambda^*).$$

In particular, we see that

$$\rho_s(\Lambda) \geq \det(\Lambda)^{-1} s^n. \tag{8.4}$$

We may use again (8.3) to show that $\rho_s(\Lambda + a) \leq \rho_s(\Lambda)$ for all $a \in \mathbb{R}^n$. Indeed, using $\widehat{\rho}_s(y) = s^n \rho_{1/s}(y) > 0$, we get

$$\rho_s(\Lambda + a) = \frac{1}{\det(\Lambda)} \sum_{y \in \Lambda^*} \hat{\rho}_s(y) e^{-2\pi i \langle a, y \rangle} \le \frac{1}{\det(\Lambda)} \sum_{y \in \Lambda^*} \hat{\rho}_s(y) = \rho_s(\Lambda). \tag{8.5}$$

Let $Z_{A,s}$ be a random vector taking values in a countable set A, such that

$$\mathbb{P}(Z_{A,s} = x) = \frac{\rho_s(x)}{\rho_s(A)}.$$

Using (8.4) we get

$$\mathbb{P}(Z_{\Lambda,s} = 0) = \frac{1}{\rho_s(\Lambda)} \le \frac{\det(\Lambda)}{s^n}. \tag{8.6}$$

Finally, if we pick a lattice $\Lambda \subset \mathbb{R}^n$, $x \in \Lambda$ and $s > 0$, we claim that

$$\mathbb{P}\big(\langle Z_{\Lambda^*,s}, x \rangle = 0\big) \ge \mathbb{P}(Z_{\mathbb{Z}\backslash|x|,s} = 0) = \rho_{s|x|}(\mathbb{Z})^{-1} \ge c \min\{1, (s|x|)^{-1}\}. \tag{8.7}$$

We may assume that $x \ne 0$ and note that by definition $\langle Z_{\Lambda^*,s}, x \rangle$ can take only integer values. Fix some $k \in \mathbb{Z}$ and consider the set of points $y \in \Lambda^*$ such that $\langle x, y \rangle = k$. Note that if this set is empty, then its ρ_s mass is clearly zero. Now consider the case where this set is not empty. The affine subspace $\{y \in \mathbb{R}^n : \langle x, y \rangle = k\}$ is a shift of $x^\perp$ at distance $k/|x|$ in the direction of $x/|x|$. Notice that such a shift will not necessarily move points of Λ^* into itself, and thus an additional shift inside $\{y \in \mathbb{R}^n : \langle x, y \rangle = k\} \cap \Lambda^*$ may be required. Thus, using the product structure of ρ_s and property (8.5) we get

$$\rho_s(\{y \in \mathbb{R}^n : \langle x, y \rangle = k\} \cap \Lambda^*) \le \rho_s(k/|x|)\rho_s(\Lambda^* \cap x^\perp).$$

Dividing the above inequality by $\rho_s(\Lambda^*)$ and summing over $k \in \mathbb{Z}$ we get (8.7).

The following theorem is due to O. Regev [114]. Consider an origin-symmetric convex body $K \subset \mathbb{R}^n$. Then

$$\#K \le \max_{\xi \in S^{n-1}}(\#(K \cap \xi^\perp)) \max\{1, cn|K|^{\frac{1}{n-1}}\}.$$

Note that this theorem improves the bound on d_n from [1] when the volume of $|K|$ is smaller than C^{n^3} and provides a polynomial bound on d_n for bodies of volume smaller than n^{cn^2}.

For the proof, using John's theorem we see that there is a linear transformation such that $\det(T) = 1$ and $TK \subset n|TK|^{1/n}B_2^n$. Without loss of generality we will consider the body TK instead of K and the lattice $\Lambda = T\mathbb{Z}^n$ instead of $\mathbb{Z}^n$. We may also assume that $|K| \ge n^{-n}$; otherwise, $K \subset \delta B_2^n$, where $0 < \delta < 1$, and using $\det(\Lambda) = 1$ we deduce that $K \cap \Lambda$ is not full-dimensional and the statement is trivial.

Now we will select the direction $\xi \in \Lambda^* \setminus \{0\}$, using a probabilistic approach. Let $\xi = Z_{\Lambda^*,s}$, where $s \ge 1$ will be chosen later. Then we may apply (8.7) to claim that for any fixed $x \in K \cap \Lambda$ we have

$$\mathbb{P}(x \in \xi^{\perp}) \geq c \min\{1, (s|x|)^{-1}\} \geq c \min\{1, (sn|K|^{1/n})^{-1}\} = \frac{c}{sn|K|^{1/n}}.$$

Then

$$\mathbb{E}\left(\frac{\#\{x \in K \cap \Lambda : \langle x, \xi \rangle = 0\}}{\#(K \cap \Lambda)} \right) \geq \frac{c}{sn|K|^{1/n}}.$$

The above inequality allows us to select ξ for which $\#\{x \in K \cap \Lambda : \langle x, \xi \rangle = 0\}$ is large. To finish the proof, we need to choose $s \geq 1$ to guarantee that ξ can be selected not to be equal to 0. Using (8.6) we see that $\mathbb{P}(\xi = 0) \leq s^{-n}$. Thus, we need to pick $s \geq 1$ such that $s^{-n} \leq c(sn|K|^{1/n})^{-1}$, i. e., $s = C|K|^{\frac{1}{n(n-1)}} \geq 1$, where $C > 0$ is a large enough absolute constant. This completes the proof.

Another very interesting bound on the cardinality of lattice points in sections of convex bodies is inspired by Meyer's inequality (3.2). It was proposed by Gardner, Gronchi, and Zong [46] and proved by Freyer and Henk [40]. For any origin-symmetric convex body $K \subset \mathbb{R}^n$ there exists a basis $b_1, \ldots, b_n$ of $\mathbb{Z}^n$ such that

$$(\#K)^{\frac{n-1}{n}} \leq O(n^2 2^n) \prod_{i=1}^{n} (\#(K \cap b_i^{\perp}))^{1/n}$$

and there are $t_1, \ldots, t_n \in \mathbb{Z}^n$ such that

$$(\#K)^{\frac{n-1}{n}} \leq O(n^2) \prod_{i=1}^{n} (\#(K \cap (t_i + b_i^{\perp})))^{1/n}.$$

Thus, we immediately obtain the slicing inequality

$$(\#K)^{\frac{n-1}{n}} \leq O(n^2) \max_{t,b \in \mathbb{Z}^n, b \neq 0} \#(K \cap (t + b^{\perp}))$$

for any origin-symmetric convex body K in $\mathbb{R}^n$. Actually, Freyer and Henk [40] removed the condition of K being symmetric in the statement above (the condition cannot be removed in the discrete analog of Meyer's inequality). Moreover, they were able to prove that in the symmetric case the $O(n^2)$-term can be replaced by $O(n)$. Unfortunately, these results do not seem to apply directly to Question 1 due to the lack of a direct analog of Brunn's inequality.

Added in Proofs: At the time of printing, Freyer and Henk [41] obtained a polynomial estimate for Question 1. More precisely, they showed that $d_n = O(n^{10/3}(\log n)^a)$, where $a > 0$ is an absolute constant.

Bibliography

[1] M. Alexander, M. Henk and A. Zvavitch, *A discrete version of Koldobsky's slicing inequality*, Isr. J. Math. **222**(1) (2017), 261–278.

[2] D. Alonso-Gutiérrez, S. Artstein-Avidan, B. González Merino, C. H. Jiménez and R. Villa, *Rogers–Shephard and local Loomis–Whitney type inequalities*, Math. Ann. **374**(3–4) (2019), 1719–1771.

[3] S. Artstein-Avidan, A. Giannopoulos and V. D. Milman, *Asymptotic geometric analysis, Vol. I*, Mathematical surveys and monographs, vol. 202, American Mathematical Society, Providence, RI, 2015, xx+451 pp.

[4] S. Artstein-Avidan, A. Giannopoulos and V. D. Milman, *Asymptotic geometric analysis, vol. II*, Mathematical surveys and monographs, vol. 262, American Mathematical Society, Providence, RI, 2021, xxxvii+645 pp.

[5] K. M. Ball, *Some remarks on the geometry of convex sets*, in: Geometric aspects of functional analysis (1986/87), Lecture notes in math., vol. 1317, Springer-Verlag, Berlin-Heidelberg-New York, 1988, pp. 224–231.

[6] K. M. Ball, *Logarithmically concave functions and sections of convex sets in $\mathbb{R}^n$*, Stud. Math. **88** (1988), 69–84.

[7] K. M. Ball, *Shadows of convex bodies*, Trans. Am. Math. Soc. **327** (1991), 891–901.

[8] K. M. Ball, *Normed spaces with a weak-Gordon–Lewis property*, in: Functional analysis, Austin, TX, 1987/1989, Lecture notes in math., vol. 1470, Springer, Berlin, 1991, pp. 36–47.

[9] K. M. Ball, *Volume ratios and a reverse isoperimetric inequality*, J. Lond. Math. Soc. (2) **44** (1991), 351–359.

[10] K. M. Ball and V. H. Nguyen, *Entropy jumps for isotropic log-concave random vectors and spectral gap*, Stud. Math. **213** (2012), 81–96.

[11] W. Banaszczyk, *Inequalities for convex bodies and polar reciprocal lattices in $\mathbb{R}^n$*, Discrete Comput. Geom. **13**(2) (1995), 217–231.

[12] W. Banaszczyk, *Inequalities for convex bodies and polar reciprocal lattices in $\mathbb{R}^n$. II. Application of K-convexity*, Discrete Comput. Geom. **16**(3) (1996), 305–311.

[13] W. Banaszczyk, A. Litvak, A. Pajor and S. Szarek, *The flatness theorem for nonsymmetric convex bodies via the local theory of Banach spaces*, Math. Oper. Res. **24**(3) (1999), 728–750.

[14] I. Bárány and A. M. Vershik, *On the number of convex lattice polytopes*, Geom. Funct. Anal. **2**(4) (1992), 381–393.

[15] A. Barvinok, *MATH 669: Combinatorics, Geometry and Complexity of Integer Points*, Lecture notes, available from: http://www.math.lsa.umich.edu/\TU\texttildelowbarvinok/latticenotes669.pdf.

[16] S. Berg and M. Henk, *Discrete analogues of John's theorem*, Mosc. J. Comb. Number Theory **8**(4) (2019), 367–378.

[17] S. Bobkov and F. Nazarov, *On convex bodies and log-concave probability measures with unconditional basis*, in: Geometric aspects of functional analysis, V. D. Milman and G. Schechtman, eds., Lecture notes in math., vol. 1807, 2003, pp. 53–69.

[18] S. G. Bobkov, B. Klartag and A. Koldobsky, *Estimates for moments of general measures on convex bodies*, Proc. Am. Math. Soc. **146**(11) (2018), 4879–4888.

[19] B. Bollobás and A. Thomason, *Projections of bodies and hereditary properties of hypergraphs*, Bull. Lond. Math. Soc. **27** (1995), 417–424.

[20] J. Bourgain, *On high-dimensional maximal functions associated to convex bodies*, Am. J. Math. **108** (1986), 1467–1476.

[21] J. Bourgain, *Geometry of Banach spaces and harmonic analysis*, in: Proceedings of the international congress of mathematicians, Berkeley, CA, 1986, Amer. Math. Soc., Providence, RI, 1987, pp. 871–878.

[22] J. Bourgain, *On the distribution of polynomials on high-dimensional convex sets*, Lect. Notes Math. **1469** (1991), 127–137.

[23]   J. Bourgain, *On the Busemann–Petty problem for perturbations of the ball*, Geom. Funct. Anal. **1** (1991), 1–13.

[24]   J. Bourgain, B. Klartag and V. D. Milman, *Symmetrization and isotropic constants of convex bodies*, in: Geometric aspects of functional analysis, Lecture notes in math., vol. 1850, 2004, pp. 101–116.

[25]   S. Brazitikos and D.-M. Liakopoulos, *On a version of the slicing problem for the surface area of convex bodies*, Trans. Am. Math. Soc. **375** (2022), 5561–5586.

[26]   S. Brazitikos, A. Giannopoulos and D.-M. Liakopoulos, *Uniform cover inequalities for the volume of coordinate sections and projections of convex bodies*, Adv. Geom. **18**(3) (2018), 345–354.

[27]   S. Brazitikos, A. Giannopoulos, P. Valettas and B. Vritsiou, *Geometry of isotropic convex bodies*, Mathematical surveys and monographs, vol. 196, American Mathematical Society, Providence, RI, 2014, xx+594 pp.

[28]   Y. D. Burago and V. A. Zalgaller, *Geometric inequalities*, Springer series in soviet mathematics, Springer-Verlag, Berlin–New York, 1988.

[29]   H. Busemann and C. M. Petty, *Problems on convex bodies*, Math. Scand. **4** (1956), 88–94.

[30]   H. Busemann and E. G. Straus, *Area and normality*, Pac. J. Math. **10** (1960), 35–72.

[31]   G. Chasapis, A. Giannopoulos and D. Liakopoulos, *Estimates for measures of lower dimensional sections of convex bodies*, Adv. Math. **306** (2017), 880–904.

[32]   Y. Chen, *An almost constant lower bound of the isoperimetric coefficient in the KLS conjecture*, Geom. Funct. Anal. **31** (2021), 34–61.

[33]   N. Dafnis and G. Paouris, *Small ball probability estimates, ψ_2-behavior and the hyperplane conjecture*, J. Funct. Anal. **258** (2010), 1933–1964.

[34]   S. Dann, G. Paouris and P. Pivovarov, *Bounding marginal densities via affine isoperimetry*, Proc. Lond. Math. Soc. **113** (2016), 140–162.

[35]   S. Dar, *Remarks on Bourgain's problem on slicing of convex bodies*, Oper. Theory, Adv. Appl. **77** (1995), 61–66.

[36]   R. Eldan and B. Klartag, *Approximately gaussian marginals and the hyperplane conjecture*, in: Proceedings of the workshop on "Concentration, functional inequalities and isoperimetry", Contemp. math., vol. 545, 2011, pp. 55–68.

[37]   M. Fradelizi, A. Giannopoulos and M. Meyer, *Some inequalities about mixed volumes*, Isr. J. Math. **135** (2003), 157–179.

[38]   M. Fradelizi, M. Madiman and A. Zvavitch, *Sumset estimates in convex geometry*, arXiv:2206.01565.

[39]   M. Fradelizi, M. Madiman, M. Meyer and A. Zvavitch, *On the volume of the Minkowski sum of zonoids*, arXiv:2206.02123.

[40]   A. Freyer and M. Henk, *Bounds on the lattice point enumerator via slices and projections*, Discrete Comput. Geom. **67**(3) (2022), 895–918.

[41]   A. Freyer and M. Henk, *Polynomial bounds in Koldobsky's discrete slicing problem*, arXiv:2303.15976.

[42]   R. J. Gardner, *Intersection bodies and the Busemann–Petty problem*, Trans. Am. Math. Soc. **342** (1994), 435–445.

[43]   R. J. Gardner, *A positive answer to the Busemann–Petty problem in three dimensions*, Ann. Math. **140** (1994), 435–447.

[44]   R. J. Gardner, *Geometric tomography*, 2nd edn., Cambridge University Press, Cambridge, 2006, xxii+492 pp.

[45]   R. J. Gardner, *Geometric tomography: update to the second edition*, https://www.wwu.edu/faculty/gardner/research.html.

[46]   R. J. Gardner, P. Gronchi and C. Zong, *Sums, projections, and sections of lattice sets, and the discrete covariogram*, Discrete Comput. Geom. **34**(3) (2005), 391–409.

[47]   R. J. Gardner and P. Gritzmann, *Discrete tomography: determination of finite sets by X-rays*, Trans. Am. Math. Soc. **349** (1997), 2271–2295.

[48]   R. J. Gardner and P. Gritzmann, *Uniqueness and complexity in discrete tomography*, in: Discrete tomography: foundations, algorithms and application, G. T. Herman and A. Kuba, eds., Birkhäuser, Boston, MA, 1999, pp. 85–113.

[49] R. J. Gardner and P. Gronchi, *A Brunn–Minkowski inequality for the integer lattice*, Trans. Am. Math. Soc. **353** (2001), 3995–4024.

[50] R. J. Gardner, A. Koldobsky and Th. Schlumprecht, *An analytic solution to the Busemann–Petty problem on sections of convex bodies*, Ann. Math. **149** (1999), 691–703.

[51] A. Giannopoulos, *A note on a problem of H. Busemann and C. M. Petty concerning sections of symmetric convex bodies*, Mathematika **37** (1990), 239–244.

[52] A. Giannopoulos and A. Koldobsky, *Variants of the Busemann–Petty problem and of the Shephard problem*, Int. Math. Res. Not. **3** (2017), 921–943.

[53] A. Giannopoulos and A. Koldobsky, *Volume difference inequalities*, Trans. Am. Math. Soc. **370** (2018), 4351–4372.

[54] A. Giannopoulos and M. Papadimitrakis, *Isotropic surface area measures*, Mathematika **46** (1999), 1–13.

[55] A. Giannopoulos, M. Hartzoulaki and G. Paouris, *On a local version of the Aleksandrov–Fenchel inequality for the quermassintegrals of a convex body*, Proc. Am. Math. Soc. **130** (2002), 2403–2412.

[56] A. Giannopoulos, A. Koldobsky and P. Valettas, *Inequalities for the surface area of projections of convex bodies*, Can. J. Math. **70**(4) (2018), 804–823.

[57] A. Giannopoulos, A. Koldobsky and A. Zvavitch, *Inequalities for the Radon transform on convex sets*, Int. Math. Res. Not. **18** (2022), 13984–14007.

[58] A. Giannopoulos, G. Paouris and B.-H. Vritsiou, *A remark on the slicing problem*, J. Funct. Anal. **262** (2012), 1062–1086.

[59] P. Goodey and G. Zhang, *Inequalities between projection functions of convex bodies*, Am. J. Math. **120** (1998), 345–367.

[60] E. L. Grinberg, *Isoperimetric inequalities and identities for k-dimensional cross-sections of convex bodies*, Math. Ann. **291** (1991), 75–86.

[61] J. Hosle, *On the comparison of measures of convex bodies via projections and sections*, Int. Math. Res. Not. **17** (2021), 13046–13074.

[62] D. Iglesias, J. Yepes Nicolás and A. Zvavitch, *Brunn–Minkowski inequalities for the lattice point enumerator*, Adv. Math. **370** (2020), 107193, 25 pp.

[63] A. Jambulapati, Y. T. Lee and S. Vempala, *A slightly improved bound for the KLS constant*, arXiv:2208.11644v2.

[64] F. John, *Extremum problems with inequalities as subsidiary conditions*, in: Courant anniversary volume, Interscience, New York, 1948, pp. 187–204.

[65] M. Junge, *On the hyperplane conjecture for quotient spaces of L_p*, Forum Math. **6** (1994), 617–635.

[66] M. Junge, *Proportional subspaces of spaces with unconditional basis have good volume properties*, in: Geometric aspects of functional analysis, Israel Seminar, 1992–1994, Oper. theory adv. appl., vol. 77, Birkhauser, Basel, 1995, pp. 121–129.

[67] R. Kannan and L. Lovász, *Covering minima and lattice-point-free convex bodies*, Ann. Math. (2) **128**(3) (1988), 577–602.

[68] B. Klartag, *An isomorphic version of the slicing problem*, J. Funct. Anal. **218** (2005), 372–394.

[69] B. Klartag, *On convex perturbations with a bounded isotropic constant*, Geom. Funct. Anal. **16** (2006), 1274–1290.

[70] B. Klartag, *Logarithmic bounds for isoperimetry and slices of convex sets*, arXiv:2303.14938.

[71] B. Klartag and A. Koldobsky, *An example related to the slicing inequality for general measures*, J. Funct. Anal. **274** (2018), 2089–2112.

[72] B. Klartag and G. Kozma, *On the hyperplane conjecture for random convex sets*, Isr. J. Math. **170** (2009), 253–268.

[73] B. Klartag and J. Lehec, *Bourgain's slicing problem and KLS isoperimetry up to polylog*, Geom. Funct. Anal. **32**(5) (2022), 1134–1159.

[74] B. Klartag and G. V. Livshyts, *The lower bound for Koldobsky's slicing inequality via random rounding*, in: Geometric aspects of functional analysis. Vol. II, Lecture notes in math., vol. 2266, Springer, Cham, 2020, pp. 43–63.

[75] B. Klartag and V. D. Milman, *The slicing problem of Bourgain*, in: Analysis at Large, a collection of articles in memory of Jean Bourgain, Springer, 2022, pp. 203–232.

[76] A. Koldobsky, *Intersection bodies, positive definite distributions and the Busemann–Petty problem*, Am. J. Math. **120** (1998), 827–840.

[77] A. Koldobsky, *Intersection bodies in* $\mathbb{R}^4$, Adv. Math. **136** (1998), 1–14.

[78] A. Koldobsky, *Fourier analysis in convex geometry*, Mathematical surveys and monographs, vol. 116, American Mathematical Society, Providence, RI, 2005, vi+170 pp.

[79] A. Koldobsky, *Stability in the Busemann–Petty and Shephard problems*, Adv. Math. **228** (2011), 2145–2161.

[80] A. Koldobsky, *A hyperplane inequality for measures of convex bodies in* $\mathbb{R}^n$, $n \leq 4$, Discrete Comput. Geom. **47** (2012), 538–547.

[81] A. Koldobsky, *Stability and separation in volume comparison problems*, Math. Model. Nat. Phenom. **8** (2013), 156–169.

[82] A. Koldobsky, *A* $\sqrt{n}$ *estimate for measures of hyperplane sections of convex bodies*, Adv. Math. **254** (2014), 33–40.

[83] A. Koldobsky, *Estimates for measures of sections of convex bodies*, in: GAFA seminar volume, B. Klartag and E. Milman, eds., Lect. notes in math., vol. 2116, 2014, pp. 261–271.

[84] A. Koldobsky, *Stability inequalities for projections of convex bodies*, Discrete Comput. Geom. **57** (2017), 152–163.

[85] A. Koldobsky, *Slicing inequalities for measures of convex bodies*, Adv. Math. **283** (2015), 473–488.

[86] A. Koldobsky, *Slicing inequalities for subspaces of* L_p, Proc. Am. Math. Soc. **144**(2) (2016), 787–795.

[87] A. Koldobsky and H. König, *On the maximal perimeter of sections of the cube*, Adv. Math. **346** (2019), 773–804.

[88] A. Koldobsky and D. Ma, *Stability and slicing inequalities for intersection bodies*, Geom. Dedic. **162** (2013), 325–335.

[89] A. Koldobsky and A. Pajor, *A remark on measures of sections of* L_p*-balls*, in: Geometric aspects of functional analysis, Lecture notes in math., vol. 2169, Springer, Cham, 2017, pp. 213–220.

[90] A. Koldobsky, A. Pajor and V. Yaskin, *Inequalities of the Kahane–Khinchin type and sections of* L_p*-balls*, Stud. Math. **184** (2008), 217–231.

[91] A. Koldobsky, G. Paouris and M. Zymonopoulou, *Isomorphic properties of intersection bodies*, J. Funct. Anal. **261** (2011), 2697–2716.

[92] A. Koldobsky, G. Paouris and A. Zvavitch, *Measure comparison and distance inequalities for convex bodies*, Indiana Univ. Math. J. **71**(1) (2022), 391–407.

[93] A. Koldobsky, D. Ryabogin and A. Zvavitch, *Projections of convex bodies and the Fourier transform*, Isr. J. Math. **139** (2004), 361–380.

[94] A. Koldobsky and D. Wu, *Extensions of reverse volume difference inequalities*, in: Analytic aspects of convexity, Springer INdAM ser., vol. 25, Springer, Cham, 2018, pp. 61–71.

[95] A. Koldobsky and V. Yaskin, *The interface between convex geometry and harmonic analysis*, CBMS regional conference series in mathematics, vol. 108, American Mathematical Society, Providence, RI, 2008, x+107 pp. Published for the Conference Board of the Mathematical Sciences, Washington, DC.

[96] A. Koldobsky and A. Zvavitch, *An isomorphic version of the Busemann–Petty problem for arbitrary measures*, Geom. Dedic. **174** (2015), 261–277.

[97] H. König, M. Meyer and A. Pajor, *The isotropy constants of Schatten classes are bounded*, Math. Ann. **312** (1998), 773–783.

[98] L. Kryvonos and D. Langharst, *Measure theoretic Minkowski's existence theorem and projection bodies*, arXiv:2111.10923.

[99] D. G. Larman and C. A. Rogers, *The existence of a centrally symmetric convex body with central sections that are unexpectedly small*, Mathematika **22** (1975), 164–175.

[100] A.-J. Li and Q. Huang, *The dual Loomis–Whitney inequality*, Bull. Lond. Math. Soc. **48** (2016), 676–690.

[101] G. V. Livshyts, *An extension of Minkowski's theorem and its applications to questions about projections for measures*, Adv. Math. **356** (2019), 106803.

[102] D.-M. Liakopoulos, *Reverse Brascamp–Lieb inequality and the dual Bollobás–Thomason inequality*, Arch. Math. (Basel) **112**(3) (2019), 293–304.

[103] D.-M. Liakopoulos, *Inequalities for the quermassintegrals of sections of convex bodies*, arXiv:2302.10568.

[104] L. H. Loomis and H. Whitney, *An inequality related to the isoperimetric inequality*, Bull. Am. Math. Soc. **55** (1949), 961–962.

[105] E. Lutwak, *Intersection bodies and dual mixed volumes*, Adv. Math. **71** (1988), 232–261.

[106] M. Meyer, *A volume inequality concerning sections of convex sets*, Bull. Lond. Math. Soc. **20**(2) (1988), 151–155.

[107] E. Milman, *Dual mixed volumes and the slicing problem*, Adv. Math. **207** (2006), 566–598.

[108] V. D. Milman and A. Pajor, *Isotropic position and inertia ellipsoids and zonoids of the unit ball of a normed n-dimensional space*, in: Geometric aspects of functional analysis, J. Lindenstrauss and V. D. Milman, eds., Lecture notes in mathematics, vol. 1376, Springer, Heidelberg, 1989, pp. 64–104.

[109] G. Paouris, *On the isotropic constant of non-symmetric convex bodies*, in: Geom. aspects of funct. analysis. Israel seminar 1996–2000, Lect. notes in math., vol. 1745, 2000, pp. 239–244.

[110] G. Paouris and P. Pivovarov, *Small-ball probabilities for the volume of random convex sets*, Discrete Comput. Geom. **49** (2013), 601–646.

[111] M. Papadimitrakis, *On the Busemann–Petty problem about convex, centrally symmetric bodies in* $\mathbb{R}^n$, Mathematika **39** (1992), 258–266.

[112] C. M. Petty, *Projection bodies*, in: Proc. coll. convexity, Copenhagen, 1965, Kobenhavns Univ. Mat. Inst., 1967, pp. 234–241.

[113] I. Rivin, *Surface area and other measures of ellipsoids*, Adv. Appl. Math. **39**(4) (2007), 409–427.

[114] O. Regev, *A note on Koldobsky's lattice slicing inequality*, arXiv:1608.04945.

[115] D. Ryabogin, V. Yaskin and N. Zhang, *Unique determination of convex lattice sets*, Discrete Comput. Geom. **57** (2017), 582–589.

[116] R. Schneider, *Zu einem problem von Shephard über die projektionen konvexer Körper*, Math. Z. **101** (1967), 71–82.

[117] R. Schneider, *Convex bodies: the Brunn–Minkowski theory*, Second expanded edn., Encyclopedia of mathematics and its applications, vol. 151, Cambridge University Press, Cambridge, 2014.

[118] R. Schneider and W. Weil, *Stochastic and integral geometry*, Probability and its applications, Springer-Verlag, Berlin, 2008.

[119] G. C. Shephard, *Shadow systems of convex bodies*, Isr. J. Math. **2** (1964), 229–306.

[120] I. Soprunov and A. Zvavitch, *Bezout inequality for mixed volumes*, Int. Math. Res. Not. **23** (2016), 7230–7252.

[121] T. Tao and V. Vu, *John-type theorems for generalized arithmetic progressions and iterated sumsets*, Adv. Math. **219**(2) (2008), 428–449.

[122] T. Tao and V. Vu, *Additive combinatorics*, Cambridge studies in advanced mathematics, vol. 105, Cambridge University Press, Cambridge, 2010.

[123] G. Zhang, *Intersection bodies and Busemann–Petty inequalities in* $\mathbb{R}^4$, Ann. Math. **140** (1994), 331–346.

[124] G. Zhang, *A positive answer to the Busemann–Petty problem in four dimensions*, Ann. Math. **149** (1999), 535–543.

[125] G. Zhang, *Sections of convex bodies*, Am. J. Math. **118** (1996), 319–340.

[126] N. Zhang, *An analogue of the Aleksandrov projection theorem for convex lattice polygons*, Proc. Am. Math. Soc. **145**(6) (2017), 2305–2310.

[127] A. Zvavitch, *The Busemann–Petty problem for arbitrary measures*, Math. Ann. **331** (2005), 867–887.

Irina Holmes Fay and Alexander Volberg

Borderline estimates for weighted singular operators and concavity

Abstract: The borderline estimates play an important part in the theory of Calderón–Zygmund operators and other types of singular integrals. It is generally true that the borderline behavior dictates the behavior of singular integrals in a large scale of spaces. But the end-of-scale behavior often exhibits certain "blow-up" near the end of scale. This blow-up is not easy to gauge. This chapter is devoted to presenting an overview of certain methods that are well suited to catch the borderline behavior. We consider only two very simple singular operators: the martingale transform and the martingale square function. These operators, being very simple dyadic singular operators, are considered in weighted spaces. We seek to find the borderline behavior in terms of the weight. The methods range from stopping time constructions and sparse domination to the PDE approach, which completely avoids stopping time arguments. The PDE approach reduces the singular integral estimate to various concavity properties of the corresponding Bellman function. This is the Bellman function technique, and the Bellman function knows how to stop time, but it rarely reveals how it does that.

Keywords: Hunt–Muckenhoupt–Wheeden weights, singular integrals estimates, stopping time, sparse domination, Bellman PDE

MSC 2020: 46B09, 46B07, 60E15

1 Introduction

The endpoint estimates play an important part in the theory of singular integrals (weighted or unweighted). They are usually the most difficult estimates in the theory, and the most interesting of course. It is a general principle that one can extrapolate the estimate from the endpoint situation to all other situations. We refer the reader to the book [11], which treats this subject of extrapolation in depth.

On the other hand, it happens quite often that the singular integral estimates exhibit a certain "blow-up" near the endpoint. To catch this blow-up can be a difficult task. We demonstrate this hunt for blow-ups by the examples of weighted dyadic singular integrals and their behavior in $L^p(w)$. The endpoint p will be naturally 1 (and sometimes slightly unnaturally 2) depending on the martingale singular operator. The singular in-

Irina Holmes Fay, Dept. Math., Texas A and M University, College Station, TX, USA, e-mail: irinaholmes@tamu.edu

Alexander Volberg, Department of Mathematics, MSU, East Lansing, MI 48823, USA; and Hausdorff Center for Mathematics, Bonn, Germany, e-mail: volberg@math.msu.edu

https://doi.org/10.1515/9783110775389-007

tegrals in this chapter are the easiest possible. They are dyadic martingale operators on the σ-algebra generated by a usual homogeneous dyadic lattice on the real line. We do not consider any non-homogeneous situation, and this standard σ-algebra generated by a dyadic lattice $\mathcal{D}$ will be provided with Lebesgue measure. The non-homogeneous case has its own surprises, but this is outside the scope of this survey. The reader can look at [13] for the non-homogeneous case.

In fact, we consider here only two dyadic martingale singular operators: the weighted martingale transform and the (weighted) square function. For the former we will explain the behavior of its norm $L^1(w) \to L^{1,\infty}(w)$ with $w \in A_1$, and for the latter we will explain the behavior of its norm $L^2(w) \to L^{2,\infty}(w)$ with $w \in A_2$.

Our goal will be to show how various techniques – from the Bellman function technique to the sparse domination technique – measure the sharp behavior of the weighted estimates of the corresponding weighted dyadic singular operators. This blow-up will be demonstrated by certain estimates from below of the Bellman function of a dyadic problem.

The Bellman function part will be reduced to the task of finding the lower estimate for the solutions of concrete Monge–Ampère differential equations with concrete first-order terms (drift). The PDE approach to harmonic analysis was pioneered by Burkholder; see, e. g., [5, 6], and his solution of Pelczynski's problem [7].

We will get a logarithmic blow-up not only for the martingale transform but also for a concrete dyadic shift; see our main result, Theorem 6.3.

Let us explain the historic background. We present it first for the dyadic martingale transform. Maria Reguera and Christoph Thiele [45, 46] disproved the Muckenhoupt–Wheeden conjecture by providing a counterexample to the following inequality:

$$\lambda \cdot \omega\{x : T\phi(x) > \lambda\} \le C \int |\phi| M\omega \, dx. \tag{1.1}$$

(Here M is the Hardy–Littlewood maximal function.) This inequality turns out to be wrong when T is a martingale transform [45], and it is also wrong when T is the Hilbert transform (see Reguera and Thiele [46]). All this shows is that weak estimates hide some blow-up. But the constructions in [45, 46] provided only a very irregular weight ω; it was practically the sum of delta-measures.

Therefore, another question of Muckenhoupt stayed open. It was called the weak A_1-conjecture or *weak Muckenhoupt conjecture*. Namely, let ω above be quite regular, $\omega \in A_1$, that is, $M\omega \le C \cdot \omega$ pointwise. The smallest C in this inequality is called the A_1-norm of ω and is sometimes denoted by $[w]_{A_1}$. The counterexamples of Reguera–Thiele did not provide such a regular weight, so it was still feasible that

$$\omega \in A_1 \quad \Rightarrow \quad \lambda \cdot \omega\{x : T\phi(x) > \lambda\} \le C[w]_{A_1} \int |\phi| \omega \, dx. \tag{1.2}$$

This was the weak Muckenhoupt conjecture.

With a logarithmically worse estimate this was in fact known. Namely, in [42, 34] the following beautiful result was proved:

$$\omega \in A_1 \quad \Rightarrow \quad \lambda \cdot \omega\{x : T\phi(x) > \lambda\} \le C[w]_{A_1} \log(1 + [w]_{A_1}) \int |\phi|\omega\,dx.$$

It has been suggested in Pérez' paper [42] that there should exist a counterexample to the weak Muckenhoupt conjecture; also, the paper [42] has several very interesting positive results, where Mw in (1.1) is replaced by a slightly bigger maximal function, in particular by M^2w (which is equivalent to a certain Orlicz maximal function).

Whether the linear in weight estimate (1.2) is true is actually important. For example, if the logarithmic blow-up were an artifact, that is, if (1.2) were true, then the A_2-conjecture would follow immediately by a standard extrapolation; see, e. g., the preprint [43].

Let us recall that by the A_p-norm, $1 < p < \infty$, of the weight ω we understand the quantity

$$[\omega]_{A_p} := \sup_I \left(\frac{1}{|I|} \int_I \omega\,dx\right)\left(\frac{1}{|I|} \int_I \omega^{-\frac{1}{p-1}}\,dx\right)^{p-1}.$$

The famous A_2-conjecture asked to give an estimate of any Calderón–Zygmund operator in norm $L^2(\omega)$, linear in the A_2-norm $[\omega]_{A_2}$ of ω. By an extrapolation argument, such an estimate can be extended to the following "phase transition"-type estimate for any strong norm Calderón–Zygmund operator T (see [15]):

$$\|T\|_{L^p(w)} \le C[w]_{A_p}^{\max(1,\frac{1}{p-1})}, \quad 1 < p < \infty. \tag{1.3}$$

There is a long story related to the solution of this problem. Briefly, for the Ahlfors–Beurling transform it was given by Petermichl–Volberg in 2002 in [41], for the Hilbert transform it was given by Petermichl in 2007 [40], for all dyadic Calderón–Zygmund operators (which are localized) it was given by Nazarov–Treil–Volberg in 2008 [38], and the final solution was given by Hytönen [19]. Then there appeared many beautiful proofs of the A_2-conjecture by different authors. The one based on the paper [38] was given by Hytönen–Pérez–Treil–Volberg [22], and the shortest and most elegant proof is due to Lerner [30].

Here we survey how to disprove this weak Muckenhoupt conjecture (which is also called the weak A_1-conjecture). The reader can also get acquainted with the results on the A_1-conjecture in the papers [34, 33].

While the endpoint estimate of the dyadic martingale transform is now fully understood due to [34, 33, 49], the $L^2(w) \to L^{2,\infty}(w)$, $w \in A_2$, estimate of the dyadic square function (our second example of a dyadic singular operator) is still mysterious.

2 Weak estimates for the square function: the Bollobás function

Recall that the symbol $\mathrm{ch}(J)$ denotes the dyadic children of J and that the martingale difference operator Δ_J was defined as follows:

$$\Delta_J f := \sum_{I \in \mathrm{ch}(J)} \mathbf{1}_I (\langle f \rangle_I - \langle f \rangle_J).$$

For our case of a dyadic lattice on the line, $|\Delta_J f|$ is constant on J and

$$\Delta_J f = \frac{1}{2}[(\langle f \rangle_{J_+} - \langle f \rangle_{J_-})\mathbf{1}_{J_+} + (\langle f \rangle_{J_-} - \langle f \rangle_{J_+})\mathbf{1}_{J_-}],$$

where J_- and J_+ denote the left and right halves of J, respectively. The square function operator is

$$(Sf)^2(x) = \sum_{J \in \mathcal{D}} |\Delta_J f|^2 \mathbf{1}_J(x).$$

An equivalent formulation in terms of Haar functions is

$$(Sf)^2(x) = \sum_{J \in \mathcal{D}} (f, h_J)^2 \frac{\mathbf{1}_J(x)}{|J|},$$

where $(\cdot, \cdot)$ denotes the usual inner product in $L^2(\mathbf{R})$ and $\{h_J\}_{J \in \mathcal{D}}$ are the Haar functions:

$$h_J(x) := \frac{1}{\sqrt{|J|}}(\mathbf{1}_{J_-}(x) - \mathbf{1}_{J_+}(x)).$$

Before even bringing in weights, it is interesting to look at the behavior of S acting from $L^1(dx)$ to $L^{1,\infty}(dx)$. To do so, we localize S and work mainly with the truncated square function:

$$S_J \varphi(x) := \sum_{I \subseteq J} (\varphi, h_I)^2 \frac{\mathbf{1}_I(x)}{|I|},$$

defined by only looking at the Haar coefficients inside a given $J \in \mathcal{D}$. The best constant C in

$$|\{x \in J : S_J^2 \varphi(x) \geq \lambda\}| \leq C \frac{1}{\sqrt{\lambda}} \int_J |\varphi|$$

(for all $\varphi \in L^1(J)$ and $J \in \mathcal{D}$) was first conjectured by Bollobás in [4] to be

$$C = \Psi(1), \quad \text{where } \Psi(\tau) = \tau\Phi(\tau) + e^{-\tau^2/2} \text{ and } \Phi(\tau) = \int_0^\tau e^{-x^2/2}\, dx. \tag{2.1}$$

In [39] it was proved using stochastic methods that this is indeed the sharp constant.

We will explain an interesting Bellman approach to this problem, based on [17]. First, let us see what the "standard Bellman" approach to this problem would be and how far it will take us. As this is an optimization of a maximal bound, the natural Bellman function of the problem is of course

$$\mathbf{M}(f,F,\lambda) := \sup \frac{1}{|J|}\left|\{x \in J : S_J^2\varphi(x) \geq \lambda\}\right|,$$

where the supremum is over all functions φ supported in $J \in \mathcal{D}$, such that $\langle\varphi\rangle_J = f$ and $\langle|\varphi|\rangle_J = F$. The obvious $|\langle\varphi\rangle_J| \leq \langle|\varphi|\rangle_J$ inequality imposes the domain restriction $|f| \leq F$, so the domain of our function $\mathbf{M}$ is then

$$\Omega_\mathbf{M} := \{(f,F,\lambda) : |f| \leq F;\ \lambda > 0\}.$$

Our problem then becomes to find the best C in

$$\mathbf{M}(f,F,\lambda) \leq C\frac{F}{\sqrt{\lambda}}. \tag{2.2}$$

Next, we would obtain some of the standard Bellman-type properties of the function $\mathbf{M}$, starting with obvious things like homogeneity, obstacle condition, or the main inequality, and culminating with the differential form of the main inequality. Usually, this would take us close to the end of the story: Ideally we will find *some* solution to the resulting PDE which satisfies the boundary conditions and the obstacle condition, and in many situations any such solution will give us the desired result – even if the solution we found was not the "true" Bellman function but a so-called *supersolution*. While this problem seemed at first quite amenable to such an approach, given the small number of variables (for a Bellman problem!) it turned out that the PDE was very difficult and we still do not know a solution for it.

Even more confounding was the following: We were able to find $\mathbf{M}$ exactly along the boundary $|f| = F$, but the only function from (2.1) appearing there was Φ, and Ψ – the one we needed! – was nowhere to be found. As it turns out, and as we will explain below, Ψ was hiding as the boundary solution of another Bellman function associated with the function, this one called $\mathbf{L}$ and defined as an infimum. The key to proving (2.1) was then a series of interactions between $\mathbf{M}$ and $\mathbf{L}$, especially the so-called Bollobás inequality. Most conveniently, we never needed to solve the overall PDEs for either one; the boundary functions (which were easy to find) turned out to be enough to obtain our desired bound.

Let us begin with the aforementioned standard properties of **M**:

- **M** *is independent of the choice of interval J*. This is a simple property, easily justified by an affine transformation which can translate any situation on one $J \in \mathcal{D}$ to any other. It is, however, crucial for the main inequality.
- **M** *is even in f*. This is easy to see once we notice that if we take any admissible function φ for $\mathbf{M}(f, F, \lambda)$ with $\langle \varphi \rangle_J = f$ and $\langle |\varphi| \rangle_J = F$, then $-\varphi$ will be admissible for $\mathbf{M}(-f, F, \lambda)$ and the square function will not see the minus sign.
- **M** *is decreasing in* λ (obvious).
- *Homogeneity:* We have

$$\mathbf{M}(f, F, \lambda) = \mathbf{M}(tf, |t|F, t^2\lambda), \quad \forall t \neq 0. \tag{2.3}$$

Here we take an admissible function φ and $t \neq 0$ and note that $t\varphi$ is admissible for $\mathbf{M}(tf, |t|F, t^2\lambda)$. Moreover, $\{x \in J : S_J^2 \varphi(x) > \lambda\}$ is the same set as $\{x \in J : S_J^2(t\varphi)(x) > t^2\lambda\}$.

- *Main inequality:* For all triplets (f, F, λ), $(f_\pm, F_\pm, \lambda_\pm)$ in the domain with $f = \frac{1}{2}(f_- + f_+)$, $F = \frac{1}{2}(F_- + F_+)$, and $\lambda = \min(\lambda_-, \lambda_+)$, we have

$$\mathbf{M}\left(f, F, \lambda + \left(\frac{f_+ - f_-}{2}\right)^2\right) \geq \frac{1}{2}(\mathbf{M}(f_+, F_+, \lambda_+) + \mathbf{M}(f_-, F_-, \lambda_-)). \tag{2.4}$$

As mentioned before, this property rests on the independence of **M** from the choice of interval J. It allows us to run the Bellman machine on J_- and J_+ completely separately and then put the results together. Specifically, we take $\varphi_\pm$ on $J_\pm$, respectively, admissible for $\mathbf{M}(f_\pm, F_\pm, \lambda)$:

$$\langle \varphi_\pm \rangle_{J_\pm} = f_\pm; \quad \langle |\varphi_\pm| \rangle_{J_\pm} = F_\pm,$$

which "almost give the supremum":

$$\frac{1}{|J_\pm|}|\{x \in J_\pm : S_{J_\pm}^2 \varphi_\pm(x) > \lambda\}| \geq \mathbf{M}(f_\pm, F_\pm, \lambda) - \epsilon,$$

for some $\epsilon > 0$. We form φ on J by concatenation, and have

$$\langle \varphi \rangle_J = f; \quad \langle |\varphi| \rangle_J = F,$$

and $\frac{1}{|J|}|\{x \in J : S_J \varphi(x) > \lambda + [(f_+ - f_-)/2]^2\}|$ is the same as

$$\frac{1}{|J|}|\{x \in J_- : S_{J_-}^2 \varphi_-(x) > \lambda\}| + \frac{1}{|J|}|\{x \in J_+ : S_{J_+}^2 \varphi_+(x) > \lambda\}|.$$

Then (2.4) follows.

– **M** *is concave and continuous in f and F.* We can obtain this by rewriting the main
 inequality as

$$\frac{1}{2}\big(\mathbf{M}(f+a,F+b,\lambda)+\mathbf{M}(f-a,F-b,\lambda)\big) \le \mathbf{M}(f,F,\lambda+a^2) \le \mathbf{M}(f,F,\lambda),$$

 for all $a \in \mathbf{R}$ and $|b| \le F$, and letting $a = 0$ or $b = 0$.
– **M** *is maximal at $f = 0$.* This will be important later, and it follows easily from the
 fact that **M** is concave and even in f.
– *Obstacle condition:* We have

$$\mathbf{M}(f,F,\lambda) = 1, \quad \forall \lambda \le F^2. \tag{2.5}$$

It is clear to see that **M** is always trapped between 0 and 1. The obstacle condition
is an easy way to obtain a condition for **M** to attain its maximum value of 1. For
every point (f,F,λ) in the domain, we consider the function $\varphi = f\mathbf{1}_J + F\sqrt{|J|}h_J$, which
satisfies $\langle\varphi\rangle_J = f$, $\langle|\varphi|\rangle_J = F$, and $S_J^2\varphi = F^2\mathbf{1}_J$. Obviously, φ is admissible for $\mathbf{M}(f,F,\lambda)$,
and if $\lambda \le F^2$, immediately $\mathbf{M}(f,F,\lambda) = 1$. As expected, the obstacle condition also
functions as a stopping condition in the so-called "Bellman induction," which we
explain next. Essentially, this is the property of **M** being the "least supersolution."

Consider the main inequality for **M** in more generality:

$$m(f,F,\lambda+a^2) \ge \frac{1}{2}\big(m(f+a,F+b,\lambda)+m(f-a,F-b,\lambda)\big). \tag{2.6}$$

We say that a function $m(f,F,\lambda)$ defined on $\Omega_\mathbf{M}$ is a *supersolution* of the main inequality (2.6) provided that m is non-negative and continuous and satisfies:
(1) the main inequality (2.6);
(2) the obstacle condition $m(f,F,\lambda) = 1$, whenever $\lambda \le F^2$.

Theorem 2.1. *If m is any supersolution as defined above, then* $\mathbf{M} \le m$.

For details of the proof, see [17].

Next, we use Taylor's formula to express the main inequality in its differential form:

$$\begin{pmatrix} \mathbf{M}_{ff} - 2\mathbf{M}_\lambda & \mathbf{M}_{fF} \\ \mathbf{M}_{fF} & \mathbf{M}_{FF} \end{pmatrix} \le 0.$$

As previously mentioned, this did not take us any further. We could however solve along
the boundary $|f| = F$, where this becomes an ODE. If $|f| = F$, any admissible φ must have
equality in $|\langle\varphi\rangle_J| = \langle|\varphi|\rangle_J$ and must therefore be either almost everywhere positive or
almost everywhere negative. Since **M** is even in f, it suffices to work with $f > 0$ and
$\varphi > 0$ a. e.

If we let $M(f,\lambda) := \mathbf{M}(f,|f|,\lambda)$, the differential form of the main inequality here becomes simply

$$M_{ff} - 2M_\lambda \le 0.$$

Using homogeneity, we note that $M(f,\lambda) = M(f/\sqrt{\lambda},1) =: a(\tau)$, where $\tau = \frac{f}{\sqrt{\lambda}}$, and the condition becomes

$$a''(\tau) + \tau a'(\tau) \le 0. \tag{2.7}$$

We see Φ emerging as part of the general solution to

$$y''(\tau) + \tau y'(\tau) = 0; \quad y(\tau) = C\Phi(\tau) + D, \quad \forall \tau \ge 0.$$

Imposing our boundary conditions $y(0) = 0$ and $y(1) = 1$, we get

$$y(\tau) = \begin{cases} \frac{\Phi(\tau)}{\Phi(1)}, & 0 \le \tau \le 1, \\ 1, & \tau \ge 1. \end{cases}$$

Now, we have to be a little careful, because this is not necessarily the function M itself, but a "candidate":

$$m(f,\lambda) = \begin{cases} \frac{\Phi(|f|/\lambda)}{\Phi(1)}, & 0 \le |f| \le \sqrt{\lambda}, \\ 1, & |f| \ge \lambda. \end{cases}$$

To see that this is indeed $M(f,\lambda)$, we first check that $m(f,\lambda)$ satisfies the main inequality for M – see the details of this calculation in [17] – which would make m a supersolution, and gives us

$$M(f,\lambda) \le m(f,\lambda).$$

For the reverse inequality, we consider a new variable $S := \Phi(\tau)$ and observe that for a function g,

$$(\tau g'(\tau) + g''(\tau))e^{\tau^2} = \frac{d^2 g}{dS^2} = g_{SS}. \tag{2.8}$$

So (2.7) is equivalent to $a_{SS} \le 0$, or a being concave in the variable S. It is easy to see that if $g(S)$ is a concave non-negative function for $S \ge 0$, then the ratio $\frac{g(S)}{S}$ is non-increasing. So, if we put $a(\tau) := g(S)$, we have for all $0 \le \tau \le 1$

$$\frac{g(S)}{S} = \frac{a(\tau)}{\Phi(\tau)} \ge \frac{g(\Phi(1))}{\Phi(1)} = \frac{a(1)}{\Phi(1)} = \frac{1}{\Phi(1)},$$

which gives exactly $M(f,\lambda) \ge m(f,\lambda)$. Therefore,

$$M(f,\lambda) = \mathbf{M}(f,|f|,\lambda) = \begin{cases} \dfrac{\Phi(\frac{|f|}{\sqrt{\lambda}})}{\Phi(1)}, & |f| < \sqrt{\lambda}, \\ 1, & |f| \geq \sqrt{\lambda}. \end{cases} \tag{2.9}$$

This also does not get us the bound, as the desired function Ψ is nowhere to be found!

So let us bring in the other Bellman function $\mathbf{L}$, which we define as follows. Given $f \in \mathbf{R}$, $0 \leq p \leq 1$, and $\lambda > 0$, define

$$\mathbf{L}(f,p,\lambda) := \inf\langle|\varphi|\rangle_J,$$

where the infimum is over all functions φ, supported in $J \in \mathcal{D}$, such that

$$\langle\varphi\rangle_J = f \quad \text{and} \quad \frac{1}{|J|}\left|\{x \in J : S_J^2\varphi(x) \geq \lambda\}\right| \geq p.$$

We say that any such φ is an admissible function for $\mathbf{L}(f,p,\lambda)$.

Interestingly, $\mathbf{L}$ has many properties which mirror those of $\mathbf{M}$: For instance, the main inequality is the same, only reversed, and concavity is replaced by convexity.

– *Domain and range:* $\mathbf{L}$ has convex domain $\Omega_\mathbf{L} := \{(f,p,\lambda) : f \in \mathbf{R}; p \in [0,1], \lambda > 0\}$ and range

$$|f| \leq \mathbf{L}(f,p,\lambda) \leq (1-p)|f| + p\max(|f|,\sqrt{\lambda}). \tag{2.10}$$

– *$\mathbf{L}$ is independent of the choice of interval $J \in \mathcal{D}$ in its definition.*
– *$\mathbf{L}$ is even in f.*
– *$\mathbf{L}$ is increasing in λ.*
– *Homogeneity:* We have

$$\mathbf{L}(tf,p,t^2\lambda) = |t|\mathbf{L}(f,p,\lambda), \quad \forall t \neq 0. \tag{2.11}$$

– *Main inequality:* For all triplets (f,p,λ), $(f_\pm,p_\pm,\lambda_\pm)$ in the domain with $f = \frac{1}{2}(f_- + f_+)$, $p = \frac{1}{2}(p_- + p_+)$, and $\lambda = \min(\lambda_-,\lambda_+)$, we have

$$\mathbf{L}\left(f,p,\lambda + \left(\frac{f_+ - f_-}{2}\right)^2\right) \leq \frac{1}{2}(\mathbf{L}(f_+,p_+,\lambda_+) + \mathbf{L}(f_-,p_-,\lambda_-)). \tag{2.12}$$

– *$\mathbf{L}$ is convex in the variables f and p.*
– *$\mathbf{L}$ is minimal at $f = 0$.*
– *Obstacle condition:* We have

$$\mathbf{L}(f,p,\lambda) = |f|, \quad \forall|f| \geq \sqrt{\lambda}. \tag{2.13}$$

Even the "least supersolution" property of $\mathbf{M}$ is mirrored by the property of $\mathbf{L}$ being the "greatest subsolution": Let us also consider the main inequality for $\mathbf{L}$ in more generality:

$$\ell(f, p, \lambda + a^2) \le \frac{1}{2}(\ell(f + a, p + b, \lambda) + \ell(f - a, p - b, \lambda)). \qquad (2.14)$$

We say that a function $\ell(f, p, \lambda)$ defined on Ω_L is a *subsolution* for the main inequality (2.14) provided that ℓ is non-negative and continuous and satisfies:
(1) the main inequality (2.14);
(2) *the range/obstacle condition:* $|f| \le \ell(f, p, \lambda) \le \max\{|f|, \sqrt{\lambda}\}$;
(3) *the boundary condition:* $\ell(f, 0, \lambda) = |f|$.

Theorem 2.2. *If ℓ is any subsolution as defined above, then $\ell \le L$.*

Now, let us explore the relationships between M and L.

Theorem 2.3. $L(f, p, \lambda)$ *is the smallest value of F for which* $M(f, F, \lambda) \ge p$:

$$L(f, p, \lambda) = \inf\{F \ge |f| : M(f, F, \lambda) \ge p\}. \qquad (2.15)$$

Moreover, $M(f, F, \lambda)$ *is the largest value of p such that* $L(f, p, \lambda) \le F$:

$$M(f, F, \lambda) = \sup\{p \in [0, 1] : L(f, p, \lambda) \le F\}. \qquad (2.16)$$

The proof is straightforward and can be found in [17]. In a sense, this means that M and L are each other's inverses in the second variable.

Looking back at the obstacle condition (2.5) for M, namely $M(f, F, \lambda) = 1$ whenever $F \ge \sqrt{\lambda}$, there is no reason to think this condition is optimal. That is, there could be values of $F < \sqrt{\lambda}$ where still M is 1. As it turns out, the optimal obstacle condition for M can be obtained from information about L. Since $M \le 1$, taking $p = 1$ in (2.15), we obtain exactly this:

$$L(f, \lambda) = L(f, 1, \lambda) = \inf\{F \ge |f| : M(f, F, \lambda) = 1\}. \qquad (2.17)$$

On the other hand, the obstacle condition for L really comes from its range, $|f| \le L(f, p, \lambda) \le \max\{|f|, \sqrt{\lambda}\}$, which clearly shows that $L = |f|$ whenever $|f| \ge \lambda^2$. However, this says nothing about p, and we do know that, for example, $L(f, 0, \lambda) = |f|$ regardless of the behavior of f and λ. What other values of p could this hold for? This is again obtained precisely from information about M, by letting $F = |f|$ in (2.16):

$$M(f, \lambda) = M(f, |f|, \lambda) = \sup\{p \in [0, 1] : L(f, p, \lambda) = |f|\}. \qquad (2.18)$$

So, if we find the expressions for L and M along these boundaries of their domains, we also obtain the optimal obstacle conditions for M and L, respectively. We have already found M, and L can be found similarly:

$$L(f, \lambda) = L(f, 1, \lambda) = \begin{cases} \dfrac{\sqrt{\lambda}\Psi(\frac{|f|}{\sqrt{\lambda}})}{\Psi(1)}, & 0 \le |f| < \sqrt{\lambda}, \\ |f|, & |f| \ge \sqrt{\lambda}. \end{cases}$$

Here is the function Ψ, finally, arisen as a solution to

$$z''(\tau) + \tau z'(\tau) - z(\tau) = 0; \quad z(\tau) = C\Psi(\tau) + D\tau.$$

By homogeneity, we can again write

$$L(f,\lambda) = \mathbf{L}(f,1,\lambda) = \sqrt{\lambda}\mathbf{L}(f/\sqrt{\lambda},1) =: \sqrt{\lambda}\beta(\tau),$$

and the main inequality in differential form here is

$$\beta''(\tau) + \tau\beta'(\tau) - \beta(\tau) \geq 0.$$

It is interesting to visualize exactly how M and L induce the optimal obstacle conditions for $\mathbf{L}$ and $\mathbf{M}$, respectively. It is easiest to do so by reducing the domains to two-variable domains, using homogeneity. Specifically, we write

$$\mathbf{M}(f,F,\lambda) = \mathbf{M}(f/\sqrt{\lambda},F/\sqrt{\lambda},1) =: \theta(\tau,\gamma) \quad \text{and} \quad \mathbf{L}(f,p,\lambda) =: \sqrt{\lambda}\eta(\tau,p),$$

where $\tau = f/\sqrt{\lambda}$ and $\gamma = F/\sqrt{\lambda}$. Thus, θ is defined on $\Omega_\theta := \{0 \leq |\tau| \leq \gamma\}$ with values in $[0,1]$, and η is defined on $\Omega_\eta := \{0 \leq p \leq 1;\ \tau \in \mathbb{R}\}$ with values satisfying $|\tau| \leq \eta(\tau,p) \leq (1-p)|\tau| + p\max(|\tau|,1)$. It is also clear that θ and η are even in τ, so we often restrict our attention to the domains Ω_θ^+ and Ω_η^+ where $\tau \geq 0$. Other properties that θ and η inherit from $\mathbf{M}$ and $\mathbf{L}$ are easy to check:

- $\theta(0,0) = 0$ and $\eta(\tau,0) = \tau$;
- θ is maximal at $\tau = 0$ and η is minimal at $\tau = 0$:

$$\theta(|\tau|,\gamma) \leq \theta(0,\gamma); \quad \eta(0,p) \leq \eta(|\tau|,p);$$

- θ is decreasing in τ for $\tau \geq 0$ and is increasing in γ; η is increasing in both $\tau \geq 0$ and p;
- θ is concave in both τ and γ, and η is convex in both τ and p;
- the original obstacle conditions (2.5) and (2.13) for $\mathbf{M}$ and $\mathbf{L}$ translate to

$$\theta(\tau,\gamma) = 1, \quad \forall\gamma \geq 1 \quad \text{and} \quad \eta(\tau,p) = |\tau|, \quad \forall|\tau| \geq 1.$$

Moreover, (2.17) and (2.18) become

$$\eta(\tau,1) = \inf\{\gamma \geq |\tau| : \theta(\tau,\gamma) = 1\} \quad \text{and} \quad \theta(\tau,|\tau|) = \sup\{p : \eta(|\tau|,p) = |\tau|\}.$$

The expression for L gives

$$\eta(\tau,1) = \begin{cases} \dfrac{\Psi(|\tau|)}{\Psi(1)}, & 0 \leq |\tau| < 1, \\ |\tau|, & |\tau| \geq 1, \end{cases}$$

which yields the optimal obstacle condition for θ (see Figure 1). Similarly, M gives

$$\theta(\tau, \tau) = \begin{cases} \dfrac{\Phi(|\tau|)}{\Phi(1)}, & 0 \le |\tau| \le 1, \\ 1, & \tau \ge 1, \end{cases}$$

which yields the optimal obstacle condition for η (see Figure 2).

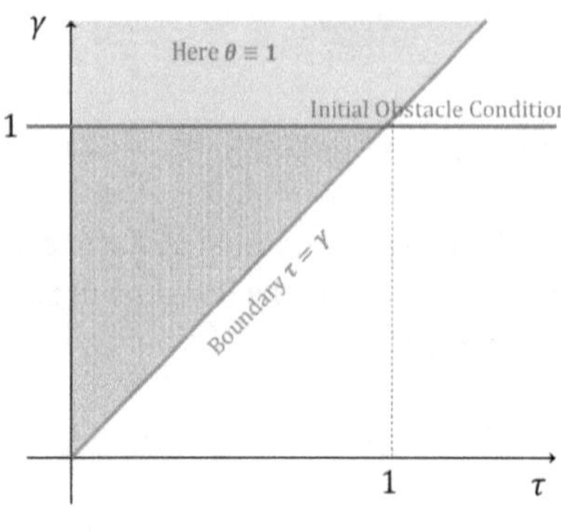

Domain of $\boldsymbol{\theta}(\tau, \gamma)$: Initial Knowledge

Knowledge of $\boldsymbol{\theta}$ after finding $\eta(\tau, 1)$.

Figure 1: Initial and optimal obstacle conditions for θ.

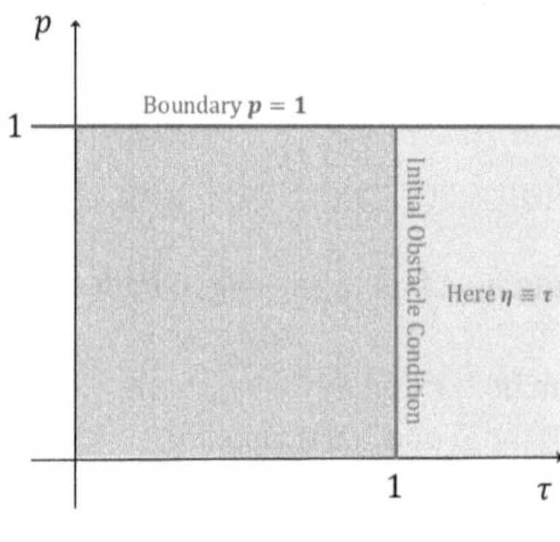

Domain of $\boldsymbol{\eta}(\tau, p)$: Initial Knowledge

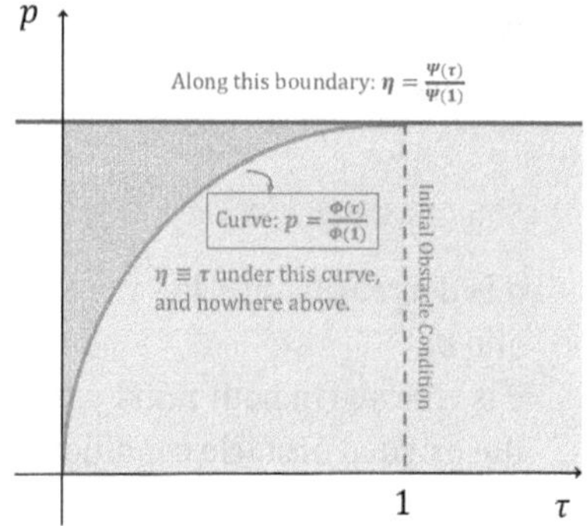

Knowledge of $\boldsymbol{\eta}$ after finding $\theta(\tau, \tau)$.

Figure 2: Initial and optimal obstacle conditions for η.

The final piece in the puzzle is the Bollobás inequality:

$$\mathbf{M}(0, F, \lambda) \le \frac{F}{\mathbf{L}(0, 1, \lambda)}. \tag{2.19}$$

First appearing in [4], this inequality ties in the maximal values of $\mathbf{M}$ (which, recall, has a maximum at $f = 0$) with the minimal value of $\mathbf{L}$. The clever proof (written in detail in

[17]) relies on the idea that one can start with a function which has $S_J^2\varphi(x) > \lambda$ on only a fraction of $|J|$, but "fill in" the parts where $S_J^2\varphi \le \lambda$ with copies of the function adapted to smaller intervals. The new function will have a square function larger than λ on a larger portion of $|J|$, and in the limit we obtain a function admissible for $\mathbf{L}(0, 1, \lambda)$.

Combined with the knowledge we already have of the boundary values for $\mathbf{M}$ and $\mathbf{L}$, this inequality can actually help us find $\mathbf{M}(0, F, \lambda)$ and $\mathbf{L}(0, p, \lambda)$ – and, since $\mathbf{M}$ is maximal at $f = 0$, these will in turn give us our desired bound.

Notice that, in terms of θ and η, the Bollobás inequality says that

$$\theta(0, y) \le \frac{F}{\sqrt{\lambda}\,\eta(0, 1)},$$

but since we actually know $\eta(\tau, 1)$ exactly, we can further improve this to

$$\mathbf{M}(f, F, \lambda) \le \mathbf{M}(0, F, \lambda) = \theta(0, y) \le \frac{F}{\sqrt{\lambda}}\Psi(1).$$

This gives us that the best constant C in (2.2),

$$C = \sup_{f, F, \lambda} \frac{\mathbf{M}(f, F, \lambda)\sqrt{\lambda}}{F},$$

must indeed satisfy $C \le \Psi(1)$.

However, to truly finish we need to show this is the best possible constant. Notice that the constant above can be expressed as

$$C = \sup_{f, F, \lambda} \frac{\mathbf{M}(f, F, \lambda)\sqrt{\lambda}}{F} = \sup_{F, \lambda} \frac{\mathbf{M}(0, F, \lambda)\sqrt{\lambda}}{F},$$

and, as mentioned before, we can find $\mathbf{M}(0, F, \lambda)$ exactly:

$$\mathbf{M}(0, F, \lambda) = \begin{cases} \frac{F}{\sqrt{\lambda}}\Psi(1), & F \le \frac{\sqrt{\lambda}}{\Psi(1)}, \\ 1, & F > \frac{\sqrt{\lambda}}{\Psi(1)}. \end{cases}$$

From this it actually follows immediately that C is *equal* to $\Psi(1)$. So let us see how to obtain this $\mathbf{M}(0, F, \lambda)$.

We know about $\theta(0, y)$ that $\theta(0, 0) = 0$ and $\theta(0, y) = 1$ for all $y \ge 1/\Psi(1)$. Moreover, we know θ is concave in y! Therefore, it must lie above its secant line determined by $(0, 0)$ and $(1/\Psi(1), 1)$, a line with equation $y(y) = \Psi(1)y$. So, $\theta(0, y) \ge \Psi(1)y$ for all $0 \le y \le 1/\Psi(1)$. The Bollobás inequality shows exactly the reverse inequality, and thus we have the above expression for $\mathbf{M}(0, F, \lambda)$.

3 Weak weighted estimates for the square function: testing

Our measure space will be $(X, \mathfrak{A}, dx)$, where the σ-algebra $\mathfrak{A}$ is generated by a standard dyadic filtration $\mathcal{D} = \bigcup_k \mathcal{D}_k$ on $\mathbf{R}$. We considered the weak weighted estimate for the martingale transform. In [36] the endpoint exponent was $p = 1$, and critical weights belong to the A_1-class.

Now we are going to consider weighted weak estimates for the dyadic square function. The endpoint exponent is now $p = 2$, and critical weights belong to the A_2-class. See the estimates for subcritical and supercritical exponents in [27, 14].

The weak estimate of the square function in $L^2(wdx)$, $w \in A_1$, is well known; see, e. g., [50]. The fact that the endpoint exponent is now at $p = 2$ is explainable by the bilinear nature of the square function transform.

Recall that the symbol $\mathrm{ch}(J)$ denotes the dyadic children of J. Recall that the martingale difference operator Δ_J was defined as follows:

$$\Delta_J f := \sum_{I \in \mathrm{ch}(J)} \mathbf{1}_I(\langle f \rangle_I - \langle f \rangle_J).$$

For our case of a dyadic lattice on the line, $|\Delta_J f|$ is constant on J, and

$$\Delta_J f = \frac{1}{2}[(\langle f \rangle_{J_+} - \langle f \rangle_{J_-})\mathbf{1}_{J_+} + (\langle f \rangle_{J_-} - \langle f \rangle_{J_+})\mathbf{1}_{J_-}].$$

The square function operator is

$$(Sf)^2(x) = \sum_{J \in \mathcal{D}} |\Delta_J f|^2 \mathbf{1}_J(x).$$

In this section we work only with the dyadic A_2-classes of weights, but we skip the word dyadic, because we consider here only dyadic operators. We consider a positive function $w(x)$, and as before we call it A_2-weight if

$$Q := [w]_{A_2} := \sup_{J \in \mathcal{D}} \langle w \rangle_J \langle w^{-1} \rangle_J < \infty. \tag{3.1}$$

We are going to consider the restricted weak estimate when the operator is applied to $f = \mathbf{1}_E$, but only for set E being itself a dyadic interval:

$$\frac{1}{|I|} w\left\{ x \in I : \sum_{J \in D(I)} |\Delta_J w^{-1}|^2 \mathbf{1}_J(x) > \lambda \right\} \leq C_T([w]_{A_2}) \frac{\langle w^{-1} \rangle_I}{\lambda}. \tag{3.2}$$

We give some information on the full weak estimate for the square function operator:

$$\frac{1}{|I|} w \left\{ x \in I : \sum_{J \in D(I)} |\Delta_J(\varphi w^{-1})|^2 \mathbf{1}_J(x) > \lambda \right\} \le C([w]_{A_2}) \frac{\langle \varphi^2 w^{-1} \rangle_I}{\lambda}. \tag{3.3}$$

Here φ runs over all functions such that $\operatorname{supp} \varphi \subset I$ and $\varphi \in L^2(I, w\, dx)$, $w \in A_2$.

We wish to understand the sharp order of magnitude of the weak estimates' constants $C_T([w]_{A_2})$ and $C([w]_{A_2})$ from (4.4) and (3.3), respectively ($[w]_{A_2}$ is supposed to be large). We wish to compare them to the similar estimates of strong type: (9.1) and (1.3) from Section 9.

Constant $C([w]_{A_2})$ gives the dependence on $[w]_{A_2}$ of the norm of the sublinear operator $S\varphi := (\sum_{J \in D} |\Delta_J \varphi|^2 \mathbf{1}_J(x))^{1/2}$ from $L^2(w)$ to $L^{2,\infty}(w)$. Constant $C_T([w]_{A_2})$ gives the dependence on $[w]_{A_2}$ of the same weak norm but measured only on test functions φ that are just characteristic functions of dyadic intervals: $\varphi = \mathbf{1}_I, I \in D$.

Let M_a be an operator of multiplication by function a. Notice a convenient change of variables $\varphi \to \varphi w^{-1}$, which reduces the problem of estimating the square function operator $S : L^2(w) \to L^{2,\infty}(w)$ to the estimate of $S_{w^{-1}} := S M_{w^{-1}}$ from $L^2(w^{-1})$ to $L^{2,\infty}(w)$.

Of course, there are several obvious estimates: (1) Weak constants are smaller than strong constants:

$$C_T([w]_{A_2}) \le C_{s,T}([w]_{A_2}), \quad C([w]_{A_2}) \le C_s([w]_{A_2}),$$

just by the Chebyshev inequality; and (2) test function estimates are trivially at least as good as the full estimates:

$$C_T([w]_{A_2}) \le C([w]_{A_2}), \quad C_{s,T}([w]_{A_2}) \le C_s([w]_{A_2}).$$

It is quite well known (although far from being trivial) that there is at least one converse inequality:

$$C_s([w]_{A_2}) \le A(d) C_{s,T}([w]_{A_2}). \tag{3.4}$$

This is an instance of the celebrated $T1$ theorem, on this occasion applied to the square function transform in the weighted situation.

Remark. Weak-type estimates $L^2(w) \to L^{2,\infty}(w)$ of a square function with sharp dependence on $[w]_{A_2}$ are rather difficult, in part because their sharp constants are much smaller than the sharp constants in the corresponding strong-type estimates.

Remark. This is in big contrast to the usual singular integrals T of Calderón–Zygmund type. We know that for $w \in A_2$ the strong norm $\|T : L^2(w) \to L^2(w)\|$ is exactly equivalent to $\|T : L^2(w) \to L^{2,\infty}(w)\| + \|T^* : L^2(w^{-1}) \to L^{2,\infty}(w^{-1})\|$; see [43, 22] for example.

In view of this remark it is important to review the strong sharp weighted estimate $S : L^2(w) \to L^2(w)$ in Section 9.

Remark. The reader can see that the (4.4) estimate is a particular case of the (3.3) estimate for a special choice of test function $f = \mathbf{1}_J$. The same remark holds for (9.1) and (1.3) in Section 9. But in Section 9 we deal with strong-type estimates, and theorem $T1$ ideology works for strong-type estimates. This is far from being true for weak estimates. In the weak case scenario there is no $T1$ theorem. However, certain extrapolation results do exist in the weak estimates world; see [10].

Remark. Unfortunately there is no $T1$ principle in general for weak-type estimates. See the papers [9, 1, 10].

3.1 Sharp constant in weak testing estimate

Theorem 3.1. *We have $C_{w,T} \leq AQ$ and this estimate is sharp.*

We will prove the estimate; the sharpness is well known just for one-point-singularity weights.

As always, in this section A_2 means dyadic A_2. We introduce the following function of three real variables:

$$B_Q(u, v, \lambda) := \sup \frac{1}{|J|} w\left\{ x \in J : \sum_{I \in D(J)} |\Delta_I w^{-1}|^2 \chi_I(x) > \lambda \right\}, \tag{3.5}$$

where the supremum is taken over all $w \in A_2$, $[w]_{A_2} \leq Q$, such that

$$\langle w \rangle_J = u, \quad \langle w^{-1} \rangle_J = v.$$

Notice that by a scaling argument our function does not depend on J but depends on $Q = [w]_{A_2}$. For brevity we can skip Q: $B := B_Q$.

Remark. Ideally we want to find the formula for this function. Notice that this is similar to solving a problem of "isoperimetric" type, where the solution of certain non-linear PDEs is a common tool; see, e. g., [2].

3.1.1 Properties of *B* and the main inequality

Notice several properties of B:
- B is defined in $\Omega := \{(u, v, \lambda) : 1 \leq uv \leq Q, u > 0, v > 0, 0 \leq \lambda < \infty\}$.
- If $P = (u, v, \lambda)$, $P_+ = (u_+, v_+, \lambda_+)$, and $P_- = (u_-, v_-, \lambda_-)$ belong to Ω and $u = \frac{1}{2}(u_+ + u_-)$, $v = \frac{1}{2}(v_+ + v_-)$, $\lambda = \min(\lambda_+, \lambda_-)$, then *the main inequality* holds with constant $c = 1$:

$$B(u, v, \lambda + c(v_+ - v_-)^2) - \frac{B(P_+) + B(P_-)}{2} \geq 0.$$

- B is decreasing in λ.
- Homogeneity: We have $B(ut, v/t, \lambda/t^2) = tB(u, v, \lambda)$, $t > 0$.
- Obstacle condition: For all points (u, v, λ) such that $10 \le uv \le Q$, $\lambda \ge 0$, if $\lambda \le \delta v^2$ for a positive absolute constant δ, then one has $B(u, v, \lambda) = u$.
- The boundary condition $B(u, v, \lambda) = 0$ if $uv = 1$.

All these properties are very simple consequences of the definition of B. However, let us explain a bit the second and the fifth bullet. The second bullet is the consequence of the scale invariance of B. We consider data P_+ and find weight w_+ that almost supremizes $B(P_+)$. By definition of B, we have it on J. But by the scale invariance we can think that w_+ lives on J_+. Then we consider data P_- and find weight w_- that almost supremizes $B(P_-)$. Again we are supposed to have it on J. But by the scale invariance we can think that w_- lives on J_-. The next step is to consider the concatenation of w_+ and w_-:

$$w_c := \begin{cases} w_+, & \text{on } J_+, \\ w_-, & \text{on } J_-. \end{cases}$$

Clearly this new weight is a competitor for giving the supremum for data P on J. But it is only a competitor; the real supremum in (3.5) is bigger. This implies the second bullet above (*the main inequality*).

Now let us explain the fifth bullet above; we call it *the obstacle condition*. Let us consider a special weight w_s in J: it is one constant on J_- and just another constant on J_+. Moreover, we wish to have $\langle w_s^{-1} \rangle_{J_+} = 4 \langle w_s^{-1} \rangle_{J_-}$. Notice that then $b \langle w_s^{-1} \rangle_J \le |\Delta_J w_s^{-1}|$ with some positive absolute constant b.

Now it is obvious that if $\lambda \le \delta^2 \langle w_s^{-1} \rangle_J^2$, then $\{x \in J : S_{w_s^{-1}}^2(\mathbf{1}_J) \ge \lambda\} = J$, so $\frac{1}{|J|} w_s \{x \in J : S_{w_s^{-1}}^2(\mathbf{1}_J) \ge \lambda\} = \langle w_s \rangle_J$. Notice now that w_s is just one admissible weight and that we have to take the supremum over all such admissible weights.

We get the fifth bullet above (the obstacle condition): $B(u, v, \lambda) = u$ for those points (u, v, λ) in the domain of definition of B where the corresponding w_s with $\langle w_s \rangle = u$, $\langle w_s^{-1} \rangle_J = v$ exists. It is obvious that for all sufficiently large Q and for any pair (u, v) such that $10 \le uv \le Q$ one can construct a just "two-valued" w_s as above with $[w_s]_{A_2} \le Q$ (we recall that we deal only with dyadic A_2-weights).

Notice that the main inequality above transforms into a *partial differential inequality* if considered infinitesimally (and if we tacitly assume that B is smooth):

$$-\frac{1}{2} d^2_{u,v} B + c \frac{\partial B}{\partial \lambda} (dv)^2 \ge 0. \tag{3.6}$$

We get it with $c = 1$ for the function B defined above (if B happens to be smooth).

We are not going to find B defined in (3.5), but instead we will construct a smooth $\mathcal{B}$ that satisfies all the properties above (and of course (3.6)) except for the boundary condition (the last bullet above). It will satisfy even slightly stronger properties, for example, the obstacle condition (the fifth bullet) will be satisfied with 1 instead of 10:

$$\forall (u, v, \lambda) \quad \text{such that} 1 \le uv \le Q, \ \lambda \ge 0,$$
$$\text{if } \lambda \le \delta v^2 \quad \text{for some } \delta > 0, \ \text{then } B(u, v, \lambda) = u. \tag{3.7}$$

Here a will be some positive absolute constant (which will not depend on Q).

Using our usual telescopic sums consideration it will be very easy to prove the following theorem.

Theorem 3.2. *Suppose we have a smooth function B satisfying all the conditions above except the boundary condition, but satisfying the obstacle condition in the form (3.7). We also allow c to be a small positive constant (say, $c = \frac{1}{8}$). Suppose it also satisfies*

$$B(u, v, \lambda) \le A\, Q \frac{v}{\lambda}. \tag{3.8}$$

Then the constant $C_{w,T}$ in (4.4) is at most $A\, Q$.

Proof. We prove this by stopping time reasoning. It is enough to think that w is constant on some very small dyadic intervals and to prove the estimate on $C_{w,T}$ uniformly. Then we start with any such w, $[w]_{A_2} \le Q$, and we use the main inequality with $u = \langle w \rangle_J$, $v = \langle w^{-1} \rangle_J$, $u_{\pm} = \langle w \rangle_{J_{\pm}}$, $v_{\pm} = \langle w^{-1} \rangle_{J_{\pm}}$,

$$\lambda - c(\langle w^{-1} \rangle_{J_+} - \langle w^{-1} \rangle_{J_-})^2 =: \lambda_{J_{\pm}},$$

to obtain

$$|J_+| B(\langle w \rangle_{J_+}, \langle w^{-1} \rangle_{J_+}, \lambda_{J_+}) + |J_-| B(\langle w \rangle_{J_-}, \langle w^{-1} \rangle_{J_-}, \lambda_{J_-})$$
$$\le B(\langle w \rangle_J, \langle w^{-1} \rangle_J, \lambda)|J| \le \frac{A\, Q \langle w^{-1} \rangle_J}{\lambda} |J|. \tag{3.9}$$

We continue to use the main inequality (because $J_{\pm}$ are not at all different from J) and finally after a large but finite number of steps on a certain collection $\mathcal{I}$ of small intervals $I = J_{\pm\pm\cdots\pm}$ we reach the situation that

$$\lambda_{J_{\pm\pm\cdots\pm}} < c(\langle w^{-1} \rangle_{J_{\pm\pm\cdots+}} - \langle w^{-1} \rangle_{J_{\pm\pm\cdots-}})^2. \tag{3.10}$$

Collection $\mathcal{I}$ may be empty of course, but we know that $I \in \mathcal{I}$ if on I the following holds for $x \in I$:

$$c \sum_{L \in D(J), I \subset L} |\Delta_I w^{-1}|^2 \mathbf{1}_L(x) > \lambda.$$

Let us combine (3.10) with an obvious inequality,

$$c(\langle w^{-1} \rangle_{J_{\pm\pm\cdots+}} - \langle w^{-1} \rangle_{J_{\pm\pm\cdots-}})^2 \le \delta \langle w^{-1} \rangle_{J_{\pm\pm\cdots}}^2.$$

At this moment we use property 5 of B called the obstacle condition. On intervals $I \in \mathcal{I}$ the obstacle condition will provide us with $B(\langle w \rangle_I, \langle w^{-1} \rangle_I, \lambda_I) = \langle w \rangle_I$. So on a certain large finite step N we get from the iteration of (3.9) N times the following estimate:

$$\sum_{I \in D_N(J): I \in \mathcal{I}} |I| \langle w \rangle_I \le \frac{A\, Q \langle w^{-1} \rangle_J}{\lambda} |J|.$$

Therefore, we proved

$$\frac{1}{|J|} w \left\{ x \in J : \sum_{I \in D(J)} |\Delta_I w^{-1}|^2 \chi_I(x) > \lambda \right\} \le A\, Q \frac{\langle w^{-1} \rangle_J}{\lambda},$$

which is (4.4). $\qquad\qquad\square$

3.1.2 Formula for the function B. Monge–Ampère equation with a drift

Here is the formula for B that satisfies all the properties in Section 3.1.1 (except for the last one, the boundary condition):

$$B(u, v, \lambda) = \frac{1}{\sqrt{\lambda}} \Theta\left(u \sqrt{\lambda}, \frac{v}{\sqrt{\lambda}} \right), \quad \text{where } \Theta(y, \tau) := \min\left(y, Q e^{-\tau^2/2} \int_0^\tau e^{s^2/2} ds \right). \tag{3.11}$$

Notice that the fact that B has the form $B(u, v, \lambda) = \frac{1}{\sqrt{\lambda}} \Theta(u \sqrt{\lambda}, \frac{v}{\sqrt{\lambda}})$ is trivial; this follows from property 4 (homogeneity).

Notice also that function Θ is given in the domain enclosed by two hyperbolas

$$H := \{ (y, \tau) > 0 : 1 \le y\tau \le Q \}.$$

All properties listed at the beginning of Section 3.1.1 (except for the sixth bullet, which is the boundary condition, but we do not use it anywhere) follow by direct computation. In the next section we explain how to get this formula.

3.1.3 Explanation of how to find such a function Θ

The main inequality (with $c = \frac{1}{8}$) in terms of Θ becomes a "drift concavity condition":

$$\frac{1}{\sqrt{1 + \frac{(\Delta\tau)^2}{8}}} \Theta\left(\sqrt{1 + \frac{(\Delta\tau)^2}{8}}\, \frac{y_- + y_+}{2}, \frac{1}{\sqrt{1 + \frac{(\Delta\tau)^2}{8}}}\, \frac{\tau_- + \tau_+}{2} \right) \ge \frac{\Theta(y_-, \tau_-) + \Theta(y_+, \tau_+)}{2}, \tag{3.12}$$

where $(y_-, \tau_-), (y_+, \tau_+) \in H$, $0 < \tau_- < \tau_+$, $\Delta\tau := \tau_+ - \tau_-$.

Assuming that Θ is smooth (we will find a smooth function), the infinitesimal version appears; it is a sort of Monge–Ampère relationship with a drift. Namely, the following matrix relationship must hold:

$$\begin{bmatrix} \Theta_{yy} & \Theta_{y\tau} \\ \Theta_{y\tau} & \Theta_{\tau\tau} + \Theta + \tau\Theta_\tau - y\Theta_y \end{bmatrix} \le 0. \tag{3.13}$$

A direct calculation shows that this property is equivalent to the following one. On any curve $y = \phi(\tau)$ lying in the domain H

such that $\phi'' + \tau\phi' + \phi = 0$

we have $\left(\Theta(\phi(\tau),\tau)\right)'' + \tau\left(\Theta(\phi(\tau),\tau)\right)' + \Theta(\phi(\tau),\tau) \le 0.$
$\tag{3.14}$

This hints at the possibility to have a change of variables $(y,\tau) \to (\Gamma, T)$ such that condition (3.13) transforms to a simple concavity. To some extent this is what happens. See the following simple.

Lemma 3.3. *Consider the following change of variables:* $T = \int_0^\tau e^{s^2/2}ds$. *Then* $\phi''(\tau) + \tau\phi'(\tau) + \phi(\tau) \le 0$ *if and only if* $(e^{\tau^2/2}\phi(\tau))_{TT} \le 0$ *and* $\phi''(\tau) + \tau\phi'(\tau) + \phi(\tau) = 0$ *if and only if* $(e^{\tau^2/2}\phi(\tau))_{TT} = 0.$

Proof. Consider $\Phi(T) := \phi(\tau)e^{\tau^2/2}$. Then $\phi(\tau) = \Phi(T)e^{-\tau^2/2}$. So

$$\phi'(\tau) = \Phi(T)T'(\tau)e^{-\tau^2/2} - \tau e^{-\tau^2/2}\Phi(T).$$

But $T'(\tau) = e^{\tau^2/2}$. Hence

$$\phi'(\tau) = \Phi'(T) - \tau\phi(\tau) \quad \Rightarrow \quad \phi''(\tau) = \Phi''(T)e^{-\tau^2/2} - \tau\phi'(\tau) - \phi(\tau).$$

Therefore,

$$\phi''(\tau) + \tau\phi'(\tau) + \phi(\tau) = \Phi''(T)e^{-\tau^2/2},$$

which means that the lemma is proved as both sides are negative simultaneously. $\square$

This lemma hints that the right change of variables should look like

$$\begin{cases} \Gamma := ye^{\tau^2/2}, \\ T = \int_0^\tau e^{s^2/2}ds, \quad (y,\tau) \in \mathbf{R}_+^1 \times \mathbf{R}_+^1. \end{cases} \tag{3.15}$$

Then in the new coordinates the family of curves $y = \phi(\tau)$ such that $\phi'' + \tau\phi' + \phi = 0$ becomes a family of all straight lines $\Gamma = CT + D$. (Notice that both families depend on two arbitrary constants.)

Denote

$$O := \{(\Gamma, T) : (\gamma, \tau) \in G\}.$$

The condition $(\Theta(\phi(\tau), \tau))'' + \tau(\Theta(\phi(\tau), \tau))' + \Theta(\phi(\tau), \tau) \leq 0$ on any of these curves becomes

$$(e^{\tau^2/2}\Theta(\phi(\tau), \tau))_{TT} \leq 0, \quad \Gamma = CT + D, \quad (\gamma, \tau) \in G, \tag{3.16}$$

which is the concavity of $e^{\tau^2/2}\Theta(\gamma, \tau)$ in a new coordinate T along the line $\Gamma = CT + D$. Let us rewrite two functions in the new coordinates:

$$\Phi(\Gamma, T) := \Theta(\gamma, \tau), \quad U(T) := e^{\tau^2/2}.$$

Then (3.16) transforms into

$$\forall C, D \in \mathbf{R}, \quad (U(T)\Phi(CT + D, T))_{TT} \leq 0, \quad (\Gamma, T) \in O. \tag{3.17}$$

This is just a concavity of $U(T)\Phi(\Gamma, T)$ on O of course. Notice that neither H nor O is convex, so we should understand (3.17) as a local concavity in O: just the negativity of its second differential form

$$d^2_{\Gamma, T}(U(T)\Phi(\Gamma, T)) \leq 0, \quad (\Gamma, T) \in O.$$

So we reduce the question to finding a concave function in new coordinates. Now we choose the simplest possible concave function:

$$U(T)\Phi(\Gamma, T) := \min(\Gamma, KT),$$

where the constant $K = K(Q)$ will be chosen momentarily.

If we write down now $\Theta(\gamma, \tau) = \Phi(\Gamma, T)$ in the old coordinates, we get exactly function Θ from (3.11) (we need to define constant K yet), namely,

$$\Theta(\gamma, \tau) := \min\left(\gamma, Ke^{-\tau^2/2}\int_0^\tau e^{s^2/2}ds\right). \tag{3.18}$$

Recall that now we can consider

$$B(u, v, \lambda) = \frac{1}{\sqrt{\lambda}}\Theta\left(u\sqrt{\lambda}, \frac{v}{\sqrt{\lambda}}\right) \tag{3.19}$$

and we are going to apply Theorem 3.1 to it. But we need to choose K to satisfy all the conditions (except the last one) at the beginning of Section 3.1.1.

First of all, it is now very easy to understand why the form of the domain $H = \{1 \le \gamma\tau \le Q\}$ plays a role. In fact, by choosing

$$K = AQ$$

with some absolute constant A, we guarantee that in this domain our function Θ satisfies the obstacle condition

$$\Theta(\gamma, \tau) = \gamma \quad \text{as soon as} \quad \tau \ge a_0 > 0, \tag{3.20}$$

where a_0 is an absolute positive constant. In fact, for all sufficiently small τ, $e^{-\tau^2/2} \int_0^\tau e^{s^2/2} ds \asymp \tau$, and therefore, for all sufficiently small τ (smaller than a certain absolute constant)

$$\Theta(\gamma, \tau) := \min(\gamma, K\tau).$$

The fifth condition at the beginning of Section 3.1.1 (the obstacle condition) requires then that $\min(\gamma, K\tau) = \gamma$ if $\tau \ge a_0 > 0$. But on the upper hyperbola then $\gamma = Q/a_0$ for $\tau = a_0$. We see that the smallest possible K we can choose to satisfy the obstacle condition is $K \asymp Q$.

Secondly, function Θ satisfies the infinitesimal condition (3.13) by construction. But we need to check that the main inequality (3.12) is satisfied as well.

This can be done by the following lemma.

Lemma 3.4. *Inequality (3.12) for function Θ built above holds if and only if the following inequality is satisfied for $\phi(\tau) := e^{-\tau^2/2} \int_0^\tau e^{s^2/2} ds$:*

$$\frac{1}{\sqrt{1 + \frac{(\Delta\tau)^2}{100}}} \phi\left(\frac{\tau_1 + \tau_2}{2\sqrt{1 + \frac{(\Delta\tau)^2}{100}}}\right) \ge \frac{\phi(\tau_1) + \phi(\tau_2)}{2}, \quad \forall 0 < \tau_1 \le \tau_2 \le \tau_0, \tag{3.21}$$

with some absolute positive small constant τ_0.

Proof. Proving the lemma is easy, because we can immediately see that the main inequality (3.12) commutes with the operation of minimum. $\qquad\square$

Now we are going to prove (3.21). To prove it we use Brownian motion in a way that is close to the idea from [3].

Proof. Put

$$U(x, y) := \frac{1}{y}\phi\left(\frac{x}{y}\right), \quad x \ge 0, \, 0 \le \frac{x}{y} \le \tau_0,$$

where τ_0 is a small positive absolute constant. It is easy to see that the claim of (3.21) follows from the following one:

$$U(x,y) \geq \frac{1}{2}\left[U\left(x+a, \sqrt{y^2 - \delta a^2}\right) + U\left(x-a, \sqrt{y^2 - \delta a^2}\right)\right] \qquad (3.22)$$

for a certain small positive absolute constant δ and with $a \leq x \leq \tau_0 y = \frac{1}{1000}y$. Let $W(t)$ be a standard Brownian motion. We consider the stochastic process

$$X(t) := U\left(x + W(t), \sqrt{y^2 - \delta t}\right).$$

By the definition of U we can easily see that

$$\frac{1}{2}U_{xx} - \frac{1}{100}\frac{U_y}{y} \leq 0.$$

Indeed this follows easily from $\phi''(\tau) + \tau\phi'(\tau) + \phi(\tau) \leq 0$. We combine this with Itô's formula to conclude that X_t is supermartingale.

Now consider the stopping time

$$T := \frac{y^2}{4} \wedge \min\{t : W(t) \notin (-a, a)\}.$$

Using the supermartingale property we can write

$$\begin{aligned}
U(x,y) = X(0) \geq\ & \mathbf{E}X(T) \\
=\ & \mathbf{P}(W(T) = -a)\mathbf{E}\left(U\left(x-a, \sqrt{y^2 - \delta T}\right)\Big|W(T) = -a\right) \\
& + \mathbf{P}(W(T) = a)\mathbf{E}\left(U\left(x+a, \sqrt{y^2 - \delta T}\right)\Big|W(T) = a\right) \\
& + \mathbf{P}(|W(T)| < a)\mathbf{E}\left(U\left(x + W(T), \sqrt{y^2 - \delta T}\right)\Big|W(T) < a\right). \qquad (3.23)
\end{aligned}$$

The crucial step is to notice that

$$\left(\frac{y^2}{4}\right)^2 \mathbf{P}(|W(T)| < a) = \left(\frac{y^2}{4}\right)^2 \mathbf{P}\left(|W(T)| < a, T = \frac{y^2}{4}\right) \leq \left(\frac{y^2}{4}\right)^2 \mathbf{P}\left(T \geq \frac{y^2}{4}\right)$$
$$\leq \mathbf{E}T^2 \leq C_4\mathbf{E}W(T)^4 \leq c_4 a^4,$$

where we used the L^4-estimate of $T^{1/2}$ by Burgess Davis [12].

We get

$$\mathbf{P}(|W(T)| < a) \leq \frac{a^4}{y^4} \quad \Rightarrow \quad \mathbf{P}(W(T) = \pm a) \geq \frac{1}{2}\left(1 - c_0\frac{a^4}{y^4}\right).$$

Combining this with (3.23) we get

$$\frac{1}{\sqrt{1 + c_1\frac{a^4}{y^4}}}U(x,y)$$

$$\geq \frac{1}{2}\Big[\mathbf{E}\big(U\big(x-a,\sqrt{y^2-\delta T}\big)|W(T)=-a\big) + \mathbf{E}\big(U\big(x-a,\sqrt{y^2-\delta T}\big)|W(T)=a\big)\Big]$$

$$\geq \frac{1}{2}\Big[\big(U\big(x-a,\sqrt{y^2-\delta\mathbf{E}T}\big)|W(T)=-a\big) + \big(U\big(x+a,\sqrt{y^2-\delta\mathbf{E}T}\big)|W(T)=a\big)\Big]$$

$$= \frac{1}{2}\Big[\big(U\big(x-a,\sqrt{y^2-\delta a^2}\big)\big) + \big(U\big(x+a,\sqrt{y^2-\delta a^2}\big)\big)\Big],$$

because

$$\mathbf{E}(T|W(T)=\pm a)=a^2.$$

Notice that the second inequality, where we move $\mathbf{E}$ from the outside to the inside of the function U, is due to the fact that $t \to \frac{1}{\sqrt{t}}\phi(\frac{x}{\sqrt{t}})$ is convex if $t \in [y^2/4, y^2]$ and x/y is sufficiently small. This is our case as a is seriously smaller than y.

Putting things together we get

$$\frac{1}{\sqrt{1+c_1\frac{a^4}{y^4}}}U(x,y) \geq \frac{1}{2}\Big[\big(U(x-a,\sqrt{y^2-\delta a^2})\big) + \big(U(x+a,\sqrt{y^2-\delta a^2})\big)\Big]. \qquad (3.24)$$

Notice that

$$a \leq \tau_0 y \quad \Rightarrow \quad \frac{\sqrt{y^2-\delta a^2}}{\sqrt{y^2-\frac{1}{2}\delta a^2}} \leq \sqrt{1-\frac{c\delta^2}{a}}. \qquad (3.25)$$

Now we remember that

$$U\big(x-a,\sqrt{y^2-\delta a^2}\big) = \frac{1}{\sqrt{y^2-\delta a^2}}\phi\bigg(\frac{x-a}{\sqrt{y^2-\delta a^2}}\bigg).$$

Thus, if we denote

$$Y_1 := \frac{x-a}{\sqrt{y^2-\delta a^2}}, \quad Y_2 := \frac{x+a}{\sqrt{y^2-\delta a^2}},$$

$$(x-a)U\big(x-a,\sqrt{y^2-\delta a^2}\big) = Y_1\phi(Y_1), \quad (x+a)U\big(x+a,\sqrt{y^2-\delta a^2}\big) = Y_2\phi(Y_2).$$

Let us denote

$$Y_3 := \frac{x-a}{\sqrt{y^2-\frac{1}{2}\delta a^2}}, \quad Y_4 := \frac{x+a}{\sqrt{y^2-\frac{1}{2}\delta a^2}}.$$

Then $Y_1 > Y_3$, $Y_2 > Y_4$. Hence, as ϕ increases near 0, we get $Y_1\phi(Y_1) > \frac{Y_1}{Y_3}Y_3\phi(Y_3)$, $Y_2\phi(Y_2) > \frac{Y_2}{Y_4}Y_4\phi(Y_4)$. We can write this as follows:

$$U\Big(x - a, \sqrt{y^2 - \delta a^2}\Big) \geq \Big(\sqrt{y^2 - \tfrac{1}{2}\delta a^2} \big/ \sqrt{y^2 - \delta a^2}\Big)U\Big(x - a, \sqrt{y^2 - \tfrac{1}{2}\delta a^2}\Big),$$

$$U\Big(x + a, \sqrt{y^2 - \delta a^2}\Big) \geq \Big(\sqrt{y^2 - \tfrac{1}{2}\delta a^2} \big/ \sqrt{y^2 - \delta a^2}\Big)U\Big(x + a, \sqrt{y^2 - \tfrac{1}{2}\delta a^2}\Big).$$

Let us combine these inequalities with (3.24) to get

$$\frac{1}{\sqrt{1 + c_1 \tfrac{a^4}{y^4}}} U(x,y) \geq \frac{\sqrt{y^2 - \tfrac{1}{2}\delta a^2}}{\sqrt{y^2 - \delta a^2}} \frac{1}{2}\Big[U\Big(x - a, \sqrt{y^2 - \tfrac{1}{2}\delta a^2}\Big) + U\Big(x + a, \sqrt{y^2 - \tfrac{1}{2}\delta a^2}\Big)\Big].$$

Using (3.25) we obtain

$$\frac{1}{\sqrt{1 + c_1 \tfrac{a^4}{y^4}}} U(x,y) \geq \frac{1}{\sqrt{1 + c \tfrac{a^2}{y^2}}} \frac{1}{2}\Big[U\Big(x - a, \sqrt{y^2 - \tfrac{1}{2}\delta a^2}\Big) + U\Big(x + a, \sqrt{y^2 - \tfrac{1}{2}\delta a^2}\Big)\Big].$$

As $a/y \leq \tau_0$ with small τ_0 we conclude that $(a/y)^2$ trumps $(a/y)^4$, and as a result we obtain

$$U(x,y) \geq \frac{1}{2}\Big[U\Big(x - a, \sqrt{y^2 - \tfrac{1}{2}\delta a^2}\Big) + U\Big(x + a, \sqrt{y^2 - \tfrac{1}{2}\delta a^2}\Big)\Big].$$

Inequality (3.21) is fully proved. $\qquad\square$

4 Restricted weak weighted estimates for the square function

Estimating the weighted square function on functions $\mathbf{1}_I$ still does not give us the restricted weak weighted estimate. To get the restricted weak weighted estimate we need to consider the action of the weighted square function on $\mathbf{1}_E$, where E is a measurable set. This is what we will be doing now. The approach below is quite different from the Bellman function approach we used above. We follow the exposition in [23].

Our measure space will be $(X, \mathfrak{A}, dx)$, where the σ-algebra $\mathfrak{A}$ is generated by a standard dyadic filtration $\mathcal{D} = \bigcup_k \mathcal{D}_k$ on $\mathbf{R}$. We considered the weak weighted estimate for the martingale transform. In [36] the endpoint exponent was $p = 1$, and critical weights belong to the A_1-class.

We are going to consider weighted weak estimates for the dyadic square function. The critical exponent is now $p = 2$, and critical weights belong to the A_2-class. See the estimates for subcritical and supercritical exponents in [27, 14].

Recall that the symbol $\mathrm{ch}(J)$ denotes the dyadic children of J. Recall that the martingale difference operator Δ_J is defined as follows:

$$\Delta_J f := \sum_{I \in ch(J)} \mathbf{1}_I (\langle f \rangle_I - \langle f \rangle_J).$$

For our case of a dyadic lattice on the line, $|\Delta_J f|$ is constant on J and

$$\Delta_J f = \frac{1}{2}[(\langle f \rangle_{J_+} - \langle f \rangle_{J_-})\mathbf{1}_{J_+} + (\langle f \rangle_{J_-} - \langle f \rangle_{J_+})\mathbf{1}_{J_-}].$$

The square function operator is

$$(Sf)^2(x) = \sum_{J \in \mathcal{D}} |\Delta_J f|^2 \mathbf{1}_J(x).$$

In this section we work only with the dyadic A_2-classes of weights, but we skip the word dyadic, because we consider here only dyadic operators. We consider a positive function $w(x)$, and as before we call it A_2-weight if

$$Q := [w]_{A_2} := \sup_{J \in \mathcal{D}} \langle w \rangle_J \langle w^{-1} \rangle_J < \infty. \tag{4.1}$$

We already considered the restricted weak estimate when the operator is applied to $f = \mathbf{1}_E$, but only for set E being itself a dyadic interval:

$$\frac{1}{|I|} w\left\{ x \in I : \left(\sum_{J \in D(I)} |\Delta_J(w^{-1})|^2 \mathbf{1}_J(x) \right)^{1/2} > \lambda \right\} \le C_{\text{weak},T}([w]_{A_2}) \frac{\langle w^{-1} \rangle_I}{\lambda}. \tag{4.2}$$

In Section 8 we give some information on the full weak estimate for the square function operator:

$$\frac{1}{|I|} w\left\{ x \in I : \left(\sum_{J \in D(I)} |\Delta_J(\varphi w^{-1})|^2 \mathbf{1}_J(x) \right)^{1/2} > \lambda \right\} \le C_{\text{weak}}([w]_{A_2}) \frac{\langle \varphi^2 w^{-1} \rangle_I}{\lambda^2}. \tag{4.3}$$

Here φ runs over all functions such that $\operatorname{supp} \varphi \subset I$ and $\varphi \in L^2(I, w\,dx)$, $w \in A_2$.

The constant $[C_{\text{weak}}([w]_{A_2})]^{1/2}$ gives the dependence on $[w]_{A_2}$ of the norm of the sublinear operator $S\varphi := (\sum_{J \in \mathcal{D}} |\Delta_J \varphi|^2 \mathbf{1}_J(x))^{1/2}$ from $L^2(w)$ to $L^{2,\infty}(w)$. The constant $[C_{\text{weak},T}([w]_{A_2})]^{1/2}$ gives the dependence on $[w]_{A_2}$ of the same weak norm but measured only on test functions φ that are just characteristic functions of dyadic intervals: $\varphi = \mathbf{1}_I$, $I \in \mathcal{D}$. The constant $[C_{\text{weak},R}([w]_{A_2})]^{1/2}$ gives the dependence on $[w]_{A_2}$ of the same weak norm but measured only on test functions $\varphi = \mathbf{1}_E$ for all measurable sets E (in particular any union of disjoint intervals).

In Section 3 we found the (sharp) estimate

$$C_{\text{weak},T}([w]_{A_2}) \lesssim [w]_{A_2}.$$

In this section we find the sharp dependence on $[w]_{A_2}$ of the restricted weak type:

$$w\left\{x \in I : \left(\sum_{J \in D(I)} |\Delta_J(\mathbf{1}_E w^{-1})|^2 \mathbf{1}_J(x)\right)^{1/2} > \lambda\right\} \leq C_{\mathrm{weak},R}([w]_{A_2}) \frac{\int \mathbf{1}_E w^{-1}}{\lambda^2}. \qquad (4.4)$$

Obviously,

$$C_{\mathrm{weak},T}([w]_{A_2}) \leq C_{\mathrm{weak},R}([w]_{A_2}) \leq C_{\mathrm{weak}}([w]_{A_2}).$$

Below we prove that $C_{\mathrm{weak},R}([w]_{A_2}) \lesssim [w]_{A_2}$. It is a sharp estimate. This is a stronger result than what we had in Section 3. In Section 3 we used the PDE approach, now we will use the stopping time/sparse domination approach.

Multiplying w by λ we do not change $[w]_{A_2}$, but we can renormalize the problem and think that $\lambda = 1$.

4.0.1 Sparse square function operators

The method of sparse domination was introduced by Andrei Lerner [28, 31, 30, 29, 32] and many applications were found; for example, a very short proof of the A_2-conjecture was found in [30, 25] by using this method. We use this method here and in Section 8. We start with the following definition.

Definition. A family $\mathcal{S}$ of intervals of $\mathcal{D}$ is called ε-sparse if the following condition is satisfied:

$$\sum_{\substack{I \in \mathcal{S}, \\ I \subsetneq J}} |I| \leq \varepsilon|J|, \quad \forall J \in \mathcal{S}. \qquad (4.5)$$

Definition. The sparse square function operator is defined for each sparse family $\mathcal{S}$ as follows:

$$S^{\mathrm{sp}}\varphi \stackrel{\mathrm{def}}{=} S^{\mathrm{sp}}_{\mathcal{S}}\varphi \stackrel{\mathrm{def}}{=} \left(\sum_{I \in \mathcal{S}} \langle\varphi\rangle_I^2 \mathbf{1}_I\right)^{1/2}.$$

Theorem 4.1. *For any $\varepsilon > 0$ and any $\varphi \in L^1$ there exist a constant $C = C(\varepsilon)$ independent of φ and a sparse family $\mathcal{S}$ (depending on ε and on φ) such that pointwise almost everywhere*

$$S\varphi \leq CS^{\mathrm{sp}}\varphi.$$

Proof. It is well known (see, e.g., [50]) that the square function operator is weakly bounded in unweighted L^1. Let us call A the norm of the operator S from L^1 to $L^{1,\infty}$. Fix ε and let $C = 100A/\varepsilon$. We start with interval $I_0 = (0,1)$, put $\mathcal{S}_0 \stackrel{\mathrm{def}}{=} \{I_0\}$, and define the first generation of stopping intervals $\mathcal{S}_1$ as follows: $Q \in \mathcal{S}_1$ if it is the maximal interval in I_0 such that

$$S_Q^{I_0} \varphi \overset{\text{def}}{=} \left(\sum_{\substack{I \in \mathcal{D}, \\ Q \subset I \subset I_0}} (\Delta_I \varphi)^2 \right)^{1/2} > C \langle |\varphi| \rangle_{I_0}.$$

The second generation of stopping intervals $\mathcal{S}_2$ will be nested inside the first generation $\mathcal{S}_1$. For every $I \in \mathcal{S}_1$ we define its subintervals from $\mathcal{D}$ by the same rule as before, but with I playing the role of I_0. Namely, we define the first generation of stopping intervals $\mathcal{S}_2$ inside $I \in \mathcal{S}_1$ as follows: $Q \in \mathcal{S}_2$ if it is the maximal interval in I such that

$$S_Q^{I} \varphi \overset{\text{def}}{=} \left(\sum_{\substack{J \in \mathcal{D}, \\ Q \subset J \subset I}} (\Delta_J \varphi)^2 \right)^{1/2} > C \langle |\varphi| \rangle_{I}.$$

We continue the construction of generations of intervals $\mathcal{S}_3, \mathcal{S}_4, \dots$ recursively, and we put $\mathcal{S} \overset{\text{def}}{=} \bigcup_{k=0}^{\infty} \mathcal{S}_k$.

Notice that by the fact that the operator S and the dyadic maximal operator M are weakly bounded in unweighted L^1, from our choice of constant C at the beginning of the proof we get $\sum_{Q \in \mathcal{S}_1} |Q| \leq \frac{\varepsilon}{50}|I_0|$, and similarly,

$$\sum_{\substack{Q \in \mathcal{S}_{k+1}, \\ Q \subset I}} |Q| \leq \frac{\varepsilon}{50}|I|, \quad \forall I \in \mathcal{S}_k.$$

Obviously, and with a good margin, we found that $\mathcal{S}$ is ε-sparse.

Now to see the pointwise estimate of the theorem, let us notice that given $x \in I_0$, which is not an endpoint of any dyadic interval, we will be able to find the tower of intervals $\cdots \subsetneq I_k \subsetneq I_1 \subsetneq I_0$ such that x is contained in all of them and such that $I_k \in \mathcal{S}_k$. This tower may degenerate to just one interval I_0, or it can be an infinite tower. But the set of points for which the tower is infinite has Lebesgue measure zero. This is clear from the fact that $\mathcal{S}$ is sparse. In any case,

$$S^2 \varphi(x) = \sum_{\substack{I \in \mathcal{D}, \\ I_1 \subsetneq I \subset I_0}} (\Delta_I \varphi)^2 + \sum_{\substack{I \in \mathcal{D}, \\ I_2 \subsetneq I \subset I_1}} (\Delta_I \varphi)^2 + \sum_{\substack{I \in \mathcal{D}, \\ I_3 \subsetneq I \subset I_2}} (\Delta_I \varphi)^2 + \cdots.$$

But then, using our stopping criterion we see that the last expression is bounded by

$$C^2 \left(\langle |\varphi| \rangle_{I_0}^2 + \langle |\varphi| \rangle_{I_1}^2 + \langle |\varphi| \rangle_{I_2}^2 + \cdots \right),$$

which proves the theorem. $\qquad\square$

Remark. The following estimate was proved in [14]:

$$w\{x \in I_0 : S^{\mathrm{sp}}(\varphi w^{-1}) > 3\} \le A[w]_{A_2} \log(1 + [w]_{A_2}) \int_{I_0} \varphi^2 w^{-1}\, dx. \tag{4.6}$$

It can be that $\log(1 + [w]_{A_2})$ can be deleted. It is a very good estimate but the problem is that nobody knows whether it is sharp. The proof that the logarithm cannot be deleted probably requires a very sophisticated weight concrete w or a very sophisticated existence theorem for such weight. We will present the reasoning from [14] in Section 8.

Remark. Now we will see what can be changed in the reasoning of [14] in order to obtain an estimate better than (4.6) for a special choice of $\varphi = w^{-1}$. In fact, the Bellman function technique of the previous section gave us a better estimate in this particular case $\varphi = w^{-1}$:

$$w\{x \in I_0 : S^{\mathrm{sp}} w^{-1} > 3\} \le A[w]_{A_2} \int_{I_0} w^{-1}\, dx. \tag{4.7}$$

Theorem 4.2. *Let $w \in A_2$ and S be a sparse collection of dyadic intervals in $\mathcal{D}(I_0)$. Let S^{sp} be the sparse square function operator built on this collection. Then the restricted weak type of the operator $S^{\mathrm{sp}}_{w^{-1}}$ from $L^2(w^{-1})$ to $L^{2,\infty}(w)$ is bounded by $A[w]^{1/2}_{A_2}$, where A is an absolute constant.*

We need the following well-known result.

Lemma 4.3. *Let M be the dyadic maximal operator and let w be in dyadic A_2. Then the norm of M from $L^2(w)$ to $L^{2,\infty}(w)$ is bounded by $[w]^{1/2}_{A_2}$.*

Proof. Given a test function $\varphi \ge 0$, let $\{I\}$ be the maximal dyadic intervals for which $M\varphi > 1$. Then

$$\sum_I w(I) \le \int \sum_I \frac{w(I)}{|I|} \varphi \mathbf{1}_I\, dx = \int \sum_I \frac{w(I)}{|I|} \varphi \mathbf{1}_I\, w^{-1/2} w^{1/2} dx$$

$$\le \left(\sum_I \left(\frac{w(I)}{|I|} \right)^2 w^{-1}(I) \right)^{1/2} \|f\|_w \le [w]^{1/2}_{A_2} \left(\sum_I w(I) \right)^{1/2} \|\varphi\|_w.$$

Hence,

$$(w\{x : M\varphi > 1\})^{1/2} = \left(\sum_I w(I) \right)^{1/2} \le [w]^{1/2}_{A_2} \|\varphi\|_w,$$

which is precisely what the lemma claims. $\qquad\Box$

Remark. One can skip the word "dyadic" everywhere in the statement of this lemma. Also, one can generalize the statement to $\mathbf{R}^n$. Then the lemma remains trues, only the estimate becomes $C_n [w]^{1/2}_{A_2}$. See [50].

From now on let us assume that $0 \le \varphi \le w^{-1}\mathbf{1}_{I_0}$.

As in [14], let $\mathcal{S}$ be an ε-sparse system of dyadic subintervals of I_0 and let $\mathcal{S}^{\mathrm{sp}}$ be a corresponding sparse square function. We split $\mathcal{S} = \mathcal{S}^0 \cup \bigcup_{m=0}^{\infty} \mathcal{S}_m$, where $\mathcal{S}^0$ consists of intervals of $\mathcal{S}$ such that $\langle \varphi \rangle_I > 1$ and $\mathcal{S}_{m+1}$ consists of intervals of $\mathcal{S}$ such that

$$2^{-m-1} < \langle |\varphi| \rangle_I \le 2^{-m}, \quad m = 0, 1, \ldots. \tag{4.8}$$

We denote

$$S_m^{\mathrm{sp}}\varphi \stackrel{\mathrm{def}}{=} \left(\sum_{I \in \mathcal{S}_m} \langle \varphi \rangle_I^2 \mathbf{1}_I \right)^{1/2}.$$

We are going to estimate these measures:

$$W_0 \stackrel{\mathrm{def}}{=} w\{x \in I_0 : (S_0^{\mathrm{sp}}\varphi)^2 > 1\},$$

$$W \stackrel{\mathrm{def}}{=} w\left\{ x \in I_0 : \sum_{m=0}^{\infty} (S_m^{\mathrm{sp}}\varphi)^2 > 4 \right\}.$$

The estimate of W_0 is easy. The sum $(S_0^{\mathrm{sp}}\varphi)^2$ is supported on intervals where the dyadic maximal function of φ is bigger than 1. The set where the dyadic maximal function is bigger than 1 has w-measure bounded by $[w]_{A_2}\|\varphi\|_w^2$ by Lemma 4.3.

Now we work with W, and we start exactly as in the estimate of W above. Namely, the support of $S_m^{\mathrm{sp}}\varphi$ is in $\cup Q_i^{(m)}$, where $Q_i^{(m)} \in \mathcal{S}_m$, which are the maximal dyadic intervals Q with the property

$$\sum_{I \in \mathcal{S}_m, Q \subset I \subset I_0} \langle |\varphi| \rangle_I^2 \mathbf{1}_I \ge 2^{-m}. \tag{4.9}$$

By maximality and by (4.8),

$$\sum_{I \in \mathcal{S}_m, Q_i^{(m)} \subset I \subset I_0} \langle |\varphi| \rangle_I^2 \mathbf{1}_I \le 2^{-m} + 2^{-2m}. \tag{4.10}$$

To estimate W we use a union bound:

$$w\left\{ x \in I_0 : \sum_{m=0}^{\infty} (S_m^{\mathrm{sp}}\varphi)^2 > 4 \right\}$$

$$\le w\left\{ x \in I_0 : \sum_{m=0}^{\infty} (S_m^{\mathrm{sp}}\varphi)^2 > \sum_{m=0}^{\infty} 2^{-m/2} \right\}$$

$$\le \sum_{m=0}^{\infty} w\left\{ x \in \bigcup_i Q_i^{(m)} : (S_m^{\mathrm{sp}}\varphi)^2 > 2^{-m/2} \right\}$$

$$\leq \sum_{m=0}^{\infty} \sum_{i} w\{x \in Q_i^{(m)} : (S_m^{\mathrm{sp}}\varphi)^2 > 2^{-m/2}\}.$$

Now let us fix $Q_i^{(m)}$. Then

$$\sum_{I \in \mathcal{S}_m,\, Q_i^{(m)} \subset I \subset I_0} \langle |\varphi| \rangle_I^2 \mathbf{1}_I \geq 2^{-m}.$$

On this interval $\langle |\varphi| \rangle_I^2 \leq 2^{2m}$ by the definition of $\mathcal{S}_m$. Hence,

$$\sum_{I \in \mathcal{S}_m,\, Q_i^{(m)} \subset I \subset I_0} \mathbf{1}_I \geq 2^m.$$

Therefore, every $Q_i^{(m)}$ lies at least 2^m generations down in $\mathcal{S}$. By sparse choice

$$\left| \bigcup_i Q_i^{(m)} \right| \leq e^{-a2^m}. \tag{4.11}$$

Consider maximal dyadic intervals $\ell \in \mathcal{S}_m, \ell \subset Q_i^{(m)}$, on which $(S_m^{\mathrm{sp}}\varphi)^2 > 2^{-m/2}$. Call them $\mathcal{L}(Q_i^{(m)})$. Let us estimate $\sum_{\ell \in \mathcal{L}(Q_i^{(m)})} |\ell|$.

Notice that for each $\ell \in \mathcal{L}(Q_i^{(m)})$

$$\sum_{I \in \mathcal{S}_m,\, \ell \subset I \subset Q_i^{(m)}} \langle |\varphi| \rangle_I^2 \mathbf{1}_I \geq 2^{-m/2} - 2^{-m} - 2^{-2m} \geq \frac{1}{2} 2^{-m/2}.$$

This is by (4.10). Now we use again the fact that $\langle |\varphi| \rangle_I^2 \leq 2^{2m}$ for all terms of this sum. Hence,

$$\sum_{I \in \mathcal{S}_m,\, \ell \subset I \subset Q_i^{(m)}} \mathbf{1}_I \geq 2^{-m/2} - 2^{-m} - 2^{-2m} \geq 2^{-m}.$$

Therefore, every ℓ from $\mathcal{L}(Q_i^{(m)})$ lies at least 2^m generations down from $Q_i^{(m)}$ in $\mathcal{S}_m$, and, thus, in $\mathcal{S}$. By sparse construction

$$\sum_{\ell \in \mathcal{L}(Q_i^{(m)})} |\ell| \leq e^{-a2^m} |Q_i^{(m)}|. \tag{4.12}$$

Recall that we assumed

$$\varphi \leq w^{-1}.$$

Hence,

$$\langle w^{-1} \rangle_\ell^{-1} \leq \langle \varphi \rangle_\ell^{-1}.$$

By definition $\langle \varphi \rangle_\ell \geq 2^{-m-1}$; hence,

$$\langle w \rangle_\ell \leq [w]_{A_2} \langle w^{-1} \rangle_\ell^{-1} \leq [w]_{A_2} \langle \varphi \rangle_\ell^{-1} \leq 2^{m+1}[w]_{A_2}, \quad \forall \ell \in \mathcal{L}(Q_i^{(m)}).$$

Therefore, by (4.12) we obtain

$$w\{x \in Q_i^{(m)} : (S_m^{\mathrm{sp}}\varphi)^2 > 2^{-m/2}\} = \sum_{\ell \in \mathcal{L}(Q_i^{(m)})} w(\ell)$$

$$= \sum_{\ell \in \mathcal{L}(Q_i^{(m)})} \langle w \rangle_\ell |\ell| \leq 2^{m+1}[w]_{A_2} \sum_{\ell \in \mathcal{L}(Q_i^{(m)})} |\ell| = 2^{m+1}[w]_{A_2} e^{-a2^m} |Q_i^{(m)}|.$$

We can conclude now that

$$w\{x \in Q_i^{(m)} : (S_m^{\mathrm{sp}}\varphi)^2 > 2^{-m/2}\} \leq 2^{m+1}[w]_{A_2} e^{-a2^m} |Q_i^{(m)}|$$

$$\leq 2^{2m+2}[w]_{A_2} e^{-a2^m} \langle \varphi \rangle_{Q_i^{(m)}} |Q_i^{(m)}|$$

$$= 2^{2m+2}[w]_{A_2} e^{-a2^m} \int_{Q_i^{(m)}} \varphi \, dx, \tag{4.13}$$

where the last inequality follows from the definition of S_m and the fact that $Q_i^{(m)} \in S_m$. Gathering things together, we obtain

$$w\left\{x \in I_0 : \sum_{m=0}^{\infty} (S_m^{\mathrm{sp}}\varphi)^2 > 4\right\}$$

$$\leq \sum_{m=0}^{\infty} \sum_i w\{x \in Q_i^{(m)} : (S_m^{\mathrm{sp}}\varphi)^2 > 2^{-m/2}\}$$

$$\overset{(4.13)}{\leq} [w]_{A_2} \sum_{m=0}^{\infty} \sum_i 2^{2m+2} e^{-a2^m} \int_{Q_i^{(m)}} \varphi \, dx$$

$$\leq [w]_{A_2} \sum_{m=0}^{\infty} 2^{2m+2} e^{-a2^m} \int_{\cup Q_i^{(m)}} \varphi \, dx$$

$$\leq [w]_{A_2} \int_{I_0} \varphi \, dx \sum_{m=0}^{\infty} 2^{2m+2} e^{-a2^m} \leq A[w]_{A_2} \int_{I_0} \varphi \, dx.$$

Estimate

$$w\left\{x \in Q_i^{(m)} : \sum_{m=0}^{\infty} (S_m^{\mathrm{sp}}\varphi)^2 > 4\right\} \leq A[w]_{A_2} \int_{I_0} \varphi \, dx \tag{4.14}$$

was just obtained by using one assumption: $\varphi \le w^{-1}$. We want to exchange in the right-hand side of (4.14) the integral $\int_{I_0} \varphi \, dx$ for the integral $\int_{I_0} \varphi^2 w \, dx$. This is trivial if one more property of φ holds, namely, if pointwise

$$\varphi \le C\varphi^2 w.$$

This is compatible with $\varphi \le w^{-1}$ if for a. e. point $x \in I_0$ one of the following properties holds: for every $x \in I_0$ either (1) $\frac{1}{C} w^{-1}(x) \le \varphi(x) \le w^{-1}(x)$ or (2) $\varphi(x) = 0$.

We conclude that (4.14) holds for every φ of the form $\varphi = w^{-1} \mathbf{1}_E$, where E is a measurable subset of I_0. In particular, Theorem 4.2 is proved.

Now we can use the sparse domination theorem, Theorem 4.1. This theorem gives us the following restricted weak type for weak estimates of square functions.

Theorem 4.4. *Let $w \in A_2$ and S be a sparse collection of dyadic intervals. Let S be the dyadic square function operator. Then the restricted weak type of the operator $S_{w^{-1}}$ from $L^2(w^{-1})$ to $L^{2,\infty}(w)$ is bounded by $A[w]_{A_2}^{1/2}$, where A is an absolute constant.*

5 Sharp weak weighted estimate for martingale transform

In [36] and [37] and in Section 6 we approach the estimate from below of the weak norm $L^1(w) \to L^{1,\infty}(w)$ of the martingale transform via the Bellman function technique. This approach, however, does not give the sharp estimate $[w]_{A_1} \log(1 + [w]_{A_1})$ from below; instead, the estimate from below had the form $[w]_{A_1} (\log(1 + [w]_{A_1}))^{1/3}$.

But the Bellman approach has a small advantage. It proves $[w]_{A_1} \log^{1/3}(1 + [w]_{A_1})$ not just for the martingale transform, but for a concrete dyadic shift; see [36].

In [34] the estimate

$$\|H : L^1(w) \to L^{1,\infty}(w)\| \le C[w]_{A_1} \log(1 + [w]_{A_1})$$

was proved. The proof of sharpness of this estimate, namely, the construction of the weight w such that

$$\|H : L^1(w) \to L^{1,\infty}(w)\| \ge c[w]_{A_1} \log(1 + [w]_{A_1}),$$

is very tough; it can be found in [33].

In the present section we transfer this construction of A. Lerner, F. Nazarov, and S. Ombrosi [33] from the Hilbert transform to the martingale transform. The Rubio de Francia technique is used in proving the existence of weights that give the sharp asymptotic estimate $[w]_{A_1} \log(1 + [w]_{A_1})$ from below. As a result of the application of the Rubio de Francia extrapolation technique, the weights are given by rather implicit construction.

The reader is warned not to think that the weights w_n below provide the desired sharp estimate of the type $[w]_{A_1} \log(1 + [w]_{A_1})$. This is not the case. The constructive weights w_n below provide us with the starting point for a certain iterative procedure, which, after being finished, gives us the desired "worst possible" weights. We follow here the exposition of [24].

In this section, A_1 and A_2 denote dyadic classes of weights, and all operators below, including the maximal operator, are dyadic too. We omit the superscript d having this in mind.

5.1 Construction of special weights $w_n \in A_2$

In this section we adapt the proof of [33] for the case of the martingale transform. The main issue is to choose the signs of the martingale transform. This should be done consistently simultaneously for all points, where we estimate the transform from below.

For a dyadic interval I we denote by I^- and I^+ its left and right children, respectively.

Definition 5.1. We also denote (see Figure 3)

$$I_0 = I, \; I_1 = I^{++}, \; I_2 = I_1^{++}, \ldots, \; I_{m-1} = I_{m-2}^{++}, \quad m = 2, \ldots, k,$$

and $C(I) \overset{\text{def}}{=} \{I_m\}_{m=0}^{k-2}$.

We put

$$\varepsilon = 4^{-k}.$$

We fix a large number p (which will be $\asymp 1/\varepsilon$) and we build the sequence of weights by the following rule. Let ω, σ be two numbers such that $\omega\sigma = p$. We put

$$w_0(\omega, \sigma, I) = \frac{\omega}{\sqrt{p}}\left(\left(\sqrt{p} - \sqrt{p-1}\right)\chi_{I_-} + \left(\sqrt{p} + \sqrt{p-1}\right)\chi_{I_+}\right),$$

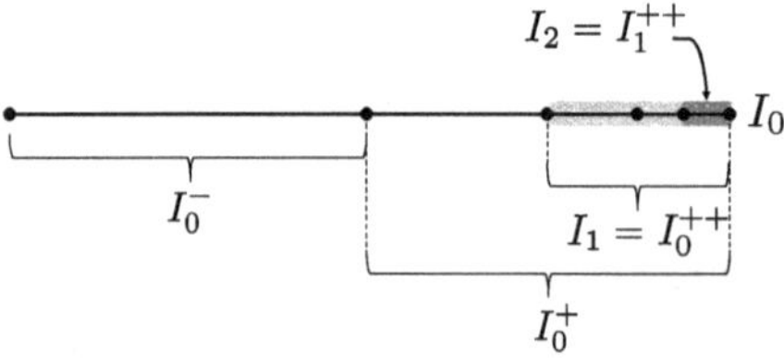

Figure 3: Intervals I_m.

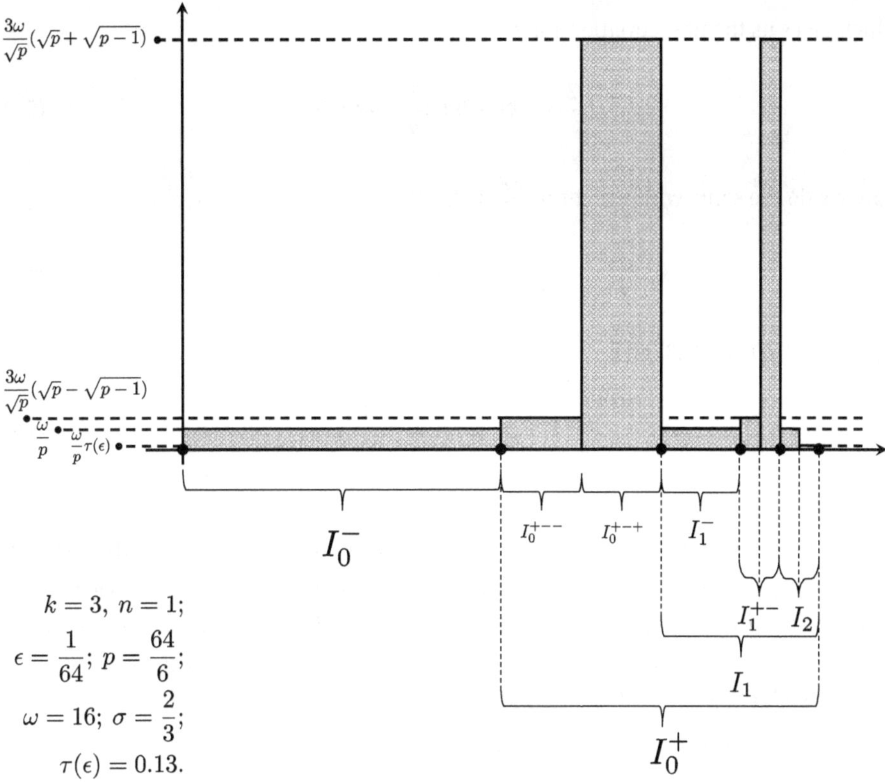

Figure 4: Weight w_1.

$$w_n(\omega,\sigma,I) = \sum_{m=0}^{k-2} w_{n-1}(3\omega,\sigma/3,I_m^{+-}) + \frac{\omega}{p}\left(\sum_{m=0}^{k-2} \chi_{I_m^-} + \chi_{I_{k-1}^-} + \tau(\varepsilon)\chi_{I_{k-1}^+} \right), \qquad (5.1)$$

where $\tau(\varepsilon) = \frac{9\varepsilon}{1+5\varepsilon}$ is chosen to satisfy Lemma 5.2 below. We illustrate weight w_1 by Figure 4.

Lemma 5.2. *We have* $\langle w_n(\omega,\sigma,I)\rangle_I = \omega$, $\langle w_n^{-1}(\omega,\sigma,I)\rangle_I = \sigma$.

Proof. The proof is by induction. For $n = 0$ the statement is clearly valid. Assume the statement is proved for $n-1$. To prove it for n, we notice that the value $\frac{\omega}{p}$ of w_n happens in different places: (1) it happens on measure $\frac{1}{2}(|I_0|+\cdots|I_{k-2}|) = \frac{1}{2}(1+1/4+\cdots+1/4^{k-2})|I|$, that is, on $1/2\frac{1-4\varepsilon}{3/4}|I|$, which is $\frac{2}{3}(1-4\varepsilon)|I|$; (2) $\frac{\omega}{p}$ happens also on measure $\frac{1}{2}|I_{k-1}| = 2\varepsilon|I|$. So, in total, the value $\frac{\omega}{p}$ is assigned to measure $\frac{2}{3}(1-4\varepsilon+3\varepsilon)|I| = \frac{2}{3}(1-\varepsilon)|I|$. The value $\frac{\omega\tau}{p}$ is assigned to measure $2\varepsilon|I|$. The total amount of measure left after these assignments is $(\frac{1}{3}-\frac{4}{3}\varepsilon)|I|$. Using the induction hypothesis, we find that the average of w_n over I is

$$3\left(\frac{1}{3}-\frac{4}{3}\varepsilon\right) + \frac{2}{3}(1-\varepsilon)\frac{1}{p} + 2\varepsilon\frac{\tau}{p} = 1,$$

which gives us the first equation for τ, p:

$$\left(\frac{2}{3}(1-\varepsilon)+2\varepsilon\tau\right)\frac{1}{p}-4\varepsilon=0. \tag{5.2}$$

Now we do the same with weight $\sigma_n \overset{\text{def}}{=} w_n^{-1}$:

$$\frac{\sigma}{3}\left(\frac{1}{3}-\frac{4}{3}\varepsilon\right)+\sigma\frac{2}{3}(1-\varepsilon)+\frac{\sigma}{\tau}2\varepsilon=\sigma. \tag{5.3}$$

Equations (5.2) and (5.3) give

$$\tau=\frac{9\varepsilon}{1+5\varepsilon}, \quad p\approx\frac{1}{6\varepsilon}. \tag{5.4}$$

$\square$

Definition 5.3. Consider $w_n(\omega,\sigma,I)$ introduced above. Interval I is called the forming interval of w_n, $\Phi_0(I) \overset{\text{def}}{=} \{I\}$. We denote by $\Phi_1(I)$ the collection $\{I_m^{+-}\}_{I_m \in C(I)}$. We call this collection the intervals forming w_{n-1}. Now we define $\Phi_k(I)$ as the collection $\Phi_1(J)$, where J runs over the collection $\Phi_{k-1}(I)$. We call $\Phi_k(I)$ the collection forming w_{n-k}. Also, we denote by $\text{supp}(w_{n-k})$ the set $\bigcup_{J \in \Phi_k(I)} J$, that is, the union of intervals forming w_{n-k}.

In what follows n will be chosen as

$$n=4^k. \tag{5.5}$$

Notice that

$$p \asymp 4^k. \tag{5.6}$$

Lemma 5.4. *We have* $[w_n(\omega,\sigma,I)]_{A_2^d(I)} \approx p^2$.

Proof. Without loss of generality we can assume that $I=[0,1)$. Take any dyadic subinterval J of I. First we consider the case when the right endpoint of J coincides with the right endpoint of $[0,1)$. Suppose that $|J|=2^{-\ell}$, where $\ell \geq 1$ is even, say $\ell=2r$. If $r=k-1$, then

$$\langle w_n\rangle_J\langle w_n^{-1}\rangle_J=\frac{1}{4}(1+\tau)\left(1+\frac{1}{\tau}\right)\approx\frac{1}{\varepsilon}\approx p.$$

Further assume $r<k-1$. Then the only m's which participate in (5.1) are $m \geq r$. Let $c \overset{\text{def}}{=} \varepsilon 4^r$. Then $\varepsilon \leq c \leq 1/16$. Clearly,

$$\langle w_n\rangle_J=\frac{\omega}{p}\cdot\frac{1}{2}\cdot\frac{1-\varepsilon 4^{r+1}}{3/4}+\frac{\omega}{p}\cdot 2\varepsilon 4^r+\frac{\omega\tau}{p}\cdot 2\varepsilon 4^r+3\omega\cdot\frac{1}{4}\cdot\frac{1-\varepsilon 4^{r+1}}{3/4}\leq 100\frac{\omega}{p}(1+c),$$

$$\langle w_n^{-1}\rangle_J = \frac{p}{\omega}\cdot\frac{1}{2}\cdot\frac{1-\varepsilon 4^{r+1}}{3/4} + \frac{p}{\omega}\cdot 2\varepsilon 4^r + \frac{p}{\omega\tau}\cdot 2\varepsilon 4^r + \frac{\sigma}{3}\cdot\frac{1}{4}\cdot\frac{1-\varepsilon 4^{r+1}}{3/4} \le 100\frac{p}{\omega}\Big(1+\frac{c}{\tau}\Big),$$

and we find that $\langle w_n\rangle_J\langle w_n^{-1}\rangle_J$ is at most of the order of $\varepsilon^{-1}\asymp p$.

Set $J = 2^{-\ell}$, where ℓ is odd. Choose the smallest integer r so that $2r > \ell$. We can estimate

$$\langle w_n\rangle_J\langle w_n^{-1}\rangle_J \le \frac{1}{2}\Big(3\omega + 100\frac{\omega}{p}(1+c)\Big)\frac{1}{2}\Big(\frac{\sigma}{3} + 100\frac{p}{\omega}\Big(1+\frac{c}{\tau}\Big)\Big)$$

and we see that in the cross product there is a term $p(1+\frac{c}{\tau})$. It is at most Cp^2. $\qquad\square$

Intervals of type I_{k-1}^+ play a special role. We call them special. Assume that I is an interval involved in forming $\omega_{n-\ell}$. Then there is only one special interval in $I\setminus\mathrm{supp}\,w_{n-\ell-1}$. Its length is $2\frac{1}{4^k}|I|$. But there are $k-2$ such special intervals in $I\cap(\mathrm{supp}\,w_{n-\ell-1}\setminus \mathrm{supp}\,w_{n-\ell-2})$. Their total length is

$$2\frac{1}{4^k}\sum_{m=1}^{k-2}\frac{1}{4^m}|I| = \frac{1}{3}\Big(1-\frac{1}{4^{k-2}}\Big)\frac{2}{4^k}|I|.$$

Similarly, the length of the union of special intervals in

$$I\cap(\mathrm{supp}\,w_{n-\ell-2}\setminus\mathrm{supp}\,w_{n-\ell-3})$$

is

$$\Big(\frac{1}{3}\Big(1-\frac{1}{4^{k-2}}\Big)\Big)^2\frac{2}{4^k}|I|,$$

etc.

If we denote the family of such special intervals in

$$[0,1)\cap(\mathrm{supp}\,w_{n-\ell}\setminus\mathrm{supp}\,w_{n-\ell-1})$$

by $\mathcal{A}_\ell$ and their union by A_ℓ, then we have

$$|A_\ell| = \Big(\frac{1}{3}\Big(1-\frac{1}{4^{k-2}}\Big)\Big)^\ell\frac{2}{4^k}. \tag{5.7}$$

5.2 Martingale transform estimate

We are going to find A_1-weights such that

$$\|T\colon L^1(w)\to L^{1,\infty}(w)\| \ge c[w]_{A_1}\log[w]_{A_1}.$$

It turns out that it is enough to construct weights $w \in A_2$ such that the dyadic maximal function has the norm in $L^2(w^{-1})$ bounded as

$$\|M\|_{w^{-1}} \le A\, p \tag{5.8}$$

and such that at the same time

$$\|T(\mathbf{1}_{I_0} w)\|_{w^{-1}}^2 \ge cp^2 (\log p)^2 \|\mathbf{1}_{I_0}\|_w. \tag{5.9}$$

The above is explained in [33] by using the extrapolation of Rubio de Francia. This explanation is repeated below during the proof of Theorem 5.6.

Take the sequence w_n of weights from Section 5.1 above. Now consider a collection of special intervals J as was introduced above. This family splits into $\mathcal{A}_\ell$ collections, $A_\ell = \bigcup_{J \in \mathcal{A}_\ell} J$. Let $J \in \mathcal{A}_\ell$, $x \in J$. First we want to estimate from below the following:

$$T_J w_n(x) = \sum_{R \in \mathcal{D}:\, R \in \mathrm{row}(J)} \varepsilon_R(w_n, h_R) h_R(x), \quad x \in J.$$

Let us explain what is $\mathrm{row}(J)$. Let the interval forming $w_{n-\ell}$ (see Definition 5.3) and containing J be called K. Then $J = K_{k-1}^+$. Consider also $K = K_0, K_1 = K^{++}, \ldots, K_m, \ldots, K_{k-2}$. J is the right child of K_{k-1}. In the sum forming $T w_n(x)$, $x \in J$, above we choose first $R = K^+, K_1^+, \ldots, K_m^+, \ldots, K_{k-2}^+$.

Call them $\mathrm{row}(J) = \{K^+(J), K_1^+(J), \ldots, K_m^+(J), \ldots, K_{k-2}^+(J)\}$. We call this collection the row of special interval J. In terms of the collection $C(K)$ (see Definition 5.1) the row of J is defined as the right children of intervals from $C(K)$.

We did not choose yet the signs ε_R. Here is the choice:

$$\begin{aligned} &\forall J \in \mathcal{A}_\ell \quad \text{special and for even } \ell, \\ &\varepsilon_R = -1 \quad \text{if } R \in \mathrm{row}(J); \text{ otherwise, } \varepsilon_R = 0. \end{aligned} \tag{5.10}$$

In other words, intervals I in the union of all rows of all special intervals in $\mathcal{A}_\ell$ with even ℓ have $\varepsilon_I = -1$, and all other intervals $I \in \mathcal{D}$ have $\varepsilon_I = 0$.

Recall that we fixed an interval $J \in \mathcal{A}_\ell$, with ℓ even and $x \in J$. We have

$$(w_n, h_{K_m^+(J)}) h_{K_m^+(J)}(x) = \left(\langle w_n \rangle_{K_m^{+-}(J)} - \langle w_n \rangle_{K_m^{++}(J)} \right), \quad x \in J.$$

By construction the interval K_m^{+-} is a forming interval of $w_{n-\ell-1}$; thus, by (5.1) the average over it is $3^{\ell+1}\omega$. On the other hand, the average $\langle w \rangle_{K_m^{++}}$ will be some average of $3^{\ell+1}\omega$ on $k - 2 - m$ intervals forming $w_{n-\ell-1}$ and lying in K to the right of K_m^{+-} and of $\frac{\omega}{p}$ on the rest of K_m^{++}. The total mass of $k - 2 - m$ intervals forming $w_{n-\ell-1}$ and lying in K to the right of K_m^{+-} is at most $\frac{1}{3}|K_m^{+-}|$. Thus, this second average is a fixed small constant smaller than $3^{\ell+1}\omega$. So, with positive absolute constant c_1,

$$(w, h_{K_m^+(J)}) h_{K_m^+(J)}(x) \ge c_1 3^{\ell+1}\omega. \tag{5.11}$$

Hence, if ℓ is even we will have positive contributions of order $3^{\ell+1}$ from its row row(J). Therefore,

$$x \in J, \quad J \in \mathcal{A}_\ell \quad\Rightarrow\quad T_J w_n(x) \geq c_1 k 3^{\ell+1}\omega, \tag{5.12}$$

where one can see that $c_1 = \frac{2}{3} - O(\frac{1}{k})$.

Now we need to bookkeep the contribution of $T_{\tilde{J}} w_n(x)$ at the same point $x \in J$, where we need to take into account all special $\tilde{J} \neq J$. That contribution is formed by intervals from row($\tilde{J}$), $\tilde{\in}\mathcal{A}_{\ell'}$, ℓ' is even, $\ell' = \ell - 2, \ell - 4, \ldots, 0$. All other contributions are zero (x in the special interval $J \in \mathcal{A}_\ell$).

As an example, consider the tower of intervals $J \subset K_{m_e}^{+-}(\tilde{J}_\ell) \subset \cdots \subset K_{m_0}^{+-}(\tilde{J}_0)$. Interval $K_{m_{\ell'}}^{+-}(\tilde{J}_{\ell'})$ is a forming interval of $w_{n-\ell'-1}$, $\ell' < \ell$, ℓ' is even. But the contribution will be not only from this tower, but also from all the intervals lying in the same rows as the intervals in the tower above.

The contribution to $T w_n(x)$, $x \in J$, of the rows assigned to other special intervals is zero.

We need to consider only the contribution of the rows of intervals to which the intervals in the tower above belong. That contribution will be (by absolute value) at most

$$k(3^{\ell+1-2} + 3^{\ell+1-4} + \cdots) \leq \frac{1}{2}k 3^\ell,$$

and, therefore, the total contribution of those $\tilde{J}$'s can be estimated by

$$\left| \sum_{\substack{i=0,\, i\,\text{even}}}^{\ell-2} T_{\tilde{J}_i} w_n(x) \right| \leq \frac{1}{2}k 3^\ell, \quad x \in J,$$

and cannot spoil the number $c_1 k 3^{\ell+1}\omega$ from (5.12) too much.

Also, the σ_n-measure of such a $J \in \mathcal{A}_\ell$ as above is $\frac{1+\varepsilon}{9\varepsilon}\frac{\sigma}{3^\ell}|J|$. Combining this with (5.12) and the estimate of $|\mathcal{A}_\ell|$ from (5.7), we get (ℓ is even)

$$\int_{\mathcal{A}_\ell} (T w_n)^2 \sigma_n\, dx \geq c\omega^2 k^2 3^{2\ell}\frac{1+\varepsilon}{9\varepsilon}\frac{\sigma}{3^\ell}\left(\frac{1}{3}\left(1 - \frac{1}{4^{k-2}}\right)\right)^\ell \frac{1}{4^k}.$$

Using the fact that $\omega\sigma = p$, $4^k = 1/\varepsilon$, we get

$$\int_{\mathcal{A}_\ell} (T w_n)^2 \sigma_n\, dx \geq c\omega\, k^2\, p\,\frac{1+\varepsilon}{9\varepsilon}\left(\left(1 - \frac{1}{4^{k-2}}\right)\right)^\ell \frac{1}{4^k} \geq c\omega\, k^2\, p\left(\left(1 - \frac{1}{4^{k-2}}\right)\right)^\ell.$$

Now,

$$\int_0^1 (Tw_n)^2 \sigma_n \, dx \geq c\omega \, k^2 \, p \sum_{\ell=0,\, \ell \text{ even}}^{4^k} \left(\left(1 - \frac{1}{4^{k-2}} \right) \right)^{\ell} \geq c\omega \, k^2 \, p 4^k.$$

But $4^k \approx p$, $k \approx \log p$. So we get

$$\int_0^1 (Tw_n)^2 \sigma_n \, dx \geq c\omega \, p^2 (\log p)^2 = c \, p^2 (\log p)^2 \int_0^1 w_n \, dx. \tag{5.13}$$

Theorem 5.5. *We have the following estimate of the dyadic maximal operator M:* $\|M\|_{w_n^{-1}} \leq A\,p$.

We prove this theorem in Section 5.3; the proof is the same as in [33]. It is a dyadic situation, so the proof is somewhat easier than in [33].

Given Theorem 5.5 and what was done above, we can now prove the following theorem, the analog of the main result of [33], but for the martingale transform instead of the Hilbert transform.

Theorem 5.6. *There is a sequence of weights $W \in A_1$ such that their dyadic A_1-norms $[W]_{A_1} \to \infty$ and martingale transforms T such that $\|T: L^1(W) \to L^{1,\infty}(W)\| \geq c[W]_{A_1} \log[W]_{A_1}$ with an absolute positive c.*

The proof of Theorem 5.6 is ad verbatim the same as in [33]; the corresponding weights $W_n \in A_1$ are obtained from $w_n \in A_2$ constructed above by the method of Rubio de Francia. For these weights one uses the same dyadic martingale transforms as above. We repeat the proof for the convenience of the reader. The proof also shows how complicated are the weights W_n built with the help of w_n constructed above.

Proof. We think of T as a linear operator with nice kernel; in practice T will be, say, a martingale transform with only finitely many ε_I's non-zero, but the estimates will not depend on how many non-zeros T has.

We always think that T has a symmetric or anti-symmetric kernel.

By T_w we understand the operator which acts on test functions as follows:

$$T_w f = T(wf).$$

Let w be a weight (for our goals it will be one of w_n built above) and let $\alpha > 0$. We let g be a function from $L^2(w^{-1})$ to be chosen soon, $\|g\|_{w^{-1}} = 1$, and we use the Rubio de Francia function

$$\mathcal{R}g = \sum_{k=0}^{\infty} \frac{M^k g}{2\|M^k\|_{w^{-1}}}.$$

Then

$$\|\mathcal{R}g\|_{w^{-1}} \le 2, \quad g \le \mathcal{R}g,$$

$$[\mathcal{R}g]_{A_1} \le \|M\|_{w_n^{-1}}. \tag{5.14}$$

Then, choosing appropriate g, $\|g\|_{w^{-1}} = 1$, and denoting $F \overset{\text{def}}{=} wf$, we can write

$$a\big(w\{T_{w^{-1}}F > a\}\big)^{1/2} = a\big(w\{Tf > a\}\big)^{1/2}$$

$$\le 2a \int\limits_{Tf>a} g\,dx \le 2a \int\limits_{Tf>a} \mathcal{R}g\,dx$$

$$\le 2N([\mathcal{R}g]_{A_1}) \int |f|\mathcal{R}g\,dx$$

$$\le 2N([\mathcal{R}g]_{A_1})\|f\|_w\|\mathcal{R}g\|_{w^{-1}}$$

$$\le 4N([\mathcal{R}g]_{A_1})\|f\|_w = 4N([\mathcal{R}g]_{A_1})\|F\|_{w^{-1}}. \tag{5.15}$$

Here $N([\mathcal{R}g]_{A_1})$ denotes the estimate from above of the weak norm $T\colon L^1(\mathcal{R}g) \to L^{1,\infty}(\mathcal{R}g)$.

This weight $\mathcal{R}g$ is the future weight W_n mentioned before the start of the proof. Henceforth, one obtains the estimate

$$N([\mathcal{R}g]_{A_1}) \ge \frac{1}{4}\big\|T_{w^{-1}}\colon L^2(w^{-1}) \to L^{2,\infty}(w)\big\|. \tag{5.16}$$

By the duality and (anti)symmetry of T, the latter norm is

$$\big\|T_{w^{-1}}\colon L^2(w^{-1}) \to L^{2,\infty}(w)\big\| = \big\|T_w\colon L^{2,1}(w) \to L^2(w^{-1})\big\| \ge \frac{\|T(\mathbf{1}_{I_0}w)\|_{w^{-1}}}{\|\mathbf{1}_{I_0}\|_w}. \tag{5.17}$$

In the last inequality we used the fact that the norms of characteristic functions in $L^{2,1}(w)$ and in $L^2(w)$ are the same.

We use (5.14), (5.16), and (5.17) to obtain

$$\frac{\|T(\mathbf{1}_{I_0}w)\|_{w^{-1}}}{\|\mathbf{1}_{I_0}\|_w} \le 4N(\|M\|_{w^{-1}}). \tag{5.18}$$

Now we plug into this inequality (with $w = w_n$) inequality (5.9) and the result of Theorem 5.5. Then we find that for all large p

$$p \log p \le CN(p).$$

This is what we wanted. $\qquad\square$

5.3 The proof of Theorem 5.5

We put initial numbers ω, σ to be

$$\omega = 1, \quad \sigma = p. \tag{5.19}$$

It is well known that for the maximal function one has the $T1$ theorem. Hence, it is sufficient to check that with finite absolute constant C

$$\forall J \in \mathcal{D}(I) \quad \int_J \big(M(w\chi_J)\big)^2 w^{-1} dx \le C p^2 w(J). \tag{5.20}$$

Following [33] we define the following function (of course we put $\omega = 1$, but it is convenient to keep writing it):

$$\tilde{w}(x) \overset{\text{def}}{=} \omega \sum_{\ell=1}^{n} 3^{\ell-1} \chi_{\operatorname{supp} w_{n-\ell+1}\setminus \operatorname{supp} w_{n-\ell}} + \omega 3^n \chi_{\operatorname{supp} w_0}. \tag{5.21}$$

Lemma 5.7. *With an absolute constant C, $Mw \le C\tilde{w}$.*

Proof. If $x \in \operatorname{supp} w_0$, we have (taking into account the normalization in (5.19))

$$w(x) = w_0\Big(3^n\omega, \frac{\sigma}{3^n}; x\Big) \le \frac{3^n}{\sqrt{p}}\Big(\big(\sqrt{p} - \sqrt{p-1}\big)\chi_{I_-} + \big(\sqrt{p} + \sqrt{p-1}\big)\chi_{I_+}\Big) \le 2 \cdot 3^n.$$

On the complement of $\operatorname{supp} w_0$, $w \le \frac{3^n}{p}$. So the claim of the lemma is obvious for $x \in \operatorname{supp} w_0$.

If $x \in \operatorname{supp} w_{n-\ell+1} \setminus \operatorname{supp} w_{n-\ell}$, $\ell = 1, 2, \ldots, n$, then

$$w(x) \le \frac{3^{\ell-1}\omega}{p},$$

and outside of $\operatorname{supp} w_{n-\ell+1}$, $w(x) \le \frac{3^{\ell-2}\omega}{p}$. If we average w over a dyadic interval J centered at $x \in \operatorname{supp} w_{n-\ell+1} \setminus \operatorname{supp} w_{n-\ell}$ and this J intersects dyadic intervals forming $\operatorname{supp} w_{n-\ell}$, then each of this dyadic interval I is inside J. Then, by Lemma 5.2, for each such I we have $w(I) = 3^\ell |I|$. Combining all this we get

$$w(J) \le 2 \cdot 3^\ell |J|,$$

which proves the lemma. $\qquad\square$

Lemma 5.8. *Let $I_0 = [0,1]$ and let $w = w_n$, as in Section 5.1. Then with a finite absolute constant C we have*

$$\int_{I_0} \tilde{w}^2 w^{-1}\, dx \le C p^2 w(I_0).$$

Proof. Let us consider first $m = 0$. Denote

$$F_j = [0,1] \cap (\operatorname{supp} w_{n-j} \setminus \operatorname{supp} w_{n-j-1}).$$

Let A_j be the union of all special intervals (see Section 5.1) contained in F_j.

On the complement of $\operatorname{supp} w_{n-j-1}$, $w(x) \leq \frac{3^j}{p}$. If in addition $x \notin A_j$, we have $w = \frac{3^j}{p}$. Hence, here $\tilde{w}(x) = pw(x)$. This is convenient, because together with $w(x)w^{-1}(x) = 1$, this implies

$$\int_{F_j \setminus A_j} \tilde{w}^2 w^{-1} dx = p^2 \int_{F_j \setminus A_j} w(x)dx = p^2 w(F_j \setminus A_j).$$

On A_j, $w^{-1}(x) = \frac{p}{3^j}\frac{1+5\varepsilon}{9\varepsilon}$. We saw in (5.7) that

$$|A_j| = \left(\frac{1}{3}\left(1 - \frac{1}{4^{k-2}}\right)\right)^j \frac{2}{4^k} \leq \frac{1}{3^j}\frac{2}{4^k}.$$

By the definition of $\tilde{w}$ we conclude now that

$$\int_{\bigcup_{j=0}^{n-1} A_j} \tilde{w}^2 w^{-1} dx \leq p \sum_{j=0}^{n-1} 3^{2j} \frac{1}{3^{2j}} \frac{2}{4^k} \frac{1+5\varepsilon}{9\varepsilon} \leq \frac{np}{4^k \varepsilon}.$$

As $n = 4^k$ and $p \asymp \varepsilon^{-1}$, by using Lemma 5.2 in the last inequality we get

$$\int_{\bigcup_{j=0}^{n-1} A_j} \tilde{w}^2 w^{-1} dx \leq \frac{np}{4^k \varepsilon} \leq Cp^2 w(I).$$

Therefore,

$$\int_I \tilde{w}^2 w^{-1} dx \leq \int_{\bigcup_{j=0}^{n-1} F_j \setminus A_j} \tilde{w}^2 w^{-1} dx + \int_{\bigcup_{j=0}^{n-1} A_j} \tilde{w}^2 w^{-1} dx + \int_{\operatorname{supp} w_0} \tilde{w}^2 w^{-1} dx$$

$$\leq p^2 \sum_{j=0}^{n-1} w(F_j \setminus A_j) + Cp^2 w(I) + \int_{\operatorname{supp} w_0} \tilde{w}^2 w^{-1} dx$$

$$\leq Cp^2 w(I) + \int_{\operatorname{supp} w_0} \tilde{w}^2 w^{-1} dx.$$

On $\operatorname{supp} w_0$, $\tilde{w}(x) = 3^n = w(x)$; hence, the last integral is just at most $w(I)$. Finally we get $(I_0 = I = [0,1])$

$$\int_{I_0} \tilde{w}^2 w^{-1} dx \leq Cp^2 w(I_0) = Cp^2 w(I). \tag{5.22}$$

$\square$

Lemma 5.9. *Let $I = [0,1]$ and $I_0 = I$, $I_1 = I^{++}$, $I_2 = I_1^{++},\ldots,I_{k-2} = I_{k-3}^{++}$ as before. Let $w = w_n$. Then with a finite absolute constant C we have for $m = 1,\ldots,k-2$*

$$\int_{I_m} \tilde{w}^2 w^{-1}\, dx \le Cp^2 w(I_m).$$

Proof. Now we consider the case of I_m, $I = [0,1]$, $m = 1,\ldots,k-2$. Notice that $I_m \setminus I_{m+1}$ consists of an interval $F \overset{\text{def}}{=} I_m^{+-}$, which is one of the intervals forming w_{n-1}, and interval G, such that G belongs to supp $w_n \setminus$ supp w_{n-1}. On such intervals, by the definition (5.21) of $\tilde{w}$, $\tilde{w} = \omega$, and $w = \frac{\omega}{p}$ (of course we can remember that ω is normalized in (5.19), but it does not matter in the calculations below). Thus, on G, $\tilde{w} = pw$.

Hence,

$$\int_{I_m \setminus I_{m+1}} \tilde{w}^2 w^{-1} dx \le p^2 \int_{G} w^2 w^{-1} dx + \int_{F} \tilde{w}^2 w^{-1} dx \le p^2 w(I) + \int_{F} \tilde{w}^2 w^{-1} dx.$$

Notice that we can estimate the last integral by Lemma 5.8. In fact, the interval F plays the role of I, and the weight w on F (and hence $\tilde{w}$) is constructed exactly as $w = w_n$ on I, only it starts not with ω but with 3ω and takes $n - 1$ steps to be constructed. By the scale invariance, using Lemma 5.8 we get

$$\int_{F} \tilde{w}^2 w^{-1} dx \le Cp^2 w(F) \le Cp^2 w(I).$$

The two last displayed inequalities together give the following:

$$\int_{I_m \setminus I_{m+1}} \tilde{w}^2 w^{-1} dx \le (C+1)p^2 w(I_m \setminus I_{m+1}). \tag{5.23}$$

Using (5.23) we can now write ($m = 1,\ldots,k-2$)

$$\int_{I_m} \tilde{w}^2 w^{-1} dx = \sum_{j=m}^{k-2} \int_{I_j \setminus I_{j+1}} \tilde{w}^2 w^{-1} dx + \int_{I_{k-1}} \tilde{w}^2 w^{-1} dx \le (C+1)p^2 w(I_m) + \int_{I_{k-1}} \tilde{w}^2 w^{-1} dx.$$

But on I_{k-1} we have $\tilde{w} = \omega$ (see (5.21)), so

$$\int_{I_{k-1}} \tilde{w}^2 w^{-1} dx \le C\omega^2 \frac{(1+5\varepsilon)p}{9\varepsilon\omega} |I_{k-1}| \le C\omega|I_{k-1}|p^2. \tag{5.24}$$

On the other hand, as $m \le k - 2$, by virtue of Lemma 5.2 we have

$$w(I_m) \ge \omega|I_m| \ge \omega|I_{k-1}|. \tag{5.25}$$

Combining (5.24) and (5.25) we get (xor $m \leq k - 2$)

$$\int_{I_{k-1}} \tilde{w}^2 w^{-1} dx \leq Cp^2 w(I_m).$$

Finally, for $m = 1, \ldots, k - 2$ we have

$$\int_{I_m} \tilde{w}^2 w^{-1} dx \leq Cp^2 w(I_m). \tag{5.26}$$

$\square$

In the next lemma we stop working with $\tilde{w}$.

Lemma 5.10. *Let $I = [0,1]$ and $I_0 = I, I_1 = I_0^{++}, \ldots, I_{k-1} = I_{k-2}^{++}$ as before. Let $w = w_n$. With a finite absolute constant C we now have the following:*

$$\int_{I_{k-1}} \left(M(\chi_{I_{k-1}} w)\right)^2 w^{-1} dx \leq Cpw(I_{k-1}).$$

Proof. Clearly, by construction of w, we have

$$M(\chi_{I_{k-1}} w) \leq \frac{\omega}{p}.$$

Hence,

$$\int_{I_{k-1}} \left(M(\chi_{I_{k-1}} w)\right)^2 w^{-1} dx \leq \frac{\omega^2}{p^2} \frac{(1 + 5\varepsilon)p}{9\varepsilon\omega} |I_{k-1}| \leq C\omega |I_{k-1}|.$$

On the other hand,

$$w(I_{k-1}) \geq \frac{1}{2} \frac{\omega}{p} |I_{k-1}|.$$

Therefore,

$$\int_{I_{k-1}} \left(M(\chi_{I_{k-1}} w)\right)^2 w^{-1} dx \leq Cpw(I_{k-1}).$$

$\square$

Lemma 5.11. *Let $I = [0,1]$ and $I_0 = I, I_1 = I_0^{++}, \ldots, I_{k-1} = I_{k-2}^{++}$ as before. Let J be the dyadic father of I_{k-1} and let $w = w_n$. Then with a finite absolute constant C we have*

$$\int_J \left(M(\chi_J w)\right)^2 w^{-1} dx \leq Cp^2 w(J).$$

Proof. Interval J consists of J_- and $J_+ = I_{k-1}$. By Lemma 5.2 and $w \leq \frac{\omega}{p}$ on I_{k-1} (see (5.1)) we have

$$x \in J_+ = I_{k-1} \quad \Rightarrow \quad M(\chi_J w) \leq \omega.$$

We also know by Lemma 5.2 that $w(J) \geq w(J_-) = \frac{1}{2}\omega|J|$. So

$$\int_{J_+} (M(\chi_J w))^2 w^{-1} dx \leq \omega^2 \frac{(1 + 5\varepsilon)p}{9\varepsilon\omega} |J| \leq Cp^2 w(J).$$

Now notice that we can estimate the integral $\int_{J_-} (M(w\chi_J))^2 w^{-1} dx$ by Lemma 5.8. In fact, interval J_- plays the role of I, and the weight w on J_- (and hence $\tilde{w}$) is constructed exactly as $w = w_n$ on I, only it starts not with ω but with 3ω and takes $n-1$ steps to be constructed. By the scale invariance, using Lemma 5.8 we get

$$\int_{J_-} \tilde{w}^2 w^{-1} dx \leq Cp^2 w(J_-) \leq Cp^2 w(J).$$

Combining the two last displayed inequalities we get the lemma's claim. $\qquad\square$

Now we combine the lemmas of this section with the fact that for $x \in \operatorname{supp} w_{n-\ell+1} \setminus \operatorname{supp} w_{n-\ell}$, $w(x) \leq \omega\frac{3^{\ell-1}}{p}$ to see that we obtained the following theorem.

Theorem 5.12. *Let $I = [0,1]$, $w = w_n$. Let J be any dyadic subinterval of I, such that it is not contained in any dyadic interval forming $\operatorname{supp} w_{n-1}$. There exists a finite absolute constant C such that*

$$\int_J (M(\chi_J w))^2 w^{-1} dx \leq Cp^2 w(J). \tag{5.27}$$

Now we just notice that if interval J is inside or equal to one of the intervals forming $\operatorname{supp} w_{n-1}$, say K, then we can just notice that K plays the role of $I = [0,1]$ and w_{n-1} is just the same type of weight as w_n, only starting with 3ω instead of ω. But the value of ω was immaterial in the above considerations. Therefore, we can extend the claim (5.27) of Theorem 5.12 to dyadic intervals that are not contained in any dyadic interval forming $\operatorname{supp} w_{n-2}$. The constant C is exactly the same. We can continue to reason this way, and we obtain (5.20).

6 PDE approach to logarithmic blow-up of the dyadic martingale transform

Below w is a dyadic A_1-weight.

By Mw we will denote the martingale maximal function of w, that is, $Mw(x) = \sup_{x \in J, J \in \mathcal{D}} \langle w \rangle_J$. Then $w \in A_1$ with "norm" Q means that

$$Mw \le Q \cdot w \quad \text{a. e.,}$$

and $Q = [w]_{A_1}$ is the best constant in this inequality.

Recall that the martingale transform is the operator given by $T\varphi = \sum_{J \in \mathcal{D}} \varepsilon_J \Delta_J \varphi$. It is convenient to use the Haar function h_J associated with dyadic interval J,

$$h_J(x) := \begin{cases} \frac{1}{|J|^{1/2}}, & x \in J_+, \\ -\frac{1}{|J|^{1/2}}, & x \in J_-. \end{cases}$$

In this notation, the martingale transform is

$$T\varphi = \sum_{J \in cD} \varepsilon_J (\varphi, h_J) h_J,$$

where (1) we always think the sum has only unspecified but a finite number of terms and (2) $|\varepsilon_J| \le 1$.

We are interested in several weak-type estimates.

We first consider the weak estimate for the martingale transform T in the weighted space $L^1(\mathbf{R}, w\,dx)$, where $w \in A_1$. The endpoint exponent is naturally $p = 1$, and we wish to understand the order of magnitude of the constant $A([w]_{A_1})$ in the weak-type inequality for the dyadic martingale transform:

$$\frac{1}{|I|} w \left\{ x \in I : \sum_{J \in \mathcal{D}(I)} \varepsilon_J (\varphi, h_J) h_J(x) > \lambda \right\} \le C_{[w]_{A_1}} \frac{\langle |\varphi| w \rangle_I}{\lambda}. \tag{6.1}$$

Here φ runs over all functions such that $\operatorname{supp} \varphi \subset I$ and $\varphi \in L^1(I, w\,dx)$, $w \in A_1$. This section will be devoted to the study of the "sharp" order of magnitude of constants $C_{[w]_{A_1}}$ in terms of $[w]_{A_1}$ if $[w]_{A_1}$ is large. We are primarily interested in the estimate of $C_{[w]_{A_1}}$ from below, that is, in finding the worst possible A_1-weight in terms of weak-type estimates (of course this involves also finding the worst test function φ).

We will prove the following result.

Theorem 6.1. *There is a positive absolute constant c and a weight $w \in A_1$ such that constant $C_{[w]_{A_1}}$ from (6.1) satisfies*

$$C_{[w]_{A_1}} \ge c[w]_{A_1} \log^{1/3}(1 + [w]_{A_1}).$$

Remark 6.2. In fact, the attentive reader will see that we prove a sharper result; see [36]. We will consider a particular dyadic shift, and we will prove the estimate

$\geq c[w]_{A_1}(\log[w]_{A_1})^{1/3}$ for one particular dyadic shift. Our dyadic shift is the following dyadic singular operator on $L^1(I, wdx)$, $I = [0, 1]$:

$$S\mathbf{1}_I = 0, \quad Sh_J = h_{J_-} - h_{J_+}, \quad J \in D(I).$$

Our main result is the following theorem.

Theorem 6.3. *There are a positive absolute constant c and a weight $w \in A_1$ such that*

$$\|S\|_{L^1(w) \to L^{1,\infty}(w)} \geq c[w]_{A_1} \log^{1/3}(1 + [w]_{A_1}).$$

In [34] the following estimate from above has been proved.

Theorem 6.4. *There is a positive absolute constant C such that for any weight $w \in A_1$ constant $C_{[w]_{A_1}}$ from (6.1) satisfies*

$$C_{[w]_{A_1}} \leq c[w]_{A_1} \log(1 + [w]_{A_1}).$$

Remark 6.5. This note is based on two preprints [36, 37], but Theorem 6.3 was not formulated in these preprints; however, as the attentive reader can notice, it was proved there.

6.1 Bellman approach to weak weighted estimate of the martingale transform

To find the "optimal" $C_{[w]_{A_1}}$ we use again the Bellman function technique. The idea is to reformulate the infinite-dimensional problem of optimization of $C_{[w]_{A_1}}$, that is, finding the "smallest" $C_{[w]_{A_1}}$ that works for all inequalities in (6.1), in terms of the growth estimate on a certain function of only a finite number of variables (five in this case).

Here it is (it will depend on the number $Q \geq 1$):

$$\mathbf{B}(F, w, m, f, \lambda) := \mathbf{B}_Q(F, w, m, f, \lambda) := \sup \frac{1}{|I|} \omega \left\{ x \in I : \sum_{J \subseteq I, J \in D} \varepsilon_J(\varphi, h_J) h_J(x) > \lambda \right\}, \quad (6.2)$$

where the sup is taken over all $\varepsilon_J, |\varepsilon_J| \leq 1, J \in D(I)$, and over all $\varphi \in L^1(I, \omega\, dx)$ such that $F := \langle |\varphi| \omega \rangle_I, f := \langle \varphi \rangle_I, w = \langle \omega \rangle_I, m \leq \inf_I \omega$, and ω are all dyadic A_1-weights, such that $[w]_{A_1} \leq Q$. This function is obviously defined in the convex subdomain of $\mathbf{R}^5$:

$$\Omega := \{(F, w, m, f, \lambda) \in \mathbf{R}^5 : F \geq |f| m, \ m \leq w \leq Q m\}. \quad (6.3)$$

Remark 6.6. We warn the reader that emotional attachment to notations F, f, w for functions should be dismissed. These symbols in this and the next sections stand for usual numbers.

6.1.1 The properties of B_Q

The first property: homogeneity
By definition, it is clear that

$$s\mathbf{B}\left(\frac{F}{s},\frac{w}{s},\frac{m}{s},f,\lambda\right) = \mathbf{B}(F,w,m,f,\lambda),$$

$$\mathbf{B}(tF,w,m,tf,t\lambda) = \mathbf{B}(F,w,m,f,\lambda).$$

Choosing $s = m$ and $t = \lambda^{-1}$ and introducing new variables

$$\alpha = \frac{F}{m\lambda}, \quad \beta = \frac{w}{m}, \quad \gamma = \frac{f}{\lambda},$$

we can see that

$$\frac{1}{m}\mathbf{B}(F,w,m,f,\lambda) = B\left(\frac{F}{m\lambda},\frac{w}{m},\frac{f}{\lambda}\right) =: B(\alpha,\beta,\gamma), \tag{6.4}$$

where function $B(\alpha,\beta,\gamma) = \mathbf{B}(\alpha,\beta,1,\gamma,1)$.
    Obviously B is defined in the domain

$$G := \{(\alpha,\beta,\gamma) : |\gamma| \le \alpha, 1 \le \beta \le Q\}. \tag{6.5}$$

The second property: special form of concavity
We formulate this property as the following theorem.

Theorem 6.7. *Let $P, P_+, P_- \in \Omega$, $P = (F, w, \min(m_+, m_-), f, \lambda)$, $P_+ = (F + A, w + u, m_+, f + a, \lambda + ta)$, $P_- = (F - A, w - u, m_-, f - a, \lambda - ta)$, $0 \le t \le 1$. Then*

$$\mathbf{B}(P) - \frac{1}{2}(\mathbf{B}(P_+) + \mathbf{B}(P_-)) \ge 0. \tag{6.6}$$

At the same time, if $P, P_+, P_- \in \Omega$, $P = (F, w, \min(m_+, m_-), f, \lambda)$, $P_+ = (F + A, w + u, m_+, f + a, \lambda - ta)$, $P_- = (F - A, w - u, m_-, f - a, \lambda + ta)$, $0 \le t \le 1$, then

$$\mathbf{B}(P) - \frac{1}{2}(\mathbf{B}(P_+) + \mathbf{B}(P_-)) \ge 0. \tag{6.7}$$

In particular, with fixed m and with all points being inside Ω, we get for all $t \in [0,1]$

$$\mathbf{B}(F, w, m, f, \lambda) \ge \frac{1}{4}(\mathbf{B}(F - dF, w - dw, m, f - d\lambda, \lambda - td\lambda)$$

$$+ \mathbf{B}(F - dF, w - dw, m, f - d\lambda, \lambda + td\lambda)$$

$$+ \mathbf{B}(F + dF, w + dw, m, f + d\lambda, \lambda - td\lambda)$$

$$+ \mathbf{B}(F + dF, w + dw, m, f + d\lambda, \lambda + td\lambda)). \tag{6.8}$$

Remark 6.8. (1) Differential notations dF, dw, $d\lambda$ just mean small numbers. (2) In (6.8) we lose a bit of information (in comparison with (6.6) and (6.7)), but this is exactly (6.8), which we are going to use in the future.

Before proving this theorem, let us explain a bit more what kind of concavity is represented by inequalities (6.6) and (6.7) (and thus by their consequence (6.8)). We can use different notations for coordinates $P_+, P_-, P_\pm := (F_\pm, w_\pm, m_\pm, f_\pm, \lambda_\pm)$. We require all $P, P_\pm$ to belong to Ω and it is evident that

$$F = \frac{F_+ + F_-}{2}, \quad w = \frac{w_+ + w_-}{2}, \quad m = m_+ \wedge m_-, \quad f = \frac{f_+ + f_-}{2}, \quad \lambda = \frac{\lambda_+ + \lambda_-}{2},$$

but also "jumps" in the fourth and fifth coordinates must be dependent on each other, namely,

$$t\Delta f := t(f_+ - f_-) = (\lambda_+ - \lambda_-) =: \Delta\lambda, \quad \text{or} \quad t\Delta f = -\Delta\lambda, \quad 0 \le t \le 1.$$

So function $\mathbf{B}$ (as we will now see) possesses such sophisticated concavity as encoded by jumps from any point $P \in \Omega$ to $P_+, P_- \in \Omega$, where P is almost the average of $P_\pm$, but not quite; the difference is that (1) the third coordinate is not an arithmetic average of the third coordinates of $P_\pm$, but their minimum, and (2) the jumps in the fourth and fifth coordinates are interdependent as above.

Proof. Fix $P, P_+, P_- \in \Omega$ as in (6.6). Let $\varphi_+, \varphi_-, \omega_+, \omega_-$ be functions and weights giving the supremum in $\mathbf{B}(P_+)$, $\mathbf{B}(P_-)$, respectively, up to a small number $\eta > 0$. Using the fact that $\mathbf{B}$ does not depend on I, we think that φ_+, ω_+ is on I_+ and φ_-, ω_- is on I_-. Consider

$$\varphi(x) := \begin{cases} \varphi_+(x), & x \in I_+, \\ \varphi_-(x), & x \in I_-, \end{cases}$$

$$\omega(x) := \begin{cases} \omega_+(x), & x \in I_+, \\ \omega_-(x), & x \in I_-. \end{cases}$$

Notice that then

$$(\varphi, h_I) \cdot \frac{1}{\sqrt{|I|}} = \Delta_I \varphi = \frac{1}{2}(P_{+,4} - P_{-,4}) =: a. \tag{6.9}$$

We denote the i-th coordinate of a point P by P_i. Then it is easy to see that $P_3 = \min(P_{3,-}, P_{3,+}) = \min(\min_{I_-} \omega_-, \min_{I_+} \omega_+)$, $P_5 = \lambda$,

$$\langle |\varphi| \omega \rangle_I = F = P_1, \quad \langle \omega \rangle_I = w = P_2, \quad \langle \varphi \rangle_I = f = P_4. \tag{6.10}$$

Notice that for $x \in I_+$, using (6.9), if $\varepsilon_I = -t$, $0 \le t \le 1$, then

$$\frac{1}{|I|}\omega_+\left\{x\in I_+:\sum_{J\subseteq I_+, J\in D}\varepsilon_J(\varphi,h_J)h_J(x)>\lambda\right\}$$

$$=\frac{1}{|I|}\omega_+\left\{x\in I_+:\sum_{J\subseteq I_+, J\in D}\varepsilon_J(\varphi,h_J)h_J(x)>\lambda+ta\right\}$$

$$=\frac{1}{2|I_+|}\omega_+\left\{x\in I_+:\sum_{J\subseteq I_+, J\in D}\varepsilon_J(\varphi_+,h_J)h_J(x)>P_{+,5}\right\}\geq\frac{1}{2}B(P_+)-\eta.$$

Similarly, for $x\in I_-$, using (6.9), if $\varepsilon_I=-t$, $0\leq t\leq 1$, then

$$\frac{1}{|I|}\omega_-\left\{x\in I_-:\sum_{J\subseteq I, J\in D}\varepsilon_J(\varphi,h_J)h_J(x)>\lambda\right\}$$

$$=\frac{1}{|I|}\omega_-\left\{x\in I_-:\sum_{J\subseteq I_-, J\in D}\varepsilon_J(\varphi,h_J)h_J(x)>\lambda-ta\right\}$$

$$=\frac{1}{2|I_-|}\omega_-\left\{x\in I_-:\sum_{J\subseteq I_-, J\in D}\varepsilon_J(\varphi_-,h_J)h_J(x)>P_{-,5}\right\}\geq\frac{1}{2}B(P_-)-\eta.$$

Combining the two left-hand sides we obtain for $\varepsilon_I=-1$

$$\frac{1}{|I|}\omega\left\{x\in I_+:\sum_{J\subseteq I, J\in D}\varepsilon_J(\varphi,h_J)h_J(x)>\lambda\right\}\geq\frac{1}{2}(B(P_+)+B(P_-))-2\eta.$$

Let us now use the simple information (6.10). If we take the supremum in the left-hand side over all functions φ, such that $\langle|\varphi|\,\omega\rangle_I=F$, $\langle\varphi\rangle_I=f$, $\langle\omega\rangle_I=w$, and for weights ω in dyadic A_1 with A_1-norm at most Q, and supremum over all $\varepsilon_J=\pm s$, $s\in[0,1]$ (only $\varepsilon_I=-1$ stays fixed), we get a quantity smaller than or equal to the one where we have the supremum over all functions φ, such that $\langle|\varphi|\,\omega\rangle_I=F$, $\langle\varphi\rangle_I=f$, $\langle\omega\rangle_I=w$, and for weights ω in dyadic A_1 with A_1-norm at most Q, and an unrestricted supremum over all $\varepsilon_J=\pm s$, $s\in[0,1]$, $\varepsilon_I=-t$, $0\leq t\leq 1$. The latter quantity is of course $\mathbf{B}(F,w,m,f,\lambda)$. So we proved (6.6).

To prove (6.7) we repeat the same reasoning ad verbatim, only keeping now $\varepsilon_I=t$, $0\leq t\leq 1$. We are done. $\qquad\square$

Remark 6.9. This theorem is a sort of "fancy" concavity property; the attentive reader will see that (6.6) and (6.7) include the biconcavity property entirely similar to the one demonstrated by the celebrated Burkholder function. We will use the consequence of biconcavity encompassed by (6.8). This is still another concavity. Let us also remark that it can be shown that $\mathbf{B}$ is a supersolution of a certain degenerate elliptic equation (but this fact does not help us in estimating $\mathbf{B}$ below).

The third property: B decreases in *m*

Function **B** is obviously decreasing in *m*. In fact, if *m* decreases (all other coordinates being fixed), then the collection of weights increases, and the supremum increases. It is not difficult to see that **B** is also continuous.

The fourth property: function *B* from (6.4) is concave

Recall that by (6.4)

$$B\left(\frac{F}{\lambda}, w, \frac{f}{\lambda}\right) = \mathbf{B}(F, w, 1, f, \lambda). \tag{6.11}$$

Choosing $t = 0$ in Theorem 6.7 we see that $\mathbf{B}(F, w, 1, f, \lambda)$ is concave when λ is fixed. This proves the fourth property, which we formulated intentionally in terms of B and not $\mathbf{B}$.

The fifth property: function $t \rightarrow \frac{1}{t}B(t\alpha, t\beta, \gamma)$ is increasing

This is the combination of (6.4) and the third property above.

The sixth property: the domain of definition of *B* is

$G = \{(\alpha, \beta, \gamma) \in \mathbf{R}^3 : 1 \le \beta \le Q, |\gamma| \le \alpha\}$

The seventh property: the symmetry and monotonicity in *γ*

It is easy to see from the definition of **B** that it is even in its variable f. Therefore,

$$B(\alpha, \beta, \gamma) = B(\alpha, \beta, -\gamma).$$

Notice that the concavity of B (in γ) and this symmetry together imply that $\gamma \rightarrow B(\cdot, \cdot, \gamma)$ is decreasing on $\gamma \in [0, \alpha]$.

6.2 The goal and the idea of the proof

In this section we are going to prove the following estimate from below on function B.

Theorem 6.10. *There is an absolute positive constant c such that for some point* $(\alpha, \beta, \gamma) \in G$

$$B(\alpha, \beta, \gamma) \ge cQ(\log Q)^{1/3}\alpha. \tag{6.12}$$

Remark 6.11. It is a subtle result and it will take some space below to prove. Recall that Muckenhoupt conjectured that for the Hilbert transform H and any weight $w \in A_1$ the following two estimates hold on a unit interval I:

$$w\{x \in I : |Hf(x)| > \lambda\} \le \frac{C}{\lambda} \int_I |f| M w \, dx, \tag{6.13}$$

$$w\{x \in I : |Hf(x)| > \lambda\} \le \frac{C[w]_{A_1}}{\lambda} \int_I |f| w \, dx. \tag{6.14}$$

Obviously, if (6.13) holds, then (6.14) is valid as well. It took many years to *disprove* (6.13). This was done by Maria Reguera and Christoph Thiele [44, 46]. The constructions involve a very irregular (almost a sum of delta-measures) weight w, so there was hope that such an effect cannot appear when the weight is regular in the sense that $w \in A_1$. Theorem 6.10 gives a counterexample to this hope for the case when the Hilbert transform is replaced by the martingale transform on a usual homogeneous dyadic filtration. The reader can consult [36] to see that for the Hilbert transform a counterexample also exists, so (6.14) fails as well. The counterexample for the Hilbert transform is the transference of a counterexample we build here for the martingale transform. Notice that Theorem 6.10 implicitly gives a certain counterexample for the Hilbert transform. We will explain in a separate note how to make this transference.

Now we will say a couple of words about the idea of the proof of Theorem 6.10. Ideally we would like to find the formula for B (and therefore for $\mathbf{B}$ because of (6.4)). To proceed we rewrite the second property of $\mathbf{B}$ as a PDE on B. Then we try to find the boundary conditions on B on ∂G, and then we may hope to solve this PDE. Unfortunately there are many roadblocks on this path, starting with the fact that the second property of $\mathbf{B}$ is not a PDE; it is rather a partial differential inequality in discrete form. We will write it down as a pointwise partial differential inequality, but for that we will need a subtle result of Aleksandrov. We also can find boundary values of B; see some of them in Section 6.2.1 below. However, the main difficulty is that our partial differential expression is in 3D.

6.2.1 Unweighted case

We first consider the simplest case of $m = \omega = 1$ identically. Then we are left with function $Bel(F, f, \lambda) = \mathbf{B}(F, 1, 1, f, \lambda)$, which is defined in a convex domain $\Omega_0 \subset \mathbf{R}^3$, $\Omega_0 := \{(F, f, \lambda) \in \mathbf{R}^3 : |f| \le F\}$, and whose concavity properties are described below.

Theorem 6.12. *Let $P, P_+, P_- \in \Omega_0$, $P = (F, f, \lambda)$, $P_+ = (F + A, f + a, \lambda + a)$, $P_- = (F - A, f - a, \lambda - a)$. Then*

$$Bel(P) - \frac{1}{2}(Bel(P_+) + Bel(P_-)) \ge 0. \tag{6.15}$$

At the same time, if $P, P_+, P_- \in \Omega_0$, $P = (F, f, \lambda)$, $P_+ = (F + A, f + a, \lambda - a)$, $P_- = (F - A, f - a, \lambda + a)$. Then

$$\mathcal{Bel}(P) - \frac{1}{2}(\mathcal{Bel}(P_+) + \mathcal{Bel}(P_-)) \geq 0. \tag{6.16}$$

Let us make the change of variables $(F, f, \lambda) \to (F, y_1, y_2)$. Then

$$y_1 := \frac{1}{2}(\lambda + f), \quad y_2 := \frac{1}{2}(\lambda - f).$$

Denote

$$M(F, y_1, y_2) := B(F, y_1 - y_2, y_1 + y_2) = \mathcal{Bel}(F, f, \lambda).$$

In terms of function M Theorem 6.12 reads as follows.

Theorem 6.13. *The function M is defined in the domain $G := \{(F, y_1, y_2) : |y_1 - y_2| \leq F\}$, and for each fixed y_2, $M(F, y_1, y_2)$ is concave in (F, y_1) and for each fixed y_1, $M(F, y_1, y_2)$ is concave in (F, y_2).*

The properties of M strongly remind of the properties of the Burkholder function.

In the unweighted situation we can find **B** (or M) precisely. Here is the result proved in [47].

Theorem 6.14.

$$\mathcal{Bel}(F, f, \lambda) = \begin{cases} 1, & \text{if } \lambda \leq F, \\ 1 - \frac{(\lambda - F)^2}{\lambda^2 - f^2}, & \text{if } \lambda > F. \end{cases} \tag{6.17}$$

This result means that we found a boundary value of the Bellman function $\mathbf{B}(F, w, m, f, \lambda)$ of the weighted problem on the part of its boundary, namely, we found this function of five variables on $\{P \in \partial\Omega : w = P_2 = P_3 = m\}$:

$$\mathbf{B}(F, m, m, f, \lambda) = m \begin{cases} 1, & \text{if } \lambda \leq F, \\ 1 - \frac{(\lambda - F)^2}{\lambda^2 - f^2}, & \text{if } \lambda > F. \end{cases} \tag{6.18}$$

In terms of function B from (6.4), we have the following boundary values of B:

$$B(\alpha, 1, \gamma) = \begin{cases} 1, & \text{if } \alpha \geq 1, \\ 1 - \frac{(1 - \alpha)^2}{1 - \gamma^2}, & \text{if } 0 \leq |\gamma| \leq \alpha < 1. \end{cases} \tag{6.19}$$

6.3 From discrete to differential inequality via Aleksandrov's theorem

By the fourth property of Section 6.1.1, function B is concave on its domain of definition G. By the result of Aleksandrov (see [16, Theorem 6.9]), B has all second derivatives almost everywhere; this means that for a. e. $x \in G^\circ$ and all small vectors $h \in \mathbf{R}^3$,

$$B(x + h) = B(x) + \nabla B(x) \cdot h + \langle H_B(x) \cdot h, h \rangle + o(|h|^2), \tag{6.20}$$

where H_B is the Hessian matrix of B. On the other hand, the second property of Section 6.1.1 can be rewritten in terms of B as follows:

$$\begin{aligned}
B\left(\frac{F}{\lambda}, \beta, \frac{f}{\lambda}\right) - \frac{1}{4}\Bigg[&B\left(\frac{F - dF}{\lambda - d\lambda}, \beta - d\beta, \frac{f - d\lambda}{\lambda - d\lambda}\right) \\
&+ B\left(\frac{F - dF}{\lambda - d\lambda}, \beta - d\beta, \frac{f + d\lambda}{\lambda - d\lambda}\right) \\
&+ B\left(\frac{F + dF}{\lambda + d\lambda}, \beta + d\beta, \frac{f - d\lambda}{\lambda + d\lambda}\right) \\
&+ B\left(\frac{F + dF}{\lambda + d\lambda}, \beta + d\beta, \frac{f + d\lambda}{\lambda + d\lambda}\right)\Bigg] \geq 0.
\end{aligned} \tag{6.21}$$

Here $(\frac{F}{\lambda}, \beta, \frac{f}{\lambda}) \in G^\circ$ and $(dF, d\beta, d\lambda)$ is just any small vector in $\mathbf{R}^3$.

Theorem 6.15. *For almost every point* $P = (\alpha, \beta, \gamma) =: (\frac{F}{\lambda}, \beta, \frac{f}{\lambda}) \in G^\circ$ *and every vector* $(dF, d\beta, d\lambda) \in \mathbf{R}^3$ *we have*

$$\begin{aligned}
&- \alpha^2 B_{\alpha\alpha}(P)\left(\frac{dF}{F} - \frac{d\lambda}{\lambda}\right)^2 - \beta^2 B_{\beta\beta}(P)\left(\frac{d\beta}{\beta}\right)^2 \\
&- (1 + \gamma^2) B_{\gamma\gamma}(P)\left(\frac{d\lambda}{\lambda}\right)^2 - 2\alpha\beta B_{\alpha\beta}(P)\left(\frac{dF}{F} - \frac{d\lambda}{\lambda}\right)\frac{d\beta}{\beta} \\
&+ 2\beta\gamma B_{\beta\gamma}(P)\frac{d\beta}{\beta}\frac{d\lambda}{\lambda} + 2\alpha\gamma B_{\alpha\gamma}(P)\left(\frac{dF}{F} - \frac{d\lambda}{\lambda}\right)\frac{d\lambda}{\lambda} \\
&+ 2\alpha B_{\alpha}(P)\left(\frac{dF}{F} - \frac{d\lambda}{\lambda}\right)\frac{d\lambda}{\lambda} - 2\gamma B_{\gamma}(P)\left(\frac{d\lambda}{\lambda}\right)^2 \geq 0.
\end{aligned} \tag{6.22}$$

Remark 6.16. We can mollify $\mathbf{B}$ to make it smooth and still have its "fancy concavity properties." But then we lose homogeneity and we cannot reduce $\mathbf{B}$ to B. We can mollify $\mathbf{B}$ to keep its homogeneity – just choose the mollifier depending on the point – but then we lose its "fancy concavity property." In short, we have a problem with the mollification. This is why Aleksandrov's theorem is very useful now.

Proof. Fix a point $P \in G^\circ$, where Aleksandrov's identity (6.20) holds. Fix an arbitrary $(dx, dy, d\lambda) \in \mathbf{R}^3$. Let us use (6.21) by expanding the fraction $\frac{x \pm \varepsilon dx}{\lambda \pm \varepsilon d\lambda}, \frac{f \pm \varepsilon d\lambda}{\lambda \pm \varepsilon d\lambda}$ up to the second order in small parameter ε and combining with the identity (6.20) after that. All terms with $\varepsilon^0, \varepsilon^1$ will disappear identically. Only the terms with ε^2 and smaller stay. After division by ε^2 we bring ε to zero and get (6.22) for a. e. point $P \in G^\circ$. $\qquad\square$

Of course we need something else from positive B to be able to prove that B satisfying this partial differential inequality (6.22) in domain $G^\circ = \{P = (\alpha, \beta, \gamma) : 1 < \beta < Q, 0 < |\gamma| < \alpha\}$ has the estimate (6.12) from below. We actually have this "something else" in the form of the obstacle condition, which we will meet in Section 6.5.

But let us first simplify (6.22). Let us denote by $\mathcal{N}$ the matrix of the quadratic form in (6.22). After a rather straightforward operation $\mathcal{N} \to \mathcal{M}_1 := A^*\mathcal{N}A$ with a certain invertible matrix A we can write down the non-negativity of the differential form in (6.22) as the a. e. in G° non-negativity of the following matrix:

$$\mathcal{M}_1 := \begin{bmatrix} -\alpha^2 B_{\alpha\alpha}, & -\alpha\beta B_{\alpha\beta}, & \alpha\gamma B_{\alpha\gamma} + \alpha B_\alpha \\ -\alpha\beta B_{\alpha\beta}, & -\beta^2 B_{\beta\beta}, & \beta\gamma B_{\beta\gamma} \\ \alpha\gamma B_{\alpha\gamma} + \alpha B_\alpha, & \beta\gamma B_{\beta\gamma}, & -(1+\gamma^2)B_{\gamma\gamma} - 2\gamma B_\gamma \end{bmatrix} \geq 0. \tag{6.23}$$

However, we saw already that $B(\alpha,\beta,\gamma)$ is concave, which implies the non-negativity of yet another matrix:

$$\mathcal{M}_2 := \begin{bmatrix} -\alpha^2 B_{\alpha\alpha}, & -\alpha\beta B_{\alpha\beta}, & -\alpha\gamma B_{\alpha\gamma} \\ -\alpha\beta B_{\alpha\beta}, & -\beta^2 B_{\beta\beta}, & -\beta\gamma B_{\beta\gamma} \\ -\alpha\gamma B_{\alpha\gamma}, & -\beta\gamma B_{\beta\gamma}, & -\gamma^2 B_{\gamma\gamma} \end{bmatrix} \geq 0. \tag{6.24}$$

Taking the half-sum of (6.23) and (6.24), we obtain the following non-negativity:

$$\mathcal{M} := \begin{bmatrix} -\alpha^2 B_{\alpha\alpha}, & -\alpha\beta B_{\alpha\beta}, & \frac{1}{2}\alpha B_\alpha \\ -\alpha\beta B_{\alpha\beta}, & -\beta^2 B_{\beta\beta}, & 0 \\ \frac{1}{2}\alpha B_\alpha, & 0, & -(\frac{1}{2}+\gamma^2)B_{\gamma\gamma} - \gamma B_\gamma \end{bmatrix} \geq 0. \tag{6.25}$$

It is now natural to restrict the quadratic form of this matrix on certain 2D hyperplanes in the 3D tangent space Tan_p of the graph $\Gamma := \{p := (P, B(P)), P \in G^\circ\}$ at a given point p. Namely, let us consider the quadratic form of matrix $\mathcal{M}$ in (6.23) on vectors of the form

$$(\xi, \xi, \eta). \tag{6.26}$$

Then, using the notation

$$\psi(\alpha,\beta,\gamma) := \psi_B(\alpha,\beta,\gamma) := -\alpha^2 B_{\alpha\alpha} - 2\alpha\beta B_{\alpha\beta} - \beta^2 B_{\beta\beta}, \tag{6.27}$$

we get the a. e. in G° non-negativity of the following matrix:

$$\begin{bmatrix} \psi(\alpha,\beta,\gamma), & \frac{1}{2}\alpha B_\alpha \\ \frac{1}{2}\alpha B_\alpha, & -(\frac{1}{2}+\gamma^2)B_{\gamma\gamma} - \gamma B_\gamma \end{bmatrix} \geq 0. \tag{6.28}$$

Then for any $(\xi,\eta) \in \mathbf{R}^2$

$$\mathcal{M}_{11}(\alpha,\beta,\gamma)\xi^2 + 2\xi\eta\mathcal{M}_{12}(\alpha,\beta,\gamma) + \eta^2\mathcal{M}_{22}(\alpha,\beta,\gamma) \geq 0.$$

Definition 6.17. Consider a subdomain of G,

$$G_1 := \left\{ (\alpha, \beta, \gamma) \in G : |\gamma| < \frac{1}{2}\alpha, 2 < \beta < Q \right\}.$$

Fix now $(\alpha, \beta, \gamma) \in G_1$ and a parameter $t \in [1/2, 1]$. Replace in the previous inequality (α, β, γ) by $(t\alpha, t\beta, \gamma)$. Denote temporarily

$$P_t := (t\alpha, t\beta, \gamma), \quad (\alpha, \beta, \gamma) \in G_1, \; 1/2 \le t \le 1.$$

Then we get for every such t and every point P_t the following inequality for all $(\xi, \eta) \in \mathbf{R}^2$:

$$\xi^2[\psi(P_t)] + \xi\eta(\alpha t B_\alpha(P_t)) + \eta^2(-\gamma B_\gamma(P_t) - (1/2 + \gamma^2)B_{\gamma\gamma}(P_t)) \ge 0. \tag{6.29}$$

Consider a new function H, which is a certain averaging of B, namely, for any $P = (\alpha, \beta, \gamma) \in G_1$, let

$$H(P) = 2 \int_{1/2}^{1} B(P_t)\,dt.$$

Notice several simple facts. First of all,

$$\alpha H_\alpha = 2 \int_{1/2}^{1} \alpha t B(t\alpha, t\beta, \gamma)\,dt, \quad \alpha^2 H_{\alpha\alpha} = 2 \int_{1/2}^{1} (\alpha t)^2 B_{\alpha\alpha}(t\alpha, t\beta, \gamma)\,dt.$$

Similarly, if for every function F we introduce the notation

$$\psi_F(\alpha, \beta, \gamma) := -\alpha^2 F_{\alpha\alpha} - 2\alpha\beta F_{\alpha\beta} - \beta^2 F_{\beta\beta}, \tag{6.30}$$

we get

$$\psi_H = 2 \int_{1/2}^{1} \psi_B(t\alpha, t\beta, \gamma)\,dt.$$

Now integrate (6.29) on the interval $t \in [1/2, 1]$. The previous simple observations allow us now to rewrite this as a pointwise inequality for function H on domain G_1 introduced in Definition 6.17:

$$\xi^2[\psi_H(P)] + \xi\eta(\alpha H_\alpha(P)) + \eta^2(-\gamma H_\gamma(P) - (1/2 + \gamma^2)H_{\gamma\gamma}(P)) \ge 0. \tag{6.31}$$

The reader may wonder why we are so keen to replace (6.29) by virtually the same (6.31). The answer is because we can give a very good pointwise estimate on $\psi_H(P)$, $P \in G_1$. Unfortunately we cannot give any pointwise estimate on $\psi(P)$, $P \in G$.

Now we deduce the desired pointwise estimate on ψ_H. Below we will use its consequences. First, let us denote

$$R := \sup \frac{B(P)}{a}, \quad P = (a, \beta, \gamma) \in G. \tag{6.32}$$

Our goal formulated in (6.12) is to prove $R \geq cQ(\log Q)^{\varepsilon}$. We are still not too close, but notice that automatically $B(P) \leq Ra$, $P = (a, \beta, \gamma) \in G$.

Lemma 6.18. *If $P = (a, \beta, \gamma)$ is such that $|\gamma| \leq \frac{1}{8}a$ and $\beta > 100$, then*

$$\psi_H(P) = 2 \int_{1/2}^{1} \psi(ta, t\beta, \gamma)dt \leq CR\left(|\gamma| + \frac{a}{\beta}\right),$$

where C is an absolute constant.

Proof. Consider the function

$$\varphi(t) := B(ta, t\beta, \gamma) \tag{6.33}$$

for a. e. $(a, \beta, \gamma) \in G_1$. It is concave.

Let us first prove that

$$\int_{1/2}^{1} -\varphi''(t)dt \leq CR\left(|\gamma| + \frac{a}{\beta}\right). \tag{6.34}$$

This would imply

$$\int_{1/2}^{1} \psi(ta, t\beta, \gamma)dt \leq CR\left(|\gamma| + \frac{a}{\beta}\right),$$

because by the definitions (6.27) and (6.33) of ψ and φ we have

$$\psi(ta, t\beta, \gamma) = -t^2\varphi''(t).$$

To prove (6.34), let us consider an auxiliary function $r(t) := \varphi(1)t - \varphi(t)$. It is defined for $t \in [\max(\frac{|\gamma|}{a}, \frac{1}{\beta}), 1]$. At 1 it vanishes, it is convex, and it attains its maximum on its left endpoint $t_0 = \max(\frac{|\gamma|}{a}, \frac{1}{\beta})$. The last statement follows from the fact that $\varphi(t)/t$ is increasing; this is property five of Section 6.1.1.

So on $[t_0, 1]$,

$$r(t) \leq r(t_0) \leq \varphi(1)t_0 \leq Rat_0 \leq Ra\left(\frac{|\gamma|}{a} + \frac{1}{\beta}\right). \tag{6.35}$$

As $\varphi(t)/t$ is increasing, we have $t\varphi'(t) - \varphi(t) \geq 0$, and thus $r'(1) \leq 0$. Let us write down the Taylor formula for convex function $r(t)$ in the integral form, keeping in mind that $r(1) = 0, r'(1) \leq 0$:

$$r(t_0) = (t_0 - 1)r'(1) + \int_{t_0}^{1} dt \int_{t}^{1} r''(s)ds.$$

Fubini's theorem, (6.35), and $r'(1) \leq 0$ imply

$$\int_{t_0}^{1} (s - t_0)r''(s)ds \leq R\alpha\left(\frac{|\gamma|}{\alpha} + \frac{1}{\beta}\right).$$

But $t_0 \leq \frac{1}{8}$ by the assumptions of the lemma. So $\int_{1/2}^{1} r''(s)ds \leq \frac{8}{3}R\alpha(\frac{|\gamma|}{\alpha} + \frac{1}{\beta})$. Hence, as $r'' = -\varphi''$,

$$\int_{1/2}^{1} -\varphi''(s)ds \leq \frac{8}{3}R\alpha\left(\frac{|\gamma|}{\alpha} + \frac{1}{\beta}\right).$$

The proof of (6.34) is finished and this, as we saw at the beginning of the proof, gives Lemma 6.18. $\qquad\square$

6.4 Logarithmic blow-up

Recall that

$$G_3 = \left\{P \in G : |\gamma| \leq \frac{1}{1000}\alpha, \beta > 100\right\}.$$

By Lemma 6.18 we conclude that for any $P = (\alpha, \beta, \gamma) \in G_3$

$$[\psi_H] \cdot [-\gamma H_\gamma - (1/2 + \gamma^2)H_{\gamma\gamma}] \geq \frac{1}{4}\alpha^2 H_\alpha^2. \tag{6.36}$$

We will consider only points P such that

$$0 < \gamma \ll \alpha \ll \beta, \quad \alpha \leq 1.$$

The absolute constants C, c will vary from line to line.

Let us temporarily take for granted the following inequality, where c_1, c_2 are absolute positive constants:

$$\alpha \leq c_2\frac{\beta}{R} \quad \Rightarrow \quad H_\alpha(\alpha, \beta, \gamma) \geq c_1\beta, \quad \beta \in (1, Q/2]. \tag{6.37}$$

Using Lemma 6.18 we obtain

$$\psi_H \le CR\left(\gamma + \frac{\alpha}{\beta}\right).$$

Now we combine this inequality with (6.37) and (6.36) to obtain

$$-\gamma H_\gamma - (1/2 + \gamma^2)H_{\gamma\gamma} \ge c_3 \frac{\alpha^2\beta^2}{R(\frac{\alpha}{\beta} + \gamma)}. \tag{6.38}$$

Using the fact that we consider only $0 < \gamma \le \alpha \le 1$ we can rewrite (6.38) as

$$-\frac{2\gamma}{(1 + 2\gamma^2)}H_\gamma - H_{\gamma\gamma} \ge c_4 \frac{\alpha^2\beta^2}{R(\frac{\alpha}{\beta} + \gamma)}.$$

Using the integrating factor, we get

$$-[\mu(\gamma)H_\gamma]_\gamma \ge c_5 \frac{\alpha\beta^3}{R(1 + \frac{\beta}{\alpha}\gamma)}.$$

We integrate this inequality from 0 to γ to produce (we use the fact that $\mu(\gamma) \approx 1$ when γ is small)

$$-H_\gamma \ge c_6 \frac{\alpha^2\beta^2}{R} \log\left(1 + \frac{\beta}{\alpha}\gamma\right). \tag{6.39}$$

From now on let us fix α as follows:

$$\alpha = c_2 \frac{\beta}{R}, \tag{6.40}$$

where c_2 is from (6.37).

We integrate (6.39) from 0 to γ and use the positivity of H to produce

$$H(\alpha, \beta, 0) - H(\alpha, \beta, \gamma) \ge c_6 \frac{\alpha^3\beta}{R}\left[\left(1 + \frac{\beta}{\alpha}\gamma\right)\log\left(1 + \frac{\beta}{\alpha}\gamma\right) - \frac{\beta}{\alpha}\gamma\right] \ge c_7 \frac{\alpha^2\beta^2}{R}\gamma\log\left(\frac{\beta}{\alpha}\gamma\right). \tag{6.41}$$

The last inequality holds true because $\frac{\beta}{\alpha} = cR$. From now on we will fix γ and β,

$$\beta = \frac{Q}{4}, \quad \gamma = c_8 \frac{\beta}{R}, \tag{6.42}$$

where the absolute positive constant c_8 is much smaller than c_2 from (6.40). In particular, $\frac{\beta}{\alpha}\gamma \asymp \beta = \frac{Q}{4}$, so it is much bigger than 1. This justifies the last inequality in (6.41). This also gives

$$\gamma \ll \alpha.$$

We just obtained the following inequality:

$$\frac{\alpha^2\beta^2}{R}\gamma\log\left(\frac{\beta}{\alpha}\gamma\right) \le C(H(\alpha,\beta,0) - H(\alpha,\beta,\gamma)). \tag{6.43}$$

Let us use the fact that $B(\alpha,\beta,\gamma)$ is concave in γ (it is concave in all three variables) and that by its definition it is even in γ (see property 7 of Section 6.1.1). The same then holds for function H, which is just some averaging of B in the first two variables. Being even in γ on $\gamma \in [-\alpha,\alpha]$ and concave, it automatically decreases for $\gamma \in [0,\alpha]$, and the concavity and non-negativity of H give $H(\alpha,\beta,\gamma) \ge (1 - \frac{\gamma}{\alpha})H(\alpha,\beta,0)$. This allows us to estimate the right-hand side of (6.43), and we have

$$\frac{\alpha^2\beta^2}{R}\gamma\log\left(\frac{\beta}{\alpha}\gamma\right) \le C(H(\alpha,\beta,0) - H(\alpha,\beta,\gamma)) \le C\frac{\gamma}{\alpha}H(\alpha,\beta,0).$$

Taking into consideration one more time that $H(\alpha,\beta,\gamma) \le R\alpha$ by the definition of R in (6.32) and by the construction of H, we get

$$\frac{\alpha^2\beta^2}{R}\gamma\log\left(\frac{\beta}{\alpha}\gamma\right) \le C(H(\alpha,\beta,0) - H(\alpha,\beta,\gamma)) \le CR\gamma, \tag{6.44}$$

or

$$\frac{Q^4}{R^4}\log\left(\frac{\beta}{\alpha}\gamma\right) \le C. \tag{6.45}$$

As $\frac{\beta}{\alpha} = cR$ and $\gamma \asymp \frac{Q}{R}$, we can see that $\log(\frac{\beta}{\alpha}\gamma) \ge \log(cQ)$, from which it follows that

$$R \ge cQ(\log Q)^{\frac{1}{4}} \tag{6.46}$$

with a positive absolute c. Theorem 6.10 is proved with $\delta = \frac{1}{4}$.

We are left to prove (6.37).

Lemma 6.19. *Suppose $H(1,\beta,\gamma) \ge A$. Then the following holds:*

$$H_\alpha\left(\frac{A}{2R},\beta,\gamma\right) \ge \frac{1}{2}A.$$

Proof. Suppose it does not. Then $H_\alpha(\frac{A}{2R},\beta,\gamma) \le \frac{1}{2}A$. Then $H_\alpha(\alpha,\beta,\gamma) \le \frac{1}{2}A$ for all $\alpha \in [\frac{A}{2R},1]$ by the fact that H_α decreases in α as H is concave.

But $H(1,\beta,\gamma) - H(\frac{A}{2R},\beta,\gamma) \ge A - R\frac{A}{2R} = \frac{1}{2}A$ by the definition of R in (6.32) and the fact that H is a certain averaging of B.

On the other hand, $H(1,\beta,\gamma) - H(\frac{A}{2R},\beta,\gamma) = H_\alpha(\theta,\beta,\gamma)(1 - \frac{A}{2R})$, $\theta \in [\frac{A}{2R},1]$. We obtain (combining the last inequalities)

$$\frac{1}{2}A \le H(1,\beta,\gamma) - H\left(\frac{A}{2R},\beta,\gamma\right) < H_\alpha(\theta,\beta,\gamma) \le \frac{1}{2}A.$$

We come to contradiction, so the lemma is proved. $\qquad\qquad\square$

The combination of Lemma 6.19 and (6.51) proves inequality (6.37).

6.5 An obstacle condition on functions B and H

Now we want to show the following obstacle condition for B, which we already used:

$$\text{if } |y| < \frac{1}{4}, \quad \text{then } B(1, \beta, y) \geq \frac{\beta}{3}. \tag{6.47}$$

Let $I := [0, 1]$. Given numbers $(F, \beta, m, f, \lambda)$ such that $|f| < \lambda/4$, $\frac{F}{m} = \lambda$, $m \leq \beta \leq Qm$, it is enough to construct functions φ, ψ, w on I such that:

- (1) each of these functions has constant values on grandchildren of I,
- (2) if $\varphi = \langle \varphi \rangle_I + (\varphi, h_{I_-})h_{I_-} + (\varphi, h_{I_+})h_{I_+}$, then

$$\psi = -\lambda + (\varphi, h_{I_-})h_{I_-} - (\varphi, h_{I_+})h_{I_+},$$

- (3) $\langle w \rangle_I = \beta$, $\min_I w = m$,
- (4) the w-measure of the subset of I where

$$\psi \geq 0 \tag{6.48}$$

is at least $c\beta$, where c is an absolute positive constant. Notice that (6.48) is the same as $(\varphi, h_{I_-})h_{I_-} - (\varphi, h_{I_+})h_{I_+} \geq \lambda$.

We will construct such a triple (φ, ψ, w). Fix $\beta \in (1, Q]$. Put $\varphi = -a$ on I_{--}, $\varphi = b$ on I_{++}, and $\varphi = 0$ otherwise. Also, $w = 1$ on $I_{--} \cup I_{++}$, and $w = \beta$ otherwise. Then put

$$\psi := -\lambda + (\varphi, h_{I_-})h_{I_-} - (\varphi, h_{I_+})h_{I_+}.$$

Let $0 < a < b$ and let a be close to b. Put $\lambda = (a + b)/4$. Then the average of φ is $\frac{b-a}{4}$. It is small with respect to λ and we can prescribe it to be any number smaller than $\lambda/4$. $F = (a + b)/4$, $m = 1$.

On the other hand, function $\lambda + \psi$ (which is a martingale transform of $\varphi - \langle \varphi \rangle_I$) is at least $-(\varphi, h_{I_+})h_{I_+} \geq \frac{1}{2}b \geq \lambda$ on I_{+-}, whose w-measure is more than $\frac{1}{3}w(I)$. So

$$B(1, (1 + \beta)/2, y) \geq \frac{1}{3}\beta, \tag{6.49}$$

for all sufficiently small y.

By the concavity and positivity of B we see immediately that

$$B(\alpha, \beta, y) \geq c\beta, \quad \alpha \geq \frac{1}{100}, \tag{6.50}$$

with absolute positive c and for $\beta \in (1, Q/2]$.

Now, from the definition of functions H we conclude that the following obstacle condition holds for the function H:

$$H(1, \beta, \gamma) \geq \frac{1}{3}\beta, \tag{6.51}$$

for all sufficiently small γ and for $\beta \in (1, Q/2]$.

6.6 Improving the exponent

From (6.51) we know that (for all γ, $0 \leq \gamma \leq 1$)

$$H\left(1, \frac{Q}{4}, \gamma\right) \geq \frac{Q}{12}.$$

As $H(\alpha, \beta, \gamma) \leq R\alpha$ we immediately conclude that $H(\frac{Q}{24R}, \frac{Q}{4}, \gamma) \leq \frac{Q}{24}$ (for all γ, $0 \leq \gamma \leq \alpha :=$ $\frac{Q}{24R}$). This, combined with the inequality displayed above, gives us

$$H_\alpha\left(\frac{Q}{24R}, \frac{Q}{4}, \gamma\right) \geq \frac{1}{24}Q. \tag{6.52}$$

But maybe there is a better point $\tilde{\alpha} \gg \alpha := \frac{Q}{24R}$, where $H(\tilde{\alpha}, \frac{Q}{4}, \gamma) \leq \frac{Q}{24}$. Then automatically we have the same estimate for H_α at this point:

$$H_\alpha\left(\tilde{\alpha}, \frac{Q}{4}, \gamma\right) \geq \frac{1}{24}Q. \tag{6.53}$$

So let us consider the largest $\tilde{\alpha} \in [\alpha, 1]$ where $\alpha = \frac{Q}{24R}$ such that the following holds:

$$\text{if } H\left(\tilde{\alpha}, \frac{Q}{4}, 0\right) = \frac{Q}{24}, \quad \text{then } H\left(\tilde{\alpha}, \frac{Q}{4}, \gamma\right) \leq \frac{Q}{24}, \ \gamma \in [0, \tilde{\alpha}]. \tag{6.54}$$

Two cases may occur.

Case 1. $\tilde{\alpha} \geq \frac{Q^{1/2}}{24R^{1/2}}$. Then in (6.38) we can use these $\tilde{\alpha} \geq \frac{Q^{1/2}}{24R^{1/2}}$ and $\beta = \frac{Q}{4}$. We just follow (6.43) and (6.44) with these new data, but without any other changes, only γ in (6.44) can be between 0 and $\tilde{\alpha}$, so in particular, we can choose $\gamma = \frac{Q^{1/2}}{24R^{1/2}}$. Then instead of (6.45) we get

$$c\frac{Q^3}{R^3}\log\left(\frac{cQ}{\tilde{\alpha}}\gamma\right) = c\frac{Q^3}{R^3}\log\left(\frac{cQR^{1/2}}{Q^{1/2}} \cdot \frac{cQ^{1/2}}{R^{1/2}}\right) \leq C. \tag{6.55}$$

This implies

$$R \geq cQ\log^{1/3} Q. \tag{6.56}$$

Case 2. $\tilde{\alpha} \le \frac{Q^{1/2}}{24R^{1/2}}$. At $\alpha_1 := \min(\frac{Q}{48R}, \frac{2}{3}\tilde{\alpha})$ we have

$$H\left(\alpha_1, \frac{Q}{4}, \gamma\right) \le \frac{Q}{48}.$$

But we saw that $\tilde{\alpha} \ge \frac{Q}{24R}$ by its definition. Hence, $\alpha_1 = \frac{Q}{48\alpha}$. Comparing the last displayed inequality with (6.54) we conclude that

$$\tilde{\alpha} H_\alpha\left(\alpha_1, \frac{Q}{4}, \gamma\right) \ge (\tilde{\alpha} - \alpha_1) H_\alpha\left(\alpha_1, \frac{Q}{4}, \gamma\right)$$

$$\ge H\left(\tilde{\alpha}, \frac{Q}{4}, \gamma\right) - H\left(\alpha_1, \frac{Q}{4}, \gamma\right) \ge \left(1 - \frac{\gamma}{\tilde{\alpha}}\right) H\left(\tilde{\alpha}, \frac{Q}{4}, 0\right) - \frac{Q}{48}$$

$$\ge \left(1 - \frac{\gamma}{\tilde{\alpha}}\right) H\left(\tilde{\alpha}, \frac{Q}{4}, 0\right) - \frac{Q}{48} \ge \left(1 - \frac{\gamma}{\tilde{\alpha}}\right)\frac{Q}{24} - \frac{Q}{48} = \frac{Q}{144}$$

if $\gamma \in [0, \frac{2}{3}\alpha_1]$. Hence, using the fact that $\tilde{\alpha} \le \frac{Q^{1/2}}{24R^{1/2}}$, we obtain the improved estimate on the derivative

$$\forall \gamma \in \left[0, \frac{2}{3}\alpha_1\right], \quad H_\alpha\left(\alpha_1, \frac{Q}{4}, \gamma\right) \ge cQ^{1/2}R^{1/2}. \tag{6.57}$$

Then in (6.38) we can use this $\alpha := \alpha_1 = \frac{Q}{48}$ and $\beta = cQ^{1/2}R^{1/2}$, $\gamma = \frac{2}{3}\alpha_1$. We have the latter because in (6.38) β was the estimate from below by $H_\alpha(\alpha, \dots)$.

Now we have a new estimate from below, namely (6.57). We just follow (6.43) and (6.44) with these new data, but without any other changes, only γ in (6.44) can be between 0 and $\alpha_1 = \frac{Q}{48R}$. Then instead of (6.45) we get the following inequality:

$$c\frac{Q^2}{R^2}\frac{QR}{R} \log\left(\frac{cQ}{\alpha_1}\gamma\right) \le CR,$$

so again, having $\gamma = \frac{2}{3}\alpha_1$, we obtain

$$R \ge cQ \log^{1/3} Q.$$

6.7 Bellman function: viscosity solution of a degenerate elliptic equation

In fact, it is a viscosity supersolution of a certain degenerate elliptic equation that will be considered now. Let us remind the reader that we defined in (6.2) function **B** on domain Ω introduced in (6.3). We want to demonstrate in this short subsection that **B** is a supersolution in the viscosity sense of a certain degenerate elliptic equation.

We never used this before, but this knowledge might happen to be important. In particular, it may happen to be true that the reader more familiar with viscosity (super)solutions can simplify a bit our proof of Theorem 6.10, which we just finished proving. In this section $D^2 u$ denotes the Hessian matrix of u.

Definition 6.20. An equation $H(x, u, Du, D^2 u) = 0$, $x \in \Omega \subset \mathbf{R}^d$, on a function u defined in a domain Ω is called degenerate elliptic if function H satisfies the following condition: For any point $(x, u, p) \in \Omega \times \mathbf{R} \times \mathbf{R}^d$ and any two $d \times d$ real symmetric matrices X and Y, from $Y \geq X$ it follows that $H(x, u, p, X) \geq H(x, u, p, Y)$.

For example, $H(x, u, p, X) = -\operatorname{trace} X$ gives a degenerate elliptic equation $-\Delta u = 0$. Many examples of degenerate elliptic operators can be found in the first sections of [35] (see also [8]; our example below can be found there too).

Definition 6.21. A lower semicontinuous function u is called a viscosity supersolution of $H(x, u, Du, D^2 u) = 0$ (a degenerate elliptic equation) if for every point $x_0 \in \Omega$ and every C^2-function φ such that (1) $\varphi(x_0) = u(x_0)$ and (2) $\varphi(x) \leq u(x)$ for x in a small neighborhood of x_0 inside Ω, one has $H(x_0, \varphi(x_0), D\varphi(x_0), D^2\varphi(x_0)) \geq 0$.

To define viscosity subsolutions, one changes lower to upper semicontinuous, requires $\varphi(x) \geq u(x)$ for x in a small neighborhood of x_0 inside Ω, and concludes that $H(x_0, \varphi(x_0), D\varphi(x_0), D^2\varphi(x_0)) \leq 0$.

To define the degenerate elliptic equation whose viscosity supersolution is $\mathbf{B}$ in Ω from (6.3), we consult Theorem 6.7 and especially inequality (6.8).

Our function $H(x, u, p, X)$ will depend only on matrices X that run over 5×5 real symmetric matrices. A vector v in $\mathbf{R}^5$ is called adapted if $v = (v_1, v_2, 0, 0, v_5)$, $\|v\| = 1$. The set of adapted vectors is called $\mathcal{A}$. Let us consider the following H_{wmt}, where the subscript stands for "weak martingale transform":

$$H_{\mathrm{wmt}}(X) := -\sup_{v \in \mathcal{A}}\left[(Xv, v) + X_{44} v_5^2\right].$$

It is very easy to check that if $Y \geq X$ are two real symmetric matrices, then $H(X) \geq H(Y)$.

Let us see that $\mathbf{B}$ from (6.2) satisfies all viscosity supersolution conditions of $H_{\mathrm{wmt}}(D^2 u) = 0$ in Ω from (6.3). The lower semicontinuity of $\mathbf{B}$ follows easily from its definition. Now let us fix $x_0 = (F, w, m, f, \lambda)$. If a smooth φ satisfies $\varphi(x) \leq \mathbf{B}(x)$ in a neighborhood of this x_0, then

$$\varphi(F \pm dF, w \pm dw, m, f \pm df, \lambda + \pm d\lambda) \leq \mathbf{B}(F \pm dF, w \pm dw, m, f \pm df, \lambda + \pm d\lambda)$$

for all sufficiently small real numbers dF, dw, df, $d\lambda$. Of course we also have $\varphi(F, w, m, f, \lambda) = \mathbf{B}(F, w, m, f, \lambda)$. Now it automatically follows from (6.8) that for all sufficiently small real numbers dF, dw, df, $d\lambda$, the following holds:

$$\varphi(F, w, m, f, \lambda) \geq \frac{1}{4}\left(\varphi(F - dF, w - dw, m, f - d\lambda, \lambda - d\lambda)\right.$$

$$+ \varphi(F - dF, w - dw, m, f + d\lambda, \lambda - d\lambda)$$
$$+ \varphi(F + dF, w + dw, m, f - d\lambda, \lambda + d\lambda)$$
$$+ \varphi(F + dF, w + dw, m, f + d\lambda, \lambda + d\lambda)). \tag{6.58}$$

The function φ is smooth. Let us use Taylor's formula for all terms on the right-hand side of (6.58). We can easily see that $\varphi(F, w, m, f, \lambda)$ will disappear together with all terms having the first derivatives of φ. After some simple algebra, which we leave to the reader, we can see that (6.58) implies an "infinitesimal" version of itself, which holds for any triple $(dF, dw, d\lambda)$:

$$-(\varphi_{FF}(dF)^2 + 2\varphi_{Fw}dFdw + \varphi_{ww}(dw)^2 + \varphi_{\lambda\lambda}(d\lambda)^2 + 2\varphi_{F\lambda}dFd\lambda + 2\varphi_{w\lambda}dwd\lambda + \varphi_{ff}(d\lambda)^2) \geq 0. \tag{6.59}$$

The reader can immediately see by the definition of H_{wmt} that we just proved that

$$H_{\mathrm{wmt}}(D^2\varphi(x_0)) \geq 0.$$

This means exactly that **B** is a viscosity supersolution of a degenerate elliptic equation $H_{\mathrm{wmt}}(D^2u) = 0$.

7 Random walk interpretation

In this section we want to prepare the ground for proving our main result, Theorem 6.3. We consider again the domain (6.3), namely,

$$\Omega^s := \{(F, w, m, f, \lambda) \in \mathbf{R}^5 : F \geq |f| \, m, \ m \leq w \leq Q\,m\}. \tag{7.1}$$

We consider special random walks in this domain. From point (F, w, m, f, λ) in Ω^s we move with equal probability 1/4 to the following four points (they have to be in this same domain Ω^s):

$$(F - dF, w - dw, m_1, f - d\lambda, \lambda - td\lambda), \quad (F - dF, w - dw, m_2, f - d\lambda, \lambda + td\lambda),$$
$$(F + dF, w + dw, m_3, f + d\lambda, \lambda - td\lambda), \quad (F + dF, w + dw, m_4, f + d\lambda, \lambda + td\lambda),$$

where $\min(m_1, m_2, m_3, m_4) = m$. Symbols $dF, dw, d\lambda$ are just real numbers and $t \in [0, 1]$. The only condition on them is that we stay in Ω^s after performing this one-step random walk.

Now we make another random step, each from the four points listed above. We consider such random walks, which also satisfy two conditions: (a) there are only a finite number of steps and (b) the last step brings us to the following part of the boundary of Ω^s:

$$\partial_w\Omega^s = \{(F, w, m, f, \lambda) : w = m\}. \tag{7.2}$$

Let us call such random walks $\mathcal{W}^s$.

Every random walk is the collection of four martingales $\mathbf{F}$, $\mathbf{w}$, $\mathbf{f}$, Λ and the "martingale-with-respect-to-minimum" M. Martingales $\mathbf{f}$ and Λ are strongly dependent. The random vector $(\mathbf{F}, \mathbf{w}, M, \mathbf{f}, \Lambda)$ should stay in Ω^s and should finish at $\partial_w \Omega^s$. We have a natural probability measure on $\mathcal{W}^s$; the expectation will be called $\mathbf{E}$.

If we start with the point (F, w, m, f, λ) in Ω^s, let us denote by

$$(\mathbf{F}(\omega), \mathbf{w}(\omega), M(\omega), \mathbf{f}(\omega), \Lambda(\omega))$$

a vector function we get after the walk ends at $\partial_w \Omega^s$. In particular, $\mathbf{w}(\omega) = M(\omega)$ identically.

We introduce

$$\mathbb{V}(F, w, m, f, \lambda) := \mathbb{V}_Q(F, w, m, f, \lambda) := \sup \mathbf{E} \mathbf{1}_{\Lambda(\omega) \leq 0},$$

where supremum is taken over all walks in $\mathcal{W}^s$ starting at (F, w, m, f, λ).

Theorem 7.1. *The function* $V = V_Q$ *satisfies all the same properties as* $\mathbf{B}_Q$ *from Section 6.1.1 with one change; instead of properties (6.6) and (6.7) it satisfies the analog of (6.8), namely,*

$$
\begin{aligned}
\mathbb{V}(F, w, m, f, \lambda) \geq \frac{1}{4}\big(& \mathbb{V}(F - dF, w - dw, m, f - d\lambda, \lambda - td\lambda) \\
& + \mathbb{V}(F - dF, w - dw, m, f - d\lambda, \lambda + td\lambda) \\
& + \mathbb{V}(F + dF, w + dw, m, f + d\lambda, \lambda - td\lambda) \\
& + \mathbb{V}(F + dF, w + dw, m, f + d\lambda, \lambda + td\lambda)\big).
\end{aligned}
\tag{7.3}
$$

The proof is ad verbatim the same as the proof of Theorem 6.7; it is based on the same trick of concatenation.

We can introduce function V starting with function $\mathbb{V}$ in ad verbatim the same manner as in (6.4), namely,

$$\frac{1}{m}\mathbb{V}(F, w, m, f, \lambda) = V\left(\frac{F}{m\lambda}, \frac{w}{m}, \frac{f}{\lambda}\right) =: V(\alpha, \beta, \gamma), \tag{7.4}$$

defined in the same domain $G = \{(\alpha, \beta, \gamma) : |\gamma| \leq \alpha, 1 \leq \beta \leq Q\}$.

Theorem 7.2.

$$\text{If } |\gamma| < \frac{1}{4}, \quad \text{then } V(1, \beta, \gamma) \geq \frac{\beta}{3}. \tag{7.5}$$

To show this we just notice that in Section 6.5 we constructed a one-step random walk from $\mathcal{W}^s$ such that (7.5) is ensured by Item (4) of Section 6.5.

In the proof of Theorem 6.10 we used only the properties of $\mathbf{B}$ and B that were listed in Section 6.1.1 and the obstacle condition (6.47). More precisely, we never used properties (6.6) and (6.7); only (6.8) was used.

But we have all those ingredients now ready for $\mathbb{V}$ and V. Therefore, we have already proved the following result.

Theorem 7.3. *There exists an absolute positive constant c_V such that*

$$\sup_{(F,w,m,f,\lambda)\in\mathfrak{Q}^s} \frac{|\lambda|\mathbb{V}_Q(F,w,m,f,\lambda)}{F} = \sup_{(\alpha,\beta,\gamma)\in G} \frac{V(\alpha,\beta,\gamma)}{\alpha} \geq c_V Q(\log Q)^{1/3}. \tag{7.6}$$

8 Full weak estimates of square functions, possibly with extra logarithm

The sharp weak weighted estimate for square functions of various types was given considerable attention in [27, 26, 20, 14] and [49, Section 5.10]. We already studied the sharp testing condition in Section 3 and the sharp restricted condition in Section 4. We saw that those sharp estimates are of order $[w]_{A_2}^{1/2}$. It is time to research the full problem of sharp weak weighted estimates.

An interesting point is that this full estimate is the only one left without fully understanding what is sharp: is it $[w]_{A_2}^{1/2}$ or $[w]_{A_2}^{1/2}\log^{1/2}(1+[w]_{A_2})$?

All other possible sharp weak estimates are known. Namely, for $p \in (1,2)$ the estimate by $[w]_{A_p}^{1/p}$ is proved in [27]; if $p \in [2,\infty)$, then the estimate of the type $[w]_{A_p}^{1/2}\log^{1/2}(1+[w]_{A_\infty})$ is proved in [14]. Finally, for $p \in (2,\infty)$ in [20] Hytönen–Li removed the logarithm; the estimate is $[w]_{A_p}^{1/p}[w]_{A_\infty}^{1/2-1/p} \leq [w]_{A_p}^{1/2}$. But this is for $p > 2$. So for both $p < 2$ and $p > 2$ there is a satisfactory sharp estimate. The borderline case $p = 2$ remains unclear. Sections 3 and 4 indicate that the sharp estimate might be $[w]_{A_2}^{1/2}$. So does [20]. On the other hand, [27, 14] point to $[w]_{A_p}^{1/2}\log^{1/2}(1+[w]_{A_\infty})$.

We follow [27] to prove here this latter estimate for the weak norm $L^2(w) \to L^{2,\infty}$ of the square function.

For the convenience of the reader we repeat some definitions from Section 4.0.1.

Definition. A family $\mathcal{S}$ of intervals of $\mathcal{D}$ is called ε-sparse if the following condition is satisfied:

$$\sum_{\substack{I\in\mathcal{S},\\ I\subsetneq J}} |I| \leq \varepsilon|J|, \quad \forall J \in \mathcal{S}. \tag{8.1}$$

Definition. The sparse square function operator is defined for each sparse family $\mathcal{S}$ as follows:

$$S^{\mathrm{sp}}\varphi \overset{\text{def}}{=} S_{\mathcal{S}}^{\mathrm{sp}}\varphi \overset{\text{def}}{=} \left(\sum_{I\in\mathcal{S}} \langle\varphi\rangle_I^2 \mathbf{1}_I\right)^{1/2}.$$

Theorem 8.1. *For any $\varepsilon > 0$ and any $\varphi \in L^1$ there exist a constant $C = C(\varepsilon)$ independent of φ and a sparse family S (depending on ε and φ) such that pointwise almost everywhere*

$$S\varphi \le CS^{\mathrm{sp}}\varphi.$$

Theorem 8.2. *The norm $\|S : L^2(w) \to L^{2,\infty}(w)\|$ is bounded by $C[w]_{A_2}^{1/2}\log^{1/2}(1 + [w]_{A_\infty})$.*

The combination of the following inequality with the sparse domination theorem, Theorem 8.1, gives us Theorem 8.2:

$$\lambda^2 w\{x \in (0,1): S^{\mathrm{sp}}\varphi(x) > \lambda\} \le C[w]_{A_2}\log(1 + [w]_{A_\infty})\,\|\varphi\|_w^2. \tag{8.2}$$

To prove this inequality we need several lemmas. Everywhere below we deal with dyadic classes A_p and we omit the superscript d in A_p^d; also, everywhere below all intervals are dyadic.

Lemma 8.3. *Let $w \in A_2$. Consider the operator of averaging over one interval: $f \to \langle f\rangle_I \mathbf{1}_I$. Its norm as operator in $L^2(w)$ is bounded by $[w]_{A_2}^{1/2}$.*

Proof. Proving this lemma is the same as proving that for any non-negative function f we have

$$\langle f\rangle_I^2 w(I) \le [w]_{A_2}\|f\|_w^2. \tag{8.3}$$

In the next chain we use the Cauchy inequality:

$$\langle f\rangle_I^2 w(I) = \langle fww^{-1}\rangle_I^2 w(I) = \left(\frac{1}{w^{-1}(I)}\int_I fww^{-1}\right)^2 \frac{w^{-1}(I)w(I)}{|I|^2}w^{-1}(I)$$

$$\le \frac{w^{-1}(I)w(I)}{|I|^2}w^{-1}(I)\frac{1}{w^{-1}(I)}\int_I f^2w^2w^{-1} = \frac{w^{-1}(I)w(I)}{|I|^2}\int_I f^2w.$$

We are done proving (8.3). $\qquad\square$

Lemma 8.4. *Let $w \in A_2$ and let $\{f_I\}_{I\in\mathcal{F}}$ be an arbitrary collection of non-negative measurable functions where f_I is supported on interval Q belonging to L^2. Then*

$$\left\|\left[\sum_{I\in\mathcal{F}}\langle g_I\rangle_I^2\mathbf{1}_I\right]^{1/2}\right\|_{L^2(w)} \le A[w]_{A_2}^{1/2}\left\|\left[\sum_{I\in\mathcal{F}}g_I^2\right]^{1/2}\right\|_{L^2(w)}.$$

Proof. One just has to apply Lemma 8.3 to each function f_I. $\qquad\square$

Lemma 8.5. *Let $w \in A_\infty$. Then there exists an absolute constant a, $a > 0$, such that for $\delta = a/[w]_{A_\infty}$ the following reverse Hölder inequality holds:*

$$\langle w^{1+\delta}\rangle_I^{\frac{1}{1+\delta}} \le 2\langle w\rangle_I.$$

Proof. The Bellman function proof is due to V. Vasyunin [48]. The stopping time proof can be found in the paper of T. Hytönen and C. Pérez [21]. $\qquad\square$

Lemma 8.6. *Let G be a measurable subset of interval I. Let $w \in A_\infty$. Then there exists an absolute constant b, $b > 0$, such that*

$$\frac{w(G)}{w(I)} \le 2\left(\frac{|G|}{|I|}\right)^{\frac{1}{b[w]_\infty}}.$$

Proof. Denote $r = 1 + \delta$, where δ is from Lemma 8.5. As always r' denotes the dual exponent. We have

$$\frac{w(G)}{|I|} = \langle w\mathbf{1}_G\rangle_I \le \langle w^r\rangle_I^{1/r} \frac{|G|^{1/r'}}{|I|^{1/r'}} \le 2\langle w\rangle_I \cdot \frac{|G|^{1/r'}}{|I|^{1/r'}} = 2\frac{|G|^{1/r'}}{|I|^{1/r'}} \cdot \frac{w(I)}{|I|},$$

which implies the lemma because clearly $r' \asymp [w]_{A_\infty}$. $\qquad\square$

Now let $\mathcal{S}$ be an ε-sparse system of dyadic intervals and let $\mathcal{S}^{\mathrm{sp}}$ be a corresponding sparse square function. We split $\mathcal{S} = \mathcal{S}^0 \cup \bigcup_{m=0}^\infty \mathcal{S}_m$, where $\mathcal{S}^0$ consists of intervals of $\mathcal{S}$ such that $\langle \varphi \rangle_I > 1$ and $\mathcal{S}_{m+1}$ consists of intervals of $\mathcal{S}$ such that

$$2^{-m-1} < \langle \varphi \rangle_I \le 2^{-m}, \quad m = 0, 1, \ldots.$$

We denote

$$S_m^{\mathrm{sp}} \varphi \stackrel{\mathrm{def}}{=} \left(\sum_{I \in \mathcal{S}_m} \langle \varphi \rangle_I^2 \mathbf{1}_I \right)^{1/2}.$$

We are going to estimate these measures:

$$W_0 \stackrel{\mathrm{def}}{=} w\{x \in I_0 : (S_0^{\mathrm{sp}}\varphi)^2 > 1\},$$

$$W_1 \stackrel{\mathrm{def}}{=} w\left\{x \in I_0 : \sum_{m=1}^{m_0} (S_m^{\mathrm{sp}}\varphi)^2 > 1\right\},$$

$$W_2 \stackrel{\mathrm{def}}{=} w\left\{x \in I_0 : \sum_{m=m_0+1}^{\infty} (S_m^{\mathrm{sp}}\varphi)^2 > 1\right\}.$$

We will choose m_0 a bit later.

Finding the estimate of W_0 is easy. The sum $(S_0^{\mathrm{sp}}\varphi)^2$ is supported on intervals where the dyadic maximal function of φ is bigger than 1. The set where the dyadic maximal function is bigger than one has w-measure bounded by $[w]_{A_2}\|\varphi\|_w^2$ by Lemma 4.3.

To estimate W_1, we first introduce for each $I \in \mathcal{S}_m$, $1 \le m \le m_0$,

$$E_m(I) \stackrel{\mathrm{def}}{=} I \setminus \bigcup_{\substack{J \subsetneq I \\ J \in \mathcal{S}_m}} J.$$

Notice that

$$\langle \varphi \mathbf{1}_{E_m(I)} \rangle_I \geq \frac{1}{4} \langle \varphi \rangle_I \tag{8.4}$$

for any $I \in \mathcal{S}_m$ if $\varepsilon \leq \frac{1}{4}$. In fact, if $\{J^*\}$ mean the maximal intervals from $\mathcal{S}_m$ lying strictly inside I, then

$$\int_I \varphi \mathbf{1}_{E_m(I)} = \int_I \varphi - \sum_{J^*} \int_{J^*} \varphi \geq 2^{-m-1}|I| - 2^{-m} \sum_{J^*} |J^*| \geq 2^{-m}|I| \left(\frac{1}{2} - \varepsilon \right) \geq 2^{-m-2}|I|,$$

if $\varepsilon \leq \frac{1}{4}$. Thus, (8.4) holds.

Now we use the Chebyshev inequality. For that we first estimate

$$\int_{I_0} (S_m^{\mathrm{sp}}\varphi)^2 \, wdx = \int_{I_0} \sum_{I \in \mathcal{S}_m} [\langle \varphi \rangle_I^2 \mathbf{1}_I] \, wdx$$

$$\leq 16 \int_{I_0} \sum_{I \in \mathcal{S}_m} [\langle \varphi \mathbf{1}_{E_m(I)} \rangle_I^2 \mathbf{1}_I] \, wdx$$

$$\leq 16[w]_{A_2} \int_{I_0} \sum_{I \in \mathcal{S}_m} [(\varphi \mathbf{1}_{E_m(I)})^2] \, wdx$$

$$\leq 16[w]_{A_2} \int_{I_0} \varphi^2 \, wdx = 16[w]_{A_2} \|\varphi\|_w^2.$$

The first inequality uses (8.4), the second one is just Lemma 8.4, and the last inequality follows because by construction the sets $E_m(I)$ are disjoint when I runs over $\mathcal{S}_m$. Now the Chebyshev inequality gives us

$$W_1 \leq 16m_0[w]_{A_2} \|\varphi\|_w^2.$$

We are left to estimate W_2. Notice that the support of $S_m^{\mathrm{sp}}\varphi$ is in $\cup Q^*$, where we denote by Q^* the maximal dyadic intervals in the family $\mathcal{S}_m$ (we drop the index m in the notation of these intervals).

We use the notation

$$b_m \overset{\text{def}}{=} 2^{-2m} (S_m^{\mathrm{sp}}\varphi)^2.$$

Notice that b_m is supported on $\cup Q^*$ and that

$$b_m \leq \sum_{I \in \mathcal{S}_m} \mathbf{1}_I. \tag{8.5}$$

For every such Q^* we denote by $G_j(Q^*)$ the subset of Q^*, where $b_m \geq j$.

On $G_j(Q^*)$ we have $\sum_{I \in \mathcal{S}_m} 1_I \geq j$. This means that maximal intervals of $G_j(Q^*)$ lie at least j generations below Q^*. From the ε-sparseness of $\mathcal{S}_m$ one immediately concludes that

$$\frac{|G_j(Q^*)|}{|Q^*|} \leq e^{-Cj}, \tag{8.6}$$

where $C = \log \frac{1}{\varepsilon}$. By Lemma 8.6 we get

$$\frac{w(G_j(Q^*))}{w(Q^*)} \leq 2e^{-cj/[w]_{A_\infty}}. \tag{8.7}$$

To estimate W_2 we will use (8.7) and a rather crude union estimate:

$$w\left\{ x \in Q^*: \sum_{m=m_0+1}^{\infty} (S_m^{\mathrm{sp}} \varphi)^2 > 1 \right\}$$

$$= w\left\{ x \in Q^*: \sum_{m=m_0+1}^{\infty} (S_m^{\mathrm{sp}} \varphi)^2 > \sum_{m=m_0+1}^{\infty} 2^{m_0-m} \right\}$$

$$= w\left\{ x \in Q^*: \sum_{m=m_0+1}^{\infty} b_m > \sum_{m=m_0+1}^{\infty} 2^{m_0+m} \right\}$$

$$\leq \sum_{m=m_0+1}^{\infty} w\{ x \in Q^*: b_m > 2^{m_0+m} \} \leq 2w(Q^*) \sum_{m=m_0+1}^{\infty} e^{-\frac{c2^{m_0}}{[w]_{A_\infty}} \cdot 2^m}.$$

On the other hand,

$$w(\cup Q^*) \leq A[w]_{A_2} 2^{2m} \|\varphi\|_w^2.$$

This is the sharp weak-type estimate for the maximal function because $\cup Q^*$ is contained in the set where the (dyadic) maximal operator $M\varphi$ is bigger than 2^{-m-1}. In the next chain of inequalities, A denote absolute constants that can change from line to line. We have

$$W_2 = w\left\{ x \in \cup Q^*: \sum_{m=m_0+1}^{\infty} (S_m^{\mathrm{sp}} \varphi)^2 > 1 \right\}$$

$$\leq A[w]_{A_2} \sum_{m=m_0+1}^{\infty} 2^{2m} e^{-\frac{c2^{m_0}}{[w]_{A_\infty}} \cdot 2^m}$$

$$\leq A[w]_{A_2} \int_{m_0}^{\infty} 2^{2x} e^{-c\frac{2^{m_0}}{[w]_{A_\infty}} \cdot 2^x} dx$$

$$\leq A[w]_{A_2} \int_{2^{m_0}}^{\infty} y^2 e^{-c\frac{2^{m_0}}{[w]_{A_\infty}} \cdot y} \frac{dy}{y} = A[w]_{A_2} \int_{2^{m_0}}^{\infty} y e^{-c\frac{2^{m_0}}{[w]_{A_\infty}} \cdot y} dy$$

$$= A[w]_{A_2} \left(\frac{[w]_{A_\infty}}{2m_0} \right)^2 \int_{\frac{2^{2m_0}}{[w]_{A_\infty}}}^{\infty} u e^{-u} du.$$

Gathering together the estimates for W_0, W_1, W_2 and choosing $2^{m_0} \asymp 1 + [w]_{A_\infty}$, we get

$$w\{x \in I_0 : S^{\mathrm{sp}}\varphi > 3\} \le A\left([w]_{A_2} + [w]_{A_2}\log(1 + [w]_{A_\infty}) + [w]_{A_2} \int_{[w]_{A_\infty}}^{\infty} u e^{-u} du \right) \|\varphi\|_w^2. \quad (8.8)$$

Inequality (8.2) is finally proved.

This proves that the norm of S^{sp} as the operator from $L^2(w)$ to $L^{2,\infty}(w)$ is at most $A[w]_{A_2}^{1/2}\log^{1/2}(1 + [w]_{A_\infty})$. Theorem 8.2 is completely proved.

Remark 8.7. The reader may suspect that by choosing m_0 differently, e. g., by choosing $m_0 \approx \log(1 + [w]_{A_\infty})/\log\log(e + [w]_{A_\infty})$, one can optimize the estimate of $W_1 + W_2$. However, it is easy to see that the choice of $m_0 \approx \log(1 + [w]_{A_\infty})$ above was already optimal.

Remark 8.8. In the reasoning above one can change L^2 to L^p and A_2 to A_p, $p > 2$. Then just minor modifications are needed to prove that the norm of S^{sp} as the operator from $L^p(w)$ to $L^{p,\infty}(w)$ is at most $[w]_{A_p}^{1/2}\log^{1/2}(1 + [w]_{A_\infty})$. As shown in [27] for $1 \le p < 2$, one can drop the term W_1 completely and conclude that the norm of S^{sp} as the operator from $L^p(w)$ to $L^{p,\infty}(w)$ is at most $[w]_{A_p}^{1/2}$.

Remark 8.9. We saw in Section 4 what can be changed in the above reasoning in order to obtain an estimate better than (8.8) for a special choice of $\varphi = 1_E w^{-1}$. We saw that we can drop the logarithmic correction for this particular case called restricted weak estimate:

$$w\{x \in I_0 : S^{\mathrm{sp}} w^{-1} > 3\} \le A[w]_{A_2} \int_{I_0} w^{-1}\, dx. \quad (8.9)$$

9 Addendum: PDEs for strong sharp weighted estimates for square functions

We use an approach that is quite different from the approach in [18], where such sharp estimate was first obtained. Notice that more general sharp weighted estimates for various square functions in $L^p(w)$ were obtained since then by Lerner [28], Lacey–Li [26], and Hytönen–Li [20]. However, the estimate we establish below is paradigmatic and also the simplest of those in [28, 26, 20]. We follow the exposition in [23]. What we pursue here is (a) to demonstrate a method that completely avoids stopping time arguments and (b) to show how different estimates of strong type are compared to the estimates of weak type worked out above. The sharp dependence on a weight's norm is of different magnitude. We have

$$\frac{1}{|J|} \sum_{I \in D(J)} |\Delta_I w^{-1}|^2 \langle w \rangle_I |I| \le C_{s,T}([w]_{A_2}) \langle w^{-1} \rangle_J, \tag{9.1}$$

$$\frac{1}{|J|} \sum_{I \in D(J)} |\Delta_I (\varphi w^{-1})|^2 \langle w \rangle_I |I| \le C_s([w]_{A_2}) \langle \varphi^2 w^{-1} \rangle_J. \tag{9.2}$$

Remark. We are working with operator $S : L^2(w) \to L^{2,\infty}(w)$ or $S : L^2(w) \to L^2(w)$. However, it is more convenient to work with isomorphic objects: $S_{w^{-1}} : L^2(w^{-1}) \to L^{2,\infty}(w)$ or $S_{w^{-1}} : L^2(w^{-1}) \to L^2(w)$, where $S_{w^{-1}}$ denotes the product $SM_{w^{-1}}$, where $M_{w^{-1}}$ is the operator of multiplication.

9.1 Sharp estimates for testing conditions

Definition. For a smooth function B of d real variables $(x_1, \ldots, x_d)$ we denote by $d^2 B(x)$ the second differential form of B, namely,

$$d^2 B(x) = (H_B(x)dx, dx)_{\mathbf{R}^d},$$

where vector $dx = (dx_1, \ldots, dx_d)$ is an arbitrary vector in $\mathbf{R}^d$ and $H_B(x)$ is the $d \times d$ matrix of the second derivatives of B (Hessian matrix) at point $x \in \mathbf{R}^d$.

For brevity we write A_2 in this section, but we mean the dyadic class A_2.

Theorem 9.1. *We have $C_{s,T} \le A[w]_{A_2}^2$ and this estimate is sharp.*

We introduce the following function of two real variables:

$$\mathbf{B}(u,v) := \mathbf{B}_Q(u,v) := \sup \frac{1}{|J|} \sum_{I \in D(J)} |\Delta_I w^{-1}|^2 \langle w \rangle_I |I|, \tag{9.3}$$

where supremum is taken over all $w \in A_2$, $[w]_{A_2} \le Q$, such that

$$\langle w \rangle_J = u, \quad \langle w^{-1} \rangle_J = v.$$

Notice that by a scaling argument our function does not depend on J but it does depend on $Q = [w]_{A_2}$. Notice also that function $\mathbf{B}_Q$ is defined in the domain O_Q, where

$$O_Q := \{(u,v) \in \mathbf{R}^2 : 1 < uv \le Q\}.$$

Function $\mathbf{B}_Q$ is the Bellman function of our problem. In particular, it is very easy to observe that proving the estimate in Theorem 9.1 is equivalent to proving $\mathbf{B}_Q(u,v) \le A\,Q^2$, and the sharpness in Theorem 9.1 is just the claim that $\sup_{(u,v)\in O_Q} \frac{1}{v}\mathbf{B}_Q(u,v) \ge c > 0$.

Remark. Therefore, Theorem 9.1 can be proved by finding the explicit formula for $\mathbf{B}_Q$. To do that we obviously need to solve an infinite-dimensional optimization problem of finding the (almost) best possible $w \in A_2$ (the reader is reminded that it is a dyadic class) such that $\langle w\rangle_J = u$, $\langle w^{-1}\rangle_J = v$. This can be done, and we give this formula in the Addendum. But here we adopt a slightly different approach to prove Theorem 9.1.

Proof. Below, A, a are positive absolute constants. Instead of finding precisely $\mathbf{B}_Q$, we will find another function B_Q such that the following properties are satisfied:

- B_Q is defined in O_{4Q}.
- $0 \le B_Q(u,v) \le A\,Q^2 v$.
- $B_Q(u,v) - \frac{B_Q(u_+,v_+)+B_Q(u_-,v_-)}{2} \ge a(v_+ - v_-)^2 u$ if $x = (u,v) \in O_Q$, $x_\pm = (u_\pm, v_\pm) \in O_Q$, and $x = \frac{x_+ + x_-}{2}$.

For example, it is not difficult to check that $\mathbf{B}_{4Q}$ satisfies the first and third properties. We leave it as an exercise to the reader. However, as we said, the second property is not easy to verify; it can be observed when the complicated formula for $\mathbf{B}$ is written down (see Addendum). Here we will write down an explicit (and rather easy) form of *some B_Q* that satisfies all three properties.

First let us observe that if the existence of such a B_Q is proved, the inequality in Theorem 9.1 gets proved. In fact, fix $I \in D(J)$ and introduce $x_I = (\langle w\rangle_I, \langle w^{-1}\rangle_I)$, $x_{I_\pm} = (\langle w\rangle_{I_\pm}, \langle w^{-1}\rangle_{I_\pm})$. Of course $\{x_I\}_{I\in D(J)}$ is a martingale. Now we compose this martingale with B_Q (notice that $x_I \in O_Q$, so $B_Q(x_I)$ is well-defined). The resulting object is not a martingale anymore, but it is a supermartingale; moreover, by the third property,

$$\langle w\rangle_I \big|\Delta_I w^{-1}\big|^2 |I| \le |I|B_Q(x_I) - |I_+|B_Q(x_{I_+}) - |I_-|B_Q(x_{I_-}).$$

Now the reader knows what happens next: We use the telescopic nature of the sum on the right-hand side to observe that the summation in all $I \in D(J)$ cancels all the terms except $|J|B_Q(x_J)$, which by the second property is at most $A\,Q^2\langle w^{-1}\rangle_J|J|$. Hence, we obtain

$$\sum_{I\in D(J)} \big|\Delta_I w^{-1}\big|^2 \langle w\rangle_I |I| \le A\,Q^2 \langle w^{-1}\rangle_J |J|.$$

We leave it as an exercise for the reader to explain where we used the positivity of B in this reasoning.

The last estimate is precisely inequality (9.1), and Theorem 9.1 gets proved (apart from the sharpness) as soon as any function B_Q as above is proved to exist.

The construction of a certain B_Q with the above-mentioned three properties is split into two steps.

Step 1. The reduction to non-linear ODEs.

First of all we wish to find a smooth $B(u, v)$ in the domain

$$O := O_Q := \{(u, v) > 0 : 1 \le uv \le Q\},$$

such that the following quadratic form inequality holds in O_Q:

$$-\frac{1}{2} d^2 B \ge u(dv)^2. \tag{9.4}$$

We will be searching for homogeneous B: $B(u/t, tv) = tB(u, v)$. Hence,

$$B(u, v) = \frac{1}{u} \phi(uv).$$

Then (9.4) becomes

$$\begin{bmatrix} x^2 \phi''(x) - 2x\phi'(x) + 2\phi & x\phi''(x) \\ x\phi''(x) & \phi''(x) + 2 \end{bmatrix} \le 0.$$

To have this it is enough to satisfy for all $x \in [1, Q]$

$$\phi''(x) + 2 \le 0, \quad -x\phi'(x) + \phi(x) \le 0,$$
$$\phi'' \cdot (-x\phi' + \phi) + x^2 \phi'' - 2x\phi' + 2\phi = 0. \tag{9.5}$$

The last equation just makes the determinant of our matrix vanish. Let us start with this equation and put $g = \phi(x)/x$. Then we know that $-x^2 g' = \phi - x\phi' \le 0$, so g is increasing. Also, $xg'' + 2g' = \phi'' \le -2$; hence, $g'' \le 0$ as g was noticed to be increasing.

In terms of g we have

$$x(-g'g'' + g'') - 2(g')^2 = 0.$$

This is a first-order non-linear ODE on $h := g'$ for which we know that $h \ge 0, h' \le 0$:

$$x(-hh' + h') - 2h^2 = 0.$$

Variables can be separated; we get

$$\frac{1 - h}{h^2} h' = \frac{2}{x}. \tag{9.6}$$

We know that $h' = g''$ is negative and x here is positive, so $h \ge 1$, and the condition $\phi - x\phi' \le 0$ is the same as $h \ge 0$. Thus, any solution $h \ge 1$ of (9.6) gives the desired result.

We want to solve this for $x \in [1, Q]$:

$$-\log h - \frac{1}{h} = 2\log x + c \quad \Leftrightarrow \quad -\frac{1}{h} e^{-\frac{1}{h}} = -x^2 C, \quad C > 0.$$

Notice that the Lambert W function (which is multi-valued) solves the equation $z = W(z)e^{W(z)}$. Thus, we must have $W(-x^2C) = -\frac{1}{h(x)}$. The condition $-1 \le -\frac{1}{h(x)} \le 0$ requires that $-1 \le W \le 0$, and this gives the single-valued function $W_0(y)$ defined on the interval $[-1/e, 0]$ such that $W_0(-1/e) = -1$, $W_0(0) = 0$, and $W_0(y)$ is increasing. So $h(x) = -\frac{1}{W_0(-x^2C)}$. The condition $-1/e \le -x^2C \le 0$ for $x \in [1, Q]$ gives the range for constant C, i. e., $0 < C \le \frac{1}{Q^2e}$. Going back to the functions ϕ and B we obtain

$$\varphi(x) = -x \int_1^x \frac{dt}{W_0(-t^2C)} + x\varphi(1) \quad \text{and thus} \quad B(u, v) = -v \int_1^{uv} \frac{dt}{W_0(-t^2C)} + v\varphi(1).$$

Let us also see how bounded is B. Choosing $C = \frac{1}{Q^2e}$ gives minimal B. $B(u, v)$ can be assumed to be 0 if $uv = 1$. Hence, $\phi(1) = 0$. Then

$$B^Q(u, v) := B(u, v) = -v \int_1^{uv} \frac{dt}{W_0(-\frac{t^2}{Q^2e})} \le v \int_1^{uv} \frac{Q^2e}{t^2} = eQ^2v\left(1 - \frac{1}{uv}\right), \quad 1 \le uv \le Q. \quad (9.7)$$

Here we used the fact that the Lambert function $W_0(x) \le x$ for $x \in [-1/e, 0]$. Actually one can get better estimates by using the series expansion for W_0, i. e.,

$$W_0(x) = \sum_{n=1}^{\infty} \frac{(-n)^{n-1}}{n!} x^n, \quad |x| < \frac{1}{e}.$$

Step 2. From the infinitesimal inequality on d^2B^Q to the global concavity property of B^Q.

The function B^Q defined in (9.7) is not function B_Q with three properties formulated at the beginning of the proof of this theorem. However, let us prove that $B_Q := B^{4Q}$ has all these three properties. The first property is just by definition, and the second property is because we just proved in (9.7) that

$$B_Q(u, v) = B^{4Q}(u, v) \le 4eQ^2v.$$

To prove the third property let us fix $x = (u, v) \in O_Q$, $x_\pm = (u_\pm, v_\pm) \in O_Q$ such that $x = \frac{x_+ + x_-}{2}$. Introduce two functions defined on $[-1, 1]$:

$$U(t) = \frac{1+t}{2}u_+ + \frac{1-t}{2}u_-, \quad V(t) = \frac{1+t}{2}v_+ + \frac{1-t}{2}v_-.$$

Compose the vector function $X(t) = (U(t), V(t))$ and function $B_Q = B^{4Q}$, namely, put

$$b(t) := B_Q(X(t)).$$

Then $X(\pm 1) = x_\pm$, $X(0) = x$. It is important to notice that b is well-defined because

$$\forall t \in [-1, 1], \quad X(t) \in O_{4Q}.$$

The latter is an elementary geometric observation saying that if three points X_+, X belong to O_Q and $X = \frac{X_+ + X_-}{2}$, then the whole segment with endpoints $X_\pm$ lies in O_{4Q} (but not, in general, on $O_{Q'}$ with $Q' < 4Q$). Now we differentiate twice function b. The chain rule gives us immediately

$$b''(t) = (H_{B_Q}(X(t))(x_+ - x_-), (x_+ - x_-))_{\mathbf{R}^2},$$

where H_{B_Q} denotes as always the Hessian matrix of function B_Q. Therefore, the use of (9.4) gives us

$$-b''(t) \geq 2U(t)(v_+ - v_-)^2.$$

On the other hand,

$$B_Q(x) - \frac{B_Q(x_+) + B_Q(x_-)}{2} = b(0) - \frac{b(1) + b(-1)}{2} = \frac{1}{2} \int_{-1}^{1} -b''(t)(1 - |t|) \, dt.$$

Notice that the integrand is always non-negative by the previously displayed formula. By the same formula, the integrand is at least $u = U(0)$ for $t \in [0, 1/2]$ because on this interval $U(t) \geq \frac{1}{2} U(0)$ by the obvious geometric reason. We obtain

$$B_Q(x) - \frac{B_Q(x_+) + B_Q(x_-)}{2} \geq \frac{3}{8} u(v_+ - v_-)^2.$$

We established all three properties for B_Q, and we have already shown that this is enough to prove the inequality in Theorem 9.1. The sharpness is not difficult to see for a weight with one singular point; see [18] for example. □

9.2 Proving the instance of the $T1$ theorem using its Bellman function

Theorem 9.2. *We have $C_s \leq AQ^2$ and this estimate is sharp.*

Let us deduce this result from Theorem 9.1. This is the occasion of the so-called weighted $T1$ theorem.

We use the notation h_I for a standard Haar function supported on a dyadic interval I; it is given by

$$h_I = \begin{cases} \frac{1}{\sqrt{I}}, & \text{on } I+, \\ -\frac{1}{\sqrt{I}}, & \text{on } I - . \end{cases}$$

It is an orthonormal basis in unweighted L^2. Now consider the same type of Haar basis but in weighted $L^2(w^{-1})$: Functions $h_I^{w^{-1}}$ are orthogonal to constants in $L^2(w^{-1})$, are normalized in $L^2(w^{-1})$, assume a constant value on each child of I, and are supported on I. For dyadic intervals on the line we get the following.

Lemma 9.3. *The following holds:* $h_I = \alpha_I^{w^{-1}} h_I^{w^{-1}} + \beta_I^{w^{-1}} \frac{1_I}{\sqrt{|I|}}$, *with*

$$\alpha_I^{w^{-1}} = \frac{\langle w^{-1}\rangle_{I+}^{1/2}\langle w^{-1}\rangle_{I-}^{1/2}}{\langle w^{-1}\rangle_I^{1/2}}, \qquad \beta_I^{w^{-1}} = \frac{\langle w^{-1}\rangle_{I+} - \langle w^{-1}\rangle_{I-}}{\langle w^{-1}\rangle_I}. \tag{9.8}$$

Proof. This is proved by a direct calculation. We need to define two constants, $\alpha_I^{w^{-1}}$ and $\beta_I^{w^{-1}}$, and we have two conditions, $\|h_I^{w^{-1}}\|_{L^2(w^{-1})} = 1$ and $(h_I^{w^{-1}}, 1)_{L^2(w^{-1})} = 0$: $\qquad\square$

The following lemma is just an instance of the chain rule.

Lemma 9.4. *Let* $\Phi(x')$, $B(x'')$ *be smooth functions of* $x' = (x_1,\ldots,x_n,x_0)$, $x'' = (x_{n+1},\ldots,x_m)$. *Then we compute the second differential form of the composition function*

$$B(x_1,\ldots,x_n,x_{n+1},\ldots,x_m) = \Phi(x_1,\ldots,x_n,B(x_{n+1},\ldots,x_m))$$

by the following formula:

$$d^2B = d^2\Phi + \frac{\partial\Phi}{\partial x_0}d^2B.$$

Remark. We understand the left-hand side as $(H_B(x)dx, dx)_{\mathbf{R}^m}$, where

$$x := (x_1,\ldots,x_n,x_{n+1},\ldots,x_m), \quad dx = (dx_1,\ldots,dx_n,dx_{n+1},\ldots,dx_m).$$

We understand $d^2\Phi$ on the right-hand side as $(H_\Phi(x_1,\ldots,x_n,B(x''))dy, dy)_{\mathbf{R}^{n+1}}$, where

$$dy = (dx_1,\ldots,dx_n,dB), \quad dB = (\nabla B(x''), dx'')_{\mathbf{R}^{m-n}}.$$

Now we are ready to prove Theorem 9.2.

Proof. The sum we want to estimate in (1.3),

$$\frac{1}{|J|}\sum_{I\in D(J)}\left(\langle\varphi w^{-1}\rangle_{I+} - \langle\varphi w^{-1}\rangle_{I-}\right)^2\langle w\rangle_I|I|,$$

is of course

$$\Sigma = \frac{2}{|J|}\sum_{I\in D(J)}\left(\varphi w^{-1}, h_I\right)^2\langle w\rangle_I.$$

We can plug the decomposition of Lemma 9.3 and take into account the fact that for a dyadic lattice obviously $\alpha_I^{w^{-1}} \leq 2\langle w^{-1}\rangle_I^{1/2}$. Then we obtain

$$\Sigma \le \frac{8}{|J|} \sum_{I \in D(J)} (\varphi w^{-1}, h_I^{w^{-1}})^2 \langle w^{-1}\rangle_I \langle w\rangle_I + \frac{2}{|J|} \sum_{I \in D(J)} \langle \varphi w^{-1}\rangle_I^2 (\beta_I^{w^{-1}})^2 \langle w\rangle_I |I| =: \Sigma_1 + \Sigma_2.$$

The system $\{h_I^{w^{-1}}\}_{I \in D(J)}$ is orthonormal in $L^2(w^{-1})$, and $\langle w^{-1}\rangle_I \langle w\rangle_I \le Q$. Hence, we immediately have

$$\Sigma_1 \le 8Q\|f\|_{L^2(w^{-1})}^2. \tag{9.9}$$

We are left to estimate Σ_2. To do that let us rewrite Σ_2:

$$\Sigma_2 = \frac{2}{|J|} \sum_{I \in D(J)} \left(\frac{\langle \varphi w^{-1}\rangle_I}{\langle w^{-1}\rangle_I} \right)^2 \gamma_I |I| = \frac{2}{|J|} \sum_{I \in D(J)} (\langle \varphi w^{-1}\rangle_{I,w^{-1}})^2 \gamma_I |I|,$$

where $\langle \cdot \rangle_{I,w^{-1}}$ means the average with respect to measure $\mu := w^{-1}(x)dx$ and

$$\gamma_I := (\langle w^{-1}\rangle_{I+} - \langle w^{-1}\rangle_{I-})^2 \langle w\rangle_I.$$

We are going to prove now that with some absolute constant A,

$$\frac{1}{|J|} \sum_{I \in D(J)} (\langle \varphi w^{-1}\rangle_{I,w^{-1}})^2 \gamma_I |I| \le AQ^2 \langle \varphi^2 w^{-1}\rangle_J. \tag{9.10}$$

This of course finishes the proof of Theorem 9.2.

To prove (9.10) we will construct a special function of four real variables $\mathcal{B}(X) = \mathcal{B}_Q(X), X = (F, f, u, v)$, that possesses the following properties:
(1) $\mathcal{B}_Q$ is defined in a non-convex domain O_{4Q}, where $O_Q := \{(F, f, u, v) \in \mathbf{R}_+^4 : f^2 < Fv, 1 < uv < Q\}$,
(2) $0 \le \mathcal{B}_Q \le F$,
(3) $\mathcal{B}_Q(X) - \frac{\mathcal{B}_Q(X_+) + \mathcal{B}_Q(X_-)}{2} \ge aQ^{-2}\frac{f^2}{v^2}u(v_+ - v_-)^2$, where $X = (F, f, u, v), X_\pm = (F_\pm, f_\pm, u_\pm, v_\pm)$ belong to O_Q and $X = \frac{X_+ + X_-}{2}$.

As soon as such a function is constructed, (9.10) and Theorem 9.2 follow immediately. In fact we repeat our telescopic consideration. We set the vector martingale $X_I := (F_I, f_I, u_I, v_I)$, where

$$F_I = \langle \varphi w^{-1}\rangle_I, \quad f_I = \langle \varphi w^{-1}\rangle_I, \quad u_I = \langle w\rangle_I, \quad v_I = \langle w^{-1}\rangle_I.$$

It is obvious that the vector martingale $\{X_I\}_{I \in D(J)}$ is always inside O_Q, so the superposition of this martingale and $\mathcal{B}_Q$ is well-defined: $\mathcal{B}_Q(X_I)$. Then, property (3) claims that $\{\mathcal{B}_Q(X_I)\}_{I \in D(J)}$ is a supermartingale, and moreover

$$|I|\frac{f_I^2}{v_I^2}u_I(v_{I_+} - v_{I_-})^2 \le AQ^2(|I|\mathcal{B}(X_I) - |I_+|\mathcal{B}(X_{I_+}) - |I_+|\mathcal{B}(X_{I_-})).$$

We use the telescopic nature of the term on the right-hand side, and summing these terms for all $I \in D(J)$, we then notice that all of them will cancel each other, except $AQ^2|J|\mathcal{B}(X_J)$, which is bounded by $AQ^2|J|F_J = AQ^2|J|\langle \varphi^2 w^{-1} \rangle_J$. We proved (9.10) provided that the existence of $\mathcal{B}_Q$ is validated.

Now we will write the explicit formula for $\mathcal{B}_Q$. Exactly as in Theorem 9.1 we first construct, by an explicit formula, an auxiliary function $\mathcal{B}^Q$. Here it is

$$\mathcal{B}^Q(F, f, u, v) := F - \frac{f^2}{v + aQ^{-2}B^Q(u, v)}, \tag{9.11}$$

where B^Q was defined in (9.7). It is clear that it satisfies property 2. It "almost" satisfies property 1, but it is defined only in O_Q, not in a larger domain O_{4Q}. Property 3 does satisfy its infinitesimal version: At the point $X = (F, f, u, v) \in O_Q$,

$$-d^2 \mathcal{B}^Q \geq aQ^{-2} \frac{f^2}{v} u(dv)^2. \tag{9.12}$$

This follows from Lemma 9.4. In fact, consider $\Phi(x_1, x_2, x_3, x_0) := x_1 - \frac{x_2^2}{x_3 + x_0}$. By a direct simple calculation one can see that it is concave in $\mathbf{R}_+^4$, so $d^2\Phi \leq 0$. Now we see that

$$\mathcal{B}^Q(F, f, u, v) = \Phi(F, f, v, aQ^{-2}B^Q(u, v)),$$

and by Lemma 9.4,

$$d^2 \mathcal{B}^Q = d^2 \Phi + aQ^{-2} \frac{\partial \Phi}{\partial x_0} \cdot (d^2 B^Q) \leq \frac{a}{Q^2} \frac{f^2}{(v + aQ^{-2}B^Q(u, v))^2} (d^2 B^Q).$$

But in Theorem 9.1 we proved that $B^Q \leq AQ^2 v$; hence, choosing a small absolute constant a we guarantee that $v + aQ^{-2}B^Q(u, v) \leq 2v$. We also proved in Theorem 9.1 that $-d^2 B^Q \geq u(dv)^2$. Combining these facts with the last displayed inequality we obtain

$$-d^2 \mathcal{B}^Q \geq \frac{a}{4Q^2} \frac{f^2}{v^2} u(dv)^2,$$

which is precisely (9.12).

We need a function with property (3) defined in the domain O_{4Q}. So let us put $\mathcal{B}_Q := \mathcal{B}^{4Q}$. Exactly as in Step 2 of Theorem 9.1 we can not only prove now that infinitesimal special concavity holds in the form $-d^2 \mathcal{B}_Q \geq \frac{a}{64Q^2} \frac{f^2}{v^2} u(dv)^2$, but we also have with some small positive a_0

$$\mathcal{B}_Q(X) - \frac{\mathcal{B}_Q(X_+) + \mathcal{B}_Q(X_-)}{2} \geq a_0 Q^{-2} \frac{f^2}{v^2} u(v_+ - v_-)^2, \tag{9.13}$$

for all triple of points $(X, X_+, X_-) \in O_Q^3$ such that $X = (F, f, u, v)$, $X_\pm = (F_\pm, f_\pm, u_\pm, v_\pm)$, and $X = \frac{X_+ + X_-}{2}$. $\qquad\qquad\square$

Remark. The constant on the right-hand side of the main inequality (9.10) is AQ^2. It is important to notice that the constant on the right-hand side of (9.10) depends only on the constant in Theorem 9.1.

In fact, these two constants are equal up to an absolute constant. So, if for example the constant in Theorem 9.1 is denoted by K, then the constant in (9.10) becomes AK, and the constant in Theorem 9.2 would also become AK (with another absolute constant A). This principle of passing from uniform testing conditions (Theorem 9.1) to full boundedness of the operator is quite general, and, as we said already, is often called the $T1$ theorem.

Here this principle was proved by showing that *the formula* that proves Theorem 9.2 can be obtained by totally formal manipulations from *the formula* that proves the testing condition (Theorem 9.1). See the next remark.

Remark. The reader should pay attention to the following curious formula:

$$\mathcal{B}_Q(F, f, u, v) = F - \frac{f^2}{v + aQ^{-2}B^Q(u, v)}. \qquad (9.14)$$

This is the function $\mathcal{B}_Q$ that proves Theorem 9.2, and B^Q is the function that proved Theorem 9.1.

So we see another instance of transference by use of the Bellman function. By formula (9.14) we transfer the claim of Theorem 9.1 to Theorem 9.2. A Bellman function of Theorem 9.1 was used as "a lego piece" to construct a Bellman function for Theorem 9.2. In this instance this "lego construction" proved for us the $T1$ theorem for the weighted square function operator.

9.3 A small sharpening of the $T1$ theorem for dyadic square functions

Recall that

$$\gamma_I = \left(\langle w^{-1} \rangle_{I_+} - \langle w^{-1} \rangle_{I_-} \right)^2 \langle w \rangle_I.$$

A natural question arises: Letting $[w]_{A_2} = Q \gg 1$ and at the same time

$$\frac{1}{|J|} \sum_{I \in \mathcal{D}(J)} \gamma_I |I| \le q^2 \langle w^{-1} \rangle_J, \quad \forall J \in \mathcal{D}, \qquad (9.15)$$

where $q \ll Q$, does this mean that the norm $\|S_{w^{-1}} : L^2(w^{-1}) \to L^2(w)\|$ is bounded by Cq? That would be quite expected, because this statement reminds us of the statement of the so-called $T1$ theorems. In fact, (9.15) is precisely the testing condition and can be rewritten as

$$\|S_{w^{-1}}\chi_J\|_w^2 \le q^2\|\chi_J\|_{w^{-1}}^2, \quad \forall J \in \mathcal{D}. \tag{9.16}$$

We are quite sure that this is wrong. If so, the next natural question is the following. If, however, $q \ll Q$, is it true that one can give a better estimate than the one we proved in the previous section,

$$\|S_{w^1} : L^2(w^{-1}) \to L^2(w)\| \le Q?$$

The answer to this question is positive. We have the following result.

Theorem 9.5. *Let $w \in A_2$ (dyadic as always), let $[w]_{A_2} = Q$, and let (9.15) hold with $q \ll Q$. Then we have the following improved estimate on $\|S_{w^{-1}} : L^2(w^{-1}) \to L^2(w)\|$:*

$$\|S_{w^{-1}} : L^2(w^{-1}) \to L^2(w)\| \le C(Q^{1/2} + q).$$

Proof. Let us consider the following aggregate, which is quite akin to the function from (9.11) (constant a is again a small positive constant):

$$\mathcal{B}(F, f, A, v) := F - \frac{f^2}{v + aq^{-2}A}. \tag{9.17}$$

Given $I \in \mathcal{D}$, we put

$$A_I := \frac{1}{|I|} \sum_{\ell \in \mathcal{D}(I)} \gamma_\ell |\ell|.$$

As before we set $Y_I := (F_I, f_I, A_I, v_I)$, where A_I is defined above and and

$$F_I = \langle \varphi w^{-1} \rangle_I, \quad f_I = \langle \varphi w^{-1} \rangle_I, \quad v_I = \langle w^{-1} \rangle_I.$$

We also denote $y_I := (F_I, f_I, \frac{1}{2}(A_{I_+} + A_{I_-}), v_I)$. We can estimate

$$|I|\mathcal{B}(Y_I) - |I_+|\mathcal{B}(Y_{I_+}) - |I_-|\mathcal{B}(Y_{I_-})$$

$$= |I|\left((\mathcal{B}(Y_I) - \mathcal{B}(y_I)) + \left(\mathcal{B}(y_I) - \frac{1}{2}\mathcal{B}(Y_{I_+}) - \frac{1}{2}\mathcal{B}(Y_{I_-}) \right) \right)$$

$$\ge \frac{|I|}{4}\frac{f_I^2}{v_I^2}\left(A_I - \frac{1}{2}(A_{I_+} + A_{I_-}) \right)\frac{a}{q^2} = \frac{a}{4q^2}\frac{f_I^2}{v_I^2}\gamma_I |I|$$

$$= \frac{a}{4q^2}\frac{\langle \varphi w^{-1} \rangle_I^2}{\langle w^{-1} \rangle_I^2}\left(\langle w^{-1} \rangle_{I_+} - \langle w^{-1} \rangle_{I_-} \right)^2 \langle w \rangle_I |I|.$$

Summing up over all $I \in \mathcal{D}(J)$ and using the telescopic nature of terms, we get

$$\frac{1}{|J|} \sum_{I \in D(J)} \left(\langle \varphi w^{-1}\rangle_{I,w^{-1}}\right)^2 \gamma_I |I| \le Aq^2 \langle \varphi^2 w^{-1}\rangle_J, \tag{9.18}$$

which is like (9.10), but with q replacing Q. In the inequality above we used the concavity of function $\mathcal{B}(F, f, A, v) = F - \frac{f^2}{v + a\,q^{-2}A}$, which, in particular, gives us

$$\mathcal{B}(y_I) - \frac{1}{2}\mathcal{B}(Y_{I_+}) - \frac{1}{2}\mathcal{B}(Y_{I_-}) \ge 0,$$

and we use the estimate from below for $\frac{\partial}{\partial A}\mathcal{B}(F, f, A, v)$:

$$\min_{\frac{1}{2}(A_{I_+} + A_{I_-}) \le A \le A_I} \frac{\partial}{\partial A}\mathcal{B}(F, f, A, v) \ge \frac{f^2}{(v + a\,q^{-2}A_I)^2} \ge \frac{f^2}{4v^2}.$$

The last inequality is clear if we use (9.15).

Now combining estimate (9.9) of the previous section and (9.18) we get the claim of Theorem 9.5. $\qquad\square$

Bibliography

[1]   I. Assani, Z. Buczolich and R. D. Mauldin, *An L^1 counting problem in ergodic theory*, J. Anal. Math. **95** (2005), 221–241.

[2]   F. Barthe and N. Huet, *On Gaussian Brunn–Minkowskii inequalities*, Stud. Math. **191** (2009), 283–304.

[3]   F. Barthe and B. Maurey, *Some remarks on isoperimetry of Gaussian type*, Ann. Inst. Henri Poincaré B, Probab. Stat. **36**(4) (2000), 419–434.

[4]   B. Bollobás, *Martingale inequalities*, Math. Proc. Camb. Philos. Soc. **87**(3) (1980), 377–382.

[5]   D. Burkholder, *Boundary value problems and sharp estimates for the martingale transforms*, Ann. Probab. **12** (1984), 647–702.

[6]   D. Burkholder, *Sharp inequalities for martingales and stochastic integrals*, in: Colloque Paul Lévy sur les processus stochastiques, Palaiseau, 1987, Astérisque, vols. 157–158, 1988, pp. 75–94.

[7]   D. Burkholder, *A proof of the Pelczynski's conjecture for the Haar system*, Stud. Math. **91** (1988), 79–83.

[8]   L. Caffarelli and X. Cabré, *Fully nonlinear elliptic equations*, Amer. math. soc. colloq. publ., vol. 43, American Mathematical Society, 1995.

[9]   M. J. Carro, *From restricted weak type to strong type estimates*, J. Lond. Math. Soc. **70**(3) (2004), 750–762.

[10]   M. Carro and L. Grafakos, *Weighted weak type $(1,1)$ estimates via Rubio de Francia extrapolation*, J. Funct. Anal. **269**(5) (2015), 1203–1233.

[11]   D. V. Cruz-Uribe, J. M. Martell and C. Pérez, *Weights, extrapolation and the theory of Rubio de Francia*, Operator theory: advances and applications, vol. 215, Birkhäuser-Springer Basel AG, Basel, 2011, xiv+280 pp.

[12]   B. Davis, *On the L^p norms of stochastic integrals and other martingales*, Duke Math. J. **43**(4) (1976), 697–704.

[13]   K. Domelevo, P. Ivanisvili, S. Petermichl, S. Treil and A. Volberg, *On the failure of lower square function estimates in the non-homogeneous weighted setting*, Math. Ann. **374**(3–4) (2019), 1923–1952.

[14] C. Domingo-Salazar, M. Lacey and G. Rey, *Borderline weak type estimates for singular integrals and square functions*, Bull. Lond. Math. Soc. **48**(1) (2016), 63–73, arXiv:1505.01804.

[15] O. Dragicevic, L. Grafakos, M. C. Pereyra and S. Petermichl, *Extrapolation and sharp norm estimates for classical operators on weighted Lebesgue spaces*, Publ. Mat. **49**(1) (2005), 73–91.

[16] L. C. Evans and R. F. Gariepy, *Measure theory and fine properties of functions*, 1992.

[17] I. Holmes, P. Ivanisvili and A. Volberg, *The sharp constant in the weak (1, 1) inequality for the square function: a new proof*, Rev. Mat. Iberoam. **36**(3) (2020), 741–770.

[18] S. Hukovic, S. Treil and A. Volberg, *The Bellman functions and sharp weighted inequalities for square functions*, in: Complex analysis, operators, and related topics, Oper. theory adv. appl., vol. 113, Birkhäuser, Basel, 2000, pp. 97–113.

[19] T. Hytönen, *The sharp weighted bound for general Calderón–Zygmund operators*, Ann. Math. (2) **175**(3) (2012), 1473–1506.

[20] T. Hytönen and K. Li, *Weak and strong $A_p - A_\infty$ estimates for square functions and related topics*, Proc. Am. Math. Soc. **146**(6) (June 2018), 2497–2507. http://dx.doi.org/10.1090/proc/13908.

[21] T. Hytönen and C. Pérez, *Sharp weighted bounds involving A_∞*, Anal. PDE **6**(4) (2013), 777–818.

[22] T. Hytönen, C. Pérez, S. Treil and A. Volberg, *A sharp estimate of dyadic shifts, which gives the proof of A_2 conjecture*, J. Reine Angew. Math. **687** (2014), 43–86, arXiv:1010.0755.

[23] P. Ivanisvili, P. Mozolyako and A. Volberg, *Strong weighted and restricted weak weighted estimates of the square function*, arXiv:1711.10578, 30 pp.

[24] P. Ivanisvili and A. Volberg, *Martingale transform and Square function: some weak and restricted weak sharp weighted estimates*, arXiv:1711.10578, 20 pp.

[25] M. T. Lacey, *An elementary proof of the A_2 bound*, Isr. J. Math. **287** (2017), 181–195.

[26] M. Lacey and K. Li, *On $A_p - A_\infty$ type estimates for square functions*, Math. Z. **284** (2016), 1211–1222.

[27] M. Lacey and J. Scurry, *Weighted weak type estimates for square function*, arXiv:1211.4219.

[28] A. K. Lerner, *Sharp weighted norm inequalities for Littlewood–Paley operators and singular integrals*, Adv. Math. **226** (2011), 3912–3926.

[29] A. K. Lerner, *Mixed $A_p - A_r$ inequalities for classical singular integrals and Littlewood–Paley operators*, J. Geom. Anal. **23**(3) (2013), 1343–1354.

[30] A. K. Lerner, *A simple proof of the A_2 conjecture*, Int. Math. Res. Not. **14** (2013), 3159–3170.

[31] A. K. Lerner, *On an estimate of Calderón–Zygmund operators by dyadic positive operators*, J. Anal. Math. **121**(1) (2013), 141–161.

[32] A. K. Lerner, *On sharp aperture-weighted estimates for square functions*, J. Fourier Anal. Appl. **20**(4) (2014), 784–800.

[33] A. K. Lerner, F. Nazarov and S. Ombrosi, *On the sharp upper bound related to the weak Muckenhoupt–Wheeden conjecture*, Anal. PDE **13**(6) (2020), 1939–1954.

[34] A. K. Lerner, S. Ombrosi and C. Pérez, *A_1 bounds for Calderón–Zygmund operators related to a problem of Muckenhoupt and Wheeden*, Math. Res. Lett. **16**(1) (2009), 149–156.

[35] N. Nadirashvili, V. Tkachev and S. Vladut, *Nonlinear elliptic equations and nonassociative algebras*, AMS math. surveys and monographs, vol. 200, 2014.

[36] F. Nazarov, A. Reznikov, V. Vasyunin and A. Volberg, *On weak weighted estimates of the martingale transform and a dyadic shift*, Anal. PDE **11**(8) (2018), 2089–2109.

[37] F. Nazarov, A. Reznikov, V. Vasyunin and A. Volberg, *A Bellman function counterexample to the A_1 conjecture: the blow-up of weak norm estimates of weighted singular operators*, arXiv:1506.04710v1, 23 pp.

[38] F. Nazarov, S. Treil and A. Volberg, *Two weight estimate for the individual Haar multipliers and other well localized operators*, Math. Res. Lett. **15**(3) (2008), 583–597, arXiv:math/0702758.

[39] A. Osekowski, *On the best constant in the weak type inequality for the square function of a conditionally symmetric martingale*, Stat. Probab. Lett. **79** (2009), 1536–1538.

[40] S. Petermichl, *The sharp bound for the Hilbert transform on weighted Lebesgue spaces in terms of the classical A_p characteristic*, Am. J. Math. **129**(5) (2007), 1355–1375.

[41] S. Petermichl and A. Volberg, *Heating of the Ahlfors–Beurling operator: weakly quasiregular maps on the plane are quasiregular*, Duke Math. J. **112**(2) (2002), 281–305.

[42] C. Pérez, *Weighted norm inequalities for singular integral operators*, J. Lond. Math. Soc. (2) **49**(2) (1994), 296–308.

[43] C. Pérez, S. Treil and A. Volberg, *On A_2 conjecture and corona decomposition of weights*, arXiv:1006.2630.

[44] M. C. Reguera, *On Muckenhoupt–Wheeden conjecture*, Adv. Math. **227**(4) (2011), 1436–1450.

[45] M. C. Reguera and C. Thiele, *The Hilbert transform does not map $L^1(Mw)$ to $L^{1,\infty(w)}$*, Math. Res. Lett. **19**(1) (2012), 1–7.

[46] M. C. Reguera and C. Thiele, *The Hilbert transform does not map $L^1(Mw)$ to $L^{1,\infty}(w)$*, Math. Res. Lett. **19**(1) (2012), 1–7, arXiv:1011.1767.

[47] A. Reznikov and A. Volberg, *The weak type estimates for two different martingale transforms coincide*, in: Analysis of operators on function spaces, Trends math., Birkhäuser/Springer, Cham, 2019, pp. 259–274.

[48] V. Vasyunin, *The sharp constant in the reverse Hölder inequality for Muckenhoupt weights*, Algebra Anal. **15**(1) (2003), 73–117 (Russian); English transl. in: St. Petersburg. Math. J., 15 (2004), no. 1, 49–79.

[49] V. Vasyunin and A. Volberg, *Bellman function technique in harmonic analysis*, Cambridge University Press, 2020, 452 pp.

[50] M. Wilson, *Weighted Littlewood–Paley theory and exponential-square integrability*, Lecture notes in math, vol. 1924, Springer-Verlag, 2007.

Piotr Nayar and Tomasz Tkocz

Extremal sections and projections of certain convex bodies: a survey

Abstract: We survey results concerning sharp estimates on volumes of sections and projections of certain convex bodies, mainly ℓ_p-balls, by and onto lower-dimensional subspaces. This subject emerged from geometry of numbers several decades ago and since then has seen the development of a variety of probabilistic and analytic methods, showcased in this chapter.

Keywords: Volume, convex sets, sections, projections, p-norms, cube, simplex

MSC 2020: Primary 52A38, Secondary 60E15, 42B10

1 Introduction

1.1 Prologue

How small can the volume of a slice of the unit cube be? This question, asked by Good in the 1970s in the context of its applications in geometry of numbers, has turned out to be rather influential, prompting the development of several important methods, as well as spurring the community to solve further problems and enter research directions of independent interest in convex geometry, with strong ties to probability. Those most notably include the *dual* question of extremal volume projections, which in the simplest non-trivial case of hyperplane projections naturally translates into probabilistic Khinchin-type inequalities. Intriguingly, questions on extremal volume sections can be similarly translated into the same probabilistic language.

The purpose of this survey is thus twofold: In addition to striving to give a systematic account of the known results, our second goal is to illustrate intertwined Fourier analytic, geometric, and probabilistic methods underpinning the old and recent approaches.

Acknowledgement: PN's research is supported in part by the National Science Centre, Poland, grant 2018/31/D/ST1/0135. TT's research is supported in part by NSF grant DMS-2246484.

Piotr Nayar, University of Warsaw, 02-097 Warsaw, Poland, e-mail: nayar@mimuw.edu.pl
Tomasz Tkocz, Carnegie Mellon University, Pittsburgh, PA 15213, USA, e-mail: ttkocz@andrew.cmu.edu

https://doi.org/10.1515/9783110775389-008

1.2 The motivating example

We begin with recalling Good's question (following [14, 132]). Suppose we are given n linear forms $L_i(x) = \sum_{j=1}^{k} a_{ij}x_j$, $i = 1,\ldots,n$, in k variables. When does the system $|L_i(x)| \leq 1$, $i \leq n$, admit a non-trivial integral solution? The cornerstone result in geometry of numbers, Minkowski's (first) theorem, provides a link to volume: If K is a symmetric convex body in $\mathbb{R}^d$ of volume at least 2^d, then it contains a non-trivial lattice point (see, e. g., [98, Chapter 2]). Let $A = [a_{ij}]_{i \leq n, j \leq k}$ be the $n \times k$ matrix whose i-th row determines L_i. Thus, immediately, if $k \geq n$ and $\det(A) \leq 1$ when $k = n$, then the answer to Good's question is affirmative because the set

$$K = \{x \in \mathbb{R}^k,\ |L_i(x)| \leq 1, i \leq n\} = \{x \in \mathbb{R}^k,\ Ax \in [-1,1]^n\}$$

is the preimage of the cube $[-1,1]^n$ under the linear map $A \colon \mathbb{R}^k \to \mathbb{R}^n$ (unbounded if A is singular and of volume exactly $2^k \det(A)^{-1}$ otherwise when $k = n$). The case $k < n$ is more interesting. Suppose A is of full rank k. Then the image of K under A is the section of the cube $[-1,1]^n$ by the k-dimensional linear subspace $A(\mathbb{R}^k)$. How small can its volume be? Good's conjecture confirmed later by Vaaler in [132] says that it is at least 2^k (the volume of the k-dimensional subcube $[-1,1]^k \times \{0\}^{n-k}$). Thus, if $\det(A^\top A) \leq 1$, we obtain

$$\mathrm{vol}(K) \geq \sqrt{\det(A^\top A)}\,\mathrm{vol}(K) = \mathrm{vol}(A(K)) \geq 2^k,$$

also asserting in view of Minkowski's theorem that the initial system of inequalities admits a non-trivial integral solution, provided the convenient sufficient condition $\det(A^\top A) \leq 1$. From a geometric point of view, it now seems natural and interesting to ask further questions about the maximal volume sections for the cube, as well as other sets.

1.3 Preliminaries and overview

We endow $\mathbb{R}^n$ with the standard inner product $\langle x,y \rangle = \sum_{j=1}^{n} x_j y_j$ between two vectors $x = (x_1,\ldots,x_n)$ and $y = (y_1,\ldots,y_n)$ in $\mathbb{R}^n$ and denote by $|x| = \sqrt{\langle x,x \rangle}$ the induced standard Euclidean norm. Its closed centered unit ball is denoted B_2^n, and for the unit sphere we write $S^{n-1} = \partial B_2^n$. Moreover, we write $e_1,\ldots,e_n$ for the standard basis vectors, $e_1 = (1,0,\ldots,0)$, $e_2 = (0,1,0,\ldots,0)$, etc. As usual, for a set A in $\mathbb{R}^n$, $A^\perp = \{x \in \mathbb{R}^n, \langle x,a \rangle = 0\ \forall a \in A\}$ is its orthogonal complement, with the convention that for a vector u in $\mathbb{R}^n$, $u^\perp = \{u\}^\perp$ is the hyperplane perpendicular to u. Dilates are denoted by $\lambda A = \{\lambda a,\ a \in A\}$ for a scalar λ. In particular, if $-A = A$, the set A is called (origin-)symmetric. The Minkowski or algebraic sum of two sets is $A + B = \{a + b,\ a \in A, b \in B\}$. The orthogonal projection onto an affine or linear subspace H in $\mathbb{R}^n$ is denoted by Proj_H. Volume, i. e., the k-dimensional Lebesgue measure in $\mathbb{R}^n$, is denoted by $\mathrm{vol}_k(\cdot)$, identified with k-dimensional Hausdorff measure (normalized so that cubes with side length

1 have volume 1). Recall that a body in $\mathbb{R}^n$ is a compact set with non-empty interior. For a symmetric convex body K in $\mathbb{R}^n$, its Minkowski functional is $\|x\|_K = \sup\{t \geq 0, x \in tK\}$, $x \in \mathbb{R}^n$, the norm whose unit ball is K. A function $f \colon \mathbb{R}^n \to \mathbb{R}_+$ is called log-concave if it is of the form e^{-V} for a convex function $V \colon \mathbb{R}^n \to (-\infty, +\infty]$. We refer for instance to the monographs [4, 31].

To put it fairly generally, given a body B in $\mathbb{R}^n$ and $1 \leq k \leq n$, the two questions of our main interest will be:

(I) What are the minimal and maximal volume *sections* $\mathrm{vol}_k(B \cap H)$ among all k-dimensional subspaces H in $\mathbb{R}^n$?	(II) What are the minimal and maximal volume *projections* $\mathrm{vol}_k(\mathrm{Proj}_H(B))$ among all k-dimensional subspaces H in $\mathbb{R}^n$?

We note the obvious fact that in contrast to Question (I), Question (II) does not change if we translate the body B.

It is worth recalling two classical convexity-type results allowing to compare such volumes in the codimension 1 case, $k = n-1$ (despite not yielding direct answers to these questions).

Theorem 1 (Busemann [33]). *Let K be a symmetric convex body in $\mathbb{R}^n$. Then the function*

$$x \longmapsto \frac{|x|}{\mathrm{vol}_{n-1}(K \cap x^\perp)}, \quad x \neq 0,$$

extended by 0 at $x = 0$ defines a norm on $\mathbb{R}^n$.

The surface area measure σ_K of a convex body K in $\mathbb{R}^n$ is a Borel measure on the unit sphere S^{n-1} defined as follows: For $E \subset S^{n-1}$, $\sigma_K(E)$ equals the volume of the part of the boundary ∂K where normal vectors belong to E (in other words, σ_K is the pushforward of the $(n-1)$-dimensional Hausdorff measure on ∂K via the Gauss map $\nu_K \colon \partial K \to S^{n-1}$).

Theorem 2 (Cauchy–Minkowski). *Let K be a convex body in $\mathbb{R}^n$. Then for every unit vector $\theta \in S^{n-1}$, we have*

$$\mathrm{vol}_{n-1}(\mathrm{Proj}_{\theta^\perp}(K)) = \frac{1}{2} \int_{S^{n-1}} |\langle \theta, \xi \rangle| \, d\sigma_K(\xi).$$

In particular, the function $x \mapsto |x| \, \mathrm{vol}_{n-1}(\mathrm{Proj}_{x^\perp}(K))$, $x \neq 0$, extended by 0 at $x = 0$, defines a norm on $\mathbb{R}^n$.

Let us explain this formula in the case of polytopes. Suppose we are given a convex polytope P in $\mathbb{R}^n$ and we want to project it onto a hyperplane $\theta^\perp$, where θ is a unit vector. Let $\mathcal{F}_P$ be the set of faces of P. If $F \in \mathcal{F}_P$, then $\mathrm{vol}_{n-1}(\mathrm{Proj}_{\theta^\perp}(F)) = \mathrm{vol}_{n-1}(F) \cdot |\langle \theta, n(F) \rangle|$, where $n(F)$ is the unit outer-normal vector to F. Note that in $\mathrm{Proj}_{\theta^\perp}(P)$ every point is *covered* two times, so one gets the following expression for the volume of projection:

$$\mathrm{vol}_{n-1}(\mathrm{Proj}_{\theta^\perp} P) = \frac{1}{2} \sum_{F \in \mathcal{F}_P} \mathrm{vol}_{n-1}(F) \cdot |\langle \theta, n(F) \rangle|.$$

The Cauchy–Minkowski formula is a straightforward generalization of this formula to general convex bodies. For further background and proofs, we refer for instance to [56, Theorem 8.1.10 and (A.49)].

For $p > 0$ and a vector $x = (x_1, \ldots, x_n)$ in $\mathbb{R}^n$, we define the ℓ_p-norm of x (quasinorm when $0 < p < 1$) and its (closed) unit ball by

$$\|x\|_p = \left(\sum_{j=1}^n |x_j|^p \right)^{1/p}, \quad \|x\|_\infty = \max_{j \leq n} |x_j|, \quad B_p^n = \{x \in \mathbb{R}^n, \|x\|_p \leq 1\}.$$

The cube $B_\infty^n = [-1, 1]^n$ often warrants the more convenient volume 1 normalization

$$Q_n = \frac{1}{2} B_\infty^n = \left[-\frac{1}{2}, \frac{1}{2} \right]^n.$$

The known results about extremal volume hyperplane sections and projections of ℓ_p-balls are summarized in Tables 1 and 2 (that is, the known answers to Questions (I) and (II) when $B = B_p^n$ and $k = n - 1$). We shall discuss them and many more in detail in the next sections.

Table 1: Extremal volume hyperplane sections of ℓ_p-balls: min / $\max_{a \in S^{n-1}} \mathrm{vol}_{n-1}(B_p^n \cap a^\perp)$.

	$0 < p < 2$	$2 < p < \infty$	$p = \infty$
min	$a = (\frac{1}{\sqrt{n}}, \ldots, \frac{1}{\sqrt{n}})$ [74]	$a = (1, 0, \ldots, 0)$ [102]	$a = (1, 0, \ldots, 0)$ [61, 62]
max	$a = (1, 0, \ldots, 0)$ [102]	?	$a = (\frac{1}{\sqrt{2}}, \frac{1}{\sqrt{2}}, 0, \ldots, 0)$ [7]

Table 2: Extremal volume hyperplane projections of ℓ_p-balls: min / $\max_{a \in S^{n-1}} \mathrm{vol}_{n-1}(\mathrm{Proj}_{a^\perp}(B_p^n))$.

	$p = 1$	$1 < p < 2$	$2 < p \leq \infty$
min	$a = (\frac{1}{\sqrt{2}}, \frac{1}{\sqrt{2}}, 0, \ldots, 0)$ [13, 130]	?	$a = (1, 0, \ldots, 0)$ [24]
max	$a = (1, 0, \ldots, 0)$ [folklore]	$a = (1, 0, , \ldots, 0)$ [24]	$a = (\frac{1}{\sqrt{n}}, \ldots, \frac{1}{\sqrt{n}})$ [24]

1.4 Existing literature and our aim

There is of course a vast body of literature on the subject. Ball's survey [14] presents stochastic comparison methods and applications of the celebrated Brascamp–Lieb inequalities to derive sharp bounds on sections. Koldobsky, Ryabogin, and Zvavich's survey [78] and Koldobsky's monograph [75] bring a common Fourier analytic treatment to bounds on both sections and projections. The aim of this chapter is to update on these and gather in one place what we know to date, as well as highlight what we would like to know around Questions (I) and (II), presenting 11 conjectures. We also showcase a unifying probabilistic point of view (via Khinchin-type inequalities) which goes hand in hand with the Fourier analytic methods, allowing to obtain additional insights and sharper results.

2 Sections

The goal here is to give a comprehensive account of known results concerning Question (I), with some indication of methods to which we come back in Section 4. We begin with some general remarks. A convex body K in $\mathbb{R}^n$ is called *isotropic* if it has volume 1, its barycenter is at the origin, and its covariance matrix is proportional to the identity matrix, that is,

$$\mathrm{vol}_n(K) = 1, \quad \int_K x\,\mathrm{d}x = 0, \quad \int_K x_i x_j \,\mathrm{d}x = L_K^2 \delta_{ij}.$$

The positive proportionality constant L_K is called the isotropic constant of K. Every convex body admits an affine image which is isotropic (diagonalizing the covariance matrix). It turns out that for symmetric isotropic convex bodies, volumes of all sections of a fixed dimension are comparable.

Theorem 3 (Hensley [63]). *Fix $1 \le l \le n$. There are positive constants c_l, c_l' which depend only on l such that for every symmetric convex body K in $\mathbb{R}^n$ which is isotropic and every l-codimensional subspace H in $\mathbb{R}^n$, we have*

$$\frac{c_l}{L_K} \le \mathrm{vol}_{n-l}(K \cap H)^{1/l} \le \frac{c_l'}{L_K}.$$

To illustrate the key insight of Hensley's argument, let us consider the hyperplane case: We take $H = a^\perp$ for a unit vector a and let

$$f(t) = \mathrm{vol}_{n-1}(K \cap (H + ta)), \quad t \in \mathbb{R}. \tag{2.1}$$

By the Brunn–Minkowski inequality, this defines a log-concave function. By the assumptions on K, it is even and integrates to 1. We claim that $f(t)$ is the probability density function of the random variable $\langle a, X \rangle$, where X is uniform on K. Indeed,

$$\mathbb{P}\left(\sum_{i=1}^n a_i X_i \le s\right) = \mathbb{P}(\langle a, X \rangle \le s) = \mathrm{vol}_{n-1}(\{x \in K : \langle a, x \rangle \le s\}) = \int_{-\infty}^s f(t)\,\mathrm{d}t,$$

by Fubini's theorem. Crucially,

$$\mathrm{vol}_{n-1}(K \cap a^\perp) = f(0). \tag{2.2}$$

Since by isotropicity $\mathbb{E}|\langle a, X\rangle|^2 = L_K^2$, we have $\int_{\mathbb{R}} t^2 f(t)\mathrm{d}t = L_K^2$. It then remains to extremize the value of $f(0)$ among all such densities. Using a "moving mass to where it is beneficial" type of argument (see the proof of Theorem 4 below), a sharp lower bound is obtained by considering a uniform density (in higher dimensions, i. e., $l > 1$, isotropicity naturally dictates a uniform density on a Euclidean ball), whilst for a sharp upper bound, using convexity, the comparison is made against a symmetric exponential density (in higher dimensions, the argument is more complicated and not sharp anymore – see [63, Lemma 3]).

Bourgain in [29] used the property of hyperplane sections having comparable volume to obtain bounds on maximal functions. He asked whether the isotropic constants L_K over all K in all dimensions are uniformly bounded by a universal constant and, equivalently, whether every (symmetric) convex body of volume 1 admits a hyperplane section of volume at least a universal constant. This has become one of the central questions in asymptotic convex geometry, the hyperplane or slicing conjecture; see, e. g., [31] for a comprehensive monograph, [73] for a recent survey, and [67, 72] for the best results to date.

By Theorem 3, for two arbitrary subspaces H_1, H_2 of codimension k, we have

$$\left(\frac{\mathrm{vol}_k(K \cap H_1)}{\mathrm{vol}_k(K \cap H_2)}\right)^{1/k} \le C_k$$

with $C_k = \frac{c_k'}{c_k}$. Hensley's proof gives an upper bound on C_k of order $k!$, which was improved to $\sqrt{k}$ by Ball in [8], who conjectured an optimal bound to be in fact of constant order, which remains open and turns out to be equivalent to the slicing conjecture. Implicit in his work and elucidated by V. Milman and Pajor in their seminal work [104] is the following reason for that equivalence: For a symmetric isotropic convex body K in $\mathbb{R}^n$ and a k-codimensional subspace H in $\mathbb{R}^n$, there is a symmetric k-dimensional convex body C in $H^\perp$ such that

$$c_1 \frac{L_C}{L_K} \le \mathrm{vol}_{n-k}(K \cap H)^{1/k} \le c_2 \frac{L_C}{L_K},$$

where $c_1, c_2 > 0$ are universal constants. The body C emerges from a generalization of Busemann's Theorem 1 to higher codimensions (see [8, 104]). Since $L_K \ge L_{B_2^n}$, if the slicing conjecture is true, then $L_C \le c_3$ for a universal constant $c_3 > 0$; thus, it in particular implies the existence of a universal constant $M > 0$ such that for all k-codimensional subspaces H, we have

$$\mathrm{vol}_{n-k}(K \cap H) \le M^k. \tag{2.3}$$

2.1 Cubes

Recall that $Q_n = [-\frac{1}{2}, \frac{1}{2}]^n$. As highlighted in the introduction, in the context of extremal volume sections, it has always been the cube sparking most interest and attention. The first sharp result concerned minimum volume hyperplane sections and was obtained independently by Hadwiger in [61] and by Hensley in [62].

Theorem 4 (Hadwiger [61], Hensley [62]). *For every unit vector a in $\mathbb{R}^n$, we have*

$$\mathrm{vol}_{n-1}(Q_n \cap a^\perp) \geq 1.$$

Equality holds if and only if $a = \pm e_j$ for some $1 \leq j \leq n$.

Proof. Let $K = Q_n$ and let us consider the function f from (2.1). Since Q_n is isotropic, the value of $\int t^2 f(t)\mathrm{d}t = \int_{Q_n} \langle x, a\rangle^2 \mathrm{d}x = \frac{1}{12}$ does not depend on a (easily found by taking $a = e_1$). Moreover, $\|f\|_\infty = f(0)$, for f is even and log-concave. It is therefore enough to show that for *every* probability density f, we have

$$\|f\|_\infty^2 \int t^2 f(t)\mathrm{d}t \geq \frac{1}{12}.$$

This goes back to Moriguti's work [106] (rederived by Ball in [7] in a slightly more general case of p-norms). For the proof, we can assume that f is even, as otherwise we consider $g(t) = \frac{1}{2}(f(t) + f(-t))$ and $\|g\|_\infty \leq \|f\|_\infty$, whereas the second moments of f and g are the same. Then we move mass towards the origin, as this is beneficial: formally, take $f_0 = \|f\|_\infty \mathbf{1}_{[-c,c]}$, where $c = (2\|f\|_\infty)^{-1}$. Clearly, $\|f_0\|_\infty = \|f\|_\infty$. We have

$$\int t^2(f(t) - f_0(t))\mathrm{d}t = \int (t^2 - c^2)(f(t) - f_0(t))\mathrm{d}t \geq 0,$$

as the integrant is non-negative. $\square$

In words, the canonical coordinate subspaces uniquely minimize the volume of hyperplane sections of the cube. Soon after, this was extended to sections of arbitrary dimension by Vaaler [132], confirming Good's conjecture.

Theorem 5 (Vaaler [132]). *Fix $1 \leq k \leq n$. For every k-dimensional subspace H in $\mathbb{R}^n$, we have*

$$\mathrm{vol}_k(Q_n \cap H) \geq 1.$$

Equality holds if and only if H is spanned by some k standard basis vectors.

Thus, every section of the cube has *large* volume. It is also "fat in all directions," in terms of quadratic forms; see Ball and Prodromou's work [16]. Vaaler used log-concavity and the notion of peakedness (introduced by Kanter in [69]) to make such comparison, generalizing Hensley's argument. Recently, Akopyan, Hubard, and Karasev [1] gave a different proof based on topological methods.

Thus, Vaaler's theorem gives a complete answer to Question (I) for minimal volume sections of the cube. Turning to the maximal ones, the first general upper bound for hyperplane sections was given by Hensley in [62], viz. $\mathrm{vol}_{n-1}(Q_n \cap a^{\perp}) \le 5$ for every unit vector a in $\mathbb{R}^n$, who also conjectured that the sharp bound would be with 5 replaced by $\sqrt{2}$ attained at $a = (\frac{1}{\sqrt{2}}, \frac{1}{\sqrt{2}}, 0, \dots, 0)$. This was later confirmed by Ball in his seminal work [7].

Theorem 6 (Ball [7]). *For every unit vector a in $\mathbb{R}^n$, we have*

$$\mathrm{vol}_{n-1}(Q_n \cap a^{\perp}) \le \sqrt{2}.$$

Equality holds if and only if $a = (\pm e_i \pm e_j)/\sqrt{2}$ for some $1 \le i < j \le n$.

Sketch of the proof. The starting point of Ball's approach was Fourier analytic: If we fix a unit vector a and let f be the probability density of $\langle a, X \rangle$, where X is a random vector uniform on Q_n (thus having i. i. d. components X_j which are uniform on $[-\frac{1}{2}, \frac{1}{2}]$), then by (2.2) and the standard Fourier inversion formula,

$$\mathrm{vol}_{n-1}(Q_n \cap a^{\perp}) = f(0) = \frac{1}{2\pi} \int_{-\infty}^{\infty} \hat{f}(t)dt = \frac{1}{2\pi} \int_{-\infty}^{\infty} \prod_{j=1}^{n} \frac{\sin(\frac{1}{2}a_j t)}{\frac{1}{2}a_j t} dt = \frac{1}{\pi} \int_{-\infty}^{\infty} \prod_{j=1}^{n} \frac{\sin(a_j t)}{a_j t} dt.$$

(This formula can perhaps be traced back to Pólya's work [119], and was also used by Hensley.) The next crucial idea is to apply Hölder's inequality with the weights $p_j = a_j^{-2}$ to get the bound

$$\int_{-\infty}^{\infty} \prod_{j=1}^{n} \frac{\sin(a_j t)}{a_j t} dt \le \prod_{j=1}^{n} \left(\int_{-\infty}^{\infty} \left| \frac{\sin(a_j t)}{a_j t} \right|^{a_j^{-2}} dt \right)^{a_j^2} = \prod_{j=1}^{n} \Psi(a_j^{-2})^{a_j^2} \tag{2.4}$$

with $\Psi(p) = \sqrt{p} \int_{-\infty}^{\infty} |\frac{\sin t}{t}|^p dt, p \ge 1$ (a similar trick was also used in Haagerup's seminal work [60] on sharp constants in Khinchin inequalities). The most technically challenging and rather intricate is the problem of maximization of Ψ. The so-called Ball integral inequality which he established to finish his proof asserts that

$$\frac{1}{\pi} \int_{-\infty}^{\infty} \left| \frac{\sin t}{t} \right|^{p} dt \le \sqrt{\frac{2}{p}}, \quad p \ge 2, \tag{2.5}$$

with equality if and only if $p = 2$. This completes the proof in the case where all $|a_j| \le \frac{1}{\sqrt{2}}$. The complimentary case is dispensed with by a geometric argument justifying that $\mathrm{vol}_{n-1}(Q_n \cap a^{\perp}) \le \frac{1}{|a_j|}$ for each j. Indeed, projecting the section onto $e_j^{\perp}$ changes its volume by the factor $|\langle a, e_j \rangle|$ and it is contained in the projection of the entire cube,

$$\mathrm{vol}_{n-1}(Q_n \cap a^{\perp}) = \frac{1}{|\langle a, e_j \rangle|} \mathrm{vol}_{n-1}(\mathrm{Proj}_{e_j^{\perp}}(Q_n \cap a^{\perp})) \le \frac{1}{|\langle a, e_j \rangle|} \mathrm{vol}_{n-1}(\mathrm{Proj}_{e_j^{\perp}}(Q_n)) = \frac{1}{|a_j|}.$$

$\square$

We mention in passing that this integral inequality has been quite influential, with a very powerful method developed by Nazarov and Podkorytov in [109] to give a "simple" proof, as well as many extensions, generalizations, discrete versions, or even stability properties (see [5, 47, 70, 95, 100, 101]).

Quite remarkably and unexpectedly, the $\sqrt{2}$ bound allows to produce a very simple counterexample to the famous Busemann–Petty problem posed in [34]: *If for two symmetric convex bodies K and L in $\mathbb{R}^n$, we have* $\mathrm{vol}_{n-1}(K \cap a^\perp) \leq \mathrm{vol}_{n-1}(L \cap a^\perp)$ *for every vector a, does it follow that* $\mathrm{vol}_n(K) \leq \mathrm{vol}_n(L)$? Indeed, Ball observed in [9] that since the volume of the hyperplane sections of the unit volume Euclidean ball in high dimensions is roughly $\sqrt{e}$ and $\sqrt{2} < \sqrt{e}$, it suffices to take $K = Q_n$, and for L, it suffices to take a ball of a slightly smaller radius. This argument in fact works in all dimensions $n \geq 10$. Later, in [58], Giannopoulos used similar ideas involving cylinders to produce such elegant and simple counterexamples in dimensions $n \geq 7$. The answer to the Busemann–Petty problem is negative for $n \geq 5$ and positive for $n \leq 4$. This is the result of significant work involving deep Fourier-analytic insights, see [57]. We refer for instance to Koldobsky's comprehensive monograph [75] for a full account.

The situation for upper bounds on volume of sections of more than one codimension is not fully understood. Ball obtained two general bounds.

Theorem 7 (Ball [10]). *Fix $1 \leq k \leq n$. For every k-dimensional subspace H in $\mathbb{R}^n$, we have*

$$\mathrm{vol}_{n-1}(Q_n \cap H) \leq \min\left\{ \sqrt{\frac{n}{k}}^k, \sqrt{2}^{n-k} \right\}.$$

The first bound $\sqrt{\frac{n}{k}}^k$ is optimal when k divides n. Rogalski's question asks for the symmetric convex body of largest volume ratio (see [10]). It turns out that the bound $\sqrt{\frac{n}{k}}^k$ is equivalent to the fact that the cube is such a maximizer. The second bound $\sqrt{2}^{n-k}$ is better than the first one for $k \geq \frac{n}{2}$ and turns out to be sharp in this case. Both bounds rely heavily on Ball's ingenious geometric version of the Brascamp–Lieb inequality (from [30]), which provides a multi-dimensional analog of Hölder's inequality. The first bound uses it in a direct way (applied to indicator functions of intervals), whereas the second bound applies it to the Fourier analytic formula. As already mentioned, the exponential bound $\sqrt{2}^{n-k}$ in codimension $n - k$ is largely motivated by the slicing problem; see (2.3), providing an explicit constant for the cube.

These bounds, although sharp for many values of k and n, leave many other cases open. A sort of folklore conjecture (see, e. g., [65, 115]) states that for arbitrary k and n, the maximal volume section of the cube is attained at an affine cube. Specifically, given $1 \leq k \leq n$, let $n = k\ell + r$ with r being the remainder from the division of n by k. We define the following k orthogonal vectors in $\mathbb{R}^n$:

$$u_{j+1} = e_{j\ell+1} + e_{j\ell+2} + \cdots + e_{(j+1)\ell}, \quad 0 \leq j < k - r,$$

$$u_{k-r+j} = e_{(d-r)\ell+(\ell+1)j+1} + e_{(d-r)\ell+(\ell+1)j+2} + \cdots + e_{(d-r)\ell+(\ell+1)(j+1)}, \quad 0 \leq j < r.$$

Let H^* be the k-dimensional subspace spanned by them. Then

$$Q_n \cap H^* = \left\{ \sum_{j=1}^{k} t_k u_k, \ |t_1|, \dots, |t_k| \le \frac{1}{2} \right\},$$

which is an affine cube of volume

$$\mathrm{vol}_k(Q_n \cap H^*) = \prod_{j=1}^{k} |u_j| = \sqrt{\ell}^{\,k-r} \sqrt{\ell+1}^{\,r}.$$

Note that this becomes $\sqrt{n/k}^{\,k}$ when k divides n and $\sqrt{2}^{\,n-k}$ when $k \ge n/2$.

Conjecture 1. *Let $1 \le k \le n$. For every k-dimensional subspace H in $\mathbb{R}^n$, we have*

$$\mathrm{vol}_k(Q_n \cap H) \le \mathrm{vol}_k(Q_n \cap H^*).$$

In addition to Ball's results of Theorem 7, this conjecture has recently been confirmed for planar sections, i. e., when $k = 2$, by Ivanov and Tsiutsiurupa in [66], who developed local conditions for extremal subspaces.

At the end of this subsection, we mention several loosely related extensions of these fundamental results.

Other measures

Let γ_n denote the standard Gaussian measure on $\mathbb{R}^n$, that is, the Borel probability measure on $\mathbb{R}^n$ with density $(2\pi)^{-n/2} e^{-|x|^2/2}$, whereas for a subspace H, let γ_H be its counterpart on H, that is, the Borel probability measure supported on H with density $(2\pi)^{-\dim H/2} e^{-|x|^2/2}$ (with respect to Lebesgue measure on H). Due to the lack of homogeneity, now of course the cube's side lengths may play a role. For the lower bounds, Barthe, Guédon, Mendelson, and Naor [22] established an analog of Vaaler's theorem.

Theorem 8 (Barthe–Guédon–Mendelson–Naor [22]). *Fix $1 \le k \le n$. For every k-dimensional subspace H in $\mathbb{R}^n$, the function*

$$t \mapsto \frac{\gamma_H(tQ_n \cap H)}{\gamma_k(tQ_k)}$$

is non-increasing on $[0, +\infty)$. In particular (letting $t \to \infty$), for every $t > 0$, we have

$$\gamma_H(tQ_n \cap H) \ge \gamma_k(tQ_k).$$

Their argument follows Vaaler's approach, crucially using the product structure of Gaussian measure. Using Ball's geometric form of the Brascamp–Lieb inequality, they also obtain an upper bound, similar to his bound for volume.

Theorem 9 (Barthe–Guédon–Mendelson–Naor [22]). *Fix $1 \le k \le n$. For every k-dimensional subspace H in $\mathbb{R}^n$ and every $t > 0$, we have*

$$\gamma_H(tQ_n \cap H) \le \gamma_k\left(t\sqrt{\frac{n}{k}}Q_k\right).$$

Again, this is sharp whenever k divides n. The maximal-Gaussian volume hyperplane sections of cubes are not known for all values of t. Zvavitch has shown in [135] that the hyperplane $(\frac{1}{\sqrt{2}}, \frac{1}{\sqrt{2}}, 0, \dots, 0)^\perp$ cannot be extremal for all dilates because the bound from Theorem 9 in the case $k = n - 1$ is tight as $t \to \infty$. König and Koldobsky [85] found conditions on product measures ensuring that the hyperplane $(\frac{1}{\sqrt{2}}, \frac{1}{\sqrt{2}}, 0, \dots, 0)^\perp$ gives the maximal volume among all hyperplanes $a^\perp$ with $\max_j |a_j| \le \frac{1}{\sqrt{2}}$. When specialized to the standard Gaussian measure, they additionally obtained that the hyperplane $(\frac{1}{\sqrt{2}}, \frac{1}{\sqrt{2}}, 0, \dots, 0)^\perp$ yields maximal volume (among *all* hyperplanes) if and only if $t < t_0 = 1.253\dots$. Sharp upper bounds for $t > t_0$ are not known.

Cylinders

Dirksen [46] studied the extremal central sections of the generalized cylinders $Z_r = Q_n \times (rB_2^m)$, $r > 0$, $m, n \ge 1$. He found sharp upper bounds in the 3-dimensional case of an ordinary cylinder, i. e., $m = 2$, $n = 1$, as well as upper bounds in the general case, sharp for large radii, developing Fourier analytic formulas and delicate integral inequalities involving Bessel functions.

Perimeter

Answering a question of Pełczyński about hyperplane sections of maximal *perimeter* (i. e., sections with the boundary of the cube), König and Koldobsky [86] have shown that the extremal direction is the same as for the volume.

Theorem 10 (König–Koldobsky [86]). *Let $n \ge 3$. For every unit vector a in $\mathbb{R}^n$, we have*

$$\mathrm{vol}_{n-2}(\partial Q_n \cap a^\perp) \le 2((n-2)\sqrt{2} + 1).$$

This bound is attained if $a = (\pm e_i \pm e_j)/\sqrt{2}$ for some $1 \le i < j \le n$. This theorem also leads to counterexamples to a perimeter version of the Busemann–Petty problem in $n \ge 14$ dimensions. For the proof, they derive a Fourier analytic formula for the perimeter; its analysis involves new ingredients, most notably local conditions for constrained extrema, as well as subtle technical estimates around Ball's integral inequality.

Diagonal sections

Here we consider the volume of the section by the hyperplane perpendicular to the main diagonal

$$a_n = \mathrm{vol}_{n-1}(Q_n \cap (\underbrace{1,\ldots,1}_{n})^{\perp}), \quad n \geq 1.$$

Perhaps a more natural interpretation of the sequence $a_1,\ldots,a_n$ is as the volumes of the sections of Q_n by hyperplanes perpendicular to the diagonals of subcubes of growing dimension, for $1 \leq k \leq n$, where we have

$$\mathrm{vol}_{n-1}(Q_n \cap (\underbrace{1,\ldots,1}_{k}, \underbrace{0,\ldots,0}_{n-k})^{\perp}) = a_k.$$

Theorems 4 and 6 in particular assert that $a_1 \leq a_i \leq a_2$. Interestingly, the volumes of the diagonal sections form a (strictly) increasing sequence.

Theorem 11 (Bartha–Fodor–González [18]). *We have $a_1 < a_3 < a_4 < a_5 < \cdots < a_2$.*

Their approach starts with Pólya's formula $a_n = \frac{\sqrt{n}}{\pi} \int_{-\infty}^{\infty} (\frac{\sin t}{t})^n dt$ and is based on an intricate asymptotic analysis by means of the Laplace method. They first argue that the sequence (a_n) increases for all $n \geq n_0$ for some n_0. Then, using numerical estimates, they bound n_0 and deal with $n \leq n_0$ by computer assisted calculations. Their arguments also show that the sequence (a_n) is eventually concave.

It is tempting to believe that critical hyperplane sections must be diagonal, that is, if $a \mapsto \mathrm{vol}_{n-1}(Q_n \cap a^{\perp})$ has an extremum at a unit vector a^*, then a^* is proportional to a diagonal $(1,\ldots,1,0,\ldots,0)$. Ambrus [3] and Ivanov and Tsiutsiurupa [66] recently independently found an elegant local condition (with vastly different methods). Moreover, Ambrus confirmed this for $n \leq 3$ and disproved it for $n = 4$.

Discrete version

Melbourne and Roberto [101] have derived a sharp discrete analog of Ball's upper bound for hyperplane sections.

Theorem 12 (Melbourne–Roberto [101]). *Let $n, \ell_1,\ldots,\ell_n \geq 1$ and $t, k_1,\ldots,k_n$ be integers. Then*

$$\left| \left\{ z \in \mathbb{Z}^n \cap \prod_{j=1}^{n} [k_j, k_j + \ell_j - 1], \sum_{j=1}^{n} z_j = t \right\} \right| < \sqrt{2} \frac{\prod_{j=1}^{n} \ell_j}{\sqrt{\sum_{j=1}^{n} (\ell_j^2 - 1)}}.$$

The constant $\sqrt{2}$ is the best possible, as can be seen by discretizing Ball's extremizer (by taking $\ell_1 = \ell_2 = m$, $\ell_3 = \cdots = \ell_n = 1$ and letting $m \to \infty$). Mimicking Ball's approach, the following integral inequality lies at the heart of the argument:

$$\int_{-1/2}^{1/2} \left| \frac{\sin(n\pi t)}{n\sin(\pi t)} \right|^p dt < \sqrt{\frac{2}{p(n^2-1)}}, \quad p \geq 2, \ n = 2, 3, \ldots.$$

This is in fact stronger than Ball's inequality (2.5) and recovers it by letting $n \to \infty$. Melbourne and Roberto developed a new viewpoint on establishing such delicate bounds for oscillatory integrands, borrowing and combining ideas from majorization and optimal transport.

Chessboard cutting

It is folklore that a line can meet the interiors of no more than $2N-1$ squares of the usual $N \times N$ chessboard and this bound is tight (consider the diagonal pushed down a bit). We refer to Bárány and Frenkel's work [35] for a short argument as well as precise estimates for a 3-dimensional analog. To tackle the problem in higher dimensions, in [36] they introduced the following quantity involving volumes of hyperplane sections of the cube:

$$V_n = \max_{v \in \mathbb{R}^n} \frac{\|u\|_1}{|v|} \, \mathrm{vol}_{n-1}(Q_n \cap v^\perp).$$

They have shown that if the cube $[0, N]^n$ is divided into N^n unit cubes in the usual way, then the maximal number of unit cubes that a hyperplane can intersect equals

$$(1 + o(1))V_n N^{n-1}$$

for a fixed $n \geq 1$ as $N \to \infty$. Confirming a conjecture from [36], Aliev recently found the constant V_n [2].

Theorem 13 (Aliev [2]). *Let $n \geq 1$. We have $V_n = \sqrt{n} \, \mathrm{vol}_{n-1}(Q_n \cap (1, \ldots, 1)^\perp)$.*

In words, it is the diagonal section that maximizes V_n; thus, $\sqrt{n} \leq V_n \leq \sqrt{2}\sqrt{n}$ and $V_n \sim \sqrt{\frac{6}{\pi}} \sqrt{n}$ for large n. Aliev's argument is purely geometric with the main observation being that the hyperplane parallel to $(1, \ldots, 1)^\perp$ supports the intersection body of the cube.

Stability

With additional insights gained from a certain probabilistic point of view (see Section 4), Chasapis and the authors recently obtained [42] a dimension-free stability result for both lower and upper bounds for hyperplane sections.

Theorem 14 (Chasapis–Nayar–Tkocz [42]). *There are universal constants $c_1, c_2 > 0$ such that for every unit vector a in $\mathbb{R}^n$ with $a_1 \geq \cdots \geq a_n \geq 0$, we have*

$$1 + c_1|a - e_1|^2 \leq \mathrm{vol}_{n-1}(Q_n \cap a^\perp) \leq \sqrt{2} - c_2\left|a - \frac{e_1 + e_2}{\sqrt{2}}\right|.$$

The exponents 2 and 1 on the left- and right-hand sides, respectively, are the best possible, as can be explicitly verified for $n = 2$. In an independent work [100], Melbourne and Roberto obtained a similar result.

2.2 Balls of p-norms

We begin with a monotonicity result for the parameter p discovered by Mayer and Pajor [102].

Theorem 15 (Meyer–Pajor [102]). *Fix $1 \leq k \leq n$. For every k-dimensional subspace H in $\mathbb{R}^n$, the function*

$$p \mapsto \frac{\mathrm{vol}_k(B_p^n \cap H)}{\mathrm{vol}_k(B_p^k)}$$

is non-decreasing on $[0, +\infty)$. In particular, comparison with the Euclidean ball yields

$$\mathrm{vol}_k(B_p^n \cap H) \leq \mathrm{vol}_k(B_p^k), \quad 0 < p < 2,$$
$$\mathrm{vol}_k(B_p^n \cap H) \geq \mathrm{vol}_k(B_p^k), \quad p > 2.$$

In each inequality, equality holds if and only if H is spanned by some k standard basis vectors.

Meyer and Pajor established this theorem for $p \geq 1$, which was extended later to $p < 1$ independently by Barthe [19] and Caetano [39]. Letting $p \to \infty$ recovers Vaaler's theorem, Theorem 5, for the cube sections. Vaaler's argument uses Kanter's peakedness to make a comparison between uniform and Gaussian distributions. The key point in [102] was that the same comparison holds across the whole family of probability measures with densities $\{e^{-c_p|x|^p}\}_{p>0}$. We will present this crucial idea in a probabilistic setting in Section 4.

More is known for hyperplane sections when $0 < p < 2$. Meyer and Pajor [102] found that the minimal volume hyperplane sections of the cross-polytope B_1^n are attained by the diagonal directions and conjectured the same for the entire range $0 < p < 2$, confirmed later by Koldobsky in [74] in a strong Schur convexity-type result.

Theorem 16 (Koldobsky [74]). *Let $0 < p < 2$. For every two unit vectors a and b in $\mathbb{R}^n$ such that $(b_1^2, \ldots, b_n^2)$ majorizes $(a_1^2, \ldots, a_n^2)$, we have*

$$\mathrm{vol}_{n-1}(B_p^n \cap a^\perp) \leq \mathrm{vol}_{n-1}(B_p^n \cap b^\perp).$$

For background on majorization and Schur convexity, we refer for instance to [25]. In particular, since

$$\left(\frac{1}{n}, \ldots, \frac{1}{n} \right) \prec (a_1^2, \ldots, a_n^2) \prec (1, 0, \ldots, 0),$$

for an arbitrary unit vector a in $\mathbb{R}^n$, the minimal and maximal volume sections follow. What makes the range $0 < p < 2$ so much more tractable compared to $p > 2$ is the fact that the Fourier transform of $e^{-|x|^p}$ is a non-negative function of the form $t \mapsto \int_0^\infty e^{-ut^2} d\mu(u)$, a Gaussian mixture. In fact, the same also holds for $e^{-|x|^p}$, which allowed the authors of [49] to bypass the Fourier analytic arguments entirely. We return to this in Section 4.

The maximal volume hyperplane sections of B_p^n-balls for $2 < p < \infty$ are unknown. Oleszkiewicz established in [113] that Ball's upper bound for the cube, Theorem 6, does not extend to all $p > 2$, as it fails for all $p < 26.265\ldots$ and large enough dimensions (by comparing the cube's extremizing hyperplane $(1, 1, \ldots, 0)^\perp$ to the diagonal one $(1, 1 \ldots, 1)^\perp$ in the limit $n \to \infty$). We conjecture that in each dimension there is a unique phase transition point.

Conjecture 2. *For every $n \geq 3$, there is a unique $p_0(n)$ such that*

$$\max_{a \in S^{n-1}} \mathrm{vol}_{n-1}(B_p^n \cap a^\perp) = \begin{cases} \mathrm{vol}_{n-1}(B_p^n \cap (1, \ldots, 1)^\perp), & 2 < p \leq p_0(n), \\ \mathrm{vol}_{n-1}(B_p^n \cap (1, 1, 0, \ldots, 0)^\perp), & p \geq p_0(n). \end{cases}$$

For lower-dimensional sections, there is a general bound of Barthe which extends a corresponding result for the cube from Theorem 7. The argument also crucially relies on the Brascamp–Lieb inequalities.

Theorem 17 (Barthe [19]). *Let $p \geq 2$. Fix $1 \leq k \leq n$. For every k-dimensional subspace H in $\mathbb{R}^n$, we have*

$$\mathrm{vol}_k(B_p^n \cap H) \leq \left(\frac{n}{k} \right)^{k(1/2 - 1/p)} \mathrm{vol}_k(B_p^k).$$

As for the cube, this is sharp when k divides n with the same extremizing subspace.

Using a direct argument involving triangulation and convexity of certain functions, Nazarov has shown that planar sections of the cross-polytope of minimal area are attained at regular polygons.

Theorem 18 (Nazarov [42]). *Let $n \geq 3$. For every 2-dimensional subspace H in $\mathbb{R}^n$, we have*

$$\mathrm{vol}_2(B_1^n \cap H) \geq \frac{n^2 \sin^3(\frac{\pi}{2n})}{\cos(\frac{\pi}{2n})},$$

which is optimal, attained when $B_1^n \cap H$ is a regular $2n$-gon.

All known results from Table 1 on extremal volume hyperplane sections for ℓ_p-balls admit robust versions (recall also Theorem 14).

Theorem 19 (Chasapis–Nayar–Tkocz [42]). *For every $p > 0$, there is a positive constant c_p such that for every $n \geq 1$ and every unit vector $a = (a_1, \ldots, a_n)$ in $\mathbb{R}^n$ with $a_1 \geq a_2 \geq \cdots \geq a_n \geq 0$, we have*

$$\frac{\mathrm{vol}_{n-1}(B_p^n \cap a^\perp)}{\mathrm{vol}_{n-1}(B_p^n \cap e_1^\perp)} \geq 1 + c_p |a - e_1|^2, \quad 2 < p \leq \infty,$$

$$\frac{\mathrm{vol}_{n-1}(B_p^n \cap a^\perp)}{\mathrm{vol}_{n-1}(B_p^n \cap (\frac{e_1 + \cdots + e_n}{\sqrt{n}})^\perp)} \geq 1 + c_p \sum_{j=1}^{n} (a_j^2 - 1/n)^2, \quad 0 < p < 2,$$

$$\frac{\mathrm{vol}_{n-1}(B_p^n \cap a^\perp)}{\mathrm{vol}_{n-1}(B_p^n \cap e_1^\perp)} \leq \left(a_1^p + (1 - a_1^2)^{p/2}\right)^{-1/p}, \quad 0 < p < 2.$$

We finish this subsection with Vaaler's conjecture on general rather precise lower bounds which have been verified to a large extent for ℓ_p-balls.

Conjecture 3 (Vaaler [132]). *Let K be a symmetric isotropic convex body in $\mathbb{R}^n$. Then for every non-zero subspace H in $\mathbb{R}^n$ of dimension $1 \leq k \leq n$, we have*

$$\mathrm{vol}_k(K \cap H) \geq 1.$$

Noteworthily, if true, it implies the slicing conjecture (made independently of it); see Hensley's theorem, Theorem 3. Vaaler's theorem confirms this inequality for the cube (which is tight). Meyer and Pajor's sharp lower bound gives this inequality for $K = B_p^n$ with $2 < p < \infty$ and all subspaces (see [102]), as well as $1 < p < 2$ and all hyperplanes (see Schmuckenschläger's note [127]); however, these are not tight anymore.

2.3 Simplices

Here we discuss results concerning sections of regular simplices. It will be most convenient to consider a regular n-dimensional simplex of side length $\sqrt{2}$ embedded in $\mathbb{R}^{n+1}$,

$$\Delta_n = \left\{ x \in \mathbb{R}^{n+1}, \ x_1, \ldots, x_{n+1} \geq 0, \ \sum_{j=1}^{n+1} x_j = 1 \right\}.$$

Central sections will refer to those by (affine) subspaces passing through the barycenter $(\frac{1}{n+1}, \dots, \frac{1}{n+1})$ of Δ_n. In particular, if a is a unit vector in $\mathbb{R}^{n+1}$ with $\sum_{j=1}^{n+1} a_j = 0$ (so parallel to the hyperplane containing Δ_n), then $\Delta_n \cap a^\perp$ is a central hyperplane section of Δ_n. Such sections of maximal volume have been determined by Webb in [134].

Theorem 20 (Webb [134]). *For every unit vector a in $\mathbb{R}^{n+1}$ with $\sum_{j=1}^{n+1} a_j = 0$, we have*

$$\mathrm{vol}_{n-1}(\Delta_n \cap a^\perp) \leq \frac{1}{\sqrt{2}} \frac{\sqrt{n+1}}{(n-1)!}.$$

This is attained if and only if $a^\perp$ passes through some $n-1$ vertices of Δ_n.

Webb gave two proofs, both based on an elegant probabilistic formula,

$$\mathrm{vol}_{n-1}(\Delta_n \cap a^\perp) = \frac{\sqrt{n+1}}{(n-1)!} f_a(0),$$

where f_a is the probability density of $\sum_{j=1}^{n+1} a_j X_j$ with X_j being i. i. d. standard exponential random variables, with density e^{-x} supported on $(0, +\infty)$. Thus, his result becomes $f_a(0) \leq \frac{1}{\sqrt{2}}$ with equality if and only if $n-1$ of the a_j vanish. His first proof mimicked Ball's Fourier analytic approach with the crucial bound coming from Hölder's inequality and an integral inequality. His second proof was probabilistic, exploiting log-concavity.

Webb also found that the 1- and 2-dimensional central sections of Δ_n of maximal volume are attained at lines and planes passing through a vertex and an edge of Δ_n, respectively (see his PhD thesis [133], as well as [94] for a different argument in the line case).

For general upper bounds on central sections, following the approach involving Ball's geometric form of the Brascamp–Lieb inequality, Dirksen [45] obtained the following result.

Theorem 21 (Dirksen [45]). *For every k-dimensional subspace of $\mathbb{R}^{n+1}$ passing through the barycenter of the simplex Δ_n, we have*

$$\mathrm{vol}_{k-1}(\Delta_n \cap H) \leq \frac{k^{\frac{k}{2(n+1)}}}{(k-1)!}.$$

Moreover, if $\mathrm{dist}(H, e_j) \leq \sqrt{\frac{n+1-k}{n+2-k}}$ *for each $j \leq n+1$, then*

$$\mathrm{vol}_{k-1}(\Delta_n \cap H) \leq \frac{1}{(k-1)!} \sqrt{\frac{n+1}{n+2-k}},$$

which is sharp, attained when H contains $k-1$ vertices of Δ_n.

As opposed to symmetric convex bodies for which maximum volume sections by all affine subspaces of a fixed dimension always occur when they pass through the barycenter (by the Brunn–Minkowski inequality), for the simplex such a question becomes nontrivial. Webb pointed out in [134] that combining two results of Ball yields that for fixed $1 \le k \le n$, we have

$$\mathrm{vol}_k(\Delta_n \cap H) \le \mathrm{vol}_k(F_k),$$

for all $(k + 1)$-dimensional affine subspaces H in $\mathbb{R}^{n+1}$, where F_k is a k-dimensional face of Δ_n, that is, the k-dimensional slices of Δ_n of maximal volume are exactly the k-dimensional faces. To explain this, fix H and consider the maximum volume ellipsoid, say $\mathcal{E}^*$ contained in the convex body $K = \Delta_n \cap H$. Ball found [11] that the n-simplex has maximal volume ratio among all convex bodies in $\mathbb{R}^n$. The volume ratio of a convex body C in $\mathbb{R}^n$ is $\mathrm{vr}(C) = (\mathrm{vol}_n(C)/\mathrm{vol}_n(\mathcal{E}))^{1/n}$, where $\mathcal{E}$ is the maximum volume ellipsoid in C. Thus,

$$\mathrm{vol}_k(\Delta_n \cap H) = \mathrm{vr}(K)^k \, \mathrm{vol}_k(\mathcal{E}^*) \le \mathrm{vr}(F_k)^k \, \mathrm{vol}_k(\mathcal{E}^*).$$

Moreover, Ball has shown in [12] that among all k-dimensional ellipsoids in Δ_n, the Euclidean balls inscribed in k-faces have maximal volume; thus, they are the maximal volume ellipsoids in F_k. Therefore,

$$\mathrm{vr}(F_k)^k \, \mathrm{vol}_k(\mathcal{E}^*) \le \mathrm{vol}_k(F_k).$$

In [55], Fradelizi has given a different argument, deriving this fact from a more general result for cones in isotropic position.

Lower bounds are much less understood.

Conjecture 4. *For every unit vector a in $\mathbb{R}^{n+1}$ with $\sum_{j=1}^{n+1} a_j = 0$, we have*

$$\mathrm{vol}_{n-1}(\Delta_n \cap a^{\perp}) \ge \left(\frac{n}{n+1}\right)^{n-1/2} \frac{\sqrt{n+1}}{(n-1)!},$$

which is attained when $a^{\perp}$ is parallel to a face of Δ_n.

This has been confirmed in low dimensions ($n \le 4$) by Brzezinski [32]. He also noticed that a bound of the correct order but off by a multiplicative constant follows by applying Fradelizi's theorem from [55] to Webb's result stated above.

2.4 Complex analogs

If we consider $\mathbb{C}^n$ as a Hilbert space equipped with the standard (complex) inner product and volume (Lebesgue measure after the natural identification $\mathbb{C}^n \simeq \mathbb{R}^{2n}$), most of the results about extremal volume sections (of real spaces) considered thus far beg for their natural complex counterparts. Vaaler's theorem and its generalization of Meyer and Pajor admit such extensions, with almost the same proofs, as was pointed out in their papers.

Theorem 22 (Vaaler [132], Meyer–Pajor [102]). *Let $1 \leq k \leq n$ and let H be a (complex) k-dimensional subspace in $\mathbb{C}^n$. Then*

$$\mathrm{vol}_{2k}(B^n_{p,\mathbb{C}} \cap H) \geq \mathrm{vol}_{2k}(B^k_{p,\mathbb{C}}),$$

when $2 \leq p \leq \infty$. The reverse inequality holds when $0 < p \leq 2$.

Here,

$$B^n_{p,\mathbb{C}} = \left\{ z \in \mathbb{C}^n, \ \left(\sum_{j=1}^{n} |z_j|^p \right)^{1/p} \leq 1 \right\}$$

is the unit ball of the complex $\ell_p(\mathbb{C}^n)$-space; in particular, $B^\infty_{\infty,\mathbb{C}}$ is the polydisc (the Cartesian product of the unit discs in $\mathbb{C}$). In fact, their proofs yield a further extension from $B^n_{p,\mathbb{C}}$ to bodies which are ℓ_p-sums of Euclidean spaces of arbitrary dimensions, which has been in turn significantly generalized by Eskanazis in [48] (see Theorem 26 below).

Ball's cube slicing result of Theorem 6 has been extended to the complex setting by Oleszkiewicz and Pełczyński in [114], who proved the following sharp polydisc slicing bound.

Theorem 23 (Oleszkiewicz–Pełczyński [114]). *For every unit vector a in $\mathbb{C}^n$, we have*

$$\mathrm{vol}_{2n-2}(B^n_{\infty,\mathbb{C}} \cap a^\perp) \leq 2\pi^{2n-2}.$$

Equality holds if and only if $a = (\xi e_i + \eta e_j)/\sqrt{2}$ for some $1 \leq i < j \leq n$ and $\xi, \eta \in \mathbb{C}$ with $|\xi|, |\eta| = 1$.

The proof strategy follows the same path of the Fourier analytic formula and defactorization by means of Hölder's inequality; however, new technical challenges arise. The heart of the proof is the following analytical inequality:

$$\int_0^\infty \left| \frac{2J_1(t)}{t} \right|^p t\,dt \leq \frac{4}{p}, \quad p \geq 2 \tag{2.6}$$

(cf. (2.5)), where J_1 is the Bessel function of the first kind of order 1. Its proof rests on precise pointwise bounds on J_1 as well as an interpolation argument. A new different proof has been very recently given in [101]. Moreover, the upper bounds for higher codimensions of Theorem 7 can be transferred almost ad verbatim to the complex case as well (as was remarked by Barthe and Koldobsky; see [114]).

The exact analog of the sharp upper bound on the perimeter from Theorem 10 also holds, as shown by König and Koldobsky in [86].

Sharp upper bounds even on hyperplane (complex codimension 1) sections in the range $2 < p < \infty$ remain open. For the same reasons as in the real case, the range $0 < p < 2$ is more tractable and we have the following analog of Koldobsky's theorem, Theorem 16.

Theorem 24 (Koldobsky–Zymonopoulou [80]). *Let $0 < p < 2$. For every two unit vectors a and b in $\mathbb{C}^n$ such that $(|b_1|^2, \ldots, |b_n|^2)$ majorizes $(|a_1|^2, \ldots, |a_n|^2)$, we have*

$$\mathrm{vol}_{2n-2}(B_{p,\mathbb{C}}^n \cap a^\perp) \le \mathrm{vol}_{2n-2}(B_{p,\mathbb{C}}^n \cap b^\perp).$$

Finally, a complex version of Busemann's theorem, Theorem 1, has been developed by Koldobsky, Paouris, and Zymonopoulou in [77], whereas a full solution to the complex Busemann–Petty problem is due to Koldobsky, König, and Zymonopoulou [76].

2.5 Miscellanea

We finish this section with a brief account of various results related to and motivated by sharp bounds on volumes of sections.

Slabs

For a unit vector a in $\mathbb{R}^n$ and $t > 0$, we set

$$H_{a,t} = \{x \in \mathbb{R}^n, \; |\langle x, a \rangle| \le t\}$$

to be the (symmetric) slab of width $2t$ orthogonal to the direction a (in other words, a thickening/enlargement $a^\perp + tB_2^n$ of the hyperplane $a^\perp$). Answering a question of V. Milman, Barthe and Koldobsky in [23] have established the following extension of Hadwiger and Hensley's Theorem 4.

Theorem 25 (Barthe-Koldobsky [23]). *For every unit vector a in $\mathbb{R}^n$ and $0 \le t \le \frac{3}{8}$, we have*

$$\mathrm{vol}_n(Q_n \cap H_{a,t}) \ge \mathrm{vol}_n(Q_n \cap H_{e_1,t}).$$

They derived this from a sharp inequality for unimodal log-concave densities in one dimension, expanding on Hensley's approach.

In words, Hadwiger and Hensley's result is stable in that, independent of the dimension, *coordinate slabs* contain the least volume of the unit cube among all symmetric slabs of fixed width at most 3/4. This bound is in the spirit of the concentration of measure (see [28, 89, 90]), providing a sharp lower bound of *small* enlargements on the volume 1/2 half-spaces $\{x \in \mathbb{R}^n, \langle x, a \rangle \ge 0\}$ in Q_n. The threshold $\frac{3}{8}$ is suboptimal: in the 2-dimensional case, a direct calculation from [23] shows that at $t = \sqrt{2}-1$ the extremizing slab changes from the coordinate one to the diagonal one. The sharp behavior in higher dimensions is not clear. The paper [23] provides asymptotic results that the slabs orthogonal to the main diagonal are optimal for *large t* of the order $\sqrt{n}$ as $n \to \infty$ (developing en route very interesting conditions for convexity properties of Laplace transforms), with a precise non-asymptotic result for the range $\frac{1}{2}\sqrt{n-1} \le t \le \frac{1}{2}\sqrt{n}$ obtained recently by Moody, Stone, Zach, and Zvavitch [105].

A detailed analysis of the (local as well as global) extremal slabs in the 2- and 3-dimensional cases has been conducted by König and Koldobsky [83], whereas in [84], they obtained a complex analog of Theorem 25.

Block subspaces

Eskenazis [48] gathered under one umbrella the results on slicing ℓ_p-balls, both real and complex, when $0 < p < 2$, thus significantly generalizing and unifying Theorems 16, 22, and 24.

Theorem 26 (Eskenazis [48]). *Let m, n be positive integers and let $0 < p < 2$. Suppose $X = (\mathbb{R}^m, \|\cdot\|)$ is a quasinormed space which admits an isometric embedding into L_p. For every two unit vectors a and b in $\mathbb{R}^n$ such that $(b_1^2, \ldots, b_n^2)$ majorizes $(a_1^2, \ldots, a_n^2)$, we have*

$$\mathrm{vol}_{mn-m}(B_p^n(X) \cap H_a) \le \mathrm{vol}_{mn-m}(B_p^n \cap H_b).$$

Here,

$$B_p^n(X) = \left\{ x = (x_1, \ldots, x_n) \in \mathbb{R}^m \times \cdots \times \mathbb{R}^m, \ \left(\sum_{j=1}^n \|x_j\|^p \right)^{1/p} \le 1 \right\}$$

is the unit ball of the ℓ_p-sum of X, whereas

$$H_a = \left\{ x = (x_1, \ldots, x_n) \in \mathbb{R}^m \times \cdots \times \mathbb{R}^m, \ \sum_{j=1}^n a_j x_j = 0 \right\}$$

is a *block* subspace of codimension m in $(\mathbb{R}^m)^n$. In particular, $X = \ell_2^m$ with $m = 1, 2$ recovers Theorems 16 (when $p < 2$), 22, and 24. The point is that there is a plethora of non-Hilbertian examples treated by this general result, most notably $X = \ell_q^m$ with $p \le q \le 2$.

Eskenazis' argument builds on [49], with the key new ingredient being Lewis' representation guaranteeing that the norm on X which embeds isometrically into $L_p, p > 0$, admits a form

$$\|x\| = \left(\int_{S^{m-1}} |\langle Ux, \theta \rangle|^p d\mu(\theta) \right)^{1/p}, \quad x \in \mathbb{R}^m,$$

for some invertible linear map $U: \mathbb{R}^m \to \mathbb{R}^m$ and an isotropic Borel measure μ on the unit sphere S^{m-1} [91, 126]. The restriction $p < 2$ is not needed here, but is included to bring about Gaussian mixtures (as highlighted after Theorem 16).

For the regime $p > 2$, only the case $p = \infty, X = \ell_2^m$ has been considered, i. e., sections of

$$B_\infty^n(\ell_2^m) = \underbrace{B_2^m \times \cdots \times B_2^m}_{n},$$

for which Brzezinski [32] obtained that for every $n, m \geq 2$ and every unit vector a in $\mathbb{R}^n$, we have

$$\mathrm{vol}_{mn-m}(B_\infty^n(\ell_2^m) \cap H_a) \leq \frac{(m+2)^{m/2}}{2^{m/2-1}m\Gamma(m/2)}. \tag{2.7}$$

This is asymptotically sharp as $n \to \infty$ because the right-hand side equals exactly $\lim_{n\to\infty} \mathrm{vol}_{mn-m}(B_\infty^n(\ell_2^m) \cap H_{(\frac{1}{\sqrt{n}},\dots,\frac{1}{\sqrt{n}})})$. The case $m = 2$ is special in that this limit also equals $A_{m,n} = \mathrm{vol}_{mn-m}(B_\infty^n(\ell_2^m) \cap H_{(\frac{1}{\sqrt{2}},\frac{1}{\sqrt{2}},0,\dots,0)})$, whilst for every $m > 2$, the limit is strictly larger than $A_{m,n}$. In other words, Ball's upper bound from Theorem 6 does *not* generalize to block subspace sections of $B_2^m \times \cdots \times B_2^m$ for any $m > 2$ (but it does when $m = 2$, as we have seen in Oleszkiewicz and Pełczyński's theorem, Theorem 23).

We finish with Eskanazis' conjecture on sharp lower bounds by block subspaces, generalizing Hadwiger and Hensley's theorem, Theorem 4.

Conjecture 5 (Eskenazis [48]). *Let $m, n \geq 1$. Let K be a symmetric convex body in $\mathbb{R}^m$. For every unit vector $a \in \mathbb{R}^n$, we have*

$$\mathrm{vol}_{mn-m}(K \times \cdots \times K \cap H_a) \geq \mathrm{vol}_m(K)^{n-1}.$$

Non-central sections

In this context, perhaps the most natural question to ask is about extremal volume sections by affine subspaces at a *fixed* distance $t > 0$ from the origin. This has arguably proved to be more difficult than the question of central sections, even for the cube. Sharp results for line sections have been found in [105] for the cube and in [94] for the cross-polytope. For hyperplane sections, we have the following conjecture of V. Milman (see [83]).

Conjecture 6 (V. Milman [83]). *The minimum and maximum of $\mathrm{vol}_{n-1}(Q_n \cap H)$ over the affine hyperplanes H at a fixed distance $t > 0$ from the origin are attained when H is orthogonal to a diagonal direction $(1, \dots, 1, 0, \dots, 0)$ with a suitable number of 1's depending on t.*

There are several partial results supporting it. König and Koldobsky verified that it holds in low dimensions ($n = 2, 3$) [83]. Moody, Stone, Zach, and Zvavitch have established that in the range $\frac{1}{2}\sqrt{n-1} < t < \frac{1}{2}\sqrt{n}$ the main diagonal direction gives the maximal section [105], later extended to all $t > \frac{1}{2}\sqrt{n-2}$ by Pournin in [116], where one of the key ideas was to employ a noteworthy combinatorial formula for sections of the cube,

$$\mathrm{vol}_{n-1}([0,1]^n \cap \{x \in \mathbb{R}^n,\ \langle x, a\rangle = b\}) = \sum_v \frac{(-1)^{\sum v_j}|a|(b - \langle v, a\rangle)^{n-1}}{(n-1)!\prod a_j},$$

where the sum is over the vertices v of the cube $[0,1]^d$ such that $\langle v, a\rangle \le b$ (see also [17]). In a recent preprint [117], Pournin also showed that the main diagonal direction is strictly locally maximal for $t = \Omega(\frac{\sqrt{n}}{\log n})$, derived from general local conditions for all diagonal directions. König and Rudelson [87] obtained dimension-free lower bounds on non-central sections of the cube as well as the polydisc. König [82] treated non-central extremal volume as well as perimeter sections of the regular simplex, cube, and cross-polytope, when the distance t is fairly large, also investigating local behavior for the entire range of t.

Probabilistic extensions

There is a natural link between the volume of sections and negative moments of linear forms, which goes back at least to Kalton and Koldobsky's work [68]. To illustrate it, first note that the value at say $x = 0$ of a probability density f on $\mathbb{R}$ which is continuous at 0 can be obtained by taking the limit of its negative moments,

$$f(0) = \lim_{q \to -1+} \frac{1+q}{2} \int |x|^q f(x)\mathrm{d}x. \tag{2.8}$$

In view of this and the basic probabilistic formula for sections (2.2), the sharp bounds for hyperplane sections of the cube from Theorems 4 and 6 can be phrased as

$$1 \le \lim_{q \to -1+} \frac{1+q}{2} \mathbb{E}\left|\sum_{j=1}^n a_j U_j\right|^q \le \sqrt{2}$$

for all unit vectors a in $\mathbb{R}^n$, where $U_1, U_2, \ldots$ are i.i.d. random variables uniform on $[-\frac{1}{2}, \frac{1}{2}]$. Do such inequalities remain true with a fixed q? The answer is known for the cube and polydisc, where a sharp phase transition of the extremizer occurs for the upper bound with diagonal directions entering the picture.

Theorem 27 (Chasapis–König–Tkocz [41]). *Let $-1 < q < 0$. Let $U_1, U_2, \ldots$ be i.i.d. random variables uniform on $[-\frac{1}{2}, \frac{1}{2}]$. For every $n \ge 1$ and unit vectors a in $\mathbb{R}^n$, we have*

$$\mathbb{E}|U_1|^q \le \mathbb{E}\left|\sum_{j=1}^n a_j U_j\right|^q \le \begin{cases} \mathbb{E}|(U_1 + U_2)/\sqrt{2}|^q, & -1 < q \le q_0, \\ \lim_{m\to\infty} \mathbb{E}|(U_1 + \cdots + U_m)/\sqrt{m}|^q, & q_0 \le q < 0. \end{cases}$$

The constant $q_0 = -0.79\ldots$ is given uniquely by equating the two expressions on the right-hand side.

A similar behavior has been established for the polydisc slicing by Chasapis, Singh, and Tkocz in [43], with the phase transition "moving to the left" where the negative moments recover volume.

3 Projections

We turn our attention to Question (II) from the introduction about projections of extremal volume of basic convex bodies such as the cube, simplex, and cross-polytope, as well as the family of ℓ_p-balls. As we will see, our understanding of hyperplane projections of ℓ_p-balls is at the same level as for sections (see Tables 1 and 2), whilst in general much less is known, particularly for lower-dimensional projections. The methods also seem to shift from analytic to more of an algebraic or combinatorial nature.

3.1 Cubes

Thanks to Cauchy's formula from Theorem 2, extremal volume projections on hyperplanes are easy to determine, for the surface area measure of the cube Q_n is the counting measure $\sum_{j=1}^{n} \delta_{\pm e_j}$ of the set of the $2n$ vectors $\{\pm e_j, \, j \le n\}$ outer normal to the facets of Q_n; thus, for every unit vector a in $\mathbb{R}^n$, we have

$$\mathrm{vol}_{n-1}(\mathrm{Proj}_{a^\perp}(Q_n)) = \sum_{j=1}^{n} |a_j|.$$

Therefore,

$$1 \le \mathrm{vol}_{n-1}(\mathrm{Proj}_{a^\perp}(Q_n)) \le \sqrt{n},$$

by squaring and neglecting the off-diagonal terms for the lower bound and simply applying the Cauchy–Schwarz inequality for the upper bound. The former is attained if and only if Q_n is projected onto a coordinate hyperplane and the latter is attained if and only if Q_n is projected onto a hyperplane orthogonal to a main diagonal.

A zonotope is the Minkowski sum of intervals. Orthogonal projections of the unit cube $Q_n = [-\frac{1}{2}, \frac{1}{2}]$ are zonotopes and, conversely, every zonotope can be obtained as such a projection (of a possibly rescaled and translated cube in a sufficiently high dimension). Shephard's decomposition of zonotopes into parallelepipeds led him in [129] to the following classical formula for volume: If $v_1, \ldots, v_n$ are vectors in $\mathbb{R}^k$, then the volume of the zonotope $Z = \sum_{j=1}^{n} [0, v_j]$ is expressed as

$$\mathrm{vol}_k(Z) = \sum_{1 \le j_1 < \cdots < j_k \le n} |\det[v_{j_1} \, \ldots \, v_{j_k}]|,$$

where $[v_{j_1} \; \ldots \; v_{j_k}]$ is the $k \times k$ matrix with columns $v_{j_1}, \ldots, v_{j_k}$. For the orthogonal projection of the cube Q_n onto a k-dimensional subspace H, the vectors v_j can be taken as columns of the $k \times n$ matrix whose rows form an orthonormal basis of H, leading to the constraint

$$\sum_{1 \le j_1 < \cdots < j_k \le n} \left| \det[v_{j_1} \; \ldots \; v_{j_k}] \right|^2 = 1,$$

by the Cauchy–Binnet formula. Then, exactly as in the codimension 1 case, we obtain upper and lower bounds on the volume. This argument goes back to Chakerian and Filliman's work [40].

Theorem 28 (Chakerian–Filliman [40]). *Fix $1 \le k \le n$. For every k-dimensional subspace H in $\mathbb{R}^n$, we have*

$$1 \le \mathrm{vol}_{n-1}(\mathrm{Proj}_H(Q_n)) \le \min\left\{ \sqrt{\binom{n}{k}}, \; \frac{\mathrm{vol}_{k-1}(B_2^{k-1})^k}{\mathrm{vol}_k(B_2^k)^{k-1}} \left(\frac{n}{k} \right)^{k/2} \right\}.$$

The lower bound is clearly sharp, attained at coordinate subspaces. It also instantly follows from Vaaler's theorem, Theorem 5, upon observing that projections contain sections. The first upper bound $\sqrt{\binom{n}{k}}$ is sharp only when $k = 1, n - 1$. The second upper bound is obtained differently, by invoking quermassintegrals (which are additive under Minkowski sums, so they go hand in hand with zonotopes), combined with Urysohn's inequality. A simpler version of the same idea is to note that every k-dimensional projection has diameter at most the diameter of the cube $\sqrt{n}$; thus, by the isodiametric inequality, its volume is at most $\mathrm{vol}_k(B_2^k)(\sqrt{n}/2)^k$. All these bounds are of the order $n^{k/2}$ for a fixed k as $n \to \infty$, which is tight. The second of the upper bounds in Theorem 28 is asymptotically better than the first one. Ivanov [64] has developed local conditions for maximizers of k-dimensional projections.

In [40], using the isoperimetric inequality for polygons, Chakerian and Filliman additionally obtained a sharp bound for 2-dimensional projections (and thus also $(n - 2)$-dimensional ones – see Theorem 31 below).

Theorem 29 (Chakerian–Filliman [40]). *For $n \ge 2$ and for every 2-dimensional subspace H in $\mathbb{R}^n$, we have*

$$\mathrm{vol}_{n-1}(\mathrm{Proj}_H(Q_n)) \le \frac{1}{\tan(\frac{\pi}{2n})}.$$

Soon after, Filliman [51] discovered a general principle that maximizing volume of the larger class of zonotopes $Z = \sum_{j=1}^{n}[-\frac{1}{2}v_j, \frac{1}{2}v_j]$ with the constraint $\sum_{j=1}^{n} |v_j|^2 = n$ on the vectors v_j in $\mathbb{R}^k$ amounts to maximizing it over all zonotopes which are k-dimensional projections of the cube. This allowed him to extend the previous estimate to 3-dimensional projections.

Theorem 30 (Filliman [51]). *For $n \geq 3$ and for every 3-dimensional subspace H in $\mathbb{R}^n$, we have*

$$\mathrm{vol}_3(\mathrm{Proj}_H(Q_n)) \leq \frac{\sqrt{n/3}}{\tan(\frac{\pi}{2n-2})}.$$

This is sharp not only for $n = 3$, but also for $n = 4$ with the extremal projection being the rhombic dodecahedron and for $n = 6$ with the extremal projection being the triacontahedron.

We finish with a striking and remarkable feature of the cube: Its projections onto orthogonal complementary subspaces have the same volume.

Theorem 31 (McMullen [99], Chakerian–Filliman [40]). *Let $1 \leq k \leq n$. For every k-dimensional subspace H in $\mathbb{R}^n$, we have*

$$\mathrm{vol}_k(\mathrm{Proj}_H(Q_n)) = \mathrm{vol}_{n-k}(\mathrm{Proj}_{H^\perp}(Q_n)).$$

This has been found by McMullen and independently by Chakerian and Filliman, using the same approach based on Shephard's formula. In particular, sharp bounds on volumes of k-dimensional projections are equivalent to those on $(n - k)$-dimensional ones.

3.2 Simplices

Recall that Δ_n is the n-dimensional regular simplex with edge length $\sqrt{2}$, assuming for convenience in this section that Δ_n is embedded in $\mathbb{R}^n$. The projections of the regular simplex onto certain orthogonal complementary subspaces are conjectured to yield minimal and maximal volume, in huge contrast to the cube, where we have seen in Theorem 31 that such projections always have the *same* volume.

Conjecture 7 (Filliman [54]). *Fix $1 \leq k \leq n$. Let H_* be a k-dimensional subspace in $\mathbb{R}^n$ such that $T_* = \mathrm{Proj}_{H_*}(\Delta_n)$ is a k-dimensional simplex, with the vertices of Δ_n projecting only onto the vertices of T_*, as evenly as possible: For each $i \leq k + 1$, letting w_i be the number of vertices of Δ_n projecting onto vertex i of T_*, we have*

$$w_i = \begin{cases} \ell + 1, & 1 \leq i \leq r, \\ \ell, & r < i \leq k + 1, \end{cases}$$

where we divide $n + 1$ by $k + 1$ with the remainder $r \in \{0, \dots, k\}$, $n + 1 = (k + 1)\ell + r$. Then

$$\min_{\substack{H \subset \mathbb{R}^n \\ \dim H = k}} \mathrm{vol}_k(\mathrm{Proj}_H(\Delta_n)) = \mathrm{vol}_k(T_*).$$

Moreover, the polytope $T^* = \operatorname{Proj}_{H_*^\perp}(\Delta_n)$ is conjectured to maximize the volume of projections onto the $(n - k)$-dimensional subspaces,

$$\max_{\substack{H \subset \mathbb{R}^n \\ \dim H = n-k}} \operatorname{vol}_k(\operatorname{Proj}_H(\Delta_n)) = \operatorname{vol}_{n-k}(T^*).$$

Filliman developed exterior algebra techniques [53] and used them [54] to confirm this conjecture in the following cases: for the minimum, $k = 1, 2, n - 1$ and n is arbitrary or $n \le 6$ and k is arbitrary; for the maximum, $k = 1, 2, n - 1$ and n is arbitrary, $k = n - 2$ and $n \le 8$, or $k = 3$ and $n = 6$.

3.3 Cross-polytopes

In view of Cauchy's formula from Theorem 2, the volume of hyperplane projections of the cross-polytope B_1^n admits a natural probabilistic expression. Since it has 2^n congruent (simplicial) facets of $(n-1)$-dimensional volume $\frac{\sqrt{n}}{(n-1)!}$ with outer normals $\frac{1}{\sqrt{n}}(\pm 1, \dots, \pm 1)$, for every unit vector in $\mathbb{R}^n$, we have

$$\operatorname{vol}_{n-1}(\operatorname{Proj}_{a^\perp}(B_1^n)) = \frac{1}{2(n-1)!} \sum_{\varepsilon \in \{-1,1\}^n} |\langle a, \varepsilon \rangle| = \frac{2^{n-1}}{(n-1)!} \mathbb{E}\left| \sum_{j=1}^n a_j \varepsilon_j \right|,$$

where the expectation is over independent random signs ε_j, $\mathbb{P}(\varepsilon_j = \pm 1) = \frac{1}{2}$. Given the constraint $|a| = 1$, the question about extremal volume projections thus becomes that of finding the best constants c, C in the homogeneous inequalities

$$c\left(\mathbb{E}\left| \sum a_j \varepsilon_j \right|^2\right)^{1/2} \le \mathbb{E}\left| \sum a_j \varepsilon_j \right| \le C\left(\mathbb{E}\left| \sum a_j \varepsilon_j \right|^2\right)^{1/2}. \tag{3.1}$$

Such L_p-moment comparison inequalities go back to Khinchin's work [71] on the law of the iterated logarithm. This motivated and should be contrasted with an analogous probabilistic viewpoint on sections from Theorem 27, where instead of the L_1-norm, we have the limit of the L_q-norm as $q \downarrow -1$. A sharp upper bound follows easily from Jensen's inequality,

$$\mathbb{E}\left| \sum a_j \varepsilon_j \right| \le \left(\mathbb{E}\left| \sum a_j \varepsilon_j \right|^2\right)^{1/2} = |a| = 1,$$

attained if and only if $a = \pm e_i$ for some $i \le n$, that is, the maximum volume projection occurs at precisely coordinate subspaces. The reverse inequality is much deeper: In a different context, a sharp lower bound was conjectured to be attained at vectors $a = \frac{\pm e_i \pm e_j}{\sqrt{2}}$, $i \ne j$, by Littlewood in [93] (cf. Ball's extremizer from Theorem 6), proved much later by Szarek in [130], with subsequently simplified and quite different proofs [60, 88, 131]. We state it here rephrased in terms of volumes of projections, together with the simple upper bound.

Theorem 32 (Szarek [130]). *Let $n \geq 2$. For every unit vector in $\mathbb{R}^n$, we have*

$$\frac{1}{\sqrt{2}} \operatorname{vol}_{n-1}(B_1^{n-1}) \leq \operatorname{vol}_{n-1}(\operatorname{Proj}_{a^\perp}(B_1^n)) \leq \operatorname{vol}_{n-1}(B_1^{n-1}).$$

The lower bound is attained if and only if $a = \frac{\pm e_i \pm e_j}{\sqrt{2}}$ for some $i \neq j$, whilst the upper bound is attained if and only if $a = \pm e_i$ for some i.

A stability version has been derived by De, Diakonikolas, and Servedio [44] (see also [100] for a local statement with explicit constants).

Much less is known in higher codimensions. In analogy to Vaaler's theorem, Theorem 5, for the cube, it is natural to conjecture that maximal volume projections of the cross-polytope B_1^n occur at coordinate subspaces.

Conjecture 8. *Fix $1 \leq k \leq n$. For every k-dimensional subspace H in $\mathbb{R}^n$, we have*

$$\operatorname{vol}_k(\operatorname{Proj}_H(B_1^n)) \leq \operatorname{vol}_k(B_1^k).$$

This conjecture has appeared in this generality in Ivanov's work [65], who has confirmed it for $k = 2, 3$ and arbitrary n, using perturbation methods for frames (see also [64]). Earlier, Filliman [52] established the same for $k = 2$, using different, more algebraic methods of his work [53], also reducing the case of $k = 3$ with arbitrary n to $n \leq 42$. For minimal volume projections, we have the following *dual* analog of Conjecture 1 for the cube.

Conjecture 9 (Ivanov [65]). *Fix $1 \leq k \leq n$. Let H^* be the k-dimensional subspace from Conjecture 1. For every k-dimensional subspace H in $\mathbb{R}^n$, we have*

$$\operatorname{vol}_k(\operatorname{Proj}_H(B_1^n)) \geq \operatorname{vol}_k(\operatorname{Proj}_{H^*}(B_1^n)).$$

This has been confirmed for $k = 2$ by Ivanov in [65]. As for the cube, the conjectured maximizer is an affine cross-polytope.

3.4 Balls of p-norms

The sharp results on hyperplane projections of the cross-polytope have been extended by Barthe and Naor in [24] to ℓ_p-balls (with $p \geq 2$), thereby bringing the knowledge on extremal volume *hyperplane* projections to the same level as for sections (see Tables 1 and 2).

Theorem 33 (Barthe–Naor [24]). *For every unit vector a in $\mathbb{R}^n$, the function*

$$p \mapsto \frac{\operatorname{vol}_{n-1}(\operatorname{Proj}_{a^\perp}(B_p^n))}{\operatorname{vol}_{n-1}(B_p^{n-1})}$$

is non-decreasing on $[1, +\infty)$.

This can be viewed as a counterpart of Meyer and Pajor's theorem, Theorem 15, for *hyperplane* projections. It is an interesting open question to find such a monotonicity result for *all* subspaces. As for the cross-polytope, it is Cauchy's formula that allows to obtain a probabilistic expression for the volume in the hyperplane case. Barthe and Naor's argument goes as follows.

First, the surface area measure is related to the cone volume measure, by a general relation of Naor and Romik [107]. To sketch this, let σ_K be the normalized surface area measure on ∂K and let S be the not normalized surface area measure, that is, $S(A) = \mathrm{vol}_{n-1}(\partial K \cap A)/\mathrm{vol}_{n-1}(\partial K)$. Let μ_K be the normalized cone volume measure, that is, for $A \subseteq \partial K$, let $\mu_K(A) = \mathrm{vol}_n(\mathrm{conv}(\{0\}\cup A))/\mathrm{vol}_n(K)$. Let C denote its not normalized version.

Lemma 34. *If K is a symmetric convex body in $\mathbb{R}^n$, then σ_K is absolutely continuous with respect to μ_K and for almost all $x \in \partial K$ one has*

$$\frac{d\sigma_K}{d\mu_K}(x) = \frac{n\,\mathrm{vol}_n(K)}{\mathrm{vol}_{n-1}(\partial K)}|\nabla(\|\cdot\|_K)(x)|.$$

Sketch of the proof. For points x such that x is perpendicular to the surface of K one has $|x|\cdot dS(x) = ndC(x)$. If the angle between the surface and x is α, then $|\cos\alpha|\cdot|x|\cdot dS(x) = ndC(x)$. We clearly have $|\cos\alpha| = |\langle n(x), x/|x|\rangle|$. Let $z = \nabla\|\cdot\|_K(x)$. If $x \in \partial K$, then $1+\varepsilon = \|x+\varepsilon x\|_K \approx \|x\|_K + \varepsilon\langle z, x\rangle = 1+\varepsilon\langle z, x\rangle$, which gives $\langle z, x\rangle = 1$. Also, z is a vector perpendicular to ∂K. Thus, $n(x) = z/|z|$. We obtain $|\cos\alpha| = \frac{1}{|x|}\cdot|\langle n(x), x\rangle| = \frac{|\langle z,x\rangle|}{|x|\cdot|z|}$. This gives

$$\frac{\mathrm{vol}_{n-1}(\partial K)d\sigma_K(x)}{|\nabla(\|\cdot\|_K)(x)|} = \frac{dS(x)}{|\nabla(\|\cdot\|_K)(x)|} = \frac{|\langle z, x\rangle|}{|z|}dS(x) = ndC(x) = n\,\mathrm{vol}_n(K)d\mu_K(x).$$

$\square$

From Lemma 34 we therefore get

$$|\mathrm{Proj}_{a^\perp} K| = \frac{n}{2}\,\mathrm{vol}_n(K)\int_{\partial K}|\langle(\nabla\|\cdot\|_K)(x), a\rangle|d\mu_K(x), \tag{3.2}$$

since $(\nabla\|\cdot\|_K)(x) = n(x)|(\nabla\|\cdot\|_K)(x)|$.

Second, the cone volume measure $\mu_{B_p^n}$ enjoys a probabilistic representation in terms of i. i. d. random variables, discovered by Rachev and Rüschendorf [120] and independently by Schechtman and Zinn [125]. We shall also later need a modification of the representation of the uniform measure on B_p^n obtained in [22] by Barthe, Guédon, Mendelson, and Naor. Let us formulate a generalization of these results discussed in [118].

Lemma 35. *Let K be a symmetric convex body and let Z be any random vector in $\mathbb{R}^n$ with density of the form $f(\|x\|_K)$ for some continuous $f : [0,\infty) \to [0,\infty)$. Let U be a random variable uniform in $[0,1]$, independent of Z. Then:*
(a) $\frac{Z}{\|Z\|_K}$ *has distribution μ_K and $U^{1/n}\frac{Z}{\|Z\|_K}$ is uniformly distributed on K,*
(b) $\frac{Z}{\|Z\|_K}$ *and $\|Z\|_K$ are independent.*

In particular, for $K = B_p^n$ one can take $Z = (Y_1, \ldots, Y_n)$, where Y_i are i. i. d. random variables having densities $(2\Gamma(1 + \frac{1}{p}))^{-1}e^{-|t|^p}$.

Proof. We first claim that for any integrable $h : \mathbb{R}^n \to \mathbb{R}$ the following identity holds:

$$\int h = n|K| \int\limits_0^\infty r^{n-1} \int\limits_{\partial K} h(rz)\mathrm{d}\mu_K(z)\mathrm{d}r. \tag{3.3}$$

To show it one can assume that $h = \mathbf{1}_A$, where $A = [a, b] \cdot A_0$, where $A_0 \subset \partial K$, as these sets generate the σ-algebra of Borel sets in $\mathbb{R}^n$. For $z \in \partial K$ and $r > 0$ we then have $h(rz) = \mathbf{1}_{[a,b]}(r)\mathbf{1}_{A_0}(z)$. Thus, (3.3) reduces to

$$|A| = |K|\left(\int\limits_a^b nr^{n-1}\mathrm{d}r\right)\mu_K(A_0) = |K|(b^n - a^n)\mu_K(A_0) = \big|[a, b]A_0\big| \tag{3.4}$$

and is therefore true. Now, let us notice that for $\phi : \mathbb{R}^n \to \mathbb{R}$ and $\psi : \mathbb{R} \to \mathbb{R}$ we have

$$\mathbb{E}\left[\phi\left(\frac{Z}{\|Z\|_K}\right)\psi(\|Z\|_K)\right] = \int\limits_{\mathbb{R}^n} \phi\left(\frac{x}{\|x\|_K}\right)\psi(\|x\|_K)f(\|x\|_K)\mathrm{d}x$$

$$= n|K| \int\limits_0^\infty \psi(r)f(r)r^{n-1}\mathrm{d}r \int\limits_{\partial K} \phi(z)\mathrm{d}\mu_K(z).$$

Taking $\phi, \psi \equiv 1$ we learn that $n|K| \int_0^\infty f(r)r^{n-1}\mathrm{d}r = 1$. Thus, taking $\psi \equiv$ and next $\phi \equiv 1$ we arrive at

$$\mathbb{E}\left[\phi\left(\frac{Z}{\|Z\|_K}\right)\right] = \int\limits_{\partial K} \phi(z)\mathrm{d}\mu_K(z), \quad \mathbb{E}[\psi(\|Z\|_K)] = n|K| \int\limits_0^\infty \psi(r)f(r)r^{n-1}\mathrm{d}r.$$

The first equation shows that $\frac{Z}{\|Z\|_K}$ has distribution μ_K. Moreover, we get

$$\mathbb{E}\left[\phi\left(\frac{Z}{\|Z\|_K}\right)\psi(\|Z\|_K)\right] = \mathbb{E}\left[\phi\left(\frac{Z}{\|Z\|_K}\right)\right]\mathbb{E}[\psi(\|Z\|_K)],$$

which shows (b). Finally, (3.4) together with the fact that $U^{1/n}$ has density nr^{n-1} on $[0, 1]$ shows that

$$\frac{|A|}{|K|} = \mathbb{P}(U^{1/n} \in [a, b])\mathbb{P}\left(\frac{Z}{\|Z\|_K} \in A_0\right) = \mathbb{P}\left(U^{1/n}\frac{Z}{\|Z\|_K} \in A\right),$$

which shows the second part of point (a). $\qquad\square$

We can now prove the probabilistic formula for the volume of the hyperplane projection of B_p^n.

Lemma 36. *For $p > 1$ and every unit vector $a \in \mathbb{R}^n$, we then have*

$$\mathrm{vol}_{n-1}(\mathrm{Proj}_{a^\perp}(B_p^n)) = \frac{\mathrm{vol}_{n-1}(B_p^{n-1})}{\mathbb{E}|X_1|} \mathbb{E}\left|\sum_{j=1}^n a_j X_j\right|, \tag{3.5}$$

where $X_1, \ldots, X_n$ are i. i. d. random variables with density $f_p(x) = \frac{p}{2(p-1)\Gamma(1/p)}|x|^{\frac{2-p}{p-1}} e^{-|x|^{\frac{p}{p-1}}}$.

Proof. By (3.2) and Lemma 35(a) for some constant $c_{p,n}$ we have

$$\mathrm{vol}_{n-1}(\mathrm{Proj}_{a^\perp} B_p^n) = C(p,n)\mathbb{E}\left|\sum_{i=1}^n a_i \left|\frac{Y_i}{S}\right|^{p-1} \mathrm{sgn}\left(\frac{Y_i}{S}\right)\right|$$

$$= C(p,n) \cdot \frac{\mathbb{E}S^{p-1}}{\mathbb{E}S^{p-1}} \cdot \mathbb{E}\left|\sum_{i=1}^n a_i \left|\frac{Y_i}{S}\right|^{p-1} \mathrm{sgn}(Y_i)\right|$$

$$= \frac{C(p,n)}{\mathbb{E}S^{p-1}} \cdot \mathbb{E}\left|\sum_{i=1}^n a_i |Y_i|^{p-1} \mathrm{sgn}(Y_i)\right|.$$

It now suffices to observe that $X_i = |Y_i|^{p-1} \mathrm{sgn}(Y_i)$ for $p > 1$ have densities f_p. We then compute $C_{p,n}$ by taking $a = e_1$. $\qquad\square$

Next, Meyer and Pajor's arguments involving peakedness are replaced by the stochastic convex (Choquet) ordering, where the independence of X_j is crucial. For $p > 2$, additional structure emerges: X_j are Gaussian mixtures. This leads to an analog of Koldobsky's theorem, Theorem 16, the proof of which was later simplified in [49] by bypassing the Fourier analytic arguments (we shall discuss the arguments in Section 4).

Theorem 37 (Barthe–Naor [24]). *Let $p > 2$. For every two unit vectors a and b in $\mathbb{R}^n$ such that $(b_1^2, \ldots, b_n^2)$ majorizes $(a_1^2, \ldots, a_n^2)$, we have*

$$\mathrm{vol}_{n-1}(\mathrm{Proj}_{a^\perp}(B_p^n)) \geq \mathrm{vol}_{n-1}(\mathrm{Proj}_{b^\perp}(B_p^n)).$$

In the range $0 < p < 1$, Cauchy's formula cannot be applied due to the lack of convexity and no non-trivial bounds are known. When $1 < p < 2$, the maximal volume hyperplane projection is onto a coordinate subspace, as follows from Theorem 33, whereas the minimal one is not known. Barthe and Naor [24] have shown that the cross-polytope minimizer $(\frac{1}{\sqrt{2}}, \frac{1}{\sqrt{2}}, 0, \ldots, 0)^\perp$ is beaten by the diagonal one for every $p > p_0 = \frac{4}{3}$ in large enough dimensions (in particular, as Oleszkiewicz has pointed out in [113], there is no "formal duality" with sections, for there is not such a phase transition at $\frac{p_0}{p_0-1} = 4$).

For higher codimensions than 1, plainly Meyer and Pajor's theorem, Theorem 15, gives a sharp lower bound: For every $p \geq 2$, $1 \leq k \leq n$, and k-dimensional subspace in $\mathbb{R}^n$, we have

$$\mathrm{vol}_k(\mathrm{Proj}_H(B_p^n)) \geq \mathrm{vol}_k(B_p^n \cap H) \geq \mathrm{vol}_k(B_p^k),$$

attained at coordinate subspaces. For $0 < p < 2$, using his reverse form of the Brascamp–Lieb inequality from [20], Barthe [21] has established the following lower bound.

Theorem 38 (Barthe [21]). *Let $0 < p < 2$. Fix $1 \leq k \leq n$. For every k-dimensional subspace H in $\mathbb{R}^n$, we have*

$$\mathrm{vol}_k(\mathrm{Proj}_H(B_p^n)) \geq \left(\frac{k}{n}\right)^{k(1/p-1/2)} \mathrm{vol}_k(B_p^k).$$

This is optimal when k divides n and $p \geq 1$ (attained at subspaces from Conjecture 1).

4 Methods

We would like to present and emphasize one particular probabilistic point of view which gathers the major results for both sections and projections under the same umbrella. The point is that as it is very natural to set up hyperplane projection problems as sharp L_1–L_2 comparison inequalities (thanks to Cauchy's formula; see, e. g., (3.5)), the same probabilistic picture captures sections upon changing the L_1-norm to L_q-norms with negative exponents q.

4.1 Sections

This is a straightforward extension to higher codimensions of Kalton and Koldobsky's observation made in [68]; recall (2.8).

Lemma 39 ([42]). *Let K be a body in $\mathbb{R}^n$ of volume 1, star-shaped with respect to the origin. Let H be a k-codimensional subspace in $\mathbb{R}^n$ and let X be a random vector uniform on K. Let $\| \cdot \|$ be a norm in $H^\perp$ with the unit ball B. Then*

$$\mathrm{vol}_{n-k}(K \cap H) = \lim_{q \to -k+} \frac{k+q}{k\,\mathrm{vol}_k(B)} \mathbb{E}\| \mathrm{Proj}_{H^\perp} X\|^q.$$

Proof. If we let $f : H^\perp \to [0, +\infty)$ be the density of $\mathrm{Proj}_{H^\perp} X$, as in (2.2), we have

$$\mathrm{vol}_{n-k}(K \cap H) = f(0).$$

The function $x \mapsto \frac{k-q}{k\,\mathrm{vol}_k(B)}\|x\|^{-q}$ as $q \to -k+$ behaves like the Dirac delta at 0: If f is continuous at 0 and integrable, then

$$\lim_{q \to -k+} \frac{k+q}{k\,\mathrm{vol}_k(B)} \int_{H^\perp} \|x\|^{-q} f(x)\mathrm{d}x = f(0),$$

and the lemma follows. To justify the last identity, for simplicity we identify $H^\perp$ with $\mathbb{R}^k$ and fix $\varepsilon > 0$. The set $\{x,\, f(x) < f(0) + \varepsilon\}$ contains a neighborhood of 0, say δB. Then

$$
\frac{k+q}{k\operatorname{vol}_k(B)} \int_{\mathbb{R}^k} \|x\|^q f(x)\,dx \le (f(0)+\varepsilon)\frac{k+q}{k\operatorname{vol}_k(B)}\int_{\delta K} \|x\|^q dx + \frac{k+q}{k\operatorname{vol}_k(B)}\delta^q \int_{\mathbb{R}^k} f
$$

$$
= (f(0)+\varepsilon)\delta^{k+q} + \frac{k+q}{k\operatorname{vol}_k(B)}\delta^q \int_{\mathbb{R}^k} f
$$

(the last equality follows by the homogeneity of volume and the layer cake representation). Taking $\limsup$ as $q \downarrow -k$ gives an upper bound of $f(0) + \varepsilon$. A lower bound is obtained similarly (the second term above can be dropped). $\qquad\square$

For hyperplane sections of the cube, the limit can be evaluated, which leads to a particularly handy expression.

Lemma 40 (König–Koldobsky [83]). *Let $\xi_1, \xi_2, \ldots$ be i. i. d. random vectors uniform on the sphere S^2 in $\mathbb{R}^3$. For a unit vector a in $\mathbb{R}^n$, we have*

$$
\operatorname{vol}_{n-1}(Q_n \cap a^\perp) = \mathbb{E}\left|\sum_{j=1}^n a_j \xi_j\right|^{-1}.
$$

Proof. Lemma 39 yields

$$
\operatorname{vol}_{n-1}(Q_n \cap a^\perp) = \lim_{q \to -1+} \frac{1+q}{2}\mathbb{E}\left|\sum_{j=1}^n a_j X_j\right|^q,
$$

where $X = (X_1, \ldots, X_n)$ is uniform on Q_n, that is, the components X_j are independent uniform on $[-\frac{1}{2}, \frac{1}{2}]$. By Archimedes' hat-box theorem, $\langle \xi_j, e_1 \rangle$ has the same distribution as $2X_j$, which allows to get for every fixed $q > -1$

$$
\frac{1+q}{2}\mathbb{E}\left|\sum_{j=1}^n a_j X_j\right|^q = 2^{-1-q}\mathbb{E}\left|\sum_{j=1}^n a_j \xi_j\right|^q
$$

(see, e. g., [41] for all details). Taking the limit finishes the proof. $\qquad\square$

Remark 41. Replacing ξ_j by i. i. d. random vectors uniform on a higher-dimensional sphere, say S^{d+1}, and the exponent -1 by $-d$ results in a formula for sections of balls in $\ell_\infty(\ell_2)$ by block subspaces (see [32, Proposition 3.2]).

To illustrate the applicability of this lemma, we sketch the proof of the lower bound of Theorem 14, the Hadwiger–Hensley bound with an optimal deficit.

Proof (Sketch). The key is to write

$$\left|\sum_{j=1}^{n} a_j \xi_j\right|^2 = \sum_{i,j} a_i a_j \langle \xi_i, \xi_j \rangle = 1 + 2 \sum_{i<j} a_i a_j \langle \xi_i, \xi_j \rangle.$$

The random variable $R = 2\sum_{i<j} a_i a_j \langle \xi_i, \xi_j \rangle$ has mean 0. Thus, by convexity,

$$\mathbb{E}(1+R)^{-1/2} \geq \mathbb{E}(1 - R/2) = 1.$$

To improve upon this, it suffices to use a more precise pointwise inequality, say

$$(1+r)^{-1/2} \geq 1 - \frac{1}{2}r + \frac{1}{3}r^2 - \frac{5}{24}r^3, \quad r > -1,$$

and estimate $\mathbb{E}R^2$ and $\mathbb{E}R^3$, which are explicitly expressed in terms of a_j. $\qquad\square$

For B_p^n-balls, a direct application of Lemma 39 leaves us with a random vector uniform on B_p^n with *mildly* dependent components. This however can be circumvented thanks to the homogeneity of L_q-norms.

Lemma 42 ([42]). *Let $p > 0$ and let $Y_1, Y_2, \ldots$ be i. i. d. random variables with density $e^{-\beta_p^p |x|^p}$, $\beta_p = 2\Gamma(1 + 1/p)$. Let H be a subspace in $\mathbb{R}^n$ of codimension k such that the rows of a $k \times n$ matrix U form an orthonormal basis of $H^\perp$. Let $v_1, \ldots, v_n \in \mathbb{R}^k$ denote the columns of U. Then*

$$\mathrm{vol}_{n-k}(B_p^n \cap H) = \mathrm{vol}_{n-k}(B_p^{n-k}) \lim_{q \to -k+} \frac{k+q}{k\,\mathrm{vol}_k(B_{\|\cdot\|})} \mathbb{E}\left\|\sum_{j=1}^{n} Y_j v_j\right\|^q,$$

where $\|\cdot\|$ is a norm on $\mathbb{R}^k$ with unit ball $B_{\|\cdot\|}$.

Proof. Let $X = (X_1, \ldots, X_n)$ be a random vector uniform on B_p^n. Lemma 39 then gives the desired formula with X_j in place of Y_j and without the factor $\mathrm{vol}_{n-k}(B_p^{n-k})$. To pass to Y we shall use Lemma 35, which ensures that for $Y = (Y_1, \ldots, Y_n)$ and $S = (\sum_{i=1}^{n} |Y_i|^p)^{1/p}$ the random vector $\frac{Y}{S}$ is independent of S and moreover $U^{1/n}\frac{Y}{S}$ is uniformly distributed in B_p^n if U is independent of Y_i and uniform on $[0, 1]$. Therefore,

$$\mathbb{E}\left\|\sum_{j=1}^{n} X_j v_j\right\|^q = \mathbb{E}\left\|\sum_{j=1}^{n} U^{1/n}\frac{Y_j}{S} v_j\right\|^q = \mathbb{E}[U^{q/n}] \cdot \frac{\mathbb{E}[S^q]}{\mathbb{E}[S^q]} \cdot \mathbb{E}\left\|\sum_{j=1}^{n} \frac{Y_j}{S} v_j\right\|^q = \frac{\mathbb{E}[U^{q/n}]}{\mathbb{E}[S^q]} \cdot \mathbb{E}\left\|\sum_{j=1}^{n} Y_j v_j\right\|^q.$$

$\qquad\square$

This has been instrumental in the proof of Theorem 19. For a simpler application, the Meyer–Pajor monotonicity result from Theorem 15 holds in fact for L_q-norms. In view of the previous lemma, this readily implies their theorem.

Theorem 43. *For $p > 0$, let $Y_1^{(p)}, Y_2^{(p)}, \ldots$ be i. i. d. random variables with density $e^{-\beta_p^p |x|^p}$, $\beta_p = 2\Gamma(1 + 1/p)$. For every vectors $v_1, \ldots, v_n$ in $\mathbb{R}^k$ and $-k < q < 0$, the function*

$$(p_1, \ldots, p_n) \mapsto \mathbb{E} \left\| \sum_{j=1}^n Y_j^{(p_j)} v_j \right\|^q$$

is non-decreasing in each variable.

Proof. Following Kanter [69], we say for two probability measures μ and ν on $\mathbb{R}^n$ that ν is more peaked than μ if $\nu(K) \geq \mu(K)$ for every symmetric convex set K in $\mathbb{R}^n$. Crucially, this is preserved by taking products and convolutions of even log-concave measures (see [69, Corollaries 3.2 and 3.3]). If $0 < p < p'$, then the density of $Y_1^{(p')}$ intersects the density of $Y_1^{(p)}$ exactly once and dominates it (pointwise) near the origin. Thus, $Y_1^{(p')}$ is more peaked than $Y_1^{(p)}$ and consequently $\sum Y_j^{(p'_j)} v_j$ is more peaked than $\sum Y_j^{(p_j)} v_j$ if $p_j \leq p'_j$. In particular, for every $t > 0$,

$$\mathbb{P}\left(\left\| \sum_{j=1}^n Y_j^{(p_j)} v_j \right\| \leq t \right) \leq \mathbb{P}\left(\left\| \sum_{j=1}^n Y_j^{(p'_j)} v_j \right\| \leq t \right)$$

and the result follows by integrating in t. $\qquad\square$

The measure with density $e^{-\beta_p^p |x|^p}$ from Lemma 39 enjoys a *Gaussian mixture* form when $0 < p < 2$. This in turn provides good convolution properties, allowing in particular to evaluate the limit from Lemma 39. We say that a random variable X is a (symmetric) Gaussian mixture if X has the same distribution as RG for some non-negative random variable R and a standard Gaussian random variable G, independent of R. Gaussian mixtures are continuous, i. e., have densities, and X is a Gaussian mixture if and only if its density f is of the form

$$f(x) = \int_0^\infty e^{-tx^2} \, d\nu(t)$$

for a Borel measure ν on $[0, +\infty)$. By Bernstein's theorem, this is equivalent to $g(x) = f(\sqrt{x})$ being completely monotone, that is, $(-1)^n g^{(n)}(x) \geq 0$ for all $n \geq 0$ and $x > 0$, which gives a practical condition. We refer to [49] for further details and more examples. Thus, if $X_1, \ldots, X_n$ are independent Gaussian mixtures, say $X_j = R_j G_j$, and $v_1, \ldots, v_n$ are vectors in $\mathbb{R}^k$, then, conditioned on the values of R_j, $\sum X_j v_j$ is a centered Gaussian random vector in $\mathbb{R}^k$ with covariance matrix $\sum R_j^2 v_j v_j^\mathsf{T}$.

Lemma 44 ([49, 108]). *Let $0 < p < 2$. There are non-negative i. i. d. random variables $R_1, R_2, \ldots$ such that for every subspace H in $\mathbb{R}^n$ of codimension k, we have*

$$\mathrm{vol}_{n-k}(B_p^n \cap H) = \mathrm{vol}_{n-k}(B_p^{n-k}) \mathbb{E}\left(\det\left[\sum_{j=1}^n R_j v_j v_j^\mathsf{T} \right] \right)^{-1/2},$$

where $v_1, \ldots, v_n$ are vectors in $\mathbb{R}^k$ such that the rows of the $k \times n$ matrix with columns $v_1, \ldots, v_n$ form an orthonormal basis of $H^\perp$.

Remark 45. To describe the distribution of R_j, for $0 < \alpha < 1$, we let g_α be the density of a standard positive α-stable random variable W_α, i.e., with the Laplace transform $\mathbb{E}e^{-uW_\alpha} = e^{-u^\alpha}$, $t > 0$, and we let $V_1, V_2, \ldots$ be i.i.d. random variables with density $\frac{\sqrt{\pi}}{2\Gamma(1+1/p)}t^{-3/2}g_{p/2}(t^{-1})$. Then $R_j = (\mathbb{E}V_j^{-1/2})^2 V_j$; see [49].

Proof of Lemma 44. Y_j from Lemma 39 are Gaussian mixtures, say $Y_j = T_j G_j$ for some non-negative random variables T_j and standard Gaussians G_j, all independent. Then, conditioned on T_j, the limit in Lemma 39 gives the density at 0 of the random variable $\sum Y_j v_j$, which, as we said, is centered Gaussian in $\mathbb{R}^k$ with covariance $\sum T_j^2 v_j v_j^\top$; thus, its density at 0 equals $(2\pi)^{-k/2}(\det[\sum_{j=1}^{n} T_j^2 v_j v_j^\top])^{-1/2}$. $\qquad\square$

For hyperplane sections, this formula directly explains Koldobsky's Schur convexity result from Theorem 16.

Proof of Theorem 16. We first observe that if $F : \mathbb{R}^n \to \mathbb{R}$ is convex and permutation-symmetric, then F is Schur convex, namely $x \prec y$ implies $F(x) \leq F(y)$. Indeed, it is a standard fact (see [25]) that there exist $(\lambda_\sigma)_{\sigma \in S_n}$, where S_n stands for the set of permutations of $\{1, \ldots, n\}$, such that $\lambda_\sigma \geq 0$, $\sum_{\sigma \in S_n} \lambda_\sigma = 1$, and $x = \sum_{\sigma \in S_n} \lambda_\sigma y_\sigma$, where $y_\sigma = (y_{\sigma(1)}, \ldots, y_{\sigma(n)})$. Thus,

$$F(x) = F\left(\sum_{\sigma \in S_n} \lambda_\sigma y_\sigma\right) \leq \sum_{\sigma \in S_n} \lambda_\sigma F(y_\sigma) = \sum_{\sigma \in S_n} \lambda_\sigma F(y) = F(y).$$

For a unit vector a in $\mathbb{R}^n$, Lemma 44 yields

$$\mathrm{vol}_{n-1}(B_p^n \cap a^\perp) = \mathrm{vol}_{n-1}(B_p^{n-1})\mathbb{E}\left(\sum_{j=1}^{n} a_j^2 R_j\right)^{-1/2}.$$

Since $(\cdot)^{-1/2}$ is convex, the right-hand side is clearly convex and permutation-symmetric (R_j are i.i.d.) as a function of $(a_1^2, \ldots, a_n^2)$ and thus it is also Schur convex. $\qquad\square$

4.2 Projections

Somewhat analogous to the Fourier analytic approach to sections, there is a formula for the volume of hyperplane projections of a convex body as the Fourier transform of its curvature function, as discovered by Koldobsky, Ryabogin, and Zvavitch [79] (see also their survey [78]). We do *not* touch upon this connection here at all. Instead, we focus on a probabilistic perspective and highlight two approaches to the L_1–L_2 moment comparison inequalities like (3.1), arising in hyperplane projections.

As we have just seen for sections, for Gaussian mixtures, thanks to their good additive structure, we readily get precise Schur majorization-type results. This proof is from [49].

Proof of Theorem 37. Recall formula (3.5) for hyperplane projections. For $p > 2$, the density $f_p(t)$ of X_i is completely monotone; thus, X_j are Gaussian mixtures, say $X_j = R_j G_j$ for some i. i. d. non-negative random variables R_j and standard Gaussians G_j, all independent. Then, adding the Gaussians first conditioning on R_j yields

$$\mathbb{E}\left|\sum_{j=1}^{n} a_j X_j\right| = \mathbb{E}\left(\sum_{j=1}^{n} a_j^2 R_j^2\right)^{1/2} \mathbb{E}|G_1|. \tag{4.1}$$

As in the proof for sections, the Schur concavity result follows from the concavity of $(\cdot)^{1/2}$. $\qquad\square$

The same argument bluntly extends to arbitrary L_q-norms, giving sharp Khinchin inequalities (see [6, 49]).

When $1 \le p < 2$, the density of X_j in (3.5) is bimodal and understanding the L_1-norm of their weighted sums is elusive, mainly due to complicated cancelations – a problem which completely disappears in (4.1). For $p = 1$, X_j become discrete (symmetric random signs). We present two completely different Fourier analytic proofs. The first proof, due to Haagerup, is in the same spirit as Ball's proof from [7] for hyperplane cube sections.

Proof of Theorem 32 (Haagerup [60]). We want to minimize $\mathbb{E}|\sum a_j \varepsilon_j|$ subject to $\sum a_j^2 = 1$. We can assume that all a_j are positive. If at least one exceeds $\frac{1}{\sqrt{2}}$, say $a_1 > \frac{1}{\sqrt{2}}$, by averaging over the other coefficients we obtain

$$\mathbb{E}\left|\sum a_j \varepsilon_j\right| \ge \mathbb{E}_{\varepsilon_1}\left|a_1 \varepsilon_1 + \mathbb{E}\sum_{j>1} a_j \varepsilon_j\right| = a_1 > \frac{1}{\sqrt{2}},$$

as desired. Now we assume that for all j, $a_j \le \frac{1}{\sqrt{2}}$. A starting point is the Fourier analytic formula

$$|x| = \frac{1}{\pi}\int_{\mathbb{R}}(1 - \cos(tx))t^{-2}dt, \quad x \in \mathbb{R}.$$

Thus, for an integrable random variable X,

$$\mathbb{E}|X| = \frac{1}{\pi}\int_{\mathbb{R}}(1 - \mathrm{Re}(\mathbb{E}e^{itX}))t^{-2}dt$$

(see also [60, Lemmas 2.3 and 4.2] and [59, Lemma 3]). In particular, using independence and $\mathbb{E}e^{it\varepsilon_j} = \cos t$, we get

$$\mathbb{E}\left|\sum a_j \varepsilon_j\right| = \frac{1}{\pi} \int_{\mathbb{R}} \left(1 - \prod \cos(ta_j)\right) t^{-2} dt.$$

By the AM-GM inequality, this gives the following bound:

$$\mathbb{E}\left|\sum a_j \varepsilon_j\right| \geq \sum a_j^2 F(a_j^{-2})$$

with

$$F(s) = \frac{1}{\pi} \int_{\mathbb{R}} \left(1 - \left|\cos\left(\frac{t}{\sqrt{s}}\right)\right|^s\right) t^{-2} dt, \quad s > 0.$$

See (2.4) and the ensuing function Ψ in Ball's proof. Here, however, function F can be expressed explicitly. Using $\sum_{n=-\infty}^{\infty} \frac{1}{(t+n\pi)^2} = \frac{1}{\sin^2 t}$, we arrive at

$$F(s) = \frac{1}{\pi\sqrt{s}} \int_{\mathbb{R}} (1 - |\cos t|^s) t^{-2} dt = \frac{1}{\pi\sqrt{s}} \sum_{n=-\infty}^{\infty} \int_{-\pi/2}^{\pi/2} (1 - (\cos t)^s)(t + n\pi)^{-2} dt$$

$$= \frac{1}{\pi\sqrt{s}} \int_{-\pi/2}^{\pi/2} (1 - (\cos t)^s) \sin^{-2} t \, dt$$

$$= \frac{2}{\sqrt{\pi s}} \frac{\Gamma(\frac{s+1}{2})}{\Gamma(\frac{s}{2})}.$$

Claim. $F(s)$ increases on $(0, +\infty)$.

Using this claim and the fact that $a_j \leq \frac{1}{\sqrt{2}}$ for all j, we finish the proof,

$$\mathbb{E}\left|\sum a_j \varepsilon_j\right| \geq \sum a_j^2 F(a_j^{-2}) \geq \sum a_j^2 F(2) = F(2) = \frac{1}{\sqrt{2}}.$$

Noteworthily, this is tight when $n = 2$ and $a_1 = a_2 = \frac{1}{\sqrt{2}}$.

To show the claim, we note that $\lim_{s\to\infty} F(s) = \sqrt{\frac{2}{\pi}}$ (e. g., by Stirling's formula) and that

$$F(s + 2) = \sqrt{\frac{s}{s+2}} \frac{s+1}{s} F(s) = (1 - 1/(s+1)^2)^{-1/2} F(s),$$

which iterated yields $F(s + 2n) = F(s) \prod_{k=0}^{n-1} (1 - 1/(s + 2k + 1)^2)^{-1/2}$, so letting $n \to \infty$,

$$F(s) = \sqrt{\frac{2}{\pi}} \prod_{k=0}^{\infty} (1 - 1/(s + 2k + 1)^2)^{1/2}. \qquad \square$$

The second proof uses the machinery of Fourier analysis on the discrete cube $\{-1, 1\}^n$. We refer for instance to [111, Chapter 1] for basic background.

Proof of Theorem 32 (Kwapień–Latała–Oleszkiewicz [81, 88, 112]).
We work with $L_2(\{-1,1\}^n, \mathbb{R})$ equipped with the product probability measure on the cube $\{-1,1\}^n$, i. e., the distribution of $(\varepsilon_1,\ldots,\varepsilon_n)$, and the inner product $\langle f, g \rangle = \mathbb{E}[f(\varepsilon)g(\varepsilon)]$, $f, g \colon \{-1,1\}^n \to \mathbb{R}$. Let

$$f(x) = \left| \sum_{j=1}^{n} a_j x_j \right|, \quad x \in \{-1,1\}^n.$$

We write its discrete Fourier expansion with respect to the orthonormal system of the Walsh functions $w_S(x) = \prod_{j \in S} x_j$ indexed by the subsets $S \subset \{1,\ldots,n\}$ with $w_\emptyset(x) \equiv 1$. We have

$$f(x) = \sum_S b_S w_S(x), \quad b_S = \langle f, w_S \rangle.$$

Since f is even, $b_S = 0$ provided $|S|$ is odd. The crux is to consider the Laplace operator $\mathcal{L}$ acting on $L_2(\{-1,1\}^n, \mathbb{R})$,

$$(\mathcal{L}g)(x) = \frac{1}{2} \sum_{y \sim x} (g(y) - g(x)),$$

where the sum is over all neighbors y of x, i. e., the points in $\{-1,1\}^n$ differing from x by one component. As can be checked, the Walsh functions are its eigenfunctions, $\mathcal{L}w_S = -|S|w_S$, and for *even* functions g, we have the following Poincaré-type inequality:

$$\langle g, -\mathcal{L}g \rangle \geq 2\,\mathrm{Var}(g).$$

Claim. We have $(-\mathcal{L}f)(x) \leq f(x)$ for every $x \in \{-1,1\}^n$.

Using this claim in the Poincaré inequality,

$$2(\mathbb{E}f^2 - (\mathbb{E}f)^2) \leq \langle f, -\mathcal{L}f \rangle \leq \langle f, f \rangle = \mathbb{E}f^2,$$

which gives $\mathbb{E}f \geq \frac{1}{\sqrt{2}}(\mathbb{E}f)^2$, as desired. The claim follows from rearranging the following consequence of the triangle inequality:

$$|-a_1x_1 + a_2x_2 + \cdots + a_nx_n| + |a_1x_1 - a_2x_2 + \cdots + a_nx_n| + \cdots + |a_1x_1 + a_2x_2 + \cdots - a_nx_n|$$
$$\geq (n-2)|a_1x_1 + \cdots + a_nx_n|. \qquad \square$$

We stress out that this proof is extremely robust: It only uses the triangle inequality and hence extends ad verbatim to the case where the coefficients a_j are vectors in an arbitrary normed vector space.

The history of this argument is a bit convoluted. Latała and Oleszkiewicz's work [88] contains all the crucial ideas of the modern proof presented above; however, it is not written in a Fourier analytic language. The proof presented here was devised by Kwapień and is based on the Walsh functions (the characters of $\{-1,1\}^n$). As we have seen, one of its main components is a strengthened Poincaré-type inequality in the presence of symmetry, the idea of which appeared first in [81] (in the continuous case), extended to the discrete case in [112] (perhaps the first place where this proof appears in print). Oleszkiewicz presented this proof in 1996 at MSRI (during a workshop in harmonic analysis and convex geometry).

We finish with a sketch of the Barthe–Naor proof from [24] of the monotonicity result from Theorem 33 featuring yet another tool, useful in proving Khinchin-type inequalities: the stochastic convex ordering. This circle of ideas was further developed in [50].

In the simplest setting sufficient for our purposes, for two symmetric random variables X and Y, we say that Y *dominates* X in the convex (or often called Choquet) stochastic ordering if $\mathbb{E}\phi(X) \leq \mathbb{E}\phi(Y)$ for every even convex function $\phi\colon \mathbb{R} \to [0,+\infty]$. It is clear that this tensorizes and is preserved by convolution: If Y dominates X and Z is a symmetric random variable, independent of them, then $Y+Z$ dominates $X+Z$. We will only need the following sufficient condition.

Lemma 46. *If random variables X and Y satisfy $\mathbb{E}|X| = \mathbb{E}|Y|$ and have even densities f and g, respectively, and there are $0 < x_1 < x_2$ such that $\{t \geq 0,\ g(t) < f(t)\}$ is the interval (x_1, x_2) (f and g intersect twice), then Y dominates X in the convex stochastic ordering.*

Proof. Let $\phi\colon \mathbb{R} \to [0,+\infty]$ be an even convex function. Thanks to the symmetry of X, Y and the constraint $\mathbb{E}|X| = \mathbb{E}|Y|$, the desired inequality $\int \phi f \leq \int \phi g$ is equivalent to

$$\int\limits_0^\infty (\phi(x) - \alpha x - \beta)(g(x) - f(x))\mathrm{d}x \geq 0$$

with some (any) $\alpha, \beta \in \mathbb{R}$. We choose α, β as the unique parameters such that the convex function $\psi(x) = \phi(x) - \alpha x - \beta$ vanishes at x_1 and x_2. Then, by convexity, $\psi \leq 0$ on (x_1, x_2) and $\psi \geq 0$ outside that interval. Thus, the integrand is pointwise non-negative. $\qquad\square$

Proof of Theorem 33. In view of (3.5), we aim at showing that the function

$$p \mapsto \frac{1}{\mathbb{E}|X_1^{(p)}|}\mathbb{E}\left|\sum a_j X_j^{(p)}\right|$$

is non-decreasing on $[1,+\infty)$, where $X_j^{(p)}$ are i. i. d. random variables with density proportional to $|x|^{\frac{2-p}{p-1}} \exp\{-|x|^{\frac{p}{p-1}}\}$. By the tensorization property, it suffices to prove that for $1 \leq p < q$, $X_1^{(q)}/\mathbb{E}|X_1^{(q)}|$ dominates $X_1^{(p)}/\mathbb{E}|X_1^{(p)}|$. This readily follows from Lemma 46. $\qquad\square$

5 Other connections

We close this survey with two tangential topics related to sections: an application of Ball's cube slicing inequality to entropy power inequalities and a reformulation of the conjectural logarithmic Brunn–Minkowski inequality in terms of sections of the cube.

5.1 Entropy power inequalities

Recall (2.2), viz. the volume of a central hyperplane section by $a^\perp$ is the maximum value of the density of the marginal $\langle a, X \rangle = \sum a_j X_j$ (there $f(0) = \|f\|_\infty$ by the symmetry and log-concavity of X). The maximum density functional

$$M(X) = \|f\|_\infty$$

of a random vector X in $\mathbb{R}^n$ with density f is closely related to classical topics in probability such as the Lévy concentration function, small ball estimates, and anticoncentration, as well as information theory, particularly the entropy power inequalities. We refer to the comprehensive surveys [96, 110]. The entropy power inequality originated in Shannon's seminal work [128] and asserts that the entropy power

$$N(X) = \exp\left\{\frac{2}{n}h(X)\right\}, \quad h(X) = -\int_{\mathbb{R}^n} f \log f,$$

is superadditive: For *independent* random vectors X and Y in $\mathbb{R}^n$, we have

$$N(X + Y) \geq N(X) + N(Y),$$

and plainly, by induction, the same is true for arbitrarily many independent summands. In analogy, we let

$$N_\infty(X) = \exp\left\{\frac{2}{n}h_\infty(X)\right\} = M(X)^{-2/n}, \quad h_\infty(X) = -\log\|f\|_\infty,$$

be the ∞-entropy power of X, sometimes called the min-entropy power (because for a fixed distribution, it is the smallest entropy power across the family of all Rényi entropies). The min-entropy power inequality in dimension 1 reads as follows.

Theorem 47 (Bobkov–Chistyakov [27], Melbourne–Roberto [100]). *Let $X_1, \ldots, X_m$ be independent random variables with bounded densities. Then*

$$N_\infty(X_1 + \cdots + X_m) \geq \frac{1}{2}\sum_{j=1}^{m} N_\infty(X_j)$$

with equality if and only if two of these variables are uniform on A and c − A, respectively, for some set A in $\mathbb{R}$ of finite measure and some $c \in \mathbb{R}$, while the other variables are constant.

Bobkov and Chistyakov proved this inequality with the sharp constant $\frac{1}{2}$ using Ball's cube slicing inequality, whereas the equality conditions have recently been established by Melbourne and Roberto using their robust version (see Theorem 14).

The argument rests on the following subtle comparison due to Rogozin.

Theorem 48 (Rogozin [122]). *Let $X_1, \ldots, X_m$ be independent random variables with bounded densities and let $U_1, \ldots, U_m$ be independent uniform random variables on intervals chosen such that $M(X_j) = M(U_j)$ for each j. Then*

$$M(X_1 + \cdots + X_m) \leq M(U_1 + \cdots + U_m).$$

Theorem 47 then follows by invoking Ball's theorem, Theorem 6, which after incorporating the variance constraint amounts to

$$M(U_1 + \cdots + U_m) \leq \sqrt{2}\left(M(U_1)^{-2} + \cdots + M(U_m)^{-2}\right)^{-1/2}.$$

In [97], Madiman, Melbourne, and Xu developed multivariate generalizations of Rogozin's result where the extremal distributions are uniform on the Euclidean ball. They have combined it with Brzeziński's bound (2.7) to obtain an extension of Theorem 47 to $\mathbb{R}^n$-valued random vectors with the sharp constant $\frac{1}{2}$ replaced by $\frac{\Gamma(1+n/2)^{2/n}}{(1+n/2)}$, which is asymptotically sharp (as $m \to \infty$). Previously, using a different argument exploiting Young's inequalities with sharp constants, Bobkov and Chistyakov in [26] obtained such an extension with a slightly worse constant $\frac{1}{e}$ ("attained" as $n \to \infty$), whereas in [121], Ram and Sason obtained constants dependent on the number of summands. Another direction, related to higher-dimensional marginals, has been explored by Livshyts, Paouris, and Pivovarov in [95].

We end this subsection with a conjectural entropic Busemann-type result.

Conjecture 10 (Ball–Nayar–Tkocz [15]). *Let X be a symmetric log-concave random vector in $\mathbb{R}^n$. Then*

$$v \mapsto \sqrt{N(\langle v, X \rangle)} = e^{h(\langle v, X \rangle)}$$

defines a norm on $\mathbb{R}^n$.

Note that Busemann's theorem, Theorem 1, is equivalent to this statement with $N_\infty(\cdot)$ in place of the entropy power $N(\cdot)$ (for uniform distributions on symmetric convex bodies which generalizes to all symmetric log-concave distributions by Ball's results from [8]). What supports this conjecture is the fact that $\sqrt{N(\langle v, X \rangle)}$ defines an e-quasinorm which is also a $\frac{1}{5}$-seminorm (see [15]), and the conjecture holds for the Rényi entropy power of order 2 (see [92]). For extensions to κ-concave measures, see [96].

5.2 The logarithmic Brunn–Minkowski conjecture

In [37], Böröczky, Lutwak, Yang, and Zhang conjectured a strengthening of the Brunn–Minkowski inequality in the presence of symmetry and convexity, namely

$$\mathrm{vol}_n\big(M_\lambda(K,L)\big) \geq \mathrm{vol}_n(K)^\lambda \, \mathrm{vol}_n(L)^{1-\lambda},$$

for all symmetric convex sets K and L in $\mathbb{R}^n$ and every $0 \leq \lambda \leq 1$, where $M_\lambda(K,L)$ is the intersection of the symmetric strips

$$S_\theta = \big\{x \in \mathbb{R}^n, |\langle x,\theta\rangle| \leq h_K(\theta)^\lambda h_L(\theta)^{1-\lambda}\big\}$$

over all unit vectors θ in S^{n-1}. Here, as usual $h_K(\theta) \sup_{y\in K}\langle\theta,y\rangle$ denotes the support functional of K. Still resisting significant efforts of many researchers over a decade, this far-reaching conjecture stems from the so-called logarithmic Minkowski problem (see [38]); we refer to E. Milman's recent work [103] for further comprehensive background, references, and the best results to date. Relevant to us is an equivalent formulation in terms of a certain convexity property of volumes of sections of rescaled cubes.

Conjecture 11. *Let $1 \leq k \leq n$. For every k-dimensional subspace H in $\mathbb{R}^n$, the function*

$$(t_1,\ldots,t_n) \mapsto \mathrm{vol}_k(\mathrm{diag}(e^{t_1},\ldots,e^{t_1})B_\infty^n \cap H)$$

is log-concave on $\mathbb{R}^n$.

For precise statements and explanations of equivalences for this and similar formulations, we refer to [108, 123, 124]. Here, as usual $\mathrm{diag}(e^{t_1},\ldots,e^{t_n})$ is the $n \times n$ diagonal matrix with the diagonal entries $e^{t_1},\ldots,e^{t_n}$, so that $\mathrm{diag}(e^{t_1},\ldots,e^{t_1})B_\infty^n = [-e^{t_1},e^{t_1}]\times\cdots\times[-e^{t_n},e^{t_n}]$. In fact, we conjecture that the conjecture remains true with B_p^n in place of the cube B_∞^n for every $p \geq 1$; we have been able to verify this for $p = 1$ in [108] using Lemma 44.

Bibliography

[1]   A. Akopyan, A. Hubard and R. Karasev, *Lower and upper bounds for the waists of different spaces*, Topol. Methods Nonlinear Anal. **53**(2) (2019), 457–490.

[2]   I. Aliev, *On the volume of hyperplane sections of a d-cube*, Acta Math. Hung. **163**(2) (2021), 547–551.

[3]   G. Ambrus, *Critical central sections of the cube*, Proc. AMS (2021), to appear, preprint, arXiv:2107.14778.

[4]   S. Artstein-Avidan, A. Giannopoulos and D. Milman V., *Asymptotic geometric analysis. Part I*, Mathematical surveys and monographs, vol. 202, American Mathematical Society, Providence, RI, 2015.

[5]   S. V. Astashkin, K. V. Lykov and M. Milman, *Majorization revisited: comparison of norms in interpolation scales*, 2021, preprint, arXiv:2107.11854.

[6]   R. Averkamp and C. Houdré, *Wavelet thresholding for non-necessarily Gaussian noise: idealism*, Ann. Stat. **31** (2003), 110–151.

[7]   K. Ball, *Cube slicing in R^n*, Proc. Am. Math. Soc. **97**(3) (1986), 465–473.

[8]   K. Ball, *Logarithmically concave functions and sections of convex sets in R^n*, Stud. Math. **88**(1) (1988), 69–84.

[9]   K. Ball, *Some remarks on the geometry of convex sets*, in: Geometric aspects of functional analysis (1986/87), Lecture notes in math., vol. 1317, Springer, Berlin, 1988, pp. 224–231.

[10]   K. Ball, *Volumes of sections of cubes and related problems*, in: Geometric aspects of functional analysis (1987–88), Lecture notes in math., vol. 1376, Springer, Berlin, 1989, pp. 251–260.

[11]   K. Ball, *Volume ratios and a reverse isoperimetric inequality*, J. Lond. Math. Soc. (2) **44**(2) (1991), 351–359.

[12]   K. Ball, *Ellipsoids of maximal volume in convex bodies*, Geom. Dedic. **41**(2) (1992), 241–250.

[13]   K. Ball, *Mahler's conjecture and wavelets*, Discrete Comput. Geom. **13**(3–4) (1995), 271–277.

[14]   K. Ball, *Convex geometry and functional analysis*, Handbook of the geometry of Banach spaces, Vol. I, North-Holland, Amsterdam, 2001, pp. 161–194.

[15]   K. Ball, P. Nayar and T. Tkocz, *A reverse entropy power inequality for log-concave random vectors*, Stud. Math. **235**(1) (2016), 17–30.

[16]   K. Ball and M. Prodromou, *A sharp combinatorial version of Vaaler's theorem*, Bull. Lond. Math. Soc. **41**(5) (2009), 853–858.

[17]   D. L. Barrow and P. W. Smith, *Classroom notes: spline notation applied to a volume problem*, Am. Math. Mon. **86**(1) (1979), 50–51.

[18]   F. Á. Bartha, F. Fodor and M. B. González, *Central diagonal sections of the n-cube*, Int. Math. Res. Not. **4** (2021), 2861–2881.

[19]   F. Barthe, *Mesures unimodales et sections des boules B_p^n*, C. R. Acad. Sci. Paris, Sér. I Math. **321**(7) (1995), 865–868.

[20]   F. Barthe, *On a reverse form of the Brascamp–Lieb inequality*, Invent. Math. **134**(2) (1998), 335–361.

[21]   F. Barthe, *Extremal properties of central half-spaces for product measures*, J. Funct. Anal. **182**(1) (2001), 81–107.

[22]   F. Barthe, O. Guédon, S. Mendelson and A. Naor, *A probabilistic approach to the geometry of the l_p^n-ball*, Ann. Probab. 33(2) (2005), 480–513.

[23]   F. Barthe and A. Koldobsky, *Extremal slabs in the cube and the Laplace transform*, Adv. Math. **174**(1) (2003), 89–114.

[24]   F. Barthe and A. Naor, *Hyperplane projections of the unit ball of ℓ_p^n*, Discrete Comput. Geom. **27**(2) (2002), 215–226.

[25]   R. Bhatia, *Matrix analysis*, Graduate texts in mathematics, vol. 169, Springer-Verlag, New York, 1997.

[26]   S. Bobkov and G. P. Chistyakov, *Entropy power inequality for the Rényi entropy*, IEEE Trans. Inf. Theory **61**(2) (2015), 708–714.

[27]   S. Bobkov and G. P. Chistyakov, *On concentration functions of random variables*, J. Theor. Probab. **28**(3) (2015), 976–988.

[28]   S. Boucheron, G. Lugosi and P. Massart, *Concentration inequalities. A nonasymptotic theory of independence. With a foreword by Michel Ledoux*, Oxford University Press, Oxford, 2013.

[29]   J. Bourgain, *On high-dimensional maximal functions associated to convex bodies*, Am. J. Math. **108**(6) (1986), 1467–1476.

[30]   H. J. Brascamp and E. H. Lieb, *Best constants in Young's inequality, its converse, and its generalization to more than three functions*, Adv. Math. **20**(2) (1976), 151–173.

[31]   S. Brazitikos, A. Giannopoulos, P. Valettas and B.-H. Vritsiou, *Geometry of isotropic convex bodies*, Mathematical surveys and monographs, vol. 196, American Mathematical Society, Providence, RI, 2014.

[32]   P. Brzezinski, *Volume estimates for sections of certain convex bodies*, Math. Nachr. **286**(17–18) (2013), 1726–1743.

[33]   H. Busemann, *A theorem on convex bodies of the Brunn–Minkowski type*, Proc. Natl. Acad. Sci. USA **35** (1949), 27–31.

[34]   H. Busemann and C. M. Petty, *Problems on convex bodies*, Math. Scand. **4** (1956), 88–94.

[35] I. Bárány and P. Frankl, *How (not) to cut your cheese*, Am. Math. Mon. **128**(6) (2021), 543–552.

[36] I. Bárány and P. Frankl, *Cells in the box and a hyperplane*, 2020, preprint, arXiv:2004.12306.

[37] K. J. Böröczky, E. Lutwak, D. Yang and G. Zhang, *The log-Brunn–Minkowski inequality*, Adv. Math. **231**(3–4) (2012), 1974–1997.

[38] K. J. Böröczky, E. Lutwak, D. Yang and G. Zhang, *The logarithmic Minkowski problem*, J. Am. Math. Soc. **26**(3) (2013), 831–852.

[39] A. M. Caetano, *Weyl numbers in sequence spaces and sections of unit balls*, J. Funct. Anal. **106**(1) (1992), 1–17.

[40] G. D. Chakerian and P. Filliman, *The measures of the projections of a cube*, Studia Sci. Math. Hung. **21**(1–2) (1986), 103–110.

[41] G. Chasapis, H. König and T. Tkocz, *From Ball's cube slicing inequality to Khinchin-type inequalities for negative moments*, J. Funct. Anal. **281**(9) (2021), 109185, 23 pp.

[42] G. Chasapis, P. Nayar and T. Tkocz, *Slicing ℓ_p-balls reloaded: stability, planar sections in ℓ_1*, Ann. Probab. (2021), to appear, preprint, arXiv:2109.05645.

[43] G. Chasapis, S. Singh and T. Tkocz, *Haagerup's phase transition at polydisc slicing*, 2022, preprint, arXiv:2206.01026.

[44] A. De, I. Diakonikolas and R. A. Servedio, *A robust Khintchine inequality, and algorithms for computing optimal constants in Fourier analysis and high-dimensional geometry*, SIAM J. Discrete Math. **30**(2) (2016), 1058–1094.

[45] H. Dirksen, *Sections of the regular simplex–volume formulas and estimates*, Math. Nachr. **290**(16) (2017), 2567–2584.

[46] H. Dirksen, *Hyperplane sections of cylinders*, Colloq. Math. **147**(1) (2017), 145–164.

[47] D. Edmunds and H. Melkonian, *Behaviour of L_q norms of the $sinc_p$ function*, Proc. Am. Math. Soc. **147**(1) (2019), 229–238.

[48] A. Eskenazis, *On extremal sections of subspaces of L_p*, Discrete Comput. Geom. **65**(2) (2021), 489–509.

[49] A. Eskenazis, P. Nayar and T. Tkocz, *Gaussian mixtures: entropy and geometric inequalities*, Ann. Probab. **46**(5) (2018), 2908–2945.

[50] A. Eskenazis, P. Nayar and T. Tkocz, *Sharp comparison of moments and the log-concave moment problem*, Adv. Math. **334** (2018), 389–416.

[51] P. Filliman, *Extremum problems for zonotopes*, Geom. Dedic. **27**(3) (1988), 251–262.

[52] P. Filliman, *The largest projections of regular polytopes*, Isr. J. Math. **64**(2) (1988), 207–228.

[53] P. Filliman, *Exterior algebra and projections of polytopes*, Discrete Comput. Geom. **5**(3) (1990), 305–322.

[54] P. Filliman, *The extreme projections of the regular simplex*, Trans. Am. Math. Soc. **317**(2) (1990), 611–629.

[55] M. Fradelizi, *Hyperplane sections of convex bodies in isotropic position*, Beitr. Algebra Geom. **40**(1) (1999), 163–183.

[56] R. J. Gardner, *Geometric tomography*, 2nd edn, Encyclopedia of mathematics and its applications, vol. 58, Cambridge University Press, New York, 2006.

[57] R. J. Gardner, A. Koldobsky and T. Schlumprecht, *An analytic solution to the Busemann–Petty problem on sections of convex bodies*, Ann. Math. (2) **149**(2) (1999), 691–703.

[58] A. Giannopoulos, *A note on a problem of H. Busemann and C. M. Petty concerning sections of symmetric convex bodies*, Mathematika **37**(2) (1990), 239–244.

[59] A. Gorin and Yu. Favorov, *Generalizations of the Khinchin inequality*, Teor. Veroâtn. Primen. **35**(4) (1990), 762–767 (Russian), translation in Theory Probab. Appl. 35 (1990), no. 4, 766–771 (1991).

[60] U. Haagerup, *The best constants in the Khintchine inequality*, Stud. Math. **70**(3) (1981), 231–283.

[61] H. Hadwiger, *Gitterperiodische Punktmengen und Isoperimetrie*, Monatshefte Math. **76** (1972), 410–418.

[62] D. Hensley, *Slicing the cube in R^n and probability (bounds for the measure of a central cube slice in R^n by probability methods)*, Proc. Am. Math. Soc. **73**(1) (1979), 95–100.

[63] D. Hensley, *Slicing convex bodies—bounds for slice area in terms of the body's covariance*, Proc. Am. Math. Soc. **79**(4) (1980), 619–625.

[64] G. Ivanov, *Tight frames and related geometric problems*, Can. Math. Bull. **64**(4) (2021), 942–963.

[65]   G. Ivanov, *On the volume of projections of the cross-polytope*, Discrete Math. **344**(5) (2021), 112312, 14 pp.

[66]   G. Ivanov and I. Tsiutsiurupa, *On the volume of sections of the cube*, Anal. Geom. Metric Spaces **9**(1) (2021), 1–18.

[67]   A. Jambulapati, Y. T. Lee and S. Vempala, *A slightly improved bound for the KLS constant*, 2022, preprint, arXiv:2208.11644.

[68]   N. J. Kalton and A. Koldobsky, *Intersection bodies and L_p-spaces*, Adv. Math. **196**(2) (2005), 257–275.

[69]   M. Kanter, *Unimodality and dominance for symmetric random vectors*, Trans. Am. Math. Soc. **229** (1977), 65–85.

[70]   R. Kerman, R. Olhava and S. Spektor, *An asymptotically sharp form of Ball's integral inequality*, Proc. Am. Math. Soc. **143**(9) (2015), 3839–3846.

[71]   A. Khintchine, *Über dyadische Brüche*, Math. Z. **18**(1) (1923), 109–116.

[72]   B. Klartag and J. Lehec, *Bourgain's slicing problem and KLS isoperimetry up to polylog*, 2022, preprint, arXiv:2203.15551.

[73]   B. Klartag and V. Milman, The slicing problem by Bourgain, 2022, to appear in the book: *Analysis at Large, a collection of articles in memory of Jean Bourgain*, preprint.

[74]   A. Koldobsky, *An application of the Fourier transform to sections of star bodies*, Isr. J. Math. **106** (1998), 157–164.

[75]   A. Koldobsky, *Fourier analysis in convex geometry*, Mathematical surveys and monographs, vol. 116, American Mathematical Society, Providence, RI, 2005.

[76]   A. Koldobsky, H. König and M. Zymonopoulou, *The complex Busemann–Petty problem on sections of convex bodies*, Adv. Math. **218**(2) (2008), 352–367.

[77]   A. Koldobsky, G. Paouris and M. Zymonopoulou, *Complex intersection bodies*, J. Lond. Math. Soc. (2) **88**(2) (2013), 538–562.

[78]   A. Koldobsky, D. Ryabogin and A. Zvavitch, *Fourier analytic methods in the study of projections and sections of convex bodies*, in: Fourier analysis and convexity, Appl. numer. harmon. anal., Birkhäuser Boston, Boston, MA, 2004, pp. 119–130.

[79]   A. Koldobsky, D. Ryabogin and A. Zvavitch, *Projections of convex bodies and the Fourier transform*, Isr. J. Math. **139** (2004), 361–380.

[80]   A. Koldobsky and M. Zymonopoulou, *Extremal sections of complex l_p-balls*, $0 < p \le 2$, Stud. Math. **159**(2) (2003), 185–194.

[81]   S. Kwapień, R. Latała and K. Oleszkiewicz, *Comparison of moments of sums of independent random variables and differential inequalities*, J. Funct. Anal. **136**(1) (1996), 258–268.

[82]   H. König, *Non-central sections of the simplex, the cross-polytope and the cube*, Adv. Math. **376** (2021), 107458, 35 pp.

[83]   H. König and A. Koldobsky, *Volumes of low-dimensional slabs and sections in the cube*, Adv. Appl. Math. **47**(4) (2011), 894–907.

[84]   H. König and A. Koldobsky, *Minimal volume of slabs in the complex cube*, Proc. Am. Math. Soc. **140**(5) (2012), 1709–1717.

[85]   H. König and A. Koldobsky, *On the maximal measure of sections of the n-cube*, in: Geometric analysis, mathematical relativity, and nonlinear partial differential equations, Contemp. math., vol. 599, Amer. Math. Soc., Providence, RI, 2013, pp. 123–155.

[86]   H. König and A. Koldobsky, *On the maximal perimeter of sections of the cube*, Adv. Math. **346** (2019), 773–804.

[87]   H. König and M. Rudelson, *On the volume of non-central sections of a cube*, Adv. Math. **360** (2020), 106929, 30 pp.

[88]   R. Latała and K. Oleszkiewicz, *On the best constant in the Khinchin–Kahane inequality*, Stud. Math. **109**(1) (1994), 101–104.

[89]   M. Ledoux, *The concentration of measure phenomenon*, Mathematical surveys and monographs, vol. 89, American Mathematical Society, Providence, RI, 2001.

[90] M. Ledoux and M. Talagrand, *Probability in Banach spaces. Isoperimetry and processes*, Springer-Verlag, Berlin, 1991.

[91] D. R. Lewis, *Finite dimensional subspaces of L_p*, Stud. Math. **63**(2) (1978), 207–212.

[92] J. Li, *Rényi entropy power inequality and a reverse*, Stud. Math. **242**(3) (2018), 303–319.

[93] J. E. Littlewood, *On a certain bilinear form*, Q. J. Math. Oxf. Ser. **1** (1930), 164–174.

[94] R. Liu and T. Tkocz, *A note on the extremal non-central sections of the cross-polytope*, Adv. Appl. Math. **118** (2020), 102031, 17 pp.

[95] G. Livshyts, G. Paouris and P. Pivovarov, *On sharp bounds for marginal densities of product measures*, Isr. J. Math. **216**(2) (2016), 877–889.

[96] M. Madiman, J. Melbourne and P. Xu, *Forward and reverse entropy power inequalities in convex geometry*, in: Convexity and concentration, IMA vol. math. appl., vol. 161, Springer, New York, 2017, pp. 427–485.

[97] M. Madiman, J. Melbourne and P. Xu, *Rogozin's convolution inequality for locally compact groups*, 2017, preprint, arXiv:1705.00642.

[98] J. Matoušek, *Lectures on discrete geometry*, Graduate texts in mathematics, vol. 212, Springer-Verlag, New York, 2002.

[99] P. McMullen, *Volumes of projections of unit cubes*, Bull. Lond. Math. Soc. **16**(3) (1984), 278–280.

[100] J. Melbourne and C. Roberto, *Quantitative form of Ball's Cube slicing in R^n and equality cases in the min-entropy power inequality*, 2021, preprint, arXiv:2109.03946.

[101] J. Melbourne and C. Roberto, *Transport-majorization to analytic and geometric inequalities*, 2021, preprint, arXiv:2110.03641.

[102] M. Meyer and A. Pajor, *Sections of the unit ball of L_p^n*, J. Funct. Anal. **80**(1) (1988), 109–123.

[103] E. Milman, Centro-affine differential geometry and the log-Minkowski problem, 2021, preprint, arXiv:2104.12408.

[104] V. D. Milman and A. Pajor, *Isotropic position and inertia ellipsoids and zonoids of the unit ball of a normed n-dimensional space*, in: Geometric aspects of functional analysis (1987–88), Lecture notes in math., vol. 1376, Springer, Berlin, 1989, pp. 64–104.

[105] J. Moody, C. Stone, D. Zach and A. Zvavitch, *A remark on the extremal non-central sections of the unit cube*, in: Asymptotic geometric analysis, Fields inst. commun., vol. 68, Springer, New York, 2013, pp. 211–228.

[106] S. Moriguti, *A lower bound for a probability moment of any absolutely continuous distribution with finite variance*, Ann. Math. Stat. **23** (1952), 286–289.

[107] A. Naor and D. Romik, *Projecting the surface measure of the sphere of l_p^n*, Ann. Inst. Henri Poincaré Probab. Stat. **39**(2) (2003), 241–261.

[108] P. Nayar and T. Tkocz, *On a convexity property of sections of the cross-polytope*, Proc. Am. Math. Soc. **148**(3) (2020), 1271–1278.

[109] F. Nazarov and A. Podkorytov, *Ball, Haagerup, and distribution functions*, in: Complex analysis, operators, and related topics, Oper. theory adv. appl., vol. 113, Birkhäuser, Basel, 2000, pp. 247–267.

[110] H. H. Nguyen and V. H. Vu, *Small ball probability, inverse theorems, and applications*, in: Erdös centennial, Bolyai soc. math. stud., vol. 25, János Bolyai Math. Soc., Budapest, 2013, pp. 409–463.

[111] R. O'Donnell, *Analysis of Boolean functions*, Cambridge University Press, New York, 2014.

[112] K. Oleszkiewicz, *Comparison of moments via Poincaré-type inequality*, in: Advances in stochastic inequalities, Atlanta, GA, 1997, Contemp. math., vol. 234, Amer. Math. Soc., Providence, RI, 1999, pp. 135–148.

[113] K. Oleszkiewicz, *On p-pseudostable random variables, Rosenthal spaces and l_p^n ball slicing*, in: Geometric aspects of functional analysis, Lecture notes in math., vol. 1807, Springer, Berlin, 2003, pp. 188–210.

[114] K. Oleszkiewicz and A. Pełczyński, *Polydisc slicing in C^n*, Stud. Math. **142**(3) (2000), 281–294.

[115] B. Pooley, *Upper bounds on the cross-sectional volumes of cubes and other problems*, MSc dissertation, University of Warwick, 2012.

[116] L. Pournin, *Shallow sections of the hypercube*, Isr. J. Math. (2022), 10.1007/s11856-022-2400-9, preprint, arXiv:2104.08484.

[117] L. Pournin, *Local extrema for hypercube sections*, 2022, preprint, arXiv:2203.15054.

[118] J. Prochno, C. Thäle and N. Turchi, *Geometry of ℓ_p^n-balls: classical results and recent developments*, in: High dimensional probability VIII – the Oaxaca volume, Progr. probab., vol. 74, Birkhäuser/Springer, Cham, 2019, pp. 121–150.

[119] G. Pólya, *Berechnung eines bestimmten Integrals*, Math. Ann. **74** (1913), 204–212.

[120] S. T. Rachev and L. Ršchendorf, *Approximate independence of distributions on spheres and their stability properties*, Ann. Probab. **19**(3) (1991), 1311–1337.

[121] E. Ram and I. Sason, *On Rényi entropy power inequalities*, IEEE Trans. Inf. Theory **62**(12) (2016), 6800–6815.

[122] B. A. Rogozin, *An estimate for the maximum of the convolution of bounded densities*, Teor. Veroâtn. Primen. **32**(1) (1987), 53–61.

[123] C. Saroglou, *Remarks on the conjectured log-Brunn–Minkowski inequality*, Geom. Dedic. **177** (2015), 353–365.

[124] C. Saroglou, *More on logarithmic sums of convex bodies*, Mathematika **62**(3) (2016), 818–841.

[125] G. Schechtman and J. Zinn, *On the volume of the intersection of two L_p^n balls*, Proc. Am. Math. Soc. **110**(1) (1990), 217–224.

[126] G. Schechtman and A. Zvavitch, *Embedding subspaces of L_p into l_p^N*, $0 < p < 1$, Math. Nachr. **227** (2001), 133–142.

[127] M. Schmuckenschläger, *Volume of intersections and sections of the unit ball of l_p^n*, Proc. Am. Math. Soc. **126**(5) (1998), 1527–1530.

[128] C. E. Shannon, *A mathematical theory of communication*, Bell Syst. Tech. J. **27** (1948), 379–423, 623–656.

[129] G. C. Shephard, *Combinatorial properties of associated zonotopes*, Can. J. Math. **26** (1974), 302–321.

[130] S. J. Szarek, *On the best constants in the Khinchin inequality*, Stud. Math. **58**(2) (1976), 197–208.

[131] B. Tomaszewski, *A simple and elementary proof of the Kchintchine inequality with the best constant*, Bull. Sci. Math. (2) **111**(1) (1987), 103–109.

[132] J. D. Vaaler, *A geometric inequality with applications to linear forms*, Pac. J. Math. **83**(2) (1979), 543–553.

[133] S. Webb, *Central slices of the regular simplex*, PhD Thesis, University, of London, University College London, United Kingdom, 1996.

[134] S. Webb, *Central slices of the regular simplex*, Geom. Dedic. **61**(1) (1996), 19–28.

[135] A. Zvavitch, *Gaussian measure of sections of dilates and translations of convex bodies*, Adv. Appl. Math. **41**(2) (2008), 247–254.

Giuseppe Negro, Diogo Oliveira e Silva, and Christoph Thiele

When does $e^{-|\tau|}$ maximize Fourier extension for a conic section?

Abstract: In the past decade, much effort has gone into understanding maximizers for Fourier restriction and extension inequalities. Nearly all of the cases in which maximizers for inequalities involving the restriction or extension operator have been successfully identified can be seen as partial answers to the question in the title. In this survey, we focus on recent developments in sharp restriction theory relevant to this question. We present results in the algebraic case for spherical and hyperbolic extension inequalities. We also discuss the use of the Penrose transform leading to some negative answers in the case of the cone.

Keywords: Sharp restriction theory, Fourier extension, maximizer, sphere, hyperboloid, cone

MSC 2020: 42B10

1 Introduction

We are interested in a family of maximization problems in Fourier extension theory for quadratic surfaces. In [42], this family of problems was formulated in a uniform manner, which we now recall.

Consider the standard cone $\mathbb{K}^{d+1}$ in $\mathbb{R}^{d+2}$, that is, the set of all (τ, η) with $\tau \in \mathbb{R}$, $\eta \in \mathbb{R}^{d+1}$, and $\tau^2 = |\eta|^2$. A conic section S is the intersection of this cone with a hyperplane W. After a rotation in the η-variables, which does not affect our object of interest, we may assume that the hyperplane W has the equation

$$\alpha\tau + \beta\rho + \gamma = 0 \tag{1.1}$$

for some real numbers α, β, γ, where $\eta = (\rho, \xi)$ with $\rho \in \mathbb{R}$ and $\xi \in \mathbb{R}^d$.

Acknowledgement: GN and DOS were supported by the EPSRC New Investigator Award "Sharp Fourier Restriction Theory", grant no. EP/T001364/1. DOS and CT acknowledge partial support from the Deutsche Forschungsgemeinschaft under Germany's Excellence Strategy – EXC-2047/1 – 390685813, and (CT) from CRC 1060. This work was initiated during a pleasant visit of CT to the Instituto Superior Técnico, Universidade de Lisboa, whose hospitality is greatly appreciated.

Giuseppe Negro, Diogo Oliveira e Silva, Departamento de Matemática, Instituto Superior Técnico, Universidade de Lisboa, Lisbon, Portugal, e-mail: diogo.oliveira.e.silva@tecnico.ulisboa.pt
Christoph Thiele, Mathematisches Institut, Rheinische Friedrich Wilhelms Universität, Bonn, Germany

https://doi.org/10.1515/9783110775389-009

A natural measure σ on the conic section S is given by

$$\int_S f \, \mathrm{d}\sigma := \int_{\mathbb{R}^{d+2}} f(\tau, \rho, \xi)\delta(\alpha\tau + \beta\rho + \gamma)\delta(\tau^2 - \rho^2 - |\xi|^2) \, \mathrm{d}\tau\mathrm{d}\rho\mathrm{d}\xi, \tag{1.2}$$

where, say, f is a continuous function on the section S. We denote by $L^2(\sigma)$ the Hilbert space closure of the space of functions satisfying

$$\|f\|_{L^2(\sigma)} := \left(\int_S |f|^2 \mathrm{d}\sigma \right)^{1/2} < \infty.$$

The Fourier extension of the function f is the function

$$\widehat{f\sigma}(x) := \int_{\mathbb{R}^{d+2}} f(\tau, \rho, \xi)e^{-i\eta\cdot x}\delta(\alpha\tau + \beta\rho + \gamma)\delta(\tau^2 - \rho^2 - |\xi|^2) \, \mathrm{d}\tau\mathrm{d}\rho\mathrm{d}\xi,$$

which we consider for $x \in W$. More precisely, we are interested in the quantity

$$\sup_f \|\widehat{f\sigma}\|_{L^p(W)}, \tag{1.3}$$

where the supremum is taken over all f in the closed unit ball B of $L^2(\sigma)$, and

$$\|\widehat{f\sigma}\|_{L^p(W)} := \left(\int_{\mathbb{R}^{d+2}} |\widehat{f\sigma}(x)|^p \delta(\alpha t + \beta r + \gamma) \, \mathrm{d}t\mathrm{d}r\mathrm{d}x \right)^{1/p}. \tag{1.4}$$

In particular, we ask to identify maximizers f in the ball B for the quantity (1.4), if they exist. Often one formulates this problem in an equivalent way, as extremizing the quotient

$$\sup_{\|f\|_{L^2(\sigma)} \neq 0} \|\widehat{f\sigma}\|_{L^p(W)}^p \|f\|_{L^2(\sigma)}^{-p}. \tag{1.5}$$

It was observed in [42] that in many instances the function $f(\tau, \rho, \xi) = e^{-|\tau|}$ maximizes (1.5). Indeed, this statement includes nearly all instances of known maximizers in sharp restriction theory, such as constant functions on spheres, maximizers which are Gaussian with respect to the projection measure in the case of paraboloids, or exponential functions on the cone. This survey summarizes some progress on the question "For which parameters of the problem is $e^{-|\tau|}$ a maximizer of (1.5)?" In particular, we emphasize progress made since [42].

Progress in identifying the maximizer attaining the supremum (1.3) was almost exclusively made in the *algebraic case*, i. e., when $p = 2k$ is an even integer. We may then write the k-th power of (1.4) as

$$\left\|(\widehat{f\sigma})^k\right\|_2 = (2\pi)^{\frac{d+1}{2}}\left\|(f\sigma)^{*k}\right\|_2,$$

where we used Plancherel's identity and wrote $(f\sigma)^{*k}$ for the successive convolution of k copies of the measure $f\sigma$. The integral power inside the norm allows for some helpful exact manipulations of the expression.

Obviously, finiteness of the supremum (1.3) is a necessary condition for it being attained by a maximizer. The range of exponents p for which (1.3) is finite is known for all conic sections in all dimensions d; see Figure 1 for the algebraic case.

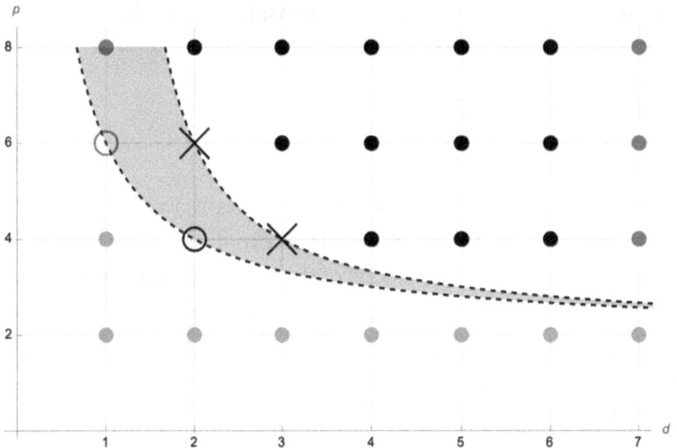

Figure 1: The horizontal axis indicates dimension d. The vertical axis indicates the Lebesgue exponent p of the $L^2(\sigma) \to L^p$ extension inequality. Circles ($\circ$) at $d \in \{1, 2\}$ correspond to endpoint Stein–Tomas for spheres and Stein–Tomas for paraboloids, while crosses ($\times$) at $d \in \{2, 3\}$ correspond to Stein–Tomas for cones. The shaded region between the two dashed curves correspond to hyperboloids. For spheres, black entries correspond to inequalities which are maximized by constants, while red entries correspond to inequalities which are conjecturally maximized by constants; orange entries correspond to situations in which Stein–Tomas does not hold, but other replacement inequalities such as Agmon–Hörmander may be available.

Letting $(\alpha, \beta, \gamma) = (1, 0, -1)$, equation (1.1) reduces to $\tau = 1$, so the section S corresponds to the unit sphere $\mathbb{S}^d$. In this case, the function $e^{-|\tau|}$ is constant. The measure on $\mathbb{S}^d$ defined by (1.2) is then nothing but the usual surface measure. The supremum in (1.3) is finite for $p \geq 2 + \frac{4}{d}$ by the Stein–Tomas inequality [82, 87] on $\mathbb{S}^d$. We call $p = 2 + \frac{4}{d}$ the endpoint Stein–Tomas exponent. The first result identifying constants as maximizers for the sphere was proved by Foschi [40] in the algebraic case $p = 4$ when $d = 2$. This corresponds to the sharp endpoint Stein–Tomas restriction theorem on $\mathbb{S}^2 \subset \mathbb{R}^3$. Remarkably, the case of the endpoint Stein–Tomas exponent for $\mathbb{S}^1 \subset \mathbb{R}^2$, the only other case where this endpoint is algebraic for the sphere, remains open. Much effort has gone into studying this case. In particular, a modification of Foschi's approach has been suggested but only led to partial progress. In Section 2, we discuss some rigorous numerical computations to verify that constants maximize (1.3) for $\mathbb{S}^1$, among those functions in the unit ball of $L^2(\sigma)$ which have Fourier modes up to degree 120. These simulations also suggest an asymptotic behavior for large Fourier modes, which in the future may help complement the numerical work with an analytic proof.

In Section 3, we address the case of higher-dimensional spheres $\mathbb{S}^d$, $d \geq 2$. We describe in some detail a common approach to all the black entries on Figure 1, corresponding to $2 \leq d \leq 6$, present a particular instance of a sharp extension inequality on $\mathbb{S}^7$, and conclude with an open problem.

On the unit sphere $\mathbb{S}^d$, the Stein–Tomas condition excludes $p = 2$ in all dimensions. Following Agmon and Hörmander [1], we therefore also consider the modified version of (1.3)

$$\sup_f R^{-1} \big\| \widehat{f\sigma} 1_{B_R} \big\|_{L^2(W)}, \tag{1.6}$$

where B_R denotes the ball in $W \equiv \mathbb{R}^{d+1}$ of radius R about the origin. Even though this functional is related to an endpoint result for the classical proof of the Stein–Tomas inequality via complex interpolation, for some values of the parameters d, R the supremum in (1.6) is maximized when f is constant, while for other values of the parameters it is maximized by other, less symmetric functions. This is an instance of *symmetry breaking*. Section 4 surveys this in more detail.

In Section 5, we specialize the hyperplane (1.1) to the limiting case $\rho = 0$, i.e., $(\alpha, \beta, \gamma) = (0, 1, 0)$. The resulting conic section $\tau^2 = |\xi|^2$, where $\xi \in \mathbb{R}^d$, is identified with the cone $\mathbb{K}^d$. The measure σ of (1.2) reduces to the Lorentz invariant measure on $\mathbb{K}^d$, which is unique up to scalar multiplication. The quantity (1.3) is finite if and only if $p = 2\frac{d+1}{d-1}$, by the Strichartz estimate for the wave equation [84]. Foschi [39] showed that $f(\tau, \xi) = e^{-|\tau|}$ maximizes this estimate when $(d, p) = (3, 4)$, and he formulated the natural conjecture that the same should be true in arbitrary dimensions. This turned out to be false for even d; we will discuss the proof, which is based on the Penrose conformal compactification of Minkowski spacetime.

Some related extension problems are also meaningful in this cone context. The aforementioned study [39] establishes that on the half-cone $\mathbb{K}_+^d = \mathbb{K}^d \cap \{\tau > 0\}$ equipped with the natural restriction σ_+ of the Lorentz invariant measure σ, the function $f(\tau, \xi) = e^{-|\tau|}$ is a maximizer for $d \in \{2, 3\}$. We will briefly discuss the conjecture that this should hold in arbitrary dimension, which is still open. Finally, we will consider the case of weighted versions of the measure σ_+, corresponding to Strichartz estimates with a different Sobolev regularity of the initial data.

Hyperboloids, corresponding to $(\alpha, \beta, \gamma) = (0, 1, 1)$ in (1.1), capture features from both spheres (Sections 2–4) and cones (Section 5), but new phenomena arise – most notably, maximizers sometimes fail to exist. In spite of this, optimal constants have been determined for all the algebraic endpoint inequalities, and concentration-compactness tools have been used together with bilinear restriction theory in order to shed some light on the behavior of arbitrary maximizing sequences in the case of non-endpoint inequalities. In Section 6, we survey both of these types of results.

While in Fourier extension theory one also studies finiteness of (1.3) when f ranges over the unit ball of $L^q(\sigma)$ for $q \neq 2$, there is little progress in identifying $e^{-|\tau|}$ as the maximizer for $q \neq 2$ and we therefore focus in this survey on $q = 2$.

1.1 Remarks and further references

Following [42] and [62, Introduction], the style of this survey is admittedly informal. In particular, some objects will not be rigorously defined, and several results will not be precisely formulated. Most of the material is not new, exceptions being some parts of Section 2, the main result in Section 5.2, and a few observations which we could not find in the literature. The subject is becoming more popular, as shown by the increasing number of works that appeared in the last decade. We have attempted to provide a rather complete set of references, which includes several interesting works on spheres [4, 9, 24, 25, 30, 32, 38, 43, 55, 81], cones [11, 20, 41, 60, 71, 75, 78], hyperboloids [28, 34, 51, 71, 72, 77], curves [17, 44, 49, 61, 63, 79], paraboloids [3, 5, 7, 10, 21, 29, 35, 45, 46, 48, 50, 54, 73, 80, 83, 89], and perturbations [33, 52, 53, 64, 66, 85, 86] that will not be discussed here. Given its young age, there are plenty of open problems in the area. We provide some more.

2 The circle

The endpoint Stein–Tomas exponent for the sphere $\mathbb{S}^d$ is an even integer in dimensions $d \in \{1, 2\}$. If $d = 2$, then the endpoint exponent is $p = 4$, and this is the only case in which constant functions are known to maximize the Fourier extension map for the endpoint Stein–Tomas exponent, thanks to the argument of Foschi [40]. Remarkably, the question remains open in dimension $d = 1$, where the exponent is $p = 6$. A modification of Foschi's argument for $d = 1$ was proposed in [22], and this approach was substantiated by extensive numerical evidence in [70, 2]. In this section, we discuss this approach to the circle problem.

We first recall Foschi's argument [40] for $d = 2$ in a form reflecting more recent insights. Let the measure σ be defined by $\sigma(x) = \delta(|x|^2 - 1)$. We aim to show

$$\|\widehat{f\sigma}\|_4^4 \leq \|\widehat{\sigma}\|_4^4 \tag{2.1}$$

for all f that have the same $L^2(\sigma)$-norm as the constant function 1.

Basic steps that are recalled in Section 3 allow to restrict our attention to f that are real and antipodally symmetric, so that also $u := \widehat{f\sigma}$ is real and even. As $\hat{u}$ is supported on the unit sphere, u satisfies the Helmholtz equation,

$$u + \Delta u = 0.$$

This equation is used in the first key step, which is sometimes referred to as the *magic identity*. It expresses the left-hand side of (2.1) in a way that later removes a singularity of $\sigma * \sigma$ at the origin. The derivation of the magic identity uses partial integration, which is justified because u and all its derivatives are in L^4; see [23, Proposition 6]. Replacing one factor of u with Helmholtz, we obtain for the left-hand side of (2.1)

$$-\int_{\mathbb{R}^3} (\Delta u)u^3 = \int_{\mathbb{R}^3} \nabla u \cdot \nabla(u^3) = 3\int_{\mathbb{R}^3} |\nabla u|^2 u^2 = \frac{3}{4}\int_{\mathbb{R}^3} |\nabla u^2|^2 = -\frac{3}{4}\int (\Delta u^2)u^2. \qquad (2.2)$$

Expressing this again in terms of f and proceeding analogously for the constant function, we reduce (2.1) to

$$\Lambda(f,f,f,f) \leq \Lambda(\mathbf{1},\mathbf{1},\mathbf{1},\mathbf{1})$$

with Λ defined by

$$\Lambda(f_1,f_2,f_3,f_4) = \int_{(\mathbb{R}^3)^4} |x_1 + x_2|^2 \delta\left(\sum_{k=1}^4 x_k\right) \prod_{j=1}^4 f_j \sigma(x_j)\mathrm{d}x_j.$$

The second key step is a positivity argument reminiscent of the Cauchy–Schwarz inequality. We use the inequality

$$2ab \leq a^2 + b^2 \qquad (2.3)$$

for real numbers a and b chosen to be $f(x_1)f(x_2)$ and $f(x_3)f(x_4)$ at every point x_1, x_2, x_3, x_4 and obtain by positivity of the integral kernel

$$2\Lambda(f,f,f,f) \leq \Lambda(f^2,f^2,\mathbf{1},\mathbf{1}) + \Lambda(\mathbf{1},\mathbf{1},f^2,f^2). \qquad (2.4)$$

Using symmetry of Λ thanks to $|x_1 + x_2| = |x_3 + x_4|$, we are reduced to showing

$$\Lambda(f^2,f^2,\mathbf{1},\mathbf{1}) \leq \Lambda(\mathbf{1},\mathbf{1},\mathbf{1},\mathbf{1}).$$

We write $f^2 = 1+g$, where $g\sigma$ has integral zero. In the integral expression for $\Lambda(\mathbf{1},\mathbf{1},\mathbf{1},g)$, the measure $g\sigma$ is integrated against a radially symmetric function and thus this integral is zero. Using expansion and symmetry of Λ again, we are reduced to showing the third key step, which is the inequality

$$\Lambda(g,g,\mathbf{1},\mathbf{1}) \leq 0. \qquad (2.5)$$

This relies on the important calculation (see, e. g., [67] or Lemma 5 below) that $(\sigma * \sigma)(x)$ is a positive scalar multiple of

$$|x|^{-1}1_{|x|\leq 2}.$$

This convolution appears in the expression for $\Lambda(g, g, \mathbf{1}, \mathbf{1})$ by integrating out the last two variables. Thanks to the support of the remaining functions, we may omit the indicator $1_{|x|\leq 2}$ and reduce (2.5) to

$$\int\limits_{(\mathbb{R}^3)^2} \frac{|x_1 + x_2|^2}{|x_1 + x_2|} g\sigma(x_1) g\sigma(x_2)\, dx_1\, dx_2 \leq 0. \tag{2.6}$$

It is here, as a result of the magic identity (2.2), that the singularity at $|x_1 + x_2| = 0$ is canceled.

Define the analytic family h_s of tempered distributions on $\mathbb{R}^3$ by

$$h_s(\phi) = \Gamma\left(\frac{s+3}{2}\right)^{-1} \int\limits_{\mathbb{R}^3} |x|^s \phi(x)\, dx. \tag{2.7}$$

The integral (2.7) is well-defined for $\mathfrak{R}(s) > -3$ and positive when ϕ is a Gaussian. Inequality (2.6) is then equivalent to

$$h_1(g\sigma * g\sigma) \leq 0. \tag{2.8}$$

The distribution h_s is rotationally symmetric and homogeneous of degree s, and it is up to positive scalar uniquely determined by these symmetries and positivity on the Gaussian. As the Fourier transform $\widehat{h}_s$ also has rotational symmetry and dilation symmetry with degree of homogeneity $-3 - s$, we have for $-3 < \mathfrak{R}(s) < 0$

$$a\widehat{h}_s = h_{-3-s} \tag{2.9}$$

for some positive constant a. Analytic continuation with (2.7) and (2.9) allows to define h_s for all complex numbers s. By unique continuation, $h_s(\phi)$ is expressed by (2.7) whenever ϕ vanishes of sufficiently high order at 0 so that the integral is absolutely convergent.

By Plancherel, we reduce (2.8) to

$$h_{-4}\left((\widehat{g\sigma})^2\right) \leq 0. \tag{2.10}$$

As $(\widehat{g\sigma})^2$ vanishes to second order at the origin, the pairing with h_{-4} is given by the expression (2.7). Inequality (2.10) follows, because $(\widehat{g\sigma})^2$ is non-negative and $\Gamma(-1/2) < 0$. This concludes our discussion of the case $d = 2$.

We turn to the case of the circle, $d = 1$. We again assume f is smooth, real-valued, and antipodally symmetric. The endpoint Stein–Tomas exponent is $p = 6$ and one may deduce a magic identity

$$\|u\|_6^6 = -\frac{5}{9} \int\limits_{\mathbb{R}^3} (\Delta u^3) u^3.$$

However, we aim to cancel a singularity of $\sigma * \sigma * \sigma$ at $|x| = 1$, so we consider the identity in the form

$$\frac{4}{5}\|u\|_6^6 = -\int_{\mathbb{R}^3} (u^3 + \Delta u^3)u^3.$$

The second step in Foschi's program suggests the inequality

$$\Lambda(f, f, f, f, f, f) \leq \Lambda(f^2, f^2, f^2, \mathbf{1}, \mathbf{1}, \mathbf{1}) \tag{2.11}$$

with

$$\Lambda(f_1, f_2, f_3, f_4, f_5, f_6) := \int_{(\mathbb{R}^2)^6} (|x_1 + x_2 + x_3|^2 - 1)\delta\left(\sum_{k=1}^{6} x_k\right)\prod_{j=1}^{6} f_j\sigma(x_j)\mathrm{d}x_j.$$

Inequality (2.11) does not follow as above from a pointwise application of the elementary pointwise inequality (2.3) because the integral kernel is not positive. Indeed, inequality (2.11) is not known to be true in the required level of generality. Assuming (2.11) nevertheless for the moment, it remains to show the third key step,

$$\Lambda(f^2, f^2, f^2, \mathbf{1}, \mathbf{1}, \mathbf{1}) \leq \Lambda(\mathbf{1}, \mathbf{1}, \mathbf{1}, \mathbf{1}, \mathbf{1}, \mathbf{1}).$$

This inequality is the main result of [22]. The technique used in [22] is to expand the function f^2 on the circle into Fourier series and use careful computations from [69] for the tensor coefficients of Λ in the Fourier basis. These arguments are very technical and go beyond the scope of this survey.

The question whether constant functions maximize the Stein–Tomas inequality on the circle is thus reduced to proving (2.11). Additional motivation for this conjecture is that thanks to the antipodal symmetry of f the values of the integrand on the negative and positive parts are correlated. We formulate an even more general conjecture.

Conjecture 1. *The quadratic form*

$$Q(f) = \Lambda(f^2, \mathbf{1}) - \Lambda(f, f)$$

with $\Lambda(f_1, f_2)$ defined as

$$\int_{(\mathbb{R}^2)^6} (1 - |x_1 + x_2 + x_3|^2)\delta\left(\sum_{k=1}^{6} x_k\right)f_1(x_1, x_2, x_3)f_2(x_4, x_5, x_6)\prod_{j=1}^{6} \sigma(x_j)\mathrm{d}x_j$$

is positive semidefinite on the space of real-valued functions $f \in L^2((\mathbb{S}^1)^3)$ that are antipodally symmetric in each variable.

Positive definiteness of Q has been verified on a large finite-dimensional subspace of $f \in L^2((\mathbb{S}^1)^3)$ in [2], using a large computing cluster and extending previous numerical results in [70]. More specifically, identifying $\mathbb{R}^2$ with the complex plane $\mathbb{C}$, we expand f in $L^2((\mathbb{S}^1)^3)$ into Fourier series

$$f(x_1, x_2, x_3) = \sum_{n \in \mathbb{Z}^3} \widehat{f}(n_1, n_2, n_3) x_1^{n_1} x_2^{n_2} x_3^{n_3}.$$

Note that antipodal symmetry translates into vanishing of Fourier coefficients with at least one odd index. The following theorem is proved in [2].

Theorem 2 ([2]). *Let $f \in L^2((\mathbb{S}^1)^3)$ so that the Fourier coefficient $\widehat{f}(n_1, n_2, n_3)$ is zero if one of the numbers n_1, n_2, n_3 is odd or larger than* 120. *Then $Q(f) \geq 0$.*

Foschi's program then shows that constants maximize the endpoint Stein–Tomas inequality on the circle among all functions with Fourier modes up to degree 120.

The numerical computations also suggest an asymptotic behavior of Q for large frequencies. This could be helpful for a resolution of Conjecture 1. We summarize some of these observations in the remainder of this section, which is best read with the displays from [2] at hand.

The quadratic form Q is symmetric under joint rotation of the variables of f, that is, $Q(f) = Q(g)$ whenever

$$g(x_1, x_2, x_3) = f(\omega x_1, \omega x_2, \omega x_3)$$

for a complex number ω of modulus 1. This rotation action decomposes $L^2((\mathbb{S}^1)^3)$ into mutually orthogonal eigenspaces, and it suffices to prove positive semidefiniteness of Q on each eigenspace separately. Numerical simulations confirm that understanding of the matter hinges on understanding of the main eigenspace V spanned by Fourier modes which satisfy

$$n_1 + n_2 + n_3 = 0. \tag{2.12}$$

The index set of the Fourier modes with even $-120 \leq n_1, n_2, n_3 \leq 120$ satisfying (2.12) is schematically depicted via black dots in the hexagon on the left in Figure 2. The colored structures in the plot show the locus of relatively large coefficients of a typical row of the symmetric quadratic form Q. There is a cluster of very large values near the diagonal element, multiplied by the sixfold permutation symmetry of the indices. Further relatively large coefficients appear near a circular structure. All remaining coefficients are quite small. The occurrence of this circular structure is observed but as of yet poorly understood.

On the right in Figure 2, the eigenvalues of Q on V are schematically depicted in increasing order. They fall into three regions: small, intermediate, and large eigenval-

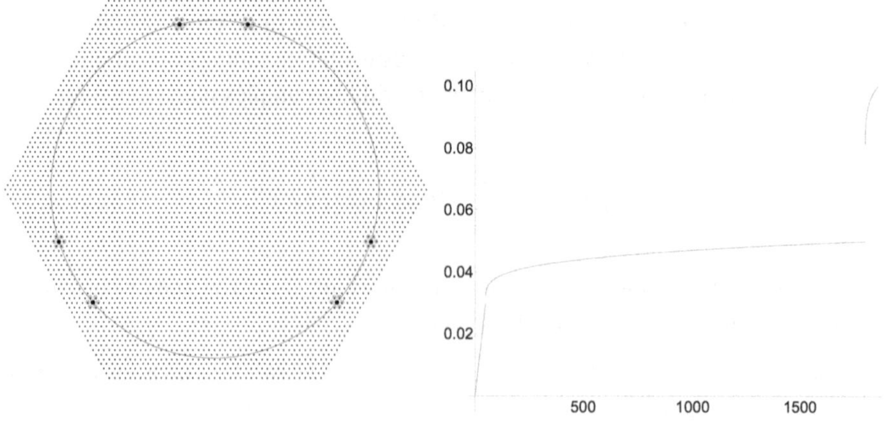

Figure 2: Fourier modes and eigenvalues; see [70] and especially [2, Figure 6].

ues. The small eigenvalues are the main difficulty in proving Conjecture 1. Numerical simulations suggest that these eigenvalues are caused by highly localized functions,

$$f_\epsilon(x_1, x_2, x_3) = \prod_{j=1}^{3} \exp(-\epsilon^{-2}\mathbb{R}(x_j)^2),$$

for small ϵ, which approximate a sum of antipodal Dirac deltas. Denote the orthogonal projection of f_ϵ to the eigenspace V by g_ϵ. Jiaxi Cheng, a graduate student at Bonn, theoretically observed the asymptotic behavior

$$\left| Q(g_\epsilon) \|g_\epsilon\|_{L^2(\sigma)}^{-2} - c \log(\epsilon)\epsilon^2 \right| = o(\log(\epsilon)\epsilon^2) \tag{2.13}$$

using Taylor expansion near the essential support of g_ϵ. This models most if not all of the small eigenvalues that were observed in the numerical simulations in [2]. An asymptotic behavior of the smallest eigenvalue was explicitly suggested in [2] and interpreted as proportional to $\epsilon^{1.74}$, which in the range of observations $1/20 < \epsilon < 1/120$ is nearly proportional to $\log(\epsilon)\epsilon^2$. The theoretically discovered constant c in (2.13) by Cheng is consistent with the numerical observations.

The functions g_ϵ inherit an approximate radial symmetry from the radial symmetry of a Dirac delta and the approximating Gaussian. This results in an approximate radial symmetry of the Fourier coefficients of the eigenfunctions with small eigenvalues, parameterized by the points in the hexagon in Figure 2. This was observed in [2]. Further computations by Jiaxi Cheng showed that these eigenfunctions are properly modeled by eigenfunctions of a Schrödinger operator in the plane with infinite hexagonal well or, approximately, circular well. The latter eigenfunctions are explicitly known; they are radial functions with radial profile given by a scaled Bessel function J_0 so that the spherical well falls on a zero of the scaled Bessel function. This is matching the numerical findings in [2, Figure 8], where the eigenfunctions with the five smallest eigenvalues are displayed.

The above discussion gives hope that one can obtain a good analytic understanding of the small eigenvalues, say on a suitably defined space of radial functions. On the orthogonal complement of such space, one mainly faces the intermediate eigenvalues displayed in Figure 2, which appear to come from a small perturbation of a multiple of the identity matrix. This is confirmed by the typical matrix row of Q, schematically depicted in Figure 2, where the blue circle and the orange dots indicate the loci of large matrix entries and are more precisely displayed in [2, Figures 3–4]. The circular structure disappears when projected to the orthogonal complement of the radial functions, while the orange clusters, up to the symmetry group of three elements, represent a nearly diagonal perturbation of the identity matrix.

The small group of large eigenvalues in Figure 2 is not well understood, but might be an artifact of the precise truncation to the hexagon in the numerical calculations and might be suppressed in a more spherically symmetric or smoothened approach.

3 Spheres

In this section, we consider the case of higher-dimensional spheres $\mathbb{S}^d \subset \mathbb{R}^{d+1}$, $d \geq 2$, equipped with the usual surface measure σ as in (1.2) or, equivalently, as in the paragraph before (2.1). The five black $L^2 \to L^4$ entries in Figure 1, corresponding to cases for which constants are global maximizers, have been previously surveyed in [42]. Above them lie infinitely many $L^2 \to L^{2k}$ estimates which have been recently put in sharp form. As in (1.5), we define the functional

$$\Phi_{d,p}[f] := \|\widehat{f\sigma}\|_p^p \|f\|_{L^2(\mathbb{S}^d)}^{-p}$$

and present the main result in [67, 68].

Theorem 3 ([67, 68]). *Let $d \in \{2, 3, 4, 5, 6\}$ and $p \geq 6$ be an even integer. Then constant functions are the unique real-valued maximizers of the functional $\Phi_{d,p}$. The same conclusion holds for $d = 1$ and even $p > 6$ if constants maximize $\Phi_{1,6}$.*

The purpose of this section is fourfold. Firstly, we briefly discuss the proof of Theorem 3 in the particular but representative case when $(d, p) = (2, 6)$; see also [65]. Secondly, we describe the extra ingredients which are needed in order to obtain sharp $L^2 \to L^{2k}$ estimates for higher $k \geq 4$. Thirdly, we characterize the class of complex-valued maximizers of $\Phi_{d,p}$, for d, p in the range covered by Theorem 3. Finally, we suggest a way to go beyond this range, by presenting a sharp extension inequality on $\mathbb{S}^7$ in the weighted setting.

3.1 The case $(d, p) = (2, 6)$

We abbreviate notation by writing $\Phi_p := \Phi_{2,p}$ and

$$\mathbf{T}_p := \sup\{\Phi_p[f]^{1/p} : 0 \neq f \in L^2(\mathbb{S}^2)\}.$$

The proof naturally splits into five steps, which use tools from the calculus of variations, symmetrization, operator theory, Lie theory, and probability. We present them next and then we will see how they all come together.

3.1.1 Calculus of variations

The existence of maximizers for Φ_6 is ensured by [36]. Let f be one such maximizer, normalized so that $\|f\|_{L^2} = 1$. Consider the extension operator $\mathcal{E}(f) := \widehat{f\sigma}$ with adjoint given by $\mathcal{E}^*(g) := g^{\vee}|_{\mathbb{S}^2}$. Then the operator norm can be estimated as follows:

$$\begin{aligned}
\|\mathcal{E}\|^6_{L^2 \to L^6} &= \|\mathcal{E}(f)\|^6_{L^6(\mathbb{R}^3)} = \langle |\mathcal{E}(f)|^4 \mathcal{E}(f), \mathcal{E}(f) \rangle = \langle \mathcal{E}^*(|\mathcal{E}(f)|^4 \mathcal{E}(f)), f \rangle_{L^2(\mathbb{S}^2)} \\
&\leq \|\mathcal{E}^*(|\mathcal{E}(f)|^4 \mathcal{E}(f))\|_{L^2(\mathbb{S}^2)} \leq \|\mathcal{E}^*\|_{L^{6/5} \to L^2} \||\mathcal{E}(f)|^4 \mathcal{E}(f)\|_{L^{6/5}(\mathbb{R}^3)} \\
&= \|\mathcal{E}^*\|_{L^{6/5} \to L^2} \|\mathcal{E}(f)\|^5_{L^6(\mathbb{R}^3)} = \|\mathcal{E}\|^6_{L^2 \to L^6}.
\end{aligned}$$

Thus, all inequalities are equalities, and in particular equality in the Cauchy–Schwarz step above yields the *Euler–Lagrange equation*,

$$\left(|\widehat{f\sigma}|^4 \widehat{f\sigma}\right)^{\vee}\Big|_{\mathbb{S}^2} = \lambda f,$$

which, in convolution form, reads as follows:

$$(f\sigma * f_*\sigma * f\sigma * f_*\sigma * f\sigma)|_{\mathbb{S}^2} = (2\pi)^{-3}\lambda f; \tag{3.1}$$

here $f_* := \overline{f(-\cdot)}$. A bootstrapping procedure can then be used to show that f, and indeed any L^2-solution of (3.1), is C^{∞}-smooth. We omit the details and refer the interested reader to [68], which extends the main result of [31] to the higher-dimensional setting of even exponents.

3.1.2 Symmetrization

Since $p = 6$ is an even integer, the problem is inherently positive, in the sense that non-negative maximizers exist. In fact,

$$\|f\sigma * f\sigma * f\sigma\|_{L^2(\mathbb{R}^3)} \leq \||f|\sigma * |f|\sigma * |f|\sigma\|_{L^2(\mathbb{R}^3)}.$$

Defining $f_\sharp := \sqrt{\frac{|f|^2+|f_*|^2}{2}}$, we also have the following monotonicity under antipodal symmetrization,[1] which can be readily verified via a creative application of the Cauchy–Schwarz inequality in the spirit of (2.4):

$$\|f\sigma * f\sigma * f\sigma\|_{L^2(\mathbb{R}^3)} \le \|f_\sharp\sigma * f_\sharp\sigma * f_\sharp\sigma\|_{L^2(\mathbb{R}^3)}. \tag{3.2}$$

We conclude that

$$\mathbf{T}_6 = \max\{\Phi_6[f]^{1/6} : 0 \ne f \in C^\infty(\mathbb{S}^2), f \text{ is non-negative and even}\},$$

an important simplification which will be crucial in the sequel.

3.1.3 Operator theory

We now explore some of the compactness inherent to the problem. Associated to a given $f \in L^2(\mathbb{S}^2)$, consider the integral operator $T_f : L^2(\mathbb{S}^2) \to L^2(\mathbb{S}^2)$ defined by

$$T_f(g)(\omega) := (g * K_f)(\omega) = \int_{\mathbb{S}^2} g(v)K_f(\omega - v)\,d\sigma(v), \tag{3.3}$$

which acts on functions $g \in L^2(\mathbb{S}^2)$ by convolution with the kernel

$$K_f(\xi) := \left(\big|\widehat{f\sigma}\big|^4\right)^{\vee}(\xi) = (2\pi)^3(f\sigma * f_*\sigma * f\sigma * f_*\sigma)(\xi).$$

Note that the Euler–Lagrange equation (3.1) can be written as the (non-linear) eigenfunction equation $T_f(f) = \lambda f$. The kernel K_f defines a bounded, continuous function on $\mathbb{R}^3$ which satisfies $K_f(\xi) = \overline{K_f(-\xi)}$, for all ξ, and crucially $K_f(0) = \|\widehat{f\sigma}\|_4^4$. Correspondingly, the operator T_f is self-adjoint and positive definite. In fact, one can check that T_f is trace-class and that its trace is given by

$$\mathrm{tr}(T_f) = 4\pi\|\widehat{f\sigma}\|_4^4. \tag{3.4}$$

This is a consequence of Mercer's theorem, the infinite-dimensional analog of the well-known statement that any positive semidefinite matrix is the Gram matrix of some set of vectors.

---

1 An analogous inequality to (3.2) holds in the non-algebraic case; see [19, Proposition 6.7] and also [47, Proposition 3.1] for the case of the cone and the underlying wave equation.

3.1.4 Lie theory

We proceed to discuss the symmetries of the problem. The set of 3×3 orthogonal matrices with unit determinant forms the special orthogonal group SO(3), with Lie algebra $\mathfrak{so}(3)$. As a preliminary observation, we note that the exponential map $\exp : \mathfrak{so}(3) \to$ SO(3), $A \mapsto e^A$, is surjective onto SO(3) and that the functional Φ_6 is rotation and modulation invariant. In other words,

$$\Phi_6[f \circ e^{tA}] = \Phi_6[f] = \Phi_6[e^{i\xi \cdot} f],$$

for all $(t, A) \in \mathbb{R} \times \mathfrak{so}(3)$ and $\xi \in \mathbb{R}^3$. As we shall now see, these symmetries naturally give rise to new eigenfunctions for the operator T_f defined in (3.3). Consider the vector field ∂_A acting on sufficiently smooth functions $f : \mathbb{S}^2 \to \mathbb{C}$ via

$$\partial_A f := \frac{\partial}{\partial t}\bigg|_{t=0} (f \circ e^{tA}).$$

We have the following key lemma, where we write $\omega = (\omega_1, \omega_2, \omega_3) \in \mathbb{S}^2$, and by $\omega_j f$ we mean the function defined via $(\omega_j f)(\omega) = \omega_j f(\omega)$.

Lemma 4. *Let* $f : \mathbb{S}^2 \to \mathbb{R}$ *be non-constant, such that* $f_\star = f \in C^1(\mathbb{S}^2)$ *and* $\|f\|_{L^2} = 1$. *Assume* $T_f(f) = \lambda f$. *Then*

$$T_f(\partial_A f) = \frac{\lambda}{5}\partial_A f, \quad \text{for every } A \in \mathfrak{so}(3),$$

$$T_f(\omega_j f) = \frac{\lambda}{5}\omega_j f, \quad \text{for every } j \in \{1, 2, 3\}.$$

Moreover, there exist $A, B \in \mathfrak{so}(3)$, *such that the set* $\{\partial_A f, \partial_B f, \omega_1 f, \omega_2 f, \omega_3 f\}$ *is linearly independent over* $\mathbb{C}$.

The proof of Lemma 4 hinges on the fact that the codimension of a proper, nontrivial subalgebra of $\mathfrak{so}(3)$ equals 2. The linear independence of the set $\{\partial_A f, \partial_B f, \omega_1 f,$ $\omega_2 f, \omega_3 f\}$ follows from the fact that $\partial_A f, \partial_B f$ are real *even* functions, whereas $\omega_1 f, \omega_2 f,$ $\omega_3 f$ are real *odd* functions.

3.1.5 Uniform random walks in $\mathbb{R}^3$

We will need explicit expressions for various convolution measures σ^{*k}. These can be interpreted in terms of random walks, and as such are sometimes available in the probability theory literature. More precisely, consider i. i. d. random variables X_1, X_2, X_3, taking values on $\mathbb{S}^2$ with uniform distribution. Then $Y_3 = X_1 + X_2 + X_3$ is known as the *uniform three-step random walk* in $\mathbb{R}^3$. If p_3 denotes the probability density of $|Y_3|$, then a straightforward computation in polar coordinates reveals that

$(\sigma * \sigma * \sigma)(r) = \sigma(\mathbb{S}^2)^2 p_3(r) r^{-2}$. Such considerations quickly lead to the following formulas for spherical convolutions.

Lemma 5 ([67]). *The following identities hold:*

$$(\sigma * \sigma)(\xi) = \frac{2\pi}{|\xi|}, \quad \textit{if } |\xi| \le 2,$$

$$(\sigma * \sigma * \sigma)(\xi) = \begin{cases} 8\pi^2, & \textit{if } |\xi| \le 1, \\ 4\pi^2(\frac{3}{|\xi|} - 1), & \textit{if } 1 \le |\xi| \le 3. \end{cases}$$

Corollary 6. *We have* $\Phi_6[\mathbf{1}] = 2\pi\Phi_4[\mathbf{1}]$.

Indeed, by Lemma 5 we have

$$\Phi_4[\mathbf{1}] = (2\pi)^3 \|\mathbf{1}\|_{L^2(\mathbb{S}^2)}^{-4} \|\sigma * \sigma\|_2^2 = 16\pi^4,$$

$$\Phi_6[\mathbf{1}] = (2\pi)^3 \|\mathbf{1}\|_{L^2(\mathbb{S}^2)}^{-6} \|\sigma * \sigma * \sigma\|_2^2 = 32\pi^5.$$

3.1.6 End of proof of Theorem 3 when $(d, p) = (2, 6)$

By Step 1, it suffices to check that any non-constant *critical point* (i. e., an L^2-solution of the Euler–Lagrange equation (3.1)) $f : \mathbb{S}^2 \to \mathbb{C} \in C^1(\mathbb{S}^2)$ of Φ_6 satisfies $\Phi_6[f] < \Phi_6[\mathbf{1}]$. By Step 2, we may further assume that $f_\star = f$ is real-valued and that $\|f\|_{L^2} = 1$. From $T_f(f) = \lambda f$, one checks that $\lambda = \Phi_6[f]$. Thus, by Steps 3 and 4,

$$\Phi_6[f] = \lambda = \frac{1}{2}\left(\lambda + 5 \times \frac{\lambda}{5}\right) < \frac{1}{2}\mathrm{tr}(T_f) = 2\pi\|\widehat{f\sigma}\|_4^4, \tag{3.5}$$

where the strict inequality is a consequence of Lemma 4 together with the fact that all eigenvalues of T_f are strictly positive, and the last identity has been observed in (3.4). But

$$2\pi\|\widehat{f\sigma}\|_4^4 = 2\pi\Phi_4[f] \le 2\pi\Phi_4[\mathbf{1}] = \Phi_6[\mathbf{1}], \tag{3.6}$$

where the inequality follows from the result (2.1) of Foschi [40] reviewed in Section 2, and the last identity was seen in Step 5 (Corollary 6). From (3.5) and (3.6) it follows that $\Phi_6[f] < \Phi_6[\mathbf{1}]$, and this concludes the sketch of the proof of the case $(d, p) = (2, 6)$ of Theorem 3.

3.2 Higher dimensions and exponents

The special case $(d, p) = (2, 6)$ of Theorem 3 which we addressed in Section 3.1, while illustrative of the general scheme, relies on several crucial simplifications which made the proof sketch fit in just a few pages. In order to deal with general even exponents and different dimensions, further ideas and techniques are needed. These turn out to be broadly connected with the following areas:

- *Non-commutative algebra.* When trying to generalize Lemma 4 to higher dimensions, one is naturally led to the following question: What is the minimal codimension of a proper subalgebra of $\mathfrak{so}(d)$? The answer is known and reveals an interesting difference that occurs in the 4-dimensional case: the minimal codimension of a proper subalgebra of $\mathfrak{so}(d)$ equals $d-1$ if $d \geq 3$, $d \neq 4$, but equals 2 if $d = 4$. In group theoretic terms, the group $SO(4)/\{\pm I\}$ is *not* simple, whereas all other groups $SO(d)$ are simple (after modding out by $\{\pm I\}$ if d is even).
- *Combinatorial geometry.* When trying to extend the relevant estimates from Corollary 6 to the multilinear setting of $(p/2)$-fold spherical convolutions, one faces certain variants of the *cube slicing problem*: Given $0 < k < d$, what is the maximal volume of the intersection of the unit cube $[-\frac{1}{2}, \frac{1}{2}]^d$ with a k-dimensional subspace of $\mathbb{R}^d$? The cube slicing problem has been intensely studied, but a complete solution remains out of reach. Fortunately, the methods that have been developed for this problem can be adapted to fulfill our needs.
- *Analytic number theory.* The rather direct approach we took in Section 3.1.5 needs to be refined in order to tackle higher dimensions. Uniform random walks in $\mathbb{R}^d$ are lurking in the background and, despite being a classic topic in probability theory, a complete answer in even dimensions remains a fascinating, largely unsolved problem, which via the theory of hypergeometric functions and modular forms exhibits some deep connections to analytic number theory [18]. In view of this, we combined known formulas for uniform random walks with rigorous numerical integration and asymptotic analysis for a certain family of weighted integrals in order to complete our task.

3.3 Complex-valued maximizers

Once real-valued maximizers have been identified, one can proceed to characterize *all* complex-valued maximizers.

Theorem 7 ([67]). *Let $d \geq 1$ and let $p \geq 2 + \frac{4}{d}$ be an even integer. Then each complex-valued maximizer of the functional $\Phi_{d,p}$ is of the form*

$$ce^{i\xi \cdot \omega} F(\omega),$$

for some $\xi \in \mathbb{R}^d$, some $c \in \mathbb{C} \setminus \{0\}$, and some non-negative maximizer F of $\Phi_{d,p}$ satisfying $F(\omega) = F(-\omega)$, for every $\omega \in \mathbb{S}^d$.

Our next result is an immediate consequence of Theorems 3 and 7.

Corollary 8 ([67]). *Let $d \in \{2, 3, 4, 5, 6\}$ and let $p \geq 4$ be an even integer. Then all complex-valued maximizers of the functional $\Phi_{d,p}$ are given by*

$$f(\omega) = ce^{i\xi \cdot \omega},$$

for some $\xi \in \mathbb{R}^d$ and $c \in \mathbb{C} \setminus \{0\}$. The same conclusion holds for $d = 1$ and even integers $p \geq 8$, provided that constants maximize $\Phi_{1,6}$.

3.4 A sharp extension inequality on $\mathbb{S}^7$

The study of sharp *weighted* spherical extension estimates is linked to the question of stability of such estimates, and was very recently inaugurated in [23]. In particular, the sharp weighted extension inequality from [23, Theorem 1] leads to the following result, which is the first instance of a sharp extension inequality on $\mathbb{S}^7$.

Theorem 9 ([23]). *For every $a > \frac{2^{25}\pi^2}{5^2 7^2 11}$, the following sharp inequality holds:*

$$\int_{\mathbb{R}^8} |\widehat{f\sigma}(x)|^4 \, dx + a \left| \int_{\mathbb{S}^7} f(\omega) \, d\sigma(\omega) \right|^4 \leq \mathbf{W}_a \left(\int_{\mathbb{S}^7} |f(\omega)|^2 \, d\sigma(\omega) \right)^2, \tag{3.7}$$

with optimal constant given by

$$\mathbf{W}_a = \int_{\mathbb{R}^8} \hat{\sigma}(x)^4 \frac{dx}{\sigma(\mathbb{S}^7)^2} + a\sigma(\mathbb{S}^7)^2.$$

Equality in (3.7) *occurs if and only if f is constant on $\mathbb{S}^7$.*

We emphasize that constants are the unique *complex-valued* maximizers for (3.7), in contrast to the situation considered in Section 3.3. An interesting open problem is to lower the value of the threshold $\frac{2^{25}\pi^2}{5^2 7^2 11}$, hopefully all the way down to 0.

4 Agmon–Hörmander-type estimates

In this section, we describe some simple estimates for the extension operator on spheres which, perhaps surprisingly, are *not always* maximized by constants. For simplicity we restrict our attention to the circle $\mathbb{S}^1 \subset \mathbb{R}^2$. However, analogous results have been recently proved in all dimensions $d \geq 1$; see [59].

Our starting point is the Agmon–Hörmander estimate on the circle,

$$\frac{1}{R} \int_{B_R} |\widehat{f\sigma}(x)|^2 \frac{dx}{(2\pi)^2} \leq \mathbf{C}_R \int_{\mathbb{S}^1} |f(\omega)|^2 d\sigma(\omega), \tag{4.1}$$

where $B_R \subset \mathbb{R}^2$ denotes a ball of arbitrary radius $R > 0$ centered at the origin and σ stands for the usual arc length measure on $\mathbb{S}^1$. Agmon and Hörmander [1] observed that (4.1) holds with a constant C_R that approaches $\frac{1}{\pi}$ as $R \to \infty$, but did not investigate its optimal value. In Theorem 10 below, such optimal value is obtained in terms of the auxiliary quantities

$$\Lambda_R^k := \frac{R}{2}J_k^2(R) - \frac{R}{2}J_{k-1}(R)J_{k+1}(R), \tag{4.2}$$

where J_n denotes the usual Bessel function; see also Figure 3.

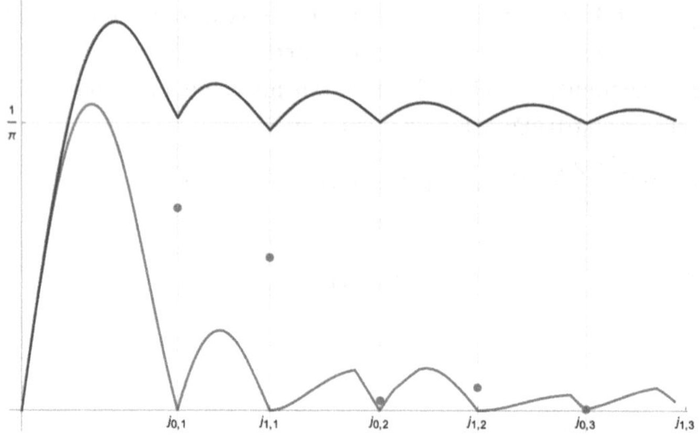

Figure 3: Optimal constant C_R (blue) and stability constant S_R (red) for the Agmon–Hörmander estimate on the circle when $0 < R < 10$.

Theorem 10 ([59]). *For each $R > 0$,*

$$C_R = \begin{cases} \Lambda_R^0, & \text{if } (J_0 J_1)(R) \geq 0, \\ \Lambda_R^1, & \text{if } (J_0 J_1)(R) \leq 0. \end{cases}$$

The corresponding space of maximizers is given by

$$\mathcal{M}_R = \begin{cases} \mathcal{H}_0, & \text{if } (J_0 J_1)(R) > 0, \\ \mathcal{H}_1, & \text{if } (J_0 J_1)(R) < 0, \\ \mathcal{H}_0 \oplus \mathcal{H}_1, & \text{if } J_0(R) = 0, \\ \mathcal{H}_0 \oplus \mathcal{H}_1 \oplus \mathcal{H}_2, & \text{if } J_1(R) = 0, \end{cases}$$

where $\mathcal{H}_k \subset L^2(\mathbb{S}^1)$ denotes the vector space of degree k circular harmonics.

One remarkable feature of Theorem 10 is that constants are seen to not always be maximizers, even though they are the unique functions which are invariant under the full rotational symmetry group of (4.1). In this way Theorem 10 identifies an instance of *symmetry breaking*, which to the best of our knowledge had not yet been observed for an estimate involving the spherical extension operator.

The next natural question concerns the *stability* of inequality (4.1), which can be phrased in terms of lower bounds for the following *deficit functional*:

$$\delta_R[f] := \mathbf{C}_R \|f\|_{L^2(\mathbb{S}^1)}^2 - \frac{1}{R} \int_{B_R} |\widehat{f\sigma}(x)|^2 \, \frac{dx}{(2\pi)^2}.$$

Clearly, $\delta_R[f] \geq 0$ for every $f \in L^2(\mathbb{S}^1)$, but, in the spirit of Bianchi–Egnell [16], more can be said; see also Figure 3 and recall the space $\mathcal{M}_R$, which has been defined in Theorem 10.

Theorem 11 ([59]). *The following sharp two-sided inequality holds:*

$$\mathbf{S}_R \operatorname{dist}^2(f, \mathcal{M}_R) \leq \delta_R[f] \leq \mathbf{C}_R \operatorname{dist}^2(f, \mathcal{M}_R). \tag{4.3}$$

Equality occurs in the right-hand side of inequality (4.3) if and only if $f \in \mathcal{M}_R$. Equality occurs in the left-hand side of inequality (4.3) if and only if $f \in \mathcal{M}_R \oplus \mathcal{E}_R$, where:

$\mathbf{S}_R =$	$\mathcal{E}_R =$	if	and
$\Lambda_R^0 - \Lambda_R^1$	$\mathcal{H}_1$		$(J_1 J_2)(R) > 0$
$=$	$\mathcal{H}_1 \oplus \mathcal{H}_2 \oplus \mathcal{H}_3$	$(J_0 J_1)(R) > 0$	$(J_1 J_2)(R) = 0$
$\Lambda_R^0 - \Lambda_R^2$	$\mathcal{H}_2$		$(J_1 J_2)(R) < 0$
$\Lambda_R^1 - \Lambda_R^0$	$\mathcal{H}_0$		$(J_0 J_1 + J_1 J_2 + J_2 J_3)(R) > 0$
$=$	$\mathcal{H}_0 \oplus \mathcal{H}_3$	$(J_0 J_1)(R) < 0$	$(J_0 J_1 + J_1 J_2 + J_2 J_3)(R) = 0$
$\Lambda_R^1 - \Lambda_R^3$	$\mathcal{H}_3$		$(J_0 J_1 + J_1 J_2 + J_2 J_3)(R) < 0$
$\Lambda_R^0 - \Lambda_R^2$	$\mathcal{H}_2$	$J_0(R) = 0$	$(J_2 J_3)(R) > 0$
$\Lambda_R^0 - \Lambda_R^3$	$\mathcal{H}_3$		$(J_2 J_3)(R) < 0$
$\Lambda_R^0 - \Lambda_R^3$	$\mathcal{H}_3$	$J_1(R) = 0$	$(J_3 J_4)(R) > 0$
$\Lambda_R^0 - \Lambda_R^4$	$\mathcal{H}_4$		$(J_3 J_4)(R) < 0$

The proofs of Theorems 10 and 11 rely on two observations. Firstly, by orthogonality of the circular harmonic decomposition $f = \sum_{k \geq 0} Y_k$, we have

$$\frac{1}{R} \int_{B_R} |\widehat{f\sigma}(x)|^2 \, \frac{dx}{(2\pi)^2} = \sum_{k \geq 0} \Lambda_R^k \|Y_k\|_{L^2}^2,$$

where Λ_R^k have been defined in (4.2). Secondly, $\mathbf{C}_R = \sup_{k \geq 0} \Lambda_R^k$, where f attains the supremum if and only if $f = \sum Y_{k_j}$, for some $k_j \in \{k \geq 0 : \Lambda_R^k = \sup_h \Lambda_R^h\}$; see also [15]. In our case of interest, we can conveniently rewrite Λ_R^k in integral form,

$$\Lambda_R^k = \frac{1}{R} \int_0^R J_k^2(r) r \, dr,$$

and invoke certain well-known Bessel recursions to start gaining control on both extremal problems corresponding to Theorems 10 and 11.

The above explicit expression for the optimal constant $\mathbf{C}_R$ leads to the following *loss-of-regularity*[2] statement: $\mathbf{C}_R$ is *not* a differentiable function of R at each positive zero of $J_0 J_1$, but it defines a Lipschitz function on $(0, \infty)$ which is real analytic between any two consecutive zeros of $J_0 J_1$. We note that such zeros precisely correspond to those values of R at which a jump in the dimension of the space of maximizers $\mathcal{M}_R$ is observed.

The behavior of the stability constant $\mathbf{S}_R$ is also interesting. Since the deficit functional $\delta_R[f]$ clearly defines a continuous function of R, the quantity $\mathbf{S}_R$ must have a jump discontinuity at the positive zeros of $J_1 J_2$, where $\mathrm{dist}(f, \mathcal{M}_R)$ likewise jumps. Moreover, the explicit expression for $\mathbf{S}_R$ can be used to establish that it defines a piecewise real analytic function of R between any two consecutive zeros of $J_0 J_1$, which fails to be differentiable at each positive zero of J_2.

5 Cones

We consider for $d \geq 2$ the two-sheeted cone

$$\mathbb{K}^d := \{(\tau, \xi) \in \mathbb{R} \times \mathbb{R}^d : \tau^2 = |\xi|^2\}.$$

It is the conical section (1.1) with $\alpha = 0$, $\beta = 1$, and $\gamma = 0$. The measure σ as in (1.2) becomes

$$\int_{\mathbb{K}^d} f \, d\sigma = \int_{\mathbb{R}^{d+1}} f(\tau, \xi) \delta(\tau^2 - |\xi|^2) d\tau d\xi.$$

We split $\sigma = \sigma_+ + \sigma_-$ via

$$\delta(\tau^2 - |\xi|^2) = \frac{\delta(\tau - |\xi|)}{2|\xi|} + \frac{\delta(\tau + |\xi|)}{2|\xi|}. \tag{5.1}$$

This follows from $\tau^2 - |\xi|^2 = (\tau - |\xi|)(\tau + |\xi|)$ by formal manipulations of delta calculus; see [42, Appendix A]. The expressions of $\sigma_\pm$ involve the singular term $1/|\xi|$, but this is singular on a set of measure zero and it is locally integrable. We conclude that $\sigma_\pm$ are well-defined, and so is σ, with explicit formulas

---

2 Similar phenomena have been recently observed in the related setting of the Brascamp–Lieb inequalities [6, 8].

$$\int_{\mathbb{K}^d} g(\tau,\xi)\,d\sigma_\pm = \int_{\mathbb{R}^d} \frac{g(\pm|\xi|,\xi)}{2|\xi|}\,d\xi.$$

By the Strichartz estimates [84], the extension operator $\mathcal{E}f := \widehat{f\sigma}$ defines a bounded linear map from $L^2(\sigma)$ into $L^p(\mathbb{R}^{1+d})$ for $p = 2\frac{d+1}{d-1}$. This is the only value of p for which such boundedness can hold, due to the scaling symmetry $\sigma_\pm(\lambda\tau,\lambda\xi) = |\lambda|^{-2}\sigma_\pm(\tau,\xi)$. Among all conic sections, only paraboloids have similar scaling symmetries as the cones. The phenomenon of such scaling does not occur for spheres or hyperboloids discussed in this survey.

We denote the conjectured maximizer of (1.3) by

$$f_\star(\tau,\xi) := e^{-|\tau|}.$$

We will study whether $f_\star$ is a maximizer for $\|\mathcal{E}f\|^p_{L^p(\mathbb{R}^{1+d})}\|f\|^{-p}_{L^2(\sigma)}$ when $p = 2\frac{d+1}{d-1}$. This is the $\mathbb{K}^d$-version of problem (1.5) from the introduction.

We will repeatedly use the property that $u(t,x) = \mathcal{E}f(t,x)$ is a solution of the wave equation. Indeed, by differentiating

$$u(t,x) = \int_{\mathbb{K}^d} f(\tau,\xi)e^{-i(t\tau+x\cdot\xi)}\,d(\sigma_+ + \sigma_-)(\tau,\xi),$$

we see that u satisfies

$$\partial_t^2 u = \Delta u, \tag{5.2}$$

with initial data

$$\widehat{u}(0,\xi) = \frac{f(|\xi|,\xi) + f(-|\xi|,\xi)}{2|\xi|}, \quad \partial_t\widehat{u}(0,\xi) = \frac{-if(|\xi|,\xi) + if(-|\xi|,\xi)}{2}, \tag{5.3}$$

where we have used the spatial Fourier transform $\widehat{v}(t,\xi) = \int_{\mathbb{R}^d} v(t,x)e^{-ix\cdot\xi}\,dx$. In particular, the conjectured maximizer extension $u_\star = \mathcal{E}f_\star$ satisfies $\widehat{u}_\star(0,\xi) = e^{-|\xi|}/|\xi|$ and $\partial_t\widehat{u}_\star(0,\xi) = 0$. An explicit computation reveals that, for some $C_d > 0$,

$$u_\star(0,x) = C_d(1+|x|^2)^{\frac{1-d}{2}}, \quad \partial_t u_\star(0,x) = 0. \tag{5.4}$$

5.1 Criticality of the conjectured maximizer via the Penrose transform

This subsection is based on [57]. Define an injective map from $(t,r) \in \mathbb{R}^2$ to $(T,R) \in \mathbb{R}^2$ by

$$T = \arctan(t+r) + \arctan(t-r), \quad R = \arctan(t+r) - \arctan(t-r).$$

This is essentially the composition of a rotation by $\pi/4$ of $\mathbb{R}^2$, a componentwise arctangent, and the inverse rotation.

We use this to define an injective map $\mathcal{P}$ from the Minkowski spacetime $\mathbb{R}^{1+d}$ into $[-\pi, \pi] \times \mathbb{S}^d$ as follows. We consider a generic $(t, x) \in \mathbb{R}^{1+d}$ in polar coordinates (t, r, ω), with $r = |x|$ and $\omega = x/r$, and then map this to $(T, \cos R, \omega \sin R)$. The last two components of the latter define a point $X = (\cos R, \omega \sin R)$ on the sphere $\mathbb{S}^d$, as claimed; see Figure 4.

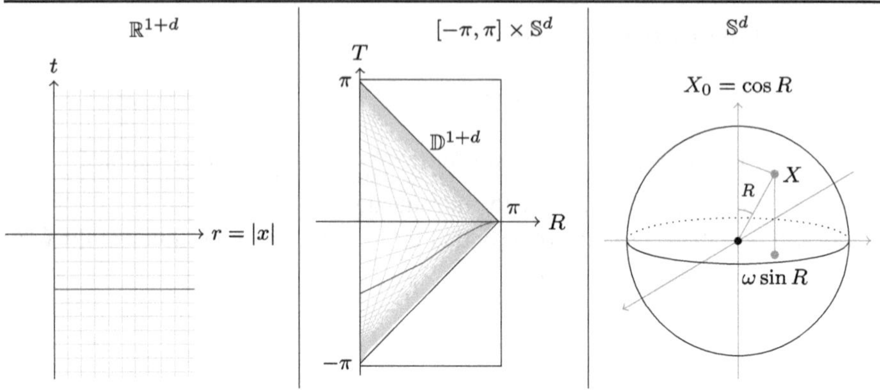

Figure 4: The map $\mathcal{P}$ of $\mathbb{R}^{1+d}$ (left) onto the Penrose diamond $\mathbb{D}^{1+d} \subset [-\pi, \pi] \times \mathbb{S}^d$ (center). Here $R \in [0, \pi]$ denotes the radial coordinate on $\mathbb{S}^d$ (right).

The map $\mathcal{P}$ was first introduced by Penrose [74], and its range is the following open submanifold of $[-\pi, \pi] \times \mathbb{S}^d$, known as the *Penrose diamond*:

$$\mathbb{D}^{1+d} := \{(T, \cos R, \omega \sin R) : \omega \in \mathbb{S}^{d-1}, R < \pi - |T|\}. \tag{5.5}$$

The $(d + 1)$-dimensional volume element of $\mathbb{D}^{1+d}$ reads $(\sin R)^{d-1}\mathrm{d}T\mathrm{d}R\mathrm{d}S(\omega)$, where $\mathrm{d}S(\omega)$ denotes the surface measure on $\mathbb{S}^{d-1}$. Similarly, the volume element of $\mathbb{R}^{1+d}$ reads $r^{d-1}\mathrm{d}t\mathrm{d}r\mathrm{d}S(\omega)$. We compute the pushforward of this volume element via $\mathcal{P}$ as

$$(\sin R)^{d-1}\mathrm{d}T\mathrm{d}R\mathrm{d}S(\omega) = 2^{d+1}\left[(1 + (t + r)^2)(1 + (t - r)^2)\right]^{-\frac{1+d}{2}} r^{d-1}\mathrm{d}t\mathrm{d}r\mathrm{d}S(\omega). \tag{5.6}$$

An important feature of $\mathcal{P}$ is that it is conformal, meaning that

$$\mathrm{d}T^2 - \mathrm{d}R^2 - (\sin R)^2\mathrm{d}\omega^2 = 4\left[(1 + (t + r)^2)(1 + (t - r)^2)\right]^{-1}(\mathrm{d}t^2 - \mathrm{d}r^2 - r^2\mathrm{d}\omega^2).$$

Therefore, we can construct solutions to the wave equation in $\mathbb{R}^{1+d}$ by pulling back solutions to the conformal wave equation

$$\partial_T^2 U = \Delta_{\mathbb{S}^d} U - \frac{(d - 1)^2}{4} U$$

in $\mathbb{D}^{1+d}$; for details we refer to [88, p. 139]. As a consequence, for each spherical harmonic $Y_n = Y_n(X)$ on $\mathbb{S}^d$ of degree n, the function

$$u(t,x) = \left[(1+(t+r)^2)(1+(t-r)^2)\right]^{\frac{1-d}{4}} e^{-iT(n+\frac{d-1}{2})} Y_n(\cos R, \omega \sin R) \tag{5.7}$$

satisfies the wave equation (5.2).

By the uniqueness of solutions to the wave equation, we see that $u = \mathcal{E}f$ for $f \in L^2(\sigma)$, uniquely identified by solving the linear initial conditions (5.3) for f. Taking a spherical harmonic of degree $n = 0$, i. e., a constant function on $\mathbb{S}^d$, and considering the real part only, we obtain a new representation for the conjectured maximizer $u_\star = \mathcal{E}f_\star$. Precisely,

$$u_\star(t,x) = C_d\left[(1+(t+r)^2)(1+(t-r)^2)\right]^{\frac{1-d}{4}} \cos\left(T\left(\frac{d-1}{2}\right)\right),$$

since this last expression is easily seen to satisfy (5.4).

Defining $C_\star$ via

$$\|\mathcal{E}f_\star\|_{L^2\frac{d+1}{d-1}(\mathbb{R}^{1+d})} = C_\star \|f_\star\|_{L^2(\sigma)}, \tag{5.8}$$

we recall that the conjecture we are studying reads as

$$\|\mathcal{E}f\|_{L^2\frac{d+1}{d-1}(\mathbb{R}^{1+d})} \le C_\star \|f\|_{L^2(\sigma)} \tag{5.9}$$

for all f. We show that this conjecture is false in even dimension by perturbing $f_\star$ into $f_\star + \epsilon f$ on each side of (5.8) and showing that for small ϵ the left-hand side of (5.9) varies of linear order in ϵ while the right-hand side varies of quadratic order in ϵ; compare with Theorem 12 below. Therefore, by choosing the sign of small enough ϵ appropriately, one can create either strict inequality between both sides, therefore contradicting (5.9).

We let $u = \mathcal{E}f$ as in (5.7) with $n > 0$. One can see that f is orthogonal to $f_\star$ using orthogonality of the spherical harmonics 1 and Y_n and the isometries[3]

$$\|f\|_{L^2(\sigma)}^2 = \|u(0,.)\|_{\dot{H}^{1/2}(\mathbb{R}^d)}^2 + \|\partial_t u(0,.)\|_{\dot{H}^{-1/2}(\mathbb{R}^d)}^2$$
$$= \|Y_n\|_{\dot{H}^{1/2}(\mathbb{S}^d)}^2 + \|Y_n\|_{\dot{H}^{-1/2}(\mathbb{S}^d)}^2,$$

where the first identity follows from (5.3) and the second one is based on a computation with fractional integrals from conformal theory; see [56, equation (2)] and (5.20) below. The orthogonality between f and $f_\star$ established the quadratic behavior in ϵ of the right-hand side of (5.9). To verify the linear behavior in ϵ of the left-hand side, we choose Y_n to be a real spherical harmonic and compute its first variation (up to a non-zero scalar constant) as

---

3 Here we define $\|F\|_{H^s(\mathbb{S}^d)}^2 = \int_{\mathbb{S}^d} |(\frac{(d-1)^2}{4} - \Delta_{\mathbb{S}^d})^{s/2} F|^2 dS$.

$$\mathbb{R} \int_{\mathbb{R}^{1+d}} |\mathcal{E}f_\star|^{p-2}\overline{\mathcal{E}f_\star}\,\mathcal{E}f. \tag{5.10}$$

With (5.6) and (5.7), we obtain for (5.10)

$$2^{d+1} \int_{\mathbb{D}^{1+d}} \left|\cos\left(\frac{d-1}{2}T\right)\right|^{p-2} \cos\left(\frac{d-1}{2}T\right) \cos\left(T\left(n+\frac{d-1}{2}\right)\right) Y_n(X). \tag{5.11}$$

Obviously but crucially, $\frac{d-1}{2}$ is an integer if and only if d is odd. If that is the case, then from the formula $Y_n(-X) = (-1)^n Y_n(X)$ we see that the integrand $U(T,X)$ of (5.11) satisfies

$$U(T+\pi,-X) = (-1)^{\frac{d-1}{2}} U(T,X). \tag{5.12}$$

Recalling the definition (5.5) of $\mathbb{D}^{1+d}$, it follows from this symmetry that

$$\int_{\mathbb{D}^{1+d}} U(T,\cos R, \omega\sin R) = \frac{(-1)^{\frac{d-1}{2}}}{2} \int_{-\pi}^{\pi}\int_{\mathbb{S}^d} U(T,\cos R, \omega\sin R)(\sin R)^{d-1}\,\mathrm{d}T\,\mathrm{d}R\,\mathrm{d}S(\omega)$$

$$= \frac{(-1)^{\frac{d-1}{2}}}{2} \int_{-\pi}^{\pi} \left|\cos\left(\frac{d-1}{2}T\right)\right|^{p-2} \cos\left(\frac{d-1}{2}T\right)\cos\left(T\left(n+\frac{d-1}{2}\right)\right)\,\mathrm{d}T \int_{\mathbb{S}^d} Y_n = 0.$$

We used the fact that the very last integral over Y_n vanishes because $n > 0$. This suggests that $f_\star$ is a critical point. A refined version of this argument even proves that $f_\star$ is a local maximizer of $\|\mathcal{E}f\|^p_{L^p(\mathbb{R}^{1+d})}\|f\|^{-p}_{L^2(\sigma)}$; see [47, Theorem 1.1].

However, when d is even, the symmetry (5.12) fails. In this case we can take an explicit spherical harmonic of degree 2 and compute the variation (5.10), which turns out to never vanish for every even d. We present a similar calculation in the next subsection. This proves that, in this case, $f_\star$ cannot be a maximizer.

Interestingly, Ramos [78] proved that maximizers exist for $\mathcal{E}$ in arbitrary $d \geq 2$. So this is an instance where maximizers exist but they are not of the type suggested by the title of this survey.

5.2 Related problems on one-sheeted cones

We consider now the one-sheeted cone

$$\mathbb{K}_+^d := \{(\tau,\xi) \in \mathbb{R}\times\mathbb{R}^d : \tau = |\xi|\}.$$

This is not a conic section in the strict sense discussed in the introduction, but it is a very natural subset of the two-sheeted cone $\mathbb{K}^d$.

We further deviate from the strict setup by considering the more general family of measures supported on $\mathbb{K}_+^d$,

$$\sigma_s(\tau, \xi) := \frac{\delta(\tau - |\xi|)}{|\xi|^{2s}}, \quad s \in \left[\frac{1}{2}, \frac{d}{2}\right). \tag{5.13}$$

The measure σ_+ introduced in (5.1) corresponds to the case $s = 1/2$ of (5.13).

By a standard interpolation of the aforementioned Strichartz estimates with Sobolev embeddings, as in [37], the Fourier extension operator $\mathcal{E}_s f := \widehat{f\sigma_s}$ is bounded from $L^2(\sigma_s)$ into $L^{p_s}(\mathbb{R}^{1+d})$, where

$$p_s := 2\frac{d+1}{d-2s}.$$

To compare with the notation of [13, 37], note that

$$\mathcal{E}_s f(-t, -x) = (2\pi)^d e^{it\sqrt{-\Delta}} u_s(x) = \int_{\mathbb{R}^d} e^{it|\xi|+ix\cdot\xi} \hat{u}_s(\xi)\, d\xi, \quad \hat{u}_s(\xi) := \frac{f(|\xi|, \xi)}{|\xi|^{2s}}.$$

In [13], it is conjectured that

$$f_\star^{(s)}(\tau, \xi) := |\tau|^{2s-1} e^{-|\tau|}$$

is a maximizer of

$$\|\mathcal{E}_s f\|_{L^{p_s}(\mathbb{R}^{1+d})}^{p_s} \|f\|_{L^2(\sigma_s)}^{-p_s} \tag{5.14}$$

if and only if

$$s \in \left\{\frac{1}{2}, \frac{d-1}{4}\right\}.$$

Note that $s = (d-1)/4$ is equivalent to $p_s = 4$. This conjecture is open, but some partial results are available.

In the case $s = 1/2$, Foschi [39] proved that $f_\star^{(1/2)}$ is a maximizer for $d \in \{2, 3\}$. In the same case, for all $d \geq 2$, in [47] it is shown that $f_\star^{(1/2)}$ is a local maximizer. This is analogous to the situation for odd d on the two-sheeted cone. However, in the present case of the one-sheeted cone there is no distinction between odd and even d. We summarize the status of the $s = 1/2$ case of this conjecture in Table 1, also for the case of the two-sheeted cone $\mathbb{K}^d$.

Table 1: Is $e^{-|\tau|} = f_\star^{(1/2)}(\tau, \xi)$ a maximizer of the Fourier extension on $\mathbb{K}^d$ and $\mathbb{K}_+^d$ when $s = 1/2$?

Spatial dimension d	$\mathbb{K}^d$	$\mathbb{K}_+^d$
2	No	Yes
3	Yes	Yes
$4, 6, 8, \ldots$	No	Local
$5, 7, 9, \ldots$	Local	Local

In the case $s = (d-1)/4$, in [14] it is proved that $f_\star^{(s)}$ is a maximizer for $d = 5$. In the same case, for $d \geq 2$, in [12] it is shown that $f_\star^{(s)}$ is a maximizer among all radially symmetric functions. We note that the $d = 5$ case of the two-sheeted cone has also been treated in [14] and further extended to a stability inequality in [58].

Finally, in the case $s \notin \{1/2, (d-1)/4\}$, in [13] it is proved that $f_\star$ is not a maximizer provided that p_s is an even integer.

We conclude with an original result, showing how the Penrose transform can be applied to settle the $s = 1$ case of the previous conjecture. This extends [13, Theorem 2.1] because we do not require p_1 to be even.

Theorem 12. *Let $d \geq 3, d \neq 5$. There exists $f \in L^2(\sigma_1)$ such that*

$$\left\| f_\star^{(1)} + \epsilon f \right\|_{L^2(\sigma_1)} - \left\| f_\star^{(1)} \right\|_{L^2(\sigma_1)} = O(\epsilon^2), \tag{5.15}$$

while, for some $C \neq 0$,

$$\left\| \mathcal{E}_1\big(f_\star^{(1)} + \epsilon f\big) \right\|_{L^{2\frac{d+1}{d-2}}(\mathbb{R}^{1+d})} - \left\| \mathcal{E}_1 f_\star^{(1)} \right\|_{L^{2\frac{d+1}{d-2}}(\mathbb{R}^{1+d})} = C\epsilon + o(\epsilon). \tag{5.16}$$

As before, it follows from Theorem 12 that $f_\star^{(1)}$ cannot be a maximizer of (5.14) for $s = 1$.

Proof of Theorem 12. As in the previous case of the two-sheeted cone, $u(t,x) = \mathcal{E}_1 f(t,x)$ satisfies the wave equation (5.2). The initial data $u(0,x), \partial_t u(0,x)$ are related by the support assumption on f on the one-sheeted cone and read as follows:

$$\hat{u}(0,\xi) = \frac{f(|\xi|,\xi)}{|\xi|^{2s}}, \tag{5.17}$$

$$\partial_t \hat{u}(0,\xi) = -i|\xi|\hat{u}(0,\xi), \tag{5.18}$$

where we used the spatial Fourier transform $\hat{v}(t,\xi) = \int_{\mathbb{R}^d} v(t,x)e^{-ix\cdot\xi}\,dx$.

As in the previous subsection, we consider $u(t,x)$ given by (5.7), which has already been seen to solve the wave equation. We claim that it also satisfies condition (5.18). Indeed, we see by direct computation that

$$u(0,x) = (1+r^2)^{\frac{1-d}{2}} Y_n(X), \quad \partial_t u(0,x) = -2i\left(n + \frac{d-1}{2}\right)(1+r^2)^{-\frac{1+d}{2}} Y_n(X),$$

where $X = (\cos R, \omega \sin R)$. Now notice that the claimed condition can be equivalently written as

$$\partial_t u(0,\cdot) = -i\sqrt{-\Delta}u(0,\cdot). \tag{5.19}$$

By the aforementioned conformal formula [56, equation (2)],

$$\sqrt{-\Delta}u(0,x) = 2(1+r^2)^{-\frac{1+d}{2}}\left(\frac{(d-1)^2}{4} - \Delta_{\mathbb{S}^d}\right)^{\frac{1}{2}} Y_n(X). \tag{5.20}$$

Recalling $-\Delta_{\mathbb{S}^d} Y_n = n(n + d - 1)Y_n$, which implies

$$\left(\frac{(d-1)^2}{4} - \Delta_{\mathbb{S}^d}\right)^{\frac{1}{2}} Y_n = \left(n + \frac{d-1}{2}\right)Y_n,$$

condition (5.19) then follows immediately.

Now we notice that the conjectured maximizer $u_\star := \mathcal{E}_1 f_\star^{(1)}$ is such that

$$\hat{u}_\star(0, \xi) = \frac{e^{-|\xi|}}{|\xi|},$$

so the same computation as in (5.4) yields

$$u_\star(0, x) = C_d(1 + |x|^2)^{\frac{1-d}{2}},$$

for some $C_d > 0$. By the uniqueness of solutions to the wave equation we conclude that

$$u_\star(t, x) = C_d[(1 + (t + r)^2)(1 + (t - r)^2)]^{\frac{1-d}{4}} e^{-iT\frac{d-1}{2}}.$$

Now consider u with $n = 2$, where we choose Y_2 in (5.7) to be

$$Y_2(X_0, \ldots, X_d) = (1 - X_0^2)^{\frac{2-d}{2}} \frac{\mathrm{d}^2}{\mathrm{d}X_0^2}[(1 - X_0)^{2+\frac{d-2}{2}}].$$

By the Rodrigues formula, this is indeed a spherical harmonic of degree 2. We let f be the corresponding function on the one-sheeted cone defined by the initial condition (5.17). The same argument as in the previous subsection shows that this f just constructed is orthogonal to $f_\star^{(1)}$, from which the quadratic variation (5.15) immediately follows.

On the other hand, the left-hand side of (5.16) reads to first order in ϵ

$$\epsilon\mathbb{R} \int_{\mathbb{R}^{1+d}} |u_\star|^{2\frac{d+1}{d-2}-2}\overline{u_\star}u \, dtdx$$

up to a non-zero scalar constant. Changing variables via $\mathcal{P}$, this is seen to equal

$$\epsilon\mathbb{R} \int_{\mathbb{D}^{1+d}} |\cos T + X_0|^{\frac{d+1}{d-2}} e^{i2T} Y_2(X)(1 - X_0^2)^{\frac{d-2}{2}} \, dTdX_0 dS(\omega). \tag{5.21}$$

Since Y_2 depends on X only via X_0, the integral in $dS(\omega)$ is constant. The integrand $U(T, X_0)$ clearly satisfies $U(T + \pi, -X_0) = U(T, X_0)$. So, arguing as in the previous subsection, we conclude that (5.21) equals

$$\frac{C}{2} \int_{-\pi}^{\pi} \int_{-1}^{1} |\cos T + X_0|^{\frac{d+1}{d-2}} \cos(2T) \frac{\mathrm{d}^2}{\mathrm{d}X_0^2}[(1 - X_0^2)^{2+\frac{d-2}{2}}] \, dX_0 dT. \tag{5.22}$$

It remains to prove that this integral does not vanish. To do so, let

$$h_d(\cos T) := \int_{-1}^{1} |\cos T + X_0|^{\frac{d+1}{d-2}} \frac{d^2}{dX_0^2}\left[(1 - X_0^2)^{2+\frac{d-2}{2}}\right] dX_0.$$

By partial integration, we see that

$$h_d(\cos T) = \int_{-1}^{1} |\cos T + X_0|^{\frac{5-d}{d-2}} (1 - X_0^2)^{2+\frac{d-2}{2}} dX_0, \tag{5.23}$$

up to a positive scalar constant.

We pause to compare this with the computation needed in the last paragraph of the previous subsection. There we did not have the weight $|\cos T + X_0|^{\frac{d+1}{d-2}}$, but we did not have the symmetry $U(T + \pi, -X_0) = U(T, X_0)$ either. Therefore, the partial integration would not result in a term like (5.23) but instead yield some boundary terms due to the more complicated region of integration.

Note that h_d is a constant function for $d = 5$, which is the only case left out of the statement of Theorem 12. In this case, (5.22) is readily seen to vanish, as it reduces to a constant multiple of the integral of $\cos(2T)$ over a full period.

We claim that $h_d = h_d(y)$ defines a *strictly increasing* function of $y \in (0,1)$ if $d \in \{3,4\}$ and a *strictly decreasing* function of $y \in (0,1)$ if $d \geq 6$. Define variables $x := X_0, t := x + y$. This yields

$$h_d(y) = \int_{y-1}^{y+1} (1 - (t - y)^2)^{\frac{d+2}{2}} |t|^{\frac{5-d}{d-2}} dt.$$

Differentiating under the integral sign and using the fact that the integrand vanishes at the boundary,

$$\frac{h_d'(y)}{d+2} = \int_{y-1}^{y+1} (t - y)(1 - (t - y)^2)^{\frac{d}{2}} |t|^{\frac{5-d}{d-2}} dt = \int_{-1}^{1} x(1 - x^2)^{\frac{d}{2}} |y + x|^{\frac{5-d}{d-2}} dx.$$

Breaking up the region of integration into two and changing x to $-x$ in one of them reveals that the last displayed formula equals

$$\int_{0}^{1} x(1 - x^2)^{\frac{d}{2}} \left(|y + x|^{\frac{5-d}{d-2}} - |y - x|^{\frac{5-d}{d-2}}\right) dx.$$

Since $|y + x| > |y - x| > 0$ for every $(y, x) \in (0,1)^2$, the bracketed term is negative for $d > 5$ and positive for $d \in \{3,4\}$. This proves the claim.

We can finally conclude that (5.22) is non-zero. Indeed, the integrand in (5.22) satisfies $U(-T,X) = U(T,X) = U(\pi - T, -X)$, and hence integration in T can be taken over the interval $[0, \pi/2]$. So (5.22) equals, with partial integration,

$$\int_0^{\pi/2} \cos(2T)h_d(\cos T)\, dT = \frac{1}{2}\int_0^{\pi/2} \sin(2T)\sin(T)h_d'(\cos T)\, dT,$$

and the right-hand integral has a definite sign. This concludes the proof of Theorem 12. $\qquad\square$

6 Hyperboloids

Hyperboloids locally look like spheres, with largest curvature at the origin, and globally resemble cones. As such, sharp restriction theory on hyperboloids shares features from both spheres and cones and serves as a natural bridge between Sections 2–4 and 5. On the other hand, genuinely new phenomena emerge, as we shall see. Consider the upper sheet of the two-sheeted hyperboloid

$$\mathbb{H}^d = \{(\tau, \xi) \in \mathbb{R} \times \mathbb{R}^d : \tau = \langle \xi \rangle\},$$

where $\langle \xi \rangle := \sqrt{1 + |\xi|^2}$, corresponding to the case $(\alpha, \beta, \gamma) = (0, 1, 1)$ of (1.1), and equip it with the Lorentz invariant measure

$$d\sigma(\tau, \xi) = \delta(\tau - \langle \xi \rangle)\frac{d\tau d\xi}{\langle \xi \rangle} \tag{6.1}$$

corresponding to (1.2). The extension operator on $\mathbb{H}^d$ is given by

$$\mathcal{E}f(t, x) = \int_{\mathbb{R}^d} e^{i(t,x)\cdot(\langle \xi \rangle, \xi)}f(\xi)\frac{d\xi}{\langle \xi \rangle},$$

and since the 1977 work of Strichartz [84] it is known that

$$\|\mathcal{E}f\|_{L^p(\mathbb{R}^{1+d})} \leq \mathbf{H}_{d,p}\|f\|_{L^2(\mathbb{H}^d)} \tag{6.2}$$

as long as[4]

$$2 + \frac{4}{d} \leq p \leq 2 + \frac{4}{d-1} \quad \text{if } d \geq 1. \tag{6.3}$$

---

4 With the caveat that the endpoint case $p = \infty$ has to be excluded when $d = 1$.

420 —— G. Negro et al.

Note that the endpoints in the latter range correspond to the endpoint exponents for the sphere and cone, respectively; recall Figure 1. $\mathbf{H}_{d,p}$ denotes the optimal constant in inequality (6.2) and, according to (6.1), $\|f\|^2_{L^2(\mathbb{H}^d)} = \int_{\mathbb{R}^d} |f(\xi)|^2 \frac{d\xi}{\langle \xi \rangle}$. The extension operator on $\mathbb{H}^d$ naturally relates to the Klein–Gordon equation, $\partial_t^2 u = \Delta u - u$. This connection comes via the Klein–Gordon propagator,

$$e^{it\sqrt{1-\Delta}}g(x) = \frac{1}{(2\pi)^d} \int_{\mathbb{R}^d} e^{i(t,x)\cdot(\langle\xi\rangle,\xi)}\widehat{g}(\xi)d\xi,$$

together with the observation that $\mathcal{E}f(t,x) = (2\pi)^d e^{it\sqrt{1-\Delta}}g(x)$ as long as $\widehat{g}(\xi) = \langle\xi\rangle^{-1}\widehat{f}(\xi)$. This relation implies that estimate (6.2) can be equivalently rewritten as

$$\left\|e^{it\sqrt{1-\Delta}}g\right\|_{L^p_{t,x}(\mathbb{R}\times\mathbb{R}^d)} \leq (2\pi)^{-d}\mathbf{H}_{d,p}\|g\|_{H^{1/2}(\mathbb{R}^d)},$$

where $\|\cdot\|_{H^{1/2}(\mathbb{R}^d)}$ denotes the usual non-homogeneous Sobolev norm.

Quilodrán [76] investigated some sharp instances of inequality (6.2), all corresponding to algebraic endpoints of the range (6.3), and proved that

$$\mathbf{H}_{2,4} = 2^{3/4}\pi, \quad \mathbf{H}_{2,6} = (2\pi)^{5/2}, \quad \mathbf{H}_{3,4} = (2\pi)^{5/4},$$

even though maximizers do not exist. He also asked about the value of $\mathbf{H}_{1,6}$, which corresponds to the last algebraic endpoint question, and whether maximizers exist in the non-endpoint case in all dimensions.[5] Both of these questions were recently answered in [26, 27].

Theorem 13 ([26]). *We have* $\mathbf{H}_{1,6} = 3^{-1/12}(2\pi)^{1/2}$, *and maximizers for* (6.2) *do not exist if* $(d,p) = (1,6)$.

Theorem 14 ([26, 27]). *Maximizers for* (6.2) *exist when* $6 < p < \infty$ *if* $d = 1$ *and* $2 + \frac{4}{d} < p < 2 + \frac{4}{d-1}$ *if* $d \geq 2$. *In fact, given any maximizing sequence* $\{f_n\}$, *there exist symmetries* S_n *such that* $\{S_n f_n\}$ *converges in* $L^2(\mathbb{H}^d)$ *to a maximizer* f, *after passing to a subsequence.*

The proof of Theorem 13 relies on two ingredients. Firstly, by Lorentz invariance it suffices to study the convolution measure $(\sigma * \sigma * \sigma)(\tau, \xi)$ along the axis $\xi = 0$, and we verify that $\tau \mapsto (\sigma * \sigma * \sigma)(\tau, 0)$ defines a continuous function on the half-line $\tau > 3$, which extends continuously to the boundary of its support, so that

$$\sup_{\tau > 3}(\sigma * \sigma * \sigma)(\tau, 0) = (\sigma * \sigma * \sigma)(3, 0) = \frac{2\pi}{\sqrt{3}}, \tag{6.4}$$

and that this global maximum is strict. By an application of the Cauchy–Schwarz inequality, which is similar to but simpler than the one for the sphere discussed in Sec-

---

5 The methods of [37] do not apply directly due to the lack of exact scaling invariance.

tion 2, it follows that $\mathbf{H}_{1,6} \leq 3^{-1/12}(2\pi)^{1/2}$. For the reverse inequality, one checks that $f_n = \exp(-n\langle\cdot\rangle)$ forms a maximizing sequence for (6.2), in the sense that

$$\lim_{n\to\infty} \frac{\|f_n\sigma * f_n\sigma * f_n\sigma\|^2_{L^2(\mathbb{R}^2)}}{\|f_n\|^6_{L^2(\mathbb{H}^1)}} = \frac{2\pi}{\sqrt{3}}.$$

This crucially relies on the fact that the strict global maximum of $\sigma * \sigma * \sigma$ occurs at the *boundary* of the support of the convolution; recall (6.4). In particular, maximizers for (6.2) do not exist if $(d, p) = (1, 6)$.

For higher-order convolutions σ^{*k}, $k \geq 4$, the global maximum occurs in the interior of the support, which already hints towards Theorem 14. The proof of this theorem is more involved and crucially relies on a *refined Strichartz estimate*. If $d \in \{1, 2\}$, corresponding to the range (6.3) whose endpoints are even integers, then this can be obtained via elementary methods, such as the Hausdorff–Young and Hardy–Littlewood–Sobolev inequalities. For instance, on $\mathbb{H}^1$ it is proved in [26, Corollary 13] that, for each $6 \leq p < \infty$, there exists $C_p < \infty$ such that

$$\|\mathcal{E}f\|_{L^p(\mathbb{R}^2)} \leq C_p \sup_{k\in\mathbb{Z}} \|f_k\|^{1/3}_{L^2(\mathbb{H}^1)} \|f\|^{2/3}_{L^2(\mathbb{H}^1)}. \tag{6.5}$$

The decomposition $f = \sum_{k\in\mathbb{Z}} f_k$ is such that $f_k = f\mathbf{1}_{\mathcal{C}_k}$, where the family of *hyperbolic caps* $\{\mathcal{C}_k\}_{k\in\mathbb{Z}} \subset \mathbb{H}^1$ is given by

$$\mathcal{C}_k := \left\{ (\tau, \xi) \in \mathbb{H}^1 : \sinh\left(k - \frac{1}{2}\right) \leq \xi \leq \sinh\left(k + \frac{1}{2}\right) \right\}.$$

Inequality (6.5) allows us to start gaining control over arbitrary maximizing sequences. In particular, it can be used to show the existence of a *distinguished cap* which contains a positive proportion of the total mass; possibly after a Lorentz boost, the distinguished cap can be assumed to coincide with $\mathcal{C}_0$. This rules out the possibility of mass concentration at infinity, which had been previously identified in [76] as the main obstruction to the precompactness of maximizing sequences modulo symmetries. If $d \geq 3$, then the refined Strichartz estimate follows from bilinear restriction theory; see [27, Theorem 5.1]. We omit the technical details, and refer the interested reader to [27, Section 5].

It would be interesting to understand whether the two-sheeted hyperboloid shares similar features to the ones described for the two-sheeted cone in Section 5.

Bibliography

[1]   S. Agmon and L. Hörmander, *Asymptotic properties of solutions of differential equations with simple characteristics*, J. Anal. Math. **30** (1976), 1–38.

[2]   J. Barker, C. Thiele and P. Zorin-Kranich, *Band-limited maximizers for a Fourier extension inequality on the circle, II*, Exp. Math., to appear, arXiv:2002.05118.

[3]   P. Bégout and A. Vargas, *Mass concentration phenomena for the L^2-critical nonlinear Schrödinger equation*, Trans. Am. Math. Soc. **359**(11) (2007), 5257–5282.

[4]   D. Beltran and L. Vega, *Bilinear identities involving the k-plane transform and Fourier extension operators*, Proc. R. Soc. Edinb. A **150**(6) (2020), 3349–3377.

[5]   J. Bennett, N. Bez, A. Carbery and D. Hundertmark, *Heat-flow monotonicity of Strichartz norms*, Anal. PDE **2**(2) (2009), 147–158.

[6]   J. Bennett, N. Bez, M. Cowling and T. Flock, *Behaviour of the Brascamp–Lieb constant*, Bull. Lond. Math. Soc. **49**(3) (2017), 512–518.

[7]   J. Bennett, N. Bez, T. Flock, S. Gutiérrez and M. Iliopoulou, *A sharp k-plane Strichartz inequality for the Schrödinger equation*, Trans. Am. Math. Soc. **370**(8) (2018), 5617–5633.

[8]   J. Bennett, N. Bez, T. Flock and S. Lee, *Stability of the Brascamp–Lieb constant and applications*, Am. J. Math. **140**(2) (2018), 543–569.

[9]   J. Bennett, N. Bez and M. Iliopoulou, *Flow monotonicity and Strichartz inequalities*, Int. Math. Res. Not. **19** (2015), 9415–9437.

[10]   J. Bennett, N. Bez, C. Jeavons and N. Pattakos, *On sharp bilinear Strichartz estimates of Ozawa–Tsutsumi type*, J. Math. Soc. Jpn. **69**(2) (2017), 459–476.

[11]   N. Bez and C. Jeavons, *A sharp Sobolev–Strichartz estimate for the wave equation*, Electron. Res. Announc. Math. Sci. **22** (2015), 46–54.

[12]   N. Bez, C. Jeavons and T. Ozawa, *Some sharp bilinear space-time estimates for the wave equation*, Mathematika **62** (2016), 719–737.

[13]   N. Bez, C. Jeavons, T. Ozawa and H. Saito, *A conjecture regarding optimal Strichartz estimates for the wave equation*, in: New trends in analysis and interdisciplinary applications, Trends math. res. perspect., 2017, pp. 293–299.

[14]   N. Bez and K. Rogers, *A sharp Strichartz estimate for the wave equation with data in the energy space*, J. Eur. Math. Soc. **15**(3) (2013), 805–823.

[15]   N. Bez and M. Sugimoto, *Optimal constants and extremisers for some smoothing estimates*, J. Anal. Math. **131** (2017), 159–187.

[16]   G. Bianchi and H. Egnell, *A note on the Sobolev inequality*, J. Funct. Anal. **100**(1) (1991), 18–24.

[17]   C. Biswas and B. Stovall, *Existence of extremizers for Fourier restriction to the moment curve*, arXiv:2012.01528.

[18]   J. Borwein, A. Straub, J. Wan and W. Zudilin, *Densities of short uniform random walks*, Can. J. Math. **64**(5) (2012), 961–990, with an appendix by Don Zagier.

[19]   G. Brocchi, D. Oliveira e Silva and R. Quilodrán, *Sharp Strichartz inequalities for fractional and higher order Schrödinger equations*, Anal. PDE **13**(2) (2020), 477–526.

[20]   A. Bulut, *Maximizers for the Strichartz inequalities for the wave equation*, Differ. Integral Equ. **23**(11–12) (2010), 1035–1072.

[21]   E. Carneiro, *A sharp inequality for the Strichartz norm*, Int. Math. Res. Not. **16** (2009), 3127–3145.

[22]   E. Carneiro, D. Foschi, D. Oliveira e Silva and C. Thiele, *A sharp trilinear inequality related to Fourier restriction on the circle*, Rev. Mat. Iberoam. **33**(4) (2017), 1463–1486.

[23]   E. Carneiro, G. Negro and D. Oliveira e Silva, *Stability of sharp Fourier restriction to spheres*, arXiv:2108.03412.

[24]   E. Carneiro and D. Oliveira e Silva, *Some sharp restriction inequalities on the sphere*, Int. Math. Res. Not. **17** (2015), 8233–8267.

[25]   E. Carneiro, D. Oliveira e Silva and M. Sousa, *Sharp mixed norm spherical restriction*, Adv. Math. **341** (2019), 583–608.

[26]   E. Carneiro, D. Oliveira e Silva and M. Sousa, *Extremizers for Fourier restriction on hyperboloids*, Ann. Inst. Henri Poincaré, Anal. Non Linéaire **36**(2) (2019), 389–415.

[27]   E. Carneiro, D. Oliveira e Silva, M. Sousa and B. Stovall, *Extremizers for adjoint Fourier restriction on hyperboloids: the higher dimensional case*, Indiana Univ. Math. J. **70**(2) (2021), 535–559.

[28] E. Carneiro, L. Oliveira and M. Sousa, *Gaussians never extremize Strichartz inequalities for hyperbolic paraboloids*, Proc. Am. Math. Soc. **150**(8) (2022), 3395–3403.

[29] M. Christ and R. Quilodrán, *Gaussians rarely extremize adjoint Fourier restriction inequalities for paraboloids*, Proc. Am. Math. Soc. **142**(3) (2014), 887–896.

[30] M. Christ and S. Shao, *Existence of extremals for a Fourier restriction inequality*, Anal. PDE **5**(2) (2012), 261–312.

[31] M. Christ and S. Shao, *On the extremizers of an adjoint Fourier restriction inequality*, Adv. Math. **230**(3) (2012), 957–977.

[32] V. Ciccone and F. Gonçalves, *Sharp Fourier extension on the circle under arithmetic constraints*, arXiv:2208.09441.

[33] B. Di and D. Yan, *Extremals for α-Strichartz inequalities*, arXiv:2204.01031.

[34] B. Dodson, J. Marzuola, B. Pausader and D. Spirn, *The profile decomposition for the hyperbolic Schrödinger equation*, Ill. J. Math. **62**(1–4) (2018), 293–320.

[35] T. Duyckaerts, F. Merle and S. Roudenko, *Maximizers for the Strichartz norm for small solutions of mass-critical NLS*, Ann. Sc. Norm. Super. Pisa, Cl. Sci. (5) **10**(2) (2011), 427–476.

[36] L. Fanelli, L. Vega and N. Visciglia, *On the existence of maximizers for a family of restriction theorems*, Bull. Lond. Math. Soc. **43**(4) (2011), 811–817.

[37] L. Fanelli, L. Vega and N. Visciglia, *Existence of maximizers for Sobolev–Strichartz inequalities*, Adv. Math. **229**(3) (2012), 1912–1923.

[38] T. Flock and B. Stovall, *On extremizing sequences for adjoint Fourier restriction to the sphere*, arXiv:2204.10361.

[39] D. Foschi, *Maximizers for the Strichartz inequality*, J. Eur. Math. Soc. **9**(4) (2007), 739–774.

[40] D. Foschi, *Global maximizers for the sphere adjoint Fourier restriction inequality*, J. Funct. Anal. **268** (2015), 690–702.

[41] D. Foschi and S. Klainerman, *Bilinear space-time estimates for homogeneous wave equations*, Ann. Sci. Éc. Norm. Supér. (4) **33**(2) (2000), 211–274.

[42] D. Foschi and D. Oliveira e Silva, *Some recent progress on sharp Fourier restriction theory*, Anal. Math. **43**(2) (2017), 241–265.

[43] R. Frank, E. H. Lieb and J. Sabin, *Maximizers for the Stein–Tomas inequality*, Geom. Funct. Anal. **26**(4) (2016), 1095–1134.

[44] R. Frank and J. Sabin, *Extremizers for the Airy–Strichartz inequality*, Math. Ann. **372**(3–4) (2018), 1121–1166.

[45] F. Gonçalves, *Orthogonal polynomials and sharp estimates for the Schrödinger equation*, Int. Math. Res. Not. **8** (2019), 2356–2383.

[46] F. Gonçalves, *A sharpened Strichartz inequality for radial functions*, J. Funct. Anal. **276**(6) (2019), 1925–1947.

[47] F. Gonçalves and G. Negro, *Local maximizers of the adjoint Fourier restriction estimate for cone, paraboloid and sphere*, Anal. PDE **15**(4) (2022), 1097–1130.

[48] F. Gonçalves and D. Zagier, *Strichartz estimates with broken symmetries*, arXiv:2011.02187.

[49] D. Hundertmark and S. Shao, *Analyticity of extremizers to the Airy–Strichartz inequality*, Bull. Lond. Math. Soc. **44**(2) (2012), 336–352.

[50] D. Hundertmark and V. Zharnitsky, *On sharp Strichartz inequalities in low dimensions*, Int. Math. Res. Not. **2006** (2006), 34080, 18 pp.

[51] C. Jeavons, *A sharp bilinear estimate for the Klein–Gordon equation in arbitrary space-time dimensions*, Differ. Integral Equ. **27**(1–2) (2014), 137–156.

[52] J.-C. Jiang, B. Pausader and S. Shao, *The linear profile decomposition for the fourth order Schrödinger equation*, J. Differ. Equ. **249**(10) (2010), 2521–2547.

[53] J.-C. Jiang, S. Shao and B. Stovall, *Linear profile decompositions for a family of fourth order Schrödinger equations*, arXiv:1410.7520.

[54] M. Kunze, *On the existence of a maximizer for the Strichartz inequality*, Commun. Math. Phys. **243**(1) (2003), 137–162.

[55] R. Mandel and D. Oliveira e Silva, *The Stein–Tomas inequality under the effect of symmetries*, J. Anal. Math., to appear, arXiv:2106.08255.

[56] C. Morpurgo, *Sharp inequalities for functional integrals and traces of conformally invariant operators*, Duke Math. J. **114**(3) (2002), 477–553.

[57] G. Negro, *A sharpened Strichartz inequality for the wave equation*, Ann. Sci. Éc. Norm. Supér., to appear, arXiv:1802.04114.

[58] G. Negro, *A sharpened energy-Strichartz inequality for the wave equation*, arXiv:2207.06350.

[59] G. Negro and D. Oliveira e Silva, *Intermittent symmetry breaking and stability of the sharp Agmon–Hörmander estimate on the sphere*, Proc. Amer. Math. Soc. **151**(1) (2023), 87–99.

[60] G. Negro, D. Oliveira e Silva, B. Stovall and J. Tautges, *Exponentials rarely maximize Fourier extension inequalities for cones*, arXiv:2302.00356.

[61] D. Oliveira e Silva, *Extremizers for Fourier restriction inequalities: convex arcs*, J. Anal. Math. **124** (2014), 337–385.

[62] D. Oliveira e Silva, *Sharp Fourier restriction theory*, Habilitation Thesis, Universität, Bonn, 2017.

[63] D. Oliveira e Silva, *Nonexistence of extremizers for certain convex curves*, Math. Res. Lett. **25**(3) (2018), 973–987.

[64] D. Oliveira e Silva and R. Quilodrán, *On extremizers for Strichartz estimates for higher order Schrödinger equations*, Trans. Am. Math. Soc. **370**(10) (2018), 6871–6907.

[65] D. Oliveira e Silva and R. Quilodrán, *A sharp inequality in Fourier restriction theory*, Bol. Soc. Port. Mat. **77** (2019), 133–150.

[66] D. Oliveira e Silva and R. Quilodrán, *A comparison principle for convolution measures with applications*, Math. Proc. Camb. Philos. Soc. **169**(2) (2020), 307–322.

[67] D. Oliveira e Silva and R. Quilodrán, *Global maximizers for adjoint Fourier restriction inequalities on low dimensional spheres*, J. Funct. Anal. **280**(7) (2021), 108825, 73 pp.

[68] D. Oliveira e Silva and R. Quilodrán, *Smoothness of solutions of a convolution equation of restricted type on the sphere*, Forum Math. Sigma **9** (2021), e12, 40 pp.

[69] D. Oliveira e Silva and C. Thiele, *Estimates for certain integrals of products of six Bessel functions*, Rev. Mat. Iberoam. **33**(4) (2017), 1423–1462.

[70] D. Oliveira e Silva, C. Thiele and P. Zorin-Kranich, *Band-limited maximizers for a Fourier extension inequality on the circle*, Exp. Math. **31**(1) (2022), 192–198.

[71] T. Ozawa and K. Rogers, *Sharp Morawetz estimates*, J. Anal. Math. **121** (2013), 163–175.

[72] T. Ozawa and K. Rogers, *A sharp bilinear estimate for the Klein–Gordon equation in* $\mathbb{R}^{1+1}$, Int. Math. Res. Not. **2014**(5) (2014), 1367–1378.

[73] T. Ozawa and Y. Tsutsumi, *Space-time estimates for null gauge forms and nonlinear Schrödinger equations*, Differ. Integral Equ. **11**(2) (1998), 201–222.

[74] R. Penrose, *Republication of: conformal treatment of infinity*, Gen. Relativ. Gravit. **43**(3) (2011), 901–922.

[75] R. Quilodrán, *On extremizing sequences for the adjoint restriction inequality on the cone*, J. Lond. Math. Soc. (2) **87**(1) (2013), 223–246.

[76] R. Quilodrán, *Nonexistence of extremals for the adjoint restriction inequality on the hyperboloid*, J. Anal. Math. **125** (2015), 37–70.

[77] R. Quilodrán, *Existence of extremals for a Fourier restriction inequality on the one-sheeted hyperboloid*, arXiv:2207.10587.

[78] J. Ramos, *A refinement of the Strichartz inequality for the wave equation with applications*, Adv. Math. **230**(2) (2012), 649–698.

[79] S. Shao, *The linear profile decomposition for the Airy equation and the existence of maximizers for the Airy-Strichartz inequality*, Anal. PDE **2**(1) (2009), 83–117.

[80] S. Shao, *Maximizers for the Strichartz and the Sobolev-Strichartz inequalities for the Schrödinger equation*, Electron. J. Differ. Equ. **2009** (2009), 3, 13 pp.

[81] S. Shao, *On existence of extremizers for the Tomas–Stein inequality for S^1*, J. Funct. Anal. **270**(10) (2016), 3996–4038.

[82] E. M. Stein, *Harmonic analysis: real-variable methods, orthogonality, and oscillatory integrals*, in: Monographs in Harmonic Analysis, III, Princeton Mathematical Series, vol. 43, Princeton University Press, Princeton, NJ, 1993.

[83] B. Stovall, *Extremizability of Fourier restriction to the paraboloid*, Adv. Math. **360** (2020), 106898, 18 pp.

[84] R. Strichartz, *Restrictions of Fourier transforms to quadratic surfaces and decay of solutions of wave equations*, Duke Math. J. **44**(3) (1977), 705–714.

[85] J. Tautges, *Extremizers for adjoint restriction to a pair of translated paraboloids*, J. Math. Anal. Appl. **513**(2) (2022), 126208, 9 pp.

[86] J. Tautges, *Extremizers for adjoint restriction to a pair of reflected paraboloids*, arXiv:2112.13130.

[87] P. Tomas, *A restriction theorem for the Fourier transform*, Bull. Am. Math. Soc. **81** (1975), 477–478.

[88] N. Tzvetkov, *Remark on the null-condition for the nonlinear wave equation*, Boll. Unione Mat. Ital., Sez. B Artic. Ric. Mat. (8) **3**(1) (2000), 135–145.

[89] C. Wayne and V. Zharnitsky, *Critical points of Strichartz functional*, Exp. Math. **30**(2) (2021), 235–257.

Carsten Schütt and Elisabeth M. Werner

Affine surface area

Abstract: We give an overview of the affine surface area, including its properties and its history.

Keywords: Affine surface area, floating body, random polytopes

MSC 2020: 52A39

1 Introduction

A convex body in $\mathbb{R}^n$ is a convex, compact subset of $\mathbb{R}^n$ with non-empty interior. The set of all convex bodies in $\mathbb{R}^n$ is denoted by $\mathcal{K}^n$. We denote the Euclidean norm of a vector $x \in \mathbb{R}^n$ by $\|x\|_2$. The Euclidean ball in $\mathbb{R}^n$ with center x and radius r is denoted $B_2^n(x, r)$. We denote $B_2^n(0, r)$ by B_2^n. We consider here $\mathcal{K}^n$ equipped with the Hausdorff metric

$$d_H(C, K) = \inf\{\rho \mid C \subseteq K + \rho B_2^n \text{ and } K \subseteq C + \rho B_2^n\}.$$

Another metric on $\mathcal{K}^n$ is the symmetric difference metric

$$d_S(C, K) = \mathrm{vol}_n(C \triangle K) = \mathrm{vol}_n((C \setminus K) \cup (K \setminus C)).$$

The support function $h_K : \mathbb{R}^n \to \mathbb{R}$ of a convex body K in $\mathbb{R}^n$ is given by

$$h_K(x) = \max_{y \in K}\langle x, y\rangle.$$

The polar body of a convex body C that contains the origin as an interior point is

$$C^* = \{y \in \mathbb{R}^n \mid \forall x \in C : \langle x, y\rangle \le 1\}.$$

The $(n-1)$-dimensional Hausdorff measure on $\mathbb{R}^n$ is denoted by $\mathcal{H}_{n-1}$. For a convex body K in $\mathbb{R}^n$ the measure $\mu_{\partial K}$ is the restriction of $\mathcal{H}_{n-1}$ to the boundary of K. We call it also the surface measure of ∂K. σ_n is the restriction of $\mathcal{H}_{n-1}$ to the Euclidean sphere S^{n-1}.

The surface area measure

$$\sigma_K \quad \text{on} \quad S^{n-1} \tag{1.1}$$

Acknowledgement: Elisabeth M. Werner was supported by NSF grant DMS-2103482.

Carsten Schütt, University of Kiel, Kiel, Germany, e-mail: schuett@math.uni-kiel.de
Elisabeth M. Werner, Case Western Reserve University, Cleveland, USA

https://doi.org/10.1515/9783110775389-010

is defined in the following way: For every Borel set A of S^{n-1} we define $\sigma_K(A)$ as the $(n-1)$-dimensional Hausdorff measure of the set of all points in ∂K that have a normal that is element in A.

The boundary of a convex body or a function mapping from $\mathbb{R}^n$ to $\mathbb{R}$ is of class C^2 if it is twice continuously differentiable and of class C_+^2 if it is twice continuously differentiable and its Gauss–Kronecker curvature is strictly positive.

For all $t_1, \dots, t_m \geq 0$ and all $K_1, \dots, K_m \in \mathcal{K}^n$ (see, e. g., Schneider [40])

$$\mathrm{vol}_n(t_1 K_1 + \cdots + t_m K_m) = \sum_{i_1,\dots,i_n=1}^{m} V(K_{i_1}, \dots, K_{i_n}) t_{i_1} \cdots t_{i_n}.$$

The coefficients $V(\cdots)$ are called *mixed volumes*. We denote

$$V_1(K, C) = V(K, \dots, K, C)$$

and we have

$$V_1(K, C) = \frac{1}{n} \lim_{\epsilon \to 0} \frac{\mathrm{vol}_n(K + \epsilon C) - \mathrm{vol}_n(K)}{\epsilon}. \tag{1.2}$$

Moreover,

$$V_1(K, C) = \frac{1}{n} \int_{S^{n-1}} h_C(\xi) d\sigma_K(\xi). \tag{1.3}$$

We introduce the affine surface area and its main properties. We discuss its history. For some results we give the main steps of their proofs. For more background on general convex geometry we refer to the books of Bonnesen and Fenchel [5], Rockefellar [39] and Schneider [40]. All details of this chapter can be found in the forthcoming book by Schütt and Werner [46].

2 Generalized Gauss–Kronecker curvature

A convex body in $\mathbb{R}^n$ may not be twice differentiable at any boundary point and therefore we do not have the classical notion of curvature. Thus, we need to introduce a more general notion of second-order differentiability.

Definition 1. Let $\mathcal{U}$ be a convex, open subset of $\mathbb{R}^n$ and let $f : \mathcal{U} \to \mathbb{R}$ be a convex function. A vector $df(x_0) \in \mathbb{R}^n$ is called subdifferential at the point $x_0 \in \mathcal{U}$ if we have for all $x \in \mathcal{U}$

$$f(x_0) + \langle df(x_0), x - x_0 \rangle \leq f(x).$$

Lemma 1. *Let $\mathcal{U}$ be a convex, open set in $\mathbb{R}^n$ and let $f : \mathcal{U} \to \mathbb{R}$ be a convex function. Then f has a subdifferential at every $x \in \mathcal{U}$ and the set of subdifferentials at a given point x is convex.*

The existence of a subdifferential follows from the theorem of Hahn–Banach.

If a function f is differentiable at a point x, then there is a unique subdifferential at x and it is equal to its gradient. By a theorem of Rademacher a convex function is almost everywhere differentiable [39, p. 246], but there are convex functions that are nowhere twice differentiable. We give an example.

Example 1. Let $q : \mathbb{N} \to \mathbb{Q} \cap [0, 1]$ be a bijection and let $g : [0, 1] \to \mathbb{R}$ be given by

$$g(x) = \sum_{q(n) \leq x} \frac{1}{2^n}.$$

Moreover, let $G : [0, 1] \to \mathbb{R}$ be the antiderivative of g,

$$G(x) = \int_0^x g(t)dt.$$

Then G is convex and differentiable at all irrational points, but not differentiable at all rational points. The derivative of G at the irrational point x is $g(x)$. In particular, G is nowhere twice differentiable in the classical sense.

Since we need a notion of curvature for convex functions, we introduce the generalized second derivative of a convex function.

Definition 2. Let $\mathcal{U}$ be a convex, open subset of $\mathbb{R}^n$ and let $f : \mathcal{U} \to \mathbb{R}$ be a convex function. The convex function f is said to be twice differentiable in the generalized sense at x_0 if there are a linear map $d^2 f(x_0) : \mathbb{R}^n \to \mathbb{R}^n$ and a neighborhood $\mathcal{V}(x_0)$ such that we have for all $x \in \mathcal{V}(x_0)$ and for all subdifferentials $df(x)$

$$\left\| df(x) - df(x_0) - d^2 f(x_0)(x - x_0) \right\|_2 \leq \Theta(\|x - x_0\|_2)\|x - x_0\|_2, \tag{2.1}$$

where $\Theta : [0, \infty) \to \mathbb{R}$ is a monotone function with $\lim_{t \to 0} \Theta(t) = 0$. The matrix of $d^2 f(x_0)$ with respect to the standard basis of $\mathbb{R}^n$ is called generalized Hesse matrix. Ignoring the difference between a linear map and its representation as a matrix we refer to $d^2 f(x_0)$ as the Hesse matrix.

Of course, if f is twice differentiable in the classical sense, then $d^2 f(x_0)$ equals the usual Hesse matrix $\nabla^2 f(x_0)$. We now give an example of a function that is not twice differentiable, but that is twice differentiable in the generalized sense.

Example 2. Let $f : [-1, 1] \to \mathbb{R}$ be defined by

$$
f(x) = \begin{cases} x^2 & \text{if } |x| = \frac{1}{n}, \\ \frac{2n+1}{n(n+1)}|x| - \frac{1}{n(n+1)} & \text{if } \frac{1}{n+1} < |x| < \frac{1}{n}, \end{cases} \qquad n \in \mathbb{N}.
$$

Then f is convex and twice differentiable at 0 in the generalized sense and $d^2 f(0) = 2$, but f is not twice differentiable at 0 in the classical sense.

Most importantly, by a theorem of Busemann–Feller–Aleksandrov a convex function is almost everywhere twice differentiable in the generalized sense (see [1, 2, 3, 6] and [19, p. 28]).

We now define the *generalized Gauss–Kronecker curvature* of a convex function. Let $\mathcal{U}$ be an open, convex set in $\mathbb{R}^n$ and let $f : \mathcal{U} \to \mathbb{R}$ be a convex function. The *epigraph* of f is

$$
\text{epi}(f) = \{(x, t) \mid f(x) \le t\}.
$$

For $x \in \mathcal{U}$ we put $z = (x, f(x))$. We assume that $\text{epi}(f)$ has a unique normal $N(z_0)$ at z_0. Let $P_{N(z_0)} : \mathbb{R}^{n+1} \to H(z_0, N(z_0))$ be the orthogonal projection. If the convex sets

$$
\frac{1}{\sqrt{2\Delta}} P_{N(z_0)}(\text{epi}(f) \cap H(z_0 - \Delta N(z_0), N(z_0))) \tag{2.2}
$$

converge pointwise for $\Delta \to 0$ to an ellipsoid or an elliptic cylinder we call its boundary the *indicatrix of Dupin* [12] and denote it by $\text{Dupin}(f, z_0)$.

We call the squares of the lengths of the principal radii $r_1, \dots, r_n$ of the indicatrix of Dupin the *generalized principal Gauss–Kronecker curvature radii*, and if $r_1, \dots, r_n > 0$ their reciprocals are called the *generalized principal Gauss–Kronecker curvatures*. Then we say that the *generalized Gauss–Kronecker curvature* of f at x_0 exists and it is equal to the product of the principal curvatures

$$
\kappa(x_0) = \left(\prod_{i=1}^{n} r_i\right)^{-2} = \left(\frac{\text{vol}_n(B_2^n)}{\text{vol}_n(\text{Dupin}(f, z_0))}\right)^{-2}. \tag{2.3}
$$

When the indicatrix of Dupin is an elliptic cylinder we define the curvature to be 0.

One can show that all eigenvalues of the generalized Hesse matrix $d^2 f(x_0)$ are nonnegative. If all eigenvalues are strictly positive, the indicatrix of Dupin is an elliptic sphere. In this case the sets (2.2) converge uniformly in the Hausdorff metric to the elliptic sphere $\text{Dupin}(f, z_0)$ because all sets are convex and bounded.

It can be shown that the generalized Gauss–Kronecker curvature of a convex function at a point x_0 at which f is twice differentiable in the generalized sense equals

$$
\kappa(x_0) = \frac{\det(d^2 f(x_0))}{(1 + \|df(x_0)\|_2^2)^{\frac{n+2}{2}}}. \tag{2.4}
$$

In order to define the curvature of a convex body K at $x_0 \in \partial K$, we parametrize the boundary in a neighborhood of x_0 by a convex function and take its curvature as the curvature of the convex body.

3 Affine surface area

Definition 3. The affine surface area of a convex body K in $\mathbb{R}^n$ is

$$as(K) = \int_{\partial K} \kappa(x)^{\frac{1}{n+1}} d\mu_{\partial K}(x), \tag{3.1}$$

where κ denotes the generalized Gauss–Kronecker curvature (2.3) and $\mu_{\partial K}$ denotes the surface measure on ∂K.

As noted above, by a theorem of Busemann–Feller–Aleksandrov the generalized curvature exists almost everywhere [1, 2, 3, 6] and thus the integrand $\kappa^{\frac{1}{n+1}}$ is well-defined, but we still have to confirm that it is an integrable function.

We discuss the history of the affine surface area [26, 27, 31] and what led to the definition of the affine surface area (3.1) and its alternative definitions. In affine differential geometry the affine surface area of a convex body whose boundary has a C_+^2-parametrization $x : U \to \mathbb{R}^n$ was defined as [4, Chapter 47]

$$\int_U \left| \det\left(\det\left(\frac{\partial x}{\partial t_1}, \dots, \frac{\partial x}{\partial t_{n-1}}, \frac{\partial^2 x}{\partial t_i \partial t_j} \right)_{i,j=1}^{n-1} \right) \right|^{\frac{1}{n+1}} d(t_1, \dots, t_{n-1}). \tag{3.2}$$

This is the simplest expression that is independent of the change of parametrization, is invariant under affine maps, depends only on the surface of a convex body, and involves only first and second derivatives of the parametrization. One can show that (3.2) equals (3.1).

Thus, (3.2) is a definition of the affine surface area for all convex bodies with a C_+^2-boundary. As the affine surface area has remarkable and useful properties – see below – it was imperative to extend it to arbitrary convex bodies. With the properties of the affine surface area for convex bodies with C_+^2-boundary in mind we expect that the extended affine surface area satisfies the following:

(i)   The extension should coincide with (3.1) and (3.2) for convex bodies with C_+^2-boundary.

(ii) For all polytopes P

$$as(P) = 0.$$

(iii) The affine surface area is not continuous with respect to the Hausdorff distance. This follows from considering a sequence of polytopes that converge in the Hausdorff metric to the Euclidean ball B_2^n. But one wants to have upper semicontinuity.

(iv) For all convex bodies K in $\mathbb{R}^n$ and all affine maps $T : \mathbb{R}^n \to \mathbb{R}^n$

$$\mathrm{as}(T(K)) = |\det T|^{\frac{n-1}{n+1}} \mathrm{as}(K).$$

(v) The affine isoperimetric inequality should hold for all convex bodies K in $\mathbb{R}^n$:

$$\mathrm{as}(K)^{n+1} \le \mathrm{as}(B_2^n)^2 n^{n-1} \mathrm{vol}_n(K)^{n-1}.$$

Since $\mathrm{as}(B_2^n) = \mathrm{vol}_{n-1}(\partial B_2^n)$, this means in particular that $\mathrm{as}(K)$ is smaller than or equal to the affine surface area of a Euclidean ball that has the same volume.

(vi) The affine surface area appears naturally in approximation of convex bodies by polytopes. This should also hold for the extended affine surface area.

In order to give the affine surface area a geometric interpretation, Blaschke introduced the floating body [4, pp. 125–126]. He introduced the *floating body* of a convex body K as the convex body $[K]_t$ whose tangent hyperplanes cut off a set of exactly volume t from K. He assumed that K has an analytic boundary to ensure that $[K]_t$ exists, at least for small $t > 0$. Then he showed that in $\mathbb{R}^3$

$$\mathrm{as}(K) = \lim_{t \to 0} \frac{\mathrm{vol}_3(K) - \mathrm{vol}_3([K]_t)}{\sqrt{t}}. \tag{3.3}$$

Leichtweiss generalized this to higher dimensions and the differentiability class C_+^2 [27]:

$$\mathrm{as}(K) = \lim_{t \to 0} \frac{\mathrm{vol}_n(K) - \mathrm{vol}_3([K]_t)}{t^{\frac{2}{n+1}}}. \tag{3.4}$$

The name affine surface area reflects the Minkowski definition of the usual surface area of a convex body:

$$\mathrm{vol}_{n-1}(\partial K) = \lim_{t \to 0} \frac{\mathrm{vol}_n(K + tB_2^n) - \mathrm{vol}_n(K)}{t}.$$

Expression (3.4) is the starting point for Leichtweiss for his extension of the affine surface area to arbitrary convex bodies: Why do we not use the right-hand side of (3.4) to define the extension? Unfortunately, the floating body may not exist in general. The simplex in $\mathbb{R}^n$ is an example. Of course, for the limit to exist in (3.4), the floating body only needs to exist for small $t \ge 0$. Again, the simplex is a counterexample.

In order to circumvent this difficulty, Leichtweiss [26] added a small Euclidean ball (or ellipsoid) to the convex body. Now the floating body exists for small t and therefore the affine surface area. Leichtweiss then took the limit of the radius of the Euclidean ball to 0 and defined this limit to be the affine surface area of the convex body K. Formally [26],

$$n \cdot c_n \lim_{\epsilon \to 0} \lim_{\delta \to 0} \frac{1}{\delta^{\frac{2}{n+1}}} \left(\mathrm{vol}_n(K + \epsilon\mathcal{E}) - V(K + \epsilon\mathcal{E}, \ldots, K + \epsilon\mathcal{E}, [K + \epsilon\mathcal{E}]_\delta) \right)$$

$$= n \cdot c_n \lim_{\epsilon \to 0} \lim_{\delta \to 0} \frac{1}{\delta^{\frac{2}{n+1}}} \left(\mathrm{vol}_n(K + \epsilon\mathcal{E}) - V_1(K + \epsilon\mathcal{E}, [K + \epsilon\mathcal{E}]_\delta) \right), \tag{3.5}$$

where $[K + \epsilon\mathcal{E}]_\delta$ is the floating body of $K + \epsilon\mathcal{E}$, $\mathcal{E}$ is an ellipsoid with $\mathrm{vol}_n(\mathcal{E}) = \mathrm{vol}_n(B_2^n)$, and

$$c_n = 2 \left(\frac{\mathrm{vol}_n(B_2^n)}{n+1} \right)^{\frac{2}{n+1}}.$$

He showed that it is an extension of the affine surface area introduced by Blaschke. A somewhat simpler expression is [28, equation (11)]

$$c_n \lim_{\epsilon \to 0} \lim_{\delta \to 0} \frac{1}{\delta^{\frac{2}{n+1}}} \left(\mathrm{vol}_n(K + \epsilon\mathcal{E}) - \mathrm{vol}_n([K + \epsilon\mathcal{E}]_\delta) \right). \tag{3.6}$$

It was shown by Schütt [41] that (3.1) and (3.6) are equal.

At the same time Lutwak gave another extension of the affine surface area to arbitrary convex bodies by quite a different approach. A compact subset L of $\mathbb{R}^n$ is star-shaped if for all $x \in L$ the line segment $[0, x]$ is also contained in L. The radial function $\rho_L : S^{n-1} \to \mathbb{R}$ of a star-shaped set L is defined by

$$\rho_L(\xi) = \max\{\lambda \geq 0 \mid \lambda\xi \in L\}.$$

If ρ_L is continuous with respect to the Euclidean norm, we call L a star body. S_c^n denotes the set of all star bodies L whose centroid

$$\frac{1}{(n+1)\,\mathrm{vol}_n(L)} \int\limits_{S^{n-1}} \rho_L(\xi)^{n+1} \xi d\sigma_n(\xi)$$

is at the origin. In [31] Lutwak introduced

$$n^{\frac{1}{n+1}} \inf_{L \in S_c^n} \left((\mathrm{vol}_n(L))^{\frac{1}{n}} \int\limits_{S^{n-1}} \frac{1}{\rho_L(\xi)} d\sigma_K(\xi) \right)^{\frac{n}{n+1}} \tag{3.7}$$

as the affine surface area. The measure σ_K on ∂B_2^n is the surface area measure (1.1).

This definition is inspired by Petty's definition of the geominimal surface area of a convex body [37]. Petty defined the geominimal surface area as

$$\mathrm{gma}(K) = \frac{n}{\mathrm{vol}_n(B_2^n)^{\frac{1}{n}}} \inf\{V_1(K, C^*)\,\mathrm{vol}_n(C)^{\frac{1}{n}} \mid C \text{ convex body with centroid at } 0\}, \tag{3.8}$$

where C^* is the polar body to C and $V_1(K, C^*)$ is the mixed volume (1.2).

Again, at the same time Schütt and Werner [43] used the convex floating body K_t of a convex body K to extend the affine surface area to arbitrary convex bodies. Let $t \geq 0$. The *convex floating body* K_t of a convex body K in $\mathbb{R}^n$ is the intersection of all half-spaces H^+ whose defining hyperplanes cut off a set of volume t, i. e.,

$$K_t = \bigcap_{\text{vol}_n(K \cap H^-) = t} H^+. \tag{3.9}$$

This notion has been independently introduced by Bárány and Larman [8] and Schütt and Werner [43].

Schütt and Werner showed that for all convex bodies K in $\mathbb{R}^n$ [43]

$$\lim_{t \to 0} \frac{\text{vol}_n(K) - \text{vol}_n(K_t)}{t^{\frac{2}{n+1}}} = \frac{1}{2} \left(\frac{n+1}{\text{vol}_{n-1}(B_2^{n-1})} \right)^{\frac{2}{n+1}} \int_{\partial K} \kappa(x)^{\frac{1}{n+1}} d\mu_{\partial K}(x), \tag{3.10}$$

where κ is the generalized Gauss–Kronecker curvature. Equation (3.10) suggests that we can use (3.1) as a definition for the extended affine surface area. In fact, it can be shown that this expression satisfies all the desired requirements.

Werner proved that an analogous formula holds for the illumination body [47].

4 Convex floating body and the rolling theorem

Blaschke said that *a Euclidean ball rolls freely in a convex body K* if it can be placed at each boundary point x of K such that it touches the boundary at x and is contained in K. Formally, a Euclidean ball with radius r rolls freely in K if for all $x \in \partial K$ we have

$$B_2^n(x - rN(x)) \subseteq K.$$

If there is $r > 0$ such that rB_2^n rolls freely in K, then K is of class C^1. Furthermore, if there is $r > 0$ such that rB_2^n rolls freely in K, then K is the r-outer parallel body of another convex body C, i. e.,

$$K = C + rB_2^n.$$

Note that there are convex bodies with a C^1-boundary such that no ball rolls freely. The convex sets

$$B_p^n = \left\{ x \in \mathbb{R}^n \;\middle|\; \sum_{i=1}^{n} |x_i|^p \leq 1 \right\}$$

for $1 < p < 2$ are examples. Furthermore, for any convex body K with a C_+^2-boundary there is a ball that rolls freely in K, but not all bodies for which there is a ball that rolls freely have a C^2-boundary. An example is $C^n + B_2^n$, where C^n is the n-dimensional cube with side length 1.

A quantitative version of *a ball rolls freely in the convex body* was introduced by Schütt and Werner [43, 46]. For $x \in \partial K$ we denote by $r(x)$ the supremum of all radii of Euclidean balls that contain x and that are contained in K, i. e., $r : \partial K \to \mathbb{R}$ is defined by

$$r(x) = \sup\{\|x - z\| \mid \exists z \in K : B_2^n(z, \|x - z\|) \subseteq K\}. \tag{4.1}$$

We call r the rolling function.

If K does not have a unique normal at x, then $r(x) = 0$. Moreover, we have

$$r(x) = \sup\{\rho \mid B_2^n(x - \rho N(x), \rho) \subseteq K\} \tag{4.2}$$

if K has a unique normal at x.

Lemma 2 ([46]). *Let K be a convex body in $\mathbb{R}^n$ and let $r : \partial K \to \mathbb{R}$ be the rolling function* (4.1). *Then we have:*
(i) *the supremum in* (4.1) *is attained;*
(ii) *r is upper semicontinuous.*

Howard [22] studied the rolling function on manifolds. There are convex bodies for which the function r is not continuous.

Example 3. Let

$$K = \mathrm{conv}\left\{ \left(\pm\frac{1}{n}, \frac{1}{n^2} \right) \,\middle|\, n \in \mathbb{N} \right\}$$

and let r be its rolling function. Then $r((0,0)) \geq \frac{1}{2}$, but for all $n \in \mathbb{N}$ we have $r((\frac{1}{n}, \frac{1}{n^2})) = 0$. In particular r is not continuous at $(0,0)$.

Schütt and Werner gave a quantitative result for *a ball rolls freely in a convex body*.

Theorem 1 ([43]). *Let K be a convex body in $\mathbb{R}^n$ such that it contains B_2^n. Then for all t with $0 \leq t \leq 1$, $\{x \in \partial K \mid r(x) \geq t\}$ is a closed set and*

$$(1 - t)^{n-1} \mathrm{vol}_{n-1}(\partial K) \leq \mathcal{H}_{n-1}(\{x \in \partial K \mid r(x) \geq t\}).$$

The inequality is optimal.

In particular, the function $r^{-\alpha} : \partial K \to \mathbb{R}$ is Lebesgue integrable for all α with $0 \leq \alpha < 1$.

Since $\kappa(x)^{\frac{1}{n-1}} \leq \frac{1}{r(x)}$, we have $\kappa(x)^{\frac{1}{n+1}} \leq r(x)^{-\frac{n-1}{n+1}}$. By Theorem 1, the function $r^{-\frac{n-1}{n+1}}$ is integrable and consequently $\kappa^{\frac{1}{n+1}}$ is smaller than an integrable function. This is used to prove (3.10). Furthermore, in order to prove (3.10) we apply Lebesgue's dominated convergence theorem and use $r^{-\frac{1}{n+1}}$ as the dominating function.

Proof of (3.10). Assume that the origin is an interior point of K. Then

$$\mathrm{vol}_n(K) - \mathrm{vol}_n(K_t) = \frac{1}{n} \int_{\partial K} \langle x, N(x) \rangle \left(1 - \left| \frac{\|x_t\|_2}{\|x\|_2} \right|^n \right) d\mu_{\partial K}(x),$$

where $x \in \partial K$ and $x_t \in \partial K_t$ is the unique point in the intersection of the line segment $[0, x]$ and ∂K_t. After dividing by $t^{\frac{2}{n+1}}$,

$$\frac{\mathrm{vol}_n(K) - \mathrm{vol}_n(K_t)}{t^{\frac{2}{n+1}}} = \frac{1}{n} \int_{\partial K} \frac{\langle x, N(x) \rangle}{t^{\frac{2}{n+1}}} \left(1 - \left| \frac{\|x_t\|_2}{\|x\|_2} \right|^n \right) d\mu_{\partial K}(x).$$

Now it remains to observe

$$\kappa(x)^{\frac{1}{n+1}} = \lim_{t \to 0} \frac{\langle x, N(x) \rangle}{t^{\frac{2}{n+1}}} \left(1 - \left| \frac{\|x_t\|_2}{\|x\|_2} \right|^n \right)$$

and apply Lebesgue's dominated convergence theorem where we use $r^{-\frac{n-1}{n+1}}$ as the dominating function. $\qquad\square$

5 Properties of the affine surface area

Theorem 2. *Expressions* (3.1), (3.5), *and* (3.7) *all coincide.*

It was shown by Dolzmann and Hug that the definitions of the affine surface area of Leichtweiss and Lutwak coincide [10].

The projection body $\Pi(K)$ of a convex body K is the convex body whose support function is given by

$$h_{\Pi(K)}(\xi) = \mathrm{vol}_{n-1}(P_\xi(K)),$$

where $P_\xi : \mathbb{R}^n \to H(0, \xi)$ is the orthogonal projection onto the hyperplane containing 0 and orthogonal to ξ.

Theorem 3. (i) *The affine surface area is affine invariant, i. e., for all affine maps* $T :$ $\mathbb{R}^n \to \mathbb{R}^n$ *and all convex bodies* K *in* $\mathbb{R}^n$

$$\mathrm{as}(T(K)) = |\det(T)|^{\frac{n-1}{n+1}} \, \mathrm{as}(K).$$

(ii) *For all polytopes P in $\mathbb{R}^n$ we have* $\mathrm{as}(P) = 0$.

(iii) *The affine surface area* $\mathrm{as} : \mathcal{K}^n \to \mathbb{R}$ *is upper semicontinuous with respect to the Hausdorff metric on* $\mathcal{K}^n$.

(iv) *For all convex bodies K in $\mathbb{R}^n$*

$$\mathrm{as}(K)^{n+1} \le \mathrm{as}(B_2^n)^2 n^{n-1} \mathrm{vol}_n(K)^{n-1}. \tag{5.1}$$

Equality holds if and only if K is an ellipsoid. The inequality is called affine isoperimetric inequality.

(v) *For all convex bodies K in $\mathbb{R}^n$*

$$\mathrm{as}(K)^{n+1} \le \frac{n^{n+1}\,\mathrm{vol}_n(B_2^n)^n}{\mathrm{vol}_{n-1}(B_2^{n-1})^n}\,\mathrm{vol}_n(\Pi(K))$$

with equality if and only if K is an ellipsoid. This inequality is called the Petty affine projection inequality.

Proof. (i) By (3.10),

$$\lim_{t\to 0} \frac{\mathrm{vol}_n(T(K)) - \mathrm{vol}_n((T(K))_t)}{t^{\frac{2}{n+1}}} = \frac{1}{2}\left(\frac{n+1}{\mathrm{vol}_{n-1}(B_2^{n-1})}\right)^{\frac{2}{n+1}} \int\limits_{\partial T(K)} \kappa(x)^{\frac{1}{n+1}}\,d\mu_{\partial T(K)}(x).$$

Since $(T(K))_t = T(K_{\frac{t}{|\det(T)|}})$,

$$\frac{1}{2}\left(\frac{n+1}{\mathrm{vol}_{n-1}(B_2^{n-1})}\right)^{\frac{2}{n+1}} \int\limits_{\partial T(K)} \kappa(x)^{\frac{1}{n+1}}\,d\mu_{\partial T(K)}(x)$$

$$= \lim_{t\to 0} \frac{\mathrm{vol}_n(T(K)) - \mathrm{vol}_n(T(K_{\frac{t}{|\det(T)|}}))}{t^{\frac{2}{n+1}}} = |\det(T)| \lim_{t\to 0} \frac{\mathrm{vol}_n(K) - \mathrm{vol}_n(K_{\frac{t}{|\det(T)|}})}{t^{\frac{2}{n+1}}}.$$

With $s = \frac{t}{|\det(T)|}$,

$$\frac{1}{2}\left(\frac{n+1}{\mathrm{vol}_{n-1}(B_2^{n-1})}\right)^{\frac{2}{n+1}} \int\limits_{\partial T(K)} \kappa(x)^{\frac{1}{n+1}}\,d\mu_{\partial T(K)}(x) = |\det(T)|^{\frac{n-1}{n+1}} \lim_{t\to 0} \frac{\mathrm{vol}_n(K) - \mathrm{vol}_n(K_s)}{s^{\frac{2}{n+1}}}$$

$$= \frac{1}{2}|\det(T)|^{\frac{n-1}{n+1}}\left(\frac{n+1}{\mathrm{vol}_{n-1}(B_2^{n-1})}\right)^{\frac{2}{n+1}} \int\limits_{\partial K} \kappa(x)^{\frac{1}{n+1}}\,d\mu_{\partial K}(x).$$

(ii) Since the curvature of a polytope is 0 almost everywhere, the equality follows immediately by (3.10). Lutwak also showed this for his definition of affine surface area (3.7) in [31].

(iii) Lutwak showed that the affine surface area is upper semicontinuous [31]. Another proof of the upper semicontinuity by Ludwig can be found in [29].

(iv) Lutwak proved this inequality [31, Corollary 7.7].

Hug gave a proof using Steiner symmetrization [23]. It is very similar to the proof of the isoperimetric inequality via Steiner symmetrization. The same idea appears in the proof of the Blaschke–Santaló inequality by Meyer and Pajor [35].

Let K be a convex body in $\mathbb{R}^n$ that contains 0 as an interior point and let $H(0,\xi)$ be the hyperplane containing 0 and orthogonal to ξ. Let P_ξ be the orthogonal projection onto the hyperplane H. We define $f_\xi^-, f_\xi^+ : P_\xi(K) \to \mathbb{R}$ by

$$f_\xi^- = \min\{t \in \mathbb{R} \mid x + t\xi \in K\},$$
$$f_\xi^+ = \max\{t \in \mathbb{R} \mid x + t\xi \in K\}.$$

By (2.4),

$$\mathrm{as}(K) = \int_{P_\xi(K)} (\det(d^2 f_\xi^-))^{\frac{1}{n+1}} \, d_H x + \int_{P_\xi(K)} (\det(d^2(-f_\xi^+)))^{\frac{1}{n+1}} \, d_H x,$$

where $d_H x$ is the Lebesgue measure on the hyperplane H. For all symmetric, positive semidefinite $(n-1) \times (n-1)$ matrices A and B

$$(\det A)^{\frac{1}{n+1}} + (\det B)^{\frac{1}{n+1}} \le 2\left(\det\left(\frac{1}{2}(A+B) \right) \right)^{\frac{1}{n+1}}.$$

This inequality follows from the standard inequality for positive semidefinite $n \times n$ matrices A and B

$$(\det A)^{\frac{1}{n}} + (\det B)^{\frac{1}{n}} \le (\det(A+B))^{\frac{1}{n}}$$

by considering the $(n+1) \times (n+1)$ matrices

$$\begin{pmatrix} A & & \\ & 1 & \\ & & 1 \end{pmatrix} \quad \text{and} \quad \begin{pmatrix} B & & \\ & 1 & \\ & & 1 \end{pmatrix}.$$

Therefore,

$$\mathrm{as}(K) \le 2 \int_{p_H(K)} \left(\det\left(\frac{1}{2} d^2 (f_\xi^- - f_\xi^+) \right) \right)^{\frac{1}{n+1}} \, d_H x = \mathrm{as}(\mathrm{St}_H(K)).$$

Now we choose a sequence of symmetrizations that transforms the convex body K into the Euclidean ball. Finally, we use the fact that the affine surface area is upper semicontinuous.

(v) Lutwak proved the Petty affine projection inequality for arbitrary convex bodies [31]. $\qquad\qquad \square$

In general, it is not easy to compute the affine surface area of a given convex body. But it is possible for the convex bodies B_p^n. For this we need the following lemma.

Let A be an $n \times n$ matrix. The (k, ℓ)-th coordinate of the cofactor matrix $\text{cof}(A)$ is $(-1)^{k+\ell}$ times the determinant of the matrix A after deleting the k-th row and ℓ-th column.

Lemma 3 ([11, 14, 17]). *Let $\mathcal{U}$ be a convex, open subset of $\mathbb{R}^{n+1}$ and let $x_0 \in \mathcal{U}$. Let $F : \mathcal{U} \to \mathbb{R}$ be twice continuously differentiable in a neighborhood of x_0. Moreover, suppose that $F(x_0) = 0$ and that $F(x) \le 0$ is a convex body contained in $\mathcal{U}$. Then*

$$\kappa(x_0) = \frac{|(\nabla F(x_0))^t \, \text{cof}(\nabla^2 F)(x_0) \nabla F(x_0)|}{\|\nabla F(x_0)\|_2^{n+2}} = \frac{|\det(\begin{smallmatrix} \nabla^2 F & \nabla F \\ (\nabla F)^t & 0 \end{smallmatrix})|}{\|\nabla F\|_2^{n+2}}, \tag{5.2}$$

where ∇F denotes the gradient and $\nabla^2 F$ denotes the Hessian of F.

Example 4. Let $1 < p < \infty$. Then we have for the Gauss–Kronecker curvature of ∂B_p^n at x with $x_i \ne 0$ for all $i = 1, \dots, n$

$$\kappa(x) = \frac{(p-1)^{n-1}(\prod_{i=1}^n |x_i|^{p-2})}{(\sum_{i=1}^n |x_i|^{2p-2})^{\frac{n+1}{2}}}.$$

For the affine surface area we get

$$\int_{\partial B_p^n} \kappa(x)^{\frac{1}{n+1}} \, d\mu_{\partial B_p^n}(x) = \frac{2^n (p-1)^{\frac{n-1}{n+1}} \Gamma(\frac{p+n-1}{(n+1)p})^n}{p^{n-1} \Gamma(\frac{n(p+n-1)}{(n+1)p})}.$$

A map $\Phi : \mathcal{K}^n \to \mathbb{R}$ is called a valuation if for all $K, C \in \mathcal{K}^n$ such that $K \cup C \in \mathcal{K}^n$

$$\Phi(K \cup C) + \Phi(K \cap C) = \Phi(K) + \Phi(C).$$

Theorem 4 ([41]). *The affine surface area is a valuation, i. e., for all $K, C \in \mathcal{K}^n$ such that $K \cup C \in \mathcal{K}^n$*

$$\text{as}(K \cup C) + \text{as}(K \cap C) = \text{as}(K) + \text{as}(C). \tag{5.3}$$

Proof. We decompose the involved boundaries into disjoint sets:

$$\partial(C \cup K) = (\partial C \cap \partial K) \cup (\partial C \cap K^c) \cup (\partial K \cap C^c),$$
$$\partial(C \cap K) = (\partial C \cap \partial K) \cup (\partial C \cap \text{int}(K)) \cup (\partial K \cap \text{int}(C)),$$
$$\partial C = (\partial C \cap \partial K) \cup (\partial C \cap K^c) \cup (\partial C \cap \text{int}(K)),$$
$$\partial K = (\partial C \cap \partial K) \cup (\partial K \cap C^c) \cup (\partial K \cap \text{int}(C)).$$

In order to show (5.3) it is enough to prove

$$\int\limits_{\partial C \cap \partial K} \kappa_{C \cup K}^{\frac{1}{n+1}} d\mu_{\partial(C \cup K)} + \int\limits_{\partial C \cap \partial K} \kappa_{C \cap K}^{\frac{1}{n+1}} d\mu_{\partial(C \cup K)} = \int\limits_{\partial C \cap \partial K} \kappa_{C}^{\frac{1}{n+1}} d\mu_{\partial(C)} + \int\limits_{\partial C \cap \partial K} \kappa_{K}^{\frac{1}{n+1}} d\mu_{\partial(C)},$$

where $\kappa_{C \cup K}$ is the curvature of $\partial(C \cup K)$ and the others accordingly. This follows from

$$\kappa_{C \cup K}(x) = \min\{\kappa_C(x), \kappa_K(x)\}, \quad \kappa_{C \cap K}(x) = \max\{\kappa_C(x), \kappa_K(x)\}.$$

This in turn is true since the indicatrix of Dupin (2.2) of $C \cup K$ is the union of the indicatrices of Dupin of C and K at x and the indicatrix of Dupin of $C \cap K$ at x is the intersection of the indicatrices of Dupin of C and K at x. Moreover, the intersection or union of two ellipsoids is again an ellipsoid if and only if one ellipsoid is contained in the other. □

By the results of Hadwiger, a map $\Phi : \mathcal{K}^n \to \mathbb{R}$ is a continuous valuation that is invariant under rigid motions if and only if it is a linear combination of quermasssintegrals. A continuous valuation that is invariant under rigid motions and that is $(n-j)$-homogeneous is a multiple of the j-th quermassintegral [21].

Ludwig and Reitzner [30] showed that aside from the Euler characteristic and the volume, there is only one non-trivial upper semicontinuous valuation that is affine invariant, namely the affine surface area.

Theorem 5 ([30]). *A functional $\Phi : \mathcal{K}^n \to \mathbb{R}$ is an upper semicontinuous and translation and $SL(n)$ invariant valuation if there are constants $c_0, c_1 \in \mathbb{R}$ and $c_2 \geq 0$ such that for all $K \in \mathcal{K}^n$*

$$\Phi(K) = c_0 V_0(K) + c_1 \operatorname{vol}_n(K) + c_2 \operatorname{as}(K),$$

where V_0 is the Euler characteristic.

6 Random polytopes and best approximation

The affine surface area appears in best approximation of convex bodies by polytopes and in approximation by random polytopes. The next theorems are examples.

Theorem 6. *Let K be a convex body in $\mathbb{R}^n$ that has a C_+^2-boundary. Then there are constants del_{n-1} and div_{n-1} depending only on the dimension n such that*

$$\lim_{N \to \infty} \left(N^{\frac{2}{n-1}} \inf_{\substack{\mathrm{vert}(P) \leq N \\ P \subseteq K}} d_S(K, P) \right) = \frac{1}{2} \mathrm{del}_{n-1} \left(\int\limits_{\partial K} \kappa(x)^{\frac{1}{n+1}} d\mu_{\partial K}(x) \right)^{\frac{n+1}{n-1}}, \tag{6.1}$$

$$\lim_{N \to \infty} \left(N^{\frac{2}{n-1}} \inf_{\substack{\mathrm{vert}(P) \leq N \\ K \subseteq P}} d_S(K, P) \right) = \frac{1}{2} \mathrm{div}_{n-1} \left(\int\limits_{\partial K} \kappa(x)^{\frac{1}{n+1}} d\mu_{\partial K}(x) \right)^{\frac{n+1}{n-1}}, \tag{6.2}$$

where P denotes polytopes.

This theorem was proved by McClure and Vitale in the 2-dimensional case [34]. Gruber proved this for arbitrary dimensions [18].

In order to estimate or compute the constants del_{n-1} it is enough to consider the case $K = B_2^n$. Kabatjanskii and Levenstein [25] estimated del_{n-1} from above, and Gordon, Reisner, and Schütt estimated del_{n-1} from below [15, 16]. The proofs were simplified and the estimates were improved by Mankiewicz and Schütt [32, 33]. We have

$$\frac{n-1}{n+1} \mathrm{vol}_{n-1}\big(B_2^{n-1}\big)^{-\frac{2}{n-1}} \le \mathrm{del}_{n-1} \le 2^{0.802}\, \mathrm{vol}_{n-1}\big(\partial B_2^n\big)^{-\frac{2}{n-1}}$$

and

$$\lim_{n\to\infty} \frac{\mathrm{del}_{n-1}}{n} = \frac{1}{2\pi e} = 0.0585498\ldots.$$

A random polytope of a convex body K is the convex hull of finitely many points that are chosen randomly from K with respect to a probability measure. The expected volume of a random polytope of N chosen points is

$$\mathbb{E}(K,N) = \left(\frac{1}{\mathrm{vol}_n(K)}\right)^N \int_K \cdots \int_K \mathrm{vol}_n([x_1,\ldots,x_N])\,dx_1\cdots dx_N, \tag{6.3}$$

where $[x_1,\ldots,x_N]$ denotes the convex hull of the points $x_1,\ldots,x_N$.

Theorem 7. *Let K be a convex body in $\mathbb{R}^n$. Then*

$$c_n \lim_{N\to\infty} \frac{\mathrm{vol}_n(K) - \mathbb{E}(K,N)}{\left(\frac{\mathrm{vol}_n(K)}{N}\right)^{\frac{2}{n+1}}} = \int_{\partial K} \kappa^{\frac{1}{n+1}}\,d\mu_{\partial K}, \tag{6.4}$$

where

$$c_n = 2\left(\frac{\mathrm{vol}_{n-1}(B_2^{n-1})}{n+1}\right)^{\frac{2}{n+1}} \frac{(n+3)(n+1)!}{(n^2+n+2)(n^2+1)\Gamma(\frac{n^2+1}{n+1})}.$$

Theorem 7 was first proved by Bárány for convex bodies with C^3-boundary [7]. For general convex bodies it was proved by Schütt [42]; see also [9].

The following theorem answers the analogous question for random polytopes whose vertices are chosen from the boundary of the convex body. Let $f : \partial K \to \mathbb{R}$ be a continuous function with $\int_{\partial K} f\,d\mu_{\partial K} = 1$. Then the expected volume of a random polytope of N points chosen randomly from the boundary of K with respect to the probability measure $f\,d\mu_{\partial K}$ is

$$\mathbb{E}(f,K) = \int_{\partial K} \cdots \int_{\partial K} \mathrm{vol}_n([x_1,\ldots,x_N]) f(x_1)\cdots f(x_N)\,d\mu_{\partial K}\cdots d\mu_{\partial K}.$$

Theorem 8 ([45]). *Let K be a convex body in $\mathbb{R}^n$ such that there are r and R in $\mathbb{R}$ with $0 < r \leq R < \infty$ so that we have for all $x \in \partial K$*

$$B_2^n(x - rN(x), r) \subseteq K \subseteq B_2^n(x - RN(x), R)$$

and let $f : \partial K \to [0, \infty)$ be a continuous function with $\int_{\partial K} f d\mu_{\partial K} = 1$. Let $\mathbb{P}_f$ be the probability measure on ∂K with $d\mathbb{P}_f(x) = f(x) d\mu_{\partial K}(x)$. Then we have

$$\lim_{N \to \infty} \frac{\mathrm{vol}_n(K) - \mathbb{E}(f, N)}{\left(\frac{1}{N}\right)^{\frac{2}{n-1}}} = c_n \int_{\partial K} \frac{\kappa(x)^{\frac{1}{n-1}}}{f(x)^{\frac{2}{n-1}}} d\mu_{\partial K}(x), \tag{6.5}$$

where κ is the generalized Gauss–Kronecker curvature and

$$c_n = \frac{(n-1)^{\frac{n+1}{n-1}} \Gamma(n + 1 + \frac{2}{n-1})}{2(n+1)!(\mathrm{vol}_{n-2}(\partial B_2^{n-1}))^{\frac{2}{n-1}}}.$$

The minimum at the right-hand side is attained for the normalized affine surface area measure with density

$$f_{\mathrm{as}}(x) = \frac{\kappa(x)^{\frac{1}{n+1}}}{\int_{\partial K} \kappa(x)^{\frac{1}{n+1}} d\mu_{\partial K}(x)}.$$

This result was proved by Schütt and Werner [45] and [44] and at the same time for convex bodies with C_+^2-boundary by Reitzner [38]. It is interesting that the minimum of (6.5) is attained for the affine surface measure.

7 Constrained convex bodies with maximal affine surface area

We introduce the analog to John's theorem [24], when volume is replaced by affine surface area. In parallel to John's maximal volume ellipsoid, Giladi, Huang, Schütt, and Werner investigated these convex bodies contained in K that have the largest affine surface areas.

The isotropic constant L_K of K is defined by

$$nL_K^2 = \min\left\{\frac{1}{\mathrm{vol}_n(TK)^{1+\frac{2}{n}}} \int_{a+TK} \|x\|^2 dx \,\Big|\, a \in \mathbb{R}^n, T \in GL(n)\right\}. \tag{7.1}$$

Theorem 9 ([13]). *There is a constant $c > 0$ such that for all $n \in \mathbb{N}$ and all convex bodies $K \subseteq \mathbb{R}^n$,*

$$\frac{1}{n^{5/6}}\left(\frac{c}{L_K}\right)^{\frac{2n}{n+1}} n\,\mathrm{vol}_n(B_2^n)^{\frac{2}{n+1}} \mathrm{vol}_n(K)^{\frac{n-1}{n+1}} \leq \sup_{C \subseteq K} \mathrm{as}(C) \leq n\,\mathrm{vol}_n(B_2^n)^{\frac{2}{n+1}} \mathrm{vol}_n(K)^{\frac{n-1}{n+1}},$$

where we take the supremum over all convex bodies C that are contained in K. Equality holds in the right inequality iff K is a centered ellipsoid.

The theorem shows that $\sup_{C \subseteq K} \operatorname{as}(C)$ is proportional to $\operatorname{vol}_n(K)^{\frac{n-1}{n+1}}$.

Proof. The right-hand side inequality follows immediately from the affine isoperimetric inequality (5.1).

We use a thin shell estimate by Guédon and E. Milman [20] (see also Paouris [36]) on concentration of volume: Given an isotropic random vector X with log-concave density in Euclidean space $\mathbb{R}^n$, we have for all $t \geq 0$

$$\mathbb{P}\big(\big|\|X\|_2 - \sqrt{n}\big| \geq t\sqrt{n}\big) \leq C \exp\big(-c\sqrt{n}\min\{t^3, t\}\big).$$

For the estimate from below we choose as the convex body

$$C = K \cap B_2^n(0, c_n),$$

where c_n is appropriately chosen and is of the order $\sqrt{n}$. By the concentration result, the volume of $K \cap B_2^n(0, c_n)$ is of the same order as K. This also implies that the surface area of $K \cap B_2^n(0, c_n)$ is of the same order as that of K. Furthermore, this implies that the part of the boundary of $K \cap B_2^n(0, c_n)$ which is also part of the boundary of $B_2^n(0, c_n)$ is big. The curvature on that part is easily computed. $\qquad\square$

Bibliography

[1] A. D. Aleksandrov, *Almost everywhere existence of the second differential of a convex function and some properties of convex surfaces connected with it*, Uchenye Zap. Leningr. Gos. Univ., Math. Ser. **6** (1939), 3–35.

[2] V. Bangert, *Analytische Eigenschaften konvexer Funktionen auf Riemannschen Mannigfaltigkeiten*, J. Reine Angew. Math. **307** (1979), 309–324.

[3] G. Bianchi, A. Colesanti and C. Pucci, *On the second differentiability of convex surfaces*, Geom. Dedic. **60** (1996), 39–48.

[4] W. Blaschke, *Differentialgeometrie II, Affine Differentialgeometrie*, Springer-Verlag, Berlin, 1923.

[5] T. Bonnesen and W. Fenchel, *Theorie der konvexen Körper*, Springer-Verlag, 1934.

[6] H. Busemann and W. Feller, *Krümmungseigenschaften konvexer Flächen*, Acta Math. **66** (1935), 1–47.

[7] I. Bárány, *Random polytopes in smooth convex bodies*, Mathematika **39** (1992), 81–92.

[8] I. Bárány and D. G. Larman, *Convex bodies, economic cap covering, random polytopes*, Mathematika **35** (1988), 274–291.

[9] K. J. Böröczky, L. M. Hoffmann and D. Hug, *Expectation of intrinsic volumes of random polytopes*, Period. Math. Hung. **57** (2008), 143–164.

[10] G. Dolzmann and D. Hug, *Equality of two representations of extended affine surface area*, Arch. Math. **65** (1995), 352–356.

[11] P. Dombrowski, *Krümmungsgrößen gleichungsdefinierter Untermannigfaltigkeiten Riemannscher Mannigfaltigkeiten*, Math. Nachr. **38** (1968), 133–180.

[12] C. Dupin, *Application de géometrie et de méchanique á la marine, aux ponts et chaussées*, Paris, 1822.

[13]   O. Giladi, H. Huang, C. Schütt and E. Werner, *Constrained convex bodies with extremal affine surface areas*, J. Funct. Anal. **279** (2020), 108531.

[14]   R. Goldman, *Curvature formulas for implicit curves and surfaces*, Comput. Aided Geom. Des. **22** (2005), 632–658.

[15]   Y. Gordon, S. Reisner and C. Schütt, *Umbrellas and polytopal approximation of the Euclidean ball*, J. Approx. Theory **90** (1997), 9–22.

[16]   Y. Gordon, S. Reisner and C. Schütt, *Erratum*, J. Approx. Theory **95** (1998), 331.

[17]   D. Gromoll, W. Klingenberg and W. Meyer, *Riemannsche Geometrie im Großen*, Lecture notes in mathematics, vol. 55, Springer-Verlag, 1975.

[18]   P. M. Gruber, *Asymptotic estimates for best and stepwise approximation of convex bodies II*, Forum Math. **5** (1993), 521–538.

[19]   P. M. Gruber, *Convex and discrete geometry*, Springer-Verlag, 2007.

[20]   O. Guédon and E. Milman, *Interpolating thin-shell and sharp large-deviation estimates for isotropic log-concave measures*, Geom. Funct. Anal. **21** (2011), 1043–1068.

[21]   H. Hadwiger, *Vorlesungen über Inhalt, Oberfläche und Isoperimetrie*, Springer-Verlag, 1957.

[22]   R. Howard, *Blaschke's rolling theorem for manifolds with boundary*, Manuscr. Math. **99** (1999), 471–483.

[23]   D. Hug, *Contributions to affine surface area*, Manuscr. Math. **91** (1996), 283–301.

[24]   F. John, *Extremum problems with inequalities as subsidiary conditions*, in: R. Courant anniversary volume, Interscience, New York, 1948, pp. 187–204.

[25]   G. A. Kabatjanskii and V. I. Levenstein, *Bounds for packings on a sphere and in space*, Probl. Inf. Transm. **14** (1978), 1–17.

[26]   K. Leichtweiss, *Zur Affinoberfläche konvexer Körper*, Manuscr. Math. **56** (1986), 429–464.

[27]   K. Leichtweiss, *Über eine Formel Blaschkes zur Affinoberfläche*, Studia Sci. Math. Hung. **21** (1987), 453–474.

[28]   K. Leichtweiss, *On the history of the affine surface area for convex bodies*, Results Math. **20** (1991), 650–656.

[29]   M. Ludwig, *On the semicontinuity of curvature integrals*, Math. Nachr. **227** (2001), 99–108.

[30]   M. Ludwig and M. Reitzner, *A characterization of affine surface area*, Adv. Math. **147** (1999), 138–172.

[31]   E. Lutwak, *Extended affine surface area*, Adv. Math. **85** (1991), 39–68.

[32]   P. Mankiewicz and C. Schütt, *A simple proof of an estimate for the approximation of the Euclidean ball and the Delone triangulation numbers*, J. Approx. Theory **107** (2000), 268–280.

[33]   P. Mankiewicz and C. Schütt, *On the Delone triangulations numbers*, J. Approx. Theory **111** (2001), 139–142.

[34]   D. McClure and R. Vitale, *Polygonal approximation of plane convex bodies*, J. Math. Anal. Appl. **51** (1975), 326–358.

[35]   M. Meyer and A. Pajor, *On the Blaschke–Santaló inequality*, Arch. Math. **55** (1990), 82–93.

[36]   G. Paouris, *Concentration of mass on convex bodies*, Geom. Funct. Anal. **16** (2006), 1021–1049.

[37]   C. Petty, *Geominimal surface area*, Geom. Dedic. **3** (1974), 77–97.

[38]   M. Reitzner, *Random points on the boundary of smooth convex bodies*, Trans. Am. Math. Soc. **354** (2002), 2243–2278.

[39]   R. T. Rockafellar, *Convex analysis*, Princeton University Press, 1970.

[40]   R. Schneider, *Convex bodies: the Brunn–Minkowski theory*, Encyclopedia of mathematics and its applications, vol. 44, Cambridge University Press, Cambridge, 1993.

[41]   C. Schütt, *On the affine surface area*, Proc. Am. Math. Soc. **118** (1993), 1213–1218.

[42]   C. Schütt, *Random polytopes and affine surface area*, Math. Nachr. **170** (1994), 227–249.

[43]   C. Schütt and E. Werner, *The convex floating body*, Math. Scand. **66** (1990), 275–290.

[44]   C. Schütt and E. Werner, *Random polytopes with vertices on the boundary of a convex body*, C. R. Acad. Sci. Paris **331** (2000), 697–701.

[45]   C. Schütt and E. Werner, *Polytopes with vertices chosen randomly from the boundary of a convex body*, in: Israel seminar 2001–2002, V. D. Milman and G. Schechtman, eds., Lecture notes in mathematics, vol. 1807, Springer-Verlag, 2003, pp. 241–422.

[46]   C. Schütt and E. Werner, *Floating body*, in preparation.

[47]   E. Werner, *Illumination bodies and affine surface area*, Stud. Math. **110** (1994), 257–269.

M. Angeles Alfonseca, Fedor Nazarov, Dmitry Ryabogin, and
Vladyslav Yaskin

Analysis and geometry near the unit ball: proofs, counterexamples, and open questions

Abstract: We present several theorems, counterexamples, and open questions related to convex bodies close to the unit ball. The techniques include spherical harmonic decomposition and some elements of perturbation theory. We hope that this short survey will attract the attention of both young and mature researchers who will be able to surpass our results and resolve some questions we left unanswered.

Keywords: Star bodies, convex bodies, sections, projections, spherical harmonics, spherical Radon transform

MSC 2020: 52A20, 52A38, 44A12, 33C55

1 Introduction

Among convex bodies, the Euclidean ball B is distinguished by several remarkable properties. The most obvious one is that it is perfectly round in the sense that any characteristic of a convex body that, generally speaking, depends on the direction (width, central cross-section area, projection area, etc.) stays constant for B. This property alone has been a source of numerous questions (some resolved and some still open) of the type "If a convex body is round in some particular sense, is it necessarily a ball?"

The second, slightly less obvious, property is that the ball is an extremizer in various minimization and maximization problems in convex geometry, the most famous of which is, probably, the isoperimetric inequality.

Finally, the unit sphere is essentially the only example of the boundary of a convex body on which harmonic analysis is not only possible in principle, but also rich and well developed. This allows one to use various tools from harmonic and functional anal-

Acknowledgement: The first author is supported in part by the Simons Foundation Grant 711907. The second and third authors are supported in part by U. S. National Science Foundation Grants DMS-1900008 and DMS-1600753. The fourth author is supported in part by NSERC.

M. Angeles Alfonseca, Department of Mathematics, North Dakota State University,Fargo, ND 58108, USA, e-mail: maria.alfonseca@ndsu.edu
Fedor Nazarov, Dmitry Ryabogin, Department of Mathematical Sciences, Kent State University, Kent, OH 44242, USA, e-mails: nazarov@math.kent.edu, ryabogin@math.kent.edu
Vladyslav Yaskin, Department of Mathematical and Statistical Sciences, University of Alberta, Edmonton, Canada, e-mail: vladyaskin@math.ualberta.ca

https://doi.org/10.1515/9783110775389-011

ysis when dealing with problems whose formulations have nothing to do with spherical harmonics, Hilbert spaces, or operator eigenvalues.

In this chapter, we will endeavor to show the reader a few tricks and techniques from perturbation theory near the unit ball. We have chosen the perturbative regime as both the easiest one to explain and the only one for which we have gained some decent understanding. In a certain sense it is quite natural: if one has some property satisfied by B or some inequality for which B is presumed to be an extremizer, it is quite tempting to ask if the same property can be preserved by a small perturbation of B or if B is at least a local extremizer. It often turns out that the local version of the question is much easier than the global one and can be answered completely. In our opinion, such investigation must be carried out every time a new conjecture is set forth, though we could not find any trace of it in the literature known to us. This curious fact served as a motivation for several recent projects of ours (often together with other people), some of which we attempt to summarize in this survey.

2 Notation and preliminary observations

Let $K \subset \mathbb{R}^n$ be a convex body containing the origin in its interior. The radial function $\varrho = \varrho_K : \mathbb{S}^{n-1} \to \mathbb{R}$ is defined by

$$\varrho(e) = \max\{t > 0 : te \in K\}.$$

The support function $\mathfrak{h} = \mathfrak{h}_K : \mathbb{S}^{n-1} \to \mathbb{R}$ is defined by

$$\mathfrak{h}(e) = \max\{\langle x, e \rangle : x \in K\}$$

(note that the same formula makes sense for all $e \in \mathbb{R}^n$ and gives a 1-homogeneous extension of $\mathfrak{h}$ to the entire space). Their geometric meanings are the length of the longest interval starting at the origin in the direction e that is contained in K and the distance from the origin to the support hyperplane of K parallel to

$$e^{\perp} = \{y \in \mathbb{R}^n : \langle y, e \rangle = 0\}$$

in the direction e.

We always have $\varrho_K \leq \mathfrak{h}_K$. For the unit ball centered at the origin, we have $\varrho = \mathfrak{h} \equiv 1$.

The closeness of a convex body K to the unit ball will be usually measured in the Hausdorff distance

$$d(K, L) = \inf\{r > 0 : K + rB \supset L, \; L + rB \supset K\}.$$

When $L = B$, the inequality $d(K, B) \leq \varepsilon < 1$ merely means that

$$(1 - \varepsilon)B \subset K \subset (1 + \varepsilon)B.$$

In the case where the formulation of the problem is invariant under linear transformations, a more natural distance to consider is the Banach–Mazur one: $d_{BM}(K,L) = \log\inf\{R > 1 : L \subset TK \subset RL$ for some linear transformation T of $\mathbb{R}^n\}$. However, if $d_{BM}(K,B) < \varepsilon$, then, replacing K by an appropriate linear image TK, we get $B \subset TK \subset e^\varepsilon B$, so $d(TK,B) \le e^\varepsilon - 1$. Thus, in the linear invariant case, the results for convex bodies close to the unit ball in the Banach–Mazur distance immediately follow from those for convex bodies close to the unit ball in the Hausdorff one.

Dealing with problems that are invariant under linear transformations presents one more difficulty: the ball is no longer going to be a unique solution here; any ellipsoid will be just as good.

To avoid this non-uniqueness, in such cases we will always consider the so-called isotropic position of K, that is, the linear image of K for which the quadratic form $x \mapsto \int_K \langle x,y\rangle^2 dy$ is a multiple of $|x|^2$ (see [4, Section 2.3.2] or [1, Section 5] for details). What is important for us here is that if $d(K,B)$ is small, then the Hausdorff distance from the isotropic position of K to B is also small. More precisely, if $d(K,B) < \varepsilon$, i. e., $(1-\varepsilon)B \subset K \subset (1+\varepsilon)B$, then for the isotropic position K' of K, we have

$$(1-\varepsilon)\left(\frac{1-\varepsilon}{1+\varepsilon}\right)^{\frac{n+2}{2}} B \subset K' \subset (1+\varepsilon)\left(\frac{1+\varepsilon}{1-\varepsilon}\right)^{\frac{n+2}{2}} B$$

(see [1, Section 5]).

2.1 Spherical harmonics

We shall now briefly remind the reader of a few basic definitions and facts from harmonic analysis on the unit sphere. More details and applications can be found in [7].

Let $\mathcal{P}$ be the linear space of polynomials of n variables, i. e., finite linear combinations of monomials x^α, where $\alpha = (\alpha_1,\ldots,\alpha_n)$, $\alpha_j \in \mathbb{Z}_+$, and $x^\alpha = x_1^{\alpha_1}\cdots x_n^{\alpha_n}$. If $P = \sum_\alpha c_\alpha x^\alpha \in \mathcal{P}$, then by $P(D)$ we shall mean the differential operator $\sum_\alpha c_\alpha (\frac{\partial}{\partial x_1})^{\alpha_1}\cdots(\frac{\partial}{\partial x_n})^{\alpha_n}$.

Let $\langle P,Q\rangle = P(D)\overline{Q}|_{x=0}$. Then a direct computation shows that $\langle x^\alpha, x^\beta\rangle = 0$ if $\alpha \ne \beta$ and $\langle x^\alpha, x^\alpha\rangle = \alpha! = \alpha_1!\cdots\alpha_n!$. It follows that $\langle \cdot,\cdot\rangle$ is a scalar product on $\mathcal{P}$ for which $\frac{x^\alpha}{\sqrt{\alpha!}}$ is an orthonormal basis.

Let $\mathcal{P}_m$ be the subspace of $\mathcal{P}$ consisting of all homogeneous polynomials of degree m ($m = 0,1,2,\ldots$), i. e., the space of linear combinations of monomials x^α with $\alpha_1 + \cdots + \alpha_n = m$. We have $\dim \mathcal{P}_m = \binom{m+n-1}{n-1}$ (see [7, pp. 65–66]). Notice that $|x|^2\mathcal{P}_{m-2}$ is a linear subspace of $\mathcal{P}_m$. If $P \in \mathcal{P}_m$ and $Q \in \mathcal{P}_{m-2}$, then

$$\langle |x|^2 Q, P\rangle = (|x|^2 Q)(D)\overline{P}|_{x=0} = Q(D)\overline{\Delta P}|_{x=0} = \langle Q, \Delta P\rangle,$$

where $\Delta = \frac{\partial^2}{\partial x_1^2} + \cdots + \frac{\partial^2}{\partial x_n^2}$ is the usual Laplace operator in $\mathbb{R}^n$. Thus, $P \in \mathcal{P}_m$ is orthogonal to $|x|^2\mathcal{P}_{m-2}$ if and only if $\Delta P = 0$, i. e., the space H_m of *harmonic* homogeneous polynomials of degree m is the orthogonal complement of $|x|^2\mathcal{P}_{m-2}$ in $\mathcal{P}_m$. We see that every $P \in \mathcal{P}_m$ can be decomposed as $P_m + |x|^2 Q_{m-2}$, where $P_m \in H_m$ and $Q_{m-2} \in \mathcal{P}_{m-2}$. Repeating this decomposition for Q_{m-2} instead of P and going all the way down, we get the representation

$$P = P_m + |x|^2 P_{m-2} + |x|^4 P_{m-4} + \cdots,$$

where $P_j \in H_j, j = m, m-2, m-4, \dots$. On $\mathbb{S}^{n-1}$, we have $|x|^2 = 1$ and, therefore,

$$P = P_m + P_{m-2} + P_{m-4} + \cdots,$$

i. e., every homogeneous polynomial $P \in \mathcal{P}_m$, as a function on $\mathbb{S}^{n-1}$, can be written as a linear combination of homogeneous harmonic polynomials of degrees $m, m-2, \dots$. Since every polynomial $P \in \mathcal{P}$ can be decomposed into a sum of homogeneous polynomials, we conclude that every $P \in \mathcal{P}$, as a function on $\mathbb{S}^{n-1}$, can be represented as a sum of finitely many $P_j \in H_j$.

Note that, for $k \neq j$, the Green formula combined with the homogeneity property yields

$$0 = \int_{B_n} [(\Delta P_k)\overline{P}_j - P_k(\Delta \overline{P}_j)]dx = \omega_{n-1} \int_{\mathbb{S}^{n-1}} \left[\left(\frac{\partial}{\partial n}P_k\right)\overline{P}_j - P_k\left(\frac{\partial}{\partial n}\overline{P}_j\right)\right]d\sigma_{n-1}$$
$$= \omega_{n-1}(k-j) \int_{\mathbb{S}^{n-1}} P_k\overline{P}_j d\sigma_{n-1},$$

where ω_{n-1} is the $(n-1)$-dimensional surface area of $\mathbb{S}^{n-1}$ and σ_{n-1} is the normalized surface area measure on $\mathbb{S}^{n-1}$. We see that P_k and P_j are orthogonal with respect to the usual scalar product in $L^2(\mathbb{S}^{n-1})$. Since $\mathcal{P}$ is dense in $L^2(\mathbb{S}^{n-1})$, it follows that we have an orthogonal decomposition

$$L^2(\mathbb{S}^{n-1}) = \bigoplus_{m=0}^{\infty} H_m,$$

i. e., every function $f \in L^2(\mathbb{S}^{n-1})$ can be uniquely written as a series $f = f_0 + f_1 + f_2 + \cdots$ with $f_m \in H_m$ and the series is orthogonal and convergent in $L^2(\mathbb{S}^{n-1})$. This decomposition is called the spherical harmonic decomposition of f.

2.2 Spherical Radon transform and spherical k-Radon transform

Recall that the Fourier transform in $\mathbb{R}^n$ is defined by

$$\widehat{f}(y) = \int_{\mathbb{R}^n} f(x)e^{-2\pi i\langle x,y\rangle} dx.$$

For reasonable (say, Schwartz class) functions $f : \mathbb{R}^n \to \mathbb{R}^n$, we have the inversion formula

$$f(x) = \int_{\mathbb{R}^n} \widehat{f}(y)e^{2\pi i\langle x,y\rangle}\,dy.$$

In particular,

$$f(0) = \int_{\mathbb{R}^n} \widehat{f}(y)\,dy. \tag{2.1}$$

Let $H \subset \mathbb{R}^n$ be a linear subspace of $\mathbb{R}^n$ and let $H^{\perp}$ be its orthogonal complement. Consider the function

$$F(x') = \int_{x'+H} f(x)\,dx, \quad x' \in H^{\perp}.$$

Notice that for $y' \in H^{\perp}$, we have

$$\int_{H^{\perp}} F(x')e^{-2\pi i\langle x',y'\rangle}\,dx' = \int_{\mathbb{R}^n} f(x)e^{-2\pi i\langle x,y'\rangle}\,dx = \widehat{f}(y'),$$

i. e., the Fourier transform of F is just $\widehat{f}|_{H^{\perp}}$. Applying (2.1), we obtain

$$\int_{H} f(x)\,dx = F(0) = \int_{H^{\perp}} \widehat{F}(y)\,dy = \int_{H^{\perp}} \widehat{f}(y)\,dy. \tag{2.2}$$

Recall also that direct integration by parts yields

$$(P(D)f)\hat{}(y) = \widehat{f}(y)P(2\pi iy)$$

for every polynomial $P \in \mathcal{P}$.

Now take $P \in H_m$ with even $m \geq 0$. A direct computation shows that

$$P(D)e^{-\pi|x|^2} = [(2\pi)^m P(-x) + Q(x)]e^{-\pi|x|^2} = [(2\pi)^m P(x) + Q(x)]e^{-\pi|x|^2},$$

where Q is a polynomial of degree $\deg Q < m$. Since $P \in H_m$ is orthogonal to all homogeneous polynomials of degree less than m in $L^2(\mathbb{S}^{n-1})$, it is also orthogonal to them in $L^2(\mathbb{R}^n, w)$ for any fast decaying radial weight w. Hence, it is orthogonal to *all* polynomials of degree less than m in $L^2(\mathbb{R}^n, w)$. In particular,

$$\int_{\mathbb{R}^n} (P(D)e^{-\pi|x|^2})\overline{Q(x)}\,dx = \int_{\mathbb{R}^n} |Q(x)|^2 e^{-\pi|x|^2}\,dx.$$

On the other hand, integration by parts yields

$$\int_{\mathbb{R}^n} (P(D)e^{-\pi|x|^2})\overline{Q(x)}\,dx = \int_{\mathbb{R}^n} e^{-\pi|x|^2}\overline{(P(D)\overline{Q})(x)}\,dx = 0$$

and we conclude that $Q \equiv 0$, so $P(D)e^{-\pi|x|^2} = (2\pi)^m P(x)e^{-\pi|x|^2}$.

Since the Fourier transform of $e^{-\pi|x|^2}$ is $e^{-\pi|y|^2}$, we have

$$(P(D)e^{-\pi|\cdot|^2})\,\hat{}\,(y) = P(2\pi i y)e^{-\pi|y|^2} = (-1)^{\frac{m}{2}}(2\pi)^m P(y)e^{-\pi|y|^2}.$$

Thus, (2.2) results in

$$\int_H P(x)e^{-\pi|x|^2}\,dx = (-1)^{\frac{m}{2}}\int_{H^\perp} P(y)e^{-\pi|y|^2}\,dy.$$

Using the m-homogeneity of P, we can rewrite this as

$$\int_{H\cap\mathbb{S}^{n-1}} P\,d\sigma_{k-1} \times \omega_{k-1} \times \int_0^\infty r^{m+k-1}e^{-\pi r^2}\,dr$$

$$= (-1)^{\frac{m}{2}}\int_{H^\perp\cap\mathbb{S}^{n-1}} P\,d\sigma_{n-k-1} \times \omega_{n-k-1} \times \int_0^\infty r^{m+n-k-1}e^{-\pi r^2}\,dr,$$

where $k = \dim H$. Taking into account that

$$\int_0^\infty r^{a-1}e^{-\pi r^2}\,dr = \frac{1}{2}\pi^{-\frac{a}{2}}\Gamma\!\left(\frac{a}{2}\right) \quad (a > 0), \qquad \omega_{l-1} = \frac{2\pi^{\frac{l}{2}}}{\Gamma(\frac{l}{2})} \quad (l = 1,2,3,\ldots),$$

we finally get

$$\frac{\Gamma(\frac{m+k}{2})}{\Gamma(\frac{k}{2})}\int_{H\cap\mathbb{S}^{n-1}} P\,d\sigma_{k-1} = (-1)^{\frac{m}{2}}\frac{\Gamma(\frac{m+n-k}{2})}{\Gamma(\frac{n-k}{2})}\int_{H^\perp\cap\mathbb{S}^{n-1}} P\,d\sigma_{n-k-1},$$

or, equivalently,

$$(-1)^{\frac{m}{2}}\frac{k(k+2)\cdots(k+m-2)}{(n-k)(n-k+2)\cdots(n-k+m-2)}\int_{H\cap\mathbb{S}^{n-1}} P\,d\sigma_{k-1} = \int_{H^\perp\cap\mathbb{S}^{n-1}} P\,d\sigma_{n-k-1}.$$

When $k = 1$ and H is just a line through a vector $e \in \mathbb{S}^{n-1}$, we get

$$\int_{H^\perp\cap\mathbb{S}^{n-1}} P\,d\sigma_{n-2} = (-1)^{\frac{m}{2}}\frac{1\cdot 3\cdots(m-1)}{(n-1)(n+1)\cdots(n+m-3)}P(e).$$

It follows that if f is an arbitrary even $L^2(\mathbb{S}^{n-1})$-function and

$$(\mathcal{R}f)(e) = \int\limits_{\mathbb{S}^{n-1}\cap e^{\perp}} f d\sigma_{n-2}$$

is the spherical Radon transform of f, then the spherical harmonic decomposition of $\mathcal{R}f$ is $\mathcal{R}f = \sum\limits_{\substack{m\geq 0 \\ m\,\text{even}}} c_m f_m$, where $f = \sum\limits_{\substack{m\geq 0 \\ m\,\text{even}}} f_m$ is the spherical harmonic decomposition of f and

$$c_m = (-1)^{\frac{m}{2}} \frac{1\cdot 3\cdots (m-1)}{(n-1)(n+1)\cdots(n+m-3)}.$$

Observe, in particular, that one can have $\mathcal{R}f \equiv 0$ only if $f \equiv 0$, i. e., the spherical Radon transform $\mathcal{R}f$ of f determines f uniquely.

When $2 \leq k \leq \frac{n}{2}$, we similarly conclude that for every even $L^2(\mathbb{S}^{n-1})$-function $f = \sum\limits_{\substack{m\geq 0 \\ m\,\text{even}}} f_m$, the function $g = \sum\limits_{\substack{m\geq 0 \\ m\,\text{even}}} c_{m,k} f_m$, where

$$c_{m,k} = (-1)^{\frac{m}{2}} \frac{k(k+2)\cdots(k+m-2)}{(n-k)(n-k+2)\cdots(n-k+m-2)},$$

is also in $L^2(\mathbb{S}^{n-1})$ and satisfies

$$\int\limits_{H\cap\mathbb{S}^{n-1}} g d\sigma_{k-1} = \int\limits_{H^{\perp}\cap\mathbb{S}^{n-1}} f d\sigma_{n-k-1} \tag{2.3}$$

for every k-dimensional plane $H \subset \mathbb{R}^n$.

Note also that if we know the averages of g over the $(k-1)$-dimensional spheres, we can average further and find the averages over $(n-2)$-dimensional spheres, i. e., we know $\mathcal{R}g$. Thus, the even function g with property (2.3) is unique. We shall call it the spherical k-Radon transform of f and denote $\mathcal{R}_k f$ (so, the usual spherical Radon transform is the same as $\mathcal{R}_1 f$).

At last, observe that the coefficients $c_{m,k}$ make sense and are bounded by 1 in absolute value for every complex k with $0 \leq \operatorname{Re} k \leq \frac{n}{2}$, so we can define $\mathcal{R}_k f$ for such k as well, despite the fact that it has no obvious geometric meaning.

Spherical harmonic decomposition is a powerful tool when we need to prove some result about all convex bodies near the unit ball (and often in a more general case), or, which is almost the same, about all reasonable functions f on $\mathbb{S}^{n-1}$ uniformly close to 1. However, when *constructing counterexamples*, one can often restrict oneself to a much narrower class of convex bodies, the so-called bodies of revolution. They are formally defined as follows. Take any not identically zero concave function $f : [a,b] \to [0,+\infty)$ with $f(a) = f(b) = 0$ and consider

$$K_f = \{x = (x_1, x') \in \mathbb{R}\times\mathbb{R}^{n-1} : |x'| \leq f(x_1)\}.$$

The unit ball corresponds to $f_o(t) = \sqrt{1-t^2}$. Note that $f_o'' \leq -1$ on the whole interval $[-1,1]$, so if h is any C^2-function on $[-1,1]$ such that $h(-1) = h(1) = 0$ and $\|h''\|_{C([-1,1])} < 1$, then $f = f_o + h$ is concave and the corresponding body of revolution K_f is convex.

The fact that the whole n-dimensional convex body K_f is completely described by a single real-valued function of one variable whose deviation from f_o can be varied almost freely often allows one to get away with elementary calculus and ODEs when building convex bodies K close to the unit ball with certain properties.

In what follows, we shall present three "local theorems" and one counterexample illustrating the usage of the above techniques. We tried to choose them so that each one has its own little twist and its own peculiar difficulty that did not appear in the previous ones. We should warn the reader that we sometimes skip the routine details in our presentations and restrict ourselves to the simplest cases of more general theorems. The reader interested in the full exposition and the highest available level of generality should follow the references to the original papers.

3 Example 1: the intersection body problem [8]

Let $K \subset \mathbb{R}^n$ be a convex body containing the origin in its interior. The intersection body IK of the body K is the star-shaped body whose radial function is given by

$$\varrho_{IK}(e) = \mathrm{vol}_{n-1}(K \cap e^{\perp}) \quad \forall e \in S^{n-1}.$$

It turns out that, for convex origin-symmetric K, IK is also convex (though we shall not use this fact in any way and the definition makes sense for any star-shaped K). Note that when K is a Euclidean ball centered at the origin, $IK = cK$ for some constant $c > 0$. A question of Lutwak (going back to the late 1980s) is whether there are any other convex, or, more generally, star-shaped bodies with this property in $\mathbb{R}^n$, $n \geq 3$ (in dimension $n = 2$, any body K that is invariant under the rotation by $90°$ gives an example).

We shall show that the answer is negative if we require in addition that K be close to B. To this end, we shall notice that

$$\mathrm{vol}_{n-1}(K \cap e^{\perp}) = c_n \mathcal{R}[\varrho_K^{n-1}](e) \quad \forall e \in S^{n-1},$$

where $c_n > 0$ is some numerical constant. Note that since σ_{n-2} is normalized by the condition that its total mass is 1, we have $\mathcal{R}1 = 1$.

The Lutwak intersection body problem can be now restated as follows: Does there exist a non-constant positive function f such that

$$\mathcal{R}[f^{n-1}] = cf \tag{3.1}$$

for some $c > 0$?

Note that the property in question is invariant under homotheties, i. e., for every $t > 0$, the condition $\mathcal{R}[f^{n-1}] = cf$ implies

$$\mathcal{R}[(tf)^{n-1}] = t^{n-1}\mathcal{R}[f^{n-1}] = c\,t^{n-2}(tf).$$

Hence, we can always normalize f so that its average over $\mathbb{S}^{n-1}$ equals 1. If f was close to 1, then it will still remain so after this normalization. Note also that if $1 - \varepsilon \le f \le 1 + \varepsilon$, then $(1 - \varepsilon)^{n-1} \le \mathcal{R}[f^{n-1}] \le (1 + \varepsilon)^{n-1}$. Now, since $(\mathcal{R}F)(-e) = (\mathcal{R}F)(e)$ for any function F and any $e \in \mathbb{S}^{n-1}$, every solution of our equation $\mathcal{R}[f^{n-1}] = cf$ must be an even function and if $1 - \varepsilon \le f \le 1 + \varepsilon$, then $(1 - \varepsilon)^{n-1}(1 + \varepsilon)^{-1} \le c \le (1 + \varepsilon)^{n-1}(1 - \varepsilon)^{-1}$.

Let now $f = 1 + f_2 + f_4 + \cdots$ be the spherical harmonic decomposition of f. Put $\varphi = f - 1 = f_2 + f_4 + \cdots$. We know that $\|\varphi\|_{L^\infty(\mathbb{S}^{n-1})} < \varepsilon < 1$. Thus,

$$\left| f^{n-1} - 1 - (n-1)\varphi \right| \le C\varepsilon|\varphi|$$

and

$$\left| \mathcal{R}[f^{n-1}] - 1 - (n-1)\mathcal{R}\varphi \right| \le C\varepsilon\mathcal{R}[|\varphi|].$$

On the other hand, as we saw above, $\mathcal{R}\varphi = c_2 f_2 + c_4 f_4 + \cdots$. The coefficients c_m ($m \ge 2$, m even) have been computed in the previous section: they are given by

$$c_m = (-1)^{\frac{m}{2}} \frac{1 \cdot 3 \cdots (m-1)}{(n-1)(n+1) \cdots (n+m-3)}.$$

Thus, $c_2 = -\frac{1}{n-1}$, $|c_m| \le c_4 < \frac{1}{n-1}$ for $m \ge 4$ (we used the fact that $n \ge 3$ to get the last property).

If $\mathcal{R}[f^{n-1}] = cf = c + c\varphi$, we get

$$\left| (c - 1) + c\varphi - (n-1)\mathcal{R}\varphi \right| \le C\varepsilon\mathcal{R}[|\varphi|]$$

and, in particular,

$$\left\| (c - 1) + c\varphi - (n-1)\mathcal{R}\varphi \right\|_{L^2(\mathbb{S}^{n-1})} \le C\varepsilon\left\| \mathcal{R}[|\varphi|] \right\|_{L^2(\mathbb{S}^{n-1})} \le C\varepsilon\|\varphi\|_{L^2(\mathbb{S}^{n-1})}.$$

Since both φ and $\mathcal{R}[\varphi]$ are orthogonal to constants, we conclude from here that

$$\left\| c\varphi - (n-1)\mathcal{R}\varphi \right\|_{L^2(\mathbb{S}^{n-1})} \le C\varepsilon\|\varphi\|_{L^2(\mathbb{S}^{n-1})}$$

as well.

However, the left-hand side squared equals

$$\left\| \sum_{\substack{m \ge 2 \\ m \text{ even}}} (c - (n-1)c_m)f_m \right\|_{L^2(\mathbb{S}^{n-1})}^2 = \sum_{\substack{m \ge 2 \\ m \text{ even}}} (c - (n-1)c_m)^2 \|f_m\|_{L^2(\mathbb{S}^{n-1})}^2$$

$$\ge \inf_{\substack{m \ge 2 \\ m \text{ even}}} (c - (n-1)c_m)^2 \sum_{\substack{m \ge 2 \\ m \text{ even}}} \|f_m\|_{L^2(\mathbb{S}^{n-1})}^2$$

$$= \inf_{\substack{m \ge 2 \\ m \text{ even}}} (c - (n-1)c_m)^2 \|\varphi\|_{L^2(\mathbb{S}^{n-1})}^2.$$

It remains to note that

$$c - (n-1)c_2 = c + 1 > 1,$$

while for $m \geq 4$,

$$|c - (n-1)c_m| \geq 1 - (n-1)|c_4| - |c - 1|,$$

so if ε is so small that

$$C\varepsilon + |c - 1| < 1 - (n-1)|c_4|,$$

we get a contradiction unless $\varphi \equiv 0$.

This simple argument illustrates the main advantage of the perturbative regime: a possibility to switch from a non-linear equation (or inequality) to its linearization. This is a trick we shall be using again and again.

It is natural to ask now what happens if we replace the intersection body operator with some of its iterations, for example, if we ask when

$$I^2 K = I(IK) = cK,$$

i. e., consider the equation

$$\mathcal{R}\big[\big(\mathcal{R}[f^{n-1}]\big)^{n-1}\big] = cf.$$

If we try to treat this equation in the same way as the previous one, after linearizing, we shall arrive at the inequality

$$\big\|(c-1) + c\varphi - (n-1)^2 \mathcal{R}^2[\varphi]\big\|_{L^2(\mathbb{S}^{n-1})} \leq C\varepsilon \|\varphi\|_{L^2(\mathbb{S}^{n-1})}.$$

We can again remove $c - 1$ on the left-hand side and write

$$c\varphi - (n-1)^2 \mathcal{R}^2[\varphi] = \sum_{\substack{m \geq 2 \\ m \text{ even}}} (c - (n-1)^2 c_m^2) f_m.$$

For $m \geq 4$, we still have $c - (n-1)^2 c_m^2$ separated from 0 if $\varepsilon > 0$ is small enough. However, $(n-1)^2 c_2^2 = 1$ now and the coefficient at f_2 can be arbitrarily small, so the previous argument fails. This is not accidental: unlike the equation $IK = cK$, which was invariant only under homotheties, the equation $I^2 K = cK$ is invariant under arbitrary linear transformations (see [6, Theorem 8.1.16]).

Therefore, any ellipsoid gives a solution and it is no longer possible to conclude that $\varphi \equiv 0$ from the equation above. Fortunately, the very same invariance under linear transformations allows us to put the body K into the isotropic position. Then, for every quadratic form $Q(e) = \sum_{i,j} a_{ij} e_i e_j$ with $\sum_i a_{ii} = 0$, we have

$$\int_{\mathbb{S}^{n-1}} \varrho_K(e)^{n+2} Q(e) d\sigma_{n-1}(e) = c_n \int_K Q(x) dx = 0,$$

which means that for the function $f = \varrho_K$, f^{n+2} is orthogonal to all spherical harmonics of degree 2. Representing $f = 1 + \varphi$, using the bound $\|\varphi\|_{L^\infty(\mathbb{S}^{n-1})} < \varepsilon$, and linearizing, we get

$$\left| f^{n+2} - 1 - (n+2)\varphi \right| \le C\varepsilon |\varphi|.$$

We see that

$$\left\| f^{n+2} - 1 - (n+2)\varphi \right\|_{L^2(\mathbb{S}^{n-1})} \le C\varepsilon \|\varphi\|_{L^2(\mathbb{S}^{n-1})}.$$

However, the second-order spherical harmonic in the expansion of $f^{n+2} - 1 - (n+2)\varphi$ is just $-(n+2)f_2$, so we conclude that

$$\|f_2\|_{L^2(\mathbb{S}^{n-1})} \le C\varepsilon \|\varphi\|_{L^2(\mathbb{S}^{n-1})},$$

and, for sufficiently small $\varepsilon > 0$, the L^2-norm of φ comes mainly from the spherical harmonics f_m with $m \ge 4$, i. e., we can write

$$\left\| c\varphi - (n-1)^2 \mathcal{R}^2[\varphi] \right\|_{L^2(\mathbb{S}^{n-1})}^2 \ge \gamma^2 \sum_{\substack{m \ge 4 \\ m \text{ even}}} \|f_m\|_{L^2(\mathbb{S}^{n-1})}^2 \ge \frac{\gamma^2}{2} \|\varphi\|_{L^2(\mathbb{S}^{n-1})}^2$$

with

$$\gamma = \inf_{\substack{m \ge 4 \\ m \text{ even}}} \left| c - (n-1)^2 c_m^2 \right| \ge 1 - (n-1)^2 c_4^2 - |c - 1|,$$

so if ε is so small that

$$C\varepsilon + |c - 1| < 1 - (n-1)^2 c_4^2,$$

we get a contradiction unless $\varphi \equiv 0$.

The reader should by now be able to show that for any $k \ge 1$, the only star-shaped bodies close to the unit ball that satisfy the equation $I^k K = cK$ are balls if k is odd and ellipsoids if k is even. The full result of [8] is stronger. Namely, it is shown there that if the body K is sufficiently close to the ball, then the iterations $I^k K$ converge to the unit ball in the Banach–Mazur distance, so IK (or $I^k K$) can be neither a homothetic image of K nor even a linear image of K, unless K is an ellipsoid. The proof of this stronger statement is more complicated, so we refer the reader to the original paper for details.

4 Example 2: Busemann's inequality [15]

The well-known Busemann intersection inequality asserts that for any star-shaped body K, we have

$$\mathrm{vol}_n(IK) \le \frac{\kappa_{n-1}^n}{\kappa_n^{n-2}}\,\mathrm{vol}_n(K)^{n-1}$$

with equality attained if and only if K is a centered ellipsoid (see [6, Corollary 9.4.5]). Here $\kappa_p = \frac{\pi^{\frac{p}{2}}}{\Gamma(1+\frac{p}{2})}$, which equals the volume of the unit ball in $\mathbb{R}^p$ when p is a positive integer.

Koldobsky introduced a generalization of the notion of an intersection body (see [9, p. 75]). Let K and L be origin-symmetric star-shaped bodies in $\mathbb{R}^n$ and let k be an integer, $1 \le k \le n-1$. We say that L is the k-intersection body of K if

$$\mathrm{vol}_k(L \cap H) = \mathrm{vol}_{n-k}(K \cap H^\perp)$$

for every k-dimensional subspace H of $\mathbb{R}^n$. Note that the 1-intersection body of K is $\frac{1}{2}IK$.

It is worth mentioning that when $k > 1$, for a given origin-symmetric star-shaped body K, its k-intersection body may not exist in general.

It has been conjectured (see [10]) that an analog of Busemann's intersection inequality holds for k-intersection bodies when $k \le \frac{n}{2}$, i. e., if L is a k-intersection body of K, then

$$\mathrm{vol}_n(L)^k \le C_{n,k}\,\mathrm{vol}_n(K)^{n-k}, \tag{4.1}$$

where $C_{n,k} > 0$ is the constant that turns (4.1) into an equality when K is a ball.

The condition that L is the k-intersection body of K can be analytically expressed in terms of the spherical k-Radon transform as

$$\varrho_L^k = b_{n,k}\mathcal{R}_k[\varrho_K^{n-k}],$$

where $b_{n,k} > 0$ is some numerical coefficient. This relation allows one to rewrite inequality (4.1) as

$$\left(\int_{\mathbb{S}^{n-1}} \mathcal{R}_k[\varrho_K^{n-k}]^{\frac{n}{k}}\,d\sigma_{n-1} \right)^k \le \left(\int_{\mathbb{S}^{n-1}} \varrho_K^n\,d\sigma_{n-1} \right)^{n-k} = \left(\int_{\mathbb{S}^{n-1}} (\varrho_K^{n-k})^{\frac{n}{n-k}}\,d\sigma_{n-1} \right)^{n-k} \tag{4.2}$$

(here one clearly has equality for $\varrho_K \equiv const$).

Raising both sides to the power $\frac{1}{n}$, we see that this inequality is almost the same as the statement that the operator norm

$$\|\mathcal{R}_k\|_{L^{\frac{n}{n-k}}(\mathbb{S}^{n-1}) \to L^{\frac{n}{k}}(\mathbb{S}^{n-1})}$$

is at most 1, except we can additionally assume that both the test function and its image are even and positive.

If $k = \frac{n}{2}$, then we have $|c_{m,k}| = 1$ for all even $m \geq 2$, so $\mathcal{R}_k$ is an isometry in $L^2_{\text{even}}(\mathbb{S}^{n-1})$ and there is nothing to prove. Assume now that $k < \frac{n}{2}$. In this case we shall prove only that the desired inequality holds if ϱ_K is close to 1, i. e., K is close to the unit ball.

The geometric meaning of the problem is useful in one more respect: it allows one to observe that inequality (4.2) is invariant under linear transformations of K (see [15]). Thus, we can assume that K is in the isotropic position, i. e., the function $\varrho_K^{n+2} = (\varrho_K^{n-k})^{\frac{n+2}{n-k}}$ has no second-order spherical harmonics. We can also assume that $\int_{\mathbb{S}^{n-1}} \varrho_K^{n-k} d\sigma_{n-1} = 1$. Then, denoting $\varrho_K^{n-k} = f = 1 + \varphi$ and arguing as in the previous example, we see that if K is close enough to the unit ball, i. e., $\|\varphi\|_{L^\infty(\mathbb{S}^{n-1})} < \varepsilon$ with sufficiently small ε, then the contribution of f_2 into the $L^2(\mathbb{S}^{n-1})$-norm of φ is negligible, so

$$\sum_{\substack{m \geq 4 \\ m \text{ even}}} \|f_m\|^2_{L^2(\mathbb{S}^{n-1})} \geq \frac{1}{2}\|\varphi\|^2_{L^2(\mathbb{S}^{n-1})},$$

say.

We thus want to show that

$$\int_{\mathbb{S}^{n-1}} (1 + \mathcal{R}_k[\varphi])^{\frac{n}{k}} d\sigma_{n-1} \leq \left(\int_{\mathbb{S}^{n-1}} (1 + \varphi)^{\frac{n}{n-k}} d\sigma_{n-1} \right)^{\frac{n-k}{k}}. \tag{4.3}$$

It is tempting to expand to the second order and write the left- and right-hand sides of (4.3) as

$$LHS \approx \int_{\mathbb{S}^{n-1}} \left(1 + \frac{n}{k}\mathcal{R}_k[\varphi] + \frac{1}{2}\frac{n}{k}\left(\frac{n}{k} - 1\right)(\mathcal{R}_k[\varphi])^2 \right) d\sigma_{n-1}$$

$$= 1 + \frac{1}{2}\frac{n}{k}\left(\frac{n}{k} - 1\right) \int_{\mathbb{S}^{n-1}} (\mathcal{R}_k[\varphi])^2 d\sigma_{n-1}$$

and

$$RHS \approx \left(\int_{\mathbb{S}^{n-1}} \left(1 + \frac{n}{n-k}\varphi + \frac{1}{2}\frac{n}{n-k}\left(\frac{n}{n-k} - 1\right)\varphi^2 \right) d\sigma_{n-1} \right)^{\frac{n-k}{k}}$$

$$= \left(1 + \frac{1}{2}\frac{n}{n-k}\left(\frac{n}{n-k} - 1\right) \int_{\mathbb{S}^{n-1}} \varphi^2 d\sigma_{n-1} \right)^{\frac{n-k}{k}}$$

$$\approx 1 + \frac{1}{2}\frac{n}{k}\left(\frac{n}{n-k} - 1\right) \int_{\mathbb{S}^{n-1}} \varphi^2 d\sigma_{n-1}.$$

Then it would remain to conclude that

$$RHS - LHS \approx \frac{1}{2}\frac{n}{k}\left(\frac{k}{n-k} \int_{\mathbb{S}^{n-1}} \varphi^2 \, d\sigma_{n-1} - \frac{n-k}{k} \int_{\mathbb{S}^{n-1}} (\mathcal{R}_k[\varphi])^2 d\sigma_{n-1} \right)$$

$$= \frac{1}{2}\frac{n}{k}\sum_{\substack{m\geq 4\\ m\ even}}\left(\frac{k}{n-k} - c_{m,k}^2\frac{n-k}{k}\right)\int_{\mathbb{S}^{n-1}} f_m^2\, d\sigma_{n-1}$$

and observe that $c_{2,k} = -\frac{k}{n-k}$ and $|c_{m,k}| \leq |c_{4,k}| < \frac{k}{n-k}$ for even $m \geq 4$, so the expression in the final line of the last formula is at least

$$\gamma \sum_{\substack{m\geq 4\\ m\ even}} \int_{\mathbb{S}^{n-1}} f_m^2\, d\sigma_{n-1} \geq \frac{\gamma}{2}\int_{\mathbb{S}^{n-1}} \varphi^2\, d\sigma_{n-1}$$

with some $\gamma = \gamma(k,n) > 0$.

To justify this approach, one needs, however, to show that the errors in the second-order Taylor approximations are small compared to $\int_{\mathbb{S}^{n-1}} \varphi^2\, d\sigma_{n-1}$ when ε is close to 0.

The right-hand side RHS presents no problem because we control φ in $L^\infty(\mathbb{S}^{n-1})$. The main difficulty with the left-hand side LHS is that unlike the usual Radon transform, $\mathcal{R}_k$ is not bounded in $L^\infty(\mathbb{S}^{n-1})$ for $k > 1$ and we cannot say that $\mathcal{R}_k[\varphi]$ is small at each individual point. What we shall use instead is the fact that $\mathcal{R}_k$ is bounded from $L^2(\mathbb{S}^{n-1})$ to $L^{\frac{n}{k}}(\mathbb{S}^{n-1})$ for all $0 \leq k \leq \frac{n}{2}$. Assuming that, we will use the inequality

$$\left| |1 + t|^p - \left(1 + pt + \frac{p(p-1)}{2}t^2\right)\right| \leq \begin{cases} c_p|t|^p, & 2 < p \leq 3,\\ c_p(|t|^p + |t|^3), & p \geq 3, \end{cases}$$

valid for all $t \in \mathbb{R}$ and $p > 2$. To prove it, just notice that the ratio of its left-hand side to $|t|^p$ or $|t|^p + |t|^3$, respectively, is a continuous function on $\mathbb{R}\setminus\{0\}$ that stays bounded as $t \to 0$ and as $t \to \pm\infty$.

This inequality allows one to estimate the error in the approximation of $\int_{\mathbb{S}^{n-1}}(1 + \mathcal{R}_k[\varphi])^{\frac{n}{k}}\, d\sigma_{n-1}$ by a constant multiple of $\int_{\mathbb{S}^{n-1}} |\mathcal{R}_k[\varphi]|^{\frac{n}{k}}\, d\sigma_{n-1}$ if $2 < \frac{n}{k} \leq 3$ and of $\int_{\mathbb{S}^{n-1}}(|\mathcal{R}_k[\varphi]|^{\frac{n}{k}} + |\mathcal{R}_k[\varphi]|^3)\, d\sigma_{n-1}$ if $\frac{n}{k} > 3$.

Using the boundedness of $\mathcal{R}_k$ from $L^2(\mathbb{S}^{n-1})$ to $L^{\frac{n}{k}}(\mathbb{S}^{n-1})$, we immediately get

$$\int_{\mathbb{S}^{n-1}} |\mathcal{R}_k[\varphi]|^{\frac{n}{k}}\, d\sigma_{n-1} = \|\mathcal{R}_k[\varphi]\|_{L^{\frac{n}{k}}(\mathbb{S}^{n-1})}^{\frac{n}{k}} \leq C\|\varphi\|_{L^2(\mathbb{S}^{n-1})}^{\frac{n}{k}}$$

$$= C\left(\int_{\mathbb{S}^{n-1}} \varphi^2\, d\sigma_{n-1}\right)^{\frac{n}{2k}} \leq C\varepsilon^{\frac{n}{k}-2}\int_{\mathbb{S}^{n-1}} \varphi^2\, d\sigma_{n-1},$$

and if $\frac{n}{k} > 3$, we also have

$$\int_{\mathbb{S}^{n-1}} |\mathcal{R}_k[\varphi]|^3\, d\sigma_{n-1} = \|\mathcal{R}_k[\varphi]\|_{L^3(\mathbb{S}^{n-1})}^3 \leq \|\mathcal{R}_k[\varphi]\|_{L^{\frac{n}{k}}(\mathbb{S}^{n-1})}^3$$

$$\leq C\|\varphi\|_{L^2(\mathbb{S}^{n-1})}^3 = C\left(\int_{\mathbb{S}^{n-1}} \varphi^2\, d\sigma_{n-1}\right)^{\frac{3}{2}} \leq C\varepsilon\int_{\mathbb{S}^{n-1}} \varphi^2\, d\sigma_{n-1}.$$

It remains only to prove the boundedness of $\mathcal{R}_k$ from $L^2(\mathbb{S}^{n-1})$ to $L^{\frac{n}{k}}(\mathbb{S}^{n-1})$ for $0 \leq k \leq \frac{n}{2}$.

Recall that the definition

$$c_{m,k} = (-1)^{\frac{m}{2}} \frac{k(k+2)\cdots(k+m-2)}{(n-k)(n-k+2)\cdots(n-k+m-2)}$$

makes sense not only for integer k but for any $k = z \in \mathbb{C}$ with $0 \le \operatorname{Re} z \le \frac{n}{2}$. Moreover, for every such z, we have $|c_{m,z}| \le 1$, so the mapping R_z defined by

$$f = f_0 + f_2 + f_4 + \cdots \quad \mapsto \quad R_z f = c_{0,z} f_0 + c_{2,z} f_2 + c_{4,z} f_4 + \cdots$$

for any even $L^2(\mathbb{S}^{n-1})$-function f is bounded in $L^2_{\text{even}}(\mathbb{S}^{n-1})$ with

$$\|R_z\|_{L^2_{\text{even}}(\mathbb{S}^{n-1}) \to L^2(\mathbb{S}^{n-1})} \le 1$$

and depends on z analytically in the strip $0 < \operatorname{Re} z < \frac{n}{2}$. Moreover, for every fixed $f \in L^2_{\text{even}}(\mathbb{S}^{n-1})$, the mapping $z \mapsto R_z f$ is continuous up to the boundary of this strip.

By the Stein interpolation theorem (see [13] or [14, Chapter 5]), it now suffices to show that when $\operatorname{Re} z = 0$, $\|R_z\|_{L^2_{\text{even}}(\mathbb{S}^{n-1}) \to L^\infty(\mathbb{S}^{n-1})}$ is finite and grows at most polynomially as $|z| \to \infty$. To this end, we first estimate $\|f_m\|_{L^\infty(\mathbb{S}^{n-1})}$ in terms of $\|f_m\|_{L^2(\mathbb{S}^{n-1})}$.

Observe that H_m is a finite-dimensional linear space of harmonic polynomials. Let $g_1, \ldots, g_N$ be an orthonormal basis in H_m with respect to the scalar product in $L^2(\mathbb{S}^{n-1})$. Then, for every $g \in H_m$, we have $g = \sum_{j=1}^{N} \langle g, g_j \rangle g_j$, so for every $x \in \mathbb{S}^{n-1}$ we have

$$g(x) = \left\langle g, \sum_{j=1}^{N} \overline{g_j(x)} g_j \right\rangle = \langle g, \mathcal{G}_x \rangle.$$

Thus, the norm of the linear functional $g \mapsto g(x)$ on H_m equals

$$\|\mathcal{G}_x\|_{L^2(\mathbb{S}^{n-1})} = \left\| \sum_{j=1}^{N} \overline{g_j(x)} g_j \right\|_{L^2(\mathbb{S}^{n-1})} = \left(\sum_{j=1}^{N} |g_j(x)|^2 \right)^{\frac{1}{2}}.$$

On the other hand, the space H_m is invariant under rotations, so this norm must be independent of x. Therefore, for every $x \in \mathbb{S}^{n-1}$, we have

$$\|\mathcal{G}_x\|_{L^2(\mathbb{S}^{n-1})}^2 = \int_{\mathbb{S}^{n-1}} \|\mathcal{G}_x\|_{L^2(\mathbb{S}^{n-1})}^2 \, d\sigma_{n-1}(x) = \int_{\mathbb{S}^{n-1}} \sum_{j=1}^{N} |g_j(x)|^2 \, d\sigma_{n-1}(x) = N.$$

Recall now that

$$\dim H_m = \dim \mathcal{P}_m - \dim \mathcal{P}_{m-2} = \binom{m+n-1}{n-1} - \binom{m-2+n-1}{n-1} = D_m = O(m^{n-2})$$

as $m \to \infty$. We see that

$$\|f_m\|_{L^\infty(\mathbb{S}^{n-1})} \le \sqrt{D_m} \|f_m\|_{L^2(\mathbb{S}^{n-1})}$$

and

$$\|R_z f\|_{L^\infty(\mathbb{S}^{n-1})} \leq \sum_{\substack{m\geq 0 \\ m\,\text{even}}} |c_{m,z}| \|f_m\|_{L^\infty(\mathbb{S}^{n-1})}$$

$$\leq \sum_{\substack{m\geq 0 \\ m\,\text{even}}} |c_{m,z}| \sqrt{D_m} \|f_m\|_{L^2(\mathbb{S}^{n-1})}$$

$$\leq \left(\sum_{\substack{m\geq 0 \\ m\,\text{even}}} |c_{m,z}|^2 D_m \right)^{\frac{1}{2}} \left(\sum_{\substack{m\geq 0 \\ m\,\text{even}}} \|f_m\|^2_{L^2(\mathbb{S}^{n-1})} \right)^{\frac{1}{2}}$$

$$= \left(\sum_{\substack{m\geq 0 \\ m\,\text{even}}} |c_{m,z}|^2 D_m \right)^{\frac{1}{2}} \|f\|_{L^2(\mathbb{S}^{n-1})}.$$

It remains to estimate $|c_{m,z}|$. When n is even, it is very easy. Since $\operatorname{Re} z = 0$, $|n - z + j| = |z + n + j|$ for every $j = 0, 2, \ldots, m - 2$, so

$$|c_{m,z}| = \frac{\prod_{\substack{0\leq j\leq m-2 \\ j\,\text{even}}} |z + j|}{\prod_{\substack{0\leq j\leq m-2 \\ j\,\text{even}}} |z + n + j|} = \frac{\prod_{\substack{0\leq j\leq n-2 \\ j\,\text{even}}} |z + j|}{\prod_{\substack{m\leq j\leq m+n-2 \\ j\,\text{even}}} |z + j|} \leq \frac{|z||z + 2| \cdots |z + n - 2|}{m^{\frac{n}{2}}}$$

for $m \geq 2$. We also have $|c_{0,z}| = 1$. Thus,

$$\sum_{\substack{m\geq 0 \\ m\,\text{even}}} |c_{m,z}|^2 D_m \leq 1 + |z|^2 |z + 2|^2 \cdots |z + n - 2|^2 \sum_{\substack{m\geq 2 \\ m\,\text{even}}} \frac{D_m}{m^n}.$$

Since $D_m = O(m^{n-2})$, the series on the right-hand side converges and we are done. A similar estimate holds for odd n too but its proof is a bit more complicated, so we refer the reader to [15] for details.

5 Example 3: a local version of the fifth Busemann–Petty problem [1]

Busemann and Petty [5] asked the following question. Let K be an origin-symmetric convex body in $\mathbb{R}^n$. Suppose that for every $(n-1)$-dimensional subspace $H \subset \mathbb{R}^n$, the volume of the cone of the largest volume with base $K \cap H$ contained in K does not depend on H. Does it follow that K is an ellipsoid?

The equivalent analytic reformulation of the question assumption is that the product $\mathfrak{h}_K(e)\mathcal{R}[\varrho_K^{n-1}](e) = const$ on $\mathbb{S}^{n-1}$, or, equivalently,

$$\mathfrak{h}_K(e) = c\left(\mathcal{R}[\varrho_K^{n-1}](e)\right)^{-1}. \tag{5.1}$$

The formulation of the problem is, clearly, invariant under linear transformations, so we can assume from the beginning that the body K is in the isotropic position.

We shall also normalize the convex body K by the condition

$$\int_{\mathbb{S}^{n-1}} \varrho_K \, d\sigma_{n-1} = 1. \tag{5.2}$$

Our task will be to show that if, under such conditions, K is close to the unit ball B, then $K = B$.

Equation (5.1) should remind the reader of equation (3.1) in the intersection body example. Just as in that example, one can conclude that since both $\mathfrak{h}_K$ and ϱ_K are close to 1, the constant c must be close to 1 as well. The key difference and the main difficulty, however, is that now we have two different functions $\mathfrak{h}_K$ and ϱ_K in the equation and while each of them determines the other one uniquely, the relation between them is highly non-linear.

What saves the day is that this non-linearity manifests itself only on high frequencies. More precisely, we have the following.

Lemma. *Let $\varrho = \varrho_K$ and $\mathfrak{h} = \mathfrak{h}_K$ be the radial and support functions of an origin-symmetric convex body K close to the unit ball, respectively. Let $\mathfrak{h} = \sum_{\substack{m \geq 0 \\ m \text{ even}}} \mathfrak{h}_m$ be the spherical harmonic decomposition of $\mathfrak{h}$. Put $\eta = \sum_{\substack{2 \leq m \leq l \\ m \text{ even}}} \mathfrak{h}_m$, $v = \sum_{\substack{m > l \\ m \text{ even}}} \mathfrak{h}_m$. Then for every $\varepsilon, l > 0$, there exists $\delta_0 = \delta_0(\varepsilon, l)$ such that whenever $\|\mathfrak{h} - 1\|_\infty \leq \delta_0$, the inequality*

$$0 \leq \mathfrak{h} - \varrho \leq \varepsilon \|\eta\|_{L^2(\mathbb{S}^{n-1})} + CMv$$

holds, where

$$Mv(e) = \max_{\theta \in (0,\pi)} \frac{1}{\sigma_{n-1}(S_\theta(e))} \int_{S_\theta(e)} |v(x)| \, d\sigma_{n-1}(x)$$

is the spherical Hardy–Littlewood maximal function, $S_\theta(e)$ denotes the set of vectors $x \in \mathbb{S}^{n-1}$ making an angle less than θ with the vector $e \in \mathbb{S}^{n-1}$, and $C > 0$ is a constant depending only on the dimension n.

Proof. We have

$$\varrho(e) = \inf\left\{ \frac{\mathfrak{h}(e')}{\langle e, e' \rangle} : e' \in \mathbb{S}^{n-1}, \langle e, e' \rangle > 0 \right\}.$$

Note that the admissible range of e' can be further restricted to $|e - e'| < \delta$ with arbitrarily small $\delta > 0$, provided that δ_0 is chosen small enough. Indeed, since $\mathfrak{h}(e') \geq \frac{1-\delta_0}{1+\delta_0}\mathfrak{h}(e)$, e' can compete with e only if $\langle e, e' \rangle \geq \frac{1-\delta_0}{1+\delta_0}$, so

$$|e - e'|^2 = 2(1 - \langle e, e' \rangle) \leq \frac{4\delta_0}{1 + \delta_0} < \delta^2$$

if $\delta_0 > 0$ is chosen appropriately. Now observe also that all norms on the finite-dimensional space of polynomials of degree not exceeding l on the unit sphere are equivalent and that any seminorm is dominated by any norm, whence

$$\|\eta\|_{C(\mathbb{S}^{n-1})} \le C(l)\|\eta\|_{L^2(\mathbb{S}^{n-1})} \quad \text{and} \quad \|\nabla\eta\|_{C(\mathbb{S}^{n-1})} \le C(l)\|\eta\|_{L^2(\mathbb{S}^{n-1})}.$$

In particular, if $|e - e''| < 2\delta$, we get

$$|\eta(e) - \eta(e'')| \le 4\|\nabla\eta\|_{C(\mathbb{S}^{n-1})}\delta \le 4C(l)\delta\|\eta\|_{L^2(\mathbb{S}^{n-1})}.$$

Before we proceed, let us prove the following claim. Let $R > \omega > 0$ and let $e \in \mathbb{S}^{n-1}$ be a unit vector. Assume that $\mathfrak{h}(e) = (R - \omega)\cos\theta$ for some $\theta \in (0, \frac{\pi}{3})$. Then

$$\frac{1}{\sigma(S_\theta(e))} \int_{S_\theta(e)} |\mathfrak{h}(e') - R|d\sigma_{n-1}(e') \ge c\,\omega \tag{5.3}$$

with some $c > 0$ depending on n only.

We will use the parametrization $e' = e'(t, v) = e\cos t + v\sin t$, where $t < \theta$ is the angle between e and e' and $v \in \mathbb{S}^{n-1} \cap e^\perp$ is the direction of the projection of e' to $e^\perp$. Note that

$$d\sigma_{n-1}(e') = c_n(\sin t)^{n-2}\, dt\, d\sigma_{n-2}(v).$$

Since $e'(\frac{t}{2}, v)\cos\frac{t}{2} = \frac{1}{2}(e'(t, v) + e)$, by the convexity and 1-homogeneity of $\mathfrak{h}$ we have

$$\mathfrak{h}\left(e'\left(\frac{t}{2}, v\right)\right)\cos\frac{t}{2} \le \frac{1}{2}[\mathfrak{h}(e'(t, v)) + \mathfrak{h}(e)],$$

so

$$2\mathfrak{h}\left(e'\left(\frac{t}{2}, v\right)\right)\cos\frac{t}{2} - \mathfrak{h}(e'(t, v)) \le \mathfrak{h}(e)$$

$$= (R - \omega)\cos\theta = R\left(2\cos^2\frac{\theta}{2} - 1\right) - \omega\cos\theta \le R\left(2\cos\frac{t}{2} - 1\right) - \frac{\omega}{2},$$

whence

$$2\left|\mathfrak{h}\left(e'\left(\frac{t}{2}, v\right)\right) - R\right| + |\mathfrak{h}(e'(t, v)) - R|$$

$$\ge 2\left|\mathfrak{h}\left(e'\left(\frac{t}{2}, v\right)\right) - R\right|\cos\frac{t}{2} + |\mathfrak{h}(e'(t, v)) - R| \ge \frac{\omega}{2}.$$

Integrating this inequality against

$$c_n(\sin t)^{n-2}\, dt\, d\sigma_{n-2}(v) \le 2^{n-1} c_n \left(\sin \frac{t}{2}\right)^{n-2} d\left(\frac{t}{2}\right) d\sigma_{n-2}(v),$$

we get

$$2^n \int_{S_{\frac{\theta}{2}}(e)} |\mathfrak{h}(e') - R| d\sigma_{n-1}(e') + \int_{S_\theta(e)} |\mathfrak{h}(e') - R| d\sigma_{n-1}(e') \ge \frac{\omega}{2} \sigma_{n-1}(S_\theta(e))$$

and the desired inequality follows with $c = \frac{1}{2(2^n+1)}$.

Let us now assume that $e' \in \mathbb{S}^{n-1}$ with $|e - e'| < \delta$ is such that

$$\frac{\mathfrak{h}(e')}{\langle e, e' \rangle} = \varrho(e) < \mathfrak{h}(e).$$

Then, if θ is the angle between e and e', we have

$$\mathfrak{h}(e') = [\mathfrak{h}(e) - (\mathfrak{h}(e) - \varrho(e))] \cos \theta$$

and we can apply (5.3) to the vector e' with $R = \mathfrak{h}(e)$ and $\omega = \mathfrak{h}(e) - \varrho(e)$ to conclude that

$$\mathfrak{h}(e) - \varrho(e) \le \frac{C}{\sigma_{n-1}(S_\theta(e'))} \int_{S_\theta(e')} |\mathfrak{h}(e) - \mathfrak{h}(e'')| d\sigma_{n-1}(e'')$$

$$\le \frac{C'}{\sigma_{n-1}(S_{2\theta}(e))} \int_{S_{2\theta}(e)} |\mathfrak{h}(e) - \mathfrak{h}(e'')| d\sigma_{n-1}(e'').$$

However,

$$|\mathfrak{h}(e) - \mathfrak{h}(e'')| \le |\eta(e) - \eta(e'')| + |v(e)| + |v(e'')|$$

and

$$|\eta(e) - \eta(e'')| \le 4C(l)\delta \|\eta\|_{L^2(\mathbb{S}^{n-1})},$$

while

$$|v(e)| \le Mv(e) \quad \text{and} \quad \frac{1}{\sigma_{n-1}(S_{2\theta}(e))} \int_{S_{2\theta}(e)} |v(e'')| d\sigma_{n-1}(e'') \le Mv(e),$$

so the desired statement follows if we choose $\delta > 0$ so that $4C'C(l)\delta < \varepsilon$. $\qquad\square$

Now fix $\varepsilon > 0$, $l > 0$ to be chosen later and assume that $\|\mathfrak{h} - 1\|_\infty < \delta_0$, where $\delta_0 > 0$ is very small. Since the body K is assumed to be in the isotropic position and normalized by (5.2), arguing as in Section 3 we can again show that the second-order spherical harmonics component of ϱ is small compared to $\varrho - 1$. Linearizing the right-hand side of the equation $\mathfrak{h} = c(\mathcal{R}[\varrho^{n-1}])^{-1}$ around 1, we get $\mathfrak{h} = c(1 - (n-1)\mathcal{R}(\varrho - 1) + \gamma)$, where $\|\gamma\|_{L^2(\mathbb{S}^{n-1})} \le \varepsilon\|\varrho - 1\|_{L^2(\mathbb{S}^{n-1})}$, provided that δ_o is small enough.

Since, for small enough δ_o, the $L^2(\mathbb{S}^{n-1})$-norm of the second-order spherical harmonics component of $\varrho - 1$ is also less than $\varepsilon\|\varrho - 1\|_{L^2(\mathbb{S}^{n-1})}$ and $c < 1 + \varepsilon$, we can incorporate the second-order spherical harmonics component into the error term γ and conclude that

$$\left\|\mathfrak{h} - c - c\,\mathfrak{M}(\varrho - 1)\right\|_{L^2(\mathbb{S}^{n-1})} \le 3\varepsilon\|\varrho - 1\|_{L^2(\mathbb{S}^{n-1})}. \tag{5.4}$$

Here $\mathfrak{M}$ is the linear operator that maps every m-th-order spherical harmonic Z_m to $\mu_m Z_m$, where

$$\mu_m = -(n-1)(-1)^{\frac{m}{2}} \frac{1 \cdot 3 \cdots (m-1)}{(n-1)(n+1) \cdots (n+m-3)}$$

for even $m \ge 4$ and $\mu_m = 0$ for other m (so $\mu_m Z_m = -(n-1)\mathcal{R}Z_m$ for even $m \ge 4$). Note that when $n \ge 3$, we have $|\mu_m| < 1$ for all m and $\mu_m \to 0$ as $m \to \infty$.

Consider the decomposition $\mathfrak{h} = \mathfrak{h}_0 + \eta + v$ and $\varrho = 1 + \varphi + \psi$, where $\mathfrak{h}_0$ is the constant term, η and φ are the parts corresponding to the harmonics of degrees 2 to l, and v, ψ are the parts corresponding to harmonics of degrees greater than l. Since the projection to any sum of spaces of spherical harmonics in $L^2(\mathbb{S}^{n-1})$ has norm 1, inequality (5.4) implies

$$\|\eta - c\,\mathfrak{M}\varphi\|_{L^2(\mathbb{S}^{n-1})} \le 3\varepsilon\|\varrho - 1\|_{L^2(\mathbb{S}^{n-1})} \le 3\varepsilon\big(\|\varphi\|_{L^2(\mathbb{S}^{n-1})} + \|\psi\|_{L^2(\mathbb{S}^{n-1})}\big)$$

and the same estimate holds for $\|v - c\,\mathfrak{M}\psi\|_{L^2(\mathbb{S}^{n-1})}$. Thus,

$$\begin{aligned}
\|v\|_{L^2(\mathbb{S}^{n-1})} &\le c\|\mathfrak{M}\psi\|_{L^2(\mathbb{S}^{n-1})} + 3\varepsilon\big(\|\varphi\|_{L^2(\mathbb{S}^{n-1})} + \|\psi\|_{L^2(\mathbb{S}^{n-1})}\big)\\
&\le c\max_{m>l}|\mu_m|\,\|\psi\|_{L^2(\mathbb{S}^{n-1})} + 3\varepsilon\big(\|\varphi\|_{L^2(\mathbb{S}^{n-1})} + \|\psi\|_{L^2(\mathbb{S}^{n-1})}\big)\\
&\le 4\varepsilon\big(\|\varphi\|_{L^2(\mathbb{S}^{n-1})} + \|\psi\|_{L^2(\mathbb{S}^{n-1})}\big),
\end{aligned}$$

provided l is chosen so large that $c\max_{m>l}|\mu_m| < \varepsilon$ and $\delta_o > 0$ is small enough.

The same computation for η, using just the crude bound $\max_{m>0}|\mu_m| < 1$, yields

$$\|\eta\|_{L^2(\mathbb{S}^{n-1})} \le 2\big(\|\varphi\|_{L^2(\mathbb{S}^{n-1})} + \|\psi\|_{L^2(\mathbb{S}^{n-1})}\big).$$

On the other hand, by Lemma 5 and the boundedness of the maximal function in $L^2(\mathbb{S}^{n-1})$, we have

$$\|\mathfrak{h} - \varrho\|_{L^2(\mathbb{S}^{n-1})} \le \varepsilon\|\eta\|_{L^2(\mathbb{S}^{n-1})} + C\|v\|_{L^2(\mathbb{S}^{n-1})} \le (2 + 4C)\varepsilon\big(\|\varphi\|_{L^2(\mathbb{S}^{n-1})} + \|\psi\|_{L^2(\mathbb{S}^{n-1})}\big),$$

which implies the same bound for both $\|\varphi - \eta\|_{L^2(\mathbb{S}^{n-1})}$ and $\|\psi - v\|_{L^2(\mathbb{S}^{n-1})}$. Combining all the above estimates, we get

$$\|\varphi - c\,\mathfrak{M}\varphi\|_{L^2(\mathbb{S}^{n-1})} + \|\psi - c\,\mathfrak{M}\psi\|_{L^2(\mathbb{S}^{n-1})}$$

$$\leq \|\varphi - \eta\|_{L^2(\mathbb{S}^{n-1})} + \|\eta - c\,\mathfrak{M}\varphi\|_{L^2(\mathbb{S}^{n-1})} + \|\psi - v\|_{L^2(\mathbb{S}^{n-1})} + \|v - c\,\mathfrak{M}\psi\|_{L^2(\mathbb{S}^{n-1})}$$
$$\leq C'\varepsilon\big(\|\varphi\|_{L^2(\mathbb{S}^{n-1})} + \|\psi\|_{L^2(\mathbb{S}^{n-1})}\big).$$

On the other hand, for any function $\chi \in L^2(\mathbb{S}^{n-1})$, we have

$$\|\chi - c\,\mathfrak{M}\chi\|_{L^2(\mathbb{S}^{n-1})} \geq \Big(1 - (1 + \varepsilon)\max_{m\geq 0}|\mu_m|\Big)\|\chi\|_{L^2(\mathbb{S}^{n-1})},$$

so we can conclude that $\varphi = 0$, $\psi = 0$ if $C'\varepsilon < 1 - (1 + \varepsilon)\max_{m\geq 0}|\mu_m|$. Thus, in this case, $\varrho = 1$ and, therefore, K is the unit ball.

6 Example 4: non-uniqueness of convex bodies with prescribed volumes of sections and projections [12]

In this section, we shall construct two essentially different convex bodies K_1 and K_2 in $\mathbb{R}^{2n}$ that have equal $(2n-1)$-dimensional volumes of central sections, maximal sections, and projections in every direction. This answers negatively an old question of Bonnesen and Klee in even dimensions. To the best of our knowledge, the case of odd dimensions still remains open. Our exposition will follow [12].

For a convex body K containing the origin in its interior and $e \in \mathbb{S}^{n-1}$, put

$$A_K(e) = \operatorname{vol}_{2n-1}(K \cap e^{\perp}) \quad \text{(the central section volume)},$$
$$M_K(e) = \max_{t\in\mathbb{R}} \operatorname{vol}_{2n-1}(K \cap (e^{\perp} + te)) \quad \text{(the maximal section volume)},$$
$$P_K(e) = \operatorname{vol}_{2n-1}(K|e^{\perp}) \quad \text{(the projection volume)}.$$

We shall construct two bodies of revolution K_1 and K_2 close to the unit ball that cannot be obtained from each other by a rigid motion but satisfy

$$A_{K_1} \equiv A_{K_2}, \quad M_{K_1} \equiv M_{K_2}, \quad \text{and} \quad P_{K_1} \equiv P_{K_2}.$$

The idea is easiest to demonstrate in $\mathbb{R}^2$. We start with the unit disc, which in the usual Cartesian coordinates x_1, x_2 is given by the inequalities $-1 \leq x_1 \leq 1$, $|x_2| \leq f_0(x_1) = \sqrt{1 - x_1^2}$. Now choose a very small $\delta > 0$ and choose two small in C^2 not identically zero functions φ and ψ supported on $[\frac{1}{2} - \delta, \frac{1}{2} + \delta]$ and $[1 - 2\delta, 1 - \delta]$, respectively.

Put

$$f_1(x_1) = f_0(x_1) + \varphi(x_1) - \varphi(-x_1) + \psi(x_1)$$

and

$$f_2(x_1) = f_0(x_1) - \varphi(x_1) + \varphi(-x_1) + \psi(x_1),$$

and put

$$K_j = \{(x_1, x_2) : -1 \le x_1 \le 1, \ |x_2| \le f_j(x_1)\}, \quad j = 1, 2.$$

If φ and ψ are small enough, K_1 and K_2 are still convex. Also, if the direction of the vector e makes with the x_1-axis an angle not close to $\frac{\pi}{6}$ or $\frac{\pi}{2}$, then $A_{K_j}(e) = M_{K_j}(e) = P_{K_j}(e) = 2$, $j = 1, 2$.

When the angle is close to $\frac{\pi}{6}$, A_{K_j}, M_{K_j}, and P_{K_j} do not feel the perturbation of f_0 by ψ in any way, so they are the same as for the convex bodies $\widetilde{K}_1$ and $\widetilde{K}_2$ corresponding to $\widetilde{f}_1(x_1) = f_0(x_1) + \varphi(x_1) - \varphi(-x_1)$ and $\widetilde{f}_2(x_1) = f_0(x_1) - \varphi(x_1) + \varphi(-x_1)$, but $\widetilde{K}_1$ and $\widetilde{K}_2$ are origin-symmetric images of each other, so we have $A_{K_1} = A_{K_2}, M_{K_1} = M_{K_2}, P_{K_1} = P_{K_2}$ again. Finally, when the angle is close to $\frac{\pi}{2}$, only ψ matters, so $A_{K_1} = A_{K_2} = A_{\overline{K}}, M_{K_1} = M_{K_2} = M_{\overline{K}}$, and $P_{K_1} = P_{K_2} = P_{\overline{K}}$, where $\overline{K}$ corresponds to $\overline{f} = f_0 + \psi$.

When $n > 1$ (i. e., we are in $\mathbb{R}^4$ and higher) one can instead consider the bodies of revolution

$$K_j = \{(x_1, x_2, \dots, x_{2n}) : -1 \le x_1 \le 1, \ x_2^2 + \cdots + x_{2n}^2 \le f_j^2(x_1)\},$$

$j = 1, 2$. The above argument holds without change except for the last part when the angle $\theta(e)$ between e and the x_1-axis is close to $\frac{\pi}{2}$ because now it is no longer true that the volumes of the corresponding sections and projections are not influenced by φ.

The projections present no problem. All one has to note is that when the angle between e and the x_1-axis is close to $\frac{\pi}{2}$, the intersection $(K|e^{\perp}) \cap \{x : |x_1| \le \frac{3}{4}\}$ is influenced by φ alone and the intersection $(K|e^{\perp}) \cap \{x : |x_1| \ge \frac{3}{4}\}$ is influenced by ψ alone.

To take care of the sections, note that for any body of revolution

$$K = \{(x_1, \dots, x_{2n}) : -1 \le x_1 \le 1, \ x_2^2 + \cdots + x_{2n}^2 \le f^2(x_1)\},$$

the volume of the section $K \cap (e^{\perp} + te)$ depends only on the angle $\theta(e)$ and on $t \in \mathbb{R}$ and can be computed as

$$c_n \sqrt{1 + s^2} \int_{x_-}^{x_+} [f^2(x_1) - (sx_1 + h)^2]^{n-1} dx_1,$$

where $s = \cot\theta(e)$, $h = \frac{t}{\sin\theta(e)}$, and $x_- < x_+$ are the x_1-coordinates of the intersections of the line $y = sx_1 + h$ with the curves $y = \pm f(x_1)$. Since both K_1 and K_2 are close to the unit ball, we always have $t \approx 0$ for both the maximal and the central section of each of them. When $\theta(e) \approx \frac{\pi}{2}$, this implies that for every choice of $\theta(e) \approx \frac{\pi}{2}$ and $t \approx 0$, the points x_- and x_+ are determined by ψ only, so they are the same for K_1 and K_2. Moreover, these points are close to -1 and 1, respectively. If we now make a choice of φ such that

$$\int\limits_{-\frac{3}{4}}^{\frac{3}{4}} [f_1^2(x_1) - (sx_1 + h)^2]^{n-1} dx_1 = \int\limits_{-\frac{3}{4}}^{\frac{3}{4}} [f_2^2(x_1) - (sx_1 + h)^2]^{n-1} dx_1,$$

for all $s, h \in \mathbb{R}$, then, since $f_1 = f_2$ outside $[-\frac{3}{4}, \frac{3}{4}]$, we will have

$$\mathrm{vol}_{2n-1}(K_1 \cap (e^\perp + te)) = \mathrm{vol}_{2n-1}(K_2 \cap (e^\perp + te))$$

as long as $\theta(e)$ is not too far from $\frac{\pi}{2}$ and t is not too far from 0. This is more than enough to conclude that $A_{K_1}(e) = A_{K_2}(e)$ and $M_{K_1}(e) = M_{K_2}(e)$ when $\theta(e) \approx \frac{\pi}{2}$, finishing the construction.

It remains to show that such choice is possible. Expanding the expressions in powers of s and h, we see that it would suffice to ensure that

$$\Gamma_{l,k}(\varphi) = \int\limits_{-\frac{3}{4}}^{\frac{3}{4}} f_1^l(x_1) x_1^k dx_1 - \int\limits_{-\frac{3}{4}}^{\frac{3}{4}} f_2^l(x_1) x_1^k dx_1 = 0$$

for all $k, l \leq 2n - 2$, say. Note that $\varphi \mapsto \Gamma(\varphi) = \{\Gamma_{l,k}(\varphi)\}_{l,k=0}^{2n-2}$ is a continuous mapping from the infinite-dimensional linear space $C_0^2([\frac{1}{2} - \delta, \frac{1}{2} + \delta])$ to $\mathbb{R}^{(2n-1)^2}$ and $\Gamma(-\varphi) = -\Gamma(\varphi)$ (when one changes φ by $-\varphi$, f_1 and f_2 swap places). The Borsuk–Ulam theorem (see [11, p. 23]) implies that for every fixed $\varepsilon > 0$, we can choose φ with $0 < \|\varphi\|_{C^2} < \varepsilon$ such that $\Gamma(\varphi) = 0$, which is exactly what we need.

7 Some open questions

We will finish this survey with five open problems that we would like to see resolved. We by no means pretend that they are "the most important" questions in the area or anything like that. What they reflect is just our personal taste and the general lack of understanding of even the most basic things about convex bodies in dimensions 2 and 3. All these problems are well known and we will accompany each of them with a reference to the earliest known to us source where it was raised. We will intentionally abstain from any comments about our own attempts to solve them.

Problem 1 (Bonnesen [3, p. 51]). Does there exist a convex body $K \subset \mathbb{R}^3$ for which all maximal sections and all projections have the same area (possibly different for sections and projections) but which is not a ball?

Problem 2 (Bonnesen [3, p. 51]). Do there exist two convex bodies $K_1, K_2 \subset \mathbb{R}^3$ such that $M_{K_1} \equiv M_{K_2}$ and $P_{K_1} \equiv P_{K_2}$ but K_1 cannot be obtained from K_2 by a rigid motion?

Problem 3 (Gardner [6, Problem 7.6]). Does there exist an origin-symmetric convex body $K \subset \mathbb{R}^3$ such that all perimeters of central sections of K have the same length but K is not a ball?

Problem 4 (Gardner [6, Problem 7.6]). Let K_1 and K_2 be two origin-symmetric convex bodies in $\mathbb{R}^3$ whose central sections have equal perimeters. Does it follow that $K_1 = K_2$?

Problem 5 ([2]). Do there exist two convex bodies $K_1, K_2 \subset \mathbb{R}^2$ containing the disc of radius 1 in their interiors such that for every $e \in S^1$,

$$\text{length}(K_1 \cap (e^\perp + e)) = \text{length}(K_2 \cap (e^\perp + e))$$

but $K_1 \neq K_2$?

We do not know the answer to any of these questions even in a small neighborhood of the unit ball.

Bibliography

[1]   M. Angeles Alfonseca, F. Nazarov, D. Ryabogin and V. Yaskin, *A solution to the fifth and the eighth Busemann–Petty problems in a small neighborhood of the Euclidean ball*, Adv. Math. **390** (2021), 107920, 28 pp., arXiv:2101.08384.

[2]   J. A. Barker and D. G. Larman, *Determination of convex bodies by certain sets of sectional volumes, selected papers in honor of Helge Tverberg*, Discrete Math. **241**(1–3) (2001), 79–96.

[3]   T. Bonnesen and W. Fenchel, *Theory of convex bodies*, BCS Associates, 1987.

[4]   S. Brazitikos, A. Giannopoulos, P. Valettas and B. Vritsiou, *Geometry of isotropic convex bodies*, Mathematical surveys and monographs, vol. 196, AMS, Providence, RI, 2014, 594 pp.

[5]   H. Busemann and C. Petty, *Problems on convex bodies*, Math. Scand. **4** (1956), 88–94.

[6]   R. J. Gardner, *Geometric tomography*, 2nd edn., Encyclopedia of mathematics and its applications, vol. 58, Cambridge University Press, Cambridge, 2006.

[7]   H. Groemer, *Geometric applications of Fourier series and spherical harmonics*, Encyclopedia of mathematics and its applications, vol. 61, Cambridge University Press, Cambridge, 1996.

[8]   A. Fish, F. Nazarov, D. Ryabogin and A. Zvavitch, *The unit ball is an attractor of the intersection body operator*, Adv. Math. **226**(3) (2011), 2629–2642.

[9]   A. Koldobsky, *Fourier analysis in convex geometry*, Mathematical surveys and monographs, vol. 116, AMS, Providence, RI, 2005.

[10] A. Koldobsky, G. Paouris and M. Zymonopoulou, *Isomorphic properties of intersection bodies*, J. Funct. Anal. **261**(1) (2011), 2697–2716.

[11] J. Matoušek, *Using the Borsuk–Ulam Theorem*, Lectures on topological methods in combinatorics and geometry, Springer Verlag, Berlin, 2008.

[12] F. Nazarov, D. Ryabogin and A. Zvavitch, *Non-uniqueness of convex bodies with prescribed volumes of sections and projections*, Mathematika **59**(1) (2013), 213–221.

[13] E. M. Stein, *Interpolation of linear operators*, Trans. Am. Math. Soc. **83** (1956), 482–492.

[14] E. M. Stein and G. Weiss, *Introduction to Fourier analysis on Euclidean spaces*, Princeton University Press, Princeton, N. J., 1971.

[15] V. Yaskin, *On a generalization of Busemann's intersection inequality*, arXiv:2208.03882v1.

Index

https://doi.org/10.1515/9783110775389-012

Advances in Analysis and Geometry

Volume 8
Suzanne Lenhart, Jie Xiao (Eds.)
Potentials and Partial Differential Equations
The Legacy of David R. Adams, 2023
ISBN 978-3-11-079265-2, e-ISBN 978-3-11-079272-0,
e-ISBN (ePUB) 978-3-11-079278-2

Volume 7
Der-Chen Chang, Jingzhi Tie
The Sub-Laplacian Operators of Some Model Domains, 2022
ISBN 978-3-11-064210-0, e-ISBN 978-3-11-064299-5,
e-ISBN (ePUB) 978-3-11-064317-6

Volume 6
Mario Milman, Jie Xiao, Boguslaw Zegarlinski (Eds.)
Geometric Potential Analysis, 2022
ISBN 978-3-11-074167-4, e-ISBN 978-3-11-074171-1,
e-ISBN (ePUB) 978-3-11-074189-6

Volume 5
Jürgen Berndt, Young Jin Suh
Real Hypersurfaces in Hermitian Symmetric Spaces, 2022
ISBN 978-3-11-068978-5, e-ISBN 978-3-11-068983-9,
e-ISBN (ePUB) 978-3-11-068991-4

Volume 4
Maurice A. de Gosson
Quantum Harmonic Analysis. An Introduction, 2021
ISBN 978-3-11-072261-1, e-ISBN 978-3-11-072277-2,
e-ISBN (ePUB) 978-3-11-072290-1

Volume 3
Alexander Grigor'yan, Yuhua Sun (Eds.)
Analysis and Partial Differential Equations on Manifolds, Fractals and Graphs, 2021
ISBN 978-3-11-070063-3, e-ISBN 978-3-11-070076-3,
e-ISBN (ePUB) 978-3-11-070085-5

www.degruyter.com